TERTIARY FAUNAS

TERTIARY FAUNAS

A Text-book
for Oilfield Palaeontologists
and Students of Geology

By

A. MORLEY DAVIES

D.SC., A.R.C.S., F.G.S., HON. F.R.G.S.

Late Reader in Palaeontology at the University of London
(Imperial College of Science and Technology)

Revised and brought up to date by

F. E. EAMES

D.SC., A.R.C.S., F.G.S.

Formerly Chief Palaeontologist of the Burmah Oil Co. Ltd
and of the British Petroleum Co. Ltd

(Section on Vertebrata contributed by R. J. G. SAVAGE, Ph.D., F.L.S.,
Department of Geology, The University of Bristol)

VOL. I

THE COMPOSITION
OF TERTIARY FAUNAS

GEORGE ALLEN & UNWIN LTD
RUSKIN HOUSE   MUSEUM STREET   LONDON

FIRST PUBLISHED BY THOMAS MURBY & CO IN 1935
SECOND EDITION PUBLISHED BY GEORGE ALLEN & UNWIN LTD 1971

ISBN 0 04 560003 1

PRINTED IN GREAT BRITAIN
in 10 point Times Roman type
by W & J Mackay & Co Ltd
Chatham

PREFACE TO VOLUME I

The prior issue of Vol. II removes the main need for a preface to Vol. I, but a few additional remarks are necessary.

In the troublesome matter of nomenclature I make no claim to authority. I have tried to place before my readers the alternative generic names they will find both in past literature and, so far as it can be foreseen, in that of the immediate future; and I have had in each case to choose one name as the most correct and enclose the rest in square brackets. In these concentration-camps are herded together many that will suffer the death-penalty (absolute synonyms), some doomed to banishment (homonyms), some that may be set free and acclaimed as the long-lost heir, and yet others that may be allowed to live as poor relations (subgenera). I have tried not to be too arbitrary a dictator in my choice of victims, though I am sure to be criticized. Where recent rectifications in nomenclature seem likely to be generally adopted I have accepted them, but where there is a conflict of opinion as to which of several forgotten names should replace a familiar one, I have retained the latter.

I may be blamed for not always quoting the genotype as an example of a genus, but for the purpose of this work I have felt that when the type is either a recent species or a rarity among fossils, a more familiar Tertiary species will better serve as an example. Doubtless, I have been inconsistent in my selections.

It has been my wish to illustrate this volume mainly by original drawings of actual fossils, but in the case of the Foraminifera and Echinoids that has proved difficult and I have had to rely largely on published figures. A few figures of Mollusca, which may be recognized by a difference of style, are direct drawings made by my wife some ten years ago, when the work was in its tentative stage, but the great majority of the figures have been redrawn by her from my own camera-lucida sketches. I have to thank her for long and patient help in this undertaking, as also in the preparation of the index.

Although I have given details of the specimens from which the drawings were made (mainly, by kind permission of the Governors, specimens in the Imperial College collection, in which those from my own collection are now incorporated), I wish them to be regarded simply as generic diagrams, not conferring on the originals the status of 'figured specimens'.

To the many helpers mentioned in the preface to Vol. II I would repeat my thanks, adding to them the name of Mr L. S. Davidashvili of Moscow, and particularly that of Dr H. Dighton Thomas, who has kindly supplied the bibliographies of Tertiary Bryozoa and Anthozoa, printed on pp. 357–359 (of the first edition).

I greatly regret that I did not make the acquaintance of Thiele's *Handbuch der Systematischen Weichtierkunde* until too late to bring my classification of the Mollusca into closer agreement with his. To keep level with the advance of science in a book the preparation of which has been spread over a fifth part of my life has not been an easy task: I trust I have not altogether failed in it.

October 1935   A. MORLEY DAVIES

PREFACE TO SECOND EDITION OF VOL I

In 1950 Morley Davies asked me if I would undertake the revision and bring-
ing up to date of *Tertiary Faunas*, but I intimated that I could not do so
because of my industrial commitments. However, when I officially retired in
1966, the British Petroleum Co. Ltd, at the instigation of Dr P. E. Kent
(Chief Geologist), made it possible for me to do so. My first thought was that
the task would be too difficult, but it became a challenge which I felt I must
accept. I am fortunate in having the collaboration of Dr R. J. G. Savage in
contributing the section on Vertebrata. I am also greatly indebted to Mr J. G.
Martin of the BP Research Centre at Sunbury-on-Thames for making adjust-
ments to some of the original illustrations and for preparing the numerous
new ones.

Many of the new illustrations are based on figures in the *Treatise on
Invertebrate Paleontology*, kindly agreed to by Professor Raymond C. Moore,
and through the courtesy of the Geological Society of America and the
University of Kansas; individual acknowledgement is given on all keys to
plates. Some of the new illustrations of gastropoda have been made after
illustrations in Wenz, and Zilch (*Handbuch der Paläozoologie*, Band 6) through
the courtesy of Gebrüder Borntraeger; individual acknowledgement is given
on the keys to plates.

I am also indebted to the following sources for permission for reproduction
of some other illustrations:

(a) Masson & Cie of Paris, for reproduction of vertebrate illustrations
from *Traité de Paléontologie* (ed. Piveteau), Tome VI, Vols 1–2 and
Tome VII; individual acknowledgement is given on all keys to plates.

(b) The Trustees of the British Museum (Natural History) for reproduction
of some vertebrate illustrations from their guides and catalogues;
individual acknowledgement is given on all keys to plates.

(c) The University of Chicago Press, for reproduction of some vertebrate
illustrations by A. S. Romer in *Vertebrate Paleontology*; individual
acknowledgement is given on all keys to plates.

(d) The Hessisches Landesamt für Bodenforschung and Professor H.
Tobien for reproduction of one vertebrate illustration (*Lagopsis*); full
acknowledgement is given on the key to plate.

(e) Springer-Verlag and Professor Dr E. Thenius for reproduction of one
vertebrate illustration (*Lantanotherium*); full acknowledgement is given
on the key to plate.

I have tried to maintain largely the original style of the book, but some
changes were inevitable. Occurrences and ranges are given in somewhat less
detail, since it would be impossible to check through all the collections at the
disposal of Morley Davies in the time available to me. Also, research work
carried out since 1935 has made necessary considerable rearrangement and

8

expansion of some chapters, and a completely new chapter on Ostracoda has been included.

In the following chapters I have, in the more important and more easily recognizable groups, carried suprageneric morphological descriptions down to subfamily level, but in other cases not. As far as genera and subgenera are concerned, I have attempted to include the following: (a) forms of widespread occurrence and long range, (b) forms of widespread occurrence and shorter range, particularly those where the upper and/or lower limits are of stratigraphical significance, (c) forms of common occurrence in a relatively restricted area, and (d) a few forms, sometimes of relatively restricted occurrence and range, which are marginal to the family description. I hope that this will give readers a reasonable coverage of the groups concerned.

During the last twenty years or so there have been many different views concerning dating and correlation within various parts of the Tertiary, and even different opinions as to where it starts and where it ends. These matters will be discussed in full in Volume II, but the following brief synopsis gives the reasons for the ranges and dating adopted in this Volume I. The faunas of the Danian are regarded as belonging to the late Mesozoic (Eames, 1969, *Geol. Soc. India Mem.*, no. 2, Seminar Vol., pp. 361–368), and I have received much international support for this opinion. The Palaeocene is regarded as the lowest of four subdivisions of the Eocene: Palaeocene, Lower Eocene, Middle Eocene and Upper Eocene; there are so many genera in common to the Palaeocene and overlying subdivisions of the Eocene (e.g. *Nummulites, Palaeonummulites, Assilina, Fasciolites, Discocyclina, Sakesaria, Operculina,* etc.) that I consider that they must be grouped together. The term Bartonian is used in the customary British sense for post-Bracklesham beds, not in the French sense which includes all beds down to the top of the Lutetian. The Lattorfian is regarded as being of early Oligocene age (not late Eocene as suggested by Krutzsch and Lotsch, 1957), this being based on the evidence of planktonic foraminifera and nannoplankton (see Eames, 1970, *Palaeogeography, Palaeoecology and Palaeoclimatology*, Vol. 8, *cum bibl.*). The Bormidian is regarded as the earliest stage of the Miocene (Lyell included the Bormidian faunas in his Miocene as originally defined), older than the stratotype Aquitanian (see Eames, 1970, loc. cit.). The Miocene/Oligocene boundary is taken within the planktonic Zone N.1 as suggested by Eames *et al.* (1962, *Fundamentals of mid-Tertiary Stratigraphical Correlation*) and by Eames (1970, loc. cit.). In the Far East (following Adams, 1970, in the press) the letter substage 'e$_5$' is agreed to be early Miocene and the substage interval 'e$_{1-4}$' is regarded as late Oligocene. The Sallomacian, which antedates the term Vindobonian (which has been variously interpreted), is, following original French interpretation, regarded as the beginning of Middle Miocene times. This involves regarding the Gaj of West Pakistan and western India as being of Middle, not Lower, Miocene age. Consequent upon this, the Tortonian is included in the Upper Miocene, where it has been placed by many palaeontologists. As far as the end of the Tertiary is concerned, I follow those (e.g. 1954, *Rept. Internat. Geol. Congr., Sess. 19, 1952*, sect. 13, fasc. 15, pt 3) who recommend that the Calabrian be regarded as the oldest stage of the Quaternary, so that the Zanclian-Astian (Upper Pliocene) marks the end of the Tertiary.

I am greatly indebted to the geological staffs of the Burmah Oil Co. Ltd and of the British Petroleum Co. Ltd, since I started my industrial career in 1927; in particular I would mention Dr W. J. Clarke, Dr F. T. Banner, Dr W. H. Blow, Mr F. C. Dilley and Dr A. H. Smout, colleagues of mine at the British Petroleum Research Centre at Sunbury-on-Thames. I am also indebted in particular to Dr C. G. Adams, M. A. Chavan, the late Dr L. R. Cox, Dr I. Crespin, Mr D. Curry, Dr J. W. Durham, Dr Myra Keen, Dr P. M. Kier, Dr R. Lagaaij, Dr N. H. Ludbrook, Dr F. Stearns MacNeil, Dr D. Mongin, Mr C. P. Nuttall, Dr D. L. Pawson, Professor G. M. Phillip, Dr A. Poignant, Dr R. M. Stainforth, Dr I. M. van der Vlerk, Dr E. Voigt, Mr D. Wilson, Dr W. P. Woodring, Dr A. Zilch, and many others for assistance and discussions from time to time. I also acknowledge the assistance, over a period of many years, of the librarians of the British Museum (Natural History) and of the Geological Society of London.

September 1971   F. E. EAMES

GENERAL ABBREVIATIONS

A.-F.-B.B.	Anglo-Franco-Belgian Basin	Medit.	Mediterranean Sea or Region
Aquit.	Aquitanian	Mesoz.	Mesozoic
Atl.	Atlantic	Mio.	Miocene
Austral.	Australian Province (see Vol. II, Chap. I)	MT.	Monotype (type and only species)
B.	Basin (e.g. Paris B.)	N.	North
Balc.	Balcombian (see Vol. II)	N.Amer.	North America
Burd.	Burdigalian	N.Z.	New Zealand
C.	Central	Olig.	Oligocene
Calif.	Californian Province (see Vol. II, Chap. I)	Ord.	Ordovician
		Pac.	Pacific
Cambr.	Cambrian	Pal.	Palaeocene
Carb.	Carboniferous	Perm.	Permian
Carib.	Caribbean Province (see Vol. II, Chap. I)	Pleist.	Pleistocene
		Plio.	Pliocene
cosmop.	Cosmopolitan	R.	Right
Cret.	Cretaceous	Rec.	Recent (Holocene)
Dev.	Devonian	R.V.	Right valve
E.	East	S.	South
Eoc.	Eocene	S.Amer.	South America
Eq. ch.	Equatorial chambers	S.Austral.	Southern Australia (part of S. Australia and Victoria with Tasmania)
esp.	Especially		
Eur.	Europe		
Ex.	Example	s.g.	subgenus
fide	on the faith of	Sib.	Siberia
I.	Isle or Island	Sil.	Silurian
I.C.Z.N.	International Commission for Zoological Nomenclature	s.l.	sensu lato (in the wide sense)
		s.s.	sensu stricto (in the restricted sense)
Indo-Pac.	Indo-Pacific Province (see Vol. II, Chap. I)	T.	Type species (of genus or subgenus)
Janj.	Janjukian (Vol. II)		
Jur.	Jurassic	Tert.	Tertiary
L.	Lower or left	Trias.	Triassic
L.V.	Left valve	U.	Upper
M.	Middle	U.S.A.	United States of America
Maastr.	Maastrichtian		
max.	maximum	W.	West

CONTENTS

13

CONTENTS

ILLUSTRATIONS

15

17

N.B. All illustrations of vertebrates are drawn with anterior to the right-hand side of the page; upper dentitions are all right side and lower dentitions left side.

Chapter I

TERTIARY FORAMINIFERIDA

REFERENCE-LETTERS on the figures, following the figure numbers (when no letter is given, 'd' may generally be understood):

a, apertural view of planispiral, biserial or triserial, etc., edge-view of trochospiral forms.

b, spiral view of planispiral, side-view of biserial or triserial, etc., apical view of trochospiral forms.

c, basal view of trochospiral forms.

d, surface view not otherwise determinable.

e, surface appearance (including decorticated surfaces), greatly enlarged.

f, equatorial or median section – (A) of megalospheric, (B) of microspheric forms.

g, the same combined with surface view or other sections.

h, axial or vertical section.

i, transverse section.

j, tangential section.

k, megalosphere.

l, diagram of chamber form.

m, septal filament enlarged.

REFERENCE-LETTERS to parts of figures:

af, astral furrow.
ap, alar prolongation.
c, chamber.
if, intercameral foramen.
mc, marginal cord.

mm, megalosphere.
nc, nucleoconch.
p, pillar.
s, septum.
sl, spiral lamina.

The Foraminiferida (foraminifera) are a class of Protozoa secreting shells, the great beauty and variety of which attracted attention as soon as magnifying glasses came into use. The majority are too small to be seen easily by the naked eye unless isolated and placed on a contrasted background, but some are larger and attain a diameter of several centimetres. Being in nearly all cases chambered shells they were naturally at first taken for small Cephalopods, a mistake corrected over a century ago when living forms were observed under the microscope to be composed of protoplasm undifferentiated into cells or tissues. Apart from this vastly simpler organic structure there are differences, both in shell texture and in the mode of shell growth, which suffice to distinguish the foraminifera.

The shell (test) may be chitinous (in which case it is almost impossible of being fossilized), or it may be calcareous, or composed of foreign bodies cemented together by a chitinous or mineral cement.

Calcareous tests may be composed of calcite, or less often aragonite, but not both. The crystal structure may be of very fine granular plates more or less parallel to the surface (porcellanous texture), similar but coarser granular structure, or of radially fibrous needles with their long axes at right angles to the surface. The calcareous wall may be hyaline in the two latter cases, but the coarser granulate walls may be opaque, and perforations can make the wall translucent. Perforations are present in most cases, except for walls of porcellanous type, but are absent at sutures, on the apertural face, and sometimes on other special areas. Lamination of the shell material arises when the shell material of one episode of chamber formation is not confined to forming a wall to the new chamber or chambers corresponding to it, but also coats part or whole of the exposed surface of the test, including the apertural face that is being converted into a septum. There is also some solution of shell material associated with the addition of chambers or with reproduction in certain species.

Cases are known where an imperforate form is perforate in its earliest growth stages; the perforate type is therefore probably the more primitive. Certain of the higher genera of perforate forms have a 'supplemental skeleton' in addition to the perforate test: this is not unlike the imperforate type of test, but is traversed by irregular cavities or canaliculi (hence 'canalicular skeleton'). This may be plastered in places on the outside of the ordinary test, or may penetrate into it as 'pillars' or otherwise (Fig. 136f).

'Arenaceous' (or more appropriately 'agglutinating') is the term more usually applied to tests formed of cemented foreign bodies, since sand grains are those most commonly used; but some species show a decided power of selection and will agglutinate sponge spicules or grains of some particular mineral in preference to quartz grains even if the latter are at hand in abundance. The cement may be organic, calcareous or ferruginous. It seems possible that this type of test is the most primitive of all, as it is that found in the most primitive order.

During the last few decades there has been much research carried out on the wall structure of foraminifera, and the following seven types of wall structure are very useful in classification:

    (i) Agglutinated tests.
    (ii) Microgranular tests.
    (iii) Hyaline calcareous tests with perforate radial walls.
    (iv) Hyaline calcareous tests with monocrystalline walls.
    (v) Hyaline calcareous tests with perforate granular walls.
    (vi) Spicular tests.
    (vii) Walls of lamellar character.

Shell growth. In a few cases growth is continuous, as in a gastropod shell without varices, but usually there is a periodicity of growth leading to a formation of chambers. The mode of formation is quite unlike what takes place in a Cephalopod. The latter shifts its body forward in the shell and secretes a new septum behind it, far back from the aperture, but the protoplasm of the foraminifer (which fills the whole of the chambers) extrudes beyond the shell and secretes around itself a new chamber, so that what had been part of the outer shell comes to be the septum between the new and the penultimate chamber (Fig. 8f).

Shapes of chambers. The different types of shape of chamber have been illustrated by Loeblich and Tappan, in Moore (**78**, p. C101), descriptive terms being: (a) spherical, (b) pyriform, (c) tubular, (d) globular, (e) ovate, (f) angular truncate, (g) hemispherical, (h) angular rhomboid, (i) angular conical, (j) radial elongate, (k) clavate, (l) tubulospinate, (m) cuneate, (n) cyclical, (o) fistulose, and (p) semicircular.

Arrangement of chambers. The different types of arrangement of chambers in foraminiferal tests have also been illustrated by Loeblich and Tappan, in Moore (**78**, p. C102), descriptive terms being (a) uniserial rectilinear, (b) trochospiral, (c) uniserial arcuate, (d) planispiral, loosely evolute, (e) planispiral involute, (f) planispiral evolute, (g) milioline, (h) streptospiral, (i) biserial enrolled, (j) biserial to uniserial, (k) biserial, (l) planispiral to biserial, (m) triserial, and (n) triserial to biserial to uniserial.

In the arrangement of chambers far more variety is shown than in the case of the Cephalopoda, where the series is linear, either straight, curved or planispiral. In foraminifera there are at least three types of spiral arrangement— plane, helicoid, and milioline (a type in which the axis is continually changing as in the winding of a ball of wool)—as well as zigzag, linear, cyclical (i.e. in rings) and acervuline (in indefinite heaps). Large chambers are often divided into chamberlets—formed simultaneously with the chamber itself.

In Cephalopoda there is reason to regard the straight or gently curved linear type as primitive, and the spiral as eventually derived from it, though racial old age may bring a return to straightness. In many foraminifera there are changes in plan during successive stages of ontogeny, but in no case does a simple spiral plan succeed any other: it is therefore probable that some form of spiral is the most primitive plan, all others being derivative. Although Cushman has suggested that the rectilinear plan may be primitive (*American Naturalist*, pp. 537–553; see also Swinnerton, *Outlines of Palaeontology*, pp. 13–15), the interpretation is a forced one: the first two chambers of a curved series can easily be imagined as in a straight line.

Shapes of test. Various types of shapes of test in the foraminifera have been illustrated by Loeblich and Tappan, in Moore (**78**, pp. C103–104); examples are: (a) tubular, (b) bifurcating, (c) radiate, (d) arborescent, (e) irregular, (f) hemispherical, (g) zigzag, (h) lanceolate, (i) conical, (j) spherical, (k) palmate, (l) discoidal, (m) fusiform, (n) biumbilicate, (o) biconvex, (p) flaring, (q) spiroconvex trochospiral, (r) umbilicoconvex trochospiral, deeply umbilicate, and (s) lenticular biumbonate.

Apertures. In addition to the nature of the test wall, the character of the aperture or apertures in the foraminifera is of very great importance in classification. The following characters have all been illustrated by Loeblich and

21

Tappan, in Moore (**78**, pp. C107–111):

   (i) *Location and form of primary apertures:*
(a) open end of tube, (b) radiate terminal, (c) slit terminal, (d) crescentic, (e) hooded, subterminal, (f) loop-shaped bulimine, (g) single, (h) multiple equatorial, interiomarginal, (i) cruciform, (j) areal, cribrate, (k) dendritic, (l) umbilical, (m) interiomarginal extraumbilical-umbilical, and (n) spiroumbilical.

   (ii) *Types of supplementary apertures:*
(a) relict, (b) areal multiple, (c) peripheral and areal, (d) multiple sutural, (e) single sutural, and (f) areal.

   (iii) *Types of accessory apertures:*
(a) sutural and umbilical canal openings, (b) infralaminal, (c) sutural (transverse), (d) sutural (parallel), and (e) intralaminal.

   (iv) *Modifications of apertures:*
(a) apertural flap, (b) bifid tooth, (c) modified tooth, (d) lateral flanges, (e) simple apertural lip, (f) pleurostomelline bifid tooth, (g) phialine lip, (h) umbilical teeth, (i) tegilla, (j) umbilical bulla, (k) areal bullae, (l) sutural bullae, and (m) umbilical-sutural bulla.

   (v) *Internal modifications:*
(a) entosolenian tube, (b) bulimine toothplate, (c) alternating hemicylindrical siphon, and (d) internal partition.

Ornamentation. Although some tests are quite smooth, many forms are beautifully ornamented; the following are some of the terms applied to different types of ornamentation: (a) acicular spines, (b) finely spinose and cancellate, (c) hispid, (d) punctate, with limbate sutures, (e) rugose, (f) costate, (g) ribbed, (h) fissured, (i) striate, (j) pitted, with chamber flanges, (k) reticulate, with elevated sutures, (l) nodose, and coarsely spinose, and (m) peripheral keel.

Canal system and intercameral foramina. Canal systems are found in the Rotaliacea and Orbitoidacea, and may be intraseptal, marginal, lateral, or umbilical; their nature, and even the presence or absence of them, are useful in classification. Intercameral foramina (sometimes referred to by the incorrect term 'stolons') are small canals connecting adjacent chambers or chamberlets.

Dimorphism and trimorphism. Just as a cephalopod starts with a protoconch quite unlike any subsequent chamber, so does a foraminifer begin by secreting an *initial chamber*, usually spherical. But foraminifera commonly show a *dimorphism* in early development, not paralleled among Cephalopods. The same species may have two forms, one starting from a much larger initial chamber (or group of chambers) than the other (Fig. 1): these are termed *megalospheric* and *microspheric*, or Forms A and B, respectively, and their differences may be tabulated as follows:

MEGALOSPHERIC (A)	MICROSPHERIC (B)
Large initial chamber, simply globular or subdivided.	Small initial chamber, simply globular, never subdivided.

22

May lack early stages of
ontogeny shown by B.
Adult shells often smaller.
Individuals more numerous.
Results from asexual repro-
duction.
Possibly several successive
generations.

May show a series of stages in
ontogeny.
Adult shells often larger.
Individuals less numerous.
Results from sexual repro-
duction.
One generation at intervals.

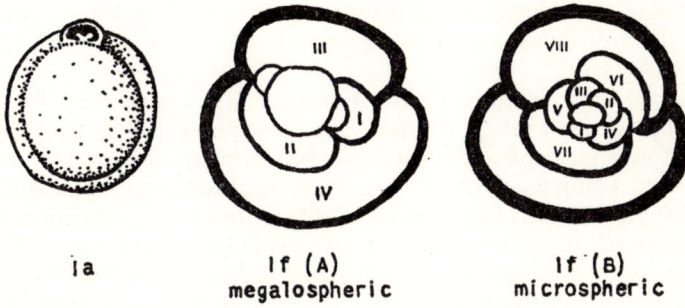

ia

If (A)
megalospheric

If (B)
microspheric

Fig. 1. *Pyrgo bulloides* (d'Orbigny), Recent. a, × 17·5; f(B) and f(A), × 125. After Schlumberger. In f(B) and f(A) only the early chambers are shown, up to attainment of adult plan: I in A corresponds to VI in B.

In the case of one living form (*Ammonia beccarii* subsp. *flevensis*) it has been shown by Hofker (**70**) that dimorphism is seasonal, the microspheric form living in winter and giving rise to megalospheric forms in spring. In this it shows close analogy with the case of certain land and freshwater arthropods (water-fleas, aphides) which multiply parthenogenetically all through the summer and only reproduce sexually in the autumn, when winter eggs are laid that only develop in spring. In these cases the seasonal control is severe, and the *Ammonia* in question lives in the less saline waters of the Zuyder Zee. It would not be safe to assume that such extinct foraminifera as the nummulites and orbitoids, which lived in a purely marine habitat and mainly in tropical and subtropical latitudes, were subject to equally great seasonal variations in their surroundings. Their dimorphism need not necessarily have been seasonal.

The same *Ammonia* has furnished an explanation of the so-called *trimorphism* that has been observed in a few fossil foraminifera. In these cases there are two kinds of megalospheric individuals, one kind having a smaller initial chamber and a larger number of whorls than the other, showing in fact a position intermediate between the latter and the microspheric form. In the Recent subspecies these are summer individuals, the most typically megalospheric individuals being those formed in spring.

When the initial chamber is simply globular, it may be called the microsphere or megalosphere as the case may be. But when (in the megalospheric form only) it is subdivided, it is termed the *nucleoconch*. The subdivision may

23

be into four chambers tetrahedrally arranged (exactly like the tetrad spores in certain plants), as in the Cretaceous *Omphalocyclus* and *Orbitoides* (*s.s.*), or into about as many as ten chambers, as in *Pliolepidina* (Fig. 208k), or into two equal (Fig. 140h) or unequal chambers (Fig. 132f), as in *Nummulites*; in the latter case the larger may more or less envelop the smaller, as in *Eulepidina* (Fig. 207k). A peculiar feature of the Miliolacea is the presence of a minute spiral tube around part of the megalosphere (Fig. 52f).

The inequality in size of the two forms of the same species may be very striking, there being cases where the ratio of the diameter of the microspheric form to that of the megalospheric form may be as much as 6:1; in other cases, however, this ratio may be slight or even 1:1.

Classification. Classifications of the Foraminiferida have been founded on different characters at different times. The great pioneer d'Orbigny (**48**) made the plan of arrangement and growth of the chambers the essential criterion, but his system, being pre-evolutionary, made no allowance for parallel development in different stocks. Later, Brady (**29**) produced a classification in which shell texture took precedence over growth plan: his scheme held the field for many years. He recognized 10 families, which Lister in 1903 raised to ordinal rank, making Brady's subfamilies into families, but not otherwise modifying the classification. Subsequently, Cushman (**39**) and Galloway (**58**) have published much more elaborate schemes, increasing the number of families to 47 and 35 respectively, but not retaining or proposing any superfamilial groups. These new classifications probably expressed the evolutionary relations of the many genera better than Brady's, but they were still tentative and presented a forbidding complexity to the student.

The more important contributions to the classification of the Foraminiferida have been listed by Loeblich and Tappan, in Moore (**78**, pp. C142–C153). Those subsequent to that of Galloway in 1933 have been by Chapman and Parr in 1936 (**33**), Glaessner in 1945 (**59**), Cushman in 1948 (**40**), Hofker in 1951 (**71**), Pokorný in 1958 (**86**), Reiss in 1958 (**89**) and Rauzer-Chernousova and Fursenko in 1959 (**87**). The suprageneric classification here adopted is close to but not identical with that of Loeblich and Tappan in 1964 (**76**), who recognized a total of 96 families; in a number of cases (for example, the Nummulitidae) the generic synonymy has not been followed since it appears that in some cases too many different forms have been grouped in synonymy, thus losing the stratigraphical value of some which have recognizable distinctive morphological characters.

Stratigraphical value. Until about the middle of the first half of this century it was generally believed that only the larger foraminifera were of stratigraphical value, the smaller species of foraminifera having far too long a time range to allow of any dating value. Brady did not hesitate to identify (in 1888) Silurian foraminifera with Recent species of *Lagena*, and many species have been listed as ranging from Cretaceous or even Jurassic to Recent. Oilfield palaeontologists had also claimed that associations of species, or even single small species,

24

served as good guides to age in particular oilfields. In most cases this appears to be a case of reliance on a succession of *facies faunules*, which may be fairly trustworthy within a single basin of deposition, but might show a different sequence in another basin (see Vol. II, Chap. II). In other cases, where the same facies has persisted for a very long time (even from Cretaceous to Pliocene as in parts of Israel and Syria) it has been claimed that a long-ranging species may show recognizable mutations of dating value.

During the last thirty years a very great deal of comprehensive and detailed research work has been carried out on foraminifera. In the case of benthonic foraminifera, very broad views have sometimes been taken as to the extent of variation in what has been considered to be one species, and consequently some reassessment may occasionally be necessary if the stratigraphical range is to be accurately known. The most important development, however, has been the carrying out of much detailed study of planktonic foraminifera, which has resulted not only in a much finer zonation of the strata in which they occur, but in much more accurate intercontinental correlation involving, in some cases, a reconsideration of the relative ages of some beds.

The list of references to foraminiferal literature at the end of this chapter is approximately double that given in the first edition. The additional references have had to be somewhat arbitrarily selected in order to give as broad a coverage as possible, and there will doubtless be authors who are disappointed that some of their works have not been included; I have tried to ensure, however, that all works will be found in the lists of references given in those contributions to which I have referred.

Suborder TEXTULARIINA

Test of agglutinated foreign material; cement of various kinds. (Cambr.-Rec.)

Superfamily AMMODISCACEA

Test irregular, spheroidal or tubular and straight, branching or enrolled, never truly chambered; wall simple or labyrinthic; aperture simple. (Cambr.-Rec.)

Family ASTRORHIZIDAE

Test free or adherent, tubular or branching; wall simple, with pseudochitinous inner layer; aperture absent, or terminal and rounded. (Cambr.-Rec.)
Astrorhiza (Fig. 2): free, flattened, with hollow central disk having numerous radiating arms. Ord.-Rec. Cosmop. *Rhabdammina* (Fig. 3): free, with long tubular arms radiating from a fairly small centre. Ord.-Rec. Cosmop. *Bathysiphon* (Fig. 4): a large, free, fairly straight, long tube. Cambr.-Rec. Cosmop. *Hyperammina*: free and cylindrical, with a bulbous proloculus of greater diameter than the tubular second chamber. Ord.-Rec. Cosmop.

Family SACCAMMINIDAE

Test free or attached, subglobular or in groups; aperture absent, single, or multiple. (Ord.-Rec.)

Saccammina: free, consisting of a single spherical chamber, the single aperture often with a short neck. Sil.-Rec. N.Amer., Eur., Atl., Pac., Antarctic. *Pelosina:* free, elongate, subcylindrical-fusiform, with fine tubular extensions at ends. Cret.-Rec. Eur., Atl., Pac., Sib., Arctic, Antarctic.

Family AMMODISCIDAE

Test free or attached; second chamber tubular, enrolled, aperture at open end. (Sil.-Rec.)

Ammodiscus: free and planispiral; it is homoeomorphic with *Cyclogyra* (Fig. 27) which has a porcellanous wall. Sil.-Rec. Cosmop. *Glomospira:* like *Ammodiscus*, but the coiling is more or less streptospiral. Sil.-Rec. Cosmop. *Turritellella:* free, elongate, close-coiled and high-spired. Sil.-Rec. Eur., N.Amer., Arctic, Antarctic.

Superfamily LITUOLACEA

Test chambered, spiral or straight, or biserial or triserial, simple or labyrinthic; aperture simple or multiple. (Carb.-Rec.)

Family LITUOLIDAE

Test free or attached; early stage coiled but later may become uncoiled, irregular or annular; wall simple to labyrinthic, epidermal layer imperforate; aperture single or multiple. (Carb.-Rec.)

Figs 2–23. FORAMINIFERIDA: AMMODISCACEA AND LITUOLACEA
Figs 2, 3, 7f, 8, 9, 15 and 17 after Brady; Fig. 22 after von Hantken; Fig. 23 after L. M. Davies; Figs 4, 5, 6, 10–14, 16, and 18–21 after Loeblich and Tappan, in Moore; Fig. 7b original.
2. *Astrorhiza limicola* Sandahl, Rec. T. × 3.
3d, f. *Rhabdammina linearis* Brady, Rec. × 15.
4. *Bathysiphon filiformis* M.Sars, in G. O. Sars, Rec. Pac. T. × 2.
5a, b. *Lituola nautiloidea* Lamarck, U.Cret. France. T. × 8.
6a, b. *Haplophragmoides canariensis* (d'Orbigny), Rec. Philippines. T. × 22.
7b, f. *Cyclammina cancellata* Brady, Mio., Trinidad (b) and Rec. (f). T. × 12.
8d, f. *Reophax nodulosa* Brady, Rec. × 9.
9a,b. *Textularia agglutinans* d'Orbigny, Rec. × 19.
10a, b. *Vulvulina pennatula* (Batsch), Rec. Italy. × 20.
11a, b. *Bigenerina nodosaria* d'Orbigny, Rec. France. T. × 17·5.
12b. *Siphotextularia wairoana* Finlay, Plio. N.Z. T. × 54·5.
13a, b. *Trochammina inflata* (Montagu), Rec. N.Atl. T. × 43.
14a, b. *Arenoparrella mexicana* (Kornfeld), Rec. Trinidad. T. × 54·5.
15b. *Verneuilina pygmaea* Egger, Rec. × 20.
16a, b. *Gaudryina rugosa* d'Orbigny, U.Cret. Germany. T. × 19·5.
17a, b. *Tritaxia lepida* Brady, Rec. × 60.
18a, b. *Dorothia bulletta* (Carsey), U.Cret. Texas. T. × 34.
19a, b. *Marssonella oxycona* (Reuss), U.Cret. Germany. T. × 31.
20b. *Karreriella siphonella* (Reuss), M.Olig. Germany. T. × 20.
21a, b. *Valvulina triangularis* d'Orbigny, Eoc. France. T. × 16·5.
22a, b, d, h. *Tritaxia szaboi* (von Hantken), U.Eoc. Hungary. × 3·75.
23. Composite diagram of abraded surface in *Lituonella* (L), *Coskinolina* (C), *Dictyoconus* (D).

26

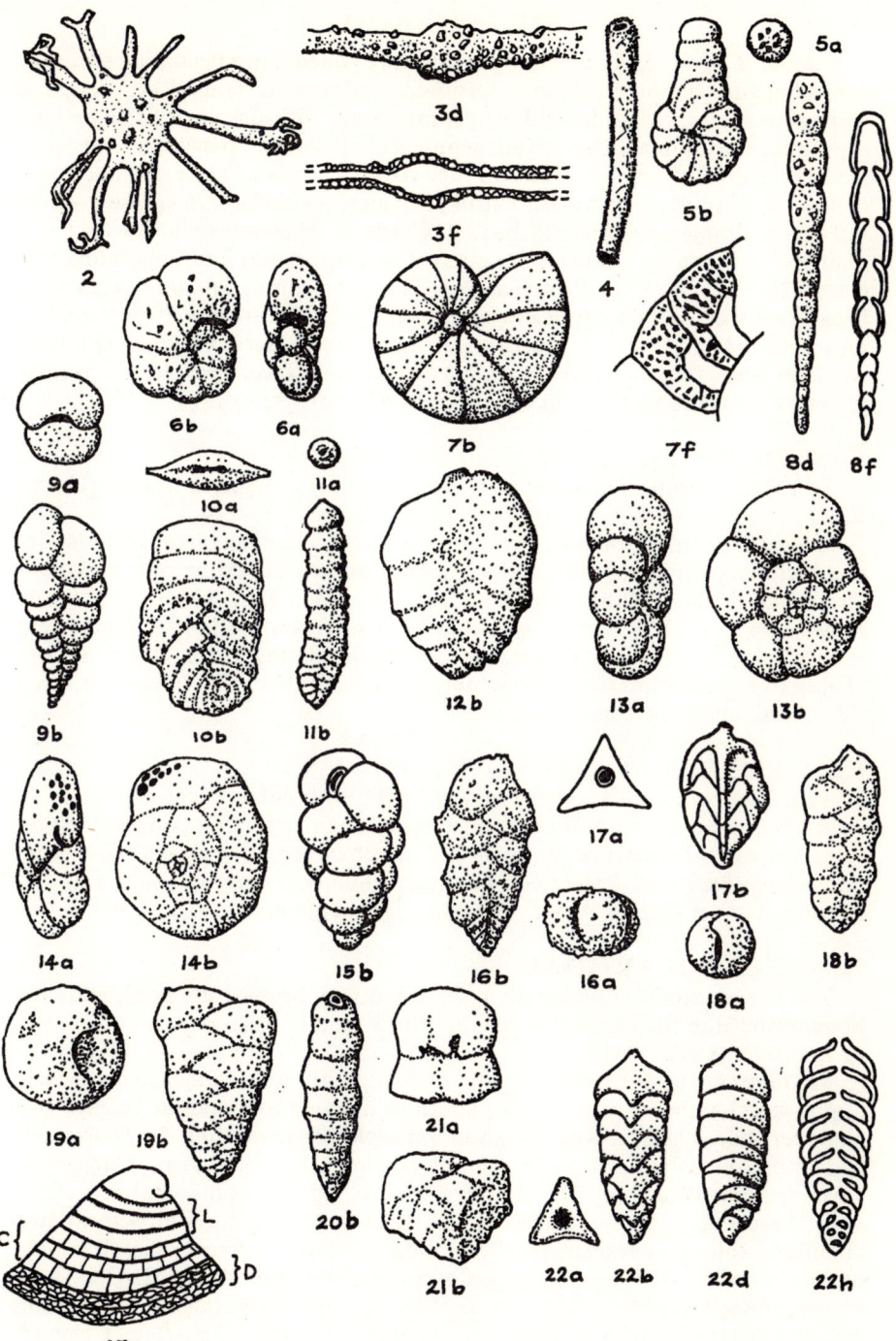

Lituola (Fig. 5): early portion planispirally coiled, later becoming straight; wall structure simple; aperture terminal, cribrate. U.Trias.-Rec. Cosmop. *Ammoastuta:* test ovate-flabelliform, compressed, with simple walls, chambers rapidly broadening and in curved, semi-enrolled series; aperture a transverse slit near middle of terminal chamber face which also has cribrate openings near its base. U.Eoc.-Rec. N.Amer., Carib., S.Amer. *Ammobaculites:* like *Lituola*, but has a simple aperture. Carb.-Rec. Cosmop. *Haplophragmoides* (Fig. 6): planispiral, involute; wall simple; aperture an equatorial interiomarginal slit. Carb.-Rec. Cosmop. *Cyclammina* (Fig. 7): planispiral, involute, with low, broad chambers; reticulate meshwork immediately beneath imperforate epidermis, interior labyrinthic; aperture an equatorial interiomarginal slit with rounded pores (having raised margins) on apertural face. Cret.-Rec. Cosmop. [Thin sections are very useful for speciation in this genus (9).]

Family HORMOSINIDAE

Test free, chambers in straight or curved series; aperture terminal. (Carb.-Rec.)

Hormosina: free, elongate, straight or arcuate; chambers increasing in size as added, large and globular, with horizontal sutures; aperture on a produced neck; wall finely agglutinating. Jur.-Rec. Atl., Pac., Eur., Medit., Antarctic. *Reophax* (Fig. 8): like *Hormosina* but lacks an apertural neck, its chambers are less globular, and the wall is more coarsely agglutinating. Carb.-Rec. Cosmop. [These two genera are homoeomorphic with *Nodosaria* (Fig. 61) which has a calcareous, finely perforate wall.]

Family RZEHAKINIDAE

Test free, coiling of various types as in the Miliolidae. (Cret.-Rec.)

Rzehakina: ovate in outline, compressed; chambers half a coil long, planispiral, involute; aperture at open end of chamber. U.Cret.-Pal. Cosmop. *Silicosigmoilina:* like *Rzehakina*, but later chambers arranged in a sigmoid manner. U.Cret.-Pal. N.Amer., S.Amer., Eur., Japan.

Family TEXTULARIIDAE

Test free or attached, generally biserial but may become uniserial, and may be planispiral in the earliest stages; aperture simple, basal or terminal, single or multiple. (Carb.-Rec.)

Textularia (Fig. 9): free, elongate, biserial, with numerous, usually closely appressed chambers; wall simple; aperture a single low arch at base of last chamber face. U.Carb.-Rec. Cosmop. *Spiroplectammina:* like *Textularia*, but early portion consists of a planispiral coil of a few chambers. Carb.-Rec. Cosmop. *Vulvulina* (Fig. 10): free, flaring or elongate, rhomboidal in section, with sharp edges; early portion coiled, then becoming biserial with arched chambers, fully grown specimens becoming uniserial; aperture in adult a long, narrow terminal slit, in earlier stages a low interiomarginal arch. U.Cret.-Rec. Cosmop. *Bigenerina* (Fig. 11): like *Textularia* in the early stages, but the adult becomes uniserial with a terminal, rounded aperture. Jur.-Rec. Cosmop.

Siphotextularia (Fig. 12): free, biserial, quadrate in section; aperture rounded and with a neck, in face of last chamber. Pal.-Rec. Cosmop.

Family TROCHAMMINIDAE

Test free or attached, trochospiral; wall simple; aperture interiomarginal or areal. (Carb.-Rec.)

Trochammina (Fig. 13): free, with rather inflated chambers increasing gradually in size; aperture a low interiomarginal extraumbilical-umbilical arch which sometimes has a narrow bordering lip. Carb.-Rec. Cosmop.
Arenoparrella (Fig. 14): primary aperture an elongate slit extending up the face of the last chamber; also a supplementary cribrate aperture near the apex of the last chamber. Mio.-Rec. N.Amer., Carib., S.Amer.

Family ATAXOPHRAGMIIDAE

Test free, trochospiral, uncoiling or uniserial; aperture a basal slit in the early chambers, but may later become terminal, cribrate or toothed. (U.Carb.-Rec.)

Verneuilina (Fig. 15): elongate, triangular with sharp angles, wholly triserial; aperture a low arch at inner face of last chamber. Jur.-Rec. Cosmop.
Gaudryina (Fig. 16): elongate, early stage triserial and usually triangular, later stage biserial; aperture interiomarginal. U.Trias.-Rec. Cosmop. *Tritaxia* (Figs. 17, 22): triserial and triangular in early stage, but later uniserial and usually triangular; aperture interiomarginal in early stage, later terminal with a thick tube connecting last one or two apertures. L.Cret.-Rec. Cosmop.
Dorothia (Fig. 18): elongate-subcylindrical, early stage slightly swollen and of four or more chambers to the whorl, later stages biserial; aperture an interiomarginal slit. L.Cret.-Rec. Cosmop. *Marssonella* (Fig. 19): like *Dorothia*, but elongate-conical, chamber faces distally usually distinctly flattened. U.Jur.-U.Eoc. Cosmop. [Pre-Cretaceous records refer to *Textilaria dumortieri* Schwager—information from Mr F. C. Dilley]. *Karreriella* (Fig. 20): elongate, early stage a trochoid spire of one or more whorls, later with well developed biserial stage; aperture rounded, in terminal face of last chamber, on a neck or with a lip. Pal.-Rec. Cosmop. *Valvulina* (Fig. 21): test sometimes triangular in section, usually triserial throughout; aperture at base of last chamber, with a large valvular tooth. U.Trias.-Rec. Cosmop. *Clavulina:* like *Valvulina*, but with final uniserial stage and terminal aperture. Pal.-Rec. Cosmop.

Family PFENDERINIDAE

Trochospiral, at least in early stage; three or more chambers to the whorl, later stages may be biserial or uniserial; wall agglutinated or microgranular; chambers may be subdivided by pillars or partitions. (U.Jur.-Rec.)

Lituonella (Fig. 23): conical, early stage an asymmetrical off-centre spire, later chambers broad, saucer-shaped and nearly circular in plan; early sutures curved, later (in uniserial portion) nearly straight; median portion of basal surface having many large apertures which are bordered by pillar-like interseptal buttresses, outer ring of buttresses external to perforations, marginal

29

area not subdivided and without apertures. Eoc. France, M.East, Pakistan.
Coskinolina (Figs. 23, 24): like *Lituonella*, with the same vertical interseptal pillars and non-perforate marginal area on base, but marginal zone subdivided by radial partial partitions. L.Cret.-basal U.Eoc. Pakistan, Eur., N.Amer., S.Amer.

In both these genera the megalospheric form tends to be more cylindrical than the microspheric. The characters of the marginal trough can be seen on abraded or acid-treated specimens if these processes are not carried too far.

Family DICYCLINIDAE
Test free, discoidal or depressed conical; chambers cyclical; wall finely agglutinated, with imperforate epidermis; aperture multiple, peripheral. (?U.Trias., Jur.-M. Eoc.)

Saudia (Fig. 25): discoidal; early stage with spiral development probably not subdivided, later flabelliform with arcuate uniserial stage, adult with cyclical chambers partially divided by secondary transverse and parallel sub-epidermal partitions projecting inward from outer wall; median plane in later, thicker parts of test also with radial interseptal pillars aligned from one cycle to the next; apertures between interseptal pillars. Pal.-M.Eoc. S.W.Asia, Libya.

Family ORBITOLINIDAE
More or less conical; test with a single series of shallow saucer-shaped

---

Figs 24–45. FORAMINIFERIDA: LITUOLACEA AND MILIOLACEA
Fig. 24 after L. M. Davies; Fig. 40 after Carpenter; Figs 30, 31, 37, 39 and 42 after Brady; Figs 25, 28, 29, 33–46, 43 and 45 after Loeblich and Tappan, in Moore; Figs 26, 27, 32, 38, 41 and 44 original.
24h, i. *Coskinolina balsilliei* L. M. Davies, L.Eoc. W. Pakistan. × 13·5.
25f, h. *Saudia discoidea* Henson, M.Eoc. M.East. T. f × 20, h × 12.
26. Diagram of 'divided scale' pattern of *Dictyoconus*.
27b. *Cyclogyra involvens* (Reuss), L.Eoc. London. × 12.
28a, c. *Nubecularia lucifuga* Defrance, M.Eoc. France. T. × 12·5.
29a, b. *Ophthalmidium acutimargo* (Brady), Rec. S.Atl. × 24.
30a, b. *Spiroloculina limbata* d'Orbigny, Rec. × 22·5.
31. *Discospirina italica* (Da Costa), Rec. T. × 15.
32b. *Quinqueloculina* aff. *seminulum* (Linné), L.Plio. Suffolk. × 12.
33a, b. *Triloculina trigonula* (Lamarck), Eoc. France. T. × 15·5.
34a, b. *Miliola saxorum* (Lamarck), Eoc. France. × 13.
35a, b. *Sigmoilina sigmoidea* (Brady), Rec. Carib. T. × 23.
36a, b. *Massilina secans* (d'Orbigny), Rec. Medit. T. × 8.
37a, b. *Hauerina compressa* d'Orbigny, Rec. T. × 28.
38a, b, d. *Pyrgo depressa* (d'Orbigny), Rec. × 12.
39. *Articulina conico-articulata* (Batsch), Rec. × 37·5.
40a, b. *Fabularia ovata* (de Roissy), M.Eoc. France. T. × 4. (Surface wrinkling omitted; two patches shown decorticated with labyrinthine structure).
41f. *Austrotrillina howchini* (Schlumberger), M.Mio. Pemba I. T. × 30.
42a, b. *Peneroplis planatus* (Fichtel and Moll), Rec. T. × 30.
43a, b. *Dendritina arbuscula* d'Orbigny, Mio. France. T. Enlarged.
44b. *Spirolina cylindracea* Lamarck, M.Eoc. France. T. × 12.
45a, b, j. *Rhapydionina urensis* Henson, M.Eoc. Iraq. a, b × 7·5, j × 9.

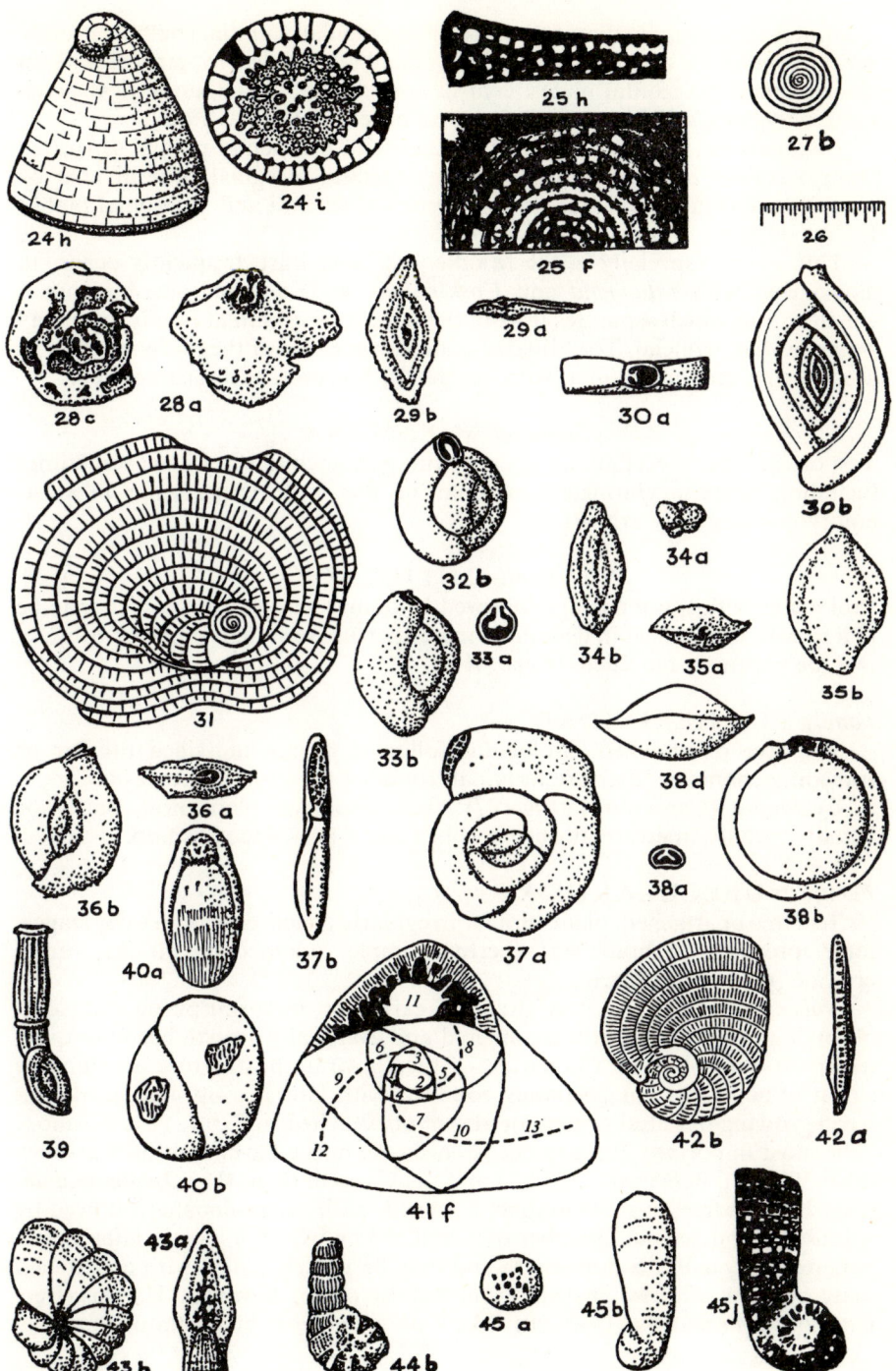

24 h
24 i
25 h
25 f
27 b
26
28 c
28 a
29 b
29 a
30 a
30 b
31
32 b
33 a
34 b
34 a
35 a
35 b
33 b
38 d
36 a
37 a
38 a
38 b
36 b
40 a
37 b
39
40 b
41 f
42 b
42 a
43 a
43 b
44 b
45 a
45 b
45 j

chambers increasing in diameter more or less regularly, initial chambers some-times forming an asymmetrical spiral; chambers divided by vertical and (in some genera) horizontal plates, central area divided by either vertical parti-tions or pillars, or both; septal apertural pores. (L.Cret.-U.Eoc.)

Dictyoconus (Figs. 23, 26): central area with interseptal pillars, separated from marginal zone by a ridge in many species; marginal zone divided by partitions and may have one or more series of vertical and horizontal plates. L.Cret.-U.Eoc., ?L.Olig. Cosmop. (**44**).

This genus, especially in the Middle and Near East, frequently occurs in association with *Lituonella* and *Coskinolina*, and it seems possible that a classification which separates it from the latter two genera at family level may be somewhat artificial. The 'divided-scale' appearance of the marginal zone as seen in thin section or on a worn surface is, however, easily recognized.

Suborder MILIOLINA

Test calcareous, porcellanous, usually with pseudochitinous lining, sometimes including some agglutinated material in the wall, imperforate in post-embryonic stages. (Carb.-Rec.)

Superfamily MILIOLACEA

Proloculus with spiral passage followed by numerous chambers which may be planispiral or arranged in definite planes; aperture terminal, single or cribrate, may be variously modified. (Carb.-Rec.)

Family FISCHERINIDAE

Test free or attached, proloculus followed by an undivided tubular or spreading chamber; aperture terminal, rounded or slit-like. (Carb.-Rec.)

Cyclogyra [*Cornuspira*] (Fig. 27): free, discoidal, planispiral, partly or wholly evolute; aperture at open end of tube. Carb.-Rec. Cosmop.

Family NUBECULARIIDAE

Test free or attached, planispiral or irregularly coiled, at least in early stages, later spreading or branched; aperture simple, rounded or slit-like, rarely cribrate. (M. Carb.-Rec.)

Nubecularia (Fig. 28): test attached, early part in cornuspirine coil, soon becoming rather irregularly chambered and eventually growth becomes quite irregular, controlled by the nature of the substratum; aperture an elongate slit at attachment, but becoming rounded with lateral tooth-like infoldings when growing free; test present or absent on attached side. Jur.-Rec. Cosmop. [The most important Tertiary occurrences are in the Sarmatian of Pressburg (near Vienna), Bessarabia, Crimea and the Caspian (Vol. II).] *Ophthalmidium* (Fig. 29): test free, ovate in outline, flattened; globular proloculus followed by spirally wound second chamber of a half to one whorl in length, later cham-bers regularly half a coil in length and may be loosely coiled with a flattened plate between whorls; aperture rounded to ovate, terminal. U.Trias.-Rec. Cosmop. *Spiroloculina* (Fig. 30): like *Ophthalmidium*, but without the early

32

cornuspirine stage and aperture with simple or bifid tooth; sides commonly flattened, outline lanceolate or fusiform. U.Cret.-Rec. Cosmop. *Discospirina* (Fig. 31): discoidal; proloculus followed by cyclogyrine-like early stage consisting of tubular chambers reducing from one and a half whorls to half a whorl at the end; chambers then become rapidly higher, flaring and flabelliform, and finally annular, with incomplete vertical partitions not reaching their roofs; a single row of apertural slits around periphery. M./U.Mio.-Rec. Eur., Medit., Atl. Superficially *Discospirina* is homoeomorphic with the Upper Cretaceous genus *Broeckina* [*Praesorites*].

Family MILIOLIDAE

Test free, forming a plane or winding spiral, usually with two chambers to a whorl, sometimes rectilinear or involute, and may be subdivided into chamberlets; aperture terminal, simple, or reduced in size by a spatulate or bifid tooth or by a cribriform plate. (Jur.-Rec.)

Quinqueloculina (Fig. 32): chambers half a coil in length, alternating regularly in five planes of coiling 72° apart, successive chambers in planes 144° apart; three chambers visible on one side of the test, four on the other; aperture terminal, flush, rounded, with a simple or short bifid tooth. Jur.-Rec. Cosmop.

A much broader view of this genus has been taken by some authors; several names have been sunk into its synonymy, and some rejected by the International Commission for Zoological Nomenclature so that the name *Quinqueloculina* could be retained. *Frumentarium* Fichtel and Moll, 1798, which has the same type species as *Quinqueloculina*, was suppressed in 1964 by I.C.Z.N. Opinion 692. However, there is a distinct group of quinqueloculine forms characterized by a much produced aperture which often has external longitudinal ridges; to this group the names *Retorta* Walker and Boys, 1784 (the juvenile of *Adelosina*, and suppressed by I.C.Z.N. Opinion 558), *Pollontes* Montford, 1808 (suppressed by I.C.Z.N. Opinion 692) and *Adelosina* d'Orbigny, 1826 apply, and it seems highly desirable now to revive the use of *Adelosina* as a genus and to regard the suppressed names *Retorta* and *Pollontes* as synonyms of it, not of *Quinqueloculina*. *Adelosina* seems to have a shorter range than *Quinqueloculina*, and is particularly common in Miocene and Pliocene deposits of the Mediterranean region. Again, the subgenus *Lachlanella* of *Quinqueloculina*, regarded as a synonym by some authors, seems worth retaining for forms in which the aperture is restricted by a very long bifid tooth.

Triloculina (Fig. 33): like *Quinqueloculina* in the early stages (at least in the microspheric form), later with successive chambers added in planes 120° apart, only three chambers being visible externally; aperture terminal, with a more or less bifid tooth. Jur.Rec. Cosmop. *Miliola* (Fig. 34): chamber arrangement quinqueloculine; aperture cribrate. Eoc. Eur., N.Amer. *Sigmoilina* (Fig. 35): ovate in outline, early stage with successive chambers added in planes 120° apart, the angle gradually changing to 180° in the adult, the changing plane of coiling forming a sigmoid curve; aperture terminal, rounded, with a tooth.

33

M.Eoc.-Rec. Cosmop. *Massilina* (Fig. 36): ovate in outline, rather flattened, early stage with chambers in quinqueloculine arrangement, later chambers added in single plane on alternate sides (as in *Spiroloculina*); aperture terminal, with bifid tooth. L.Cret.-Rec. Cosmop. *Hauerina* (Fig. 37): flattened, at first milioline, then planispiral with three or more chambers to the whorl; aperture cribate. Eoc.-Rec. Cosmop. *Pyrgo* [*Biloculina*] (Figs 1, 38): inflated, ovate; microspheric form with chambers added successively in quinqueloculine, triloculine and finally biloculine arrangement; megalospheric forms may be biloculine throughout; chambers about as wide as long, only two visible externally; aperture terminal, near the junction of the last two chambers, rounded to elongate, with a usually bifid tooth. Jur.-Rec. Cosmop. *Biloculinella:* like *Pyrgo*, but aperture nearly covered by a broad flap. Eoc.-Rec. Medit., Antarctic, N.Amer. *Articulina* (Fig. 39): early stage milioline, later rectilinear, sometimes with longitudinal costae; aperture simple, with everted margin. M.Eoc.-Rec. Cosmop.

So far the genera are recognizable by external form, but there are genera resembling one or other of these externally but internally having chambers subdivided by ingrowths producing a labyrinthine cavity; they belong to the subfamily Fabulariinae, and the three genera mentioned below all have apertures with a *trematophore* (**99, 100**).

Fabularia (Fig. 40): like *Pyrgo* in chamber arrangement; wall thick, chambers subdivided by secondary partitions; surface covered with fine wrinkles and punctations; aperture with trematophore. M.Eoc.-Rec. France, Africa, N.Amer. *Austrotrillina* (Fig. 41): chamber arrangement almost triloculine, new chambers adding layers against previous ones as well as the new chamber wall; wall thick, inner portion of outer part alveolar, floor simple; chamber cavities undivided; aperture terminal, with trematophore. Olig.-M.Mio. Austral., Pac., M.East, E.Africa, Medit. [Some of these forms were previously placed in the genus *Trillina* Munier-Chalmas, 1882, the type species of which, however, is a true *Triloculina*. Eocene records of *Austrotrillina* cannot be substantiated.] *Lacazinella* (Fig. 46): elongate-ovate in outline, circular in cross-section, each chamber completely enveloping test so that only one is visible externally, aperture of successive chambers being alternately at opposite ends of the test; chambers partially infilled by longitudinal internal ribs which more or less anastomose below apertural region; aperture terminal, with trematophore. Pal.-L.Olig. W.Pac., N.Africa, M.East. [The Palaeocene records are of unpublished species from North Africa and the Middle East; now known from the Palaeocene of Turkey. These Eocene and Lower Oligocene forms differ from the Senonian genus *Lacazina* Munier-Chalmas, 1882, to which they have previously been referred by some authors, in not being discoidal and in having an elongate axis.]

Family SORITIDAE [ORBITOLITIDAE, PENEROPLIDAE]

At first planispiral, afterwards serial, flabelliform or cyclical, the planispiral stage sometimes greatly shortened; interior simple or labyrinthic; aperture single and simple, dendritic or multiple. (U.Trias.-Rec.) (**49**)

34

The relation to the Miliolidae is shown by the presence of a spiral passage (Fig. 52) round the megalosphere. Neither the milioline nor the spiroloculine chamber arrangement is found in the Soritidae. In the Miliolidae, the genus *Hauerina* finishes with a planispiral stage, which seems to corroborate the close relationship of the two families. As demonstrated by Henson (**68, 69**), the degree of development of subepidermal partitions and of interseptal buttresses is of considerable importance in differentiating between genera in the family Soritidae.

Peneroplis (Fig. 42): compressed, planispirally coiled at first, with many chambers in a whorl, some species keeping this plan to the end, others tending to uncoil and flare; aperture a row of slits in a slight depression along the apertural face; no internal partitions in the chambers. ?U.Cret., Eoc.-Rec. Cosmop. *Dendritina* (Fig. 43): planispiral, nearly or completely involute; chambers not subdivided; aperture dendritic (narrow and branched). Eoc.-Rec. Eur., Carib., Africa, Atl. *Spirolina* (Fig. 44): at first planispiral and umbilicate, later uniserial and cylindrical; aperture terminal, round, margined by teeth. ?Pal., L. Eoc.-L./M.Mio. Cosmop. *Coscinospira:* placed in synonymy of *Spirolina* by some authors, but seems best retained as a separate genus since it has a cribrate aperture. M. Eoc.-Rec. Red Sea, Egypt, N.Amer. *Rhapydionina* (Fig. 45): elongate-subcylindrical, sometimes with an initial involute coil; chambers subdivided by transverse subepidermal partitions; aperture terminal, cribrate. M.Eoc.-U.Eoc. Istria, M.East, Carib. [The Jurassic genus *Haurania* Henson, 1948, placed in synonymy of *Rhapydionina* by some authors, has a wall structure of lituolid type, and should be regarded as a distinct homoeomorphic genus.] *Rhipidionina* (Fig. 47): compressed, flabelliform, early stage with very short, involute planispiral coil, later chambers uniserial, broad, numerous, with subepidermal partitions; aperture of numerous pores on last chamber face. M.Eoc. Istria, M.East *Praerhapydionina:* like *Rhapydionina*, but with a single terminal aperture. U.Cret.-early L.Mio. M.East, Cuba, Medit., Africa, Papua. [The species *P. delicata* is known from the early part of the Lower Miocene of Malta, Libya, East Africa, Iran, Iraq and Papua as well as from the middle and late Oligocene of the Middle East.] *Archaias* (Fig. 48): compressed, early stage planispiral and involute, later chambers flaring, becoming evolute; chambers with interseptal buttresses; aperture multiple, a double row of pores on peripheral face of last chamber. M.Eoc.-Rec. Cosmop. *Sorites* (Figs. 49, 50): discoidal, early stage with proloculus followed by tubular enrolled second chamber of nearly one coil, later chambers added simultaneously in flaring peneropline series, finally becoming annular; apertures connecting each chamber with two in the preceding and two in the succeeding series (hence no communication between chambers of single series). Mio.-Rec. Carib., Medit., Red Sea, Pac., Atl., S.Amer. *Amphisorus* (Fig. 50): like *Sorites*, but later cyclical chambers with two layers of chambers alternating in position with those of previous cycle, and those of same cycle alternating in position when viewed from periphery; apertures in two alternating rows on periphery. Mio.-Rec. Carib., N.Amer., S.Amer., Pac., Medit., Eur. *Marginopora* (Fig. 50): discoidal, biconcave, with distinct second

35

tubular chamber similar to that of *Sorites* and *Amphisorus*; annular canals connecting chambers as in *Amphisorus*, with one at each side of main chamber in all except first one or two cyclical chambers, primary chambers with intercameral foramina, smaller lateral chambers not interconnected, but with foramina leading into main chambers; the *Amphisorus*-like stage which follows the early *Sorites*-like stage is itself followed by an *Orbitolites*-like (but less regular) stage lacking the engine-turned appearance of *Orbitolites*; numerous apertural pores irregularly arranged on periphery. Mio.-Rec. Pac., Atl. **Hensoniella** (Fig. 51): spiral-pseudevolute, becoming flabelliform or cyclical, never with meandriform lateral chambers; subepidermal partitions. U.Olig.-M.Mio. M.East. **Pseudotaberina** (Fig. 52): lenticular, spiral-involute, becoming flaring pseudevolute, not cyclical; interseptal pillars and primary subepidermal partitions. M.Mio. India, M.East, E.Africa, W.Pac. **Pseudorbitolites:** discoidal-cyclical, with vestigial initial spiral stage; anastomosing interseptal pillars; primary and secondar subepidermal partitions. Pal. M.East. **Orbitolites** (Fig. 53): discoidal, similar to *Sorites*, gently biconcave; megalospheric form with large subglobular multilocular nucleoconch surrounded by numerous chambers in annual series, chambers in successive alternate series, pores connecting chambers with preceding and succeeding chambers only, there being no foramina connecting chambers of the same series; equatorial section with a readily recognizable 'engine-turned' appearance; numerous apertures in transverse rows across the periphery. L.Eoc.-basal U.Eoc. Eur., Medit., M.East, Asia, E.Africa. [It does not seem Possible to substantiate Palaeocene records of this genus; they evidently refer to specimens from the Ilerdian stage in Spain, which was originally referred

---

Figs 46–61. FORAMINIFERIDA: MILIOLACEA AND NODOSARIACEA
Figs 48, 50 and 60 after Carpenter; Figs 49 and 53 after H. Douvillé; Fig. 56 after van der Vlerk and Umbgrove; Figs 46, 47 and 61 after Loeblich and Tappan, in Moore; Figs 55 and 58 after Reichel, in Moore; Figs 51, 52, 54, 57 and 59 original.
46h, i. *Lacazinella wichmanni* (Schlumberger), U.Eoc. W.Pac. T. × 20.
47a, b. *Rhipidionina liburnica* (Stache), M.Eoc. Istria. T. a enlarged, b × 5.
48b. *Archaias aduncus* (Fichtel and Moll), Rec. × 12.
49f. Diagram of *Sorites* pattern. (Cyclical chamber walls are shown as straight, instead of curved as parts of concentric circles.)
50h. Radial section of *Marginopora* (diagrammatic) showing changes in plan from within outwards.
51i. *Hensoniella anahensis* (Henson), L.Mio. Iraq. T. × 30.
52f(A), f(B), k. *Pseudotaberina malabarica* (Carpenter), M.Mio. Ceylon. T. f(A) and k × 22·5, f(B) × 15.
53f. Diagram of *Orbitolites* pattern. (Cyclical chamber walls are shown as straight, instead of curved as parts of concentric circles.)
54. *Yaberinella jamaicensis* Vaughan, M.Eoc. Jamaica. T. × 9.
55g. *Borelis melo* (Fichtel and Moll) *curdica* (Reichel), Mio. Turkey. × 42·5.
56g. Diagram of structure of *Fasciolites*.
57. *Fasciolites ovoidea* (d'Orbigny) *globosa* Leymerie, L.Eoc. Baluchistan. × 4.
58g. *Bullalveolina bulloides* (d'Orbigny), L.Olig. France. T. × 40.
59f. Diagram of section of *Flosculinella*.
60f. Diagram of section of *Alveolinella*.
61a, b. *Nodosaria radicula* (Linné), U.Plio. Italy. T. × 15,

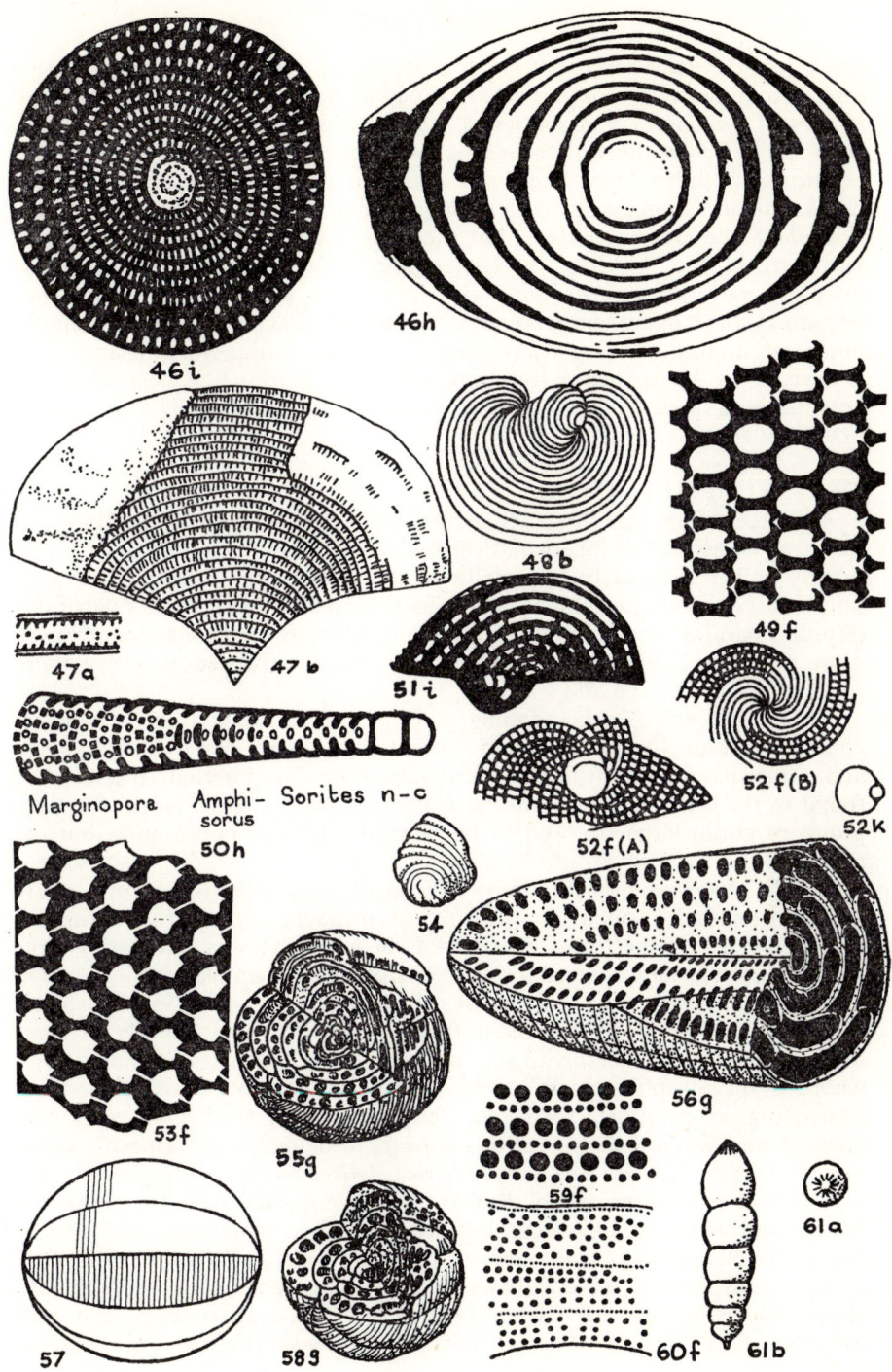

46h

46i

47a

47b

48b

49f

51i

52f(B)

52k

Marginopora   Amphi-   Sorites n-c
sorus

50h

52f(A)

54

53f

55g

56g

57

58g

59f

60f

61a

61b

37

entirely to the Palaeocene, but much of its higher part (overlying the Palaeocene planktonic microforaminiferal faunas of its lower part) contains large species of *Fasciolites* and species of *Nummulites* which have nowehere else been found below the Lower Eocene (see Vol. II).] ***Opertorbitolites*** has an equatorial chamber layer like *Orbitolites*, but possesses thick, imperforate, non-vacuolar laminae on either side. L.Eoc. Pakistan, India, M.East. ***Somalina:*** like *Opertorbiotlites*, but the thick, imperforate lateral layers are vacuolar. M.Eoc. E.Africa, M.East. ***Yaberinella*** (Fig. 54): operculine to discoidal, early stage peneropline with very broad, low chambers which may become cyclical; chambers subdivided into secondary chamberlets, the septula numerous and intersecting at low angles to form a lattice-work; chamberlets connected by intercameral foramina occurring in three planes; apertures apparently a series of pores on apertural face. M.Eoc.-U.Eoc. Carib. (?Miliolidae.)

Family ALVEOLINIDAE

Test free, often quite large, coiled about an elongate axis, subcylindrical, fusiform, ellipsoidal or spherical. Proloculus followed by spiral tube. Juveniles usually coiled irregularly, especially in the microspheric form. Chambers numerous, divided into tubular chamberlets by means of secondary partitions (septula) parallel to direction of coiling, chamberlets sometimes occurring in more than one layer; apertures numerous, usually arranged in one or more rows. (L.Cret.-Rec.) (**6, 88**)

This is the only Tertiary family of foraminifera dominantly with spiral coiling about an elongate axis, and consequent fusiform shape. Similar shape and somewhat similar (though often more complex) internal structure are found in the Carboniferous and Permian Fusulinacea, the chief pre-Tertiary group to attain large size and to be of stratigraphical importance, but the relationship does not seem to be close.

Borelis [*Neoalveolina*] (Fig. 55): very small, spheroidal to fusiform, early whorls irregularly coiled; no post-septal passage; septula in continuous arrangement, in some forms chamberlets of same chamber alternately large and small, the latter displaced toward exterior, so that septula are Y-shaped in axial section. U.Eoc.-Rec. Asia, Africa, Eur. ***Fasciolites*** [*Alveolina, Flosculina*] (Figs. 56, 57): spherical, ellipsoidal, fusiform or cylindrical; septula alternating in adjacent chambers, with pre- and post-septal passages; coiling of first whorls of megalospheric form regular (*Fasciolites s.s.*); two rows of apertures alternating in position; there may be great basal thickening of several internal whorls (*Flosculina*), but this appears to be merely intraspecific variation. Pal.-basal U.Eoc. Eur., Asia, Africa. [*Alveolina* is here regarded as a synonym of *Fasciolites*, not of *Borelis*, since the quoted type designation of Parker and Jones in 1860 does not seem to be a valid one—they merely referred to all species as synonyms of '*Nautilus melo*'.] ***Bullalveolina*** (Fig. 58): minute, subspherical; several rows of apertures, upper ones opening into alveoli which occupy rear part of chambers; septula alternating; first coils irregular. Olig. France, Italy, M.East. ***Flosculinella*** (Fig. 59): globular to

38

fusiform, septula alternating; apertures in two rows, upper ones small, leading into narrow chamberlets; early whorls irregularly coiled. L.Mio.-M.Mio. E.Indies, Austral., E.Africa. *Alveolinella* (Fig. 60); fusiform, elongate, septa continuously arranged; several layers of chamberlets; preseptal passages on floor of chambers; apertures in several rows. M. Mio.-Rec. Indo-Pac.

Suborder ROTALIINA

Wall calcareous, perforate. (Perm.-Rec.)

Superfamily NODOSARIACEA

Wall finely perforate, of radial laminated calcite; chambers planispirally coiled or uncoiled, or straight, or coiled about longitudinal axis; aperture terminal or peripheral, usually radiate, sometimes slit-like or rounded. (Perm.-Rec.)

Family NODOSARIIDAE

Free, of one or more chambers in planispiral, biserial, uncoiling, curved or straight series; aperture simple, slit-like or radiate, peripheral in coiled forms, terminal in straight forms, and may have an elongate neck. (Perm.-Rec.)

Nodosaria (Fig. 61): multilocular, rectilinear, rounded in cross-section, smooth or variously ornamented; aperture terminal, central, radiate. Perm.-Rec. Cosmop. *Chrysalogonium:* like *Nodosaria*, but with a cribrate aperture. U.Cret.-Rec. Pac., N.Amer., Carib., Atl., Eur. *Dentalina* (Fig. 62): like *Nodosaria*, but arcuate and with oblique septa; aperture radiate, usually eccentric. Perm.-Rec. Cosmop. *Lingulina* (Fig. 63): uniserial, stubby, compressed; aperture a long terminal slit in plane of compression. Perm.-Rec. Cosmop. *Marginulina* (Fig. 64): early portion slightly coiled but not enrolled, later rectilinear; sutures oblique, especially in early portion; aperture on dorsal angle. Trias.-Rec. Cosmop. *Vaginulina* (Fig. 65): straight to arcuate like *Dentalina*, initially sometimes slightly coiled, compressed in cross-section; aperture radiate, at dorsal margin. Trias.-Rec. Cosmop. *Lenticulina* [*Cristellaria* in part, *Robulus, Robulina*] (Figs. 66, 67): planispiral, lenticular, biumbonate, periphery sharp, smooth or ornamented; aperture radiate, at peripheral angle, sometimes (*Robulus*) with an additional slit-like aperture just below the radiate one. Trias.-Rec. Cosmop. *Saracenaria:* planispiral in early stage, later uncoiling; triangular in cross-section; aperture radiate, at peripheral angle. Jur.-Rec. Cosmop. *Frondicularia* (Fig. 68): elongate or palmate, flattened, both microspheric and megalospheric forms with low, equitant chambers throughout, sutures strongly arched or angled along midline; aperture terminal, radiate. Perm.-Rec. Cosmop. [Some species attain a large size, e.g. *F. alata*, Mio.-Rec., Italy etc., may be over 1 cm. long by 8 mm. wide.] *Neoflabellina* [*Flabellina*]: like *Frondicularia*, but early portion coiled in microspheric forms and arcuate in megalospheric forms. U.Cret.-Pal. Cosmop. *Lagena* (Fig. 69): unilocular, flask-shaped; aperture at end of a long neck which may have a phialine lip; may be variously ornamented. Jur.-Rec. Cosmop. *Parafrondicularia* (Fig. 70): elongate, compressed, early stage biserial, later uniserial with somewhat arched chambers; aperture terminal,

39

radiate. L.Eoc.-Rec. Cosmop. *Citharina* (Fig. 71): flattened, rather triangular in outline, sometimes keeled; chambers broad, extending well down inner side; aperture radiate, at outer margin. L.Jur.-Pal. Cosmop. *Planularia* (Fig. 72): flattened, ovate-triangular in outline, margins carinate; chambers broad, extending far down inner margin; aperture radiate, at outer margin. Mio.-Rec. Cosmop. *Pseudonodosaria* (Fig. 73): uniserial, rectilinear, chambers usually embracing strongly, later ones may be more inflated; sutures horizontal; aperture terminal, radiate. Perm.-Rec. Eur., Asia, Pac., Atl., N.Amer. [Some forms placed in this latter genus might be better placed in the Polymorphinidae.]

Family POLYMORPHINIDAE

Chambers in spiral or sigmoidal coil about longitudinal axis, sometimes

---

Figs 62–95. FORAMINIFERIDA: NODOSARIACEA AND
BULIMINACEA

Figs 66 and 67 after von Hantken; Figs 62, 65, 68, 69, 74, 82 and 89 after Brady; Figs 63, 64, 70–73, 75–81, 83–88 and 90–95 after Loeblich and Tappan, in Moore.

62a, b. *Dentalina communis* d'Orbigny, Rec. × 28.
63a, b. *Lingulina carinata* d'Orbigny, Rec. Medit. T. × 5.
64b. *Marginulina glabra* d'Orbigny, Plio. Italy. T. × 30.
65b. *Vaginulina legumen* (Linné), Rec. T. × 22·5.
66a, b. *Lenticulina princeps* (Reuss), U.Eoc. Hungary. × 12.
67a, b. *Lenticulina depauperatus* (Reuss), U.Eoc. Hungary. × 15.
68a, b. *Frondicularia alata* d'Orbigny, Rec. × 7·5.
69b. *Lagena sulcata* (Walker and Jacob), Rec. T. × 45.
70a, b. *Parafrondicularia japonica* Asano, Plio. Japan. T. × 24.
71b. *Citharina strigillata* (Reuss), Cret. Bohemia. T. Enlarged.
72b. *Planularia auris* (Defrance, in de Blainville), Plio. Italy. T. × 8·25.
73b. *Pseudonodosaria discreta* (Reuss), U.Tert. Java. T. × 31.
74a, b. *Polymorphina ovata* d'Orbigny, Rec. × 19.
75b. *Globulina gibba* (d'Orbigny, in de la Sagra), Mio. Austria. T. × 22·5.
76b. *Guttulina communis* (d'Orbigny), Plio. Italy. T. Enlarged.
77b. *Pyrulina gutta* (d'Orbigny), Plio. Italy. T. Enlarged.
78. *Webbinella hemisphaerica* (Jones, Parker and Brady), Plio. England. × 24.
79b. *Ramulina laevis* Jones, in Wright, U.Cret. Eire. × 8·5.
80b. *Glandulina laevigata* (d'Orbigny), Rec. Canada. T. × 24·5.
81a, b. *Tristix reesidei* Loeblich and Tappan, Jur. U.S.A. × 24.
82b. *Oolina lineata* (Williamson), Rec. × 45.
83a, b. *Fissurina marginata* (Montagu), Rec. Alaska. × 37·5.
84b. *Turrilina alsatica* Andreae, M.Olig. France. T. × 78·3.
85b. *Buliminoides williamsoniana* (Brady), Rec. Fiji. T. × 47.
86b, c. *Sphaeroidina bulloides* d'Orbigny, Rec. Italy. T. Diagrammatic.
87b, d. *Bolivinita quadrilatera* (Schwager), Rec. Philippines. T. × 32·5.
88a, b. *Bolivina plicata* d'Orbigny, Rec. Panama. T. × 49·5.
89a, b. *Brizalina hantkeniana* (Brady), Rec. × 30.
90a, b. *Loxostomoides applinae* (Plummer), Pal. Texas. T. × 25.
91a, b. *Rectobolivina bifrons* (Brady), Rec. Pac. T. × 32·5.
92a, b. *Tappanina selmensis* (Cushman), U.Cret. Tennessee. T. × 65.
93b, c. *Islandiella islandica* (Nørvang), Rec. Iceland. T. × 16·5.
94b. *Eouvigerina aculeata* (Ehrenberg), U.Cret. Texas. T. × 81.
95b. *Siphonodosaria abyssorum* (Brady), Rec. S.Pac. T. × 11.

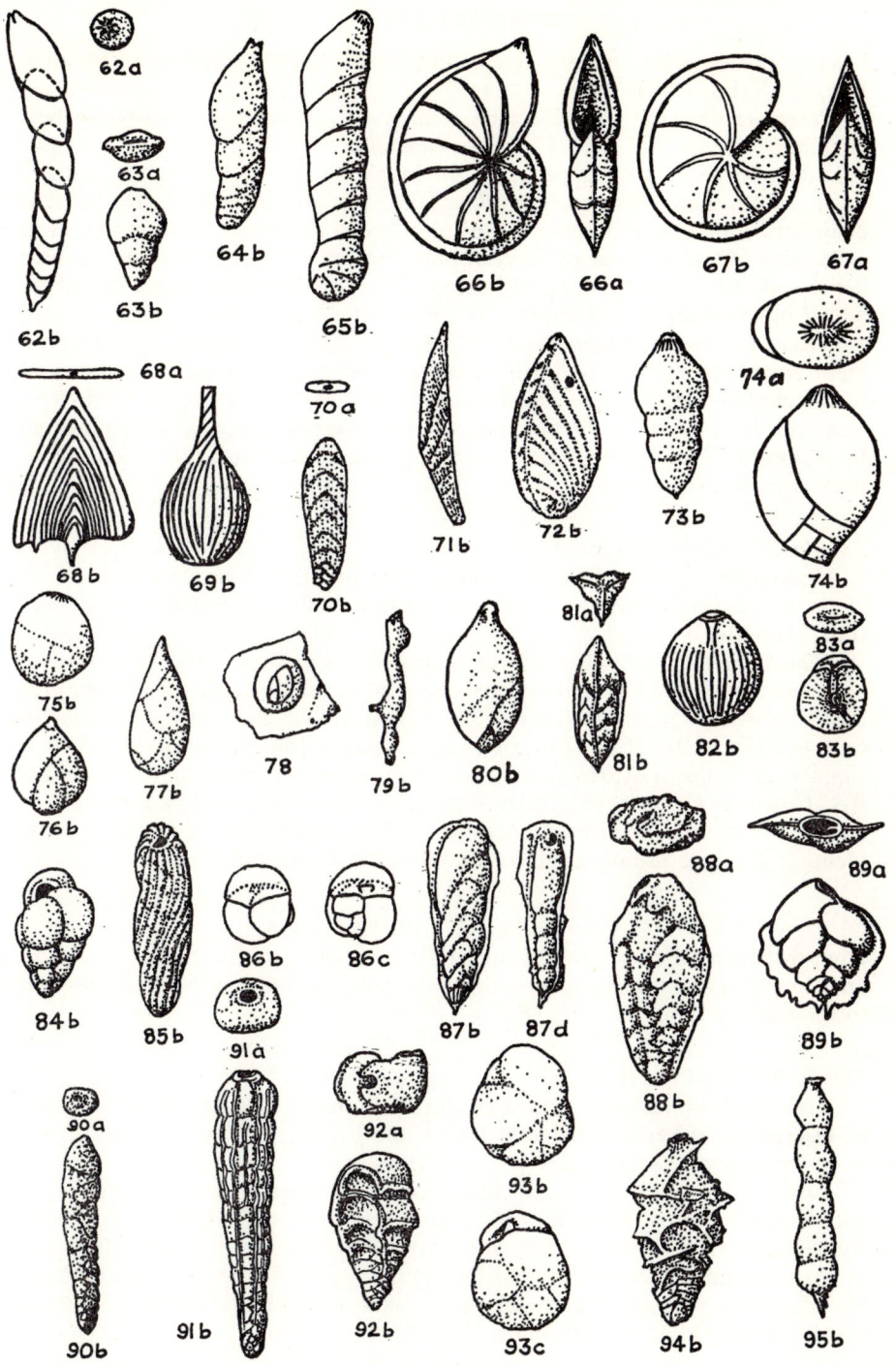

62a
62b
63a
63b
64b
65b
66b
66a
67b
67a
68a
68b
69b
70a
70b
71b
72b
73b
74a
74b
75b
76b
77b
78
79b
80b
81a
81b
82b
83a
83b
84b
85b
86b
86c
87b
87d
88a
88b
89a
89b
90a
90b
91a
91b
92a
92b
93b
93c
94b
95b

41

biserial or uniserial; mostly free, sometimes attached; aperture terminal, radiate. (Trias.-Rec.).

Polymorphina (Fig. 74): elongate, somewhat compressed, usually twisted; chambers biserial. Pal.-Rec. Cosmop. *Globulina* (Fig. 75): globular to ovate; chambers strongly overlapping, added in planes about 144° apart, sutures flush. U.Jur.-Rec. Cosmop. *Guttulina* (Fig. 76): ovate to elongate; inflated chambers added in quinqueloculine spiral series (in planes 144° apart), each successive chamber strongly overlapping but extending further up; sutures depressed. Jur.-Rec. Cosmop. *Pyrulina* (Fig. 77): fusiform; early chambers added in spiral series and about 120° apart, later stage biserial; sutures flush. Jur.-Rec. Cosmop. *Webbinella* (Fig. 78): the only attached polymorphinid mentioned here; early polymorphine or pyruline stage surrounded by a flange-like chamber attached to substratum; aperture not visible. L.Cret.-Rec. Cosmop. *Ramulina* (Fig. 79): globular or irregular chambers loosely connected by stolon-like necks; sometimes branching; aperture rounded, at open end of neck. Jur.-Rec. Cosmop.

Family GLANDULINIDAE

Unilocular, or with biserial, uniserial or polymorphine arrangement of chambers; aperture terminal, radial or slit-like, with an internal entosolenian tube. (Jur.-Rec.)

Glandulina (Fig. 80): elongate, circular in cross-section, early stage biserial, later uniserial; chambers strongly overlap and increase rapidly in size; sutures flush. Pal.-Rec. Cosmop. *Tristix* (Fig. 81): uniserial, normally triangular in cross-section. L.Jur.-Eoc. Eur., N.Amer. *Oolina* [*Entosolenia*] (Fig. 82): a singular globular to ovate chamber, often ornamented; aperture radiate. Jur.-Rec. Cosmop. *Fissurina* (Fig. 83): like *Oolina*, but compressed and trigonal or tetragonal in cross-section, sometimes keeled, sometimes beautifully ornamented; aperture usually transversely elongate. Cret.-Rec. Cosmop.

Superfamily BULIMINACEA

High trochospiral or biserial or uniserial; aperture a primary basal slit, or in apertural face, or terminal, sometimes on a neck, and may have an internal tooth-plate or tube. (U.Trias.-Rec.)

Family TURRILINIDAE

High trochospiral with more than three chambers to a whorl, or modified to biserial or uniserial; apertural face without pores, formed by outgrowth from tooth-plate. (M.Jur.-Rec.)

Turrilina (Fig. 84): elongate, high-spired, with three or more chambers per whorl; aperture a small basal arch in last chamber face. L.Eoc.-U.Olig. Eur. *Buliminoides* (Fig. 85): elongate, early stage a low trochospiral coil, then spire increases rapidly in height and coiling round an open umbilicus, with about five chambers per whorl and oblique to axis; longitudinal costae; aperture umbilical, with simple tooth-plate. Plio.-Rec. Indo-Pac., W.Atl.

Family SPHAEROIDINIDAE

Early stage trochospiral, later streptospiral and with chambers largely embracing previous ones; aperture interio-marginal, with rounded tooth, sometimes with later secondary sutural openings. (U.Cret.-Rec.)

Sphaeroidina (Fig. 86): subglobular, coiling variable; chambers hemispherical, few, each placed centrally above previous aperture; aperture a crescentic slit near suture, with a lip, and also may have a simple or bifid tooth; no secondary sutural openings. U.Eoc.-Rec. Cosmop.

Family BOLIVINITIDAE

Biserial, at least in early part; aperture comma-shaped, parallel to compression of test, basal or terminal, with internal tooth-plate. (U.Trias.-Rec.)

Bolivinita (Fig. 87): compressed, rectangular in cross-section, angles carinate; biserial; not retral processes; aperture basal, short-elliptical, with bordering lip. Mio.-Rec. Cosmop. *Bolivina* (Fig. 88): elongate, biserial, basal margins of chambers with retral processes; smooth, striate, or costate, sometimes with marginal keel or flange; aperture a long, narrow loop. U.Cret.-Rec. Cosmop. *Brizalina* (Fig. 89): like *Bolivina*, but without retral processes; sometimes flanged. U.Trias.-Rec. Cosmop. *Loxostomoides* (Fig. 90): like *Bolivina*, but oval in cross-section and with retral processes or crenulations at base of chambers; tends to become uniserial in later stages. U.Cret.-Pal. N.Amer., M.East. *Rectobolivina* (Fig. 91): elongate, subcircular in cross-section, early stage biserial (much reduced in megalospheric form), later uniserial and with terminal, rounded or elongate aperture. M.Eoc.-Rec. Cosmop. *Tappanina* (Fig. 92): biserial, slightly flaring, quadrate in cross-section; chambers cuneiform, concave on broad sides and convex laterally; sutures sunk, margined by ridges which continue down median zigzag line. U.Cret.-Pal. N.Amer., Eur.

Family ISLANDIELLIDAE

Biserially arranged enrolled chambers at least in early stage, may uncoil at late stage; like Cassidulinidae, but with calcareous, perforate, radiate fibrous wall and primary aperture with internal tooth-plate. (?U.Cret., Pal.-Rec.)

Islandiella (Fig. 93): like *Cassidulina* in morphology, but aperture (elongate, interiomarginal) with internal tooth-plate. ?U.Cret. Pal.-Rec. Cosmop. *Cassidulinoides:* early stage morphologically like *Cassidulina*, later stage uncoiling but remaining biserial; aperture in adult loop-shaped, extending up from base of chamber to a rounded terminal opening. U.Eoc.-Rec. Cosmop.

Family EOUVIGERINIDAE

Biserial, biserial becoming uniserial, or uniserial; aperture terminal, with internal tube, and may have phialine lip. (L.Cret.-Rec.)

Eouvigerina (Fig. 94): biserial, sutures depressed; smooth, carinate or hispid; aperture terminal, with neck and phialine lip, usually with crenulated margin. L.Cret.-U.Eoc. N.Amer., Eur., N.Z., M.East. *Siphonodosaria* (Fig. 95): elongate, uniserial, straight or curved; chambers inflated; aperture

43

rounded, on a neck with a phialine lip, with teeth projecting inwards. Eoc.-Rec. Cosmop.

Family BULIMINIDAE

High trochospiral, with three or two chambers to the whorl; aperture a loop in apertural face, with internal tooth, or may be cribrate. (Pal.-Rec.)

Bulimina (Fig. 96): triserial, sometimes tending toward uniserial in later stage; aperture a loop extending up from base of chamber face, sometimes with a rim. Pal-Rec. Cosmop. *Pavonina* (Fig. 97): Small triserial stage initially, then biserial, finally uniserial and flabelliform with chambers arched well down; aperture cribrate. Mio.-Rec. Indo-Pac., Atl., Africa, N.Amer. *Reussella* (Fig. 98): triserial, triangular throughout, with carinate margins; aperture basal. M.Eoc.-Rec. Cosmop.

Family UVIGERINIDAE

Triserial to biserial in early stage, may become biserial or uniserial later; aperture terminal, with neck and internal tooth-plate. (U.Cret.Rec.)

Uvigerina (Fig. 99): triserial, rounded in cross-section, with sunken sutures; smooth, hispid or costate; aperture terminal, round, on a neck which may have a phialine lip. L.Eoc.-Rec. Cosmop. *Hopkinsina:* like *Uvigerina*, but later stage becomes biserial. L.Eoc.-Rec. N.Amer., Eur. *Rectuvigerina* (Fig. 100): triserial initially, becoming uniserial in adult. M.Eoc.-Rec. Cosmop. *Siphogenerina:* like *Rectuvigerina*, but biserial initially, uniserial later. L.Eoc.-Rec. Cosmop. *Trifarina* (Fig. 101); triangular in cross-section, triserial, later chambers more loosely appressed and tending to become uniserial; aperture terminal, ovate, on a neck with a thickened rim. L.Eoc.-Rec. Cosmop.

Figs 96–113. FORAMINIFERIDA: BULIMINACEA, DISCORBACEA, SPIRILLINACEA AND ROTALIACEA

Figs 96 and 97 after Brady; Figs 98–111 after Loeblich and Tappan, in Moore; Fig. 112 after L. M. Davies; Fig. 113 after Macfadyen.

96a, b. *Bulimina marginata* d'Orbigny, Rec. T. × 60.
97b. *Pavonina flabelliformis* d'Orbigny, Rec. T. × 26.
98b. *Reussella spinulosa* (Reuss), Mio. Austria. T. × 50.
99b. *Uvigerina pigmea* d'Orbigny, Plio. Italy. T. × 47.
100b. *Rectuvigerina multicostata* (Cushman and Jarvis), Mio. Trinidad. T. × 22.
101a, b. *Trifarina bradyi* Cushman, Rec. Indonesia. T. ×. 47.
102a, b. c. *Discorbis colliculus* (Bandy), Eoc. Oregon. × 37.
103b, c. *Laticarinina pauperata* (Parker and Jones), Rec. Carib. T. × 9·5.
104b, c. *Baggina californica* Cushman, Mio. Calif. T. × 28.
105a, b, c. *Cancris auriculus* (Fichtel and Moll), Plio. Italy. T. × 22·5.
106a, b, c. *Valvulineria californica* Cushman, Mio. Calif. T. × 24·5.
107a, b, c. *Siphonina reticulata* (Cžjžek), Mio. Austria. T. × 47 (approx.).
108a, b, c. *Asterigerina carinata* d'Orbigny, in de la Sagra, Rec. Carib. T. × 30.
109a, b, c. *Epistomaria rimosa* (Parker and Jones), M.Eoc. France. T. × 20.
110b, c. *Spirillina vivipara* Ehrenberg, Rec. Off Florida. T. × 75.
111a, b. *Patellina corrugata* Williamson, Rec. Canada. T. × 50.
112 h, l. Diagrams of *Rotalia*.
113a, b, c. *Ammonia beccarii* (Linné), Pleist. Norfolk. T. × 34.

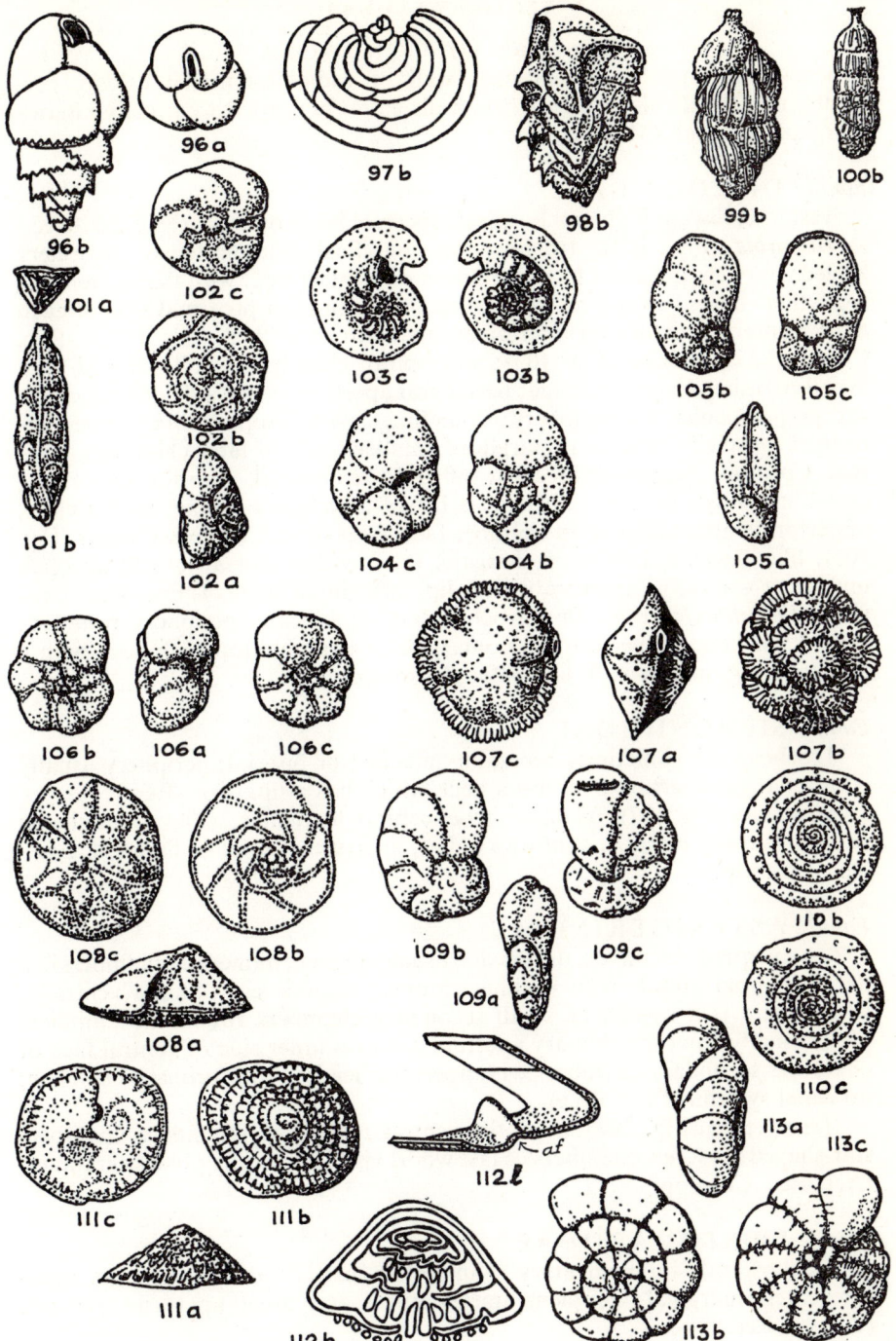

Superfamily DISCORBACEA

Trochospiral, sometimes modified; wall of radial laminated calcite, perforate, not canaliculate, with single walls and septa; aperture usually interiomarginal or areal. (M.Trias.-Rec.)

Family DISCORBIDAE

Test free, low- or high-trochospiral; aperture basal or areal. (M.Trias.-Rec.)
Discorbis (Fig. 102): trochospiral, umbilical side flattened, periphery keeled; only chambers of last whorl seen on ventral side, with flaps extending from their bases toward centre leaving an opening on proximal side of each flap; primary aperture an interiomarginal, extraumbilical arch. L.Eoc.-Rec. Cosmop. **Laticarinina** (Fig. 103): planispiral, chambers saddle-shaped with broad peripheral keel or flange; peripheral aperture at one side of keel, a low slit perpendicular to periphery; sometimes with supplementary openings beneath posterior umbilical margin of smaller lobes of later chambers. Pal.-Rec. Cosmop. **Baggina** (Fig. 104): inflated-trochospiral, with few chambers; closed umbilicus on base; aperture a broad umbilical opening, with a clear, imperforate area above it on chamber face. Cret.-Rec. Cosmop. **Cancris** (Fig. 105): like *Baggina*, but more elongate, fully evolute dorsally, with an open umbilicus ventrally; aperture with a lip; margin carinate. L.Eoc.-Rec. Cosmop. **Valvulineria** (Fig. 106): trochospiral, umbilicate, periphery rounded; sutures thickened; aperture interiomarginal, extraumbilical-umbilical, with a flap covering umbilicus. L.Cret.-Rec. Cosmop.

Family SIPHONINIDAE

Trochospiral, sometimes becoming uncoiled or biserial; periphery usually with a flange; aperture oval, on a neck which has a lip. (Eoc.-Rec.)
Siphonina (Fig. 107): biconvex, trochospiral, lenticular, without umbilicus; sutures dorsally oblique, ventrally radial; aperture at level of flange. L.Eoc.-Rec. Cosmop.

Family ASTERIGERINIDAE

Test unequally biconvex, dorsal side usually higher; numerous chambers in a flat, turbinoid spiral, with oblique sutures; sutures less oblique ventrally, where there is a rosette of small secondary chambers round the umbilical plug; usually smooth; primary aperture a slit on inner side of ventral face of chamber; secondary chambers with apertures leading into primary chamber; no canal system. (Cret.-Rec.)
Asterigerina (Fig. 108): smooth; ventral rosette of secondary chambers star-shaped; first few chambers of last whorl visible ventrally possess granules. Cret.-Rec. Cosmop.

Family EPISTOMARIIDAE

Trochospiral; supplementary chamberlets on ventral side; primary aperture interiomarginal; supplementary sutural and areal apertures present. (U.Cret.-Rec.)

Epistomaria (Fig. 109): biconvex, early whorls visible dorsally, chambers enlarging rather rapidly, sutures depressed, curved; complex series of internal partitions causing appearance of supplementary chambers around umbilicus; primary aperture a low interiomarginal slit extending from periphery nearly to umbilicus, a second aperture in last chamber face, and a series of slit-like accessory apertures parallel to periphery ventrally. L.Eoc.-U.Eoc. France.

Superfamily SPIRILLINACEA

Planispiral to conical; simple forms with an enrolled, tubular second chamber, non-septate or with septa in later stages; advanced forms with septa throughout, becoming biserial or even annular; wall perforate, calcareous. (?Trias., Jur.-Rec.)

Family SPIRILLINIDAE

Non-septate enrolled tubular second chamber, may become septate in later stages, biserial, or terminally annular; wall optically a single calcite crystal. (?Trias., Jur.-Rec.)

Spirillina (Fig. 110): free, more or less planispiral; second chamber undivided, close-coiled; aperture terminal, peripheral, somewhat crescentic. ?Trias., Jur.-Rec. Cosmop. *Patellina* (Fig. 111): conical, ventral side flat and involute; second chamber of a few whorls, later stages biserial; numerous incomplete secondary septa of one or two orders, not crossing umbilical portion of chambers; aperture a low arch under sigmoid edge of median septum of last chamber in median portion of base. L.Cret.-Rec. Cosmop.

Superfamily ROTALIACEA

Canaliculate, double walls and septa of radial laminated calcite; no primary aperture or large pores, or with pores on apertural face or elsewhere; interiomarginal intercameral foramina sometimes present. (U.Cret.-Rec.) (**19**)

Family ROTALIIDAE

Trochospiral; radial canals or fissures and intraseptal and subsutural canals present. (U.Cret.-Rec.)

Rotalia (Fig. 112): lenticular to planoconvex, all whorls visible dorsally; spire multilocular, single; chambers simple, septa double; wall coarsely perforate; dorsally smooth; ventral side with plug split by anastomosing fissures into numerous tubercles and pillars that crowd central portion of test and are not continuous from one whorl to the next; umbilical canal beneath cortical chamber layer receiving tributary canals from umbilical slit-like apertures at inner side of chambers. U.Cret.-Rec. Cosmop. [L. M. Davies (**45**) has given a good account of the genera *Rotalia*, *Lockhartia* and *Dictyoconoides*, but it should be noted that *Lockhartia newboldi*, which is not the type species of *Lockhartia*, is the juvenile of *Rotalia trochidiformis*, which is the type species of *Rotalia*. See also Smout (**106**).] *Ammonia* [*Streblus*] (Fig. 113): biconvex; sutures thickened, depressed on ventral side; septa double; wall finely perforate; young stage with open umbilical fissures and plug which in adult is

47

broken up into numerous fused pillars and bosses which extend upward to proloculus; no umbilical canal; umbilical area and ventral sutures usually with granules. Eoc.-Rec. Cosmop. [Although Loeblich and Tappan, in Moore (**78**), give the range as Miocene to recent, an unpublished species of *Ammonia* occurs together with *Nummulites hormoensis* in a sample from East Africa, indicating that the genus ranges down at least to the Upper Eocene; evidence from elsewhere suggests that it may range down to the Palaeocene.] *Dictyoconoides* (Fig. 114): low- to high-conical, chambers on dorsal side arranged in multiple spiral beneath a thin imperforate lamina; ventral side with radiating pillars extending down from apex, the intervals between them about the same size and divided by horizontal partitions; septa double; ventral side with granules; aperture multiple, consisting of pores between pillars on ventral side. Pal.-M.Eoc. Asia, M.East, Africa. *Lockhartia* (Fig. 115): like *Dictyoconoides*, but smaller; dorsal chambers in a simple spiral, aperture an interiomarginal slit; pillars in umbilical region irregular in some species, the granules on ventral surface partly anastomosing. Pal.-M.Eoc. Asia, M.East, Africa, Carib. (**45**). [Although the wall of the type species, *L. haimei*, is coarsely perforate, that of some other species—e.g. *L. conditi* and *L. tipperi*—is only finely perforate.] *Dictyokathina* (Fig. 116): multispiral dorsal side as in *Dictyoconoides*, and umbilical region with strong radial vertical canals as in *Kathina*; wall finely perforate; intercameral foramen an interiomarginal slit. Pal. M.East, India. *Kathina* (Fig. 117): like *Dictyoconoides*, but dorsal chambers in simple spire; ventral side may have central plug with strong vertical canals; chambers without the umbilical extensions present in *Lockhartia* and *Sakesaria*. U.Cret.-Pal. M.East, E.Africa, Carib. *Sakesaria* (Fig. 118): like *Lockhartia*, but axis of coiling elongate, whorls more numerous,

---

Figs 114–130. FORAMINIFERIDA: ROTALIACEA
Figs 114, 115 after L. M. Davies; Figs 116–121, 122b, c, 123–129 after Loeblich and Tappan, in Moore; Figs 130f, h after Cole, in Moore; Fig. 122f after Carpenter; Fig. 126 after Brady; Fig. 130b after Nuttall.
114c, h. *Dictyoconoides vredenburgi* (L. M. Davies), L.Eoc. Hindu Bagh. × 7·5.
115h, 1. Diagrams of *Lockhartia*.
116b, h. *Dictyokathina simplex* Smout, Pal. Qatar. T. × 12·5.
117h. *Kathina delseota* Smout, Pal. Qatar. T. × 12·5.
118h. *Sakesaria cotteri* L. M. Davies, in Davies and Pinfold, L.Eoc. Qatar. T. × 8·3.
119a, b. *Cuvillierina vallensis* (Ruiz de Gaona), L.Eoc. France. T. × 41.
120b, c, h. *Daviesina khatiyahi* Smout, Pal. Qatar. T. a, b × 8·5, h × 14.
121a, c. *Chapmanina gassinensis* (A. Silvestri), U.Eoc. Italy. T. × 17·5.
122b, c, f. *Calcarina spengleri* (Gmelin), Rec. Pac. T. b, c × 10, f × 9.
123. *Baculogypsina sphaerulata* (Parker and Jones), Rec. Pac. T. × 15.
124i. *Silvestriella tetraedra* (Gümbel), U.Eoc. Italy. T. × 7·5.
125. *Baculogypsinoides spinosus* Yabe and Hanzawa, Rec. Pac. T. × 16·5.
126b. *Elphidium macellum* (Fichtel and Moll), Rec. T. × 30.
127b. *Cribrononion heteroporum* (Egger), Mio. Bavaria. T. × 30.
128a, b. *Elphidiella arctica* (Parker and Jones), Rec. Arctic. T. × 11·5.
129a, b. *Laffitteina bibensis* Marie, Pal. France. T. × 7.
130b, f, h. *Miscellanea miscella* (d'Archiac and Haime), Pal. W.Pakistan. T. b × 6, f(A) × 6·25, h × 10.

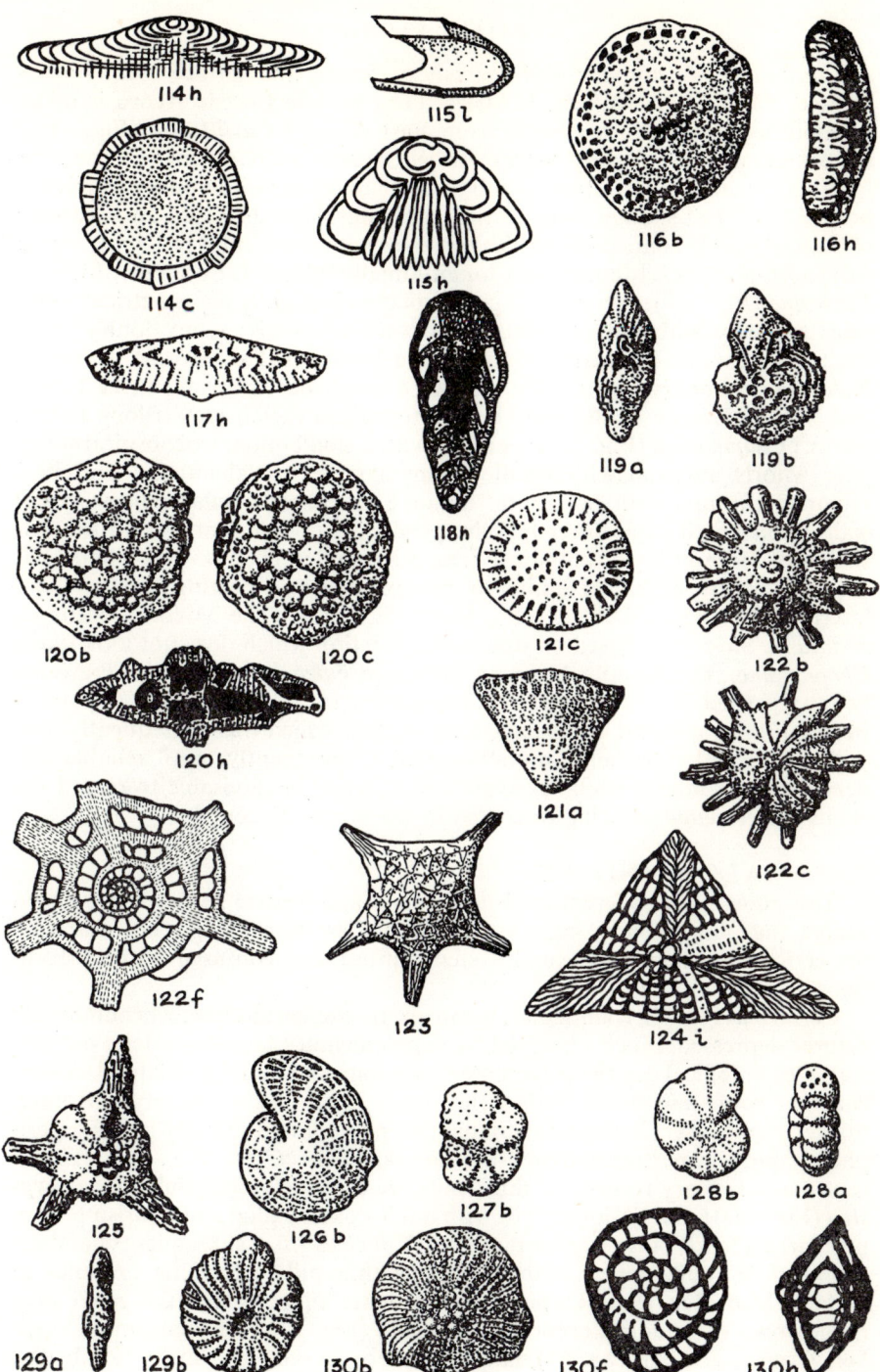

114h

115l

116b

116h

114c

115h

117h

119a

119b

118h

120b

120c

121c

122b

120h

121a

122c

122f

123

124i

125

126b

127b

128b

128a

129a

129b

130b

130f

130h

49

and umbilical side convex; wall coarsely perforate, and surface usually orna-mented. Pal.-M.Eoc. M.East, E.Africa. [The Middle Eocene record is of an unpublished species in a sample from East Africa.] *Cuvillierina* (Fig. 119): almost planispiral, slightly asymmetrical, usually with reticulate ornament forming chevrons over sutures; umbilical region open, with numerous pillars; both sides of test spongy, with vertical and lateral canals; septa double, rows of sutural canals connecting vertical grooves with intraseptal passages; septal flap tooth-plate nearly equatorial, longitudinally folded. L.Eoc. Eur., M.East. *Daviesina* (Fig. 120): biconvex to concavoconvex, slightly asymmetrical; both sides medially with pillars, fissures and vertical canals; septa double, with intraseptal canals; intercameral foramen a basal slit. Pal. M.East, E.Africa, N.Africa. [Some imperfectly known and unpublished forms from the Lower and Middle Eocene of the eastern Mediterranean region may belong to this genus.] *Chapmanina* (Fig. 121): conical, with a small initial trochospiral stage, later whorls uniserial with small rectangular cortical chambers; umbilical region perforated with horizontal laminae and interlamellar pillars; septal walls invaginated from lower margin, producing double septa; aperture con-sisting of large pores in umbilical area, surrounded by tube-like pillars that extend from one umbilical lamina to the next, pores connecting chambers and interlamellar spaces. U.Eoc. Eur., Medit. E.Africa. [The Middle Eocene *C. sertata* A. Silvestri is based on one sagittal section which does not look like a *Chapmanina*. The Middle Miocene *Preverina galea* (A. Silvestri) (*Preverina* being regarded as a synonym of *Chapmanina* by Loeblich and Tappan, in Moore (**78**)) was based on a single axial section which was subsequently lost, and was recorded as being monolamellar. Consequently, until reliable evi-dence to the contrary becomes available, it would be advisable to regard the genus *Chapmanina* as being restricted to the Upper Eocene.]

Family CALCARINIDAE

Test coiled, not differentiated into dorsal and ventral surfaces; advanced genera may be globose; large spines formed by thickenings, not marginal projections of chambers; canal system diffuse and confused with perfora-tions. (U.Cret.-Rec.)

Calcarina (Fig. 122): lenticular, biconvex, trochospiral, chambers numerous; sutures depressed, much obscured by supplementary lamellar calcite on ven-tral side; umbilical cavities interrupted by pillars and radial and lateral canals; chamber roofs and floors with a thin inner layer and a coarsely perforate, thicker outer layer; surface tubercular; elongate longitudinally striated peri-pheral spines; aperture narrow, interio-marginal. ?Cret., Rec. Pac. [Creta-ceous and Tertiary records of this genus are extremely doubtful.] *Baculogyp-sina* (Fig. 123): margin lobulated, with a few coarse radial spines arising from juvenarium; early stage trochospiral and '*Cibicides*'-like, subsequent chambers arranged in radial layers with numerous thin pillars forming granules at surface. Mio.-Rec. Pac. [Unpublished evidence suggests that the genus may range down to Middle Eocene.] *Silvestriella* (Fig. 124): juvenarium of 'rasp-berry' type (see Hanzawa, **64**), later chambers acervuline; usually tetrahedral and with four spines. M.Eoc.-U.Eoc. Italy. **Baculogypsinoides** (Fig. 125): the large *Calcarina*-like trochospiral juvenarium is followed by trochospiral chambers and then by acervuline chambers. Mio.-Rec. Pac. [See Hanzawa, **64**, for a full account of the morphology of these forms. It seems that *Baculogypsinoides bonarellii* Osimo, originally stated to be of Upper Eocene age, and later suggested by Hanzawa to be of Oligocene age, is more likely to be of Lower Miocene age. The Upper Cretaceous genus *Siderolites*, also having spines, is planispiral.]

Family ELPHIDIIDAE

Planispiral or trochospiral, rarely uncoiling; single or double row of sutural pores; wall calcareous, perforate, of radial structure; aperture interiomarginal, single or multiple, or areal. (Pal.-Rec.)

Elphidium [*Polystomella*] (Fig. 126): planispiral, involute; each of the numerous chambers with retral processes or internal chamber projections along septal borders, each with a pore so that there are numerous perforations connecting the chambers; complex canal system; aperture a V-shaped row of pores at base of septal face. L.Eoc.-Rec. Cosmop. **Cellanthus:** large, planispiral, with numerous chambers; large, porous umbilical plug on each side; canal system even more complex than in *Elphidium*; surface without retral processes, only with perforations of the canal system; aperture as in *Elphidium*. Plio.-Rec. Indo-Pac. **Cribrononion** (Fig. 127): planispiral, involute; no retral processes, sutures sunk and with a row of pores; aperture as in *Elphidium*. Mio.-Rec. Cosmop. **Elphidiella** (Fig. 128): planispiral, involute, chambers equitant; umbilical plug without canals; double row of pores along sutures, no retral processes; aperture areal, a group of pores. U.Cret. (Danian)-Rec. Cosmop. **Laffitteina** (Fig. 129): lenticular, planispiral, internally asymmetrical; double row of sutural pores; vertical umbilical canals; aperture a basal slit. L.Pal. France, W.Africa.

Family MISCELLANEIDAE

Planispiral, bilaterally symmetrical, involute; median chambers moderately numerous, simple; spiral lamina coarsely perforate, composed of pillars; no marginal cord. (Pal.)

Miscellanea (Fig. 130): lenticular to subspherical. Pal. Asia, M.East, Medit.

Family PELLATISPIRIDAE

Planispiral, bilaterally symmetrical, evolute; median chambers moderately numerous, simple; spiral lamina thick, coarsely perforate, in one genus developing vacuoles; marginal cord present. (U.Eoc.)

Pellatispira (Fig. 131): lenticular; median chamber layer single; spiral lamina without vacuoles. U.Eoc. Eur., Asia, Pac. **Biplanispira:** like *Pellatispira*, but median chamber layer double peripherally. U.Eoc. Indo-Pac. **Vacuolispira:** more inflated than *Pellatispira*; spiral lamina with coarser canals and developing vacuoles. U.Eoc. Indo-Pac.

Family NUMMULITIDAE

Planispiral, bilaterally symmetrical, rarely becoming annular; involute or evolute; median chambers numerous, simple or subdivided into chamberlets; lateral chambers present or absent; marginal cord present; aperture believed to be an arched slit at the base of the septal face. (Pal.-Rec.) (**1, 28, 43, 46, 47, 50, 51, 93–96**)

Nummulites [*Camerina, Nummulina, Nummularia*] (Figs. 132–143): involute; spiral lamina and median chambers simple; proloculus of megalospheric form usually not less than 0·2 mm. in diameter, and may attain over 1·0 mm.; microspheric form noticeably larger than the megalospheric form, sometimes very much so. Pal.-M.Olig. (warm seas). [These are the typical forms which gave their name to the 'Nummulitic', which constitutes practically the whole of the Palaeogene.] *Palaeonummulites* [*Operculinella, Operculinoides*] (Fig. 144): like *Nummulites* in general form, but a miniature, never attaining the large size of so many species of that genus, the microspheric form being very little larger than the megalospheric form, if at all; the diameter of the proloculus rarely attains 0·2 mm., and is usually considerably smaller. Pal.-Rec. Cosmop. [Compared with *Nummulites*, *Palaeonummulites* is a miniature, and it is only such miniature forms that survived the Middle Oligocene. It seems that the species *N. fabianii* should be regarded as a reticulate *Palaeonummulites*—see Fig. 152.] *Assilina* (Fig. 145): usually more or less flatly lenticular, evolute but with thickened lateral layers, multispiral; microspheric form usually distinctly larger than the megalospheric form; septa upright; spiral lamina slightly equitant but without long alar prolongations. Pal.-M.Eoc. Eur., Asia, Africa, Indo-Pac. *Operculina* (Fig. 146): flat, thin, slightly thicker at centre, evolute (except for the first whorl or two in a few cases), paucispiral and laxly coiled, chambers rapidly increasing in height; septa upright for most of their length, occasionally strongly swept back, often

---

Figs 131–143. FORAMINIFERIDA: ROTALIACEA

Figs 132 and 141 after de la Harpe; Fig. 134 after Nuttall; Figs 135 and 138 after Douvillé; Fig. 136e after Boussac; Fig. 139 after d'Archiac and Haime; Fig. 131 after Cole, in Moore; Figs 133, 136f, 137, 140, 142 and 143 original.

131b, f. *Pellatispira madaraszi* (Hantken), U.Eoc. Italy. T. b × 3·5, f × 5·5.
132a, b, f(A). *Nummulites globulus* Leymerie, Pal. (L. Libyan). Egypt. a × 1, b × 1·5, f(A) × 7·5.
133b. *Nummulites atacicus* Leymerie, M.Eoc. Mortola. × 3.
134j. *Nummulites clipeus* Nuttall (local variant of *N. intermedius* (d'Archiac)), M.Olig. Cutch. × 7·5. (The arrow indicates the radial direction.)
135b, m. *Nummulites planulatus* (Lamarck), L.Eoc. Paris B. b × 3, m × 7·5.
136e, f. *Nummulites laevigatus* (Bruguière), M.Eoc. T. e, Paris B. × 3·75; f, Hampshire × 17.
137f, g. *Nummulites gizensis* (Forskål), M.Eoc. Egypt. × 5·3.
138b. *Nummulites lucasanus* (Defrance), M.Eoc. Bordeaux boring. × 3·75.
139h. *Nummulites aturicus* Joly and Leymerie, M.Eoc. Mortola. × 1.
140h. *Nummulites perforatus* (Montfort), M.Eoc. Sind. × 10.
141f(A). *Nummulites irregularis* Deshayes, M.Eoc. Steinbach, Switzerland. × 3.
142f(A). *Nummulites brongniarti* d'Archiac and Haime, M.Eoc. Hungary. × 15.
143a, b. *Nummulites vascus* Joly and Leymerie, L.Olig. Schio, N.Italy. × 5·6.

52

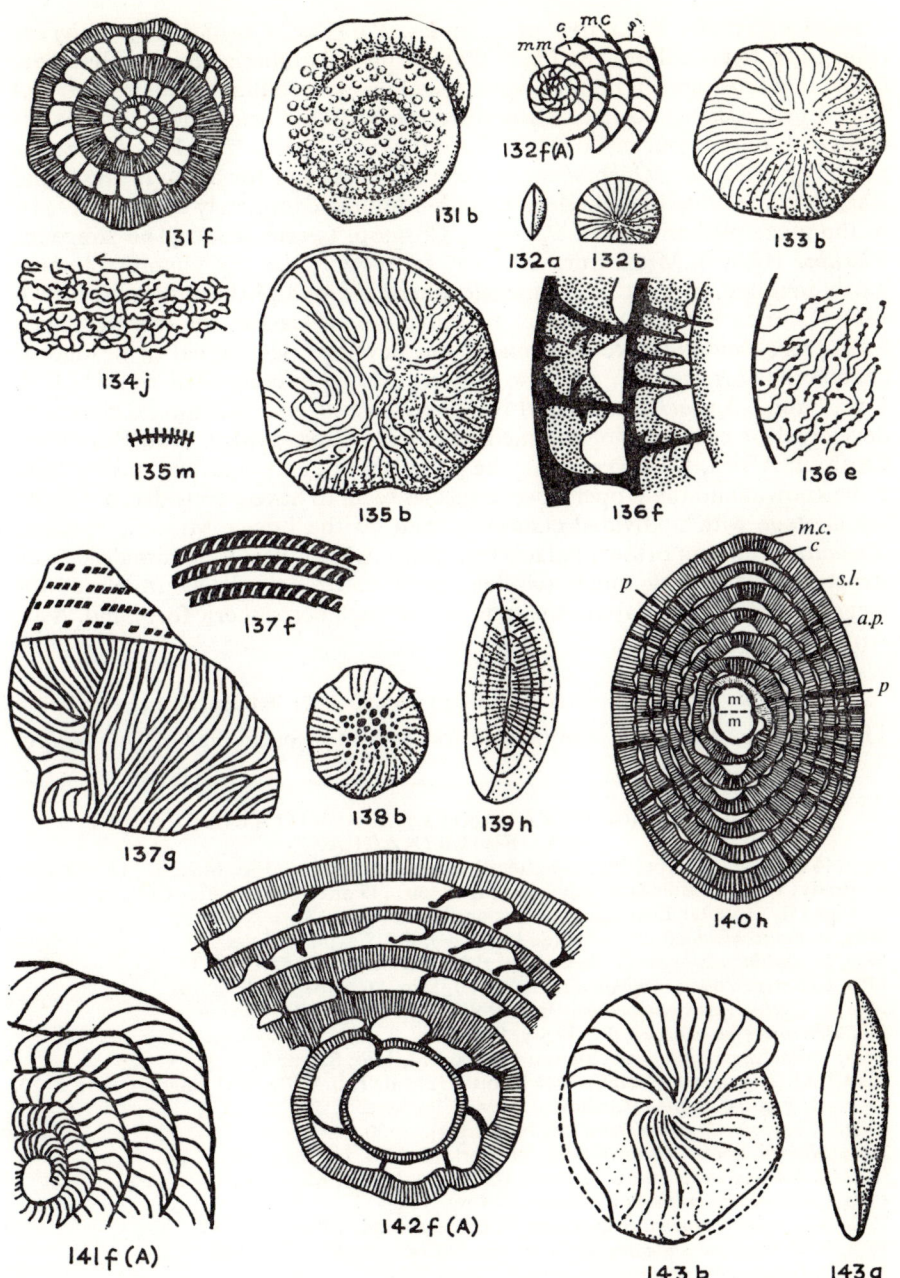

131 f

131 b

132 f(A)

mm *c* *m.c.* *s.*

132 a 132 b

133 b

134 j

135 m

135 b

136 f

136 e

137 f

137 g

138 b

139 h

140 h

m.c.
c
s.l.
a.p.
p
p
m | m

141 f (A)

142 f (A)

143 b

143 a

marked externally by a row of granules. Pal.-Rec. Cosmop. *Cycloclypeus* (Fig. 147): flat, evolute, chambers divided into chamberlets; early chambers coiled in a loose spiral (heterostegine stage), later becoming annular. M.Olig.-Rec. Eur., Asia, Indo-Pac. *Katacycloclypeus:* like *Cycloclypeus*, but externally with raised concentric ridges. Late L.Mio.-M.Mio. Indo-Pac. *Heterostegina* (Fig. 148): like *Operculina*, but chambers (except for a few in the initial stages) divided into chamberlets; chamber lumina completely evolute, at least in the megalospheric form. Pal.-Rec. Cosmop. (warm seas). [The subgenus *Vlerkina* (Olig.-L.Mio., Eur., M.East, Indo-Pac.) has the chamber lumina partly involute, at least in the megalospheric form, and the subgenus *Vlerkinella* (U.Eoc., S.Amer.) is completely involute, at least in the megalospheric form.] *Grzybowskia:* like *Vlerkinella*, but develops pentagonal or hexagonal chamberlets arranged in a favose pattern as seen in equatorial view. U.Eoc. Eur., Papua. *Spiroclypeus* (Fig. 149): like *Heterostegina*, but lateral chambers developed on each side of the median layer. U.Eoc.-L.Mio. Indo-Pac., Eur., Carib., E.Africa. [This genus may be polyphyletic; no Oligocene species seem to be known, and the Upper Eocene species seem to have a considerably larger initial stage with undivided chambers than do the Lower Miocene species.] *Sindulites:* an important Palaeocene form which has a nummuloid wall structure but a very much swollen spiral cord; completely involute; the megalospheric form is about the same size as the microspheric form. Pal. Asia, Africa.

MORPHOLOGY AND CLASSIFICATION OF NUMMULITES

The importance of these nummuloid forms, and in particular of *Nummulites*

Figs 144–149, 153–164. FORAMINIFERIDA: ROTALIACEA AND GLOBIGERINACEA

Fig. 144 after de la Harpe; Fig. 145 after d'Archiac and Haime; Figs 146a, b, 147, 148 after Brady; Fig. 153 after Schlumberger; Figs 146h, 149 and 154–157 after Cole, in Moore; Figs 158–164 after Loeblich and Tappan, in Moore.

144a, b. *Palaeonummulites deserti* (de la Harpe), Pal. Egypt. × 6.
145g, h. *Assilina exponens* (J. de C. Sowerby), M.Eoc. g × 2·5, h × 3·3.
146a, b, h. *Operculina complanata* (Defrance), Mio., T. a, b × 9, h × 10.
147a, b. *Cycloclypeus gümbelianus* Brady, Rec. × 17·5.
148. *Heterostegina depressa* d'Orbigny, Rec. T. × 2.
149h. *Spiroclypeus tidoenganensis* van der Vlerk, L.Mio. Saipan I. × 8.
153b, f(A). *Miogypsina globulina* (Michelotti), L.Mio. T. b × 7·5, f(A) diagrammatic × 10.
154h. *Miogypsina antillea* (Cushman), L.Mio. Panama. × 20.
155h. *Miogypsinella complanata* (Schlumberger). × 30.
156h. *Miogypsinoides dehaartii* (van der Vlerk), L.Mio. Indonesia. T. × 20.
157f(A). *Miolepidocyclina burdigalensis* (Gümbel), L.Mio. Morocco. T. × 12·5.
158b. *Guembelitria cretacea* Cushman, U.Cret. Texas. T. × 156.
159b, b¹. *Chiloguembelina midwayensis* (Cushman), Pal. Texas. T. × 97.
160a, b. *Hantkenina alabamensis* Cushman, U.Eoc. Alabama. T. × 18.
161a, b. *Hastigerina pelagica* (d'Orbigny), Rec. S.Atl. T. × 18.
162a, b. *Clavigerinella akersi* Bolli, Loeblich and Tappan, M.Eoc. Trinidad. T. × 24·5.
163a, b. *Pseudohastigerina micra* (Cole), M.Eoc. Trinidad. T. × 109.
164a, b. *Cassigerinella boudecensis* Pokorný, M.Olig. Czechoslovakia. T. × 109·5.

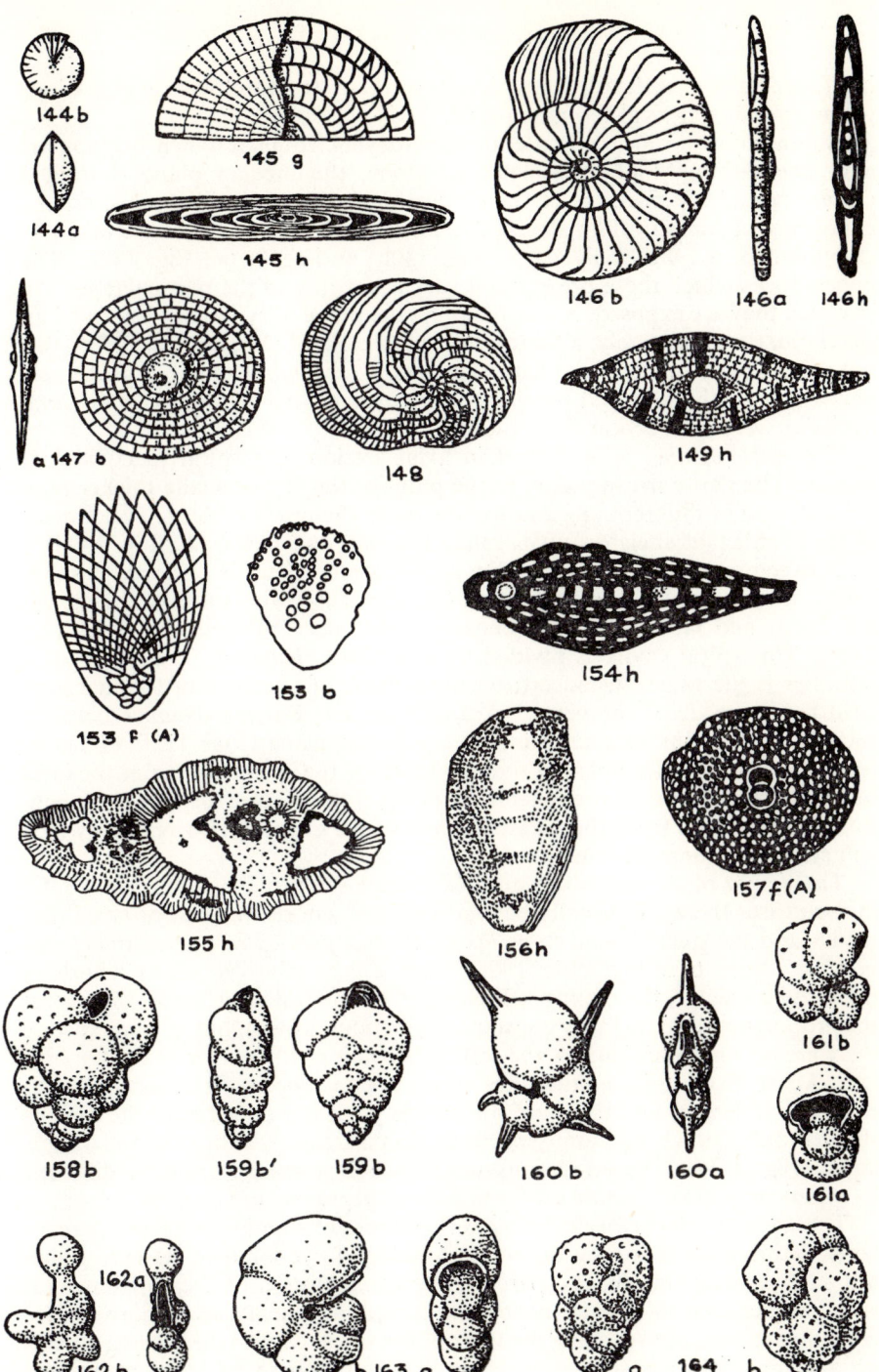

144 b

144 a

145 g

145 h

146 b

146 a   146 h

a 147 b

148

149 h

153 f (A)

153 b

154 h

155 h

156 h

157 f (A)

158 b

159 b'   159 b

160 b

160 a

161 b

161 a

162 a

162 b

b 163 a

a   164   b

55

itself, necessitates a detailed description of their structure, with explanations of the special technical terms required.

A nummulite is more or less lenticular (varying from flat to nearly globular) and, being thought of as a flattened globe, the median plane is termed *equatorial* and the portions of the surface most distant from it as *polar* (the equivalent of 'umbilical' in Rotaliidae or *Nonion*). The test may be taken as composed of a *spiral lamina* (*s.l.*, Fig. 140h) and septa (*s.*, Fig. 132f). The appearances which these present will vary according to the particular section in which they are exposed. A section along the plane of symmetry is termed an *equatorial* (less accurately, a *horizontal*) *section* (Fig. 132f); one parallel to this but nearer the surface of the lenticular test is *tangential*; one in a plane perpendicular to the first and passing through the centre of the test is an *axial* (*radial* or *vertical*) section (Fig. 140h).

The *spiral lamina* is V-shaped in axial section and spiral in equatorial section. The part corresponding to the point of the 'V' is usually thicker than the rest, and of different appearance (owing to the much greater development of the canalicular skeleton): it is called the *marginal cord* (*m.c.*, Fig. 140h).

Between successive whorls of the spiral lamina there is enclosed a spiral cavity, divided into a median portion (between successive turns of the marginal cord) and lateral portions known as the *alar prolongations* (*a.p.*, Fig. 140h). This spiral cavity is divided by *septa* into *chambers*. The *height* of a chamber is the radial distance from the outside of one turn of the marginal cord to the inside of the next (Fig. 132f); its *length* is the distance between successive septa, as measured in the equatorial plane (Figs 132f, 142f); its *breadth* is the distance apart of the two limbs of the spiral lamina at the level of the marginal cord of the inner whorl (Fig. 140h). Thus the alar prolongations are excluded from the measurements of the chamber, for reasons which will shortly become obvious.

The height of the chambers may be greater than their length (Fig. 132f); in such cases there will usually be relatively few whorls in a test of given diameter, and the spiral is said to be *lax* or the test *paucispiral*; when the chambers are longer than high (Fig. 142f), there will be relatively many whorls in the same diameter, the coiling will be *tight*, or the test *multispiral*. In general the proportions remain fairly constant throughout growth, but in some species there are ontogenetic changes, the coiling becoming either laxer or tighter with age. Apart from this, nummulites seem to have been very susceptible to external influences, and there are sometimes great irregularities in the dimensions of different whorls, some species being particularly subject to the irregularity (Fig. 141f). The conditions under which a species lived in different locations may also to some extent affect the degree of inflation of its test.

The *septa* are imperforate, being of the nature of supplemental or canalicular skeleton. Each septum is double, consisting of two laminae between which are narrow, irregular spaces, parts of the canal system of the supplemental skeleton. As seen in equatorial section, a septum appears as a more or less curved partition, not reaching the floor of the spiral cavity (there being always an aperture between successive chambers at the base of the septum and

56

nowhere else) and inclined more or less backwards (i.e. towards the earlier part of the spiral) as it approaches the roof, which it meets obliquely. The aperture is not always seen in nominally equatorial sections, since they often are not exactly in the median plane; in Fig. 142f a few of the septa are seen to stop short of the floor of the chamber, being cut more nearly in the centre than the others. The *intercameral aperture* in the equatorial plane is the only one, the lateral extensions of the septa dividing up the alar prolongations completely. These lateral extensions of each septum leave on the lateral portions of the spiral lamina a linear trace called the *septal filament*, which may be compared to the suture line of an ammonite and is of similar classificatory value. But, whereas the ammonite septum ends definitely in its suture line against the inner face of the test, the nummulite septa penetrate into the thickness of the test (Fig. 136f): consequently there is no need of an internal cast in order to see the septal filaments, since they are revealed by merely a slight decortication (Figs. 133b, 135b, 136e, 137g, 138b). The following are the chief varieties of pattern shown by them:

(1) *Radial:* filaments running straight, or with a slightly sinuous course, from poles to circumference (Figs 144b, 132b, 138b).

(2) *Sigmoidal:* radial, but with a slight twist towards the pole and a reverse curve towards the periphery (Fig. 133b); in the absence of a polar twist the term *falciform* is more appropriate (Fig. 143b).

(3) *Meandrine* (an inappropriate term): grouped in bundles running in a fairly straight course, with occasional abrupt bends, mostly ending polewards by abutting at an angle against another bundle (Fig. 137g). The beginning of this arrangement may be seen among simpler forms (e.g. in Fig. 143b, a little above the centre, where three filaments in one case, four in another, unite into one towards the pole).

(4) *Subreticulate:* irregularly radial, frequently branching or uniting with other filaments (Fig. 136e).

(5) *Reticulate:* forming a complex network in which the individuality of the filaments is completely lost (Fig. 134j). The beginning of this pattern is seen in simpler forms where transverse laminae (Fig. 135m) and other branches (Fig. 136f) are formed to root the filaments firmly in the perforate test.

Trabeculae. In some forms with more complicated septal filaments, especially the meandrine type, the filaments, especially laterally, have numerous small, fine canals on their surfaces so that in tangential section they present a frosted or moss-like appearance; these small, fine canals are called *trabeculae,* also an inappropriate term.

Granules and pillars. Some species of nummulites have on the surface a scatter of round spots, usually lighter in colour than the rest of the test, varying greatly in size and numbers in different species (Figs. 136e, 138b). These are termed *granules,* and in axial sections (Figs. 139h, 140h) they are seen to be the outer ends of more or less cylindrical or acutely conical *pillars* which extend perpendicular to the surface through several chambers; they are part of the canalicular skeleton.

A larger structure, or cluster of structures, of similar nature, polar in position, found in some species, is described as the *polar pustule*.

Classification of nummulites. The most obvious characters, other than shape, are the form of the septal filaments and the presence or absence of granules. d'Archiac, who made the first wide researches on the genus (43), divided it according to these two characters into five groups:

Group 1. *Laeves aut Sublaeves.* Surface smooth or nearly so; when abraded, septal filaments are simple, sinuous or meandrine, directed towards centre. This group includes most of the large species, e.g. *N. millecaput* [*complanatus*], *N. gizensis* (Fig. 137).

Group 2. *Reticulatae.* Filaments forming a complete network; with thin pillars and inconspicuous granules, e.g. *N. intermedius* (cf. Fig. 134).

Group 3. *Subreticulatae.* Network less pronounced, the filaments retaining individuality; with fine granules e.g. *N. laevigatus* (Fig. 136e).

Group. 4. *Punctulatae.* Granules abundant and large; shape more or less globose, e.g. *N. aturicus* (Fig. 139).

Group 5. *Plicatae vel Striatae.* Filaments more or less radial; no granules. This group includes medium and fairly small species, e.g. *N. atacicus* [*biarritzensis*] (Fig. 133), *N. striatus*, *N. vascus* (Fig. 143).

Prever (1902) modified this into a more rigid form, introducing subgeneric and sectional names (though not in accord with the rules of nomenclature, which require one of the subgenera and one of its sections to bear the same name as the genus):

Nummulites
{
Camerina filaments reticulate
{
Bruguieria, without granules = Group 2 of d'Archiac approximately.
Laharpeia, with granules = Group 3.

Lenticulina filaments radial or meandrine
{
Paronaea [*Hantkenia*], without granules = Groups 1 and 5 approximately.
Gümbelia, with granules = Group 4.

This scheme looks simple, but is really artificial, being founded on two characters both of which are probably 'progressive', i.e. which develop independently in parallel lines of descent.

Boussac (28) sought for some 'static' character which should be a more faithful guide to affinity, and believed he had found it in the curve of the septum as seen in equatorial section. He still accepted the division into granulate and non-granulate as of primary importance, though recognizing that some of the former were evolved from the latter. He indicated three main lineages, one entirely non-granulate with radial filaments, a second with mainly meandrine filaments and progressing from non-granulate to granulate: these two

58

have long sloping septa, curved throughout and becoming thinner towards their distal ends. The third lineage has straighter and more upright septa of more uniform thickness; its filaments develop from radial to reticulate, and granules are always present.

This was a great advance on previous schemes, but the slope of the septa seems also to be a 'progressive' character. Boussac's work was cut short by his premature death, and it is impossible to say how he would have improved his classification had he lived.

H. Douvillé, whose researches both preceded and followed those of Boussac, put forward a scheme in 1919 (**51**), in which Boussac's three lineages can be recognized in a modified form, the curvature of the septum not being accepted as a stable character, while granulation is recognized as a progressive character independently developed in several stocks. There can be little doubt that Douvillé's views approximate to what will eventually be recognized as the true evolutionary classification of the nummulites, though minor modifications of his scheme will prove to be necessary.

Abrard (**1**) has proposed some such alterations, and at the same time propounded certain general laws of evolution in nummulites, which may be summarized as follows:

(1) Progressive evolution (anagenesis) in coiling is from lax and irregular, to tight and regular, with change of chamber shape from high and short to low and long; it may be followed by the reverse (retrogressive or katagenetic) change. This sequence is seen in the phylogeny, and in a few cases can be recognized in the ontogeny of microspheric individuals.

(2) Each line of nummulites starts with a non-granulate form and may develop in two directions, with and without granulation. The granulate forms become extinct at their acme, hardly ever undergoing retrogression (*N. aturicus* is possibly an exception); while the same sometimes happens to the non-granulates, but these more usually end in retrogressive forms with lax coiling and high, short chambers. Pustulate species, according to Abrard, are always blind-alley side-shoots.

(3) A third principle (not specially mentioned by Abrard) is that in both ontogeny and phylogeny nummulites begin with a *lenticular* form: they may retain this or may develop in either of two directions—towards *flat* or *globose*.

The charts in the first edition of this volume, showing what was believed to be the lines of evolution in three lineages of nummulities, are not reproduced here because subsequent work has yielded much new evidence and it is obvious that some previously held concepts have to be radically revised. Except for the genus *Assilina* (which is here considered in the text only) the species concerned, together with a few additional ones, are shown on Figs 150–152, not as lineages, but as successive species or groups of species in time, grouped under different genera; in the case of the genus *Nummulites* there is a subdivision into four subgroups according to the type of septal filament. On these figures ranges are not intended to be precise in detail.

FIG 150: RANGES OF SOME SPECIES OF NUMMULITES

		Reticulate	Radial tending to sigmoidal, non-granulate

Reticulate: hormoensis, intermedius

Radial tending to sigmoidal, non-granulate: vascus, striatus, beaumonti, atacicus, globulus, exilis

OLIGOCENE	UPPER	
	MIDDLE	
	LOWER	
EOCENE	UPPER	
	MIDDLE	
	LOWER	
	PALAEOCENE	
UPPER CRETACEOUS	DANIAN	
	MAASTRICHTIAN	

FIG 151   RANGES OF SOME SPECIES OF NUMMULITES

meandrine,non-granulate; septa very oblique & bending forward to meet roof at a very oblique angle.	meandrine,granules largely between filaments.	sub-reticulate,granules on the filaments.
bolcensis, planulatus, irregularis, murchisoni, distans, millecaput	pustulosus, partschi, lucasanus, uranensis, perforatus, aturicus	nuttalli, aquitanicus, laevigatus, acutus, gizensis, brongniarti, yawensis

OLIGOCENE — UPPER, MIDDLE, LOWER
EOCENE — UPPER, MIDDLE, LOWER
PALAEOCENE — DANIAN
UPPER CRETACEOUS — MAASTRICHTIAN

61

FIG 152: RANGES OF SOME SPECIES OF PALAEONUMMULITES

			frasi	deserti	variolarius	orbignyi	garnieri	chavannesi	budensis	pulchellus	incrassatus		tabianii
							Radial tending to sigmoidal, non-granulate						Reticulate
OLIGOCENE	UPPER												
	MIDDLE											▄	▄ tabianii
	LOWER										▄ incrassatus		
EOCENE	UPPER				▄ variolarius	▄ orbignyi	▄ garnieri	▄ chavannesi	▄ budensis	▄ pulchellus	▄		
	MIDDLE				▄								
	LOWER												
	PALAEOCENE	frasi ▄	deserti ▄										
UPPER CRETACEOUS	DANIAN												
	MAASTRICHTIAN												

62

Some of the reasons for abandoning the previous hypothetical arrangements are:

(a) It is now known that no 'nummulites' (including the genera *Nummulites*, *Palaeonummulites*, *Assilina* and *Operculina*) existed prior to about the middle of the Palaeocene; they did not occur in the Upper Cretaceous (including the Danian)—it seems that rare records of the genus *Operculina*, based almost entirely on axial sections, from the Senonian cannot be substantiated.

(b) The suggestion that the genus *Assilina* might have evolved from *Nummulites exilis* of the Upper Palaeocene does not seem likely since *Assilina ranikoti* appeared at least as early; present evidence seems to show that the four genera *Nummulites*, *Palaeonummulites*, *Assilina* and *Operculina* first appeared almost simultaneously when the tremendous and sudden burst of foraminiferal evolution took place at about the middle of Palaeocene times.

(c) The existence of two or three genera within one lineage is not borne out by present evidence; for example, the suggestion that *Operculina canalifera* may have evolved from *Nummulites nuttalli* cannot be upheld because the species appeared at least as early as *N. nuttalli*.

(d) The dating or ranges of some species are now known to be different; for example, (i) *Palaeonummulites deserti* and *P. fraasi* are Palaeocene, not Danian, (ii) *Nummulites partschi* occurs in the Lower Eocene as well as the Middle Eocene, and (iii) *Nummulites globulus* ranges from Palaeocene to Middle Eocene.

(e) In some cases different names shown on the chart were actually synonyms; for example, (i) *Nummulites granifer* of the 'second lineage' is regarded by Schaub (**94**) as a synonym of *N. partschi* which was shown in the 'first lineage', and (ii) *N. djokdjokartae* is the megalospheric form of *N. acutus* which was shown at a different level in the 'second lineage'.

In West Pakistan there are three species of *Assilina* which are of considerable use in stratigraphy. *Assilina ranikoti* is restricted to the Palaeocene, *Assilina granulosa* is restricted to the Lower Eocene, and *Assilina exponens* is restricted to the Middle Eocene. The latter two species have for over a century been very badly misidentified; while these microspheric forms are somewhat similar in external appearance and equatorial section, the megalospheric forms are easily recognized in equatorial section, the megalosphere of *A. leymeriei* (the megalospheric form of *A. granulosa*) being only about one quarter of a millimetre in diameter, while the megalosphere of *A. mamillata* (the megalospheric form of *A. exponens*) is about one millimetre in diameter. Schaub (**95**, **96**) has given a fuller list of species of *Assilina* which he believed formed an evolutionary sequence, but in this connection it must be remembered that his Ilerdian species of larger foraminifera occur above the levels at which Palaeocene planktonic foraminifera were found, and that the Ilerdian stage is not entirely of Palaeocene age, but includes both Palaeocene and Lower

Eocene, the larger foraminifera, including *Assilina leymeriei* and *Nummulites bolcensis*, being of Lower Eocene, not Palaeocene, age. Schaub gave two 'lineages' under the genus *Assilina* (the species are given here in ascending stratigraphical order):

(i) *A. pustulosa*, Ilerdian (Lower Eocene, not Palaeocene), *A. reicheli*, Cuisian (i.e. Lower Eocene), and *A. exponens tenuimarginata* and *A. exponens*, Lutetian (i.e. Middle Eocene), and

(ii) *A. leymeriei*, Ilerdian (Lower Eocene, not Palaeocene), *A. placentula*, *A. laxispira*, *A. major*, Cuisian (i.e. Lower Eocene), *A. spira* subspecies, *A. spira* (*s.s.*) and *A. gigantea*, Lutetian (i.e. Middle Eocene).

Schaub (loc. cit.) recorded six 'lineages' for the genus *Nummulites*, and the reader can refer to his works for detailed information. In some cases (e.g. *Nummulites haimana*) species are not given at the same stratigraphical horizon, but, more importantly, the examples given in the 'lineages' are not drawn from information in any *one* section, but are dependent upon correlation of strata in quite different areas.

The brief outlines of Schaub's six 'lineages' of the genus *Nummulites*, in ascending stratigraphical order, are:

(i) *N. pernotus* of the Ilerdian (Lower Eocene, not Palaeocene), *N. burdigalensis*, *N. pergranulatus* of the Cuisian (i.e. Lower Eocene), *N. gallensis*, *N. verneuili*, *N. uranensis* of the Upper Lutetian (i.e. Middle Eocene), and *N. perforatus* of the Biarritzian (apparently Middle Eocene),

(ii) *N. burdigalensis* subspecies, *N. inkermanensis* of the Cuisian (i.e. Lower Eocene), *N. haimana*, *N. obesus* of the Lower Lutetian (i.e. Middle Eocene), *N. crassus* of the Middle Lutetian (i.e. Middle Eocene), *N. meneghinii* of the Upper Lutetian (i.e. Middle Eocene), and *N. perforatus* var. *C* of the Biarritzian (apparently Middle Eocene),

(iii) *N.* cf. *praecursor*, *N. bolcensis*, *N. haimana* of the Ilerdian (Lower Eocene, not Palaeocene), *N. nemkovi*, *N. kaufmanni*, *N. distans*, *N.* aff. *distans*, *N. polygyratus* of the Cuisian (i.e. Lower Eocene), and *N. millecaput*, *N. maximus*, *N. dufrenoyi* of the Lutetian (i.e. Middle Eocene),

(iv) *N. spileccensis* of the Ilerdian (Lower Eocene, not Palaeocene), *N. subdistans*, *N. archiaci*, *N. regulatus*, *N. pratti*, *N. 'formosus'* of the Cuisian (i.e. Lower Eocene),

(v) *N. laxus* of the Ilerdian (Lower Eocene, not Palaeocene), *N. nitidus*, *N. formosus* of the Cuisian (i.e. Lower Eocene), and

(vi) *N. praemurchisoni*, *N. irregularis* of the Cuisian (i.e. Lower Eocene), and '*N. murchisoni major*' of the Lutetian (i.e. Middle Eocene).

While it seems that some of the successive species listed by Schaub are closely related, the following example will indicate that apparent dissimilarity in morphological characters need not necessarily indicate distant relationship. In West Pakistan, in a Lower Eocene section of about 3,600 ft, samples collected at every few feet produced the following evidence. At the lowest horizons appeared a small species of *Nummulites* similar to *N. burdigalensis*, gently lenticular, with well developed pillars and well curved septa. As the succession was ascended, for a short time the characters of the species did not change, but then very gradually changes began to take place. The degree of inflation of the test increased, granulation became less conspicuous, the curvature of the distal ends of the septa became less, the coiling gradually became tighter, and the megalosphere gradually increased in size. These changes gradually continued until, near the top of the Lower Eocene, the end form was a strongly inflated simple radiate form with no granules, straighter septa, a larger megalosphere, and a tighter rate of coiling of the spiral lamina. Since on morphological grounds it did not seem desirable to regard the beginning and end forms as belonging to the same species, the series was regarded as being constituted of (i) an initial species by itself, (ii) the initial species accompanied by a sub-species, (iii) this subspecies accompanied by a new species, only subspecifically different from the previous subspecies, and (iv) the new species alone. Thus, while similarity of morphological characters may in some cases indicate close relationship, there probably exist examples where the closest related form is actually one which is not the most similar in morphological characters.

It seems possible that some such form as *Palaeonummulites deserti* was the root stock of nummulites. Although the genera *Palaeonummulites* and *Nummulites* may to some extent be polyphyletic, it seems to be imperative to retain the two genera since nowhere in the world is there any sign of any species conforming to the diagnosis of *Nummulites*, as here given, above the Middle Oligocene, whereas *Palaeonummulites* continues abundantly to the present day.

Geological and geographical range of nummulites. The acme of nummulites is in the Middle Eocene, which is characterized both by the largest species and by the greatest abundance of granulate species. The change from Middle to Upper Eocene, when nearly all the large species disappear, is striking. The Upper Eocene and Lower and Middle Oligocene are characterized by smaller species except that the Lower and Middle Oligocene *Nummulites intermedius* attains rather more than medium size.

The headquarters of *Nummulites* were in the Old World Tethys, from Spain to New Caledonia. In the New World the genus is represented only by two Upper Eocene species (*N. macgillivrayi* and *N. striatoreticulatus*), the latter reaching the Pacific Islands. The supposed 'nummulite' *Camerina matleyi* from the Middle Eocene of Jamaica is now placed in the genus *Pellatispirella* and regarded as an elphidiid. The genus *Assilina* is unknown in the New World. In Africa, apart from the Egypt-Tunisia-Algeria-Morocco region, which is part of the Tethys, *Nummulites* is known from the Middle Eocene of Tanzania, Madagascar and Mozambique, but not from the west coast where *Sindulites* occurs in the Palaeocene having evidently spread there through the Lake Tchad area from the Middle East. In north-west Europe very few true *Nummulites* managed to make their way, the Lower Eocene *N. planulatus* and

65

the Middle Eocene *N. laevigatus* being the best known. There are no authentic published records of *Nummulites* or *Assilina* from Australia or New Zealand (see Vol. II).

NOTE ON THE NOMENCLATURE OF NUMMULITES

By the strict law of priority the generic name *Nummulites* is a synonym of *Camerina*, but in 1945 Opinion 192 of the International Commission for Zoological Nomenclature suppressed the name *Camerina* in favour of *Nummulites*, a name which has been used in the vernacular and adjectivally ('nummulites', nummulitic) for so many years.

Prior to the recognition of dimorphism, separate names were in many cases given to the two forms (A and B) of one species. When dimorphism came to be understood, a custom grew up of hyphenating the two names, e.g. *N. planulatus-elegans, N. laevigatus-lamarcki.* Although not in accordance with the strict laws of nomenclature, as a temporary method this was convenient, although occasionally mistakes were made in coupling. Another less excusable method, introduced by Prever, was that of distinguishing the name of the megalospheric form by the prefix *sub*: thus the two forms of one species would be, e.g. *N. atacicus* and *N. subatacicus.* Unfortunately this prefix had already been used in some cases, as it is in all branches of the animal kingdom, to denote a supposedly allied species, e.g. *N. sub-brongniarti* Verbeek is not the megalospheric form of *N. brongniarti* d'Archiac and Haime, but was regarded as a close ally of, and now a synonym of the microspheric *N. intermedius* (d'Archiac).

The only sound rule is to give the same name, fixed by the rule of priority, to both forms of one species. It is sometimes urged that as it is sometimes impossible to be certain that the two forms are one species they should not be given the same name; but if it should prove that a mistaken association has been made, it will always be easier to give a new name to the divorced partner than to transfer a name beginning with *sub-* from one species to another.

The following is a list of some of the more important trivial names, with the chief synonyms, in the genera *Nummulites* and *Palaeonummulites*:

NUMMULITES;

Trivial names	Synonyms arising from dimorphism	Synonyms otherwise arising
acutus	djokdjokartae	douvillei, vredenburgi
atacicus	guettardi	biarritzensis
aturicus	rouaulti	
fichteli	?intermedius	
gizensis (often misspelt 'gizehensis')	curvispira	
globulus		ramondi
laevigatus	lamarcki	
millecaput	helveticus	
murchisoni	heeri	

nuttalli	thalicus	
perforatus	obtusus	
striatus	contortus	
vascus	boucheri	bezançoni, germanicus

PALAEONUMMULITES;

bouillei	tournoueri	
budensis	bericensis	
chavannesi	rütimeyeri	
orbignyi	prestwichianus	wemmelensis
variolarius	heberti	

Family MIOGYPSINIDAE

Trigonal or suborbicular, sometimes digitate; lenticular; equatorial layer with lateral chambers, or with solid lateral layers, or without either; megalosphere bilocular, succeeded by one to four series of spirally arranged periembryonic chambers situated apically, subapically or subcentrally (see Drooger, **53**), which are in turn followed by essentially rhombic equatorial chambers with intercameral foramina. (U.Olig.-M.Mio.) **(97)**

Miogypsina [*Flabelliporus, Lepidosemicyclina*] (Figs 153, 154): megalosphere situated apically; lateral chambers present. L.Mio.-M.Mio. Cosmop. *Miogypsinella* (Fig. 155): like *Miogyspina*, but without lateral chambers. U.Olig.-L.Mio. Eur., M.East, Africa, Asia, Indo-Pac., Carib. *Miogypsinoides* (Fig. 156): like *Miogypsinella*, but with a thick solid layer on each flank. L.Mio.-M.Mio. Asia, Africa. *Miolepidocyclina* (Fig. 157): like *Miogypsina*, but proloculus subapically to subcentrally placed. L.Mio. Eur., Indo-Pac., N.Amer., S.Amer.

Members of this family were previously grouped with the Orbitoids, but are now regarded as having been developed from some form of rotaliid. *Miogypsina irregularis* (Michelotti), a name that has been much used in stratigraphical literature, is unfortunately a homonym, and is replaced by *M. globulina* (Michelotti), the type species of the genus. Members of the family are of common occurrence in strata in the Caribbean region as well as in the Tethys proper and the Indo-Pacific region, but concepts as to the ranges of species have been confused by the different principles of dating used in the Caribbean region as compared with elsewhere.

Superfamily GLOBIGERINACEA

Test free, spirally coiled, planispiral, trochospiral or sometimes streptospiral, or modified from one of these; wall calcareous, radial hyaline in structure, with radially disposed pores, surface usually coarsely or moderately coarsely perforate, cancellate and often with distinct surface pits into which the pores open; multilocular, chambers globular, subglobular, or modified from this; aperture always interiomarginal or areal, sometimes modified by small plates across the aperture; supplementary apertures may be present; habitat planktonic. (M.Jur.-Rec.) **(10, 12, 13, 22, 22a, 24, 25, 54, 75)**

67

Family HETEROHELICIDAE

Early stage trochospiral, planispiral, biserial or triserial, later sometimes modified; aperture large, simple and interiomarginal, terminal in uniserial forms. (M.Jur.-Olig.).

Guembelitria (Fig. 158): triserial with inflated chambers and sunken sutures; aperture an interiomarginal arch at base of last chamber. L.Cret.-Eoc. Cosmop. *Chiloguembelina* (Fig. 159): flaring, biserial; smooth or hispid; aperture a low arch bordered by a flap which is better developed on one side. Pal.-Olig. Cosmop.

Family HANTKENINIDAE

Planispiral or enrolled biserial, chambers spherical to elongate or clavate; primary aperture symmetrical and equatorial, single or multiple; relict or areal secondary apertures sometimes present. (Pal.-Rec.)

Hantkenina (Fig. 160): planispiral, involute, biumbilicate, each chamber usually with a long hollow spine at forward margin on periphery; primary aperture interiomarginal, like an inverted 'Y' with the shaft bordered by portici. L.Eoc.-U.Eoc. Cosmop. *Cribrohantkenina:* like *Hantkenina*, but primary aperture simple and with a series of circular, completely enclosed small openings above it. U.Eoc. N.Amer., Carib., Africa. *Hastigerina* (Fig. 161): early stage slightly trochospiral, adult planispiral, involute or loosely coiled, biumbilicate, chambers inflated; aperture a low interiomarginal arch. L.Mio.-Rec. Cosmop. (11). *Clavigerinella* (Fig. 162): planispiral, involute, later chambers radially elongate or clavate; aperture an elongate interiomarginal slit extending up chamber face and with sides bordered by flanges. M.Eoc.-U.Eoc. Carib., N.Amer. *Pseudohastigerina* (Fig. 163): planispiral, biumbilicate, chambers inflated, smooth; aperture an equatorial arch with a porticus. L.Eoc.-Olig. Carib. *Cassigerinella* (Fig. 164): early stage planispiral, later with biserially arranged chambers continuing to spiral in the same plane; biumbilicate, chambers inflated; aperture interiomarginal-extraumbilical. U.Eoc.-Mio. Eur., N.Amer., Carib., S.Amer.

Family GLOBOROTALIIDAE

Trochospiral, chambers more or less rounded or angular; primary aperture interiomarginal, extraumbilical-umbilical; supplementary sutural apertures may occur on dorsal side. (Pal.-Rec.)

Globorotalia (*s.s.*) (Fig. 165): periphery carinate; no supplementary apertures. Pal.-Rec. Cosmop. [In the subgenus *Turborotalia* (Pal.-Rec. Cosmop.) the periphery is rounded.] *Truncorotaloides:* like *Globorotalia*, but with supplementary sutural apertures on dorsal side. L.Eoc.-M.Eoc. Carib., N.Amer., Eur.

Family GLOBIGERINIDAE

Trochospiral, streptospiral or globular, chambers spherical to clavate;

primary aperture umbilical or spiroumbilical; secondary sutural or areal apertures, bullae, and accessory infralaminal apertures sometimes present. (Maastr.-Rec.).

Globigerina (*s.s.*) (Fig. 166): trochospiral, chambers inflated, smooth, pitted, cancellate, hispid or spinose; aperture interiomarginal, umbilical. Maastr.-Rec. Cosmop. [*Subbotina* (U.Cret.(Danian)-L.Eoc. Cosmop.) may prove to be a useful subgeneric name for forms with strongly lipped apertures.] *Globigerinoides:* like *Globigerina*, but with supplementary sutural apertures on dorsal side. Aquit.-Rec. Cosmop. *Globoconusa:* like *Globigerina*, but with strongly convex dorsal side and chambers increasing rapidly in size; aperture small; a few small supplementary sutural openings on dorsal side. Maastr.-Danian-?L.Pal. Eur., N.Amer., S.Amer., Russia. *Globoquadrina* (Fig. 167): aperture initially extraumbilical in part, but later becoming solely intraumbilical, typically with strong lips. Aquit.-L.Plio. Austral., N.Z., Carib. *Globorotaloides:* trochoid; primary aperture interiomarginal umbilical-extra-umbilical during ontogeny, but covered in adult by a bulla. Maastr.-Rec. Cosmop. *Pulleniatina* (Fig. 168): trochoid, becoming strepospiral in adult; primary aperture umbilical-extraumbilical and ventral in ontogeny and adult. Plio.-Rec. Cosmop. *Sphaeroidinella* (Fig. 169): adult trochoid; primary aperture interiomarginal umbilical; primary wall covered by an additional cortex with perforations reduced or absent; lips thickened, often crenulate, symmetrical; supplementary apertures present in adult. Plio.-Rec. Cosmop. *Sphaeroidinellopsis:* like *Sphaeroidinella*, but no supplementary sutural apertures. Aquit.-Plio. S.Amer., Carib., Pac., N.Z., Eur. *Orbulina* (Fig. 170): early stage like *Globigerinoides*, but the last chamber is spherical and completely or partially envelopes all the others; adult aperture areal and in part also sometimes sutural. M.Mio.-Rec. Cosmop. (**21**). *Praeorbulina:* last chamber largely, but not completely, embracing earlier chambers; adult aperture consisting of numerous sutural openings. L.Mio.-M.Mio. Cosmop. *Biorbulina:* bilobate, last two chambers similar in size, penultimate chamber partly or completely embracing earlier ones; aperture consisting of areal and sutural openings. M.Mio.-Rec. Cosmop. *Candeina* (Fig. 171): early stage like *Globigerina*, later like *Globigerinoides*; no primary aperture, but numerous small secondary openings on both sides of test. U.Mio.-Rec. Cosmop. *Globigerapsis* [*Porticulasphaera*] (Fig. 172): early stage trochospiral, final chamber embracing and covering umbilical region; adult with a few arched secondary sutural apertures at lower margin of final chamber. M.Eoc.-U.Eoc. Carib., N.Amer., S.Amer., Eur., N.Z., Japan. *Orbulinoides:* similar to *Globigerapsis*, but adult with secondary sutural openings on dorsal side. M.Eoc. N.Amer., Carib., Eur., Japan. *Globigerinatella* (Fig. 173): early stage trochospiral, final chamber embracing and obscuring umbilical region; later chambers with secondary sutural and areal apertures which have lips and are sometimes covered by bullae. L.Mio. Carib., N.Amer., Pac. *Globigerinatheka:* like *Globigerapsis*, but with bullae. ?M.Eoc. U.Eoc., Cosmop. *Globigerinita* [*Catapsydrax*]: like *Globigerina*, but possesses a bulla. M.Eoc.-Rec. Cosmop. *Turborotalita:* like *Turborotalia*, but possesses a bulla. Mio.-Rec. Atl., Carib.

The Globigerinacea have been considered above in some detail because the considerable amount of research carried out on the superfamily during the last quarter of a century has very greatly increased their use for detailed world-wide *correlation*, although, as far as *dating* is concerned, it has to be remembered that other groups of fossils have to be considered as well. Details of zonation by planktonic foraminifera will be found in Bolli (**24, 25**), Loeblich and Tappan (**75**), Banner and Blow (**13**) and Blow (**22a**), in which further references will be found.

Superfamily ORBITOIDACEA

Test basically coiled; walls of radially laminated calcite, of two layers, outer lamella covering all previously deposited parts of test as well as forming new chamber; septa double. (Cret.-Rec.) (**61, 66, 98, 114**)

Family GLABRATELLIDAE
Trochospiral, ventral side flattened; aperture umbilical. (Eoc.-Rec.)

Glabratella (Fig. 174): hemispherical, only chambers of last whorl visible on ventral side which has radial ornament. U.Eoc.-Rec. N.Z., Pac., Atl., Austral., E.Africa, Medit.

Family PEGIDIIDAE
Modified trochospiral; chambers few, successive ones opposite to or partly enveloping previous chamber; aperture a series of tubes in umbilical region. (Mio.-Rec.)

Pegidia (Fig. 175): sublenticular, unequally biconvex, of three or four chambers arranged in apposition. Mio.-Rec. France, Indo-Pac., Africa, Carib.

Family EPONIDIDAE
Low trochospiral, rarely uncoiling; aperture basal or areal, single or multiple. (U.Cret.-Rec.)

Figs 165–178. FORAMINIFERIDA: GLOBIGERINACEA AND ORBITOIDACEA

Figs 166, 170 and 176 after Brady; Figs 165, 167–169, 171–175, 177 and 178 after Loeblich and Tappan, in Moore.

165a, b, c. *Globorotalia tumida* (Brady), post-Tert. W.Pac. T. × 22.
166a, b, c. *Globigerina bulloides* d'Orbigny, Rec. T. × 28.
167a, b, c. *Globoquadrina dehiscens* (Chapman, Parr and Collins), Mio. Austral. T. × 53·5.
168a, b, c. *Pulleniatina obliqueloculata* (Parker and Jones), Rec. S.Atl. T. × 41.
169a, c. *Sphaeroidinella dehiscens* (Parker and Jones), Rec. Pac. T. × 19.
170d, e. *Orbulina universa* d'Orbigny, in de la Sagra, Rec. T. d × 25, e × 75.
171b, c. *Candeina nitida* d'Orbigny, in de la Sagra, Rec. Atl. T. × 41.
172b, c. *Globigerapsis kugleri* Bolli, Loeblich and Tappan, Eoc. Trinidad. T. × 36.
173b, c. *Globigerinatella insueta* Cushman and Stainforth, L.Mio. Trinidad. T. × 46·5.
174a, b, c. *Glabratella crassa* Dorreen, U.Eoc. N.Z. T. × 59·5.
175a, b, c. *Pegidia dubia* (d'Orbigny), Rec. Mauritius. T. × 16·5.
176a, b, c. *Eponides karsteni* (Reuss), Rec. × 37·5.
177a, b, c. *Neoeponides schreibersii* (d'Orbigny), Mio. Austria. T. × 17.
178a, b. *Poroeponides lateralis* (Terquem), Rec. Rhode I. T. × 22.

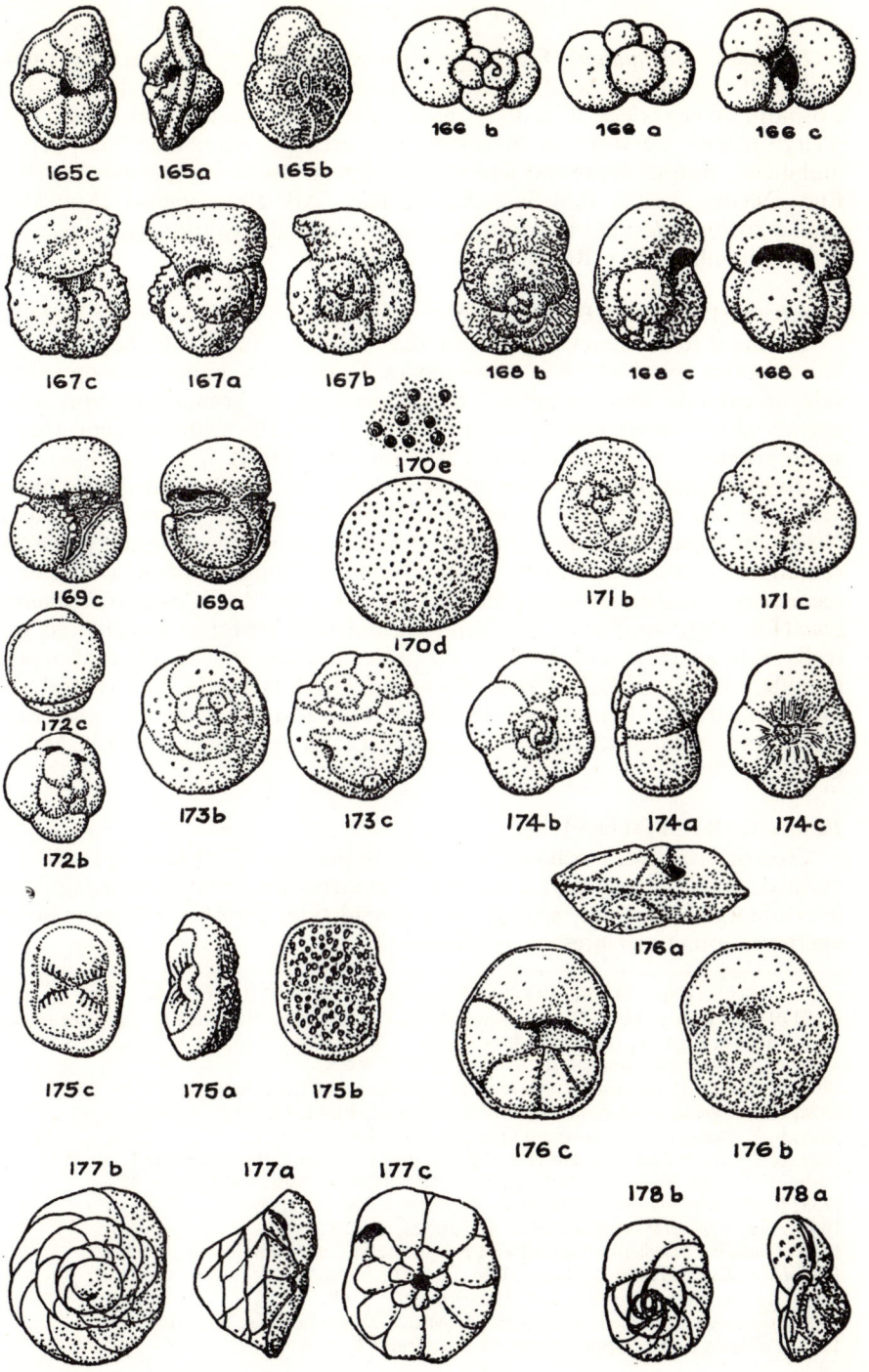

165c 165a 165b

166 b 166 a 166 c

167c 167a 167b 168 b 168 c 168 a

169c 169a 170e 171 b 171 c

170d

172c 173b 173c 174 b 174 a 174 c

172b

176 a

175c 175a 175b 176 c 176 b

177 b 177 a 177 c 178 b 178 a

Eponides [*Pulvinulina*] (Fig. 176): biconvex, periphery angular to carinate, umbilical region depressed; aperture an interiomarginal arch. U.Cret.-Rec. Cosmop. *Neoeponides* (Fig. 177): planoconvex to unequally biconvex, periphery carinate; aperture an interiomarginal arch extending from periphery to umbilicus; sutures depressed and thickened near umbilical margin. Pal.-Rec. Eur., Medit., Israel, Red Sea, Africa, Pac., Atl. *Poroeponides* (Fig. 178): similar to *Eponides*, but in addition possesses round areal pores on lower face of last chamber. Plio.-Rec. Medit., Atl., Pac.

Family AMPHISTEGINIDAE

Trochoid to asymmetrically lenticular, involute; many chambers arranged in a complex spiral which in some genera divides into chamberlets on ventral side or extends into peripheral flange; smooth to granulated; aperture a narrow slit at inner margin of last chamber, usually with adjacent area of small granules; no canal system. (?U.Cret., Eoc.-Rec.)

Amphistegina (Fig. 179): lenticular, usually more or less unequally biconvex, with low spire; many equitant chambers with alar prolongations as in *Nummulites* (but no marginal cord); dorsal septa simple, radiate, falciform, sometimes wavy near umbo; ventral septa dividing to form a rosette of secondary chambers around pole. ?U.Cret., L.Eoc.-Rec. Cosmop. *Tremastegina* (Fig. 180): similar to *Amphistegina*, but with pustules dorsally and ventrally, and aperture with backwardly projecting lips and no adjacent area of granules. L.Eoc.-U.Eoc. Carib. *Eoconuloides:* dorsal side flat, ventral side high-conical; last chambers subdivided on ventral side into small chamberlets; spiral wall in later stages thinner and with less prominent pillars. M.Eoc. Carib.

Family CIBICIDIDAE

Free or attached, trochospiral to nearly planispiral, or later spreading or cyclical; wall coarsely perforate, radial in structure; septa double; aperture interiomarginal but may spread on to dorsal side; peripheral supplementary apertures sometimes present. (Cret.-Rec.)

Figs 179–191. FORAMINIFERIDA: ORBITOIDACEA
Fig. 179 after Brady; Fig. 181 after Macfadyen; Fig. 187 after Nuttall; Fig. 180 after Barker, in Moore; Figs 182–186 and 188–191 after Loeblich and Tappan, in Moore.
179b, c. *Amphistegina lessonii* d'Orbigny, Rec. × 22·5.
180a, b. *Tremastegina senni* (Cushman, in Vaughan), M.Eoc. Barbados. T. × 25.
181a, b, c. *Cibicides lobatulus* (Walker and Jacob), Pleist. Norfolk. × 34.
182a, b, c. *Planulina ariminensis* d'Orbigny, Rec. Italy. T. × 23.
183a, b. *Cycloloculina annulata* Heron-Allen and Earland, Tert. England. T. × 35·5.
184b, c. *Dyocibicides biserialis* Cushman and Valentine, Rec. Calif. T. × 37.
185a, b. *Planorbulina mediterranensis* d'Orbigny, Rec. × 22·5.
186b. *Planorbulinella larvata* Parker and Jones, Rec. T. × 30.
187h. *Linderina buranensis* Nuttall and Brighton, M.Eoc. Somaliland. × 15.
188. *Acervulina inhaerens* Schultze, Rec. Italy. T. × 18.
189i. *Gypsina plana* (Carter), Rec. Mauritius. T. × 48.
190f. *Sphaerogypsina globulus* (Reuss), U.Mio. Czechoslovakia. T. × 10.
191h. *Fabiania cassis* (Oppenheim), M.Eoc. Italy. T. × 6.

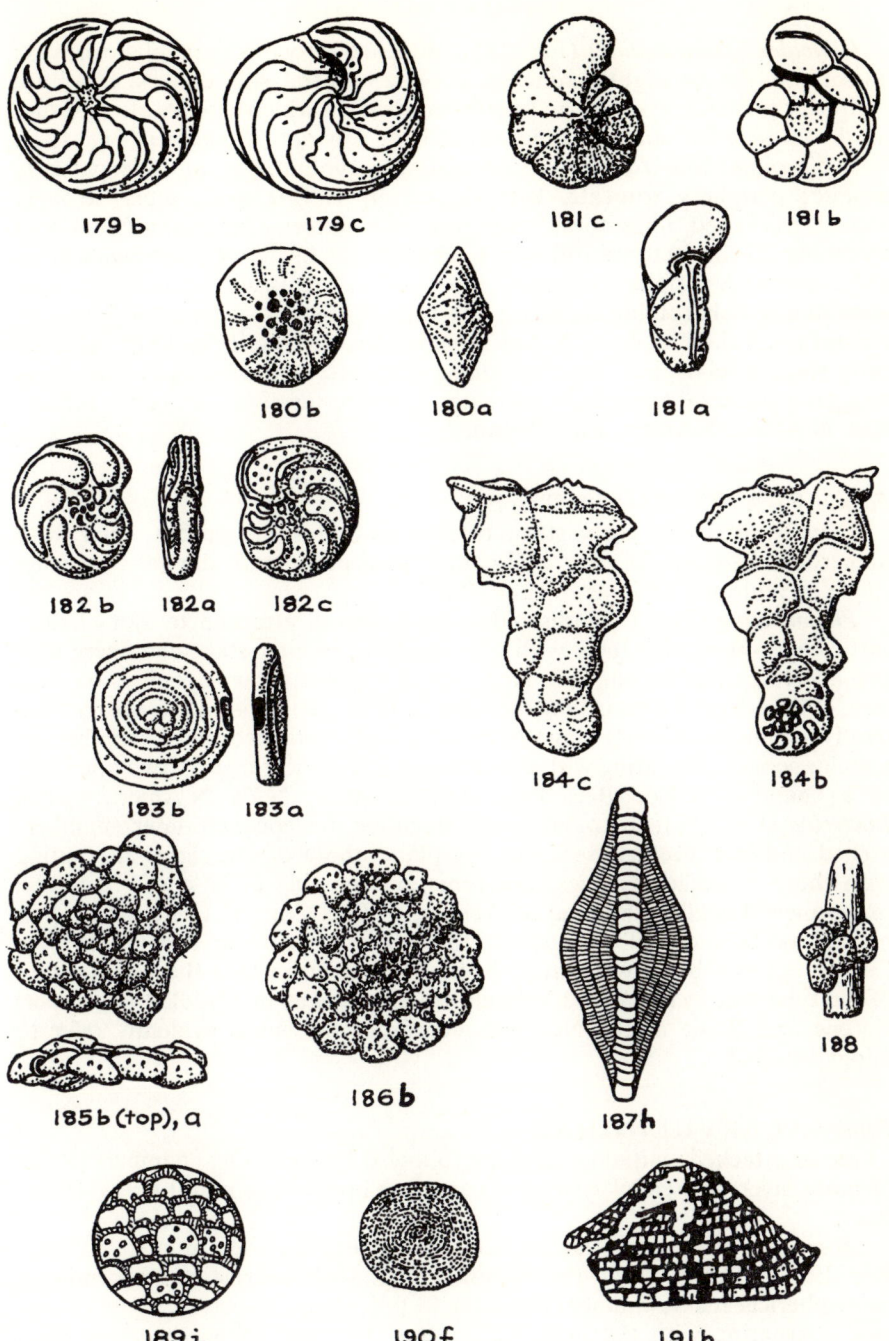

179 b   179 c   181 c   181 b

180 b   180 a   181 a

182 b   182 a   182 c   184 c   184 b

183 b   183 a

185 b (top), a   186 b   187 h   188

189 i   190 f   191 h

Cibicides [*Truncatulina*] (Fig. 181): attached, trochospiral, planoconvex; dorsal side flat to slightly concave, evolute; ventral side convex, involute; periphery carinate; aperture an interiomarginal slit which extends along suture on dorsal side for some distance. Cret-Rec. Cosmop. *Planulina* (Fig. 182): free, discoidal, low-trochospiral; dorsal side evolute, ventral side partially evolute; periphery truncate, with thick, imperforate keel; septa strongly arched, thickened, imperforate; aperture an equatorial, interiomarginal arch extending a little on to ventral side. U.Cret.-Rec. Cosmop. *Cycloloculina* (Fig. 183): discoidal, periphery rounded; more or less planispiral initial stage, later uncoiling and becoming annular; aperture consisting of large perforations on surface. Pal.-Mio. Eur., N.Amer., Asia. *Dyocibicides* (Fig. 184): attached early stage trochospiral, later uncoiling, elongate, and irregularly biserial to staggered uniserial; periphery carinate; aperture terminal, elongate. L.Eoc.-Rec. N.Amer., S.Amer., Eur., Japan.

Family PLANORBULINIDAE

Usually attached, early stage trochospiral in some genera, later with numerous chambers forming a discoidal, conical or lenticular test; aperture single or multiple, peripheral. (Eoc.-Rec.)

Planorbulina (Fig. 185): discoidal, trochospiral, attached by dorsal side; early stage spiral, chambers with one aperture; in later stage chambers with two apertures, and arranged in rings; small supplementary openings on each side. L.Eoc.-Rec. Cosmop. *Planorbulinella* (Fig. 186): discoidal, nearly bilaterally symmetrical; early stage trochoid, usually attached; later chambers in annular series, alternating with those of previous series; usually two apertures to a chamber. L.Eoc.-Rec. Pac., Atl., Austral., N.Z., N.Amer., Carib. *Linderina* (Fig. 187): discoidal, with biloculine nucleoconch, later chambers arched and in concentric series in one plane, those of one series alternating with those of adjacent series; apertures at each side of base of chambers; pronounced lamellar umbonal thickening. M.Eoc. Eur., M.East, Africa. [No normal occurrences of the genus other than those from beds of Middle Eocene age can be substantiated, Szöts having shown that the type locality of the type species (*L. brugesi*) should be regarded as of Middle (not Upper) Eocene age. Some of the Neogene records of the genus evidently refer to miogypsinids.]

Family ACERVULINIDAE

Free or attached; initial spiral stage followed by spreading chambers in one or more layers; no canal system; no aperture apart from mural pores. (Eoc.-Rec.)

Acervulina (Fig. 188): attached, later encrusting, chambers inflated. U.Tert.-Rec. Eur., N.Amer., Indo-Pac. *Gypsina* (Fig. 189): encrusting or forming a hemispherical mass; chambers circular to polygonal in section, those of one row alternating with the next. L.Eoc.-Rec. Cosmop. *Sphaerogypsina* (Fig. 190): like *Gypsina*, but globular and free. Pal.-Rec. Eur., Medit., M.East,

Asia, Africa, Pac., Carib. [Palaeocene records refer to material from East Africa.]

Family CYMBALOPORIDAE

Trochospiral, later chambers annular and in flat or conical layer; apertures numerous. (U.Cret.-Rec.)

Fabiania (Fig. 191): conical; three simple, globose chambers followed by a cyclical series of chambers with area beneath outer wall subdivided by horizontal and vertical partitions forming coarse alveoli which themselves are subdivided by shorter partitions into a few small alveoli; outer wall coarsely perforate, umbilical walls and partitions imperforate. M.Eoc.-U.Eoc. Eur., Medit., M.East, Pac., Japan, Carib. *Halkyardia* (Fig. 192): small, planoconvex to unequally biconvex with dorsal side higher; a few irregular early chambers followed by numerous small chambers in alternating annular series; thick lamellae obscuring early spire chambers; umbilical area with horizontal and connecting hollow vertical pillars; sutures radial on ventral side; inner walls non-porous; aperture consisting of small pores at periphery. U.Eoc.-Aquit. Eur., Africa, Pac., N.Amer., S.Amer. [The genus has been found in the Oligocene of Venezuela and in the early part of the Lower Miocene in Papua.]

Family HOMOTREMATIDAE

Attached; early chambers more or less trochospiral, later modified; wall coarsely perforated. (U.Cret.-Rec.)

Sporadotrema (Fig. 193): later branched, chambers at periphery of branches; centre of branches with irregular spiral tubes; septal wall non-perforate. ?L.Eoc.-Rec. Indo-Pac., Carib., Africa, Medit. *Victoriella* (Fig. 194): adult high-spired, umbilicus depressed or forming axial hollow; chambers inflated; aperture umbilical. U.Eoc.-Mio. Austral., Papua, N.Z., Eur. *Eorupertia:* similar to *Victoriella*; aperture umbilical, interiomarginal, slit-like. U.Eoc. Pac., N.Amer., Carib., S.Amer., Japan. *Carpenteria* (Fig. 195): planoconvex, flat dorsal side attached; only chambers of last whorl visible on convex ventral side; keeled; aperture a slit extending down to base of last chamber and into umbilicus. M.Eoc.-Rec. Pac., Austral., Carib., E.Africa, M.East, Medit.

Family ORBITOIDIDAE

Biconcave to spherical; embryonic chambers enclosed by thick perforate wall, or with thinner-walled bilocular chambers followed by several fairly large periembryonic chambers; equatorial and lateral chambers not differentiated, or equatorial chambers covered on each side by distinct zones of lateral chambers; equatorial chambers arcuate or short; no canal system; intercameral foramina present. (U.Cret.-Pal.)

Members of this family (e.g. *Orbitoides, Schlumbergeria, Lepidorbitoides, Omphalocyclus* and *Simplorbites*) flourished in warm seas in Upper Cretaceous times. Only one genus (tentatively placed in the family) has been suggested to exist in the Tertiary: *Actinosiphon* (Fig. 196): embryonic chambers bilocular, completely surrounded by a ring of about eleven periembryonic chambers; equatorial chambers in irregular radial rows, each with a large median inter-cameral foramen. Pal. Mexico, Indo-Pac.

Family DISCOCYCLINIDAE
Circular or stellate in outline, thin or inflated, composed of an equatorial layer and lateral chambers on each side; microspheric forms with initial coil of small chambers; megalospheric proloculus consisting of a subspherical protoconch partly or completely surrounded by a deuteroconch; equatorial chambers annular, subdivided into rectangular to subhexagonal chamberlets; radial chamberlet walls, when present, arranged in annuli; equatorial chamberlets with annular and radial intercameral foramina connecting adjacent chambers; fissural interseptal spaces (referred to by some as intraseptal and intramural canal system) present. (Eoc.)

Discocyclina [*Orthophragmina*] (Figs 197, 198, 199): circular in plan, discoidal or lenticular, sometimes saddle-shaped; proximal intercameral foramen; radial chamberlet walls of equatorial chambers in adjacent annuli usually alternating. Eoc. Cosmop. (**98**). [Forms in the group of *D. archiaci* (Schlumberger) and *D. pratti* (Michelin) are flattened lenticular, nearly always with a small central boss or mamelon, and uniformly and finely granular; those in the group of *D. sella* (d'Archiac), *D. discus* (Rütimeyer) and *D.*

Figs 192–212. FORAMINIFERIDA: ORBITOIDACEA AND CASSIDULINACEA

Figs 198 and 200 after Schlumberger; Fig. 199 after Deprat; Figs 202, 204, 206 and 207 after H. Douvillé; Fig. 203 after Lemoine and R. Douvillé; Fig. 208 after Vaughan; Figs 192–196 and 212 after Loeblich and Tappan, in Moore; Figs 201 and 209–211 after Cole, in Moore; Figs 197 and 205 original.

192a, b, c. *Halkyardia minima* (Liebus), U.Eoc. France. T. × 65.
193. *Sporadotrema cylindricum* (Carter), Rec. Indian Ocean. T. × 2.
194h. *Victoriella conoidea* (Rutten), Olig. Austral. T. × 15.
195a, b, c. *Carpenteria balaniformis* Gray, Rec. Pac. T. × 10.
196f(A). *Actinosiphon semmesi* Vaughan, Pal. Mexico. T. × 40.
197. *Discocyclina sowerbyi* Nuttall, M.Eoc. Baluchistan. × 1·5.
198f, h. *Discocyclina pratti* (Michelin), U.Eoc. Aquitaine. f × 19, h × 11.
199f, h. *Discocyclina umbilicata* (Deprat), Eoc. New Caledonia. f × 10, h × 3·5.
200. *Aktinocyclina radians* (d'Archiac), M.Eoc. Alpes Maritimes, France. T. × 3.
201. *Asterocyclina stellata* (d'Archiac), M.-U.Eoc. France. T. × 2.
202f, k. *Lepidocyclina* (*s.s.*). k × 15.
203. *Nephrolepidina morgani* (Lemoine and R. Douvillé), L.Mio. Aquitaine. × 5.
204f, k. *Nephrolepidina*. k × 15.
205h. *Nephrolepidina galliennei* (Lemoine ad R. Douvillé), Mio. Funza I., E.Africa. × 19.
206k. *Nephrolepidina*. × 15.
207f, k. *Eulepidina*. k × 15.
208k. *Pliolepidina duplicata* (Cushman), U.Eoc. Panama. × 15.
209f.(A). *Polylepidina antillea* (Cushman), M.Eoc. W.Indies. T. × 20.
210f(A). *Helicolepidina spiralis* (Tobler), U.Eoc. Trinidad. T. × 13·5.
211f(A). *Helicostegina dimorpha* Barker and Grimsdale, M.Eoc. Mexico. T. × 28.
212a, b. *Pleurostomella subnodosa* (Reuss), U.Cret. Germany. T. × 24.

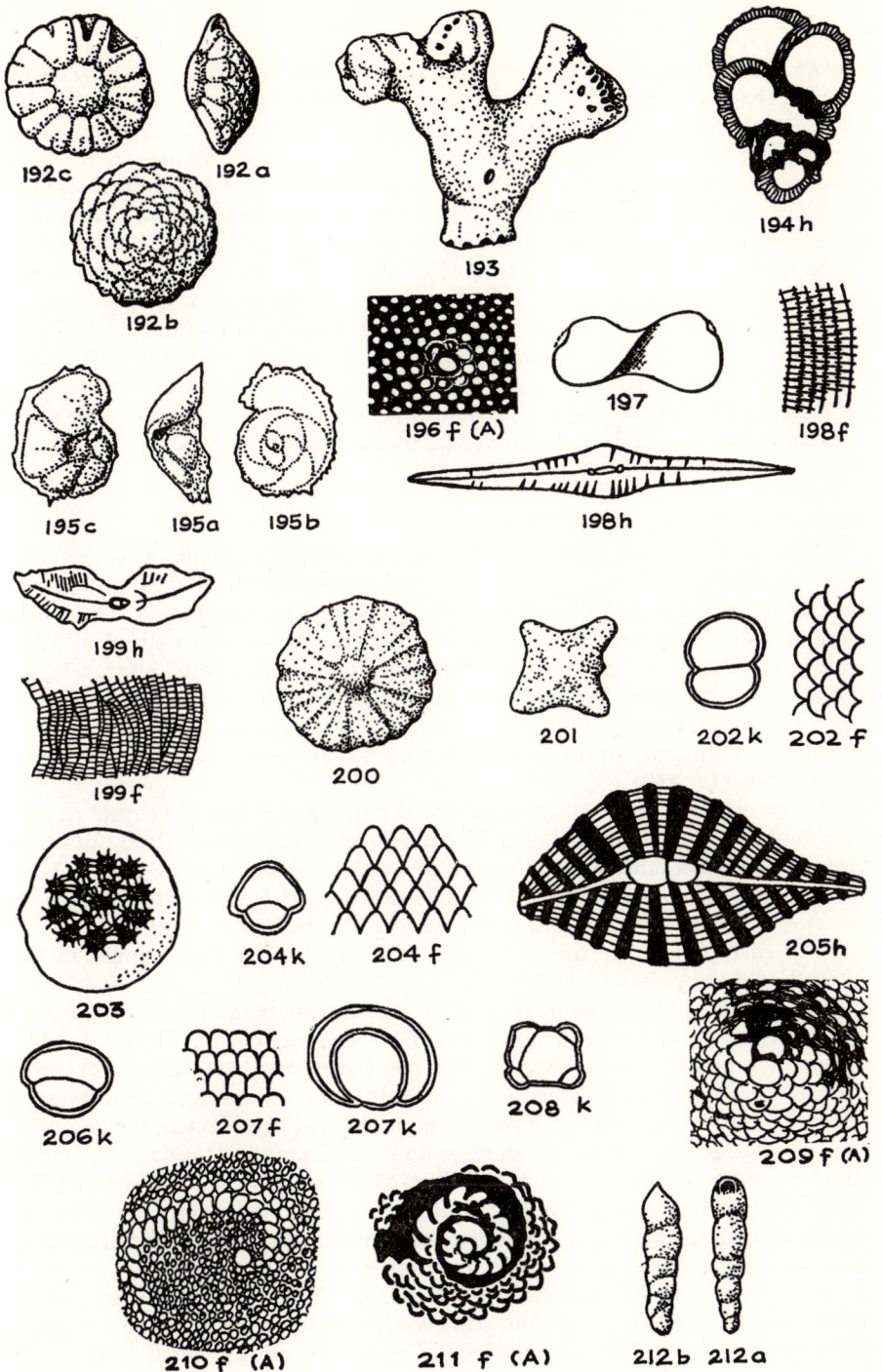

192c   192a

192b

193

194h

195c   195a   195b

196 f (A)   197   198f

198h

199 h

199 f   200   201   202k   202 f

203   204k   204 f   205h

206k   207f   207k   208 k   209 f (A)

210 f (A)   211 f (A)   212b 212a

umbilicata (Deprat) have a less pronounced boss and a tendency to be saddle-shaped, the granules are larger and more irregularly disposed, and the lateral chamberlets are rather smaller and grouped in rosettes of nine to thirteen round the granules; those in the group of *D. nummulitica* Gümbel are thicker, more lenticular, and with a thin brim, and the granules may be much larger in the centre.] *Aktinocyclina* (Fig. 200): like *Discocyclina*, but with radiating ridges which do not affect the outline. L.Eoc.-U.Eoc. Eur., Asia. *Asterocyclina* [*Asteriacites, Asterodiscus, Cisseis*] (Fig. 201): like *Discocyclina*, but with radiating ridges producing an irregular or stellate outline, number of ridges varying from 4 to 16 or more, often 5. L.Eoc.-U.Eoc. Eur., Indo-Pac., N.Amer., Carib., S.Amer. *Pseudophragmina:* like *Discocyclina*, but with distal intercameral foramen, and the distal parts of radial chamber walls incomplete or degenerate. L.Eoc.-U.Eoc. N.Amer., Carib., S.Amer. [In the subgenus *Athecocyclina* (L./M.Eoc.-U.Eoc. N.Amer., Carib., S.Amer.) the radial chamber walls are more or less absent.]

Family LEPIDOCYCLINIDAE

General form much as in *Discocyclina* except that angular forms are very few. Equatorial chamber layer on each side covered by tiers of lateral chambers or by laminated shell material with vacuoles; megalospheric embryonic chambers either (a) bilocular and followed by a long or short spiral of peri-embryonic chambers, or by reduced number of periembryonic chambers, or (b) of one large chamber and several smaller peripheral ones, all inside a thick wall; equatorial chambers arcuate, ogival, rhombic, spatulate or hexagonal; intercameral foramina present; no canal system. (M.Eoc.-M.Mio.) **(15, 52, 66, 115, 118)**

Pillars are often present at angles of the polygonal lateral chambers, but vary much in size and abundance. The family seems to have evolved out of an *Amphistegina*-like ancestor; it appeared in the Middle Eocene in the New World, but was very rare in the Middle and Upper Eocene elsewhere, only becoming common in the Mediterranean region, the Middle East and the Far East in early Oligocene times; it apparently died out a little sooner in the New World than elsewhere.

Lepidocyclina [*Isolepidina*] (Fig. 202): megalospheric embryonic chambers bilocular, more or less equal, separated by a straight wall. M.Eoc.-L.Mio. N.Amer., Carib., S.Amer., N.Afr., Pac. *Nephrolepidina* [*Amphilepidina*] (Figs. 203, 204; 205, 206 (*Amphilepidina*)): megalospheric embryonic chambers of unequal size, the deuteroconch partly embracing the protoconch, the common wall convex; equatorial chambers rhomboid, tending to hexagonal, or (*Amphilepidina*) more or less spatulate. U.Eoc.-M.Mio. Cosmop. *Eulepidina* (Fig. 207): megalospheric embryonic chambers large and bilocular, the deuteroconch almost completely surrounding the protoconch, both much flattened as seen in axial section; equatorial chambers more or less spatulate. M.Olig.-L.Mio. Cosmop. *Pliolepidina* [*Multicyclina, Multilepidina, Cyclolepidina*] (Fig. 208): like *Lepidocyclina*, but embryonic chambers consisting of one large chamber and several smaller peripheral ones, all inside a thick wall, much flattened as seen in axial section. U.Eoc.-M.Mio. Carib., S.Amer., Indo-Pac., Eur., E.Africa. [This feature has been regarded as 'teratological' (i.e. a freak) by some, but it occurs so commonly in all components of large populations of one species that it should be regarded as a normal and useful morphological character for use in classification.] *Polylepidina* (Fig. 209): megalospheric nucleoconch bilocular, followed by up to nearly one coil of distinct periembryonic chambers. M.Eoc.-U.Eoc. N.Amer., Carib., S.Amer., Burma. *Helicolepidina* (Fig. 210): megalospheric embryonic chambers bilocular, followed by an open spiral of chambers limited on their proximal side by a spiral band of one or two volutions, sometimes reaching periphery of test; equatorial chambers arcuate to subhexagonal. U.Eoc. N.Amer., Carib., S.Amer. [A few specimens of this genus have been found in the Lower Oligocene of Ecuador, but it seems possible that they might be derived.] *Helicostegina* (Fig. 211): lenticular, early chambers coiled in involute trochoid spire, later ones subdividing ventrally into chamberlets; aperture a narrow slit near inner margin of ventral face of last chamber, with backwardly directed lip as in *Tremastegina*. M.Eoc.-U.Eoc. Carib., N.Amer.

'ORBITOIDS' IN GENERAL

The vernacular term 'Orbitoids' is usually applied to members of the families Orbitoididae, Discocyclinidae and Lepidocyclinidae. They are more or less discoidal foraminifera with about the same range in size as nummulites. Their structure is usually fundamentally cyclical, not spiral (except in the earliest growth stages) and the test appears to be perforated throughout.

In external appearance the Orbitoids are discoidal, rarely of uniform thickness, usually thicker at the centre: the decrease in thickness from centre to margin may be gradual, or abrupt, giving rise to a central boss or button (termed the 'centrum') on a thin disk (Fig. 198). The disk may be flat or occasionally bent, sometimes saddle-shaped (Fig. 197). One group of species has radiating ribs which, when projecting from the margin, give a stellate outline (Fig. 201). The surface is usually, but not always, covered with very numerous tubercles (granulations) which, like those of nummulites, are the ends of vertical pillars (Figs 198, 205). The external form gives little clue to the internal structure.

It is usual to regard the Orbitoids as being flattened in a horizontal plane: sections perpendicular to this plane are then described as *vertical*, or, if they pass through the centre, *radial* (or *axial*); while those in a horizontal plane, if they divide the test quite symmetrically are *equatorial*, but if they are above or below the equatorial plane they are *tangential*. It is possible, however, to think of the test as being flattened in a vertical plane, and thus some authors (e.g. Wayland Vaughan) have used *meridional* instead of equatorial, and even those who use the latter term inconsistently call the regions above and below the equatorial plane *lateral*. As this last term cannot very well be misunderstood it is used here along with *equatorial*, in spite of the inconsistency.

Two special features characterize a large proportion of the Orbitoids: (1) their very early adoption of cyclical growth: this at once distinguishes them

79

from *Nummulites* (but not from *Cycloclypeus*); (2) the formation of external chambers, with or without pillars, on both sides of the median or equatorial plane: this distinguishes them (with the exception of the Cretaceous *Omphalocyclus*) from other cyclical genera (*Cycloclypeus, Cyclorbiculina, Orbitolites*) but not from *Spiroclypeus*. The pattern of the equatorial chambers is not seen, as in those other cyclical genera, on the surface, or even when the shell is decorticated; it can only be determined (the Cretaceous *Omphalocyclus* again excepted) by grinding away the lateral chambers and developing an equatorial section or polished surface. The patterns so exposed may be of great stratigraphical value, distinguishing broadly the Upper Cretaceous to Palaeocene Orbitoids, the Eocene Discocyclines, and the Middle Eocene to Middle Miocene Lepidocyclines. The age value of these orbitoidal groups has been disputed, mainly by Italian geologists who have asserted the presence of all three in Eocene strata in the Apennines and Sicily. So far as concerns *Lepidocyclina*, H. Douvillé, after a critical review of the evidence, came to the conclusion that Middle Eocene nummulites occur as derived fossils in post-Eocene beds with *Lepidocyclina*, and that similar mixtures may occur in other cases. There is little doubt that careful consideration of the geological history of a region from which a fossiliferous sample has been collected will in some cases help to decide whether radically to extend the ranges of some well known forms present or whether to regard them as derived fossils; regions where it is known that fossiliferous beds have been uplifted and eroded are particularly suspect.

Lepidocyclines, while flourishing in Middle and Upper Eocene times in the Central American region, were extremely rare in the Upper Eocene in the Old World. In Morocco, Bourcart and David found, in beds below those containing the Upper Eocene *Palaeonummulites fabianii, Lepidocyclina* associated with *Discocyclina* and *Asterocyclina*. In Algeria, Flandrin found a bed only half a metre thick in which *Lepidocyclina* was associated with Upper Eocene *Discocyclina* and *Palaeonummulites fabianii*, but also with *Assilina* and *Nummulites lucasanus* which are not believed to survive the Middle Eocene; hence, although Flandrin thought the association to be a normal one, it seems that some of the fossils have to be regarded as derived ones (**27, 57, 101**).

More recently, Professor van der Vlerk (e.g. **119, 120**) has used a formula concerned with the degree of enclosure of the protoconch by the deuteroconch in some lepidocyclines as a marker of geological time. Apart from disregarding the principle, which works very well in practice, that many other morphological characters are used to define species of lepidocyclines and to determine the specific name to be given to populations, the formula does result in some queer conclusions; for example, all the Far East 'letter stage' subdivisions from 'e$_1$' to 'e$_4$' were concluded to be of Rupelian (Middle Oligocene) age, whereas it is known that the genus *Nummulites* is well developed in the Middle Oligocene and does not survive into the Upper Oligocene. It therefore seems that the interval 'e$_{1-4}$' is, as Adams (in the press) suggests, of Upper (not Middle) Oligocene age and that the mathematical formula has not yet been proved to

80

be a marker of geological time. Its use by itself is hypothetical, and should be co-ordinated with other evidence.

Geographically, the Orbitoids were specially characteristic of the warm oceans—Tethys and Pacific. In late Cretaceous times the Orbitoididae spread as far north as the southern part of the Netherlands (Maastricht); in the Miocene the Lepidocyclinidae reached southern Australia and New Zealand, but at no time in the Tertiary did they extend north of southern France or the fortieth parallel in Japan. Unlike *Nummulites*, the Orbitoids were plentiful in tropical and subtropical America.

Superfamily CASSIDULINACEA

Test enrolled, planispiral, or trochospiral, or reduced to triserial or biserial or even uniserial, or biserial series may be planispiral; wall of perforate granular calcite; septa monolamellid; aperture a slit, loop-shaped, or multiple. (?Jur., L.Cret.-Rec.)

Family PLEUROSTOMELLIDAE

Early stage triserial or biserial and later uniserial, or uniserial throughout; aperture a curved slit, lateral or terminal, those of adjacent chambers joined by a siphon. (?Jur., L.Cret.-Rec.)

Pleurostomella (Fig. 212): elongate; early chambers biserial or alternating, later uniserial; aperture terminal, with projecting hood distally and two teeth proximally. L.Cret.-Rec. Cosmop. *Ellipsoglandulina* (Fig. 213): elongate, uniserial, not involute, with strongly overlapping chambers, base tapering; aperture terminal, semilunate. ?Jur., ?Cret., L.Eoc.-Rec. Eur., Carib., N.Amer., N.Z.

Family ANNULOPATELLINIDAE

Conical, proloculus followed by reniform second chamber, then uniserial, dorsal annular chambers overlapping on to ventral side, chambers subdivided by many radial tubules functioning as apertures. (Mio.-Rec.)

Annulopatellina (Fig. 214): characters of the family. Mio.-Rec. Austral., Carib.

Family CAUCASINIDAE

Elongate, early stage spiral, later may become uniserial; aperture loop-shaped, with internal tooth-plate connecting those of adjacent chambers. (U.Cret.-Rec.)

Caucasina (Fig. 215): early stage low-trochospiral with up to eight chambers per whorl, later stage becoming high-spired and with only three chambers per whorl; aperture an elongate loop at inner margin of last chamber, with distal lip. U.Cret.-Mio. Eur., Asia, N.Amer. *Fursenkoina* [*Virgulina* d'Orbigny *non* Bory de St. Vincent] (Fig. 216): elongate, rounded to ovate in cross-section; chambers higher than broad; early stage biserial and strongly twisted, later more typically biserial; aperture elongate comma-shaped, in face of last chamber. U.Cret.-Rec. Cosmop. *Sigmavirgulina:* like *Fursenkoina*, but test,

81

compressed and laterally carinate, and chambers broader than high. Mio.-Rec. Cosmop.

Family LOXOSTOMIDAE

Biserial, or later may become uniserial, usually compressed and laterally carinate; aperture interiomarginal or terminal; no tooth-plate or internal siphon. (U.Cret.-Eoc.)

Loxostomum (Fig. 217): elongate, compressed, quadrate in section, with more or less flat sides; biserial, tending to become uniserial in adults; sutures limbate, arched; aperture a terminal slit, usually with a finely crenulated lip. U.Cret.-Pal. Eur., N.Amer. *Aragonia:* biserial, compressed, flabelliform, with marginal keel; sutures limbate; aperture small, a low opening at base of last chamber. Eoc. N.Z., N.Amer., Carib., Eur.

Family CASSIDULINIDAE

Lenticular, subglobular or elongate; planispiral at least in early stage, later may uncoil; alternating chambers biserially arranged; aperture a comma-shaped slit in lower part of last chamber face. (Eoc.-Rec.)

Cassidulina (Fig. 218): lenticular, usually biumbonate; only alternate chambers on each side of periphery reach boss; usually smooth. Eoc.-Rec. Cosmop.

Superfamily NONIONACEA

Planispiral to low trochospiral; wall of perforate granular calcite; septa monolamellid; aperture interiomarginal or areal. (Jur.-Rec.)

Family NONIONIDAE

Planispiral or trochospiral; aperture interiomarginal or areal. (Jur.-Rec.)

Nonion (Fig. 219): planispiral, involute, biumbonate, smooth, periphery

---

Figs 213–229. FORAMINIFERIDA: CASSIDULINACEA, NONIONACEA AND ANOMALINACEA

Fig. 229 after Brady; Figs 213–228 after Loeblich and Tappan, in Moore.
213a, b. *Ellipsoglandulina laevigata* A. Silvestri, Plio. Sicily. T. × 22.
214a, b, c. *Annulopatellina annularis* (Parker and Jones), Rec. Austral. T. × 64.
215a, b. *Caucasina schischkinskayae* (Samoylova), Olig. Caucasus. T. × 53.
216a, b. *Fursenkoina squammosa* (d'Orbigny), Plio. Italy. T. × 22.
217a, b. *Loxostomum eleyi* (Cushman), U.Cret. Arkansas. × 52.
218a, b, c. *Cassidulina laevigata* d'Orbigny, Rec. Italy. T. × 39.
219a, b. *Nonion incrassatus* (Fichtel and Moll), Mio. Albania. T. × 50.
220a, b. *Florilus asterizans* (Fichtel and Moll), Rec. Italy. T. × 24.
221a, b, c. *Nonionella miocenica* Cushman, Mio. Calif. T. × 50·5.
222a, b. *Pullenia bulloides* (d'Orbigny), Mio. Austria. T. × 26.
223a, b. *Astrononion stelligera* (d' Orbigny), Rec. Atl. T. × 62·5.
224a, b. *Chilostomella ovoidea* Reuss, Mio. Austria. T. × 31·5.
225a, b, c. *Alabamina wilcoxensis* Toulmin, L.Eoc. Alabama. T. × 70.
226a, b, c. *Gyroidina orbicularis* d'Orbigny, Rec. Italy. T. × 37.
227a, b, c. *Cibicidoides mundula* (Brady, Parker and Jones), Rec. Brazil. T. × 54·5.
228a, b, c. *Heterolepa dutemplei* (d'Orbigny), Mio. Austria. T. × 16·5.
229a, b. *Melonis pompilioides* (Fichtel and Moll), Rec. T. × 56.

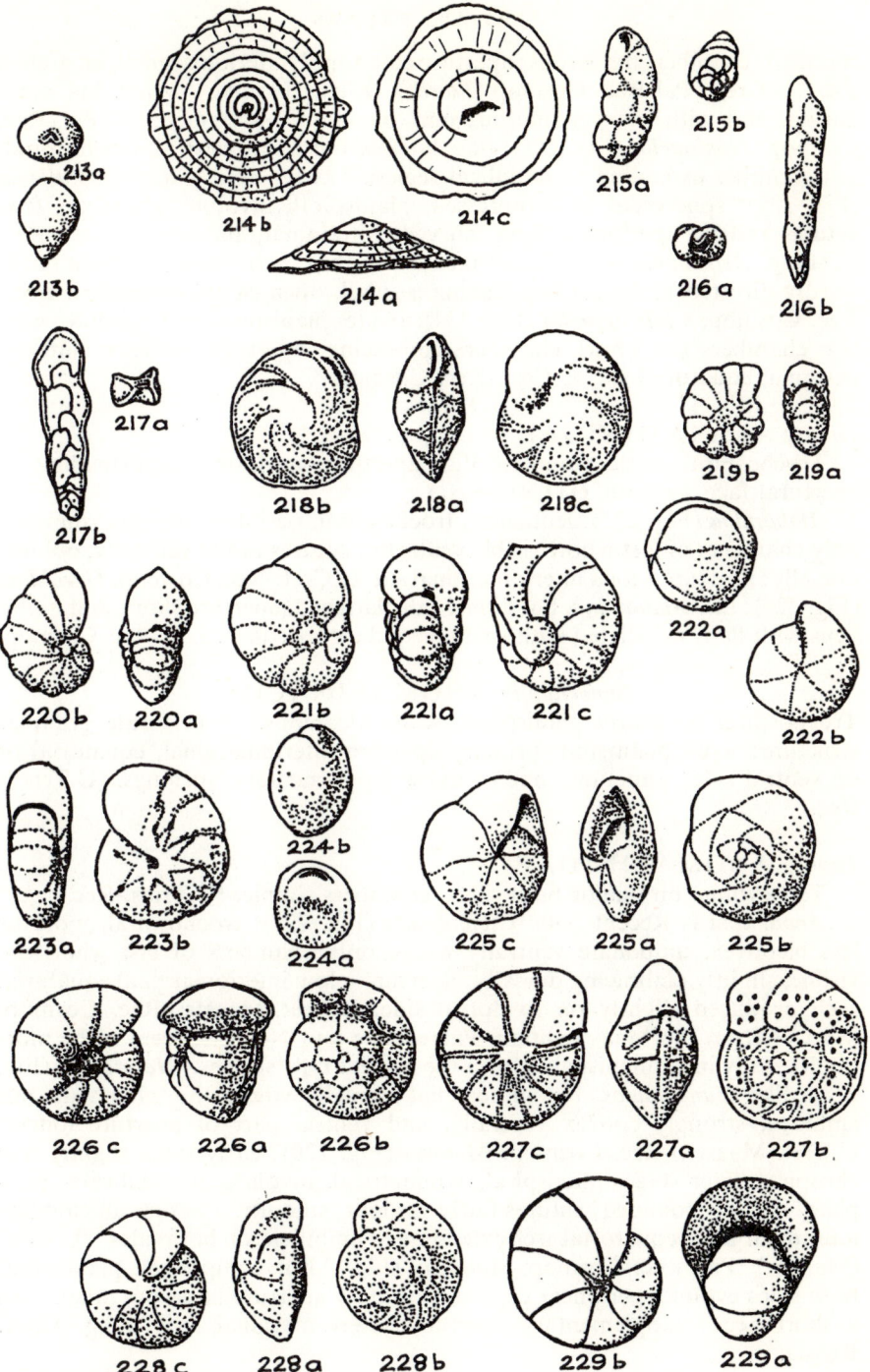

213a

213b

214b

214c

214a

215a

215b

216a

216b

217a

217b

218b

218a

218c

219b 219a

220b 220a

221b 221a 221c

222a

222b

223a 223b

224b

224a

225c 225a 225b

226c 226a 226b

227c 227a 227b

228c 228a 228b

229b 229a

rounded; chambers numerous; aperture equatorial, interiomarginal, an arched slit. ?U.Cret., Pal-Rec. Cosmop. *Florilus* (Fig. 220): like *Nonion*, but more flaring, and with sharper margin; may be slightly asymmetrical. Pal.-Rec. Cosmop. *Nonionella* (Fig. 221): like *Nonion*, but asymmetrical, trochospiral, last chamber extending to umbilical region. U.Cret.-Rec. Cosmop. *Pullenia* (Fig. 222): spheroidal to compressed, planispiral, involute; chambers few, sutures radial; aperture a long, curved, interiomarginal slit. U.Cret.-Rec. Cosmop. *Astrononion* (Fig. 223): like *Nonion*, but each chamber with a backwardly directed umbilical flap leaving a small, open cavity beneath it. Eoc.-Rec. Cosmop. *Chilostomella* (Fig. 224): ovate, planispiral and involute, with two chambers per whorl, chambers embracing; aperture a narrow, interiomarginal equatorial slit. U.Cret.-Rec. Cosmop.

Family ALABAMINIDAE

Trochospiral; septa monolamellid; aperture basal, or a slit extending up apertural face, or both. (U.Cret.-Rec.)

Alabamina (Fig. 225): lenticular, trochospiral, periphery sharply rounded, only chambers of last whorl visible ventrally; sutures radial ventrally, oblique dorsally; aperture a long interiomarginal slit. U.Cret.-Rec. Cosmop. *Gyroidina* (Fig. 226): like *Alabamina*, but dorsal side flattened, aperture short, and umbilicus with flaps associated with small secondary apertures. Eoc.-Rec. Cosmop.

Superfamily ANOMALINACEA

Trochospiral to nearly planispiral; wall calcareous, of perforate granular structure; septa bilamellid; primary aperture interiomarginal, equatorial or on ventral side; sometimes one or more supplementary openings. (U.Trias.-Rec.)

Family ANOMALINIDAE

Test evolute on one or both sides; chambers simple. (U.Trias.-Rec.)

Anomalina is Recent only. *Cibicidoides* (Fig. 227): trochospiral, more or less biconvex, umbonate ventrally where only chambers of last whorl are visible, slightly embracing dorsally; aperture a low interiomarginal equatorial arch produced slightly on to dorsal side. U.Cret.(Maastr.)-Rec. Cosmop. [Species from the Maastrichtian, Palaeogene and Neogene seem to be morphologically indistinguishable from the Recent type species.] *Heterolepa* (Fig. 228): like *Cibicidoides*, but almost flat dorsally where the whorls do not embrace, strongly convex ventrally, and ventral part of aperture longer. U.Cret.(Maastr.)-Rec. Cosmop. *Melonis* (Fig. 229): early stage slightly trochospiral, later stages planispiral, symmetrical, involute, biumbilicate; periphery broadly rounded; sutures fairly straight; smooth; aperture an elongate interiomarginal, equatorial arch extending to umbilicus on both sides. ?U.Cret. (Maastr.), Pal.-Rec. Cosmop. *Almaena* (Fig. 230): compressed planispiral, both sides evolute, periphery carinate; primary aperture basal but areal, with a short neck; supplementary apertures latero-marginal only. Olig.-Aquit. Russia.

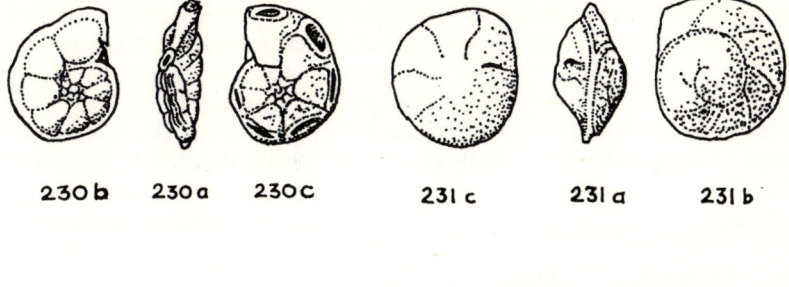

230b   230a   230c   231c   231a   231b

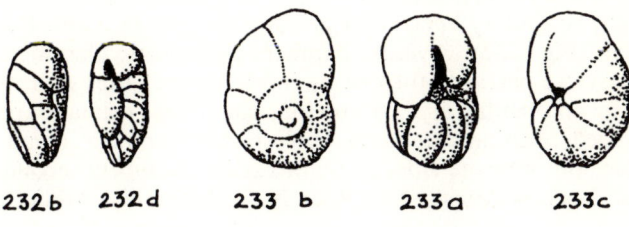

232b   232d   233 b   233a   233c

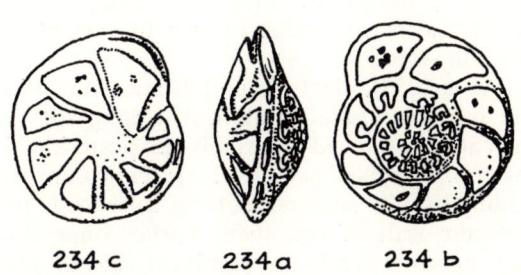

234 c   234a   234 b

Figs 230–234. FORAMINIFERIDA: ANOMALINACEA AND ROBERTINACEA

Figs 230–234 after Loeblich and Tappan, in Moore.
230a, b, c. *Almaena taurica* Samoylova, U.Eoc. Crimea. T. × 12·5.
231a, b, c. *Osangularia lens* Brotzen, Danian. Sweden. × 55·5.
232b, d. *Robertina arctica* d'Orbigny, Rec. Spitzbergen. T. × 27·5.
233a, b, c. *Ceratobulimina contraria* (Reuss), M.Olig. Germany. T. × 45.
234a, b, c. *Hoeglundina elegans* (d'Orbigny), Rec. Carib. T. × 15·5.

Family OSANGULARIIDAE

Trochospiral; aperture with interiomarginal and vertical or oblique parts which may be separate or joined, the areal part sometimes multiple. (L.Cret.-Rec.)

Osangularia (Fig. 231): lenticular, biumbonate, periphery carinate; only last whorl visible ventrally; interiomarginal and simple oblique parts of aperture separate or joined. Cret.-Rec. Cosmop. *Gyroidinoides:* dorsal side

85

flattened, ventral side very convex, periphery rounded; aperture a long, low, interiomarginal slit extending from periphery to umbilicus, the umbilical part partially obscured by a flap from each chamber. Cret.-Rec. Cosmop.

Superfamily ROBERTINACEA

Trochospiral; chambers divided internally by partitions which are more important in advanced forms; wall aragonitic, perforate-radial; aperture a slit in chamber face; secondary aperture in each septum above partition. (?Trias., Jur.-Rec.)

Family ROBERTINIDAE

Test high; secondary aperture a primary feature. (U.Cret.-Rec.)

Robertina (Fig. 232): several chambers per whorl; chambers divided by double transverse partition formed by infoldings of outer wall, the two parts connected by an opening against previous chambers; primary aperture a long comma-shaped opening extending up last chamber face; small supplementary aperture on opposite side, those of earlier chambers secondarily closed. L.Eoc.-Rec. Eur., N.Amer., Atl., Pac., N.Z., Tasmania, Arctic, Antarctic.

Family CERATOBULIMINIDAE

Trochospiral, coiling dextral or sinistral; primary aperture closed when new chamber added and new opening formed by resorption above internal partition. (?Trias., Jur.-Rec.)

Ceratobulimina (Fig. 233): dextral, umbilicate, whorls opening rapidly, smooth; aperture interiomarginal, umbilical, a long slit extending up last chamber face; partition attached internally to posterior side of aperture. U.Cret.-Rec. Cosmop. *Hoeglundina* (Fig. 234): lenticular, periphery angular, no umbilicus, coiling usually sinistral, sutures thickened; internal partitions extend from posterior wall of chambers, earlier ones always secondarily resorbed; primary aperture oval, on umbilical side of peripheral margin. M.Jur.-Rec. Cosmop.

THE MICROSCOPIC STUDY OF FORAMINIFERA

Larger foraminifera may be found in comparatively soft strata from which they can be picked or washed out, or embedded in hard limestones or occasionally in hard sandstones. When such hard rocks are examined at outcrop, it is always advisable to search the weathered surface or the overlying soil for loose specimens, but, unless the matrix matches exactly the underlying rock, the collector should note that they were '*not in situ*': this cannot, of course, be done in the case of hard rocks brought up as cores from a boring. The disadvantage of hard rocks is that it is very time-consuming to obtain properly oriented sections, but when no soft material is available the time is well spent. Even discoidal foraminifera do not usually lie along the bedding planes: they lie indiscriminately in all planes and sections seen in an ordinary rock slide are usually almost completely random sections. It is only by good luck that a section will cut an orbitoid equatorially or a nummulite tangentially, and structures essential for accurate determination can often not be seen. For ordinary rock slide work the methods of sectioning are the same as for any igneous or other rock, but owing to limestones being usually considerably softer they are ground down much more quickly, and the later stages of grinding are best carried out on glass or water-of-ayr stone.

However, in the case of hard rocks, a technician well trained in the morphology of foraminifera and who knows for what particular sections to look can be invaluable. He can take relatively small chips of rock and, by varying the plane of grinding from time to time, make an almost complete, properly oriented thin section for study.

If isolated specimens are available, the requisite oriented sections can be obtained much more easily. In general, at least two and often three oriented thin sections are required for the accurate identification of any one species of foraminifer. The grinding may, as in the case of hard rocks, be begun with carborundum or emery, but as the required plane of section is approached it is better to use glass or water-of-ayr stone to finish off to avoid over-shooting the mark by too rapid abrasion. Careful cleaning and sorting of specimens from a sample, into groups for study, is essential.

For examination in the field, where proper equipment is not available, rough sections or polished surfaces can be made by using glass or steel plates with various grades of carborundum, completing the work on a water-of-ayr stone. When isolated specimens are abundant, a view of the equatorial layer may be obtained (e.g. in nummulites) by heating the specimen in a flame and dropping it when hot into cold water when it will often split into two parts along the equatorial plane.

The prime importance of thin section work in the study of foraminifera cannot be emphasized too much. Such studies are now also applied to microforaminifera becasue it is not only the internal morphology which is useful, but also the wall structure, the type of which is extremely important in classification.

GLOSSARY OF PRINCIPAL TECHNICAL TERMS
APPLIED TO FORAMINIFERIDA

(Including some terms not used in the present work. A full Glossary will be found in Loeblich and Tappan, in Moore (78), pp. C58–C65).

Accessory apertures. Test openings that do not lead directly into primary chambers but extend beneath or through accessory structures.

Acervuline. With chambers in irregular clusters or heaps (e.g. *Acervulina*).

Adventitious. Composed of foreign particles (e.g. in agglutinated test).

Alar prolongations. The portions of the septa or of the chamber cavities which lie within the lateral or overlapping part of a whorl (e.g. *Nummulites*).

Alveolus. Minute blind cavity in shell wall or blind chamberlet opening only backwards.

Annular. Ring-shaped; cyclical.

Aperture. Opening(s) from chamber of test to exterior.

Areal aperture. Aperture in face of last chamber of test.

Axial. Lying in or containing the axis of a spiral or cyclical shell; e.g. *axial section*, a section along a plane passing through the axis.

Axis [of construction]. The imaginary line around which a spiral or cyclical shell is built up.

Bilamellar. Walls of each chamber consisting of two primarily formed layers.

Biserial. Having the chambers in two parallel series.

Bulla. Blister-like structure in planktonic foraminifera that partially or completely covers primary or secondary apertures, not closely related to primary chambers; may be umbilical, sutural or areal in position and may have one or more accessory marginal apertures.

Canalicular skeleton. Shell substance not perforated by vertical pores, but containing a branching series of fine tubular cavities (canals): in rotaliids and nummulites.

Canal system. See *Canalicular skeleton.*

Chamber. The unit of construction in foraminiferal tests: a cavity or group of cavities, within the bounding wall or walls, secreted all at one time.

Chamberlet. Subdivision of a chamber produced by presence of axial or transverse septula.

Clavate. Club-shaped.

Column. (1) The same as *pillar, q.v.;* (2) a vertical series of chambers in biserial or triserial forms.

Convolute. Evolute, enrolled (all whorls visible).

Cornuspirine. Having tube-like, planispirally coiled test.

Costa. Raised ridge.

Cribate, cribriform. Perforated like a sieve.

Cyclical. In rings; used especially of that mode of growth in which each chamber forms a complete ring encircling the previous chamber.

Deuteroconch. Chamber formed immediately subsequent to proloculus and forming part of the embryonic apparatus.

Dimorphism. The occurrence in one species of two distinct forms—megalospheric and microspheric tests.

Embryonic apparatus. Group of chambers usually at centre of some megalospheric tests, larger in size and different in shape and arrangement from other chambers; nucleoconch.

Entosolenian Having internal tube-like apertural extension.

Ephebic. Pertaining to adult stage in ontogeny.

88

Equatorial. Situated in the median plane.

Equitant. Inverted V-shaped in section, the limbs of the V being astraddle the inner whorls.

Extraumbilical aperture. One in last chamber, not connecting with umbilicus, usually midway between umbilicus and periphery.

Extraumbilical-umbilical aperture. One in last chamber that extends along its forward margin from umbilicus toward periphery, becoming extraumbilical in part.

Falciform. Sickle-shaped, i.e. consisting of a strong curve and a straight (or nearly straight) line.

Flabelliform. Shaped like a fan.

Foramen. Opening between chambers.

Frondicularian. Consisting of a column of inverted V-shaped chambers.

Granulations or *granules.* Protuberances from the surface of a lenticular test, being the ends of pillars (*q.v.*).

Helicoid. Asymmetrically spiral (like most gastropod shells).

Homoeomorphy. Similarity in appearance between forms of different descent (e.g. *Dictyoconus* and *Dictyoconoides*, *Orbitolites* and *Discospirina*).

Hyaline. Glassy clear, transparent.

Imperforate. Without the innumerable pores normal to the surface which are part of the fundamental structure of a 'perforate' test.

Infralaminal accessory aperture. Opening in planktonic foraminifera at margin of accessory structures and leading to cavity beneath.

Intercameral. Between chambers: applied to septa and (more frequently to an opening connecting adjacent chambers (*intercameral foramen*).

Interiomarginal aperture. Basal opening in test at margin of last chamber, along final suture; in coiled forms may be umbilical, extraumbilical or equatorial in position.

Interseptal. Located between septa.

Intralaminal accessory aperture. Opening in planktonic foraminiferal test leading through accessory structures into cavity beneath, not directly into chamber cavity.

Intraseptal. Located within septum; e.g. between the two layers of a septum in nummulites: applied to cavities forming part of the canal system.

Intraumbilical aperture. Opening of test located solely in umbilicus.

Involute. In enrolled forms, later whorls completely enclosing earlier ones.

Lenticuline. Lens-shaped (e.g. *Lenticulina*).

Limbate. Referring to thickened margin of chamber, usually at suture.

Lineage. A series of species following one another in time and apparently related to one another as direct (or closely collateral) ancestors and descendants.

Lip. Elevated border of aperture.

Loculus. Chamber.

Marginal cord. The portion of the spiral lamina of a nummulite lying in the equatorial plane, at the junction of the two limbs of the V-shaped section; thicker and more coarsely perforate than the rest of the spiral lamina.

Meandrine. Changing direction repeatedly: applied particularly to the septal filaments of certain types of nummulites.

Megalosphere. The large initial chamber of a megalospheric test (A Form). (When there are two or more initial chambers, the term nucleoconch is generally applied.)

Microsphere. The small initial chamber of a microspheric test (B Form).

Milioline. Spiral with a continually changing axis (e.g. as in Miliolacea).

Monolamellid. Lamellar hyaline tests with single-layered septa and wall of last formed chamber.

Multilocular. Consisting of many chambers: *multicamerate.*

89

Multispiral. Having a relatively large number of whorls in a given diameter. (Contrast *paucispiral.*)

Nautiloid. Planispiral with symmetrically equitant chambers.

Neanic. Youthful stage in ontogeny.

Nepionic. Stage immediately after embryonic stage in ontogeny.

Nodosarian. Consisting of a single straight row or column of chambers.

Nucleoconch. See *embryonic apparatus.*

Ontogeny. Life history, with special reference (in the case of foraminifera) to changes in shell structure, in the shape and size, structural details and arrangement of the chambers, as between those secreted at different periods of life.

Paucispiral. Having a relatively small number of whorls in a given diameter. (Contrast *multispiral.*)

Peneropline. Having the form of *Peneroplis.*

Periembryonic chambers. Nepionic parts of test.

Phylogeny. The history of the evolution of any lineage: believed to be indicated by the ontogeny of the later members.

Pillar. A solid, tapering mass of ordinary test (orbitoids) or canalicular skeleton (nummulites), more or less acutely conical or pyramidal, traversing several whorls or layers of a shell and ending at the surface in a *granule.*

Planispiral. Coiled in a single plane.

Polar. Situated at or near the poles of a discoidal shell, i.e. the points farthest from the equatorial plane.

Polyphyletic. Of more than one lineage. Applied to 'genera' containing homoeomorphic but unrelated species, unlike true genera which should consist only of immediately related species.

Porcellanous. Having calcareous, white, shiny, and usually imperforate wall resembling porcelain in surface appearance; usually brown in transmitted light (contrasted with *hyaline*, resembling glass).

Porticus. An asymmetrical, imperforate apertural flap.

Primary aperture. Main opening of test, may be only one or accompanied by secondary apertures.

Progressive characters. Characters which show a definite trend in evolution, so that they undergo very closely similar changes in the successive forms of different lineages.

Proloculus. Initial chamber.

Pustule. A coarser and projecting form of granulation.

Quinqueloculine. Having five externally visible chambers in milioline type of growth.

Radial. In lenticular forms the direction from pole to any part of circumference: (1) Radial section = axial section; (2) Radial filaments, running approximately in the radial direction.

Radiate aperture. An aperture with numerous radiating ridges and depressions around it.

Relict apertures. Short radial slits around umbilicus which remain open when umbilical portions of equatorial aperture are not covered by succeeding chambers.

Reticulate. Forming a network: applied especially to the most complex forms of septal filament in nummulites, which, by the irregularity of their course and the presence of transverse laminae, produce a network in which the individual filaments are not traceable.

Retral processes. Pocket-like cavities in the rear wall of a chamber (e.g. in *Elphidium*).

Rotaliform. Asymmetrically planispiral, chambers being equitant on one side and not on the other.

90

Secondary apertures. Additional or supplementary openings into main chamber cavity; position may be areal, sutural or peripheral.

Septal filament. The trace of the alar extension of a septum on the lateral part of the spiral lamina (in nummulites).

Septum. The partition between two chambers; usually composed of the same kind of shell substance as the outer wall, being in fact part of the wall of one chamber enclosed by the formation of another; may be a single layer (monolamellid), be secondarily doubled enclosing canal systems (e.g. rotaliids), or be primarily double (bilamellid).

Sieve-plate. See *trematophore.*

Sigmoidal. S-shaped.

Spatulate. Having an outline formed by a semicircle, two parallel straight sides and two re-entrant quarter-circles, thus approximating to an elongate hexagon.

Spiral canals. Part of the canal system, in the umbilical region of a spiral shell, parallel and just internal to the lateral margins of the chambers (e.g. *Elphidium*).

Spiral lamina. The main part of a nummulite shell, consisting of a V-shaped lamina spirally enrolled.

Spiroloculine. Planispiral with each chamber occupying half a whorl (e.g. *Spiroloculina*).

Spiroumbilical. Interiomarginal aperture extending from umbilicus to periphery and finally on to dorsal (spiral) side.

Stolon. An inappropriate term. See *intercameral foramen.*

Streptospiral. Coiled like a ball of twine.

Subreticulate. Applied to the network formed by the contact of irregular septal filaments, not complicated by the union of transverse laminae, in nummulites.

Supplemental skeleton. Canalicular skeleton.

Supplementary apertures. Secondary openings in test which may be additional to primary aperture; may sometimes completely replace primary aperture.

Sutural supplementary apertures. Rather small sutural openings, one or more per suture; may be restricted to dorsal side (e.g. *Truncorotaloides*), restricted to ventral side, or present on both sides (e.g. *Candeina*).

Suture. The external line of contact of successive whorls in a spiral shell.

Tangential section. One through part of test parallel to axis of coiling but not through proloculus.

Tegillum. In planktonic foraminifera, umbilical coverings formed by extensions of chambers across umbilicus and completely covering primary aperture; may have small openings along their margins or a median hole.

Test. The shell or skeletal covering.

Tooth. Projection in aperture.

Tooth-plate. An internal contorted plate that extends from the aperture to the previous chamber's septal foramen.

Trabeculae. Broad crinkles or grooves on the septal filaments of nummulites bordered by short transverse laminae.

Trematophore. Perforated plate over aperture of some milioloids.

Triloculine. Having three externally visible chambers as in *Triloculina.*

Trimorphic. Having two kinds of megalospheric individuals as well as the microspheric.

Triserial. Having the chambers arranged in three columns, high trochospiral, with three chambers in each whorl.

Trochoid. Conical with fairly flat base and sides.

Trochospiral. Spirally coiled to produce a trochoid test.

91

Umbilicus. Space formed between inner margins of umbilical walls of chambers belonging to the same whorl.

Umbo. Central elevated structure in discoidal and lenticular forms usually due to lamellar thickening (e.g. *Lenticulina*).

Umbonate. Having an umbo: if on both sides, *biumbonate.*

Unilocular. Consisting of one chamber only.

Uniserial. Having the chambers in a single column.

Vacuoles. Cavities in test wall.

Vertical. A term varying in meaning according to the position in which a shell is placed. In elongate shells (e.g. *Nodosaria, Textularia*) the long axis is thought of as vertical; in planispiral shells, it is the direction perpendicular to the axis of coiling (which in *Fasciolites* is the long axis, but in *Nummulites* is the short axis); in cyclical shells, on the contrary, it is any direction which includes the axis (i.e. it is the same as radial.)

Vitreous. Having the appearance and lustre of glass; hyaline. (Contrast *porcellanous.*)

Whorl. A single turn of a spiral test (through 360°).

SOME FRENCH TERMS

Bourrelet spiral. Marginal cord.

Cloison. Septum.

Filets cloisonnaires. Septal filaments.

Lame spirale. Spiral lamina.

Loge. Chamber.

Logette. Chamberlet.

Maille. Best translated as 'pattern'.

Méandriforme. Meandrine.

Ogivales. Pointed-arched. (*Ogive* is the Gothic arch.)

Parois. Wall or periphery.

Tourbillonant. Sigmoidal.

SOME GERMAN TERMS

Dorsalstrang. Marginal cord.

Kammer. Chamber.

Kammermündung. Aperture of chamber.

Pfeil, Pfeilerkegel. Pillar.

Röhrchen. Tubules, perforations.

Scheidewand. Septum.

Windung. Whorl.

Zwischenskelet. Supplementary skeleton.

SELECT BIBLIOGRAPHY OF TERTIARY FORAMINIFERIDA

1. ABRARD, R. 1928. 'Contribution à l'étude de l'évolution des Nummulites', *Bull. Soc. géol. Fr.*, (4), **28**, 161–182.
2. ADAMS, C. G. 1965. 'The Foraminifera and stratigraphy of the Menilau Limestone, Sarawak, and its importance in Tertiary correlation', *Q. Jl. geol. Soc. Lond.*, **121**, (3), 283–338, pl. 21–30.
3. AGIP MINERARIA. 1957. *Foraminiferi Padani* (*Terziario e Quaternario*), pl. 1–52 (AGIP Mineraria: Milano).
4. ASTRE, G. 1923. 'Étude paléontologique des Nummulites de Crétacé supérieur de Cezan-Lavardens (Gers) [*Nummulites mengaudi* n.sp.]', *Bull. Soc. géol. Fr.*, (4), **23**, 360–368, pl. 12.
5. BAGG, R. M. 1905. 'Miocene Foraminifera from the Monterey Shale of California . . .', *Bull. U.S. Geol. Surv.*, **268**, 1–78, pl. 1–11.
6. BAKX, L. A. J. 1932. 'De genera *Fasciolites* en *Neoalveolina* in het Indo-Pacifische Gebied', *Verh. geol-mijnb. Genoot. Ned.*, **9**, 205–266, pl. 1–6.
7. BANDY, O. L. 1949. 'Eocene and Oligocene Foraminifera from Little Stave Creek, Clarke County, Alabama'. *Bull. Am. Paleont.*, **32**, no. 131, 210 pp., pl. 1–27.
8. BANDY, O. L. 1960. 'The geologic significance of coiling ratios in the foraminifer *Globigerina pachyderma* (Ehrenberg)', *J. Paleont.*, **34**, 671–681.
9. BANNER, F. T. 1966. 'The Morphology, Classification and stratigraphic Value of the Spirocyclinidae', *Vop. Mikropaleont.*, **10**, 201–224, pl. 1–20.
10. BANNER, F. T. and BLOW, W. H. 1959. 'The classification and stratigraphical distribution of the Globigerinaceae', *Palaeontology*, **2**, (1), 1–27, pl. 1–3.
11. BANNER, F. T. and BLOW W. H. 1960. 'The taxonomy, morphology and affinities of the genera included in the subfamily Hastigerininae', *Micropaleontology*, **6**, (1), 19–31.
12. BANNER, F. T. and BLOW, W. H. 1960. 'Some primary types of species belonging to the superfamily Globigerinacea', *Contr. Cushman Fdn foramin. Res.*, **11**, (1), 1–41, pl. 1–8.
13. BANNER, F. T. and BLOW, W. H. 1965. 'Progress in the Planktonic Foraminiferal Biostratigraphy of the Neogene', *Nature, Lond.*, **208**, no. 5016, 1164–1166.
14. BARKER, R. W. 1939. 'Species of the foraminiferal family Camerinidae in the Tertiary and Cretaceous of Mexico', *Proc. U.S. Natn. Mus.*, **86**, no. 3052, 305–330, pl. 11–22.
15. BARKER, R. W. and GRIMSDALE, T. F. 1936. 'A contribution to the phylogeny of the orbitoidal Foraminifera, with descriptions of new forms from the Eocene of Mexico', *J. Paleont.*, **10**, 231–247, pl. 30–38.
16. BELLEN, R. C. van. 1946. 'Foraminifera from the middle Eocene in the southern part of the Netherlands Province of Limburg', *Meded. geol. Sticht.*, (C), **5**, no. 4, 1–144, pl. 1–13.
17. BERGGREN, W. A. 1966. 'Phylogenetic and taxonomic problems of some tertiary planktonic foraminiferal lineages', *Vop. Mikropaleont.*, **10**, 309–332, 4 figs.
18. BERMÚDEZ, P. J. 1949. 'Teritiary smaller Foraminifera of the Dominican Republic', *Spec. Publs Cushman Fdn*, **25**, 1–322, pl. 1–26.
19. BERMÚDEZ, P. J. 1952. 'Estudio sistemático de los Foraminíferos rotaliformes', *Boln Geol. Minist. Minas Venez.*, **2**, no. 4, 1–230, pl. 1–35.

20. BERMÚDEZ, P. J. 1961. 'Contribución al estudio de las Globigerinidea de la region Caribe-Antillana (Paleoceno-Reciente)', *3rd. Congr. Geol. Venezolano, Bol. Geol., Mem.*, 3, spec. publ. 3 (1960), 1, 119–1, 393, pl. 1–20.

21. BLOW, W. H. 1956. 'Origin and evolution of the foraminiferal genus *Orbulina* d'Orbigny', *Micropaleontology*, 2, (1), 57–70.

22. BLOW, W. H. 1959. 'Age, correlation, and biostratigraphy of the upper Tocuyo (San Lorenzo) and Pozón formations, Eastern Falcón, Venezuela', *Bull. Am. Paleont.*, 39, no. 178, 67–251, pl. 6–191.

22a. BLOW, W. H. 1969. 'Late Middle Eocene to Recent Planktonic Foraminiferal Biostratigraphy', *Proc. 1st. Internat. Conf. on Planktonic Microfossils Geneva 1967*, 1, 199–422, pl. 1–54 (E. J. Brill: Leiden).

23. BOGDANOVICH, A. K. 1952. 'Miliolidy i Peneroplidy, Iskopaemye Foraminifery SSSR', *VNIGRI, Trudy*, N.S., no. 64, 1–338, 39 pls.

24. BOLLI, H. M. 1957. 'Planktonic Foraminifera from the Oligocene-Miocene Cipero and Lengua formations of Trinidad, B.W.I,', *Bull. U.S. Natn. Mus.*, 215, 97–121, pl. 22–29.

25. BOLLI, H. M. 1957. 'Planktonic Foraminifera from the Eocene Navet and San Fernando formations of Trinidad, B.W.I.', *Bull. U.S. natn. Mus.*, 215, 155–172, pl. 35–39.

26. BORNEMANN, J. G. 1855. 'Die mikroskopische Fauna des Septarionthones von Hermsdorf bei Berlin', *Z. dt. geol. Ges.*, 7, 307–371, pl. 12–21.

27. BOURCART, J. and DAVID, E. 1933. 'Étude stratigraphique et paléontologique des grès à foraminifères d'Ouezzan au Maroc', *Mém. Soc. Sc. nat. Phys. Maroc*, no. 37, 1–55, pl. 1–14.

28. BOUSSAC, J. 1911. 'Études paléontologiques sur le Nummulitique Alpin', *Mém. Carte géol. dét. Fr.*, 1, 1–437; 2, pl. 1–22.

29. BRADY, H. B. 1884. 'Report on the Foraminifera dredged by H.M.S. *Challenger* during the years 1873–76', *Rep. scient. Results Voy. Challenger: Zoology*, 9 (2 vols), 1–814, pl. 1–115.

30. BRÖNNIMANN, P. 1940. 'Über die tertiären Orbitoididen und die Miogypsiniden von Nordwest-Marokko', *Abh. schweiz. paläont. Ges.*, 63, 1–113, pl. 1–11.

31. BROTZEN, F. 1948. 'The Swedish Paleocene and its foraminiferal fauna', *Sver. geol. Unders., Afh.*, 42, no. 2, ser. C, no. 493, 1–140, pl. 1–19.

32. CARPENTER, W. B. 1862. *Introduction to the Study of the Foraminifera*, 1–319, pl. 1–22 (Ray Soc.: London). [Pioneer work: very clear and accurate account of structure of many genera. Nomenclature largely out of date].

33. CHAPMAN, F. and PARR, W. J. 1936. 'A classification of the Foraminifera', *Proc. R. Soc. Vict.*, N.S., 49, (1), 139–151.

34. COLE, W. S. 1938. 'Stratigraphy and micropaleontology of two deep wells in Florida', *Geol. Bull. Fla*, 16, 73 pp., 12 pls.

35. COLE, W. S. 1952 [1953]. 'Eocene and Oligocene larger Foraminifera from the Panama Canal Zone and vicinity', *Prof. Pap. U.S. geol. Surv.*, 244, 41 pp., 28 pls.

36. COLE, W. S. 1954. 'Larger Foraminifera and smaller diagnostic Foraminifera from Bikini drill holes', *Prof. Pap. U.S. geol. Surv.*, 260–O, 569–608, p . 204–222.

37. COLE, W. S. 1957. 'Larger Foraminifera' in 'Geology of Saipan Mariana Islands; Pt 3. Paleontology', *Prof. Pap. U.S. geol. Surv.*, 280–1, 321–360, pl. 94–118.

38. COLOM, G. 1956. 'Los Foraminíferos del Burdigaliense de Mallorca', *Mem. R.*

Acad. Cienc. Artes Barcelona, **32**, no. 5 (tercera epoca, no. 653), 7–140, pl. 1–25.

39. CUSHMAN, J. A. 1933. 'Foraminifera: their classification and economic use', *Spec. Publs Cushman Lab.*, **4**, ed. 2, 1–349, pl. 1–40.

40. CUSHMAN, J. A. 1948. '*Foraminifera: their classification and economic use*', ed. 4, 605 pp., 55 pls (Harvard Univ. Press: Cambridge, Mass.).

41. CUSHMAN, J. A. and STAINFORTH, R. M. 1945. 'The foraminifera of the Cipero marl formation of Trinidad, British West Indies', *Spec. Publs Cushman Lab.*, no. 14, 1–74, pl. 1–16.

42. CUVILLIER, J. and SZAKALL, V. 1949. 'Foraminifères d'Aquitaine: Pt 1. Reophacidae à Nonionidae', *Soc. Nat. Petroles d'Aquitaine*, 112 pp. 32 pls, (Imprimerie F. Boisseau: Toulouse).

43. D'ARCHIAC, A. and HAIME, J. 1853–54. *Description des Animaux Fossiles du Groupe Nummulitique de l'Inde*, 1–373, pl. 1–36 (Gide et J. Baudry: Paris). [Fundamental work on Nummulites, also echinoids, for stratigraphical position of which consult in the first place: Fedden, F., 1879, *Mem. Geol. Surv. India*, **17**, 197–210, and also various papers by Vredenburg, E. W., between 1906 and 1928, in *Rec. geol. Surv. India* and *Mem. geol. Surv. India.*]

44. DAVIES, L. M. 1930. 'The genus *Dictyoconus* and its allies . . .', *Trans. R. Soc. Edinb.*, **56**, 485–505, pl. 1–2.

45. DAVIES, L. M. 1932. 'The genera *Dictyoconoides* Nuttall, *Lockhartia* nov. and *Rotalia* Lamarck . . .', *Trans. R. Soc. Edinb.*, **57**, 397–428, pl. 1–4.

46. DE LA HARPE, P. 1881–1883. 'Études des Nummulites de la Suisse', *Abh. schweiz. paläont. Ges.*, **7**, 1–140, pl. 1–2; **10**, 141–180, pl. 3–7.

47. DE LA HARPE, 1883. 'Monographie der in Ægypten und der Libyschen Wuste vorkommenden Nummuliten', *Palæontographica*, **30**, 157–216, pl. 30–35.

48. D'ORBIGNY, A. D. 1850–52. *Prodôme de Paléontologie Stratigraphique Universelle des Animaux mollusques et rayonnés*, 3 vols, (V. Masson: Paris). [A pioneer work.]

49. DOUVILLÉ, H. 1906. 'Évolution et enchâinements des Foraminifères', *Bull. Soc. géol. Fr.*, (4), **6**, 588–602.

50. DOUVILLÉ, H. 1919. 'L'Éocène Inférieur en Aquitaine et dans les Pyrénées', *Mém. Carte géol. dét. Fr.*, 1–84, pl. 1–8. [Nummulites].

51. DOUVILLÉ, H. 1919. 'Les Nummulites, Évolution et classification', *C. r. hebd. Séanc. Acad. Sci., Paris*, **168**, 651–656.

52. DOUVILLÉ, H. 1924–25. 'Révision des Lépidocyclines', *Mém. Soc. géol. Fr.*, **2**, N.S., **1**, 1–50, pl. 1–2; **2**, 51–115, pl. 3–7. [Also by the same author numerous papers in *Bull. Soc. géol. Fr.* from 1902 on; *C. r. hebd. Séanc. Acad. Sci., Paris* from 1915 on; and elsewhere..

53. DROOGER, C. W. 1952. *Study of American Miogypsinidae*, 80 pp. (Vonk & Co.: Zeist).

54. EAMES, F. E., BANNER, F. T., BLOW, W. H. and CLARKE W. J. 1962. *Fundamentals of mid-Tertiary Stratigraphical Correlation*, 1–163, pl. 1–17 (Cambridge Univ. Press).

55. EGGER, J. G. 1857. 'Die Foraminiferen der Miocän-Schichten bei Ortenburg in Nieder-Bayern', *Neues Jb. Miner Geogn. Geol. Petrefakt.*, 266–311, pl. 5–15.

56. ELLIS, B. F. and MESSINA, A. 1940. *Catalogue of Foraminifera* (Am. Mus. Nat. Hist.). [Supplements post-1940].

57. FLANDRIN, J. 1934. 'La faune de Tizi Renif près Dra el Mizan (Algérie)', *Bull. Soc. géol. Fr.*, (5), **4**, 251–272, pl. 14–16.

58. GALLOWAY, J. J. 1933. *A Manual of Foraminifera*, 1–450, pl. 1–42 (Principia Press, Inc.: Bloomington, Indiana).

59. GLAESSNER, M. F. 1945. *Principles of Micropaleontology*, 1–296, pl. 1–14 (Melbourne Univ. Press).

60. GRIMSDALE, T. F. 1952. 'Cretaceous and Tertiary Foraminifera from the Middle East', *Bull. Brit. Mus. nat. Hist., Geol.*, 1, no. 8, 221–248, pl. 20–25.

61. GÜMBEL, C. W. 1868 [1870]. 'Beiträge zur Foraminiferenfauna der nordalpinen Eocängebilde', *Abh. bayer. Akad. Wiss.*, Cl. II, 10, (2), 581–730, pl. 1–4.

62. HANTKEN, M. VON 1875 [1876]. 'A *Clavulina Szabói* rétegek Faunája, I. Foraminiferak', *Magy. kir. foldt. Intéz. Evk.*, 4, 1–82, pl. 1–16.

63. HANZAWA, S. 1935. 'Some fossil *Operculina* and *Miogypsina* from Japan and their stratigraphical significance', *Sci. Rep. Tohoku Univ.*, (2), (Geol.), 18, no. 1, 1–29, pl. 1–3.

64. HANZAWA, S. 1952. 'Notes on the Recent and fossil *Baculogypsinoides spinosus* Yabe and Hanzawa from the Ryukyu Islands and Taiwan (Formosa), with remarks on some spinose Foraminifera', *Short Pap. Inst. geol. Paleont. Tohoku Univ.*, no. 4, 1–22, pl. 1–2.

65. HANZAWA, S. 1957. 'Cenozoic Foraminifera of Micronesia', *Mem. geol. Soc. Am.*, 66, 163 pp., 38 pls.

66. HANZAWA, S. 1962. 'Upper Cretaceous and Tertiary three-layered foraminifera and their allied forms', *Micropaleontology*, 8, (2), 129–186, pl. 1–8.

67. HAQUE, A. F. M. M. 1956. 'The Foraminifera of the Ranikot and the Laki of the Nammal Gorge, Salt Range', *Mem. geol. Surv. Pakist.*, 1, 1–300, pl. 1–34.

68. HENSON, F. R. S. 1948. 'Larger imperforate Foraminifera of southwestern Asia, Families Lituolidae, Orbitolinidae and Meandropsinidae', *Brit. Mus. nat. Hist. Mon.*, 1–127, pl. 1–16.

69. HENSON, F. R. S. 1950. *Middle eastern Tertiary Peneroplidae (Foraminifera), with remarks on the phylogeny and taxonomy of the family*, 1–70, pl. 1–10 (West Yorkshire Printing Co.: Wakefield, England).

70. HOFKER, J. 1930. 'Der Generationswechsel von *Rotalia beccarii* var. *flevensis* nov.', *Z. Zellforsch. mikrosk. Anat.*, 10, 756–758.

71. HOFKER, J. 1951. 'The toothplate-Foraminifera', *Archs néerl. Zool.*, 8, (4), 353–372.

72. HOFKER, J. 1956. 'Tertiary Foraminifera of coastal Ecuador: Pt II. Additional notes on the Eocene species', *J. Paleont.*, 30, 891–958.

73. LE CALVEZ, J. 1938. 'Recherches sur les Foraminifères I. Developpement et reproduction', *Archs Zool. exp. gén.*, 80, (3), 163–333, pl. 2–7.

74. LEHMANN, R. 1962. 'Strukturanalyse einiger Gattungen der Subfamilie Orbitolitinae', *Eclog. geol. Helv.*, 54, no. 2, 597–667, pl. 1–14.

75. LOEBLICH, A. R. and TAPPAN, H. 1957. 'Planktonic Foraminifera of Paleocene and early Eocene age from the Gulf and Atlantic Coastal Plains, *Bull. U.S. natn. Mus.*, 215, 173–198, pl. 40–64.

76. LOEBLICH, A. R. and TAPPAN, H. 'Foraminiferal Classification and Evolution', *J. geol. Soc. India*, 5, 5–40.

77. MACFADYEN, W. A. 1930. *Miocene Foraminifera from the Clysmic area of Egypt and Sinai* . . ., 1–149, pl. 1–4, (Survey of Egypt: Cairo).

78. MOORE, R. C. (ed.). 1964. *Treatise on Invertebrate Paleontology: Part C, Protista 2*, 1–2, C1–C900, (Univ. Kansas Press).

79. NUTTALL, W. L. F. 1927. 'Larger Foraminifera of Middle and Lower Khirthar Series, *Rec. geol. Surv. India*, 59, 115–164, pl. 1–8.

80. NUTTALL, W. L. F. 1928. 'Tertiary Foraminifera from the Naparima Region of Trinidad', *Q. Jl geol. Soc. Lond.*, **84**, 57–115, pl. 3–8.

81. NUTTALL, W. L. F. 1930. 'Eocene Foraminifera from Mexico', *J. Paleont.*, **4**, 271–293, pl. 23–25.

82. NUTTALL, W. L. F. 1932. 'Lower Oligocene Foraminifera from Mexico', *J. Paleont.*, **6**, 3–35, pl. 1–9.

83. PFENDER, J. 1935. 'Á propos du *Siderolites Vidali* Douvillé et de quelques autres', *Bull. Soc. géol. Fr.*, (5), **4**, 225–236, pl. 11–13.

84. PIVETEAU, J. 1952. '*Traité de paléontologie*', **1**, 1–782 (Masson & Cie.: Paris).

85. PLUMMER, H. J. 1927. 'The Foraminifera of the Midway Formation in Texas', *Bull. Univ. Tex. Bur. econ. Geol. Technol.*, **2644**, 1–206, pl. 1–15.

86. POKORNÝ, V. 1958. '*Grundzüge der Zoologischen Mikropaläontologie*', **1–2**. (Diebel: Berlin).

87. RAUZER-CHERNOUSOVA, D. M. and FURSENKO, A. V. 1959. 'Osnovy Paleontologii. Obshchaya chast prosteyshie', *Akad. Nauk SSSR*, 1–368, pl. 1–13.

88. REICHEL, M. 1931. 'Sur la structure des Alvéolines', *Eclog. geol. Helv.*, **24**, 289–303, pl. 13–18.

89. REISS, Z. 1957. 'The Bilamellidea, nov. superfam., and remarks on Cretaceous globorotaliids', *Contr. Cushman Fdn. foramin. Res.*, **8**, (4), 127–145, pl. 18–20.

90. REISS, Z. 1958. 'Classification of lamellar Foraminifera', *Micropaleontology*, **4**, 51–70, pl. 1–5.

91. RENZ, H. H. 1948. 'Stratigraphy and fauna of the Agua Salada group, State of Falcon, Venezuela', *Mem. geol. Soc. Am.*, **32**, 219 pp., 12 pls.

92. REUSS, A. E. 1855. 'Beiträge zur Charakteristik der Tertiärschichten des nördlichen und mittleren Deutschlands, *Sber. Akad. Wiss. Wien*, **18**, 197–272, pl. 1–12. [Many other papers by this author].

93. ROZLOZSNIK, P. 1927. Einleitung in das Studium der Nummulinen und Assilinen', *Jahrb. Ungar. Geol. Anst.*, **26**, 1–156, pl. 1. [First published in Hungarian in 1924].

94. SCHAUB, H. 1951. 'Stratigraphie und Paläontologie des Schlierenflysches mit besonderer Berücksichtigung der paleocaenen und untereocaenen Nummuliten und Assilinen, *Schweiz. paläont. Abh.*, **68**, 1–222, 9 pls.

95. SCHAUB, H. 1963. 'Über einige Entwicklungsreihen von *Nummulites* und *Assilina* und ihre stratigraphische Bedeutung', *Evolutionary Trends in Foraminifera*, 283–297 (Elsevier Publishing Co.: Amsterdam).

96. SCHAUB, H. 1966. 'Nummulitic zones and evolutional series of *Nummulites* and *Assilina*', *Vop. Mikropaleont.*, **10**, 298–301, fig. 1–2.

97. SCHLUMBERGER, C. 1900. 'Note sur le genre *Miogypsina*,' *Bull. Soc. géol. Fr.*, (3), **28**, 327–333, pl. 2–3.

98. SCHLUMBERGER, C. 1901–4. 'Notes sur les Orbitoides', *Bull Soc. géol. Fr.*, (4), **1**, 459–467, pl. 7–9 [*Orbitoides*]; **3**, 273–289 [*Discocyclina*]; **4**, 119–135, pl. 3–6 [*Asterocyclina*]. [Many other papers in the same *Bulletin* from 1883 to 1905].

99. SCHLUMBERGER, C. 1905. 'Deuxième note sur les Miliolidées Trematophorées', *Bull. Soc. géol. Fr.*, (4), **5**, no. 2, 115–134, pl. 2–3.

100. SCHLUMBERGER, C. and MUNIER-CHALMAS, E. 1884. 'Note sur les Miliolidées trematophorées'. *Bull. Soc. géol. Fr.*, (3), **12**, no. 8, 629–630.

101. SENN, A. 1935. 'Die stratigraphische Verbreitung der tertiären Orbitoiden, mit specieller Berücksichtigung ihres Vorkommens in Nord-Venezuela und Nord-Marokko', *Eclog. geol. Helv.*, **28**, 51–113, 370–373, tables 8, 9.

102. SHERBORN, C. D. 1888. *A bibliography of the Foraminifera, Recent and fossil, from 1565–1888* . . ., 1–152, (Dulau & Co.: London).

103. SHERBORN, C. D. 1893–96. 'An index to the genera and species of the Fora-minifera', *Smithson. misc. Collns*: (a) 1893, no. 856, 1–240; (b) 1896, no. 1031, 241–485. [A further bibliography, 1888–98, was published by P. Toutkowski at Kiev in 1898. For later publications see the annual *Zoological Record* (Zool. Soc. London) under 'Protozoa', the annual bibliography (and list of new genera and species) by H. E. Thalmann in *J. Paleont*. from 1933 (for 1931); and later lists of literature by R. Todd in *Contrib. Cushman Fdn foramin. Res*.].

104. SHERBORN, C. D. and CHAPMAN, F. 1886. 'On some microzoa from the London Clay exposed in the drainage works, Piccadilly, London, 1885', *J. R. microsc. Soc*., (2), **6** 737–763, pl. 14–16. [Supplement: 1889. ibid., 483–488, pl. 11].

105. SILVESTRI, A. 1937. 'Foraminiferi dell'Oligocene e del Miocene della Somalia', *Palaeontogr. ital*., **32**, suppl. 2, 45–264, pl. 4–22.

106. SMOUT, A. H. 1954. *Lower Tertiary Foraminifera of the Qatar Peninsula*, 1–96, pl. 1–15, (Brit. Mus. (N.H.): London).

107. SUBBOTINA, N. N. 1953. 'Verkhneeotsenovye Lyagenidy i Buliminidy yuga SSSR', *Mikrofauna SSSR, Sbornik 6, VNIGRI, Trudy*, N.S., no. 69, 115–255, pl. 1–13.

108. SUBBOTINA, N. N. 1953. 'Globigerinidy, Hantkeninidy i Globorotaliidy: Iskopaemye Foraminifery SSSR', *VNIGRI, Trudy*, N.S., no. 76, 1–296, 41 pls.

109. TERQUEM, O. 1882. 'Les Foraminifères de l'Éocène des Environs de Paris', *Mém. Soc. géol. Fr*., (3), **2**, mém. 3, 1–193, pl. 9–28.

110. THALMANN, H. E. See **102** and **103**.

110a. TODD, R, See **103**.

111. TOUTKOWSKI, P. See **103**.

112. UMBGROVE, J. H. F. 1928. 'Het genus *Pellatispira* in het indo-pacifische gebied', *Nederland.-Indië, Dienst. Mijnb*., no. 10, 43–71. pl. 1–15.

113. UMBGROVE, J. H. F. 1931. 'Tertiary Foraminifera [of East Indies]', *Leid. geol. Meded*., pt 5 (Feestbundel K. Martin), 35–91. [List of species and bibliography, with introduction and stratigraphical remarks and tables in English.]

114. VAUGHAN, T. W. 1924. 'American and European Tertiary larger Foraminifera', *Bull. geol. Soc. Am*., **35**, 785–822, pl. 30–36.

115. VAUGHAN, T. W. 1933. 'Studies of American species of foraminifera of the genus *Lepidocyclina*', *Smithson. Misc. Collns*, **89**, no. 10 (publ. 3222), 1–53, pl. 1–31.

116. VAUGHAN, T. W. 1945. 'American Paleocene and Eocene larger Foraminifera', *Mem. geol. Soc. Am*., **9**, (1), 1–175, pl. 1–46.

117. VAUGHAN, T. W. and COLE, W. S. 1941. 'Preliminary report on the Cretaceous and Tertiary larger Foraminifera of Trinidad, British West Indies', *Spec. Pap. geol. Soc. Am*., **30**, 1–137, pl. 1–46.

118. VLERK, I. M. VAN DER. 1928. 'The genus *Lepidocyclina* in the Far East', *Eclog. geol. Helvet*., **21**, no. 1, 182–211, pl. 6–23. [Numerous other papers, mostly in Dutch, by the same author.]

119. VLERK, I. M. VAN DER. 1959. 'Modification de l'ontogénèse pendant l'évolution des Lépidocyclines (Foraminifères)', *Bull. Soc. géol. Fr*., (7), **1**, no. 7 669–673.

120. VLERK, I. M. VAN DER. 1966. 'Tertiary correlation based on biometrical investigation of the genus *Lepidocyclina*', *Vop. Mikropaleont*., **10**, 302–308, fig. 1–4, plate.

121. VLERK, I. M. VAN DER. 1966. 'Stratigraphie du Tertiaire des Domaines Indo-Pacifique et Mesogéen: Essai de Correlation', *Proc. K. ned. Akad. Wet*. (b), **69**, no. 3, 336–344.

122. VLERK, I. M. VAN DER and UMBGROVE, J. H. F. 1927. 'Tertiaire Gidsfora-miniferen van Nederlandsch Oost-Indië', *Wet. Meded.* (*Dienst Mijnb. Ned. Oost.-Indië*), no. 6, 3–35. [A convenient pocket-book with figures, stereo-diagrams and range tables of the principal larger Foraminifera of the East Indies.]

123. WOOD, A. 1949. 'The structure of the wall of the test in the Foraminifera; its value in classification', *Q. Jl geol. Soc. Lond.*, **104**, 229–225, pl. 13–15.

Chapter II

TERTIARY ECHINOIDEA

REFERENCE-LETTERS following figure numbers:
a, apical view.
b, oral view.
c, side view.
d, posterior view.
e, apical disk.
f, amb plate(s) enlarged.
g, interamb plate(s) enlarged.

h, amb and interamb plates enlarged.

h′ amb and interamb plates (stippled) on oral surface.
i, girdle, face view.
j, girdle, side view.
k, internal view, oral half.
l, internal view, apical half.
m, phyllode.
n, radiole.
o, profile of primary tubercle.

REFERENCE-LETTERS to parts of figures:
ad, apical disk.
af, anal fasciole.
ag, actinal groove (furrow).
am or *amb*, ambulacral area.
an, anus.
br, bourrelet.
dd, dot-and-dash pore pairs.
f, peristome funnel (external).
gc, gill cuts (branchial incisions).
gp, genital plate or pore.
i or *ia*, interambulacral area.
if, internal fasciole.
ip, interporiferous area.
lb, labrum.
lf, lateral fasciole.
md, madreporite.
mf, marginal fasciole.
oc, ocular plate.

p, pillar supporting interamb I.
pe or *pt*, petal.
pf, pore field.
pg, perignathic girdle.
ph, phyllode.
pl, plastron.
pp, simple pore pair.
ppf, peripetalous fasciole.
pr, periproct.
ps, peristome.
pt, petal.
s, perradial suture of amb I.
sf, subanal fasciole.
sg, sutural grooves.
sp, subpetal.
st, secondary tubercle.
t, primary tubercle.

The Echinoidea are the only class of Echinoderms sufficiently abundant in Tertiary rocks to be of stratigraphical importance. Like other Echinoderms they have a skeleton composed of plates and other units, each a single calcite crystal. Though really internal in origin, this skeleton lies so near the outer surface as to be apparently external, but in life it has outside it a thin layer of living tissue including muscle and nerve. The whole skeleton consists of the *test* and the *radioles* (movable articulated spines, Figs. 235, 237, 239), the test being composed of the *apical system* (or apical disk), *corona* and *peristome plates.* These last, as well as the radioles, are usually detached after death, so that the fossil echinoid consists usually of corona and apical system, or in some cases often or corona alone.

The Echinoidea were considered to fall into two main divisions, the Regularia or Endocyclica and Irregularia or Exocyclica. The former are radially symmetrical, their most usual shape is an oblate spheroid, sometimes greatly flattened both above and below, at other times tending towards hemispherical owing to flattening of the under surface only. The apical disk lies in the centre of the upper (*adapical* or aboral) surface and encloses the anus; the mouth is in the centre of the lower (oral or *adoral*) surface. The margin between adoral and adapical surfaces is termed the *ambitus.*

In the Irregularia the symmetry is bilateral, though in those nearest to the Regularia it is still almost radial. The shape varies very greatly, the apical system may be central in the adapical surface or it may be shifted forwards or backwards (*prae-* or *post-central*), may be a disk in shape, or elongate and never enclosing the anus. The mouth may be central in the adoral surface or shifted forwards (prae-central).

The *corona* in all Tertiary echinoids consists of 20 columns of plates, arranged in alternating double columns known as the 5 ambs (ambulacral areas) and 5 interambs (interambulacral areas). In the ambs each plate is typically perforated by a pair of pores (pore pair), through which pass the roots of a *podium*—a soft, hollow, movable structure which may serve in different cases for adhesion, locomotion, the seizing of food, or respiration: for the first three purposes a podium is cylindrical and the two pores of a pair are rounded and near together (*pp*, Fig. 235h), but when the podia serve as gills they become flat and leaf-like and the two pores move apart, one of them becoming slit-like (dot-and-dash type of pore pair, *dd*, Fig. 289c). When, in parts of an amb, the podia are degenerate, one of the two pores of a pair may be suppressed.

The interamb plates are not perforated. The plates of both areas bear tubercles, which give articulation to a radiole, each by a ball-and-socket joint: the tubercles are often larger on the interambs than on the ambs. According to their size and elaboration of structure, tubercles are classed as *primary, secondary* and *miliary*; the smallest and simplest protuberances are termed *granules.*

The five radiating directions corresponding to the central lines of the ambs are termed *perradial*; those of the interambs *interradial*; the ten corresponding to the margins of ambs and interambs, *adradial.* Most plates extend half way across an area, thus having both an adradial and a per- (or inter-) radial margin, and are termed *primary* plates. A plate which fails to reach the perradial line is termed a *demiplate*; one that fails to reach the adradial, an *occluded* plate; while one that is shut off from both is an *included* (or isolated) plate. These various shortened plates are mainly found as components of compound ambulacral plates in the Diadematacea and Echinacea.

The apical system (Figs 235e, 242e) consists typically of 10 principal plates —5 *ocular* plates, each at the summit of an amb, 5 *genital* plates topping the interambs and each with a pore (*gonopore*) by which the reproductive cells are shed. One of the genital plates has in addition numerous smaller pores: it is termed the madreporic plate (*madreporite*), and in fossil Regularia affords the only means of orientating the test, as it occupies a right anterior position. As may be seen on Fig. 245a, the ambs are numbered I–V, and the interambs 1–5, in a counter-clockwise direction (from below as seen in apical view), and the madreporite tops interamb 2.

In the Regularia the oculars and genitals form a ring, within which are a number of small and loose *periproct plates* which surround the minute anus. The oculars are smaller than the genitals, and the latter often meet so as to exclude the former from contact with the periproct: the oculars are then said to be *exsert*; when they touch the periproct they are *insert*. Thus in Fig. 273e, ocular I is insert, the others exsert. In most fossils the periproct plates are lost and the apparent anal opening is really the *periproct* (*pr*, many Figs).

In Irregularia, the periproct is quite outside the apical system and lies in interamb 5, either on the upper or lower surface or on the ambitus. The ring of oculars and genitals may consequently contract into a smaller disk, or may undergo elongation or other changes: the number of gonopores may be reduced to 4, 3 or 2: only exceptionally is the fifth present (some Clypeastroids). The madreporite (genital 2) tends to extend into the centre of a disk which keeps its circular shape. When it extends no farther, the disk is *ethmophract*: it becomes *ethmolytic* when the madreporite pushes its way back between genitals 1 and 4 (Fig. 354).

The mouth is always on the adoral surface and is surrounded by loose peristome plates, usually lost in fossils: the actual visible opening (*peristome*) may therefore be much larger than the true mouth. In Regularia and the less modified Irregularia there arise from the margin of the peristome a series of internal upgrowths constituting the *perignathic girdle* (or simply *girdle*) (Figs 235i, j and 293), and separate from these is a complex mechanism known as the *lantern* (of Aristotle), which works a set of five teeth converging on the centre of the mouth: the lantern is commonly lost in fossils. In the more specialized Irregularia both lantern and girdle disappear, as the mouth departs from its central position.*

The Echinoidea are exclusively marine and mainly of shallow water habitat. Many burrow in sand or nestle or bore in rock cavities. In their ontogeny they nearly all pass through a larval stage, totally unlike the adult in appearance, structure and habits, since they swim near the surface by means of ciliated bands: a startling metamorphosis precedes the beginning of adult structure. In the larval stage they may be carried far by currents, hence the geographical distribution of species is often wider than would be expected from the adult habits. In a few genera the larvae are retained in 'marsupia' formed by the sunken interporiferous areas of the upper surface of the parent.

During the Palaeozoic era the Echinoidea had experimented in the direction of varying the numbers of columns of plates in the corona and had discovered

* Other structural details are dealt with under the several Orders. See also Glossary, pp. 161–165.

that twenty (five double ambulacral and interambulacral columns alternately) was the best number. During the Mesozoic era experiment took two other directions. The first was in the compounding of ambulacral plates so as to increase greatly the number of podia without depriving them of the protection of adjacent large radioles: the families in which these experiments were tried remained 'regular', i.e. radially symmetrical. The second advance was much more venturesome and opened up a number of new possibilities in the way of corona shape, new uses for the podia and so forth: this was the escape of the periproct from the apical disk for a journey along the posterior interamb. The urchins which thus, towards the middle of the Jurassic period, became 'irregular', gave rise to a variety of lineages. Of the many characteristic families, regular and irregular, of the Mesozoic era, some attained their acme and died down before the end of that era, or left a few isolated descendants to survive for a part or the whole of the Caenozoic, while others were only initiated in the former and underwent most of their evolution in the latter.

Although until fairly recently the Echinoidea were divided into Regularia (Endocyclica) and Irregularia (Exocyclica), that part of the *Treatise on Invertebrate Paleontology* (Durham *et al.*, in Moore, **21**), which is followed here, uses two different subdivisions—Perischoechinoidea and Euechinoidea.

Subclass PERISCHOECHINOIDEA

Regular (endocyclic) echinoids with interambs of one to many columns and ambs of two to twenty columns, the latter without compound plates; perignathic girdle of apophyses only or none; teeth grooved; no gills slits, spheridia or ophiocephalous pedicellariae. (Ord.-Rec.)

Order CIDAROIDA

Test subspherical, radially symmetrical, rigid or with imbricating plates; ambs of two columns, each plate with a single pore pair, grouped in diads or triads in some forms; interambs much wider than ambs, of two or more columns, each plate with one large primary tubercle carrying a large primary radiole; areole conspicuous, usually surrounded by a ring of secondary tubercles; lantern present; teeth not keeled; no gills or gill slits; apical system enclosing periproct; no spheridia; pedicellariae of two types: globiferous and (rarely) tridentate. (U.Sil.-Rec.)

The order Cidaroida is the one group that is definitely known to pass up from the Palaeozoic to the Mesozoic. Primary interamb tubercles are always (Psychocidaridae excepted) perforate. All post-Palaeozoic cidarids (except *Tetracidaris*) have interambs with two columns of plates, and all post-Triassic ones have rigid tests.

Family CIDARIDAE

Test rigid; interamb plates in two columns; primary tubercles perforate;

amb pore pairs uniserial aborally, in some with pluriserial tendencies adorally, but not forming compound plates. (U.Trias.-Rec.)

In members of this family the apical disk is large, the oculars all being exsert. The ambs are very narrow, flexuous, consisting of a very large number of similar primary plates. The interambs are broad, of a small number (five to eleven) of very large plates, each with one large primary tubercle. Primary radioles are of varied form, often long and cylindroidal. In the Recent *Cidaris papillata* (Fig. 235i, j) the girdle is discontinuous, being formed of interamb processes only. The very simple ambs with their close-set perforations constitute a mechanical weakness, very obvious in fossils broken or crushed in along these lines, but offset in the living animal by the protective action of the long radioles. Most of the advances in the tests of other Regularia tend to do away with this weakness, enabling the animal to carry long radioles on ambs as well as interambs and increasing the efficiency of the podia.

When the apical disk is missing it may not seem easy to distinguish the adapical from the adoral surface, but there is an essential difference: the primary tubercles grow gradually smaller towards the peristome, but not so towards the apical disk.

The genus **Cidaris** [*Dorocidaris*] as now interpreted is Recent only (Fig. 235). **Eucidaris** (Fig. 236): pores non-conjugate; primary tubercles non-crenulate adorally; primary radioles usually cyclindrical, truncate, otherwise fusiform or clavate, crown with a median prominence, with regular longitudinal rows of warts. U.Eoc.-Rec. N.Z., Fiji, Calif., Carib., tropical and subtropical seas. **Goniocidaris** (*s.s.*): median part of horizontal sutures sunken or naked, forming conspicuous grooves in ambs and interambs; pores non-conjugate; primary tubercles non-crenulate; primary radioles with distal spurs,

---

Figs 235–248. ECHINOIDEA: CIDAROIDA, ECHINOTHURIOIDA, DIADEMATOIDA, PEDINOIDA, SALENIOIDA AND PHYMOSOMATOIDA

Fig. 243f after Cotteau; Fig. 242e after R. T. Jackson; Figs 236–238, 240, 241, 242f, 243c and 244 after Fell, in Moore; Figs 246–248 after Fell and Pawson, in Moore; Figs 235, 239 and 245 original.

235c, e, h, i, j, n, o. *Cidaris papillata* Leske, Rec. England. c, n × 1, e × 1·5, h, o × 2·25, i, j × 1·25.

236n. *Eucidaris strobilata* Fell, L.Olig. N.Z. × 1·4.

237n. *Rhabdocidaris orbignyana* (Agassiz), Jur. France. × 0·37.

238n. *Phyllacanthus titan* Fell, L.Mio. N.Z. × 0·35.

239n. *Porocidaris schmidelii* (Münster), U.Eoc. N.Italy. T. × 0·75.

240n. *Prionocidaris marshalli* Fell, M.Eoc. N.Z. × 0·75.

241n. *Araeosoma violaceum* Mortensen, Rec. Atl. × 4·25.

242e, f. *Diadema setosum* (Leske), Rec. e, W.Indies, Enlarged; f, Indo-Pac., × 3·1. T.

243c, f. *Echinopedina gacheti* (Desmoulins), Eoc. c, France, × 0·8; f, Blaye (Gironde), × 2·25.

244c, f. *Leiopedina tallevignesi* (Cotteau), M.Eoc. France. T. c × 0·6, f × 1·65.

245a. *Salenia petalifera* (Desmarest), Cenomanian. Wiltshire. × 1·5.

246f. *Phymosoma regulare* (Agassiz), Turonian. Switzerland. × 1·3.

247h. *Acanthechinus nodulosus* Duncan and Sladen, Pal. W.Pakistan. T. × 0·32.

248h. *Aeolopneustes delorioli* Duncan and Sladen, Pal. W.Pakistan. T. × 1·8.

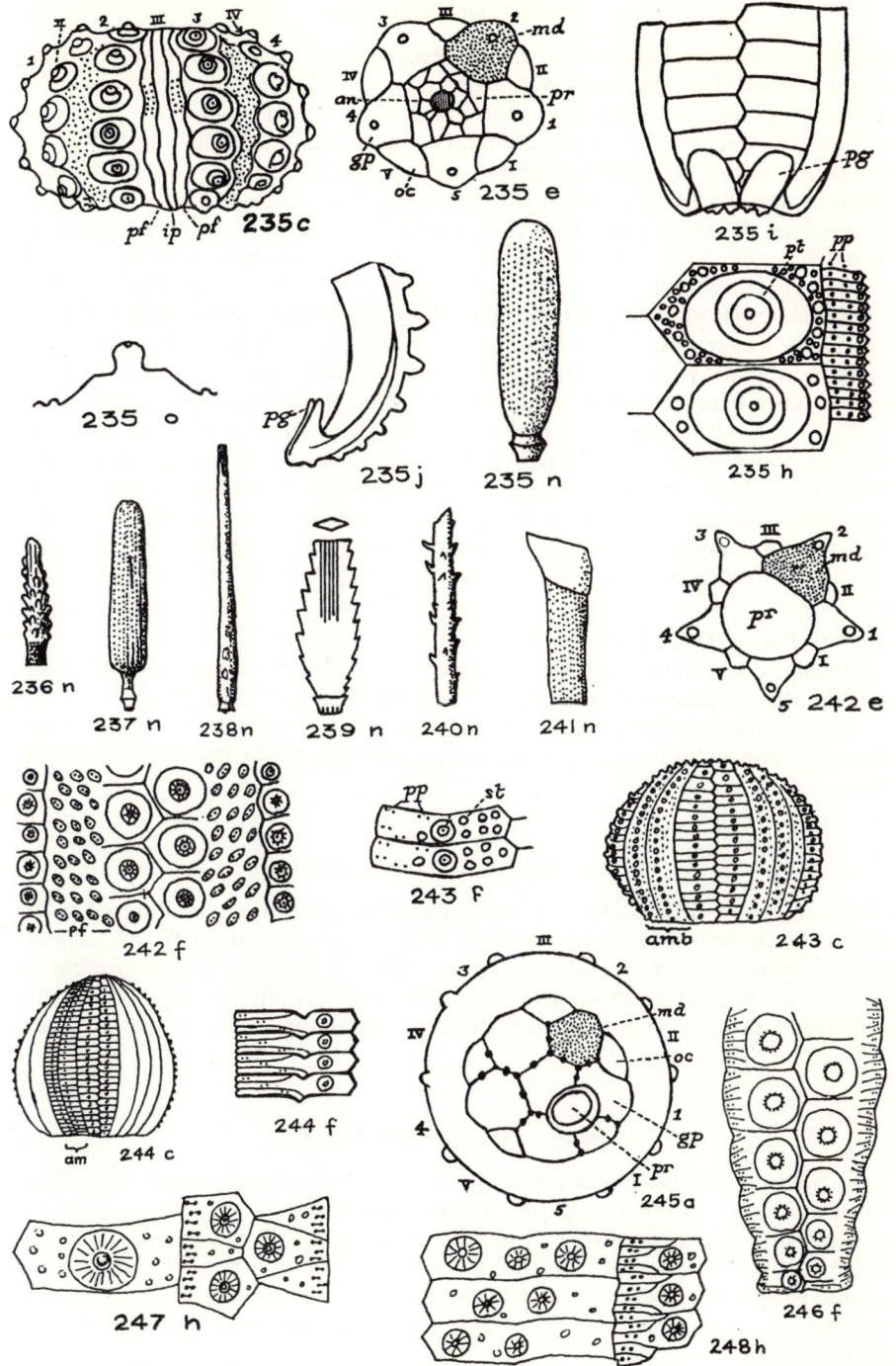

235 c

235 e

235 i

235 o

235 j

235 n

235 h

236 n

237 n

238n

239 n

240n

241 n

242 e

242 f

243 f

243 c

244 c

244 f

245 a

246 f

247 h

248 h

shaft with coarse ridges or thorns. Eoc.-Rec. Austral., N.Z., India, W.Pakistan, Iran, Indo-W.Pac. *Stereocidaris:* interambs with horizontal sutural grooves aborally, plates (especially aborally) higher than broad, uppermost ones with rudimentary sculpture; dense secondary and miliary tubercles; ambs usually sinuous; pores non-conjugate; primary tubercles non-crenulate; primary radioles usually flaring toward tip. Cret.-Rec. Eur., N.Amer., Austral., N.Z., S.E.Africa. *Rhabdocidaris* (Fig. 237): ambs sinuous, pores conjugate; primary tubercles crenulate; primary radioles long, usually depressed and fan-shaped, shaft with longitudinal rows of small spines. L.Jur.-Eoc. Eur. *Phyllacanthus* [*Leiocidaris*] (Fig. 238): primary tubercles non-crenulate; scrobicular tubercles distinctly larger than other secondaries; madreporite large, periproct small; pores conjugate, primary radioles cylindrical, with fine granules in longitudinal series. Olig.-Rec. Austral., N.Z., W.Pakistan, Indo-Pac. *Porocidaris* (Fig. 239): like *Rhabdocidaris*, but ambs straight, areoles with a marginal series of radiating, narrow grooves, and primary (adoral) radioles flattened and serrate. Eoc. Eur., N.Africa, E.Africa, M.East, Pakistan. *Prionocidaris* (Fig. 240); primary tubercles non-crenulate; pores conjugate; primary radioles long, tapering or cylindrical, with longitudinally or spirally arranged rows of coarse spines. U.Cret.-Rec. Eur., W.Pakistan, Austral., N.Z., Medit., Indo-Pac.

Family PSYCHOCIDARIDAE

Primary tubercles imperforate; pores non-conjugate; no peristomial interamb plates; peristomial amb plates in double series. (U.Jur.-Rec.)

Psychocidaris is Recent only. *Sardocidaris:* like *Psychocidaris*, but primary radioles long, cylindrical, tapering, not glandiform. Mid-Cret.-Mio. N.Africa, Eur.

Subclass EUECHINOIDEA

Test composed of five bicolumnar ambs and five alternating bicolumnar interambs. Plates imbricating, or joined by flexible integument, or (more usually) united by rigid sutures; periproct endocyclic or exocyclic; lantern present or absent, or present only in juveniles; gills and gill slits present or absent; spheridia present; pedicellariae present, including ophicephalous types. (?Carb., U.Trias.-Rec.)

Superorder DIADEMATACEA

Test rigid or flexible, plates united by sutures or by membranous interstices, or imbricating beveled margins; no distinct bilateral symmetry; primary tubercles perforate; periproct endocyclic or exocyclic. Perignathic girdle complete; lantern well developed; gills and gill slits usually present in adult; amb plates simple or (more usually) compounded in diadematoid or arbacioid groups, or in complex modifications of the diadematoid type. (?L.Carb., U.Trias.-Rec.)

Order ECHINOTHURIOIDA

Test hemispherical to flattened, flexible, with imbricating plates or interstitial membranous junctions; peristomial amb plates simple, others compounded in diadematoid or derived patterns; periproct endocyclic; apical system dicyclic in young, monocyclic in adult; five genital pores; tubercles non-crenulate; radioles striate, usually hollow; gills and gill slits inconspicuous, or lost in adult; spheridia present aborally and adorally; pedicellariae present. (U.Jur.-Rec.)

Family ECHINOTHURIIDAE
Characters of the order. (U.Jur.-Rec.)

The genus **Echinothuria** is Upper Cretaceous only. *Araeosoma* (Fig. 241): test large, depressed; conspicuous membranous interstices between plates; primary amb plates entire, much larger than demiplates; pores on oral surface in three series on either side of the interporiferous area. ?U.Cret., Plio.-Rec. ?Eur., N.Z., Indo-Pac., Atl.

Order DIADEMATOIDA [CENTRECHINOIDA]

Test subspherical, low hemispherical or pentagonal; rigid or flexible, plates usually imbricated internally; ambs composed of simple or compound diadematoid plates; gills present, conspicuously notching peristomial margin; periproct endocyclic; apical system monocyclic, or with anterior oculars (II, III, IV) exsert; five genital pores; tubercles crenulate or non-crenulate; radioles hollow, cylindrical, usually verticillate; spheridium in each compound plate (sometimes lacking adapically); pedicellariae present. (?L.Carb., U.Trias.-Rec.)

Family DIADEMATIDAE [CENTRECHINIDAE]
Test usually rather flattened and flexible; primary tubercles crenulate; primary and secondary radioles usually hollow and verticillate. (L.Jur.-Rec.)

Diadema [*Centrechinus*] (Fig. 242): test large, low-subhemispherical, rigid; primary amb tubercles conspicuous in two regular series; interambs also with two columns of primary tubercles; ambs not much narrower than interambs; pore pairs in simple arcs; primary tubercles perforate and crenulate; peristome large, decagonal, test with narrow bare spaces by the gill slits; apical system large, with oculars I, IV and V insert; primary radioles of oral surface not clavate, not expanded distally; no globiferous pedicellariae. ?U.Cret., Rec. Indo-Pac., Atl., Carib. [Rare records of radioles exist, of material from the Neogene of the Mediterranean region, but identification seems doubtful.] **Centrostephanus:** like *Diadema*, but embryonic adoral amb plate not resorbed in adult and globiferous pedicellariae present; interambs twice as broad as ambs. ?Mio., Plio.-Rec. Indo-Pac., Atl., Medit., S.Eur. [Fossil records are mainly of radioles]. **Kierechinus:** pores uniserial aborally, in trigeminate arcs of three at ambitus, irregularly uniserial adorally; twelve small primary tubercles on each interamb plate. Pal./?L.Eoc. Somalia.

107

Order PEDINOIDA

Test subspherical, high-subconical to low-hemispherical or flattened, rigid but fragile; ambs composed of simple or compounded diadematoid plates; gills present, gill slits shallow; periproct endocyclic; apical system dicyclic; five genital pores; tubercles non-crenulate; radioles finely striate, more or less spinose, primaries solid, secondaries hollow; pedicellariae present. (U.Trias.-Rec.)

Family PEDINIDAE
Characters of the order. (U.Trias.-Rec.)

Echinopedina [*Hebertia* Lambert *non* Michelin] (Fig. 243): test subspherical, flattened below; pore zones arranged in arcs of three, adapical pair slightly nearer adradial margin than others (hence obscurely inverse), but not forming three vertical series; tubercles perforate, non-crenulate; ambs and interambs with only a single vertical series of primary tubercles. Eoc. Eur., N.Africa, Carib. *Leiopedina* (Fig. 244): test large, as high as broad, or higher; pore zones broad; pore pairs in oblique arcs of three, in inverse sequence (adapical pair outermost), forming three well-defined series. M.Eoc. Eur. *Loriolipedina:* like *Leiopedina*, but pores of outermost pore pairs of each series elongate. Eoc. Eur.

Superorder ECHINACEA

Corona rigid; periproct endocyclic; radioles solid; branchial slits present in adult; perignathic girdle complete in adult; lantern present in adult; ambs simple or with various types of compound plating. (U.Trias.-Rec.)

Order SALENIOIDA

Test of cidaroid type (each interamb plate with single large primary tubercle and many much smaller secondary tubercles); ambs simple or compounded in diadematoid manner; apical system with one or several large, polygonal suranal plates; inner border of oculogenital ring angular; periproct posterior or postero-dextral, encroaching on posterior edge of suranal plate(s) which become emarginated; primary tubercles usually crenulate; lantern stirodont; primary radioles with collar and cortex layer. (?U.Trias., L.Jur.-Rec.)

Family SALENIIDAE
Primary tubercles imperforate; primary amb tubercles non-crenulate, those of interamb usually crenulate; apical disk large, usually dicyclic, but oculars I and V sometimes insert; gill slits usually present; amb plates simple, bigeminate or trigeminate, apparently of diadematoid type; pore zones usually straight; primary radioles long and slender, with a more or less spinulose cortex layer on shaft; spheridia present; pedicellariae present (but not the globiferous type). (U.Jur.-Rec.)

Salenia (Fig. 245): test hemispherical; apical disk with plates ornamented with grooves and ridges, and with pitted sutures, oculars exsert, except for I which touches the postero-dextrally displaced periproct; ambs narrow, bigeminate throughout, each compound plate with two pore pairs and one primary tubercle; interamb primary tubercles large; primary radioles usually slender and curved, spinose and verticillate. L.Cret.-Rec. Eur., Africa, Asia, N.Amer., Austral., Indo-Pac., Carib. *Salenidia:* like *Salenia*, but ambs throughout consisting of primary plates, each with tubercle and pore pair. Mid.-Cret.-U.Eoc. Eur., Pakistan, Austral.

Order PHYMOSOMATOIDA

Lantern stirodont; apical disk without large polygonal suranal plates; primary tubercles imperforate; amb plates simple throughout, or (more usually) compounded in diadematoid manner, trigeminate or polyporous. (L.Jur.-Rec.)

Family PHYMOSOMATIDAE [CYPHOSOMATIDAE]

Primary tubercles crenulate, amb tubercles usually as large as the interamb ones; amb plates simple or compound; polyporous and diplopodous in more specialized genera; apical system dicyclic or monocyclic, usually prolonged posteriorly into interamb 5; peristome large, with distinct gill slits; pedicellariae known in *Glyptocidaris;* spheridia present. (L.Jur.-Rec.)

Phymosoma [*Cyphosoma* L. Agassiz *non* Mantell] (Fig. 246): test low, flattened above; amb plates compound, polyporous, pore pairs in double series adapically; primary tubercles in regular series, without distinct radiating striae. U.Jur.-Eoc. Eur., N.Africa, Madagascar, W.Pakistan, N.Amer., S.Amer. *Acanthechinus* (Fig. 247): test small, hemispherical, flattened below; ambs polyporous, pore pairs in double series adapically but in single series adorally; primary tubercles in regular double series in both areas, sharply crenulate and with radiating ridges on the flanks. Pal. W.Pakistan. *Aeolopneustes* (Fig. 248): test large, subconical; ambs polyporous, pores adapically almost horizontal arcs of five or six, adorally narrowing to form a straight vertical series; primary and secondary tubercles small, of about the same size, both forming vertical series. Pal. W.Pakistan. *Eurypneustes* (Fig. 249): test large; ambs as broad as interambs, amb plates polyporous; poriferous zones very wide, pore pairs in three vertical series; primary tubercles in regular double series in each area, and secondary tubercles in series outside the primaries attain almost the same size. Pal. W.Pakistan. *Gauthieria* (Fig. 250): test low, flattened; amb plates polyporous throughout; poriferous zones simple, undulating; primary tubercles large, in two series in each area; apical disk large, monocyclic, pentagonal, extending into posterior interamb. U. Cret.-Pal. Eur., N.Africa, Madagascar, N.Amer. *Glyptocidaris:* test low, hemispherical, large; amb plates compound, polyporous, pore pairs in single series adapically, in double series only at ambitus; ocular I insert. Eoc.-Rec. France, N.Amer., Japan. *Porosoma* [*Coptosoma*] (Fig. 251): test low hemispherical; amb plates compound, polyporous, pores all in single wavy series;

109

apical system small, ocular I insert; ambs and interambs differing little in width and in size of primary tubercles. L.Cret.-Olig. Eur., N.Africa, S.Africa, N.Amer. *Thylechinus* (*s.s.*): test hemispherical; amb plates trigeminate; apical disk dicyclic, oculars usually widely exsert; ambs and interambs with single series of primary tubercles in each column. U.Cret.-Olig. Eur., N.Africa, W.Pakistan, Peru. [The subgenus *Orthechinus* (Fig, 252) has well developed secondary tubercles forming vertical series beside primary tubercles, and the apical disk has some oculars insert. U.Cret.-Eoc. Eur., N.Africa, Asia Minor, Iran, N.Amer.]

Family STOMECHINIDAE

Primary tubercles non-crenulate, those of ambs and interambs of about the same size; ambs compounded in diadematoid manner, with one exception trigeminiate or polyporous; diplopodous ambs may be present adapically or throughout; apical system dicyclic or monocylclic, small, rarely extending backward; peristome large, with distinct gill slits; primary radioles without cortex and collar; pedicellariae and spheridia present. (L.Jur.-Rec.)

Circopeltis (Fig. 253): test low hemispherical; ambs polyporous, pore pairs in single undulating line; two series of large primary tubercles in each area, sometimes with series of secondary tubercles alongside. U.Cret.-Eoc. Eur. *Phymotaxis:* like *Circopeltis*, but pore pairs in double series adorally. U.Cret.-Aquit. Eur. *Stomopneustes:* large; amb plates compound, each of four to six trigeminate plates covered by one large primary tubercle; poriferous zones broad, rather petaloid adorally; primary tubercles in regular series; distinct undulating median furrow in each interamb. Mio.-Rec. Java, Indo-Pac. [The type species, *Echinus variolaris* Lamarck, is the only known Recent stomechinid.]

Figs 249–263. ECHINOIDEA: PHYMOSOMATOIDA, ARBACIOIDA
AND TEMNOPLEUROIDA

Fig. 252 after Fourtau; Fig. 256 after Duncan and Sladen; Figs 251 and 257 after Cotteau Figs 249, 250, 253–255, 258 and 260–263 after Fell and Pawson, in Moore; Fig. 259 original.

249h. *Eurypneustes grandis* Duncan and Sladen, Pal. W.Pakistan. T. × 0·67.
250f. *Gauthieria radiata* (Sorignet), Senonian. France. T. × 3·3.
251b, c, f. *Porosoma cribrum* (L. Agassiz), M.Eoc. France. T. b × 0·8, c × 1, f × 3·5.
252h. *Thylechinus (Orthechinus) humei* (Fourtau), M.Eoc. Egypt. × 2·25.
253f. *Circopeltis meridanensis* (Cotteau), Turonian. France. T. × 7.
254c. *Arbacia waccamaw* Cooke, ?U.Mio. S.Carolina. × 1.
255c. *Codiopsis doma* (Desmarest), Cenomanian. France. T. × 1·1.
256c. *Coelopleurus forbesi* d'Archiac and Haime, M.Mio. (Gaj). Sind. × 0·75.
257a, e. *Coelopleurus delbosi* Desor, M.Eoc. Charente Inférieur. a × 0·75, e × 3·75.
258f. *Goniopygus major* L. Agassiz, Cenomanian. France. × 2·2.
259h. *Temnopleurus hardwicki* L. Agassiz, Plio. Zanzibar. × 2·75.
260g. *Arbacina monilis* (Desmarest), M.Mio. France. T. × 8·9.
261h. *Brochopleurus stellulatus* (Duncan and Sladen), M.Mio. W.Pakistan. T. × 8.
262h. *Grammechinus regularis* Duncan and Sladen, M.Mio W.Pakistan. T. × 3.
263h. *Leptopleurus hemisphaericus* (Duncan and Sladen), M.Mio. W.Pakistan. T. Enlarged.

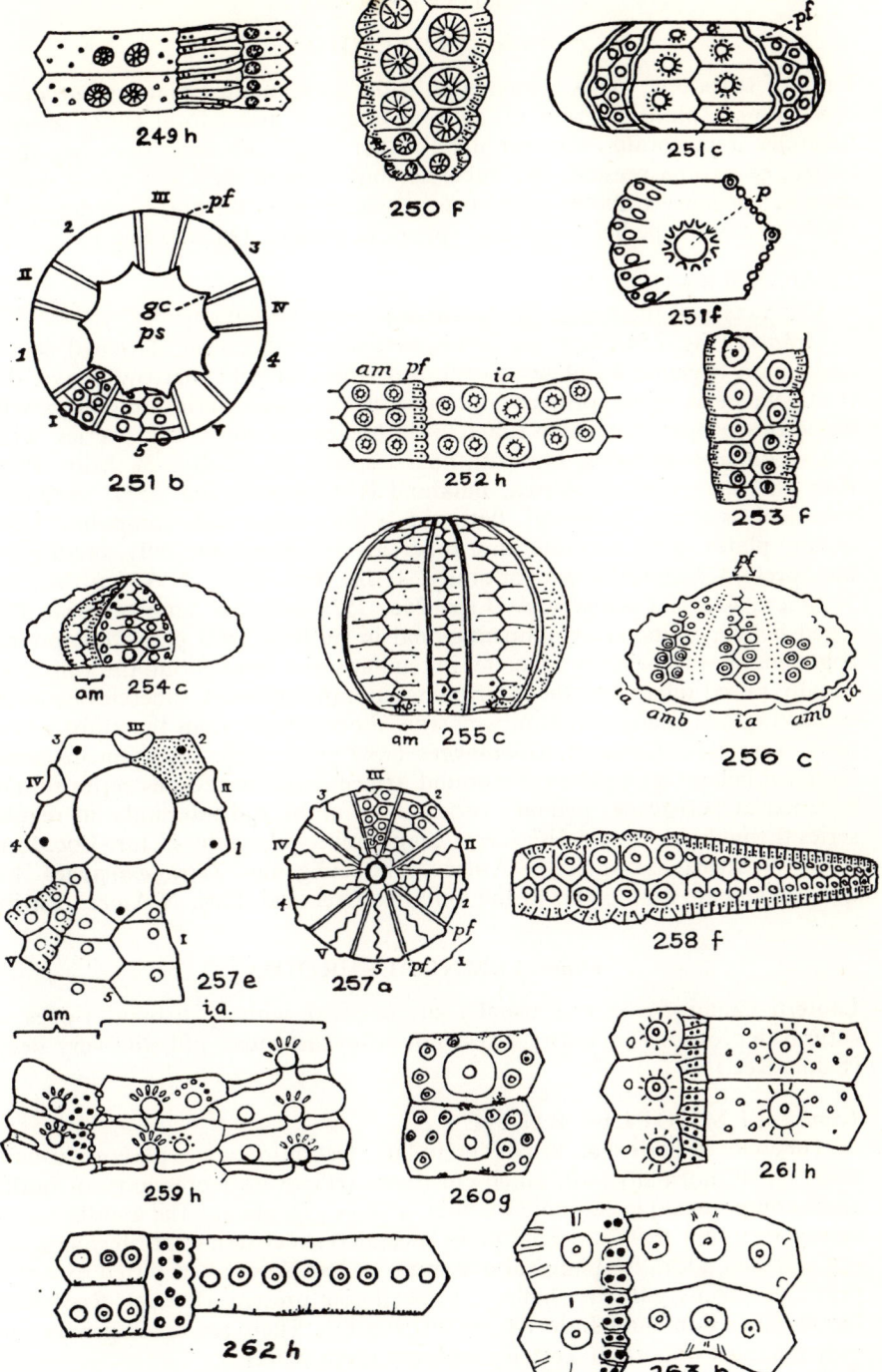

249 h

250 f

251 c

251 b

251 f

252 h

253 f

254 c

255 c

256 c

257 e

257 a

258 f

259 h

260 g

261 h

262 h

263 h

111

Order ARBACIOIDA

Lantern stirodont; ambs always with some compounded plates of *arbacioid type*; simple plates, if present, only at adapical and adoral ends; primary tubercles imperforate, non-crenulate, rather small, interamb ones becoming larger; epistroma present, like tubercles but without radioles; apical system dicyclic; primary radioles with cortex, usually smooth, but secondary ones small or absent; pedicellariae and spheridia present. (M.Jur.-Rec.)

Family ARBACIIDAE

Characters of the order. (M.Jur.-Rec.)

Arbacia (Fig. 254): test low hemispherical or subconical, flattened below; ambs with trigeminate plates, poriferous zones straight, narrow above, distinctly widened below; primary amb tubercles in regular series; interambs with numerous primary interamb tubercles in horizontal and vertical series, without secondary tubercles, but with naked spaces adapically. ?U.Mio., Plio.-Rec. N.Amer., Eur., W.Africa, Falkland Is. *Codiopsis* (Fig. 255): test hemispherical or almost spherical, flattened below; ambs with compound trigeminate plates; amb and interamb tubercles on adoral side only, small granules present adapically and adorally. U.Jur.-Eoc. Eur., N.Africa, M.East, N.Amer., Carib. *Coelopleurus* (Figs. 256, 257): test low hemispherical, flattened below, rounded or subpentagonal in outline; amb plates compound, trigeminate, with primary tubercles in regular series throughout, the amb areas usually raised relative to the interambs; interamb primary tubercles incipient or lacking adapically; interambs narrow more rapidly than the ambs adapically. Eoc.-Rec. Cosmop. *Goniopygus* (*s.s.*) (Fig. 258): test hemispherical, flattened below: amb plates compound, trigeminate; poriferous zones simple, widened at peristome; primary tubercles of ambs and interambs in regular series throughout; apical disk large, genital plates elongate. U.Jur.-Eoc. Eur., Asia, N.Africa, N.Amer., S.Amer. [The subgenus *Tetragoniopygus* has quadrigeminate amb plates at the ambitus. Cret.-Pal. Eur., N.Amer., Carib.]

Order TEMNOPLEUROIDA

Lantern camarodont; test usually ornamented with epistromal ridges or sutural depressions or both; if test not so ornamented, gill slits very deep. (L.Jur.-Rec.)

Family TEMNOPLEURIDAE

Tubercles imperforate, usually crenulate; test more or less globose, usually fairly small, normally with much epistroma (ridges or depressions, or both); ambs compounded in echinoid manner, always trigeminate, the middle plate a demiplate; pore pairs uniserial or in several vertical series; poriferous zones not widened adorally; plate sutures more or less grooved or pitted except in primitive forms; gill slits shallow; pedicellariae present. (U.Cret.-Rec.) [The family has a dominant Indo-Pacific distribution. There are many genera, but only the most important Tertiary ones are given here.]

Temnopleurus (*s.s.*) (Fig. 259): of medium or small size, low hemispherical or subconical; deep sutural pits, ambs with one pit, interambs with two pits, one median, the other adradial, pits obsolete adorally; tubercles crenulate; ambs about four-fifths the width of the interambs; pore pairs in triplet arcs; interporiferous areas wide, with two columns of primary tubercles; interambs with two columns of similar primaries, numerous secondaries and miliaries; primaries tending to be united by vertical ridges. M.Mio.-Rec. W.Pakistan, Indonesia, Iran, E.Africa, Indo-Pac. *Arbacina* (Fig. 260); no angular pores or pits, merely slight depressions in the horizontal sutures; tubercles non-crenulate; dense secondary tuberculation, those near primaries sometimes elongate. L.Mio.-Plio. Eur., Sicily, W.Africa. *Brochopleurus* (Fig. 261): no angular pores or pits; primary tubercles non-crenulate; they and some secondaries having radiating ridges around them; apical system dicyclic; gill slits small. Eoc.-M.Mio. W.Pakistan, N.Amer., N.Africa, N.Z., Austral., Egypt. *Grammechinus* (Fig. 262): depressed; tubercles non-crenulate; interamb plates transversely elongate, with median primary tubercle, and secondaries on each side attaining almost the same size; small tubercles near horizontal sutures elongate and tending to form bridges. M.Mio. W.Pakistan. *Irenechinus:* like *Brochopleurus*, but tubercles crenulate and secondary tubercles carried on ridges of epistroma forming zigzag series between primaries. L.Olig.-L.Mio. N.Z., Austral. *Leptopleurus* [*Lepidopleurus* Duncan and Sladen *non* Risso] (Fig. 263): hemispherical; primary tubercles smooth; ridges cross interporiferous zones; interamb plates broadly V-shaped; apical system dicyclic. M.Mio. W.Pakistan, Egypt. *Opechinus* (Fig. 264): four to six pits on each horizontal interamb suture; primary tubercles crenulate, interamb ones not in horizontal series adorally. Mio.-Rec. W.Pakistan, Indonesia, Japan. [The older Pakistan Eocene records are really of Miocene age.] *Paradoxechinus* (Fig. 265): like *Brochopleurus*, but more flattened, with raised ridges forming zigzag lines across interamb sutures, and lacking secondary tubercles. Eoc.-Mio. Eur., Austral. *Pseudarbacina:* primary tubercles non-crenulate; no pores or pits; granulation simple. M.Mio. Egypt. *Salmacis:* like *Temnopleurus*, but sutural pits reduced to small angular pores. Plio.-Rec. Java, Timor, E.Africa, Indo-Pac. *Temnechinus* (Fig, 266): primary tubercles non-crenulate, elevated; broad, deep angular pits; apical system dicyclic; gill cuts distinct; secondary tubercles not numerous. Plio.-Pleist. England. *Temnotrema* [*Dicoptella*] (Fig. 267): primary tubercles vaguely crenulate; horizontal interamb sutures with deep pit at each end, but horizontal amb sutures with deep pit at median end and smaller, shallow pit at abradial end; ten buccal plates. Mio.-Rec. Indonesia, Burma, Indo-Pac. *Triplacidia* (Fig. 268): large; no sutural pits; primary interamb tubercles crenulate, in horizontal and vertical series; apical system monocyclic or dicyclic. Eoc. Eur., Egypt.

Possibly also belonging to this family is *Gagaria* (Fig. 269): low hemispherical; no angular pores or pits and no distinct ornament other than tubercles which are crenulate and form regular series in both areas; apical system with ocular I insert. Eoc.-L.Mio. W. Pakistan, N.Amer.

Family GLYPHOCYPHIDAE

Tubercles perforate, crenulate; test sculptured; ambs compounded in diadematoid manner, trigeminate or polyporous. (L.Jur.-Eoc.)

Ambipleurus (Fig. 270): hemispherical; amb plates trigeminate; primary tubercles in regular series; horizontal sutures with deep pits; apical system dicyclic, ocular I insert. Eoc. Eur., Egypt, W.Pakistan. *Arachniopleurus* (Fig. 271): ambs polyporous, pore pairs in slight arcs; each primary tubercle is surrounded by a ring of secondaries connected to it by radiating ridges, the rings being in contact vertically and joined laterally by irregular horizontal ridges. Eoc. W.Pakistan, Eur. *Dictypopleurus* (Fig. 272): amb plates trigeminate; each interamb plate carries a single small, perforate, crenulate tubercle those of the same column being joined by a vertical ridge and those in adjacent columns by oblique ridges; similarly in the ambs, but tubercles are close to pore fields and the oblique ridges are more irregular. Pal. W.Pakistan. *Echinopsis* [*Herbertia*] (Fig. 273): hemispherical; amb plates trigeminate, in single series; primary tubercles of both areas small, perforate, crenulate, those of ambs close to pore pairs; small secondary tubercles. Eoc. Eur., W.Africa. *Progonechinus* (Fig. 274): hemispherical, flattened and concave adorally; ambs and interambs tumid; poriferous zones uniserial; both areas with up to four series of larger tubercles. Eoc. W.Pakistan. [In this genus the tubercles are atypically non-crenulate.]

Family TOXOPNEUSTIDAE

Tubercles imperforate, non-crenulate; test not sculptured; ambs compounded in echinoid manner, trigeminate to polyporous, usually much widened adorally; gill slits narrow and deep; pedicellariae present. (?Cret., ?Eoc., ?Olig., Mio.-Rec.)

Oligophyma (Fig. 275): small; amb plates trigeminate, pore pairs in erect arcs; single series of primary tubercles in each column; apical system with oculars I and V insert. M.-U.Mio. N.Africa. *Schizechinus* (Fig. 276): large, more or less hemispherical; amb plates trigeminate, with vertical row of primary tubercles in each column and also a vertical row of secondaries alongside; interambs with many rows of primary tubercles in vertical and hori-

---

Figs 264–273. ECHINOIDEA: TEMNOPLEUROIDA
Fig. 266 after Forbes; Fig. 272 after Duncan and Sladen; Figs 268 and 273 after Cotteau; Figs 264–267 and 269–271 after Fell and Pawson, in Moore.
264h. *Opechinus costatus* (d'Archiac and Haime), M.Mio. India. T. Enlarged.
265f, g. *Paradoxechinus bardini* (Cotteau), Mio. France. × 8.
266a, h. *Temnechinus excavatus* Forbes, Plio. Suffolk. T. a × 0·75, h × 2·25.
267g. *Temnotrema pulchellum* (Mortensen), Rec. Indonesia. × 7·3.
268c, e, f. *Triplacidia biarritzensis* Cotteau, Eoc. France. c × 1, e × 6, f × 8.
269h. *Gagaria venustula* (Duncan and Sladen), L.Eoc. W.Pakistan. T. × 5.
270h. *Ambipleurus douvillei* (Lambert), Eoc. Egypt. T. × 3·3.
271h. *Arachniopleurus reticulatus* Duncan and Sladen, Pal. W.Pakistan. T. Enlarged.
272e, h. *Dictyopleurus ziczac* Duncan and Sladen, Pal. Sind. T. × 15.
273a, c, e, f, g. *Echinopsis elegans* (Desmoulins), U.Eoc. T. a, c × 1, e × 3, f, g × 2.

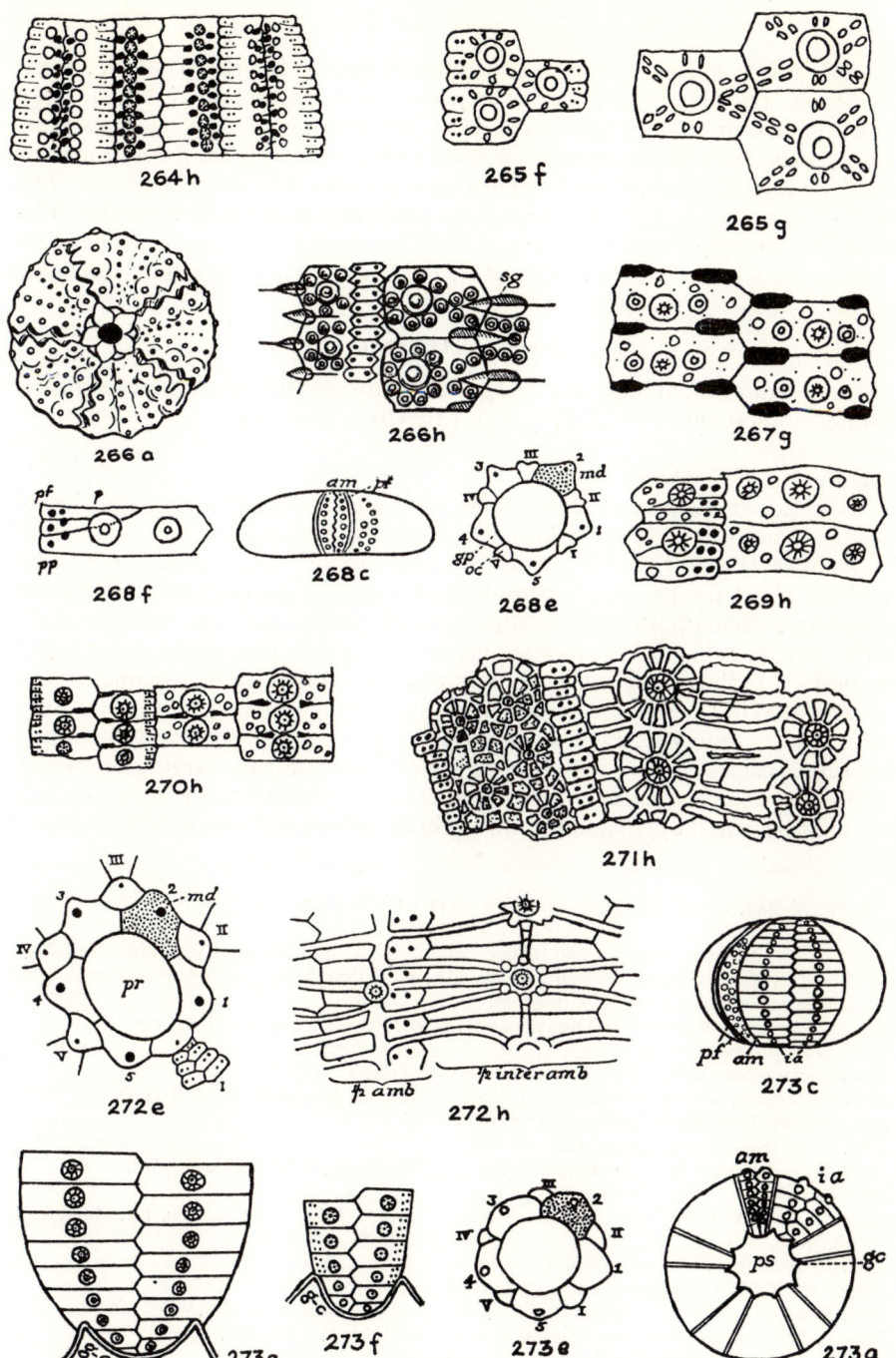

264 h

265 f

265 g

266 a

266 h

267 g

268 f

268 c

268 e

269 h

270 h

271 h

272 e

272 h

273 c

273 g

273 f

273 e

273 a

zontal series; apical system with oculars I and V insert. Mio.-Plio. Eur., N.Africa. *Tripneustes* [*Hipponoë* Gray *non* Milne-Edwards] (Fig. 277): large and high; amb plates trigeminate, with primary tubercle on every three or four plates; pore pairs arranged in three vertical series; pore fields almost as wide as interporiferous areas; plates compounded of one primary and five demiplates; interambs with two main columns of primary tubercles; distinct median naked space aborally in both areas; gill cuts deep; apical system with oculars I and V insert. Mio.-Rec. Eur., S.Amer., W.Pakistan, Carib.

Order ECHINOIDA

Lantern camarodont; test not sculptured; gill slits shallow; tubercles imperforate, non-crenulate; radioles solid. (?U.Cret., Pal.-Rec.)

Family ECHINIDAE

Globiferous pedicellariae with one or more lateral teeth on each side of blade (Mortensen's classification). (?U.Cret., Mio.-Rec.)

Echinus (Figs. 278, 279): subhemispherical or subglobose, large, with radioles short for its size; interamb primaries in numerous rather irregular columns, secondaries tending to disappear adapically; amb plates trigeminate, with primary tubercle on every second or third plate, pore fields rather wide; secondary radioles not much shorter than primaries, rather sparse; apical system dicyclic; peristome rather small and feebly notched. Plio.-Rec. Eur., Atl., Medit., Indo-Pac. *Psammechinus* (Fig. 280): subhemispherical; ambs and interambs each with two columns of primary tubercles, converging adapically in the ambs but remaining almost parallel in the interambs, one to each triad; secondaries small and passing into granules adapically; pore fields narrow;

Figs 274–288. ECHINOIDEA: TEMNOPLEUROIDA, ECHINOIDA AND HOLECTYPOIDA

Figs 277 and 280 after Duncan and Sladen; Figs 279, 282 and 283 after Hawkins; Figs 274–276, 281, 284 and 285 after Fell and Pawson, in Moore; Figs 286–288 after Wagner and Durham, in Moore; Fig. 278 original.

274h. *Progonechinus eocenicus* Duncan and Sladen, Pal. W.Pakistan. T. × 6·6.
275h. *Oligophyma cellense* Pomel, M.Mio. Algeria. Enlarged.
276c. *Schizechinus tuberculatus* (Pomel), Mio. Algeria. T. × 0·8.
277f. *Tripneustes proavia* (Duncan and Sladen), M.Mio. Nr. Karachi. × 1·7.
278. *Echinus.* Diagram of perignathic girdle. × 1.
279f. *Echinus esculentus* Linné, Rec. × 1·2.
280h. *Psammechinus*(?) *subcrenatus* (Duncan and Sladen), M.Mio. (Gaj). Sind. × 5.
281f. *Stirechinus scillae* (Desmoulins), Plio. Sicily. × 1·6.
282a, e, f. *Echinometra lucunter* (Linné), Rec. T. a × 0·75, e × 2·5, f × *c.* 1·3. (Tubercles omitted).
283f. *Heterocentrotus mammillatus* (Linné), Rec. T. × *c.* 3.
284g. *Diplosalenia gosseleti* (Cotteau), Eoc. France. T. × 3·3.
285f. *Strongylocentrotus droebachiensis* (Müller), Rec. Arctic. T. × 4·5.
286a, c. *Echinoneus cyclostomus* Leske, Rec. Lord Howe I. T. × 0·5.
287b, d. *Galeraster australiae* Cotteau, Eoc. Austral. T. × 0·5.
288b, d. *Globator nucleus* Agassiz, Senonian. Belgium. T. × 1·3.

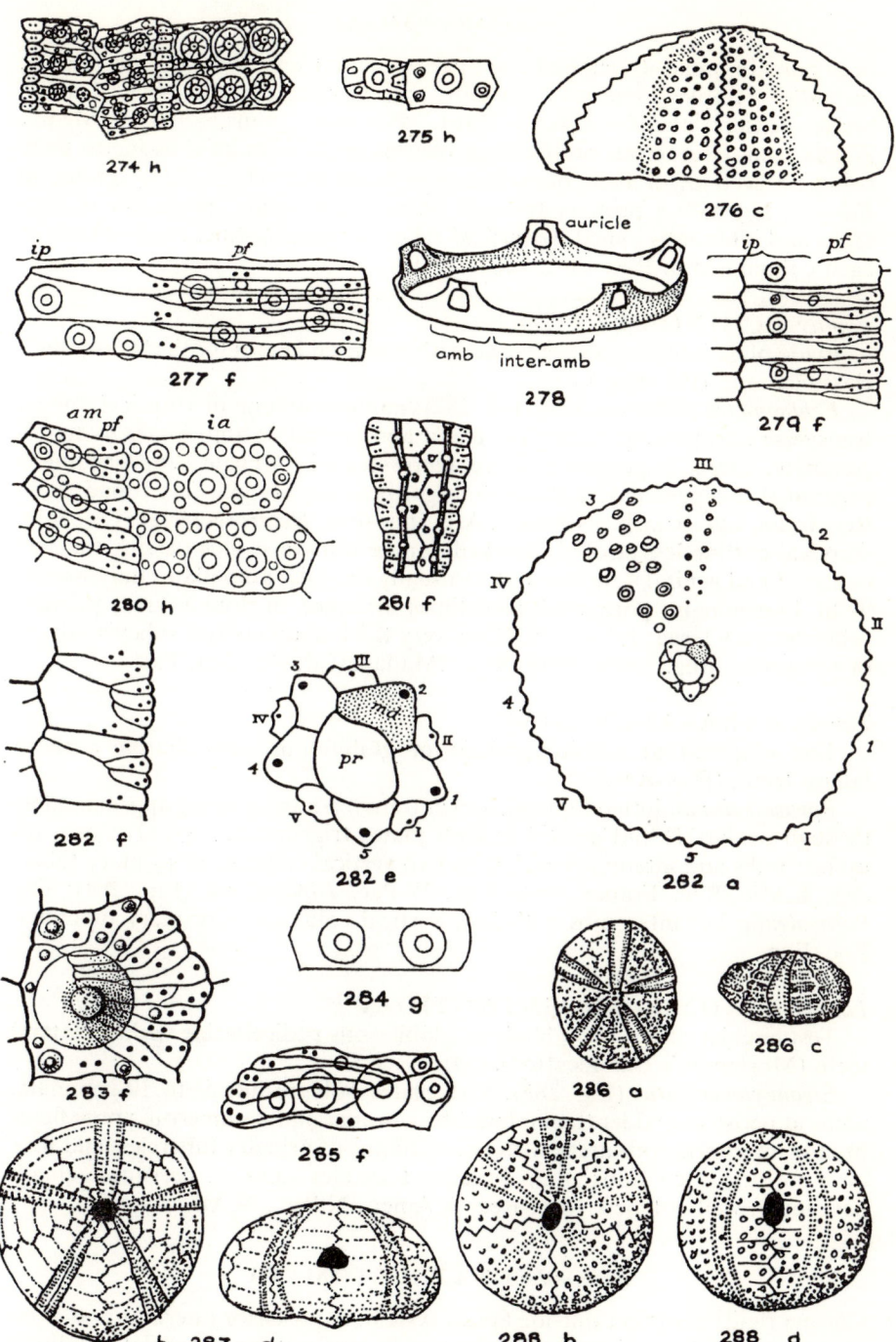

274 h

275 h

276 c

277 f

auricle

amb   inter-amb

278

279 f

280 h

281 f

282 f

282 e

282 a

283 f

284 g

285 f

286 a

286 c

b 287 d

288 b

288 d

117

pore pairs in pseudotrigeminal arcs; peristome not very wide, gill cuts feeble; secondary radioles numerous, smooth; apical system dicyclic; buccal membrane densely plated. ?Cret., ?Mio., Plio.-Rec. Eur., N.Atl., Medit., W.Pakistan. [Numerous pre-Pliocene records remain doubtful owing to there being no account of their buccal plates.] *Stirechinus* (Fig. 281): similar to *Echinus*, but with a primary tubercle on each amb plate; primaries of both ambs and interambs joined by vertical ridges forming distinct median keels in each column. Mio.-Rec. Eur., Medit., Atl.

Family ECHINOMETRIDAE

Blade of globiferous pedicellariae with unpaired lateral tooth (Mortensen's classification). (Pal.-Rec.)

Echinometra [*Ellipsechinus*] (Fig. 282): ambitus oblong or elliptical, longer transverse axis passing through ocular I and genital 3; amb plates 4- to 10-geminate, rarely trigeminate; pore pairs strongly arcuate, approaching a trigeminal arrangement; peristome large, pentagonal; gill cuts broad. Pal.-Rec. India, Eur., Carib., N.Amer., Atl., Indo-Pac. *Heterocentrotus* (Fig. 283): elliptical outline less strongly marked, longer transverse axis passing through ocular II and genital 4; apical system usually dicyclic; amb plates polyporous, 9- to 16-geminate, pore pairs sometimes arranged in double arcs; primary tubercles very large; primary radioles very thick and massive, subcyclindrical or prismatic, long. ?Mio., Plio.-Rec. ?Madagascar, Red Sea, Pac.

Family PARASALENIIDAE

Test elliptical at ambitus; blade of globiferous pedicellariae without lateral teeth. (Eoc.-Rec.)

Parasalenia: ambitus elliptical, longer transverse axis passing approximately through ocular III and genital 5; amb plates trigeminate; apical system dicyclic; ambs and interambs each with two vertical columns of primary tubercles. L.Mio.-Rec. France, Indo-Pac., W.Pac. *Diplosalenia* (Fig. 284): like *Parasalenia*, but interambs with four vertical columns of primary tubercles. Eoc. Eur.

Family STRONGYLOCENTROTIDAE

Test circular at ambitus; blade of globiferous pedicellariae without lateral teeth (Mortensen's classification). (Mio.-Rec.)

Strongylocentrotus (Fig. 285): amb plates polyporous, 5- to 10-geminate, ambs at peristome wider than interambs; coronal plates numerous; pore fields broad; interporiferous areas with two columns of primary tubercles, and four or more minor columns; amb primary tubercles smaller; peristome small, decagonal, well notched. Mio.-Rec. N.Amer., N.Pac., N.Atl., Arctic.

Superorder GNATHOSTOMATA

Corona rigid; periproct outside apical system, in posterior interamb; no compound amb plates; primary tubercles usually perforate and crenulate; radioles hollow; lantern and girdle usually present in adult; apical system and peristome usually approximately opposite. (Jur.-Rec.)

Order HOLECTYPOIDA

Test hemispherical to globular or ovoid; ambs petaloid or not, narrower than interambs; apical system monobasal or with four or five genital plates; girdle well developed or rudimentary or absent in adult; gill slits present or not; periproct supramarginal to inframarginal. (L.Jur.-Rec.)

The Holectypoida are mainly Mesozoic, with a few later survivors. The shifting of the periproct is almost the only sign of irregularity, the peristome remaining central and generally retaining lantern and girdle, though often in a more or less degenerate form. In spite of the approximate radial symmetry, there are two obvious distinctions from Regularia applicable to fragments not showing the periproct: (1) the more marked difference between upper and lower surfaces, and (2) the much less prominent character of the primary tubercles. Gradual divergence from Regularia is shown by (1) the tubercles, at first arranged in vertical columns, becoming irregularly scattered, and at the same time both by diminution in size and depression in pits tending to reduce the test surface almost to smoothness, and (2) the peristome, at first large and notched by gill cuts, becoming contracted and circular. Only four families are present in the Tertiary.

Suborder ECHINONEINA

Ambs nonpetaloid; auricles radial or lantern and girdle absent in adult; gill slits small or absent; ornament not orderly except in the Cretaceous genus *Conulus*; apical system tetrabasal or monobasal, with four genital pores. (Jur.-Rec.)

Family ECHINONEIDAE

Ambs with reduced plates in part or throughout; lantern and girdle absent in adult; peristome oblique or elongate. (U.Cret.-Rec.)

Echinoneus (Fig. 286): test ovoid; amb plates in groups of three; pore pairs uniserial adapically, in arcs of three adorally, poriferous zones slightly sunk; apical system monobasal; peristome oblique; periproct inframarginal; tubercles perforate or imperforate, non-crenulate; pedicellariae present. Olig.-Rec. Eur., Carib., Indo-Pac., Austral.

Family CONULIDAE

Ambs with reduced plates; girdle continuous; pore pairs uniserial or in arcs of three adorally; ornament not orderly except in the Cretaceous genus *Conulus*; peristome round or oblique. (Jur.-Eoc.)

Members of this family are common in the Jurassic and Cretaceous; two genera survived into the Eocene.

Galeraster (Fig. 287): test low hemispherical, rounded in outline or slightly produced posteriorly; apical system with four genital plates; periproct

119

marginal; peristome sunken; tubercles perforate, crenulate. Cret.-Eoc. Austral., France. **Globator** [*Pyrina auctt. non* Desmoulins] (Fig. 288): test usually round, orally flattened; pore pairs uniserial; poriferous zones flush to slightly sunk; four genital plates; peristome round or oblique; periproct marginal or a little supramarginal; tubercles perforate, crenulate. Cret.-Eoc. Eur., Medit., Carib., N.Amer., S.Amer., Madagascar, W.Pakistan.

Suborder CONOCLYPINA

Ambs petaloid or subpetaloid; pores of pore pairs at least partly conjugate; ornament not orderly; apical system monobasal, with four genital pores. (U.Cret.-Mio.)

Family CONOCLYPIDAE

Corona large, hemispherical; ambs petaloid, with pores of pairs widely separated, the outer one elongate; bourrelets well developed, and peristome with oral funnel. (Eoc.-Mio.)

Conoclypus [*Conoclypeus auctt.*] (Fig. 289): test high, flattened below, slightly elongated posteriorly; amb plates all primaries except near peristome, pores in single series adorally; periproct inframarginal, oval; primary tubercles perforate and crenulate. Eoc.-Mio. Eur., Medit., Madagascar, India, W.Pakistan, ?S.Amer. **Oviclypeus:** like *Conoclypus*, but lower and margin more rounded, periproct marginal, petals less petaloid, and outer pore of a pair only slightly longer than the inner one. Eoc. Eur.

Family OLIGOPYGIDAE

Test of small to medium size; periproct small; no oral funnel. (U.Cret. (Danian)-L.Mio.)

Oligopygus (Fig. 290): broadly elliptical in outline, not very high, slightly concave adorally, peristome deeply sunk, symmetrical; ambs petaloid, anterior petal usually longest; pores subequal, conjugate; apical system subcentral; periproct inframarginal, small; tubercles imperforate, non-crenulate. U.Eoc.-L.Mio. Carib. **Haimea:** similar to *Oligopygus*, but has a pentagonal peristome and numerous small scattered tubercles. Eoc. Carib., W.Africa.

Doubtfully related to the Oligopygidae, but differing in lacking a lantern and in possessing a large periproct, is the genus **Amblypygus** (Fig. 291): outline circular to ovate; low arched to high subconical, flattened below, margin tumid; ambs subpetaloid; pores of a pair adapically conjugate, the outer pore elongate, but adorally the pores are small, in simple oblique pairs; apical system tetrabasal, with four gonopores; peristome subcentral, subcircular to

---

Figs 289–292. ECHINOIDEA: HOLECTYPOIDA AND CLYPEASTEROIDA
Figs 290 and 291 after Wagner and Durham, in Moore; Figs 289 and 292 original.
289b, c. *Conoclypus delanouei* de Loriol, L.Eoc. (U.Libyan). Egypt. b × 0·5, c × 0·6.
290a, b, c. *Oligopygus wetherbyi* de Loriol, U.Eoc. Florida. T. × 0·7.
291a, b, c. *Amblypygus dilatatus* Agassiz and Desor, Eoc. France. T. × 0·75.
292a, b, c. *Clypeaster humilis* (Leske), Pleist. Red Sea. × 0·5.

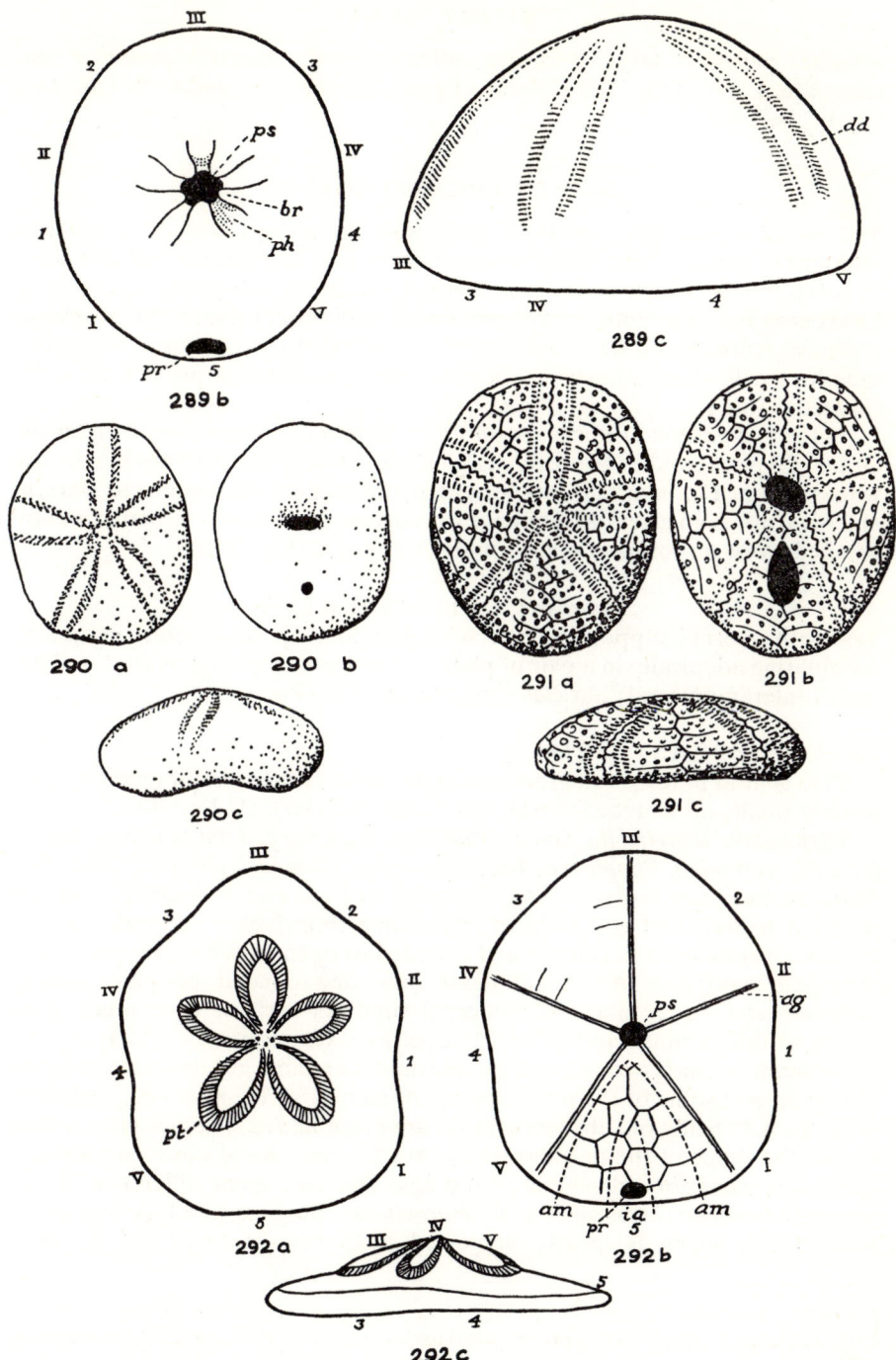

289 b

289 c

290 a 290 b

291 a 291 b

290 c

291 c

292 a 292 b

292 c

oblique; periproct large, pyriform, inframarginal, tubercles perforate and crenulate. L.Eoc.-Olig. Eur., Medit., Egypt, Madagascar, India, W.Pakistan, Carib., N.Amer.

Order CLYPEASTEROIDA

Test ovoid to flattened; petaloid ambs aborally always at least as wide as interambs; genital plates fused; primary tube feet respiratory and restricted to petals, but accessory ones are numerous, extend outside petals and in some forms even into interambs; peristome small, without gill slits; lantern without compass, flattened, girdle or auricles only; test usually with internal supports; radioles small, short, numerous, of two types; pedicellariae present. (U.Cret. (Maastr.)-Rec.)

In the first edition of this book only four families were mentioned under the Order Clypeasteroida. However, as a result of very detailed studies of the whole group, Durham (19) has increased the number of families to sixteen, and his work is an essential reference for anyone who intends to carry out any detailed work on forms classified under this order (Fig. 292A).

Suborder CLYPEASTERINA

Test with internal supports; petals with pseudocompound plates; interambs terminating adapically in a pair of plates; apical system pentagonal or stellate, apices interambulacral; auricles separate. (U.Eoc.-Rec.)

Family CLYPEASTERIDAE

Five genital pores; food grooves simple, weak; primordial interamb plates usually small; no 'combed' areas (see Arachnoididae). (U.Eoc.-Rec.)

Clypeaster [*Biarritzella, Guebhardanthus, Laganidea, Laubeanthus, Paleanthus, Platyclypeina, Pliophyma, Rhaphidoclypus, Stolonoclypus,* etc.] (Figs 292, 293): medium size to large; test flattened to high and bell-shaped, margin rounded to flattened and inflated; peristome funnel deep; adorally flat or concave; petals variable, closed and rounded to open or sublyrate; pore pairs conjugate, outer pore of a pair elongate, inner one rounded; periproct usually inframarginal, rarely marginal; internal supports variable in number, consisting of thin laminae and pillars. U.Eoc.-Rec. Cosmop.

As many as thirty-seven supraspecific names are now grouped in synonymy of *Clypeaster*; some of the more important names, which have been regarded in the past as being of subgeneric rank, are given above. There is great variation in the shape of the test and petals, but Durham could not recognize any systematic basis for subgeneric groupings. The very great difference in appearance between forms such as *C. marbellensis* Boussac (the type species of *Biarritzella*), which is greatly depressed, with thin margins and *C. altus*

Fig. 292A. DEVELOPMENT AND EVOLUTION OF FOOD GROOVES on oral surface in Clypeasteroida. Vertical spacing proportional to time except for Pleistocene. (After Durham, in Moore).

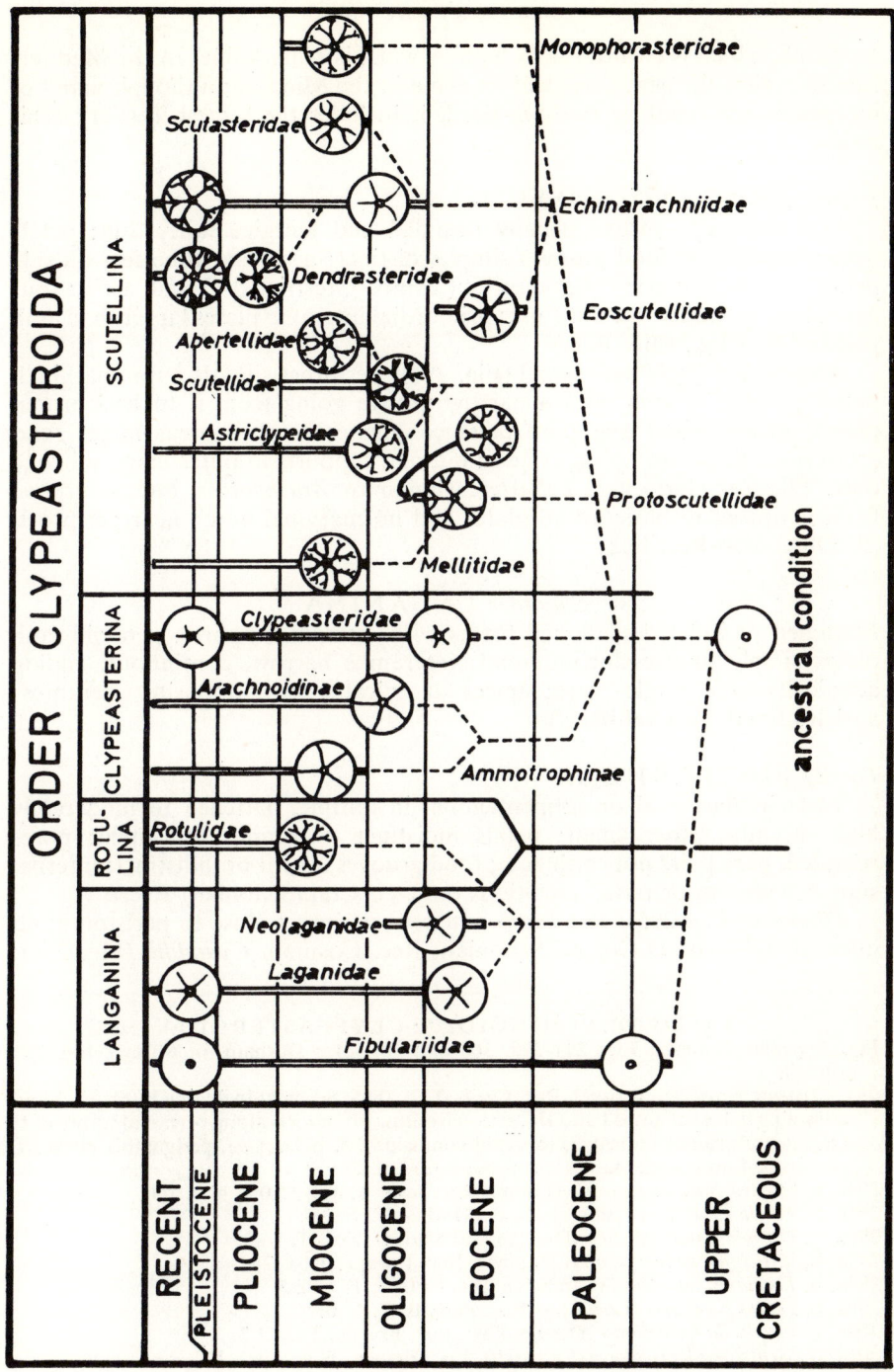

ORDER CLYPEASTEROIDA

SCUTELLINA

Monophorasteridae
Scutasteridae
Echinarachniidae
Dendrasteridae
Eoscutellidae
Abertellidae
Scutellidae
Astriclypeidae
Protoscutellidae
Mellitidae

CLYPEASTERINA

Clypeasteridae
Arachnoidinae
Ammotrophinae

ROTU-LINA

Rotulidae

LANGANINA

Neolaganidae
Laganidae
Fibulariidae

ancestral condition

RECENT
PLEISTOCENE
PLIOCENE
MIOCENE
OLIGOCENE
EOCENE
PALEOCENE
UPPER CRETACEOUS

123

Lamarck, which is two-thirds as high as wide, is remarkable. In the Mediterranean region the genus reached its acme in the Miocene; although many of
its species are small or medium-sized, it includes the largest Tertiary echinoids.

Family ARACHNOIDIDAE

Test flattened, outline usually rounded and margins fairly thin; petals
open; ambulacral food grooves simple, distinct; accessory tube feet outside
petals usually in dense oblique series ('combs'), restricted to amb areas; four
genital pores; peristome not sunk; primordial interamb plates larger than amb
plates. (?U.Olig., Mio.-Rec.)

Arachnoides [*Echinarachnius*] (Fig. 294): periproct slightly supramarginal,
notching edge of test, with a narrow groove going from it to basicoronal
plates; petals raised and pore pairs conjugate; combed areas large; food
grooves extending to apical system; internal supports in outer marginal zone
only. Plio-Rec. Indo-Pac. *Fellaster*: similar to *Arachnoides*, but no groove
from periproct to basicoronal plates and no marginal notch near periproct.
?U.Olig., Mio-Rec. N.Z.

Suborder LAGANINA

Flattened or inflated, flattened forms with internal supports; petaloid amb
plates simple or pseudocompound; interambs narrow, continuous, ending
adapically in a single plate; apices of apical system opposite interambs;
auricles fused. (U.Cret.-Rec.)

Family FIBULARIIDAE

Test circular, oval or subpentagonal in outline, flattened to moderately
high, usually rather small; petals indistinct or simple and open; pores
rounded, pore pairs not conjugate; food grooves absent or indistinct; internal
supports absent, or radial partitions only. (U.Cret.(Senonian)-Rec.)

Fibularia (Fig. 295): test ovate, inflated; periproct close to peristome; no
internal supports. U.Cret.(U.Senonian)-Rec. Cosmop. *Cyamidia* (Fig. 296):

Figs 293–301. ECHINOIDEA: CLYPEASTEROIDA
Fig. 299 after Cotteau; Figs 294–298, 300 and 301 after Durham, in Moore; Fig. 293
   original.
293. *Clypeaster rosaceus* (Linné), Rec. Carib. T. × 0·66. Segment including interamb 1 and
   adjoining halves of ambs I and II; section fronting observer is along perradial suture of II
   (zigzag surface not indicated, to avoid confusion). *p*, pillars; *pg*, perignathic girdle; *f*,
   peristome-funnel (external); *s*, perradial suture of I.
294a, b, h′. *Arachnoides placenta* (Linné), Rec. Java. T. a, b × 0·5, h′ × 0·6.
295a, b. *Fibularia ovulum* Lamarck, Rec. E.Indies. T. × 2·5.
296a, b. *Cyamidia nummulitica* (Duncan and Sladen), Eoc. T. × 3.
297a, b, k. *Echinocyamus pusillus* (Müller), Rec. Europe. T. × 2.
298a, b. *Eoscutum doncieuxi* (Lambert), Eoc. France. T. × 2·5.
299a, b. *Lenita patellaris* (Leske), M.Eoc. Paris B. T. × 6.
300a, b. *Porpitella hayesianus* (Desmoulins), Eoc. France. T. × 1·3.
301a, h′. *Scutellina lenticularis* (Lamarck), Eoc. France. T. a × 1·5, h′ × 3.

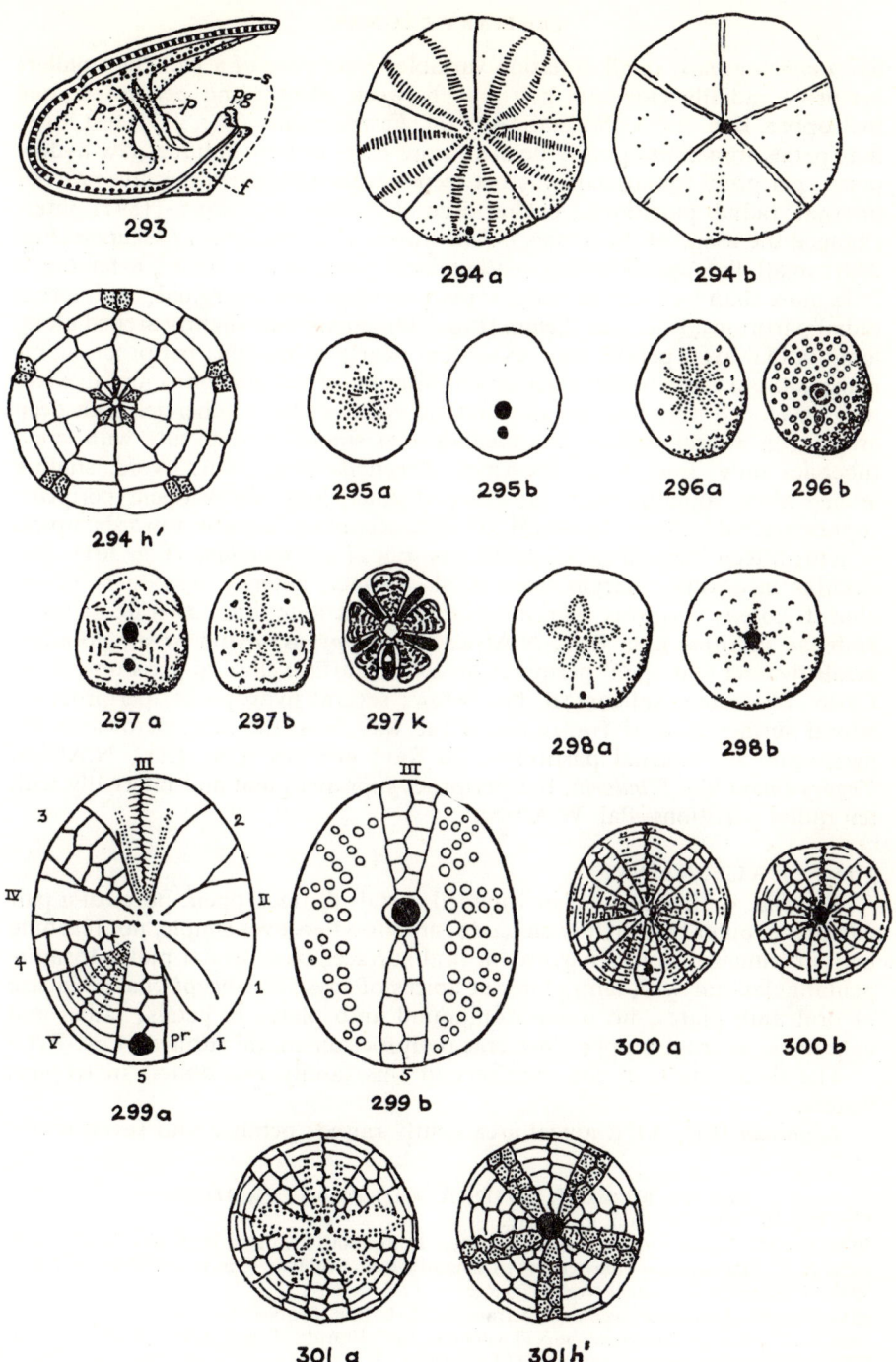

293

294 a   294 b

294 h'

295 a   295 b   296 a   296 b

297 a   297 b   297 k   298 a   298 b

299 a   299 b   300 a   300 b

301 a   301 h'

125

like *Echinocyamus*, small, inflation variable, inner pore of a pair the smaller; periproct radially elongate, half way between margin and peristome; one hydropore. Eoc. India, Pakistan, Austral. **Echinocyamus** (Fig. 297): test fairly flat; petals sometimes poorly defined, pore pairs usually oblique; few hydropores; periproct round, half way between margin and peristome; five pairs of internal radial partitions. U.Cret.-Rec. Cosmop. [Lambert (1891) interchanged the usage of the names *Echinocyamus* and *Fibularia*.] **Eoscutum** (Fig. 298): small, flat, apical system gently raised; petals nearly closed, extending a little more than half way to margin; periproct just supramarginal; ten internal radial partitions. Eoc. Eur. **Lenita** (Fig. 299): small, flat, slightly arched along longitudinal axis; petals open, extending nearly to margin; periproct supramarginal, but close to the margin; ten internal radial partitions and five less well developed ones interradially; lateral zones of large tubercles with deep areolae on adoral surface, the median area smooth; adapically with small tubercles only. Eoc. Eur., ?N.Amer. **Porpitella** (Fig. 300): ovoid, slightly arched along longitudinal axis; petals distinct, long, fairly open; periproct supramarginal, about one-third of distance from margin towards apical system; fifteen internal radial partitions. Eoc. Eur. **Scutellina** (Fig. 301): flat, circular in outline, margin thin; petals distinct, anterior one open, others almost closed; periproct marginal; ambs about four times as wide as interambs at ambitus. Eoc. Eur., N.Africa. **Tarphypygus:** ovoid to subglobular; petals distinct and open; periproct on adoral surface close to peristome. Eoc. Carib. **Thagastea:** subconical, flat below; several hydropores; periproct on adoral surface at least two-thirds of the way from the margin towards the peristome; no internal partitions; no food grooves. Eoc. Eur., N.Africa. **Togocyamus:** like *Fibularia*, but periproct supramarginal and internally with ten radial partitions. Pal. W.Africa.

Family LAGANIDAE

Test flat, outline angular or rounded; petals distinct, open; pores of a pair conjugate, outer one slightly elongate; amb food grooves simple, not reaching margin; interambs very narrow on oral surface; basicoronal plates forming pentangular star with amb plates at apices of rays; no abrupt change in size of oral amb plates; no pseudocompound amb plates in petals; radial and concentric internal supports present; periproct on adoral surface. (Eoc.-Rec.)

The distribution of the members of this family was and is in tropical waters.

Laganum (Fig. 302): apical area gently raised; petals about seven tenths

---

Figs 302–307. ECHINOIDEA: CLYPEASTEROIDA
Figs 302–307 after Durham, in Moore.
302a, b, e, h'. *Laganum laganum* (Leske), Rec. Indonesia. T. a, b × 0·35, e × 3, h' × 0·7.
303a, b, h'. *Jacksonaster conchatus* (M'Clelland), Rec. Malaya. T. a, b × 0·6, h' × 1·5.
304a, b. *Rumphia rostratum* (Agassiz), Rec. N.Z. T. × 0·35.
305a, b, c, h'. *Sismondia occitana* (Defrance), M.-U.Eoc. France. T. × 1.
306a, b, h'. *Neolaganum archerensis* (Twitchell), Eoc. Florida. T. a × 1, b × 0·65, h' × 1·3.
307h'. *Cubanaster torrei* (Lambert), U.Eoc. Cuba. T. × 1.

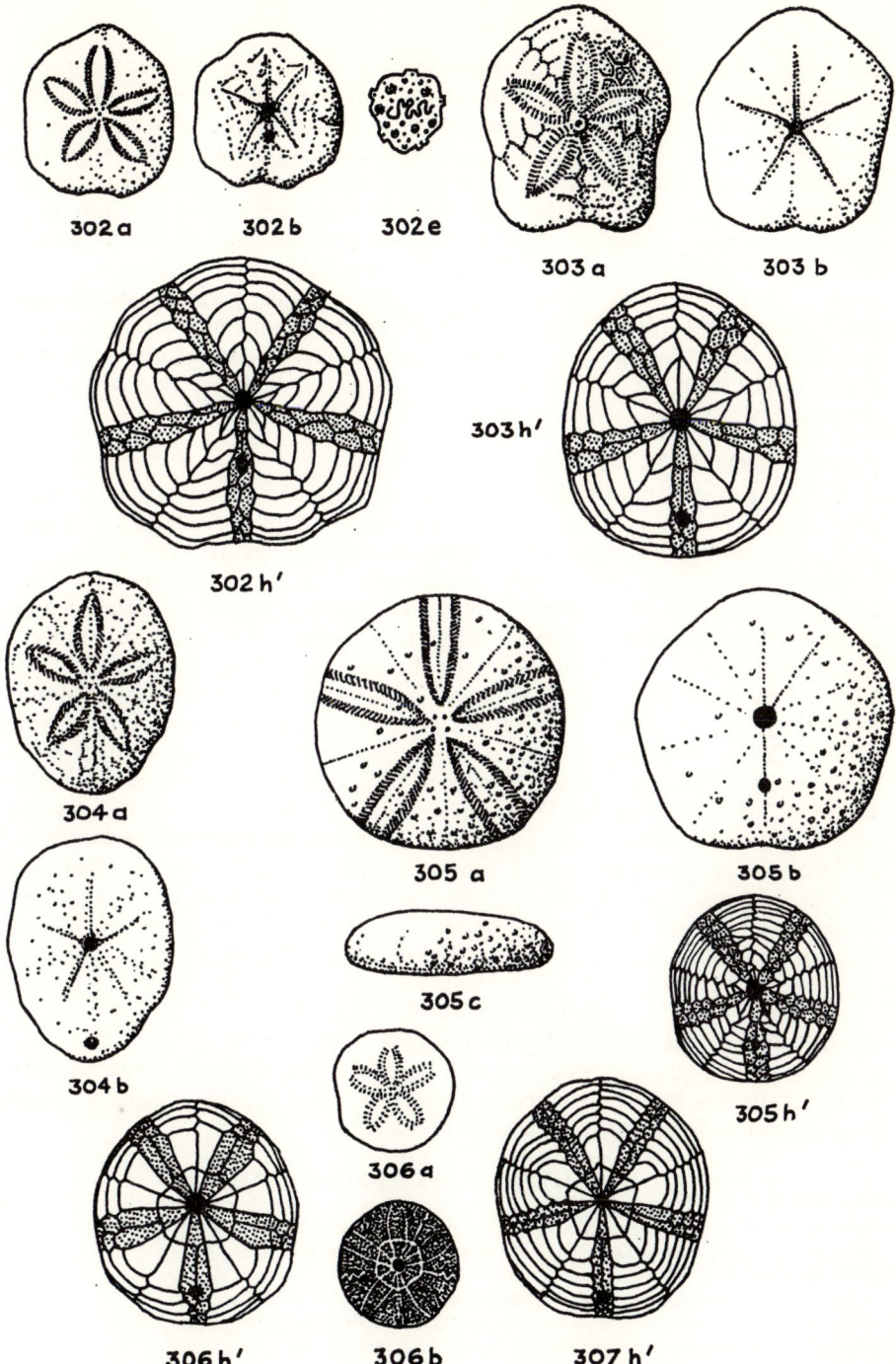

302 a 302 b 302 e

303 a 303 b

303 h'

302 h'

304 a

305 a

305 b

304 b

305 c

305 h'

306 a

306 h' 306 b 307 h'

the length of the radius; five gonopores; hydropores in a groove; periproct elongate in a radial direction, about half way between peristome and margin. Eoc.-Rec. Eur., Indo-Pac. *Jacksonaster* (Fig. 303): like *Laganum*, but periproct transversely oval and situated only one-quarter of the distance from the margin towards the peristome. Mio.-Rec. Indo-Pac. *Rumphia* (Fig. 304): elongate and with long petals; only four gonopores; hydropores not in a groove; periproct close to margin. Mio.-Rec. Indo-Pac., N.Z. *Sismondia* (Fig. 305): small, margin inflated, outline rounded to subpentagonal; petals slightly lyrate, length about three-quarters of the radius; periproct nearly half way from margin towards peristome; food grooves indistinct; internal radial partitions well developed, concentric ones incipient; basicoronal interamb plates larger than the amb ones. Eoc.-Mio. Eur., Africa, Asia, Indo-Pac., Austral.

Family NEOLAGANIDAE

Like Laganidae, but usually with pseudocompound plates in petals, outer pore of pore pairs very long, basicoronal plates in regular pentagon, and first pair of coronal plates much larger than succeeding ones. (Eoc.)

Neolaganum (Fig. 306): fairly small; petals nearly closed, nearly three quarters the length of the radius; plates of petals in dyads and triads; four gonopores; hydropores in a branching groove; periproct on adoral surface, nearly one-quarter of the way towards the peristome from the margin. Eoc. Carib. *Cubanaster* (Fig. 307): similar to *Neolaganum*, but petals longer, some dyads in plates of petals (other plates simple), and periproct nearer margin. U.Eoc. Carib. *Wythella:* similar to *Cubanaster*, but larger, with thinner margins, petals raised, and median part of interamb areas widened adorally. U.Eoc. Carib. *Neorumphia* (Fig. 308): large, posteriorly truncated; apical area raised and ambitus thick; petals pointed, plates mostly triads and tetrads; periproct near posterior adoral margin. ?U.Eoc. Carib. [Dr W. Durham (verbal information) tells me that he thinks that, although this genus was originally recorded as of post-Eocene age, it is more likely to be Upper Eocene, since all other members of the family are of Eocene age.]

Suborder SCUTELLINA

Rather flat; radial and concentric internal supports; no pseudocompound plates in petals, outer pore of pore pair elongate; apical system pentagonal or stellate; auricles fused; basicoronal interamb plates usually at least as large as amb plates; food grooves present. (Eoc.-Rec.)

Figs 308–314. ECHINOIDEA: CLYPEASTEROIDA
Figs 308–314 after Durham, in Moore.
308a, h', h. *Neorumphia elegans* (Sánchez Roig), ?U.Eoc. Cuba. T. a, h' × 0·3, h × 3.
309a, b, c. *Scutella subrotunda* (Leske), Mio. Malta. T. × 0·3.
310a, b. *Parascutella leognanensis* (Lambert), Mio. France. T. × 0·35.
311a, b. *Parmulechinus agassizi* (Oppenheim), Olig. Eur. T. × 0·35.
312a, h'. *Protoscutella mississippiensis* (Twitchell), M.Eoc. Mississippi. T. × 0·6.
313a, b, c. *Mortonella quinquefaria* (Say), U.Eoc. Georgia. T. × 0·4.
314a, c. *Periarchus alta* (Conrad), U.Eoc. N.Carolina. T. × 0·4.

128

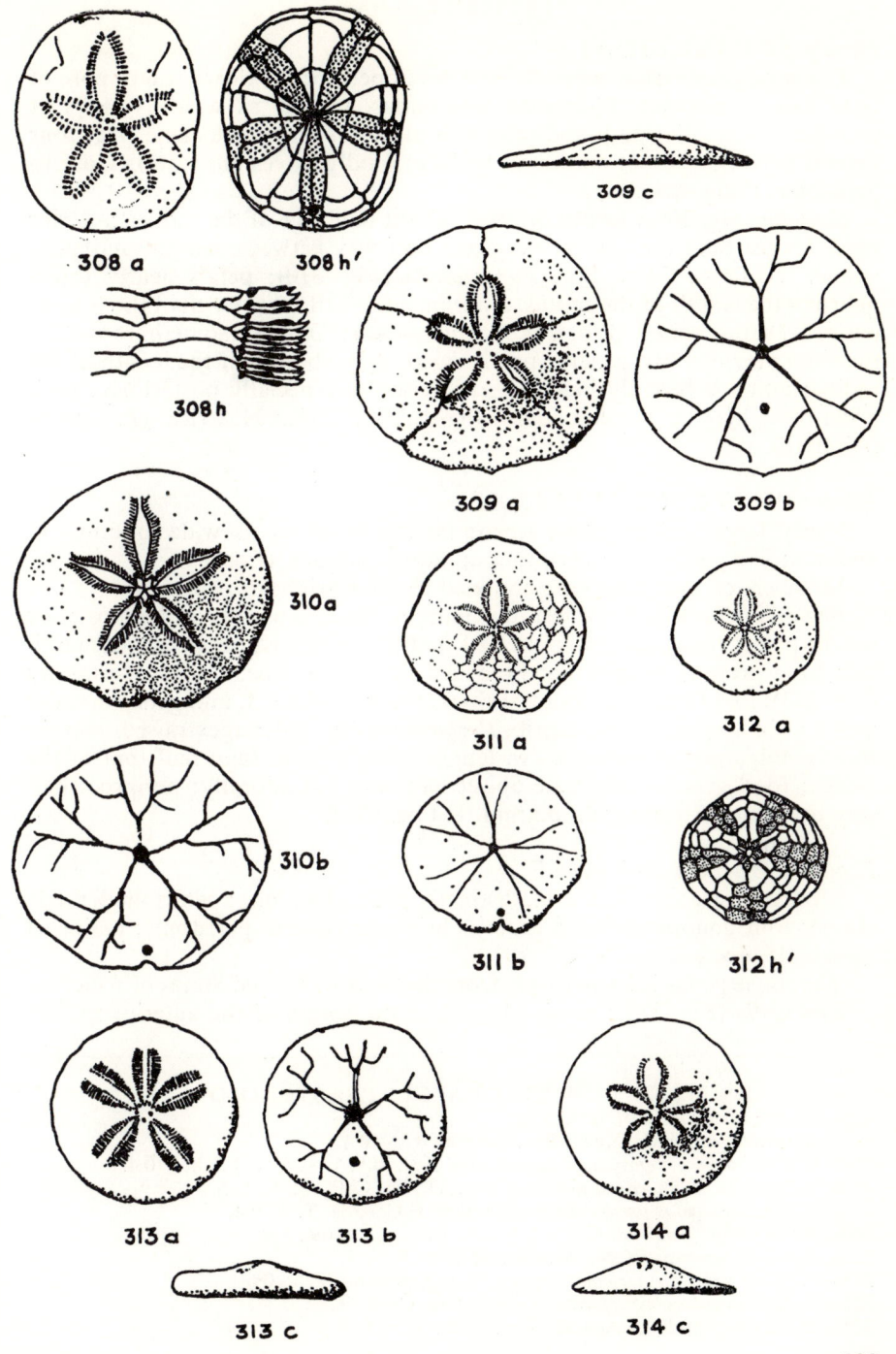

308 a   308 h′

308 h

309 c

309 a

309 b

310 a

311 a

312 a

310 b

311 b

312 h′

313 a   313 b

314 a

313 c

314 c

Family SCUTELLIDAE

Fairly large; internal supports well developed; petals closed, outer pore of pore pair subdivided; interambs continuous, usually as wide as ambs at ambitus; primordial amb and interamb plates of about the same size; four genital pores; periproct on adoral surface; food grooves bifurcating close to peristome. (Olig.-Mio.)

Scutella (Fig. 309): length of petals about half that of the radius, anterior petal longer than the others; periproct half way between margin and peristome. ?U.Olig., Mio. Eur. *Parascutella* (Fig. 310): petals nearly three-quarters the length of the radius, anterior one a little shorter than the others; periproct submarginal. Mio. Eur. *Parmulechinus* (Fig. 311): sometimes smaller than the previous two genera; petals small, only about half the length of the radius; ambitus broadly indented at interambs, especially posteriorly; periproct marginal to submarginal. Olig.-L.Mio. Eur., N.Africa (*P. chiesai* (Airaghi)).

Family PROTOSCUTELLIDAE

Fairly large; petals partly open; interambs about as wide as ambs at ambitus; five gonopores; periproct on adoral surface. (Eoc.)

Members of this family are confined to the Caribbean region.

Protoscutella (Fig. 312): test low, margin thin and usually with a posterior notch; petals equal, length about half that of the radius; periproct submarginal; food grooves simple, not branched. M.Eoc.-U.Eoc. Carib. *Mortonella* (Fig. 313): like *Periarchus*, but test thick, margin rounded, and petals broader and a little longer. U.Eoc. Carib. *Periarchus* (Fig. 314): apex raised, margin thin; petals open, rather narrow, length slightly more than half that of the radius; food grooves bifurcate about half way out adorally; periproct half way between peristome and margin. U.Eoc. Carib.

Family EOSCUTELLIDAE

Moderate size; test flat, thin, distinctly wider than long; petals moderately closed; four gonopores; food grooves bifurcate close to peristome; periproct marginal. (Eoc.)

The single genus is known only from the western United States of America.

Eoscutella (Fig. 315): petals about half the length of the anterior radius;

Figs 315–323. ECHINOIDEA: CLYPEASTEROIDA
Figs 315–323 after Durham, in Moore.
315a. *Eoscutella coosensis* (Kew), Eoc. Oregon. T. × 0·4.
316a, b. *Dendraster excentricus* (Eschscholtz), Rec. U.S.A. T. a × 0·5, b × 0·4.
317a, b. *Scutellaster interlineatus* (Stimpson), U.Plio. Calif. T. × 0·35.
318a, b. *Echinarachnius parma* (Lamarck), Rec. E.Canada. T. × 0·4.
319a, b. *Astrodapsis antiselli* Conrad, L.Plio. Calif. T. × 0·4.
320a. *Kewia blancoensis* (Kew), M.Mio. Oregon. T. × 1.
321a, b. *Monophoraster darwini* (Desor), Mio. Argentine. T. × 0·6.
322a, b. *Mellita quinquiesperforata* (Leske), Rec. Carib. T. × 0·35.
323a, b. *Encope grandis* L. Agassiz, Rec. Calif. T. × 0·3.

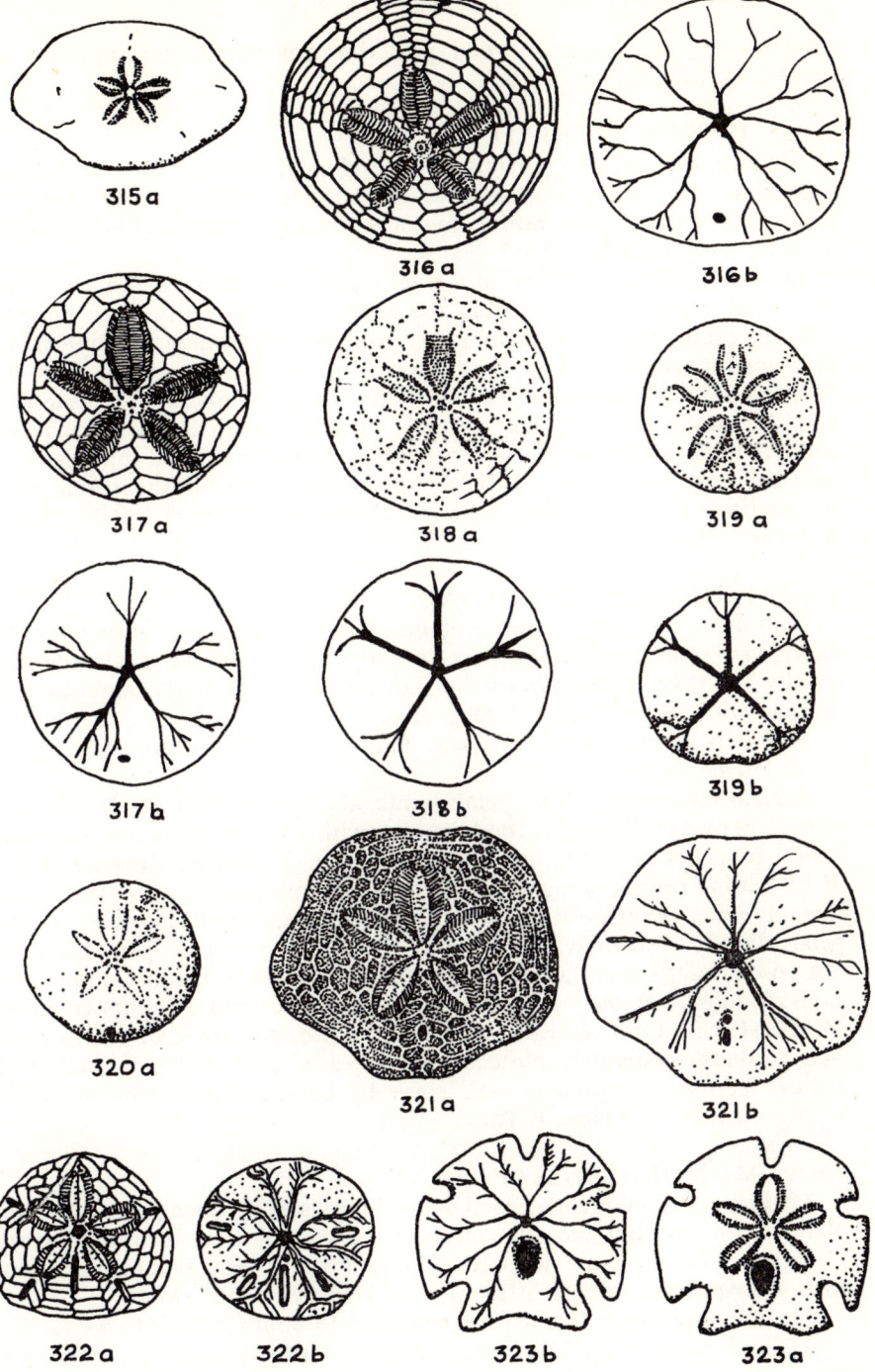

315a

316a

316b

317a

318a

319a

317b

318b

319b

320a

321a

321b

322a

322b

323b

323a

margin very thin, with a broad posterior notch; internal supports well developed. Eoc. Western U.S.A.

Family DENDRASTERIDAE

Medium size to large; petals well developed, anterior one more widely open than the others; interambs nearly as wide as ambs at ambitus; four gonopores; food grooves bifurcating or trifurcating; periproct inframarginal to supramarginal. (Plio.-Rec.)

Dendraster (Fig. 316): apical system usually eccentric posteriorly so that the anterior petal is longer than the others; margin fairly thin; periproct inframarginal, near ambitus; food grooves bifurcating, complex, best developed posteriorly, and may extend on to apical surface. Plio.-Rec. Western U.S.A. *Scutellaster* (Fig. 317): apical system only slightly posterior to middle; margin thin or thick; outline rounded or with interambs indented; petals at least three-quarters the length of the radius; periproct supramarginal, near ambitus; food grooves trifurcating about one-third of the distance from the peristome to the ambitus, poorly developed anteriorly. Plio. Western N.Amer., ?Sakhalin.

Family ECHINARACHNIIDAE

Medium size to large; petals well developed, anterior one more open than the others; four gonopores; periproct marginal to inframarginal; interambs usually less than three-quarters the width of the ambs at the ambitus; food grooves with central trunk. (L.Olig.-Rec.)

Members of this family are only known from North America and the north Pacific region.

Echinarachnius (Fig. 318): petals lyrate, about three-fifths the length of the radius; periproct marginal; food grooves with straight trunk and two equal lateral branches near margin. Mio.-Rec. N.Amer., N.Pac. *Astrodapsis* (Fig. 319): outline round, elongate or pentagonal; margin thin to inflated, more strongly indented at posterior amb; periproct marginal to just submarginal; petals slightly to strongly raised, usually broad, more or less open; apical disk not raised; adapical surface of advanced species with broad interamb depressions; food grooves as in *Echinarachnius*, but may extend on to upper surface. M.Mio.-L.Plio. Calif. *Kewia* (Fig. 320): small to medium size; anterior petal open, others moderately closed, length nearly three-quarters that of the radius; periproct supramarginal, close to margin; food grooves simple. L.Olig.-M.Mio. N.Amer., N.Pac.

Family MONOPHORASTERIDAE

Medium size to large, flat; well defined but variably open petals; interamb plates continuous but adorally narrower at the ambitus; primordial interamb plates much larger than amb plates; four gonopores; periproct inframarginal; food grooves bifurcating a little way out from peristome. (Mio.)

The members of this family are restricted to South America.

Monophoraster (Fig. 321): petals large, anterior one slightly longer than the others, length nearly three-quarters that of the radius; posterior anal lunule situated about two-fifths the length of the radius in from the margin; periproct just anterior to lunule. Mio. S.Amer.

Family MELLITIDAE

Medium size to large, flat; petals well defined, fairly closed, outer pore of pore pair very long; internal supports well developed; lunules or notches in posterior interamb and paired ambs; paired interambs not continuous; adorally interambs widest at ambitus; basicoronal plates small; periproct inframarginal, between posterior lunule and peristome; food grooves bifurcating close to peristome. (L.Mio.-Rec.)

The members of this family are confined to the Americas.

Mellita (Fig. 322): thin, flat, margin sharp; no anterior amb lunule; lunules narrow, long; anterior paired petals a little shorter than the others, reaching half way or more to the margin; peristome and apical system slightly anterior; four gonopores. Mio.-Rec. N.Amer., S.Amer., Carib. *Encope* (Fig. 323): like *Mellitella*, but apical system and peristome slightly anterior, posterior petals longer than the others, and posterior lunule more than half inside the line connecting the tips of the posterior petals. L.Mio.-Rec. N.Amer., S.Amer., Carib. *Mellitella* (Fig. 324): margin thick to thin; five amb lunules or notches; posterior interamb lunule outside a line joining the tips of the posterior petals; apical system and peristome slightly posterior; posterior pair of petals shorter than the others; five gonopores. Mio.-Rec. E.Pac., Carib.

Family ASTRICLYPEIDAE

Medium size to large, flat; margin thin; paired posterior amb lunules or notches and (in one genus) anterior amb lunules as well; petals well defined; interambs about as wide as ambs at ambitus; primordial interamb plates much larger than ambs; four gonopores; periproct on adoral side; food grooves bifurcating close to peristome; internal supports well developed. (Olig.-Rec.)

Astriclypeus (Fig. 325): five amb lunules; apical system central; periproct half way from peristome towards margin. Mio.-Rec. W.Pac. *Amphiope* (Fig. 326): like *Echinodiscus*, but lunules transversely oval (except in Oligocene species), and apical system slightly anterior. Olig.-Mio. Eur., W.Africa, Asia. *Echinodiscus* [*Tretodiscus*] (Fig. 327): two long, narrow posterior amb lunules or notches; apical system central; anterior petal longest, posterior petals shortest; periproct about one-seventh of the way in from the margin towards the peristome. Mio.-Rec. Indo-Pac.

Family ABERTELLIDAE

Medium size to large; margin with broad ambulacral and anal indentations; petals well defined, nearly closed; periproct inframarginal; four gonopores; internal supports well developed; food grooves bifurcating just outside peristome. (Mio.)

The single genus is restricted to North American and Caribbean regions.

Abertella (Fig. 328): petals about three-quarters the length of the radius;

133

posterior marginal indentations more prominent; periproct not far from posterior margin; interambs about half the width of the ambs at ambitus. Mio. N.Amer., Carib.

Family SCUTASTERIDAE

Medium size to large; three ovate anterior amb lunules or indentations; anterior petal open, the others nearly closed; four gonopores; periproct inframarginal; internal supports well developed; food grooves bifurcating fairly close to peristome. (L.Mio.)

The single genus is as yet known only from California.

Scutaster (Fig. 329): posterior marginal notch; apical system and peristome slightly posterior; periproct very close to margin; interambs very narrow adorally; primordial plates of paired interambs long. L.Mio. Calif.

Suborder ROTULINA

Test flat, posteriorly dentate or digitate; ambs petaloid adapically, petals well defined, the anterior more widely open; pore pairs usually conjugate; no pseudocompound plates; interambs about as wide as ambs at ambitus, terminating adapically in series of single plates; four gonopores in the compact apical system; periproct inframarginal; twenty small basicoronal plates of about the same size; concentric and radial internal supports; food grooves bifurcate near peristome. (Mio.-Rec.)

Family ROTULIDAE

Characters of the suborder. (Mio.-Rec.)

Rotula (Fig. 330): unequally digitate posteriorly, with paired anterior interamb lunules; pore pairs conjugate, outer pore of a pair elongate and subdivided; periproct nearly half way from peristome towards the margin. Mio.-Rec. W.Africa. *Heliophora* (Fig. 331): posteriorly equally digitate; no lunules; pore pairs conjugate, outer pore of a pair simple; periproct nearly half way to margin from peristome. Mio.-Rec. W.Africa. *Rotuloidea* (Fig. 332): posteriorly merely dentate; no lunules; margin thick; petals long, pore pairs only partly conjugate, outer pore simple; periproct half way from peristome to margin. Mio.-Plio. W.Africa, N.Africa.

Figs 324–331. ECHINOIDEA: CLYPEASTEROIDA

Figs 324–331 after Durham, in Moore.

324a, b. *Mellitella stokesii* (L. Agassiz), Rec. Ecuador. T. × 0·4.
325a, b. *Astriclypeus manni* Verrill, Rec. Japan. T. × 0·25.
326a, b. *Amphiope bioculata* (Desmoulins), Mio. France. T. × 0·4.
327a, b. *Echinodiscus bisperforatus* Leske, Rec. Indo-Pac. T. × 0·35.
328a b. *Abertella aberti* (Conrad), Mio. Maryland. T. × 0·17.
329a. *Scutaster andersoni* Pack, L.Mio. Calif. T. × 0·6.
330a, b. *Rotula octiesdigitatus* (Leske), Rec. W.Africa. T. × 0·3.
331a, b. *Heliophora orbiculus* (Linné), Rec. W.Africa. T. × 0·6.

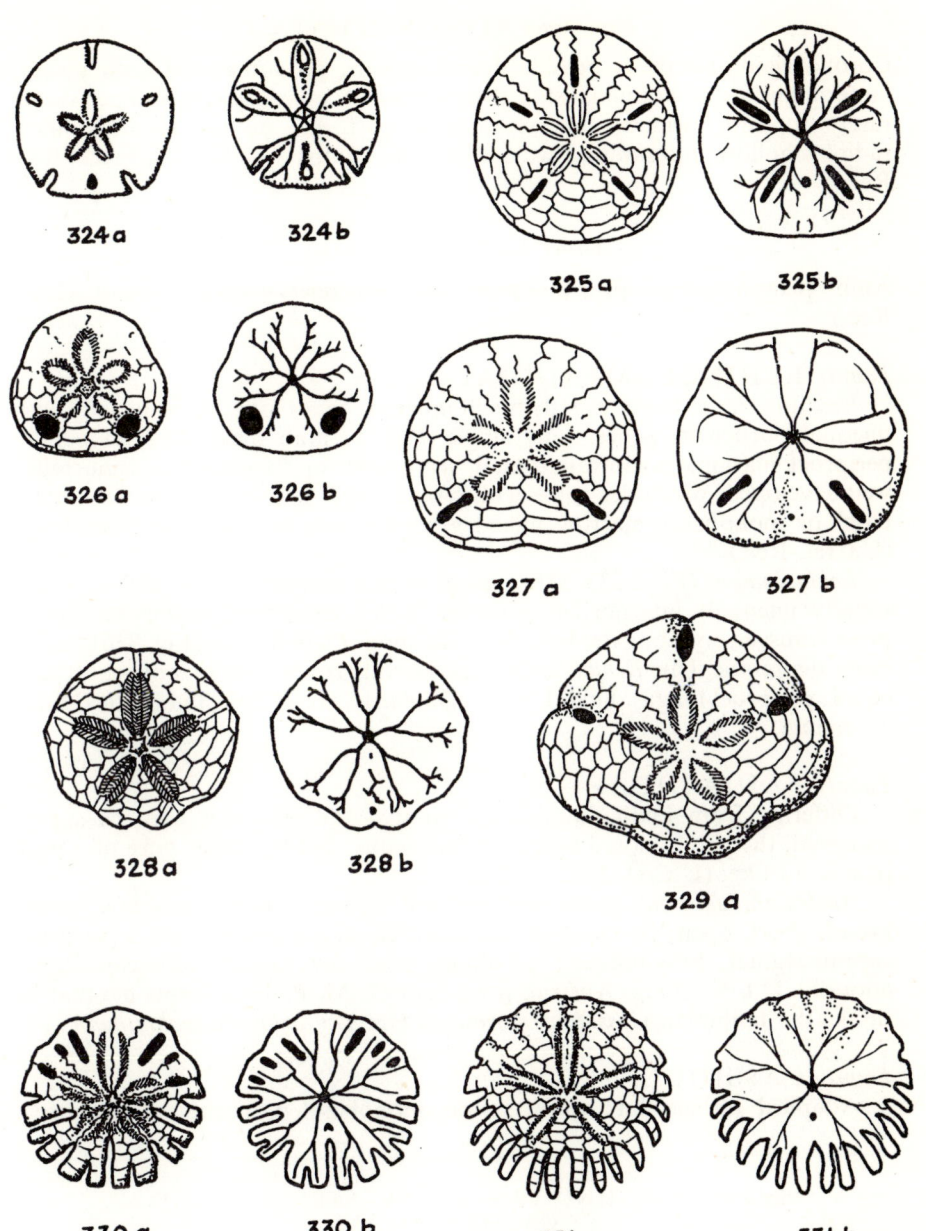

324 a

324 b

325 a

325 b

326 a

326 b

327 a

327 b

328 a

328 b

329 a

330 a

330 b

331 a

331 b

135

Superorder ATELOSTOMATA

Corona rigid; periproct outside apical system; no compound amb plates; lantern, girdle and branchial slits absent in adult; apical system and peristome rarely opposite; primary tubercles usually perforate and crenulate; primary radioles hollow; interambs always wider than ambs on adoral surface. (Jur.-Rec.)

Order CASSIDULOIDA

Ambs petaloid adapically; phyllodes and bourrelets usually present. (Jur.-Rec.)

Family ECHINOLAMPADIDAE

Medium size to large, usually much inflated; petals long, open, usually with unequal poriferous zones, only single pores in amb plates beyond petals; periproct marginal to inframarginal, transverse or longitudinal; bourrelets well developed; phyllodes wide, with few or many pores; apical system tetrabasal or monobasal; narrow naked granular zone in interamb 5 adorally. (U.Cret.-Rec.)

Echinolampas (Figs. 333–335): apical system monobasal; poriferous zones usually unequal; interporiferous zones wide; peristome pentagonal; periproct transversely elongate. Eoc.-Rec. Cosmop. *Plesiolampas* (Fig. 336): much more depressed than *Echinolampas*; poriferous zones not so noticeably unequal; periproct longitudinal. U.Cret.(Danian)-Eoc. W.Pakistan, Africa, Eur., Tasmania.

Family FAUJASIIDAE

Differing from Echinolampadidae in that the periproct may be supramarginal, the petals closed except in one genus, and the outer pore of a pore pair is slit-like. (U.Cret.-Eoc.-?Olig.)

Australanthus: oval, moderately inflated; apical system monobasal; petals broad, short, open; pores conjugate, with equal poriferous zones; periproct supramarginal, longitudinal; phyllodes with few pores; tubercles larger adorally. U.Eoc.-?Olig. Austral. [Professor G. M. Philip informs me that he thinks the Janjukian *Cassidulus florescens* Gregory is an *Australanthus*.]

Family CASSIDULIDAE

Adapical surface flat; apical system monobasal or tetrabasal; periproct

---

Figs 332–335. ECHINOIDEA: CLYPEASTEROIDA AND CASSIDULOIDA

Fig. 332 after Durham, in Moore; Fig. 333 after Fourtau; Figs 334 and 335 original.
332h′. *Rotuloidea fimbriata* Etheridge, Plio. Morocco. T. × 0·66.
333a, b. *Echinolampas proteus* Fourtau, U.Eoc. Libyan Desert. × 0·75.
334c. *Echinolampas kleini* (Goldfuss), Olig. N. Germany. × 0·5.
335b, c. *Echinolampas stellifer* Lamarck, U.Eoc. (Auversian). Blaye, Aquitaine. × 0·75.

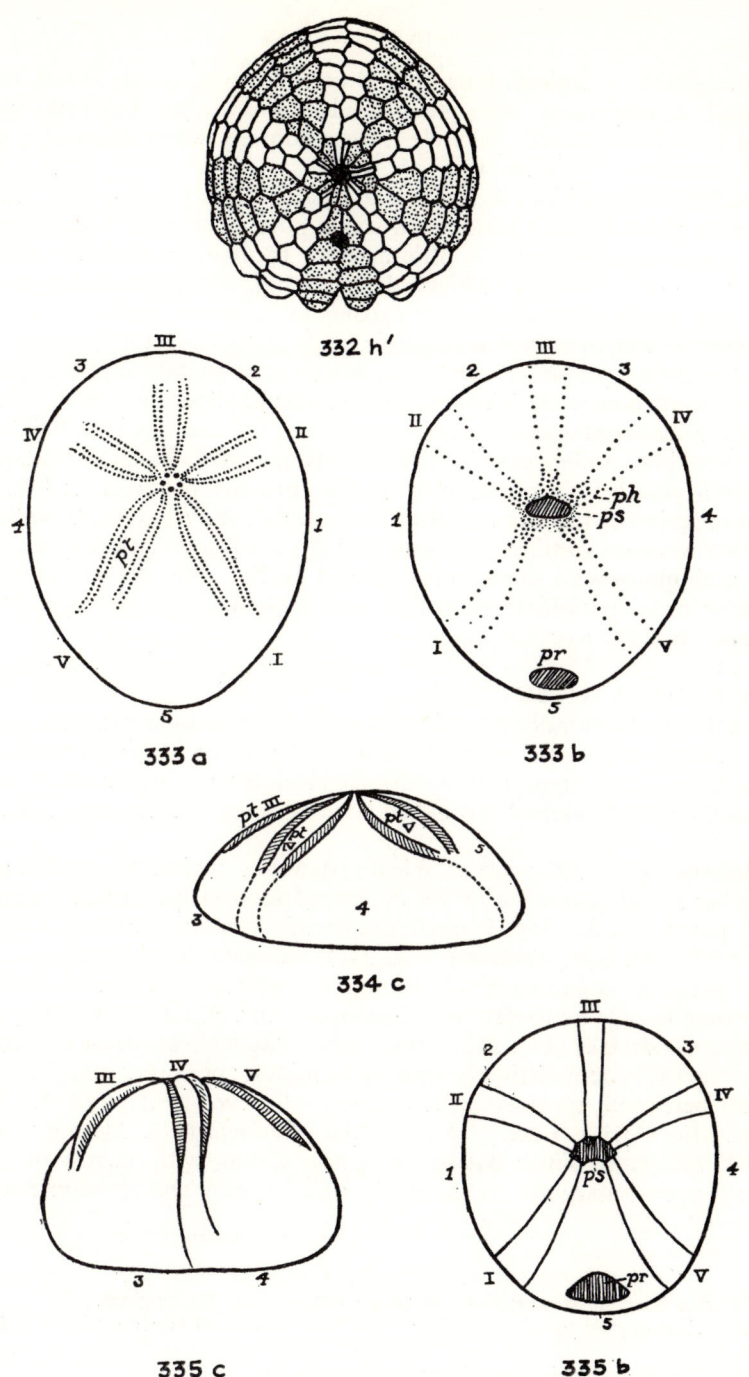

332 h'

333 a

333 b

334 c

335 c

335 b

137

supramarginal to marginal, longitudinal or transverse; petals broad, usually equal and inconspicuous, amb plates double-pored in pre-Senonian species; bourrelets well developed; tubercles larger adorally where there is a naked zone in interamb 5. (L.Cret.-Rec.)

Cassidulus (Fig. 337): oblong; apical system monobasal; petals straight, open, amb plates beyond petals with single pores; periproct supramarginal; apical system and peristome a little anterior to middle; phyllodes with few pores. Eoc.-Rec. Cosmop. *Rhyncholampas* (Fig. 338): somewhat concave below and larger compared with *Cassidulus*; greatest width posterior to centre; apical system monobasal; petals long, lanceolate, equal, with unequal poriferous zones; periproct transverse, marginal or slightly supramarginal or inframarginal; bourrelets fairly well developed; phyllodes widened. Pal.-Rec. Cosmop. *Paralampas:* small suboval, high and inflated, greatest height posterior to middle; ambs petaloid, short, subequal; pores equal, round, conjugate; peristome a little anterior to middle, broadly pentagonal; bourrelets wall-like; periproct high on posterior margin. Pal. W.Pakistan. *Rhynchopygus* (Fig. 339): greatest width posterior to middle; flat below; apical system tetrabasal; amb plates with single pores beyond petals; periproct supramarginal, transverse or longitudinal; bourrelets well developed; phyllodes wide, with few pores. U.Cret.-Rec. Cosmop.

Family PLIOLAMPADIDAE

Apical system monobasal, with three or four gonopores; pore pairs in petals strongly conjugate, amb plates beyond petals with single pores; periproct marginal, supramarginal or inframarginal, usually longitudinal; bourrelets usually well developed; no naked zone adorally in interamb 5. (U.Cret.-Rec.)

Pliolampas (Fig. 340): ovate; apical system and peristome a little anterior to middle; apical system with three or four gonopores; peristome longer than broad; petals well developed; phyllodes broad, with large pores. L.Eoc.-Plio. Eur., Africa, Malaya. *Eurhodia* (Fig. 341): elongate, low, flat below; petals equal; periproct supramarginal, transverse, with a roof; peristome much longer than broad; bourrelets well developed. Eoc. W.Pakistan, Eur., Africa, N.Amer. *Gitolampas* (Fig. 342): ovate-subpentagonal, broadest posterior to middle; apical system distinctly anterior to middle; periproct marginal, longitudinal; peristome transverse; bourrelets well developed; phyllodes broad. U.Cret.-Mio. Africa, Eur., N.Amer., Asia. *Ilarionia* (Fig. 343): moderately inflated, with steep sides; flat below; petals closed, with narrow poriferous zones; periproct marginal, longitudinal; apical system and peristome anterior

---

Figs 336–338. ECHINOIDEA: CASSIDULOIDA
Fig. 337 after Kier, in Moore; Fig. 338 after Cotteau; Fig. 336 original.
336a, b, e. *Plesiolampas paquieri* Lambert, Pal. Near Sokoto, N.Nigeria. a, b × 0·75, e × 3·25.
337a, b, c. *Cassidulus cariboearum* Lamarck, Rec. Carib. T. × 1·5.
338a, b, c, d. *Rhyncholampas grignonensis* (Defrance), M.Eoc. Paris B. × 1.

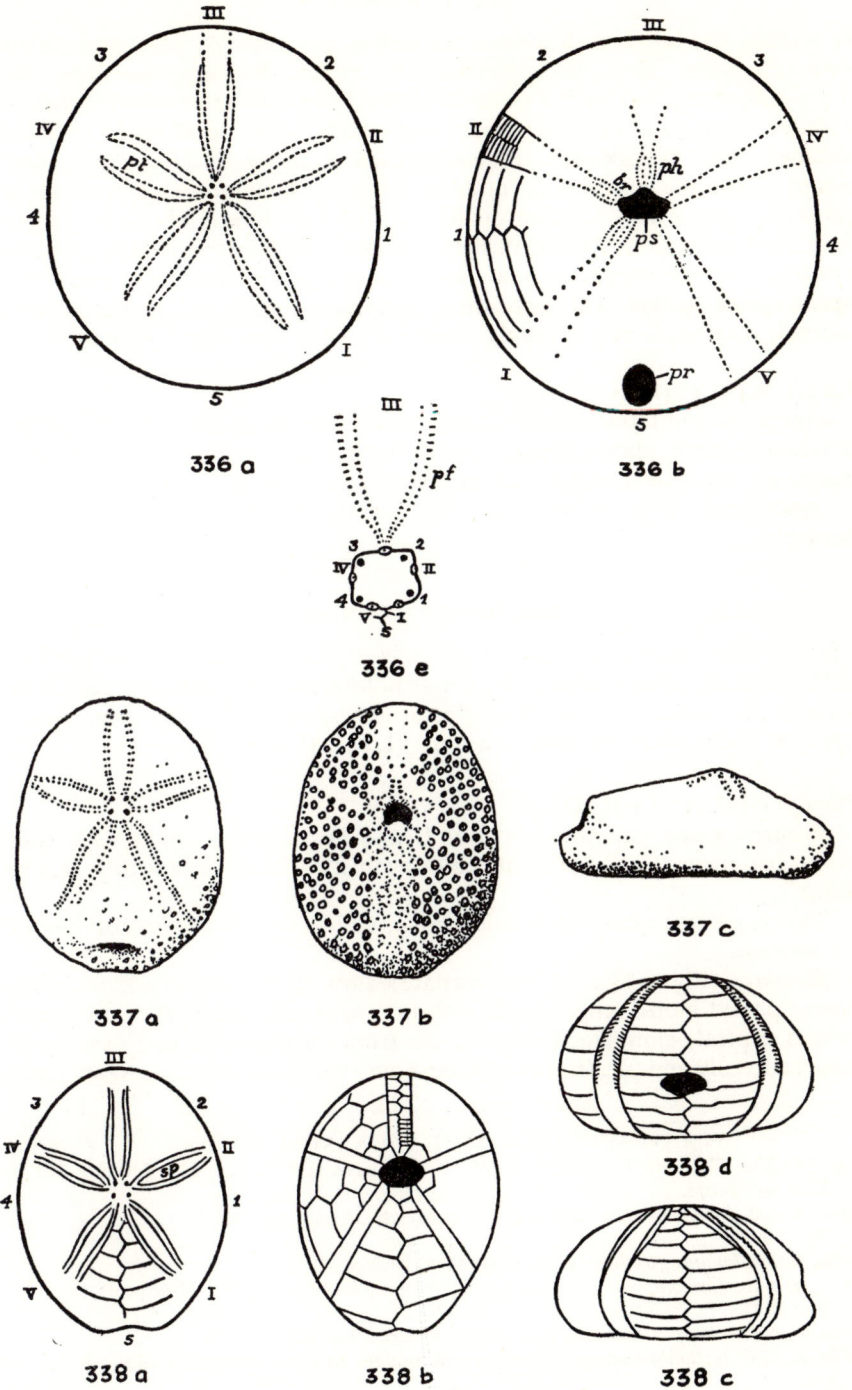

336 a

336 b

336 e

337 a

337 b

337 c

338 a

338 b

338 c

338 d

to middle, the latter with a rim; bourrelets not inflated; phyllodes narrow, with few pores. Eoc. Eur., W.Pakistan, Africa. **Neocatopygus** (Fig. 344): ovate-subpentagonal, rather inflated; apical system and peristome anterior to middle; periproct inframarginal; bourrelets well developed; phyllodes widened. Pal. W.Pakistan. **Pseudopygaulus** [*Eolampas*] (Fig. 345): small, ovate; peristome anterior to middle, apical system more so: amb III not petaloid; periproct inframarginal, transverse; bourrelets only slightly developed. Eoc. Eur., Africa, W.Pakistan. **Studeria** [*Tristomanthus*] (Fig. 346): ovate to slightly subpentagonal, a little truncated posteriorly; petals long, open; apical system with only three gonopores; periproct marginal, longitudinal; bourrelets very strongly developed. Olig.-Rec. Eur., Africa, Austral.

Family APATOPYGIDAE

Apical system tetrabasal in young, monobasal in adult; petals moderately developed; amb plates beyond petals with single pores; periproct supramarginal; bourrelets slightly developed. (U.Tert.-Rec.)

Apatopygus (Fig. 347): characters of the family. U.Tert.-Rec. N.Z., Austral.

Order HOLASTEROIDA

Apical system usually elongate or disjunct, without genital 5; plastron feebly differentiated or meridosternous; petals not always differentiated, paired ones usually not impressed; no floscelle; apical system and peristome may be opposite each other; fascioles variable. (L.Jur.-Rec.)

Family HOLASTERIDAE

Plastron meridosternous; oculars II and IV touching; ambs with double pores; basicoronal interamb plate abuts against two plates adapically. (L.Cret.-Rec.)

This is an essentially Mesozoic family, with only five genera surviving the Cretaceous.

Holaster (Fig. 348): outline cordate; amb III non-petaloid, with small pores; paired ambs subpetaloid, with non-conjugate, elongate pores; apical system central, elongate; periproct on truncate posterior end; interamb 5 raised adorally and plastron meridosternous; peristome anterior, semicircular,

---

Figs 339–343. ECHINOIDEA: CASSIDULOIDA
Fig. 339 after Cotteau; Figs 341 and 343 after Duncan and Sladen; Figs 340 and 342 after Kier, in Moore.
339a, b, c, d. *Rhynchopygus marmini* (L. Agassiz, in Agassiz and Desor), U.Cret. Normandy. T. × 0·75.
340a, b. *Pliolampas gauthieri* (Cotteau), Mio. France. T. × 1.
341a, b, c, m. *Eurhodia morrisi* Haime, in d'Archiac and Haime, Pal. (Ranikot). Sind. T. a, b, c × 0·9, m × 5·5.
342a, b, c. *Gitolampas tunetana* (Gauthier), U.Cret. Tunisia. T. × 1.
343a, b, c, d. *Ilarionia sindensis* Duncan and Sladen, M.Eoc. (Khirthar). Sind. × 1.

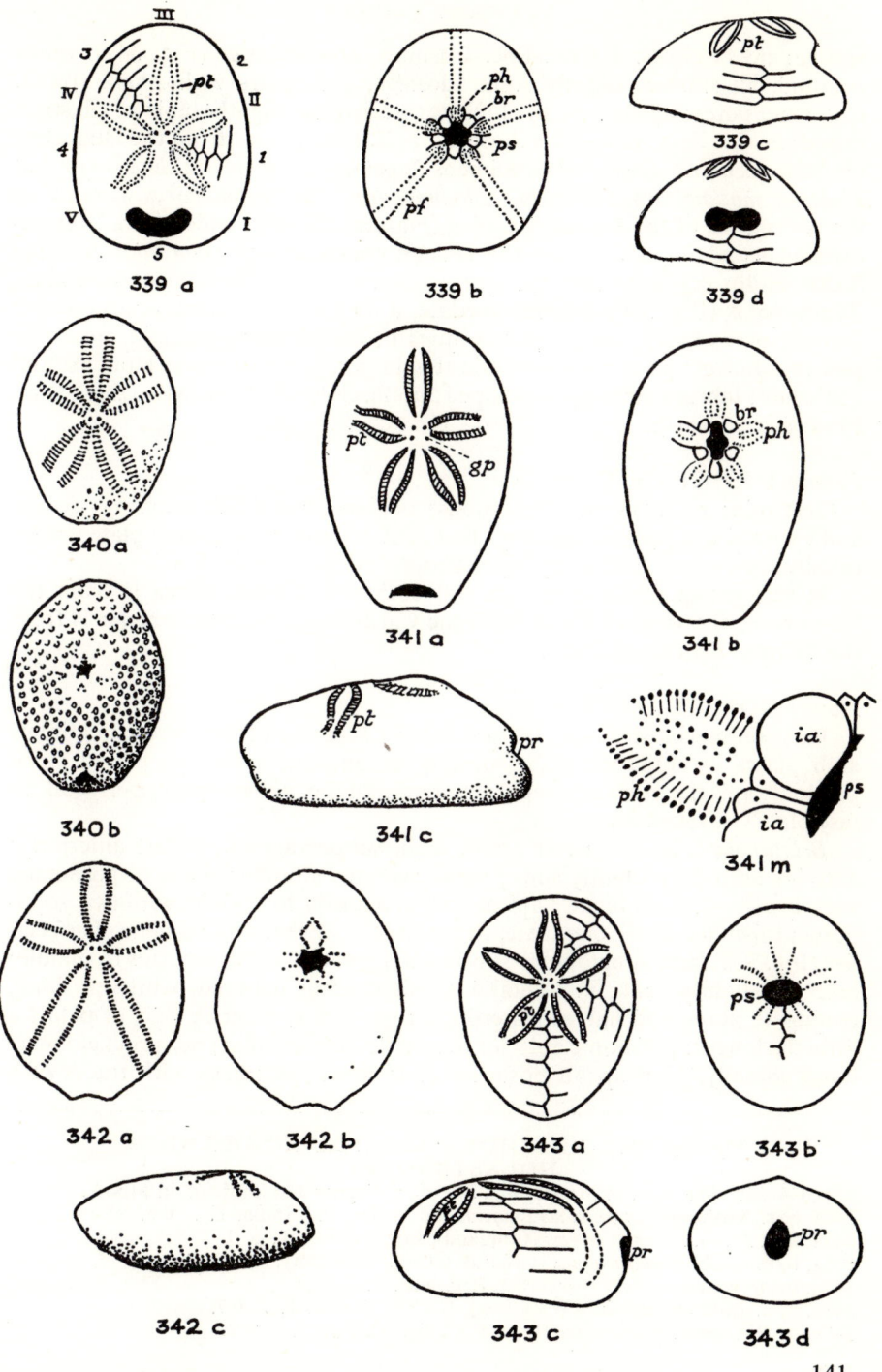

339 a

339 b

339 c

339 d

340 a

341 a

341 b

340 b

341 c

341 m

342 a

342 b

343 a

343 b

342 c

343 c

343 d

simple; no fascioles. L.Cret.-Eoc. Cosmop. *Duncaniaster* (Fig. 349): ambs petaloid, all similar, slightly sunk adorally; pore pairs with small, round, conjugate pores; apical system elongate; peristome slightly labiate; plastron raised; no fascioles. Eoc.-Mio. Austral., N.Z. [The record from the Oligocene of New Zealand refers to specimens labelled 'probably *Echinocorys* but *Holaster spatangiformis* Hutton', '*?Echinocorys*' and '*Echinocorys scutatus*' in the collections of the British Museum (Natural History) which have been re-examined and found to have the generic characters of *Duncaniaster*; Professor Philip confirms that the genus ranges up to Miocene in Australia.] *Toxopatagus* (Fig. 350): outline cordate, amb III deeply sunk and notching margin, pores small, round, not conjugate; paired ambs petaloid, posterior pair the shorter; posterior pore series the larger, pores of a pair elongate and conjugate; labrum strongly developed; no fascioles. U.Cret.-Mio. Eur., Iran, Madagascar, Carib.

Family URECHINIDAE

Test thin; plastron meridosternous; oculars II and IV touching; paired ambs with single pores adapically; first post-primordial interamb plate single; usually with subanal or marginal fasciole. (?U.Eoc., Mio.-Rec.)

In this group fossil forms are not common. *Chelonechinus* is from the Miocene of the western Pacific and the Caribbean, and *Sanchezaster* is from the ?Upper Eocene of Cuba.

Family SOMALIASTERIDAE

Plastron meridosternous, labrum usually separated from succeeding interamb plates by adjacent ambs meeting at mid-line adorally; paired ambs petaloid; apical system ethmophract or ethmolytic, not elongate; peripetalous fasciole. (U.Cret.-Pal.)

Brightonia (Fig. 351): test small, high, subpentagonal, widest anteriorly; anterior amb only slightly sunk; petals narrow, straight; pores of pore pairs slightly elongate, conjugate; apical system anterior to middle, with two gonopores; tubercles small, crenulate, irregularly arranged. Pal. Somalia. *Leviechinus* (Fig. 352): test subglobular-subcordate, greatest width anterior to middle; amb III slightly sunk, non-petaloid; paired ambs petaloid, with very long, conjugate pores and narrow interporiferous zones; anterior pair of petals a little the longer; peristome anterior to middle, semicircular; periproct elongate longitudinally, high up on posterior end; some specimens with traces of a

---

Figs 344–349. ECHINOIDEA: CASSIDULOIDA AND HOLASTEROIDA

Figs 344–348 after Kier, in Moore; Fig. 349 after Wagner and Durham, in Moore.
344a, b, c. *Neocatopygus rotundus* Duncan and Sladen, Pal. India. T. × 0·5.
345a, b, c. *Pseudopygaulus trigeri* (Coquand), Eoc. Tunisia. T. × 1.
346a, b, d. *Studeria subcarinatus* (Goldfuss), Olig. ?Germany. × 0·75.
347a, b. *Apatopygus recens* (Edwards), Rec. N.Z. T. × 0·66.
348a, b, d. *Holaster nodulosus* (Goldfuss), U.Cret. France. T. × 0·7.
349a, b. *Duncaniaster australiae* (Duncan), Olig. Austral. T. × 0·7.

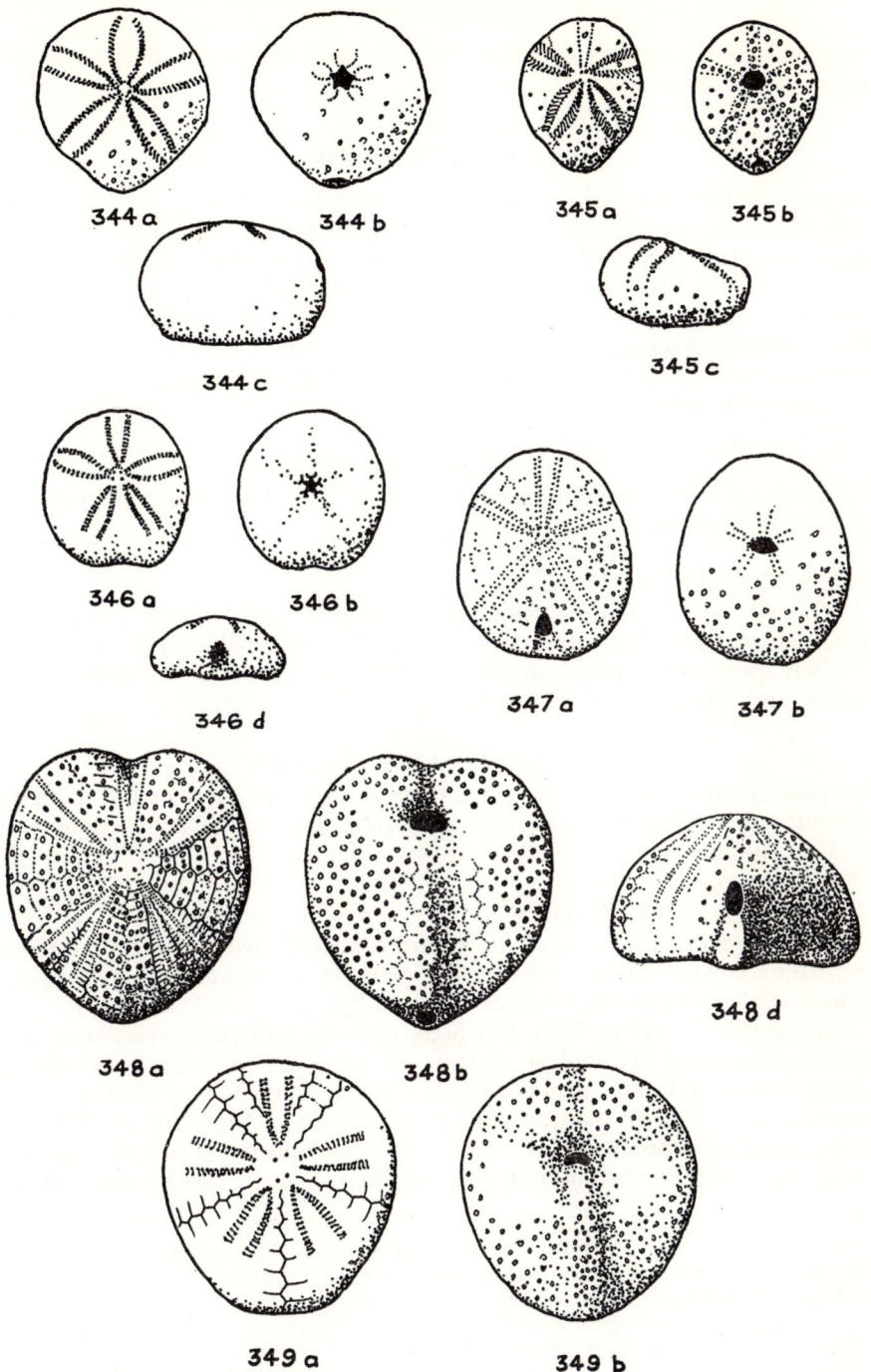

344 a 344 b 345 a 345 b

344 c 345 c

346 a 346 b 347 a 347 b

346 d

348 a 348 b 348 d

349 a 349 b

143

marginal as well as a peripetalous fasciole; tubercles small, irregularly distributed. Pal. Somalia.

Order SPATANGOIDA

Apical system ethmophract or ethmolytic, with four or fewer gonopores, not opposite peristome; amphisternous plastron; amb plates primitively with pore pairs, but many forms with only single pores in some plates; all or some amb areas aborally petaloid; peristome usually anteriorly eccentric and labiate; phyllodes usually present; no bourrelets; periproct near posterior end; most forms possess fascioles (except Toxasteridae), called peripetalous, marginal, subanal, lateroanal, anal and internal, occurring either singly or in partial combination; outline usually cordate; mostly burrowers. (L.Cret.-Rec.)

Suborder TOXASTERINA
Petaloid; usually without fascioles and primary radioles; apical system ethmophract. (L.Cret.-Rec.)

Family TOXASTERIDAE
Three or four gonopores; paired ambs petaloid, usually open, anterior amb petaloid or not; ambital and circumoral amb pores double. (L.Cret.-Rec.)

Only two rare genera are known from the Tertiary: *Adytaster* (Eoc., Spain) and *Palmeraster* (U.Eoc., Cuba).

Suborder HEMIASTERINA
Petaloid; peripetalous fasciole, some families also with lateroanal or marginal fasciole; no subanal fasciole; apical system ethmophract to ethmolytic. (L.Cret.-Rec.)

Family HEMIASTERIDAE
Apical system ethmophract or ethmolytic, with two to four gonopores; peristome labiate; paired ambs usually petaloid, fairly well closed; peripetalous fasciole; no primary radioles. (L.Cret.-Rec.)

Hemiaster (*s.s.*) (Fig. 353): test high, inflated, posteriorly truncated, with gentle anterior sinus; apical system ethmophract, with four gonopores; anterior amb not fully petaloid, with small round pores; posterior paired petals a little shorter than anterior paired petals. L.Cret.-Eoc. Eur. In the subgenus *Leymeriaster* (U.Cret.-Mio., Eur.), the posterior paired petals are much shorter than the anterior ones. In the subgenus *Trachyaster* (U.Cret,

---

Figs 350–353. ECHINOIDEA: HOLASTEROIDA AND SPATANGOIDA
Figs 350–352 after Wagner and Durham, in Moore; Fig 353 original.
350a, b, c. *Toxopatagus italicus* (Manzoni), Mio. Italy. T. × 0·5.
351a, c. *Brightonia macfadyeni* Kier, Pal. Somalia. T. × 1.
352a, b, c. *Leviechinus gregoryi* (Currie), Pal. Somalia. T. × 0·75.
353a, b, c, d. *Hemiaster bufo* (Brongniart), Cenomanian. Vaches Noires, Normandy. T. × 1.

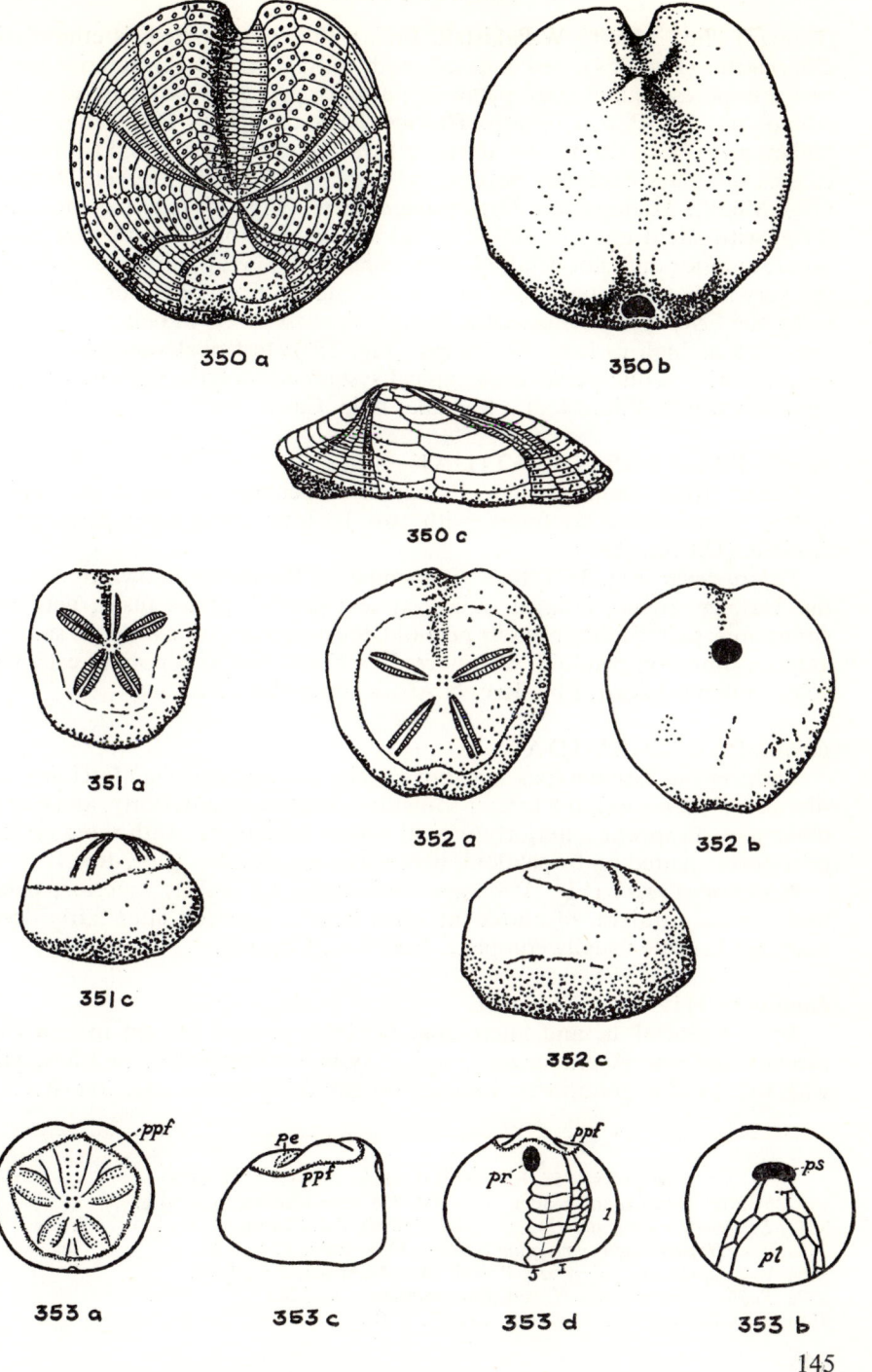

350 a

350 b

350 c

351 a

352 a

352 b

351 c

352 c

353 a

353 c

353 d

353 b

(Danian)-Plio., Medit., W.Pakistan, India), the apical system is ethmolytic. *Ditremaster* (Fig. 354): test inflated, with only a very slight anterior sinus; two gonopores; paired ambs petaloid, posterior pair very short; apical system ethmolytic. Pal.-Plio. Cosmop. *Holcopneustes* resembles *Hemiaster*, but is highest anterior to the middle, the apical system is ethmolytic, and the fasciole crosses the ambs distinctly beyond the ends of the petals. U.Cret.(Danian)-Olig. Medit., Madagascar. *Hypsopatagus* (*s.s.*) (Fig. 355): test ovate-cordiform, with slight anterior sinus; apical system ethmolytic, with four gonopores; anterior amb not petaloid, with small pores; petals narrow and closed; test very high, maximum height anterior to middle. Eoc.-Olig. Eur., Asia. The subgenus *Leiopneustes* (Eoc.-Olig. Eur., N.Amer.) differs in being only about one-third as high as long. *Opissaster* (Fig. 356): test ovate-cordiform, with deep anterior sinus; petals sunk; apical system ethmolytic, with two to four gonopores. Eoc.-Plio. Medit., Somalia, Asia, Carib.

Family PALAEOSTOMATIDAE

Differs from Hemiasteridae in having a pentagonal peristome; apical system ethmophract to fused, with two to four gonopores; peripetalous fasciole. (U.Cret.-Rec.)

Palaeostoma (Fig. 357): the only member of the family known to occur in the Tertiary; ovoid, inflated; apical system central, plates fused, with two gonopores; paired ambs broadly petaloid, the posterior pair a little the shorter; anterior amb non-petaloid, with pores in a single row, one pore of each pair being comma-shaped. Eoc.-Rec. N.Africa, Indo-Pac., Red Sea.

Family PERICOSMIDAE

Peripetalous fasciole (passing above periproct) and marginal fasciole (passing below periproct), the former sometimes branching anteriorly, and one or other may disappear anteriorly; apical system ethmolytic, with three or four gonopores; paired ambs petaloid, depressed; no radioles. (Eoc.-Rec.)

Pericosmus (*s.s.*) (Fig. 358): apex subcentral to slightly anterior; petals fairly broad, straight, of about the same length; anterior sinus fairly deep; marginal fasciole usually complete. Eoc.-Rec. Cosmop.

Family SCHIZASTERIDAE

Both peripetalous and latero-anal fascioles present (except in one Cretaceous and one Recent genus); apical system ethmophract to ethmolytic, with two to four gonopores; radioles usually fairly coarse. (U.Cret.-Rec.)

Figs 354–358. ECHINOIDEA: SPATANGOIDA
Fig. 354 after Duncan and Sladen; Figs 355–358 after Fischer, in Moore.
354a. *Ditremaster elongatus* (Duncan and Sladen), Pal. (Ranikot). Sind. × 3·75.
355a, b, c. *Hypsopatagus meneghini* (Desor), Olig. Italy. T. × 0·37.
356a, b, c. *Opissaster polygonalis* Pomel, Mio. N.Africa. T. × 0·75.
357a. *Palaeostoma mirabile* (Gray), Rec. Indo-Pac.–Red Sea. T. × 0·75.
358a, c. *Pericosmus latus* (Agassiz), Rec. T. × 0·75.

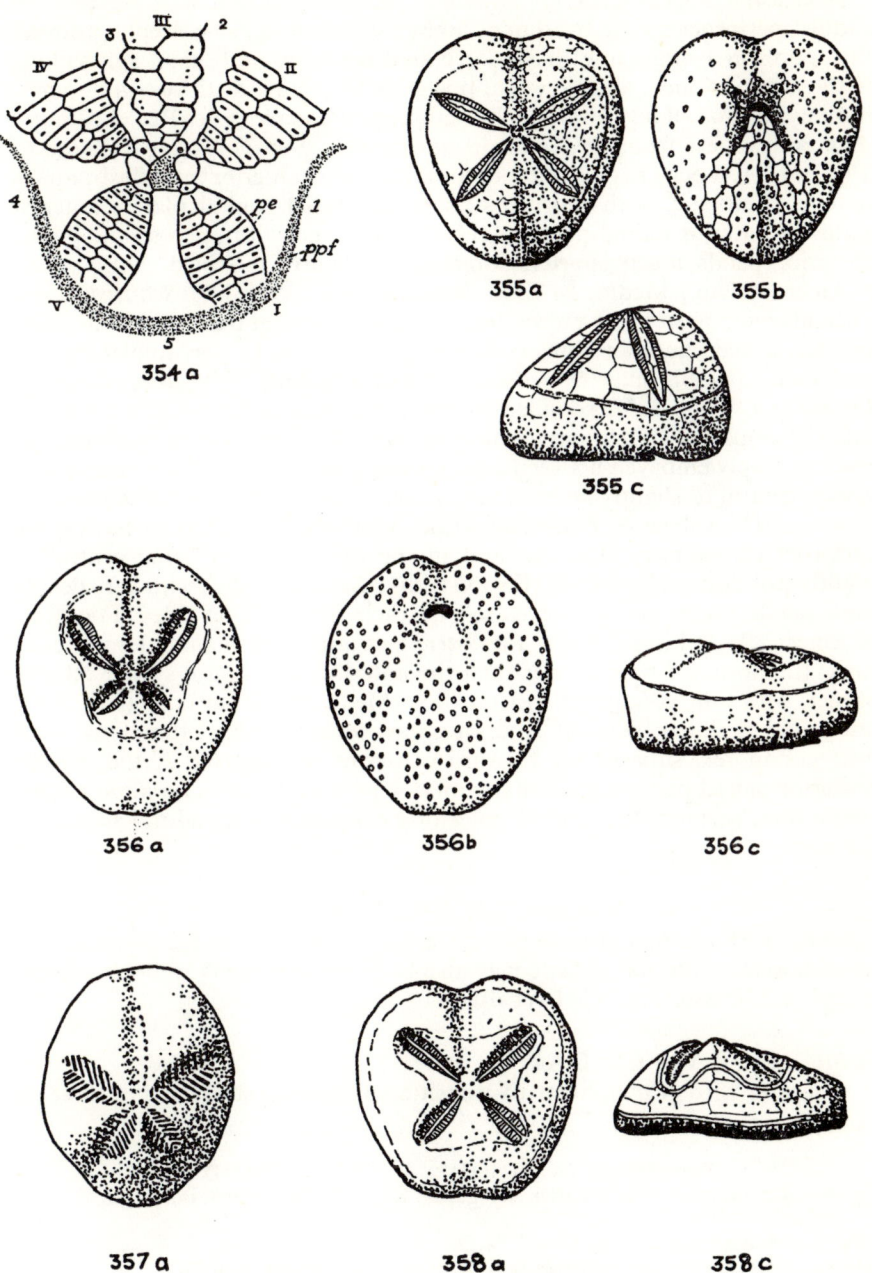

354 a

355 a   355 b

355 c

356 a   356 b   356 c

357 a   358 a   358 c

Schizaster (*s.s.*) (Fig. 359): test high, maximum height well posterior to middle, periproct a little overhung; ambs sunk, anterior one deeply grooved, with pores in a single row; posterior paired petals no more than half as long as the anterior ones; apical system posterior to middle, with two gonopores, ethmolytic. Eoc.-Rec. Cosmop. The subgenus *Paraster* (Eoc.-Rec. Cosmop.) (Fig. 360) differs from *Schizaster* in having four gonopores. *Agassizia* (*s.s.*) (Fig. 361): test egg-shaped, without anterior sinus, anterior amb flush; apical system ethmolytic, with four gonopores and fused genital plates; anterior plates of anterior paired petals reduced in size and with microscopic pores; posterior petals much shorter, sometimes similarly modified. U.Eoc.-Rec. N.Amer., Carib., Medit., M.East. *Brisaster* (Fig. 362): like *Schizaster*, but with three gonopores and a lower test which is somewhat truncated posteriorly and has a deep anterior sinus continuing to the sunken peristomial region; latero-anal fasciole may be reduced or lost in adults. Olig.-Rec. Cosmop. *Linthia* (*s.s.*) (Fig. 363): rounded-cordate, with depressed anterior amb and anterior sinus; apical system ethmolytic, with four gonopores; peripetalous fasciole deeply embayed between anterior and posterior pairs of petals; apical system central to slightly anterior; periproct vertically elongate. U.Cret.-Plio. Cosmop. The subgenus *Lutetiaster* (Eoc.-Mio. Medit.) differs in having the periproct transversely elongate, and in the apical system being central to slightly posterior. *Moira* (*s.s.*) (Fig. 364): like *Schizaster*, but with very deeply sunk petals which are almost closed by their overhanging sides. Eoc.-Rec. N.Amer., ?W.Pakistan, India. *Periaster:* test almost as high as long; petals sunk, the posterior pair shorter than the anterior pair; apical system ethmophract, with four gonopores. U.Cret.-Eoc. Medit. *Prenaster* (*s.s.*) (Fig. 365): test ovoid, inflated, without anterior sinus; apical system ethmolytic, with four gonopores, situated far forward at about one-quarter of the length; posterior paired petals longer than anterior pair which form an almost horizontal line; peripetalous fasciole anteriorly extending on to adoral side. Eoc. Cosmop.

Suborder MICRASTERINA

Petaloid, with subanal fasciole (except in some members of the Loveniidae); peripetalous or internal fasciole may also be present; primary radioles present except in Micrasteridae. (U.Cret.-Rec.)

Family MICRASTERIDAE

Cordate, with ethmophract to transitional apical system having three or

---

Figs 359–362. ECHINOIDEA: SPATANGOIDA

Fig. 359 after Fourtau; Fig. 361 after A. Agassiz; Fig. 362 after Fischer, in Moore; Fig. 360 original.

359a, c. *Schizaster africanus* de Loriol, M.Eoc. Egypt. × 0·75.

360b, d. *Schizaster* (*Paraster*) *parkinsoni* (Defrance), Mio. Malta. × 0·85.

361a, c. *Agassizia excentrica* A. Agassiz, Rec. Panama. × 2·25.

362a. *Brisaster fragilis* (Düben and Koren), Rec. T. × 0·5.

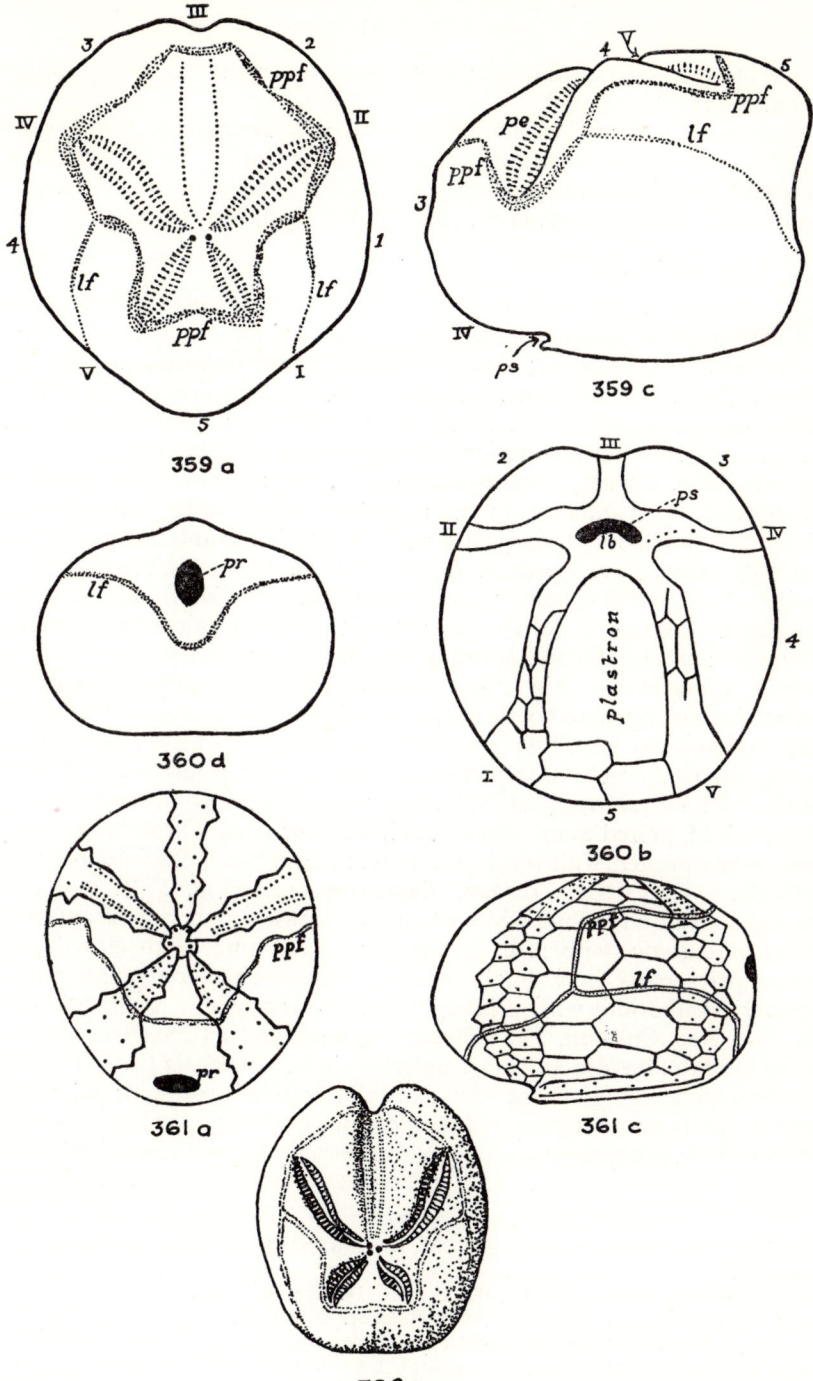

359 a

359 c

360 d

360 b

361 a

361 c

362 a

149

four gonopores; subanal fasciole; no primary tubercles or primary radioles. (U.Cret.-Eoc.)

The genus *Micraster* is restricted to the Upper Cretaceous. *Brissopneustes* (Fig. 366): subcordate, rostrate, without anterior sinus; paired petals broad, with round or elongate conjugate pores; anterior amb with small pores; three gonopores. U.Cret.(Maastr.)-Eoc. Eur., W.Pakistan, Madagascar.

Family BRISSIDAE

Test cordate, usually with both peripetalous and subanal fascioles, sometimes with anal branches as well; apical system ethmophract to ethmolytic, with two to four gonopores; usually with some large radioles. (U.Cret.-Rec.)

Brissus (Fig. 367): test ovate, with no anterior sinus; apical system distinctly anterior to middle, ethmolytic, with four gonopores; anterior amb barely sunk; paired petals sunk, anterior pair almost horizontal; peristome far forward. Eoc.-Rec. Cosmop. *Brissopatagus* (Fig. 368): test cordate, rather low, with anterior sinus; anterior amb non-petaloid, paired ambs petaloid; similar to *Eupatagus* (*q.v.*), but has large depressions anterior to anterior paired petals or to both pairs. Eoc. Cosmop. *Brissopsis* (Fig. 369): test cordiform, with anterior sulcus; apical system central, ethmolytic, with two to four gonopores; ambs slightly depressed, paired ones petaloid, anterior pair forming an angle of about 90°; proximal plates of petals may have rudimentary pores. Eoc.-Rec. Cosmop. *Cionobrissus* (Fig. 370): elongate-cordiform, inflated, with deep anterior sinus; periproct above a posterior bulge; apical system well anterior to middle, ethmolytic, with four gonopores; anterior amb flush adapically, but forming anterior notch, deeply sunk on adoral side; petals sunk. Eoc.-Rec. Iran, S.W.Pac. *Cyclaster* (Fig. 371): test cordate, with no anterior sinus; all ambs petaloid, paired ambs about equal, anterior petal a little longer; apical system ethmophract, with three gonopores; peripetalous fasciole incomplete anteriorly. U.Cret.-Rec. Cosmop. *Eupatagus* (*s.s.*) (Fig. 372): test ovoid, flattened above and below, without anterior sinus; apical system anterior to middle, ethmolytic, with four gonopores; paired ambs with closed petals, anterior pair nearly horizontal; anterior amb non-petaloid, with pores in single series; aborally with primary tubercles only within peripetalous fasciole. Eoc.-Rec. Cosmop. The subgenus *Gymnopatagus* (Eoc.-Rec. Cosmop.) is more cordate, having a distinct anterior sinus. *Gualtieria* (*s.s.*) (Fig. 373): test ovate, with slight anterior sinus; anterior amb non-petaloid; paired ambs petaloid, petals extending beyond peripetalous fasciole; apical system ethmolytic, with four gonopores; adorally with ridges or nodes on anterior amb and

---

Figs 363–367. ECHINOIDEA: SPATANGOIDA

Fig. 365 after Duncan and Sladen; Figs 364, 366 and 367 after Fischer, in Moore; Fig 363 original.

363a, c. *Linthia sudanensis* (Bather), Pal. Sokoto, N.Nigeria. × 1·6.
364a, c, d. *Moira atropos* (Lamarck), Rec. N.Amer. T. × 0·75.
365a, b, c, d. *Prenaster oviformis* Duncan and Sladen, Pal. (Ranikot). Sind. × 0·75.
366a, b, c. *Brissopneustes vilanovae* Cotteau, U.Cret. Eur. T. × 1.
367a, b, c, d. *Brissus unicolor* (Leske), Rec. Cuba. T. × 1·5.

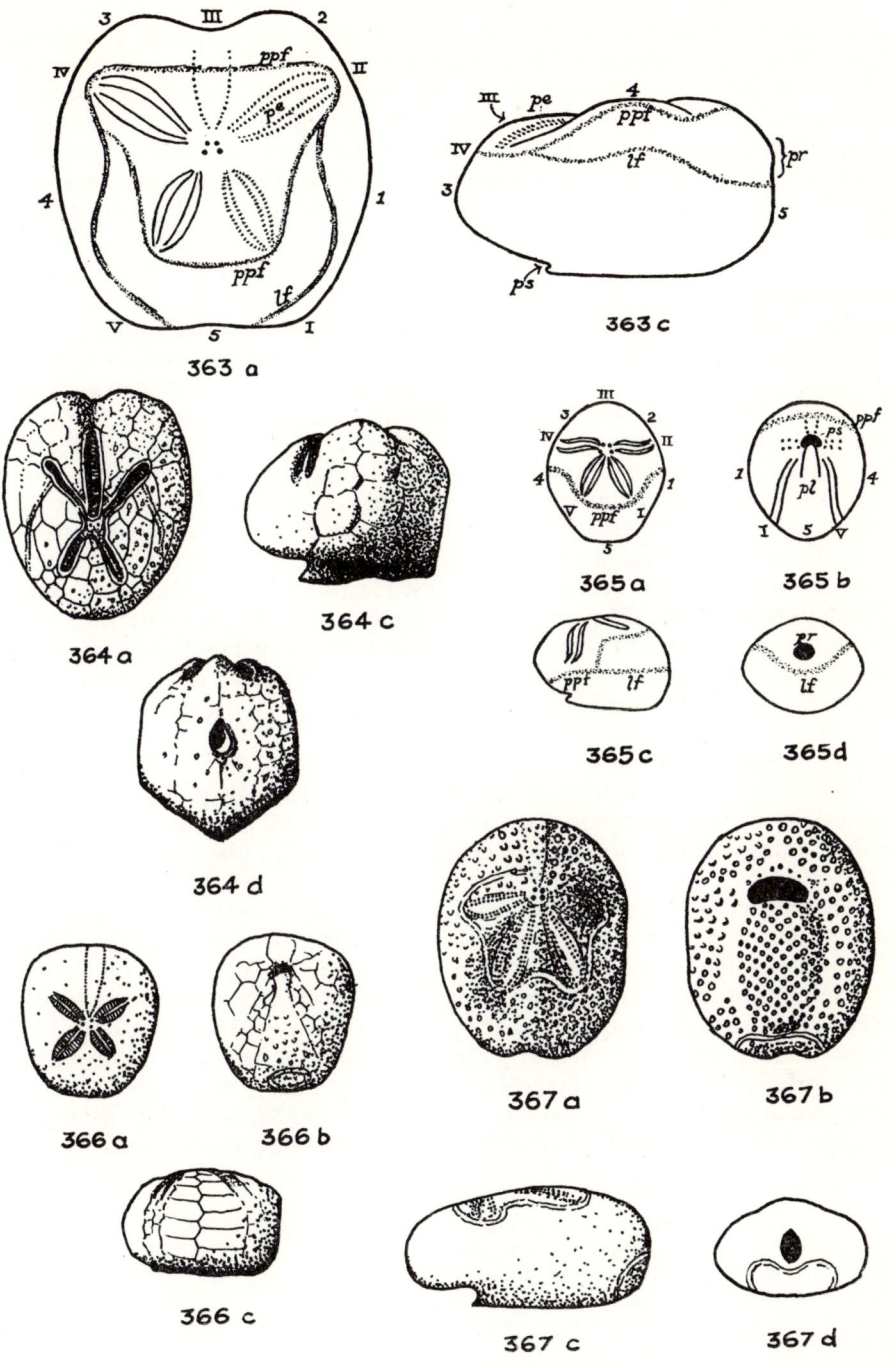

363 a

363 c

364 a

364 c

365 a

365 b

364 d

365 c

365 d

366 a

366 b

367 a

367 b

366 c

367 c

367 d

posterior interamb near peristome; anterior amb slightly sunk. Eoc.-Olig. Eur., N.Africa. *Hikelaster* [*Troschelia* Duncan and Sladen *non* Moerch] (Fig. 374): differs from *Eupatagus* in having a distinct anterior sinus, a narrow and sunken anterior amb, sunken petals with small pores proximally and large pores distally, and large tubercles outside as well as within the peripetalous fasciole. M.Mio. India. *Macropneustes* (*s.s.*) (Fig. 375): like *Eupatagus*, but petals sunk and large, anterior sinus distinct, and test broad; peripetalous fasciole without embayments between petals. Eoc.-Rec. Medit., Carib., W.Pakistan. The subgenus *Deakia* (Eoc. Eur.) is depressed and has a sharper ambitus, and the peripetalous fasciole is embayed between the petals. *Meoma* (*s.s.*) is Recent only. Its subgenus *Schizobrissus* [*Peripneustes*] (Fig. 376) is like *Macropneustes*, but has deeply sunk, narrow petals, a deep anterior sinus, a peripetalous fasciole which is deeply embayed between the petals, and a higher test. Eoc.-Mio. Carib. India. *Plagiobrissus* (*s.s.*) (Fig. 377): similar to *Eupatagus*, but has a distinct anterior sinus, long, narrow and flexed petals, anal branches on the subanal fasciole, a long plastron and a short labrum. Eoc.-Rec. Cosmop. *Spatangomorpha* (Fig. 378): like *Eupatagus*, but with anterior sinus and rather depressed anterior amb, more plates included in subanal fasciole, and posterior ambs meeting to separate labrum from sternum; primary tubercles within peripetalous fasciole rather large. Mio.-Plio. Asia. *Trachypatagus* (Fig. 379): similar to *Eupatagus* in having no anterior sinus, but is larger, higher, with less smooth course of the peripetalous fasciole which has no large tubercles within it; the peristome is situated very far forwards at almost one-fifth of the length. Eoc.-Mio. Medit.

Family SPATANGIDAE

Test oval to cordate, with subanal fasciole only; apical system ethmolytic, with three or four gonopores; anterior amb with small pores in single series; paired ambs petaloid, never much depressed; primary radioles differentiated. (Eoc.-Rec.)

Spatangus (*s.s.*) (Fig. 380): test cordate; four gonopores; primary tubercles in all interambs. Eoc.-Rec. Cosmop. The subgenus *Granopatagus* (Eoc.-Rec. Medit., Indo-Pac.) is less broad, and primary tubercles are present only along the margins of the anterior sinus and in the posterior interamb. The subgenus *Phymapatagus* (Eoc.-Mio. Eur.) has primary tubercles only in interambs 1, 2, 3 and 4, and pores in the anterior plates of the anterior petals are rudimentary. The subgenus *Platyspatus* (Eoc.-Mio. Medit.) has a very broad anterior sinus, and primary tubercles are present in all interambs. *Maretia* (Fig. 381): test cordate, with broad, shallow anterior sinus; amb III only slightly sunk; paired

---

Figs 368–371. ECHINOIDEA: SPATANGOIDA
Fig. 369 after A. Agassiz; Figs 368, 370 and 371 after Fischer, in Moore.
368a, b, c. *Brissopatagus caumonti* Cotteau, Eoc. France. T. × 0·75.
369a, b. *Brissopsis columbaris* A. Agassiz, Rec. Panama. × 0·75.
370b, c, d. *Cionobrissus revinctus* A. Agassiz, Rec. S.W.Pac. T. × 0·75.
371a, b, c, d. *Cyclaster declivus* Cotteau, Eoc. France. T. × 0·75.

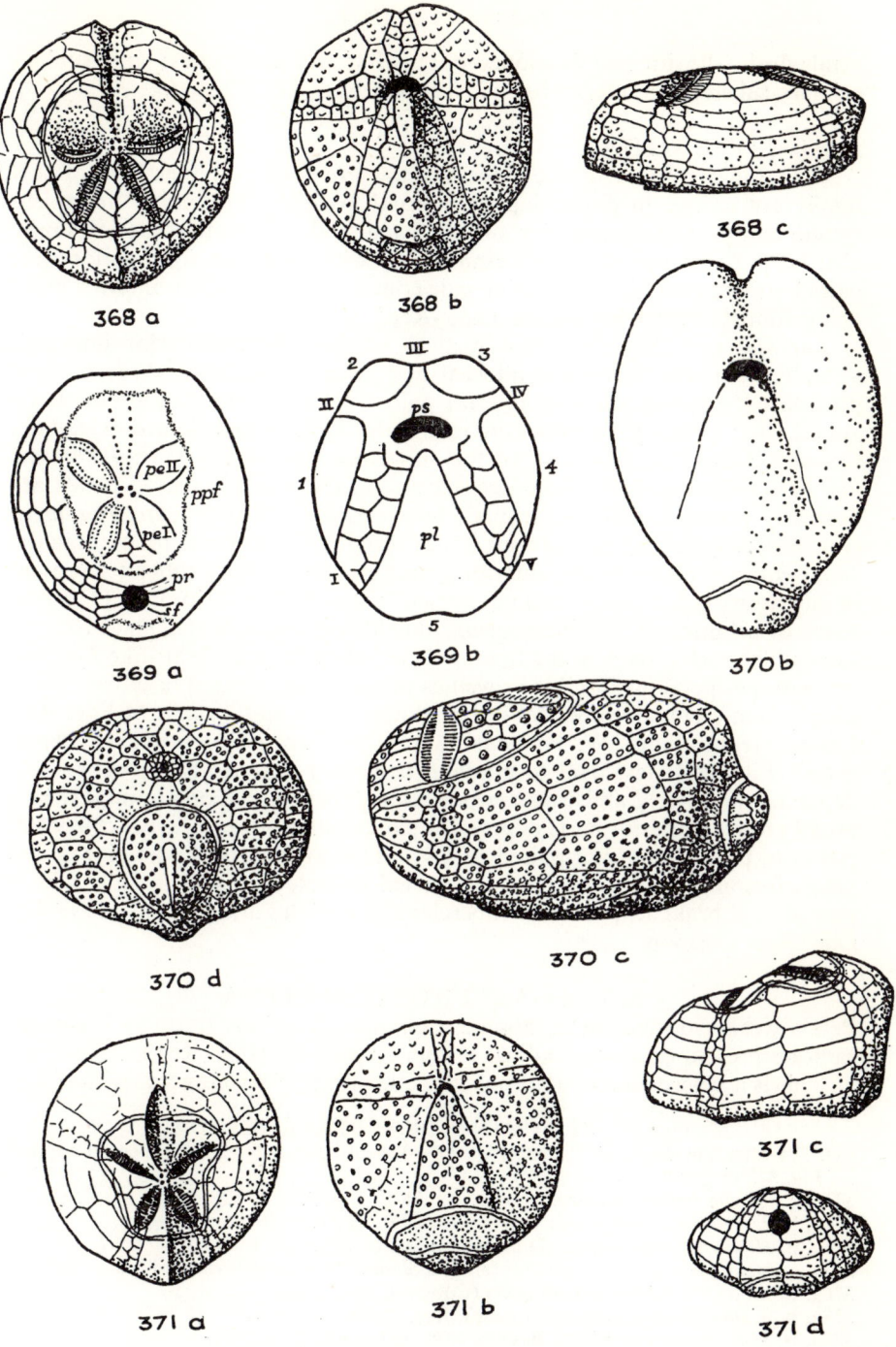

368 a

368 b

368 c

369 a

369 b

370 b

370 d

370 c

371 a

371 b

371 c

371 d

petals flush, all with well developed pore pairs; four gonopores; large sunken primary tubercles in interambs 1, 2, 3 and 4. Eoc.-Rec. Cosmop.

Family LOVENIIDAE

With the exception of the Recent genus *Homolampas*, members of this family are unique in possessing an internal fasciole surrounding the apical system and part of the anterior amb; subanal and peripetalous fascioles may also be present; apical system ethmolytic, with three or four gonopores; paired ambs petaloid, the anterior pair commonly fused, their anterior parts being almost or quite horizontal. (Eoc.-Rec.)

Lovenia (*s.s.*) (Fig. 382): test oval to cordate, low, posterior interamb aborally somewhat carinate; internal and subanal fascioles; three or four gonopores; anterior parts of anterior paired petals forming an almost horizontal line; primary tubercles in interambs 1, 2, 3 and 4, non-crenulate, recessed. Olig.-Rec. Cosmop. In the subgenus *Vasconaster* [*Sarsella* Pomel *non* Haeckel] (Fig. 383) (Eoc.-Rec. Cosmop.) the anterior paired petals are normally petaloid, and the primary tubercles are not sunk. *Atelospatangus* (Fig. 384): small, cordate, anterior sinus deep; four gonopores; no pores in the small anterior plates of the anterior paired petals; primary tubercles present in interambs 1, 2, 3 and 4; a small internal fasciole. Eoc.-Mio. Eur. [The presence of an internal fasciole suggests that this genus is better placed in the Loveniidae rather than in the Spatangidae.] *Breynia* (Fig. 385): like *Lovenia*, but with peripetalous fasciole as well as internal and subanal fascioles; large, usually non-crenulate tubercles in interambs 1, 2, 3 and 4 sunk. M.Eoc.-Rec. Libya, Medit., India, W. Pakistan, Burma, W.Pac. *Echinocardium* [*Amphidetus*] (Fig. 386): test oval to cordate, truncated posteriorly, apical surface depressed anteriorly and carinate posteriorly; four gonopores; amb III in a broad groove; anterior parts of anterior paired petals short, nearly horizontal; posterior parts of anterior paired petals and anterior parts of posterior paired petals forming continuous curves; internal fasciole; subanal fasciole with a pair of anal branches; no large tubercles and radioles and no sunken areoles. Olig.-Rec. Cosmop.

Suborder ASTEROSTOMATINA

Petals weakly developed or absent; fascioles of various types and primary radioles present or absent; apical system ethmolytic. (Eoc.-Rec.)

This is a heterogeneous and polyphyletic group of spatangoids with a ten-

Figs 372–378. ECHINOIDEA: SPATANGOIDA

Fig. 375 after Duncan and Sladen; Figs 373, 374 and 376–378 after Fischer, in Moore; Fig. 372 original.

372a. *Eupatagus hollisi* Stockley, M.Mio. Pemba I., E.Africa. × 1.
373a, b, d. *Gualtieria orbignyana* Agassiz, Eoc. France. T. × 0·75.
374a, b. *Hikelaster tuberculatus* (Duncan and Sladen), M.Mio. India. T. × 0·66.
375a, c. *Macropneustes speciosus* Duncan and Sladen, M.Eoc. (Khirthar). Sind. × 0·5.
376a. *Meoma* (*Schizobrissus*) *antillarum* (Cotteau), Eoc. Cuba. × 0·35.
377a, b, d. *Plagiobrissus grandis* (Gmelin), Rec. Carib. T. × 0·1.
378a, b. *Spatangomorpha eximia* Boehm, Mio. Indonesia. T. × 0·66.

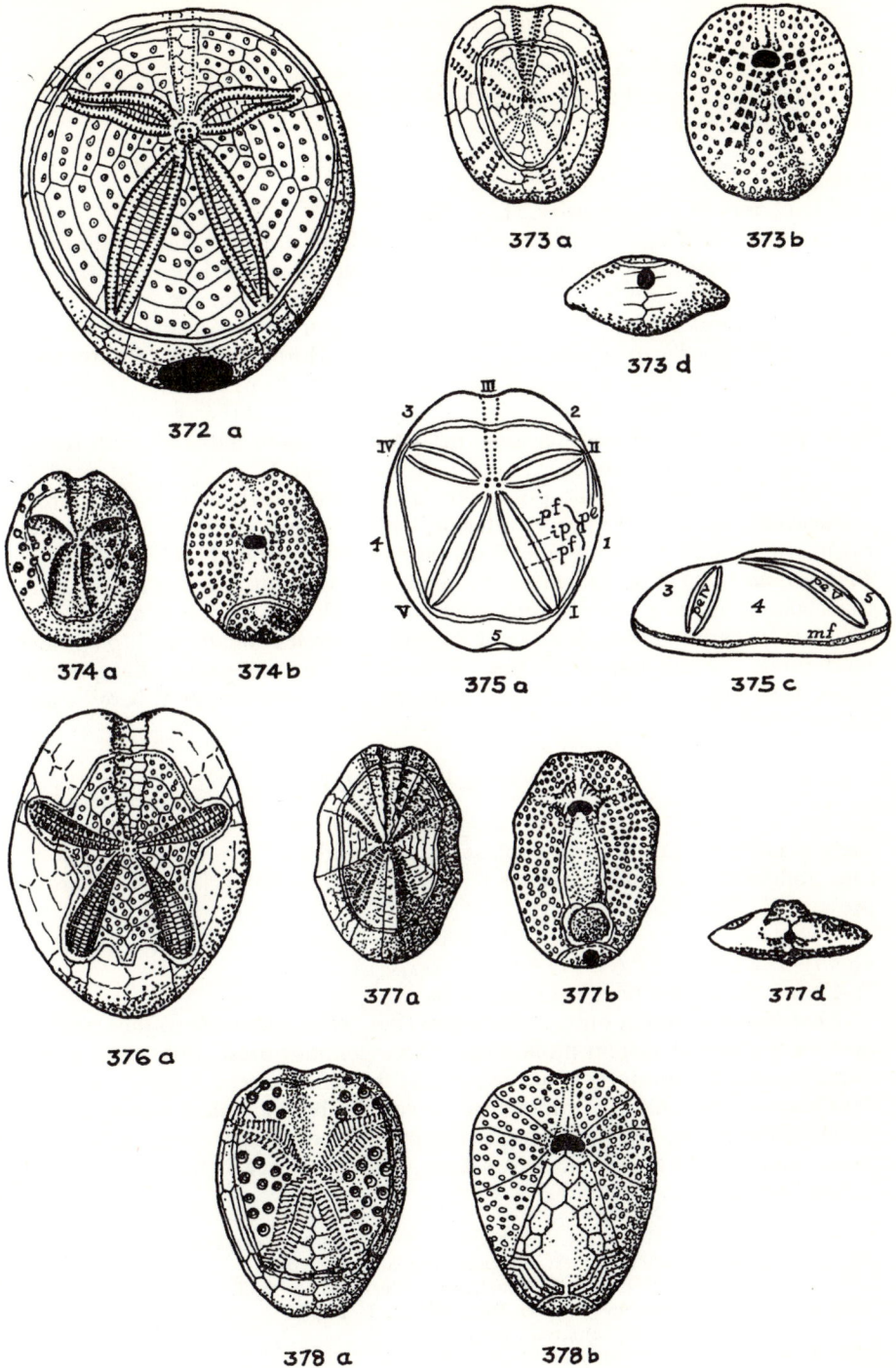

372 a

373 a

373 b

373 d

374 a

374 b

375 a

375 c

376 a

377 a

377 b

377 d

378 a

378 b

155

dency to lose fascioles and petaloid structure of ambs. The peristome is labiate, and phyllodes are well developed. The test is generally fragile, and most forms possess radioles.

Family ASTEROSTOMATIDAE
Characters of the suborder. (Eoc.-Rec.)

Asterostoma (Fig. 387): test ovoid, flat below; apical system anterior to middle, at maximum height and at about one-third the length, with four gonopores; paired ambs subpetaloid, open at ambitus; median parts of all ambs on adoral surface form grooves radiating from peristome to margin. Eoc. Carib. *Antillaster:* like *Asterostoma*, but adoral surface without amb grooves, or with them developed only near the peristome in ambs II and IV. Eoc.-Mio. Carib. *Brissolampas* (Fig. 388): test ovoid, rather pointed posteriorly, flat below, with inframarginal periproct; all ambs petaloid, with round pores; no fascioles. Mio. Eur., Carib. *Brissomorpha* (Fig. 389): test cordate but without anterior sinus, possessing a posterior beak with the periproct underneath; apical system well anterior, with four gonopores; petals narrow, open, with round pores; peripetalous fasciole. Mio. Eur., N.Africa, W. Pac. *Platybrissus* (s.s.) (Fig. 390): test elliptical, low, without anterior sinus; ambs flush, paired ambs petaloid; four gonopores; subanal fasciole tending to be obsolete in adults; peristome not broad, tubercles small, and phyllodes moderately developed. Mio.-Rec. Indo-Pac. The subgenus *Eurypatagus* (Tert.-Rec. Java, Indo-Pac.) has a broader peristome, larger tubercles and deeper phyllodes.

Order NEOLAMPADOIDA

Rather small; ambs non-petaloid, with pores simple or absent adapically; there may be a weak floscelle; apical system tetrabasal or monobasal; two to four gonopores; primary radioles few and short; pedicellariae and spheridia present. (U.Eoc.-Rec.)

Family NEOLAMPADIDAE
Characters of the order. (U.Eoc.-Rec.)

Neolampas is Recent only. *Notolampas* (Fig. 391): test ovate, rather pointed posteriorly; apical system monobasal; three genital pores; ambs with single pores adapically; periproct inframarginal; floscelle present. L.Mio. Austral. *Pisolampas* (Fig. 392): test subhemispherical, flat below; apical system monobasal; three gonopores; amb pores rudimentary or absent adapically; periproct supramarginal, on posterior end; bourrelets weak: floscelle present. U.Eoc. Austral.

Figs 379–382. ECHINOIDEA: SPATANGOIDA
Figs 379, 381 and 382 after Fischer, in Moore; Fig. 380 original.
379a, b, c. *Trachypatagus oranensis* Pomel, Mio. Algeria. T. × 0·12.
380a, b. *Spatangus purpureus* (Müller), Rec. Britain. T. × 0·75.
381a, b. *Maretia planulata* (Lamarck), Rec. T. × 0·75.
382a, b. *Lovenia elongata* (Gray), Rec. T. × 0·75.

156

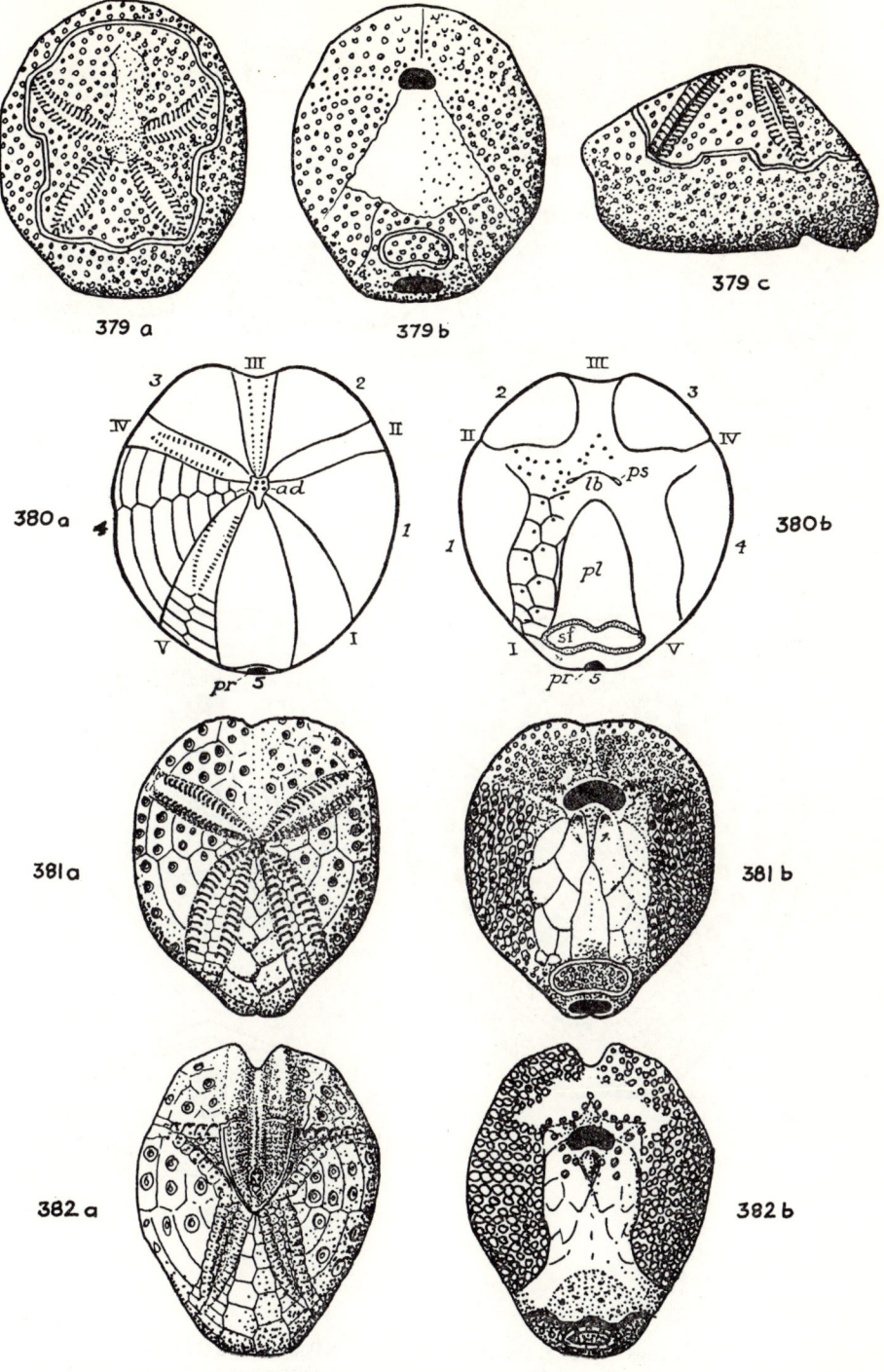

379 a

379 b

379 c

380 a

380 b

381 a

381 b

382 a

382 b

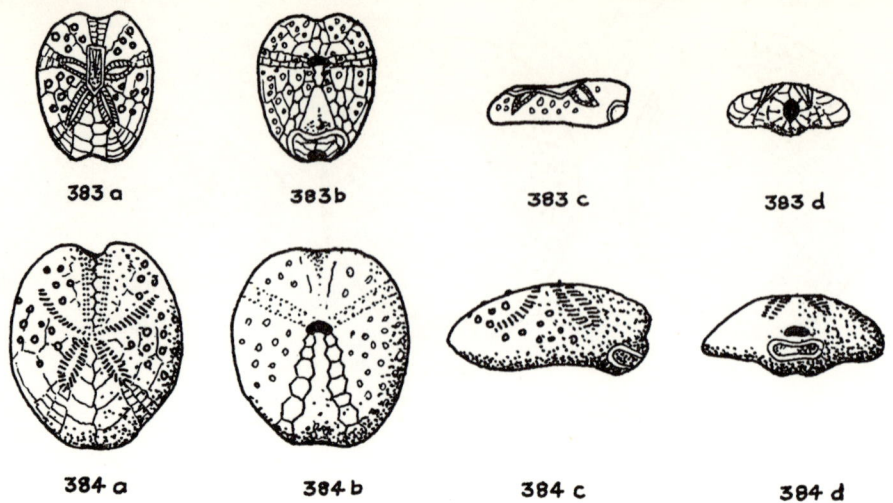

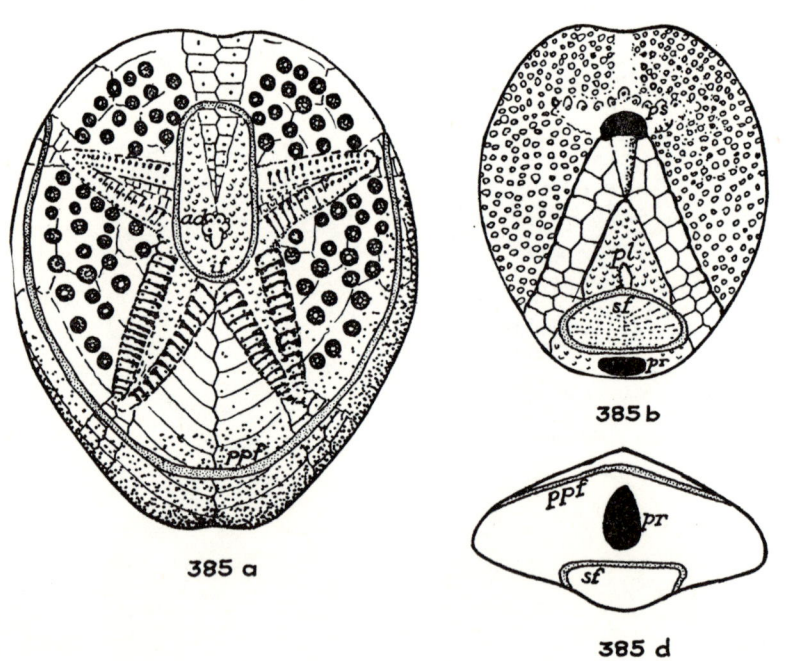

Figs 383–385. ECHINOIDEA: SPATANGOIDA
Fig. 385 after d'Archiac and Haime (corrected after Duncan and Sladen); Figs 383, 384
after Fischer, in Moore.
383a, b, c, d. *Lovenia (Vasconaster) sulcatus* (Haime), Olig. France. T. × 0·25.
384a, b, c, d. *Atelospatangus transilvanicus* Koch, Tert. Romania. T. × 1.
385a, b, d. *Breynia carinata* d'Archiac and Haime, M.Mio. (Gaj.). Sind. × 0·75.

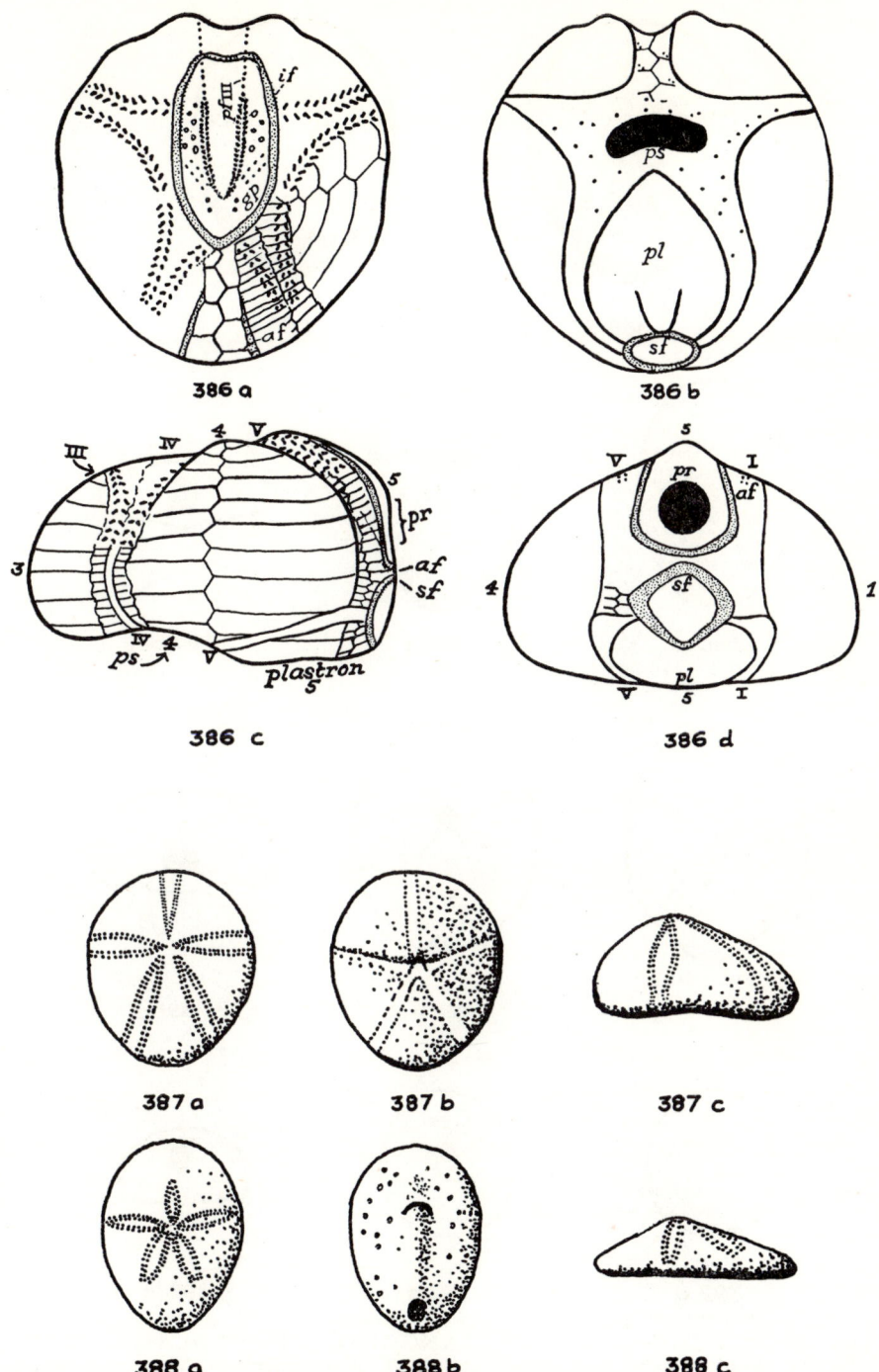

Figs 386–388. ECHINOIDEA: SPATANGOIDA

Figs 387, 388 after Fischer, in Moore; Fig. 386 original.

386a, b, c, d. *Echinocardium cordatum* (Pennant), Rec. Britain. T. × 0·9.

387a, b, c. *Asterostoma excentricum* Agassiz, Eoc. Carib. T. × 0·25.

388a, b, c. *Brissolampas conicus* (Dames), Mio. Italy. T. × 0·2.

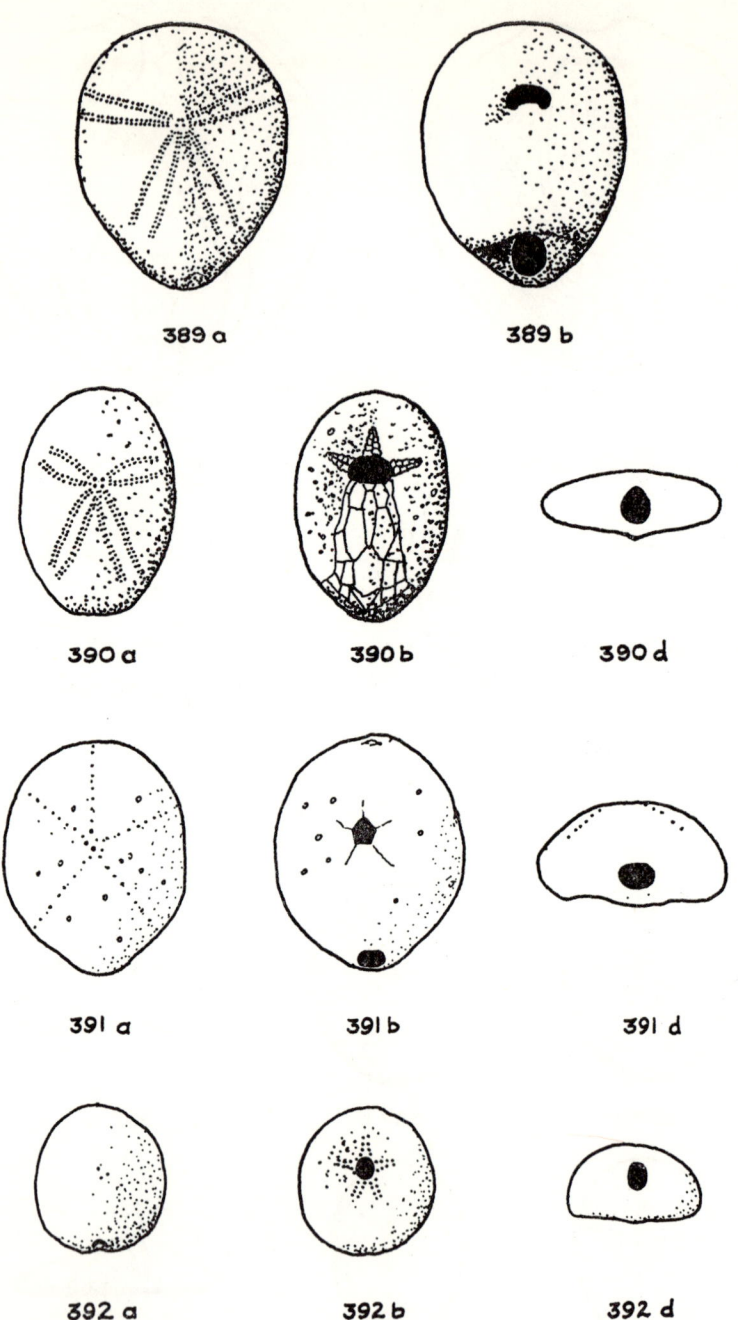

389 a 389 b

390 a 390 b 390 d

391 a 391 b 391 d

392 a 392 b 392 d

Figs 389–392. ECHINOIDEA: SPATANGOIDA AND NEOLAMPADOIDA
Figs 389, 390 after Fischer, in Moore; Figs 391, 392 after Durham and Wagner, in Moore.
389a, b. *Brissomorpha fuchsi* Laube, Mio. Austria. T. × 0·37.
390a, b, d. *Platybrissus roemeri* Grube, Rec. Indo-Pac. T. × 3.
391a, b, d. *Notolampas flosculus* Philip, L.Mio. Austral. T. × 2.
392a, b, d. *Pisolampas concinna* Philip, U.Eoc. Austral. T. × 2.

GLOSSARY OF PRINCIPAL TECHNICAL TERMS
APPLIED TO ECHINOIDEA

Abactinal. Aboral, adapical.

Aboral. Side opposite mouth, upper surface of test.

Actinal. Adoral.

Actinal furrows. Furrows radiating from the mouth in a perradial direction.

Adapical. Towards the apex, e.g. the horizontal suture of any plate on the side towards the apex. Especially, the upper surface of the test.

Adoral. Towards the mouth, e.g. the horizontal suture of any plate on the side towards the mouth. Especially, the lower surface of the test.

Adradial. The direction of the boundary between an amb and an interamb.

Amb. Abbreviation for ambulacrum or ambulacral area.

Ambitus. The widest horizontal outline of a test, forming the margin between the upper (adapical) and lower (adoral) surfaces.

Ambulacrum. A complete double column of perforated plates, touching an ocular plate at the summit and the peristome at the base.

Amphisternous. In spatangoids, with the labrum followed by two large, more or less equal sternal plates opposite one another.

Anal fasciole. Fasciole adoral and lateral to periproct.

Anterior. That direction indicated by the amb (III) which starts on the immediate left of the madreporite.

Apetaloid. In which the arrangement of the pores in the ambs is similar throughout the column.

Apical disk (*system*). A group typically of 10 plates (subject to certain variations), 5 ocular, 5 genital, from which the ambs and interambs radiate.

Areole (=scrobicule). The depression around the base of a tubercle.

Aristotle's lantern. See 'lantern'.

Auricles. Internal upgrowths from the amb plates around the peristome, each of which may unite with its fellow to form an arch; parts of the girdle.

Basicoronal. Referring to corona at edge of peristome.

Bidentate. Type of pedicellaria with head consisting of two long pointed valves.

Bigeminate. An arrangement of the pore pairs of a pore field in two vertical columns.

Bivium. The two postero-lateral ambs, when more or less separated at their origin from the other three. (See *Trivium.*)

Boss. Part of tubercle below mamelon, truncated cone-shaped.

Bourrelet. A blunt and rounded projection from each interamb in the immediate neighbourhood of the peristome.

Branchial slits. Gill slits.

Buccal plates. The plates on the peristomial membrane immediately next to the mouth (not usually preserved in fossils).

Camarodont. Type of lantern with keeled teeth and closed foramen magnum.

Compact (apical disk). Having all the plates concentrated in an approximately circular area.

Compound plate. Amb plate unit composed of two or more simple plates fused together, with only one primary tubercle.

Conjugate pores. Pores of a pair united by a groove in external surface of test.

Corona. The main part of the test, excluding the apical disk, peristome, periproct, lantern and appendages.

Cortex. Differentiated dense outer layer of a radiole.

Crenulate tubercles. Tubercles with radiating grooves on the boss.

Dactylous. Type of pedicellaria with spoon-shaped jaws mounted on individual stalks.

Demiplate. An amb plate which occupies less than the width of a single column, reaching the outer (adradial) but not the inner (perradial) boundary of its column.

Dicyclic. Type of apical system with ocular and genital plates in two concentric circles, genitals alone in contact with margin of periproct.

Disjunct (apical system). Separated into two portions, in correspondence with the separation of bivium and trivium (*q.v.*) (no genital 5).

Endocyclic. Having the periproct within the apical system.

Epistroma. Surface ornament of plates, other than tubercles and granules.

Ethmolytic (apical system). In which the madreporite extends back so as to separate the postero-lateral genitals (1 and 4), e.g. *Ditremaster* (Fig. 354).

Ethmophract (apical system). In which the madreporite does not extend back between the postero-lateral genitals (1 and 4).

Exocyclic. Having the periproct outside the apical system.

Exsert (ocular plates). Not separating their adjacent genital plates and therefore not in contact with the periproct.

Fasciole. A narrow band of closely packed, minute tubercles, the spines (clavulae) of which, in the living animal, are ciliated and set up currents of water in the direction in which the fasciole runs (most Spatangoida and some Cassiduloida).

Floscelle. The star-shaped area composed of five bourrelets and five phyllodes surrounding the peristome (some Cassiduloida)

Food groove. Narrow grooves leading to peristome in adoral amb areas supplied with specialized tube feet for food gathering and transport; may extend into interamb areas and on to aboral surface.

Genital plates. Five (or sometimes fewer) plates, each at the top of an interamb, forming part of the apical system, each perforated by a fairly large pore for discharge of genital products.

Gill slits. Ten notches in the outline of the peristome interambs for passage of stem of external branchia.

Globiferous. Type of pedicellaria with three valves containing poison glands.

Gonopore (genital pore). Pore in genital plate for discharge of reproductive products.

Imperforate. (1) plates not traversed by pores for the podia; (2) tubercles not having a pore in the centre of the mamelon.

Included plate. Amb plate which touches neither perradial nor adradial suture.

Inframarginal periproct. One below the ambitus, i.e. on the adoral surface.

Insert. With ocular plates separating the adjacent genitals so as to be in contact with the periproct.

Interamb. Abbreviation for interambulacrum.

Interambulacrum. A complete double column of imperforate plates, touching a genital plate at the summit and the peristome at the base, alternating with the ambs. (In some pre-Tertiary echinoids there are more or fewer than two columns.)

Internal fasciole. One surrounding apical system and crossing all petals.

Interporiferous zone. That portion of the amb lying between the two pore fields.

Interradial. The five secondary radial directions (or structures lying along them) alternating with the five primary (perradial) directions.

Irregular. With periproct situated outside oculogenital ring.

Isolated plate. See 'included plate'.

Labrum. A blunt, lip-like projection of the posterior interamb into the peristome (Cassiduloida and Spatangoida).

Lantern. A complicated moving framework of forty or fewer skeletal elements carrying the teeth and serving for mastication (same as 'Aristotle's lantern').

Lateral fasciole. One branching off from the peripetalous on each side and passing under the periproct behind.

Lateroanal fasciole. One formed by the union of lateral and anal fascioles.

Lovénian system. Numbering system in which amb and interamb areas of test are designated by Roman (I–V) and Arabic (1–5) numerals, respectively.

Madreporite. The right anterior genital plate (2), distinguished by its sieve-like character, being perforated by many minute pores providing access to water-vascular system (in addition to the single large genital pore).

Mamelon. The rounded top part of a tubercle on which a radiole articulates.

Marginal fasciole. One running round the test near the ambitus.

Marsupium. A deeply sunk petal in the female, serving as a cavity in which the eggs develop into larvae (some spatangoids).

Meridosternous. Type of plastron with labrum followed by single large plate. (Spatangoids.)

Miliary tubercles. The smallest, having no differentiation into areolae, mamelons, etc.

Monobasal. Type of apical system with genital plates apparently fused, no suture lines being seen.

Monocyclic. Type of apical system with genital and ocular plates arranged in a single ring around periproct.

Occluded plate. One reaching the perradial suture (central line of amb area) but not the amb (adradial) margin.

Ocular plates. Five plates forming part of the apical system, each at the summit of an amb.

Oculogenital ring. System of ocular and genital plates at summits of ambs and inter-ambs, surrounding periproct in regular echinoids.

Ophicephalous. Type of pedicellaria with jaws which lock together.

Oral side. That on which the peristome is located.

Pedicellaria. A minute stalked, specialized, grasping or defensive radiole articulated on a granule.

Perforate. Traversed by a pore or pores, applicable to (1) amb plates, (2) certain primary tubercles.

Perignathic girdle. A series of vertical arches and plates, surrounding the peristome on the inner face of the corona, for attachment of muscles supporting and controlling the lantern.

Peripetalous fasciole. One passing around petals of ambs I, II, IV and V and around or across amb petal III.

Peripodium. A raised elliptical rim around a pore pair.

Periproct. The space surrounding the anus, occupied in life by a membrane bearing plates not firmly articulated together and therefore usually lost in fossils.

Peristome. The space surrounding the mouth, having the same characters as the periproct (*q.v.*).

Perradial. The five primary radial directions (or structures lying along them).

Petal. Differentiated adapical segment of amb with tube feet more or less specialized for respiration; pores of pair in petal typically unequal or enlarged.

Petaloid (ambs). Having the pore pairs first diverging more or less as they recede from the apex and then converging again.

Phyllode. Area of enlarged pores in ambs near peristome.

163

Plastron. A protuberant region on the adoral surface of Spatangoids, composed of large plates (interamb 5) extending from labrum to periproct.

Plate. A unit of the skeleton, consisting of a single crystal of calcite.

Podium. See 'tube foot'.

Pore field. That part of an amb column occupied by the pore pairs.

Pore pair. The two associated pores through which the two roots of a typical podium emerge.

Poriferous zone. See 'pore field'.

Posterior. The direction of the odd (unpaired) interamb (5), determined in Regularia as the second interamb from the madreporite counting clockwise (as seen from above).

Primary. (1) (plates) occupying the whole breadth of a column (half the breadth of an amb or interamb); (2) (tubercles) those of the largest size and most elaborate structure; (3) (radioles) those of the largest size, articulating with primary tubercles.

Radiole. Movable elongated calcareous shaft mounted on tubercle and articulating with it.

Regular. Type of test having periproct within oculogenital ring.

Rostrum. Raised or attenuated area of interamb 5.

Scrobicule. Depressed ring around base of tubercle for attachment of muscles of radiole.

Secondary tubercle .Those next in size to the primaries, and with which secondary radioles articulate.

Spheridia. Minute spherically modified radioles on short stalks usually situated adorally in pits near perradial suture.

Spine. Commonly used for radiole, but preferably applied to immovable pointed outgrowths, e.g. on the radioles of *Prionocidaris.*

Stereom. The original solid calcite of any plate or other skeletal unit, as distinguished from the organic matter occupying its interstices or its calcite replacement in fossils.

Sternum. See 'plastron'.

Stirodont. Type of lantern with keeled teeth and open foramen magnum.

Subanal fasciole. A closed fasciole, often reniform or rhomboidal, situated below the periproct.

Subpetaloid. With pore pairs diverging but not closing again distally as in a true petal.

Suranal plate. An exceptionally large periproct plate which sometimes becomes incorporated in the apical system (e.g. *Salenia*).

Suture. The line of junction between any two plates.

Test. The whole skeleton of an Echinoid, except the radioles and lantern (corona + apical system + periproct + peristome).

Tetrabasal. Type of apical system with four separate genital plates (genital 5 absent).

Tridentate. Type of pedicellaria with three long, pointed, jaw-like valves.

Trigeminate (trigeminal, triserial). Pore pairs arranged in three vertical series in each pore field.

Triphyllous. Type of minute pedicellaria with three leaf-like jaws not hinged to one another.

Triplet arcs (of pore pairs). Arranged in a series of vertical arcs each composed of three pore pairs.

Trivium. The three anterior (i.e. one unpaired and two antero-lateral) ambs (II, III, IV), when more or less separated from the two postero-lateral (bivium; I, V).

Tube foot. A movable, soft, tubular outgrowth serving for adhesion, locomotion, respiration, or the grasping of food; connected with the interior of the test, typically through two roots corresponding with the two pores of a pair.

Tubercle. A rounded protuberance from the surface of a plate for the articulation of a radiole.

Uniserial. Referring to amb with pore pairs in single longitudinal row.

SELECT BIBLIOGRAPHY OF TERTIARY ECHINOIDEA

1. AIRAGHI, C. 1939. 'Echinidi cretacici e terziari della Sirtica', *Ann. Mus. libico Stor. nat.*, **1**, 253–286, pl. 10–12.
2. D'ARCHIAC, E. J. A. D. and HAIME, J. 1853. *Description des animaux fossiles du groupe nummulitique de l'Inde*, 1–373, pl. 1–36 (Gide et J. Baudry: Paris).
3. ARNOLD, B. W. and CLARK, H. L. 1927. 'Jamaican fossil Echini', *Mem. Harvard Univ. Mus. comp. Zool.*, **50**, no. 1, 84 pp., 22 pls.
4. CLARK, W. B. and TWITCHELL, M. W. 1915. 'The Mesozoic and Cenozoic Echinodermata of the United States', *Monogr. U.S. geol. Surv.*, **54**, 341 pp., 108 pls.
5. COOKE, C. W. 1959. 'Cenozoic echinoids of the eastern United States', *Prof. Pap. U.S. geol. Surv.*, **321**, 1–106, pl. 1–43.
6. COTTEAU, G. H. 1878. 'Description des Échinides du Calcaire Grossiere de Mons', *Mém. cour. Acad. r. Sci. Belg.*, **42**, 1–12, pl. 1.
7. COTTEAU, G. H. 1878. 'Description des Échinides tertiaires de Belgique', *Mém cour. Acad. r. Sci. Belg.* **43**, 1–90, pl. 1–6.
8. COTTEAU, G. H. 1885–89. *Paléontologie française: Description des Animaux Invertébrés: Terrains tertiaries: Éocène*, **1**, 672 pp., pl. 1–200 (G. Masson & Cie.: Paris).
9. COTTEAU, G. H. 1889–94. *Paléontologie français: Description des Animaux Invertébrés: Terrains tertiaires: Éocène*, **1**, 789 pp., pl. 210–384 (G. Masson & Cie.: Paris).
10. COTTEAU, G. H. and GAUTHIER, V. (in de Morgan, J.). 1895. *Mission scientifique en Perse: Vol. 3. Études géologiques: Pt. 2. Paléontologie: Pt. 1. Échinides fossiles*, 107 pp., 16 pls. (E. Leroux: Paris).
11. COTTEAU, G. H., PERON, P. A. and GAUTHIER, V. 1885. *Échinides fossiles de l'Algérie*, **3**, (9), (Étage Éocène), 89 pp., 8 pls. (G. Masson & Cie.: Paris).
12. COTTEAU, G. H., PERON, P. A. and GAUTHIER, V. 1891. *Échinides fossiles de l'Algérie*, **3**, (10), (Étages Miocène et Pliocène), 273 pp., 8 pls. (G. Masson & Cie.: Paris).
13. COTTREAU, J. 1913. 'Les Échinides néogènes du bassin mediterranéen', *Annls Inst. océanogr., Monaco*, **6**, (3), 1–192, pl. 1–15.
14. COTTREAU. J. 1923. 'Rotuloidea du Pliocène marocain', *Annls Paléont.*, **12**, 135–147, pl. 17–19.
15. DAMES, W. 1877. 'Die Echiniden der Vicentinischen und Veronesischen tertiaerablagerungen', *Palaeontographica*, **25**, 1–100, pl. 1–11.
16. DUNCAN, P. M. 1877. 'Echinodermata of the Australian Cainozoic (Tertiary) Deposits', *Q.Jl geol. Soc. Lond.*, **33**, 42–73, pl. 3–4. [See also a revision of the same, *Q. Jl geol. Soc. Lond.*, **43**, 411–430].
17. DUNCAN, P. M. and SLADEN, W. P. 1882–86. 'Fossil Echinoidea of Western Sind and the coast of Bilûchistan and of the Persian Gulf, from Tertiary formations', *Mem. geol. Surv. India, Palaeont. indica*, (14), **1**, (3), 392 pp., 58 pls. [To be checked stratigraphically by the Vredenburg reference.]
18. DUNCAN, P. M. and SLADEN, W. P. 1883. 'The fossil Echinoidea of Kachh and Kattywar' *Mem. geol. Surv. India, Palaeont. indica*, (14), **1**, (4), 104 pp., 13 pls.
19. DURHAM, J. W. 1955, 'Classification of clypeasteroid echinoids', *Univ. Calif. Publs geol. Sci.*, **31**, 73–198, pl. 3–4.

20. DURHAM, J. W. and MELVILLE, R. V. 1957. 'A classification of echinoids', *J. Paleont.*, **31**, no. 1, 242–272.
21. DURHAM, J. W. *et. al.* (in Moore, R. C.). 1966. *Treatise on Invertebrate Palaeontology. Part U, Echinodermata 3: Echinoidea*, U211–U640 (Kansas Univ. Press).
22. EBERT, T. 1889. 'Die Echiniden des nord- und mitteldeutschen Oligocaens', *Abh. geol. SpecKarte preuss. thur. St.*, **9**, not. 1, 1–111, 10 pls.
23. FELL, H. B. 1954. 'Tertiary and Recent Echinoidea of New Zealand: Cidaridae', *Palaeont. Bull., Wellington*, **23**, 62 pp., 15 pls.
24. FELL, H. B. 1964. 'Oligocene echinoids from Trelissic Basin, New Zealand', *Trans. R. Soc. N.Z.*, 201–205, pl. 1–4.
25. FORBES, E. 1852. 'Monograph of the Echinodermata of the British Tertiaries', *Palaeontogr. Soc. Monogr.*, 1–36, pl. 1–4.
26. FOURTAU, R. 1913. *Catalogue des invertébrés fossiles de l'Égypte . . . Terrains tertiaires: pt. 1, Échinodermes éocènes.*, 1–93, pl. 1–6 (Egypt Geol. Surv.: Cairo).
27. FOURTAU, R. 1920. 'Catalogue des invertébrés fossiles de l'Égypte . . . Terrains tertiaires: pt. 2. Échinodermes néogènes', *Egypt. geol. Surv., Paleont.* (4), 100 pp., 12 pls. (Egypt Geol. Surv.: Cairo).
28. GAGEL, C. 1903. 'Ueber einige neuen Spatangiden aus dem Norddeutschen Miocän', *J. preuss. geol. Landesanst. BergAkad.*, **23**, 525–543, pl. 24–25.
29. GERTH, H. (in Martin, K.). 1922. 'Die Fossilien van Java', *Samml. geol. Reichsmus. Leiden*, **1**, Abt. 2, Heft 4, 471–538, pl. 60–63.
30. GRANT, U. S. and HERTLEIN, L. G. 1938. 'The West American Cenozoic Echinoidea', *Univ. Calif., Los Angeles publs, Math. Phys. Sci.*, **2**, 1–226, pl. 1–30.
31. GREGORY, J. W. 1891. 'A revision of the British fossil Cainozoic Echinoidea', *Proc. Geol. Ass.*, **12**, 16–60, pl. 1–2.
32. HAWKINS, H. L. 1934. 'The Lantern and Girdle of some Recent and Fossil Echinoidea', *Phil. Trans. R. Soc.*, (B), **223**, 617–649, pl. 68–70.
33. JACKSON, R. T. 1918. 'Fossil Echinoidea of the Panama Canal zone and Costa Rica', *Bull. U.S. natn Mus.*, **103**, 103–116, pl. 46–52.
34. JACKSON, R. T. 1922. 'Fossil Echini of W. Indies', *Publs Carnegie Instn*, **306**, 1–103, pl. 1–18.
35. KEW, W. 1920. 'Cretaceous and Cenozoic Echinoidea of the Pacific Coast of N. America, *Univ. Calif. Publs Bull. Dep. Geol.*, **12**, no. 2, 23–236, pl. 3–42.
36. KIER, P. M. 1957. 'Tertiary Echinoidea from British Somaliland', *J. Paleont.*, **31**, no. 5, 839–902, pl. 103–108.
37. KIER, P. M. 1962. 'Revision of cassiduloid echinoids', *Smithson. misc. Collns*, **144**, no. 3, 262 pp., 44 pls.
38. LAMBERT, J. 1907–8. 'Description des Échinides fossiles des terrains mioceniques de la Sardaigne', *Abh. schweiz. palaont. Ges.*, **34**, 1–142, pl. 11.
39. LAMBERT, J. and THIERY, P. 1909–25. *Essai de nomenclature raisonné des Échinides*, 1–607, 15 pls. (Libraire Ferriere: Chaumont). [At the time of publication, the most complete taxonomic work on Echinoidea, listing all Recent and fossil species, but with certain defects: (1) generic nomenclature confused by acceptance of pre-Linnean names as valid: (2) classification difficult to follow, owing to subfamilies being given the same termination (-*idae*) as families, the correct suffix (-*inae*) being allotted to the minor divisions called 'tribes'; (3) the running headline of the pages gives no assistance in finding the taxonomic position of a genus, which can only be done by turning back through

many pages; (4) there is no easy way of tracking a species which has been transferred from its customary genus to a new one.]

40. LORIOL, P. DE. 1875. *Échinologie Hélvetique: Échinides de la period tertiaire*, 1–142, pl. 1–23 (Ramboz et Schuchardt: Geneve).

41. LORIOL, P. DE. 1881. 'Monographie des Échinides contenus dans les couches nummulitiques de l'Égypte', *Mém. Soc. Phys. Hist. nat. Genève*, **27**, 59–148, 11 pls.

42. MICHELIN, J. L. H. 1861. 'Monographie des Clypéastres fossiles', *Mém. Soc. géol. France*, (2), **7**, 101–147, 28 pls.

43. MORTENSEN, T. 1928–51. *A Monograph of the Echinoidea* (C. A. Reitzel: København).

44. POMEL, A. 1885–87. *Paléontologie ou description des animaux fossiles de l'Algérie: Échinodermes*, pp. 131, 344, 79 pls (A. Jourdan: Alger).

45. STEFANINI, G, 1908. 'Echini del Miocene medio dell' Emila. *Palaeontogr. ital.*, **14–15**, 65–119, pl. 13–16.

46. STEFANINI, G. 1924. 'Fossili terziari della Cirenaica', *Palaeonotgr. ital.*, **27** (1921), 101–145, pl. 1–3.

47. STEFANINI, G. 1924. 'Relations between American and European Tertiary Echinoid faunas', *Bull. geol. Soc. Am.*, **35**, 827–846.

48. VREDENBURG, E. W. 1906. 'Classification of the Tertiary system in Sind with reference to the zonal distribution of the Eocene Echinoidea described by Duncan and Sladen.' *Rec. geol. Surv. India*, **34**, 172–198.

49. WEISBORD, N. E. 1934. 'Some Cretaceous and Tertiary echinoids from Cuba', *Bull. Am. Paleont.*, **20**, no. 70C, 1–270, pl. 20–28.

Chapter III

TERTIARY BIVALVIA

REFERENCE-LETTERS following figure numbers (where there is no number 'a' may be understood):

a, R. valve, internal view (a¹, inverse form).
b, L. valve, internal view (b¹, inverse form).
c, L. valve, external view.
d, R. valve, external view.
e, dorsal view of one or both valves.
f, anterior view of one or both valves.
g, external ornament, enlarged.

REFERENCE-LETTERS to parts of figures:

aa, anterior adductor scar.	*ms*, myophoric septum.
ae, anterior ear (wing).	*n*, nymph.
b, buttress (clavicle).	*pa*, posterior adductor scar.
bs, byssal sinus or notch (*bs¹*, the same closed).	*pb*, pedal embayment.
	pe, pedal muscle scar; also posterior ear.
bt, buttress tooth.	*pl*, pallial line.
c, ctenolium.	*pm*, pedal muscle scar.
ca, cardinal area.	*ps*, pallial sinus.
e, escutcheon.	*r*, resilium or resiliifer.
f, flange,	*ra*, roughened area of hinge plate.
ht, hinge tooth (or teeth).	*rg*, resiliifer gap.
l, ligament (or its area of attachment).	*rhm*, reflected hinge margin.
lp, ligament pit or pits.	*r.v.*, right valve.
lu, lunule.	*s*, hinge socket.
l.v., left valve.	*t*, hinge tooth (or teeth).
ma, myophoric apophysis.	*u*, umbo.

Note: In general, the ligament area or resiliifer is marked by parallel shading; hinge teeth are shown white and numbered (e.g. 2, 3b); sockets are shown black and numbered (e.g. 2¹, 3b¹).

The Bivalvia (also termed Lamellibranchia and Pelecypoda) are the only mollusca (as distinct from the brachiopoda, which used to be called molluscoidea) with bivalve shells, except for a few very rare cases where gastropods become bivalved. The two valves are right and left in position, each being a curved and asymmetric cone. The apical area of the cone is the *umbo* (*u*, Figs. 463, 464), the tip of which is called the *beak*, and represents the region in which shell growth started: it may still show the first-formed shell (*prodissoconch*), though this is usually lost by wear and tear. The greatly abbreviated side of the

169

asymmetric cone faces its fellow of the other valve, and here, on the *dorsal* aspect of the shell, the two valves are united along the *hinge line*. The opposite side, where maximum growth has occurred, forms the greater part of the valve and terminates in the *ventral margin*, where the valves diverge most widely when open. The umbo is thus always near the hinge line, and the *lines of growth*, which mark the position of successive margins reached during the animal's life history, crowd together towards the side of the umbo nearest the hinge line and open out on the opposite side.

The two ends, midway between dorsal and ventral margins, are *anterior* and *posterior* (fore and hind). The distance between them is the *length* of the shell; that between dorsal and ventral margins is the *height*; while the maximum measurement across from valve to valve, at right angles to both length and height, is the *thickness*. (Some ambiguity may arise between thickness thus defined and thickness of the shell substance of each valve: in descriptions, 'shell thick' refers to the former, 'valves thick' to the latter).

The shell is composed of three layers, of which the innermost is *nacreous* in primitive forms (being composed of alternating thin laminae of aragonite and conchiolin, internal reflection between which produces interference colours), but in more advanced forms is *porcellanous*. The middle layer is more or less *prismatic*, i.e. composed of columns of calcite perpendicular to the surface; while the outer layer (*periostracum*) is purely organic (conchiolin) and therefore usually lost in fossils. This outer layer is most thickly developed in freshwater shells, as a protection against the solvent action of the water; but it is also well developed in certain primitive marine shells, such as *Solemya*.

In the living animal (Fig. 393) the valves are lined by thin extensions of the

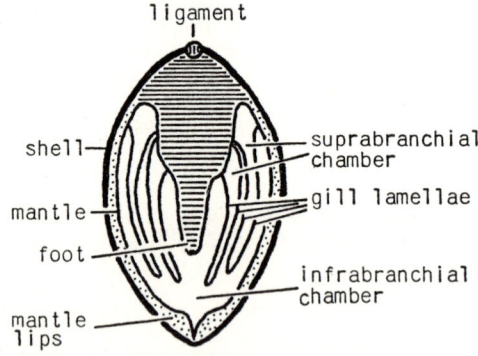

Fig. 393. DIAGRAM OF TRANSVERSE SECTION OF A BIVALVE.

body wall, the *mantle lobes*, enclosing a space around the animal's body, the *mantle cavity*, in which are suspended the *gills* (which in their typical, sacklike form divide the cavity into an upper or *suprabranchial* and a lower or *infrabranchial* chamber). The ventral mantle edges are thickened into *mantle lips* which function both in secreting new shell and in enclosing the mantle

170

cavity when the valves are not closed. The lips may close by muscular action only or may be permanently fused along a great part of their length, but there are two regions where they can always separate: (1) anteroventrally, where the muscular organ of locomotion, the *foot*, is protruded; (2) at the posterior end, where a double or 'figure of 8' opening is left for the inflowing and outflowing currents on which the animal's life depends. These latter openings may be drawn out into tubes (*siphons*) of variable length (Figs 394, 395).

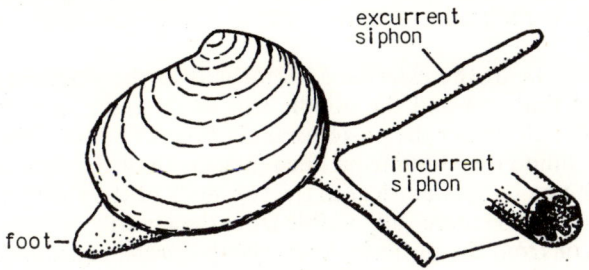

Fig. 394. *Arcopagia crassa* (Pennant). Sketched from life. × 1.

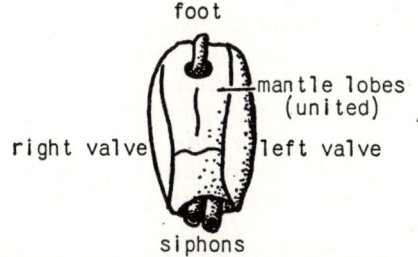

Fig. 395. *Hiatella arctica* (Linné), Rec. Sketched from life. × 1·2.

The shell is *equivalve* when one valve is approximately the mirror image of the other; *inequivalve* when one valve is larger. When a vertical line through the beak separates anterior and posterior halves with approximate symmetry, the shell is *equilateral*: such a form is rare (*Glycymeris*, Fig. 421; *Pecten*, Fig. 438). Much more usually there is a marked difference between the *pedal* and *siphonal* regions, and the shell is *inequilateral*. The general outline of each valve is most usually more or less *ovoid* (*Unio*, Fig. 462); but when length and height are more equal this becomes *orbicular* (*Linga*, Fig. 465); when the length is much greater and the ventral and dorsal margins tend to be straight and parallel, it is *oblong* (*Panopea*, Fig. 618); when the anterior end is rounded and the posterior end is drawn out to enclose the siphons, it is *rostrate* (*Lembulus*, Fig. 401); when an ovoid or orbicular outline appears to be cut off behind by a line of lesser curvature or a straight line, the shell is *truncate* (*Eucrassatella*, Fig. 501; *Oraphocardium*, Fig. 516). When length exceeds height, whatever the general shape, the shell is *elongate* (or *transverse*); in the opposite

171

case it is *high* (or *vertically elongate*). A shell is *inflated* when the ratio of thickness to length is more than one-half; *compressed* when it is less than one-sixth. In the case of an orbicular shell these terms may be replaced by *globose* and *lenticular* respectively.

If the valve margins can meet without leaving any opening, the shell is described as *closed*; if not, it is *gaping*. It may gape at the anterior end (for the permanent protrusion of the foot or of the byssus which it secretes) or at the posterior end (for the siphons), or at both. In several dysodont families the anterior gape is reduced to a narrow notch in the right valve only (*byssal notch*).

The beaks may be at the extreme anterior end of the shell (*Mytilus*, Fig. 423), or between that end and the middle (the usual case), or in the middle (equilateral shells), or nearer the posterior end (*Donax* (*Chion*), Fig. 532). The curvature of the umbo (resulting from unequal growth of the shell on its outer and inner sides) may be symmetrical to the two ends of the shell (*orthogyrate*, e.g. *Pecten*), or it may be more fully visible from the anterior end (*prosogyrate*, the usual case), or from the posterior end (*opisthogyrate*, e.g. *Eotrigonia*, Fig. 464a). Its curvature is always part of a spiral, but it may be so strong as to produce a completely *spiral umbo* (*Meiocardia*, Fig. 558). The *hinge line*, along which the dorsal margins of the two valves are in continuity, may be *straight* (*Arca*, Fig. 409), *angulated* (*Nucula* (*Leionucula*), Fig. 397) or *curved* (a usual condition). Between hinge line and umbo there may be a continuous flat *cardinal area*, facing its fellow and separated from the general surface by an abrupt change of direction (*Arca*, Fig. 409). In the absence of such a continuous area, a similar smaller area, at an angle to the general surface and usually concave, may be present in front of the umbo; this is the *lunule* (*lu*, Figs 502, 500). A similar, but often less clearly defined area behind the umbo is termed an *escutcheon* (*e*, same Figs).

Part of the periostracum uniting the two valves dorsally is usually thickened and modified to form an elastic *ligament* which ensures the opening of the valves when the closing muscles relax. This ligament may be preserved in fossils, but more usually its place is marked by a gap in the hinge line. The ligament always lies within the cardinal area, or, if that is absent, within the escutcheon or where the escutcheon would be. In the first case it is described as *amphidetic*, and in the latter as *opisthodetic*. The latter is the commoner case, and, then, the position of the ligament provides the surest external distinction between the anterior and posterior ends.

The surface of the shell is rarely perfectly smooth, but it may be termed *smooth* when the only marks are delicate *lines of growth*. It may show more or less *ornament* or *sculpture*, the result of unequal secretion of shelly matter by the mantle lips. When the mantle activity varies *in time* only, secreting alternately more or less along its whole lengths, *commarginal* (loosely called 'concentric') sculpture results—and accentuation of growth lines. When there is variation in *space*, some portions of the mantle being always more active than others, *radial* ornament is produced. By differences in strength of either of these, and their varied combination, different patterns may be produced.

There are certain cases which do not come under either of the two main types, and which imply more complicated variations in mantle activity:

(1) *Oblique* ornament consists of lines or costae which are only approximately commarginal and are crossed by the growth lines at very acute angles (some species of *Trigonia* and *Eotrigonia*, Fig. 464; *Digitaria*, Fig. 504).

(2) *Divaricate* ornament consists of two sets of lines both approximately radial, but diverging from one another at an angle so as to form an inverted V (*Divalucina*, Fig. 480).

(3) The curious *loop ornament* on the youthful shells of some freshwater mussels, on the other hand, consists of converging lines, forming an erect V, the Jurassic genus *Goniomya* has the same ornament on the adult shell.

If a line be drawn from the beak to the lower limit of the siphonal openings, it will delimit a posterior *siphonal area* which is sometimes clearly marked off from the main (or pedal) area either by a bend or fold in the shell surface, or by a raised keel, or by a difference of sculpture (*discrepant* ornament) in the two areas, or in more than one of these ways. The case of *discrepant ornament* is found in many diverse families of Bivalvia, e.g. *Trigonia*, *Nemocardium*, *Pinna*, *Brachidontes*, *Pholadomya*, *Pholas*. It is noteworthy that in certain cases (*Protocardia*, *Pinna*) the ornament is more elaborate on the siphonal area. This would seem to support the view that ornament has no original utility, but is simply a by-product of useful activities; since the mantle edge must be more sensitive where it has to control water currents than in other parts. A remarkable fact is that in the family Cardiidae, when radial ornament first appears it is confined to the siphonal area (*Protocardia*, Jurassic): later it extends to the whole surface. Something similar is seen in the Tertiary *Neotrigoniae* of Australia. In certain shells which show only a commarginal sculpture on the surface (*Venus*, *Glycymeris*), radial structure may appear in the deeper layers of the shell when the surface layer is worn away (*decorticated*),

When the valves are completely separated, internal features become visible, chief of which are (1) the hinge plate, (2) the muscle scars, and (3) the pallial line.

The *hinge plate* is a flange of shell extending downwards from the dorsal edge for a short distance, practically in contact with its fellow along the median plane separating the valves. Projections from the hinge plate beyond this median plane constitute the *hinge teeth*; depressions in the contrary direction form the *sockets*. The teeth of one valve correspond in position with the sockets of the other, into which they fit. The hinge plate may also support the ligament: sometimes it is extended upwards above the dorsal contour, forming a *nymph* (*n*, Fig. 505), to which the ligament is attached. At other times the attachment is flush with the dorsal contour. Again, it may sink below that contour, though still above the level of the teeth: the ligament is then semi-internal. When it sinks deeper it becomes an internal ligament, or *resilium*: the hollow on the hinge plate occupied by it is the ligament pit or *resilium pit* (*r*, Figs 521, 520). (Such a pit is distinguished from a tooth socket by occupying an exactly corresponding position in both valves.) To make room for the pit the outline of the hinge plate may bulge downwards: this forms the rudiment of a *resiliifer* (or *chondrophore*) which in some genera (*Corbula*, *Mya*, Fig. 609) is a projecting process.

The hinge plate varies greatly in depth. It may be broad, narrow, or even altogether wanting, the teeth in that case springing from the inner face of the umbonal cavity. Where the hinge plate is narrow or missing, *buttresses* (or 'clavicles') may be developed to support the teeth: these are pillar-like thickenings of the internal surface of the shell (*b*, Fig. 528).

The most primitive type of dentition consists of a row of numerous teeth, all very similar, on a long hinge line; this is the *taxodont* type (*Nucula*, Fig. 396; *Arca*, Fig. 409), the term also covering various modifications in which the teeth may be dissimilar but without approximating to any other type (*Cucullaea*, Fig. 415). By a diminution in the number of teeth, accompanied by a tendency for them to radiate outwards and downwards from the beak, this gave rise to the *actinodont* type, which by degeneration became *dysodont* or by association with a central resilium, *isodont* (*Spondylus*, Fig. 446). From the actinodont may also have been derived the *schizodont* or *palaeoheterodont* type (*Eotrigonia*, Fig. 464; *Unio*, Fig. 462), with a limited number of teeth, often taking the form of long laminae. The breaking up of these laminae into subumbonal (*cardinal*) and horizontal (*lateral*) portions produced the *heterodont* type.

The French palaeontologists Bernard and Munier-Chalmas devised a system of notation for the hinge teeth which facilitates description of the hinge by making it possible to refer to teeth individually (**47**). The notation applies strictly only to the type of hinge known as *heterodont*, and its application to other types (though sometimes convenient) often seems rather forced. The teeth are conceived of as laminae, set horizontally on the hinge plate, and numbered from below upwards (or from within outwards), odd numbers applying to right valve teeth, even numbers to those of the left valve. This condition is actually found in the anterior lateral teeth of many forms, which receive the symbols AI and AIII in the right valve, AII and AIV in the left. The cardinal teeth are regarded as the detached and turned up posterior ends of the anterior laterals. The innermost lamina gives rise to a single cardinal (pivotal tooth), the outer laminae may each give rise to two teeth arranged in a series of inverted V's over the pivotal tooth. In the Corbiculoid type of hinge the pivotal tooth is in the right valve and is numbered 1, the others being 2a and 2b, 3a and 3b, and so on. The usual formula for this type is

$$\frac{\text{RV. 3a, 1, 3b}}{\text{LV. 2a, 2b, 4b}},$$

4a not being developed. There are thus three cardinals in each valve.

In the Lucinoid type the pivotal tooth is in the left valve and is numbered 2, the normal formula being

$$\frac{\text{RV. 3a, 3b}}{\text{LV. 2, 4b}}.$$

Thus there are only two cardinals in each valve, and distinction from the corbiculoid type would be easy but for certain exceptional cases. On the one hand an extra tooth, 5b, may arise in a lucinoid right valve, giving it a corbiculoid aspect; on the other hand, one or more cardinals in a corbiculoid hinge may become obsolete, giving rise to a lucinoid aspect. The true interpretation of such cases is only to be reached by comparison with allied forms of more normal type.

When the umbo is orthogyrate or only slightly prosogyrate, the cardinals are short and fairly upright; but when it is displaced far forward, the posterior cardinals become long, curved and almost horizontal. The posterior laterals are a late development, independent of the anterior laterals and cardinals. They arose to strengthen the hinder region of the hinge as the ligament became shortened. There should therefore never be any confusion between them and the lengthened posterior cardinals seen in forms with the beaks displaced far forward: the latter are in front of or below the ligament; posterior laterals are behind it. They are numbered PI, PII, etc., on the same principle as the antero-laterals.

Typically there are two closing muscles or *adductors*, the attachment of which leaves a distinct *muscle scar* on the interior of each valve at each end of the hinge line, the outer margin of each scar being continuous with the corresponding end of the hinge plate. The scars are usually deep in thick shells, shallow in thin shells; but in the same shell the anterior scar may be much deeper than the posterior. Other muscle scars (*pedal* muscles) may also be present, usually behind and rather above the anterior adductor, in front of and above the posterior: they are well marked in some superfamilies (Unionacea Fig. 463; Crassatellacea, Figs 500–503).

In the majority of bivalvia the two adductors are of approximately equal size, though they may differ in shape (*Linga*, Fig. 465): the shell is then said to be *isomyarian*. In one important group (dysodont bivalves) the anterior adductor becomes progressively smaller (*anisomyarian* condition) until it finally disappears (*monomyarian*, in antithesis to both the previous cases which are *dimyarian*). Step by step with these changes the posterior adductor becomes larger and assumes a more central position with reference to the hinge line.

The inner margin of the thick mantle lip is marked on the interior of the valves as the *pallial line*: this runs from one main muscle scar to the other, keeping parallel to the valve margin for the whole or most of its course. This line is indistinct in many Pteriomorpha; it is a broad and irregular band in some Pandoracea. When the posterior mantle openings are lengthened out into siphons, some adaptation to this condition is necessary. As already mentioned, the shell may become rostrate, or gaping; but whether either or neither of these changes takes place it is generally necessary to increase the area of muscular attachment, and this is shown on the shell by an embayment of the pallial line, the *pallial sinus*. Shells are described as *integripalliate* or

175

sinupalliate according to the absence or presence of this sinus. It varies greatly in size (small in *Polymesodea* (*Geloina*), Fig. 562, large in *Tellina* (*Tellinella*), Fig. 533) and in shape may be broad or narrow, rounded or angular, while its direction also varies.

Rare in the older Palaeozoic era, the Bivalvia became increasingly common as time went on and may be regarded as at their acme in the Caenozoic era or at the present day. They are therefore important fossils in Tertiary stratigraphy, but most of the species are usually longer-lived than those of Palaeozoic and Mesozoic Cephalopoda or Brachiopoda, and do not mark horizons with such precision: it is rather on associations than on single species that the stratigrapher relies.

In 1884 Neumayr (**67**) proposed a classification based on shell characters; in 1912 an improvement on this was sketched out but never formally completed by H. Douvillé (**53**). This was a natural or evolutionary classification, based on the idea of adaptive radiation. The scheme assumes a primary divergence or radiation into three main branches, adapted primarily to the three main modes of life of bivalvia:

(1) The 'normal' life on the sea bottom, either active in the upright position or passive lying on one side;
(2) the fixed life, either suspended by the byssus from, or cemented by shell substance to, some foreign body;
(3) the burrowing or boring existence.

In each of the three branches, secondary radiations took place, so that the actual mode of life of any genus is no indication of its position in classification. For instance, the most active of all bivalvia, *Pecten*, belonged to the 'fixed' branch.

In 1906 Pelseneer (**30**) proposed a zoological classification which in its original form was based fundamentally on gill structure. In the most primitive forms (*Protobranchia*), the gills are plume-like, as are those of other Mollusca, while in others they show either a transitional structure (*Filibranchia*), or the typical sack-like form (*Eulamellibranchia*), or the rare septum-form connected with very specialized feeding (*Septibranchia*). This classification was thus based on a character which was not only indeterminable in the case of extinct forms, but belonged to the category of 'progressive' characters, i.e. characters the direction of evolution in which is in some degree inevitable, and likely to be repeated on parallel lines in a number of separate lines of descent. Thus the divisions were really *grades*, not groups of common ancestry, and it was found that *Trigoniacea* and *Arcacea* were united because they had reached the same grade in gill structure, while *Ostreacea* were separated from *Pectinacea* because the former had evolved farther than the latter, although in fundamental structure the two are obviously related. Pelseneer recognized these anomalies and later modified his classification by introducing an additional order, *Pseudolamellibranchia*, to include Pteriacea, Pectinacea and Ostreacea, but in so doing he abandoned gill structure as a defining character.

In 1965 Newell (**67a**) and in 1967 Vokes (**72b**) gave comprehensive accounts

176

of the classification of the Bivalvia. The classification Vokes recommended, which was designed for use in the *Treatise on Invertebrate Paleontology* (ed. Moore, **66a**), is followed here except for a few very minor differences. See also Cox, 1960 (**49a**). (M.Cambr.-Rec.)

Subclass PALAEOTAXODONTA

The nuculoids; shell structure nacreous or crossed lamellar; equivalve; isomyarian; ligament usually amphidetic. (Ord.-Rec.)

Order NUCULOIDA

Protobranch taxodonts. (Ord.-Rec.)

Superfamily NUCULACEA
Integripalliate. (Ord.-Rec.)

Family NUCULIDAE
Internal resilium; umbones opisthogyrate; shell closed, inequilateral, posterior end shorter; internally nacreous (**71**). (Ord.-Rec.)

Nucula (*s.s.*) (Fig. 396): ovate-trigonal; escutcheon rather swollen, smooth; beaks posterior to middle; weak to moderately strong commarginal threads and sometimes weak radial ornament; valve margins internally crenulate; radial shell-structure. U.Cret.-Rec. Cosmop. The subgenus *Leionucula* [*Ennucula, Nuculopsis* Woodring *non* Girty] (Fig. 397) resembles *Nucula*, but is quite smooth and the valve margins are internally smooth. Cret.-Rec. Cosmop. *Acila* (*s.s.*) (Fig. 398) has divaricate ornament and is somewhat restrate posteriorly. Olig.-Rec. Asia, Japan, N.Amer., Carib., S.Amer. In the subgenus *Truncacila* there is no posterior rostration and the posterior slope is rather steeper. L.Cret.-Rec. N.Amer., S.Amer., Japan, Eur., N.Africa.

Superfamily NUCULANACEA
Usually with a pallial sinus; posterior end often produced and rostrate; shell not nacreous. (Ord.-Rec.)

Family NUCULANIDAE [LEDIDAE]
Valves often gaping; valve margins internally smooth; except in the Recent subfamily Sareptinae the external ligament is not closely joined to the resilium. (Dev.-Rec.)

Nuculana [*Leda*] (*s.s.*) (Fig. 399): surface with commarginal rugae dorsally, fading to threads ventrally; umbones at about one-third of the length; posteriorly gaping and rostrate, rostrum with two ill-defined keels; shallow pallial sinus. Trias.-Rec. Cosmop. [It seems that earlier records from the Palaeozoic have to be placed elsewhere.] The subgenus *Saccella* [*Ledina* Sacco *non* Dall] (Fig. 400) has the umbones at about two-fifths of the length, so that the anterior and posterior rows of hinge teeth are approximately of the same length;

177

the rostrum is shorter, pointed, and carries only one keel; a shallow groove runs from the umbo to the ventral margin at each end. Eoc.-Rec. Cosmop. The subgenus *Lembulus* (Fig. 401) differs from *Saccella* in that both siphonal area and escutcheon are defined by keels, and the former is also marked off by a notch in the shell outline, while there is a delicate oblique ornament which may become prominent on the keels. Eoc.-Rec. Cosmop. *Jupiteria* (Fig. 402): corbuloid, with only weak rostration, beaks at about mid-length; shell inflated and with weak commarginal ornament; shallow pallial sinus. U.Cret.-Rec. Eur., Medit., N.Amer., Carib., Austral., N.Z. *Calorhadia* (*s.s.*): '*Nuculana*'-like, large (up to 4 cm. long), with inconspicuous umbones situated a little anterior to mid-length, distinctly rostrate and with two keels posteriorly, and the surface has distinct, well spaced commarginal threads. Eoc.-?Rec. N.Amer., Carib. *Yoldia* (*s.s.*) (Fig. 403): more compressed than '*Nuculana*', elongate-subelliptical, slightly rostrate posteriorly; beaks nearly central; resilium pit large, symmetrically underlapping both rows of teeth; pallial sinus extending almost to mid-length; surface nearly smooth. Cret.-Rec. Cosmop. The subgenus *Orthoyoldia* is not rostrate, the posterior end being almost as high as the anterior end; resilium pit smaller and narrower; shell moderately inflated. M.Eoc.-Rec. Carib., N.Amer. [Although *Yoldia* (*s.s.*) extends into the Arctic, Boreal and Antarctic provinces, *Orthoyoldia* is confined to tropical and sub-tropical America.]

Family MALLETIIDAE

Shell rounded or elongate; ligament external; no resilium, hinge line almost straight. (Ord.-Rec.)

Malletia (*s.s.*) (Fig. 404): shell oval, subequilateral, not inflated, smooth or

---

Figs 396–412. BIVALVIA: NUCULACEA, NUCULANACEA, SOLEMYACEA AND ARCACEA

Figs 396 and 398 after Keen, in Moore; Figs 399 and 402 after Puri, in Moore; Figs 404–406 after McAlester, in Moore; Fig. 408 after Keen and Newell, in Moore; Fig. 412 after Newell, in Moore; Figs 397, 400, 401, 403, 407 and 409–411 original.

396d. *Nucula nucleus* (Linné), Rec. France. T. × 2·5.
397. *Nucula* (*Leionucula*) *obliqua* Lamarck, Plio. (Kalimnan). Muddy Creek, Victoria. × 1
398d. *Acila divaricata* (Hinds), Rec. Korea. T. × 0·7.
399c. *Nuculana pernula* (Müller), Rec. Eur. T. × 1.
400c. *Nuculana* (*Saccella*) sp. Plio. Greece. × 4.
401c, e. *Nuculana* (*Lembulus*) *emarginata* (Lamarck), L.Mio. Aquitaine. × 1·25.
402c. *Jupiteria oculata* (Iredale), Rec. Austral. × 1·5.
403. *Yoldia oblongoides* (S. V. Wood), Pleist. (Norwich Crag). Norfolk. × 1.
404d. *Malletia chilensis* des Moulins, Rec. Chile. T. × 0·5.
405c. *Neilo cumingii* Adams, Rec. N.Z. T. × 1·5.
406c. *Tindaria arata* Bellardi, L.Plio. Italy. T. × 2.
407. *Solemya* (*Solemyarina*) *australis* Lamarck, Rec. Austral. × 1·25.
408b. *Nucinella ovalis* (Wood), Plio. England. T. × 6.
409. *Arca noae* Linné, Rec. Medit. T. × 1.
410. *Barbatia barbata* (Linné), Mio. Vienna B. T. × 1·35.
411b. *Obliquarca* sp. Diagrammatic.
412d. *Bathyarca pectunculoides* (Scacchi), Rec. × 6.

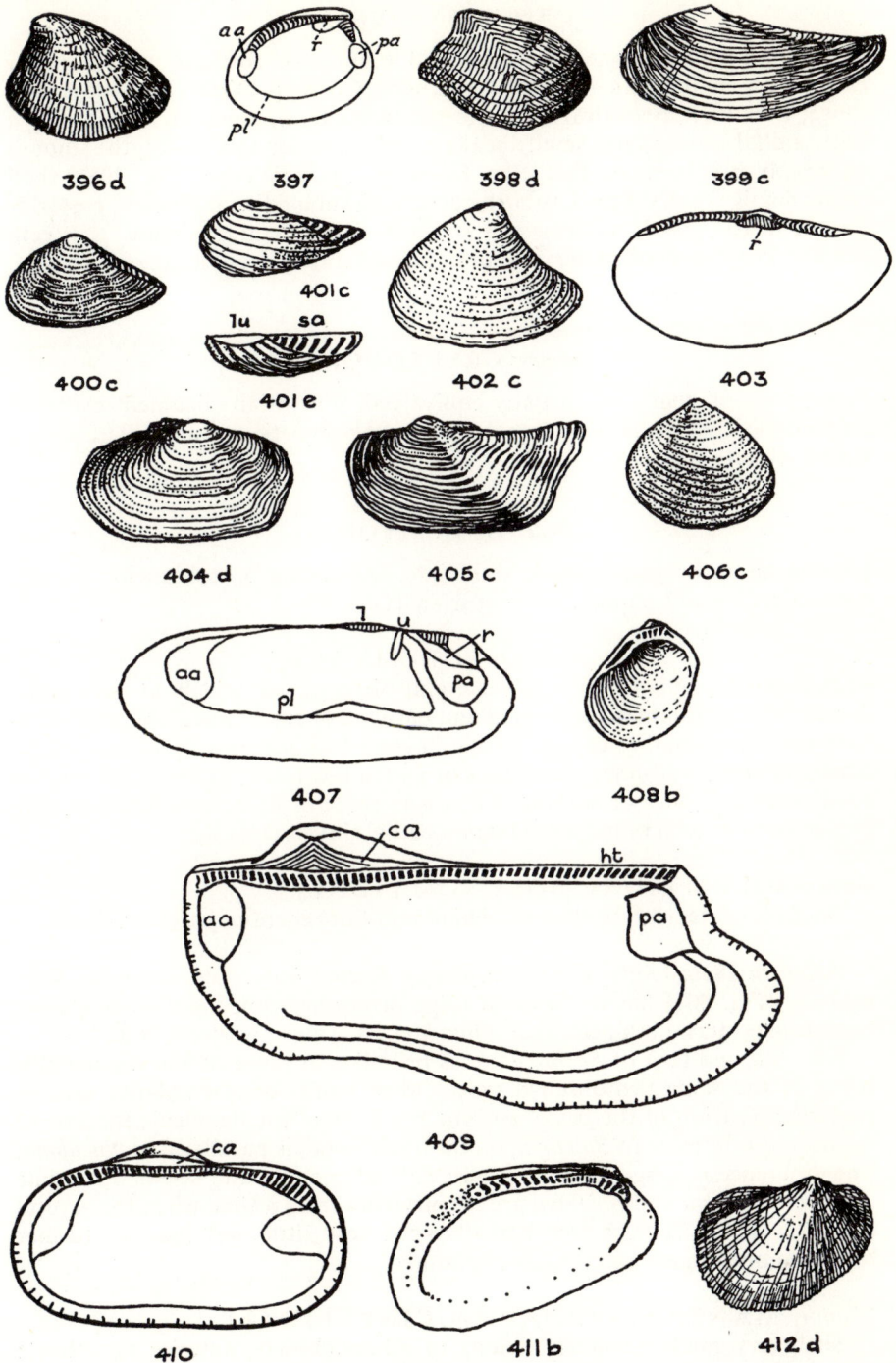

396 d   397   398 d   399 c

400 c   401 c   402 c   403

401 e

404 d   405 c   406 c

407   408 b

409

410   411 b   412 d

179

with commarginal striae; posterior end blunt, not rostrate; no lunule or escutcheon; pallial sinus large. Mesoz.-Rec. Cosmop. *Neilo* (Fig. 405): shell longer than high, posteriorly rostrate and truncate, siphonal area limited by a keel; pallial sinus rather small; beaks a little anterior to mid-length; smooth or weakly ornamented. Tert.-Rec. Cosmop. *Tindaria* (*s.s.*) (Fig. 406): shell rather small, solid, inflated, suborbicular, with submedian umbones; posterior row of teeth curved; anterior row of teeth shorter and straighter, the teeth heavier; commarginal rugae; pallial sinus moderately deep, its apex V-shaped. Tert.-Rec. Cosmop.

Subclass CRYPTODONTA

Edentulous or nearly so; usually equivalved, dimyarian; ligament external, amphidetic to opisthodetic; probably a polyphyletic group. (?U.Cambr., L.Ord.-Rec.)

Order SOLEMYOIDA

Homogeneous aragonite shell; siphonate, burrowing protobranchs; usually edentulous, usually gaping, anisomyarian. (Dev.-Rec.)

Superfamily SOLEMYACEA

Shell elongate, soleniform, equivalve and often gaping widely at both ends, dorsal and ventral margins straight and parallel, or rarely nuculoid; periostracum usually extending conspicuously beyond free margins of valves; beaks situated behind mid-line; edentulous or with a few teeth; ligament internal or partly external; with or without buttress in front of posterior adductor; surface smooth or with radial costellae; no pallial sinus. (Dev.-Rec.)

Family SOLEMYIDAE [SOLENOMYIDAE]
    Shell elongate, with beaks well behind mid-line; edentulous. (Dev.-Rec.)

Solemya [*Solenomya*] (*s.s.*) is probably Recent only. The subgenus *Solemyarina* (Fig. 407) has an internal ridge diverging from the chondrophore, bounding posterior adductor scar. Olig.-Rec. N.Amer., Austral., N.Z.
    The gills and foot of *Solemya* are as primitive as those of *Nucula*, the gills being of the same protobranch type. Other points of resemblance are the posterior position of the beaks and the relations of the ligament; the loss of the nacreous interior in *Solemya*, on the other hand, is paralleled in *Nuculana*. The difference in hinge teeth, however, is fundamental, and shows that while *Nucula*, *Nuculana* and *Solemya* are all survivors from a time when the several branches of the Bivalve tree had diverged very little, yet the two former belong to one branch and the last to another.

?*Family* MANZANELLIDAE [NUCINELLIDAE]
    Shell very small, obliquely oblong to subnuculiform, anterior side short;

180

not nacreous; ligament internal, resiliifer small and bordered by a few sub-taxodont teeth. (Perm.-Rec.) [Although previously classified in the Arcoida, the anatomy indicates relationship to *Solemya*.]

Nucinella (Fig. 408): subnuculiform, obliquely oval; surface smooth or with fine growth lines; valve margins internally smooth; resiliifer bordered on each side by a few relatively large teeth; left valve with a strong posterior lateral tooth (PII). L.Jur.-Rec. Eur., Austral., Pac., Carib., S.Amer., N.Amer., Africa.

Subclass PTERIOMORPHA

Cyrtodonts, arks, anisomyarians and monomyarians; shell structure, ligament, gills and stomach variable; usually byssate or cemented in adults, sometimes free; phyletic unity suggested by the fossil record. (L.Ord.-Rec.)

Order ARCOIDA

Isomyarian filibranchs with crossed-lamellar shells; adults free or byssate; shells usually equivalved, orbicular to trapezoidal; dorsally mostly with cardinal areas above hinge axis. (L.Ord.-Rec.)

Superfamily ARCACEA

Valves trapezoidal, usually with radial ornament, often longer than high, with long, straight or gently curved, well developed taxodont hinge, posterior teeth sometimes being few in number and very oblique; cardinal area usually relatively long compared with Limopsacea. (L.Ord.-Rec.)

Tendency to fixation is slight: *Arca* fixes itself by a very short byssal plug, but as it is not suspended by this byssus there is no tendency either to asymmetry of the valves or to forward rotation of the foot and reduction of the anterior adductor as in dysodont shells. Thus the shell is equivalve and isomyarian, without 'ears' and either with a ventral gape (marking the position of foot and byssus) or completely closed (indicating loss of all tendency to fixation).

As in primitive dysodont shells, the hinge line is long and often straight, and a large triangular cardinal area intervenes between beak and hinge line. This area may be entirely occupied by the insertion of the amphidetic ligament, or unoccupied portions may be left—lunule in front, escutcheon behind. The ligament insertion is marked by a series of grooves, which are either straight and vertical or inverted Vs (*chevrons*). The hinge is more or less taxodont in character. Although nearer the ancestral bivalve stock than dysodont shells, the Arcacea depart more from the primitive state in one respect: the interior is porcellanous, not nacreous.

Family ARCIDAE

Shell subtrapezoidal to ovate, usually equivalve, sometimes slightly inequivalve; valves equally inflated, usually longer than high; beaks nearly

181

always anterior to mid-line; ornament of commarginal lines with or without radial costae; with or without a byssal gape; ligament external, duplivincular, amphidetic, prosodetic or opisthodetic; hinge straight or nearly so; teeth numerous; adductor scars subequal; valve margins often crenulate internally. (?Trias. Jur.-Rec.)

Arca [*Navicula*] (*s.s.*) (Fig. 409): transversely subquadrate, inflated, with ventral byssal gape corresponding to a wide, shallow radial depression on the flank of the shell; beaks well anterior to mid-line, remote, slightly opistho-gyrate; posterior end rather alate, separated off from the rest of the surface by a radial carina and somewhat differently ornamented; lunule and escutcheon narrow, especially the lunule; ligament in chevrons; base of cardinal area meeting upper edge of hinge plate at an angle of 90°; cardinal area large, as long as hinge line; teeth small and numerous, becoming only very slightly oblique at the ends of the hinge plate; ornament of more or less granular radial costae; valve margins internally finely scalloped except at the byssal gape. U.Cret.-Rec. Cosmop. (warm seas). *Barbatia* (*s.s.*) (Fig. 410): differs from *Arca* in being more ovoid, more nearly equilateral, without defined siphonal area, with prosogyrate beaks, the distal teeth somewhat oblique, and the byssal gape smaller; the cardinal area is narrow, and its base meets the upper edge of the hinge plate at an angle of 150°; the costellae are rather fine. ?Trias., Jur.-Rec. Cosmop. There seem to be at least five additional valid subgenera of *Barbatia*, of which the following are the more important. The subgenus **Cucullaearca** is generally distorted and higher posteriorly than anteriorly, with a distinctly sinuous ventral margin and a large byssal opening; the hinge teeth are coarser and more oblique distally. U.Cret.-Rec. Cosmop. (warm seas). The subgenus *Acar* is rather small and solid, trapezoidal in outline and with a sharp umbonal ridge, has coarse imbricate and reticulate ornament, a narrow byssal gape, the cardinal area is narrow with the ligament grooves restricted to the posterior end, and the adductor scars are elevated and conspicuous. U.Cret. (Danian)-Rec. Eur., N.Amer., S.Amer., W.Pakistan. *Obliquarca* (Fig. 411): modioliform, with beaks near anterior end, and with fine, slightly beaded radial costae; valve margins internally smooth. Eoc.-Plio.-?Rec. Eur., Medit., N.Amer., Carib., S.Africa. *Bathyarca* (Fig. 412): usually very small and sub-globular; left valve slightly the larger; cardinal area very narrow, ligament grooves only at posterior end; fine radial and commarginal threads, discrepant on the two valves; valve margins internally finely crenulate. Eoc.-Rec. Cosmop. *Trisidos* (Fig. 413): elongate parallelepipedic, asymmetrically twisted, posteriorly carinate; cardinal area very narrow; weak radial costae; teeth not meeting medially, terminally oblique. Eoc.-Rec. Indo-Pac., M.East, Eur. *Anadara* [*Diluvarca*] (*s.s.*) (Fig. 414): valves thick; equivalve, ovate trapezoidal with commarginally beaded radial costae, ornament similar on the two valves; beaks prosogyrate, at about one-third of the length; cardinal area high, mostly covered by ligament, with chevrons; hinge line relatively short; valve margins internally scalloped. L.Mio.-Rec. Cosmop. The subgenus **Lunarca** [*Argina* Gray *non* Hübner] has the beaks situated well forward, the ligament area is narrow and occurs only posterior to the beaks, and the anterior teeth are few and irregular. Eoc.-Rec. N.Amer., Carib., S.Amer., Eur., Atl., Pac. *Scapharca:* left valve larger than right valve, its postero-ventral margin projecting beyond the edge of the right valve; valves subequilateral and subquadrate, ornament similar. Olig.-Rec. Eur., Indo-Pac., Carib., widespread. The subgenus *Cunearca* has the left valve a little larger than the right valve; the ornament is discrepant, the costae on the left valve being larger and more heavily beaded. Olig.-Rec. N.Amer., Carib., S.Amer., Eur., Austral., widespread.

Family CUCULLAEIDAE
Valves rather thick; smooth or with radial costae; umbones submedian; less transverse than most Arcidae, and with rather shorter hinge which has a smaller number of short, vertical median teeth and several very oblique to horizontal terminal teeth; ligament amphidetic, chevron-shaped grooves few. (Jur.-Rec.)
Cucullaea (*s.l.*) is the only genus of this family that survived the Mesozoic.
Cucullaea (*s.s.*) (Fig 415): trapeziform or subquadrate, never pronouncedly inequilateral, closed (byssus vestigial), ornamented with fine radial striae, well inflated and posteriorly carinate; cardinal area moderately wide, with chevron-shaped ligament grooves; valve margins internally finely scalloped; posterior adductor scar usually bounded anteriorly by a myophoric septum. Jur.-Rec. Indo-Pac., Eur., N.Amer., S.Amer., China, Japan, Austral., Burma.

Family NOETIIDAE
Subtrapezoidal to ovoid, equivalve, inequilateral; ligament prosodetic, amphidetic or opisthodetic; cardinal area with vertically arranged ligament grooves; earliest members orthogyrate or prosogyrate, later members all opisthogyrate; no byssal gape; sometimes smooth, more usually with radial costae, commarginal bands, or both; in some forms strong secondary costae alternate with small primary costae; hinge line gently to strongly arched, teeth oblique at ends; posterior adductor scar often anteriorly with a raised flange. (L.Cret.-Rec.)
Noetia (*s.s.*) (Fig. 416): anadariform, beaks opisthogyrate; subrhombic to subtrigonal, usually produced or angular postero-ventrally; cardinal area mainly anterior to beaks, with vertical ligament grooves; umbonal ridge well defined; weakly ornamented, undivided interstitial costae becoming less strong distally; valve margins internally with scalloping better developed posteriorly. U.Eoc.-Rec. Indo-Pac., Carib. *Eontia:* more ovoid; secondary costae often divided; ligament amphidetic. L.Mio.-Rec. S.Amer., N.Amer., Carib., Eur., Atl. *Protonoetia* (Fig. 417): subquadrate, inflated, with low, nearly central, orthogyrate beaks; ligament amphidetic; umbonal ridge subcarinate; costae smooth in adult stage. Low U.Eoc. Nigeria. *Arcopsis* [*Fossularca*] (Fig. 418): small, valves thick, subquadrate, beaks a little anterior to mid-line, surface with a median radial depression; cardinal area narrow, vertical ligament grooves in a small triangular area beneath beak; fine radial costae and conspicuous growth lines; hinge line with a small submedian edentulous gap;

183

valve margins internally smooth or finely crenulate; inner margin of adductor scars raised. U.Cret. (Danian)-Rec. Eur., Indo-Pac., Medit., Carib., N.Amer., S.Amer., W.Pakistan, N.Z. *Striarca:* similar to *Arcopsis*, but has no median edentulous gap in the hinge line, the outline is more barbatiiform, and the adductor scars seem larger; ligament longer. U.Cret.-Rec. Eur., Medit., N.Amer. *Trinacria* (Fig. 419): subtriangular, rostrate and sharply carinate posteriorly, practically smooth; umbones high, beaks strongly opisthogyrate; ligament usually amphidetic, situated in a deep triangular depression restricted to a small part of the cardinal area below beak; valve margins internally smooth. Eoc. Eur., N.Amer.

Family PARALLELODONTIDAE

Arciform; posterior hinge teeth elongate, tending to be parallel to hinge margin; ligament duplivincular. (L.Ord.-Rec.)

Cucullaria (Fig. 420): form and ornament of *Barbatia*; hinge line arched, with a few granular teeth in the middle and two to three horizontal teeth anteriorly and posteriorly. Eoc. Eur. *Porterius* [*Pseudogrammatodon*]: resembles *Arcopsis* or *Barbatia* in form but has a median edentulous gap on the hinge between about four short anterior teeth and about six posterior teeth the three hindmost of which are nearly horizontal. Eoc.-Rec. Eur., N.Amer., Japan. The subgenus *Notogrammatodon* is much more elongate, the dorsal and ventral margins are not parallel, and the edentulous gap on the hinge line is much longer. U.Eoc.-Mio. N.Z.

Superfamily LIMOPSACEA

Valves usually suborbicular or even higher than long; often without umbonal ridge; smooth or with radial ornament; hinge line shorter and more arched than in Arcacea, with fewer teeth, sometimes almost edentulous. (L.Perm.-Rec.)

Family GLYCYMERIDAE

Hinge line arched, sometimes with an edentulous median gap caused by the downward encroachment of the cardinal area; shell suborbicular, usually slightly truncated behind, equivalve, non-byssate, of large to moderately

---

Figs 413–423. BIVALVIA: ARCACEA, LIMOPSACEA AND MYTILACEA
Figs 413–414 and 416–420 after Newell, in Moore; Figs 415 and 421–423 original.
413c. *Trisidos tortuosa* (Linné), Rec. Philippines. T. × 0·5.
414c. *Anadara antiquata* (Linné), Rec. Madagascar. T. × c. 0·8.
415. *Cucullaea concamerata* Martini, Rec. Ceylon. × 0·56. *ml*, myophoric lamina.
416d. *Noetia reversa* (Sowerby), Rec. Panama. T. × 1.
417d. *Protonoetia nigeriensis* (Newton), U.Eoc. T. × 1.
418c. *Arcopsis limopsis* (Koenen), Danian. Denmark. T. × 3.
419c. *Trinacria crassa* (Deshayes), Eoc. Paris B. T. × 4.
420d. *Cucullaria heterodonta* (Deshayes), Eoc. France. T. × 0·5.
421. *Glycymeris siculus* (Reeve), L.Mio. Aquitaine. × 0·7.
422. *Limopsis beaumariensis* Chapman, Plio. (Kalimnan). Grange Burn, Victoria. × 1·3.
423. *Mytilus* sup., Rec. × 0·75.

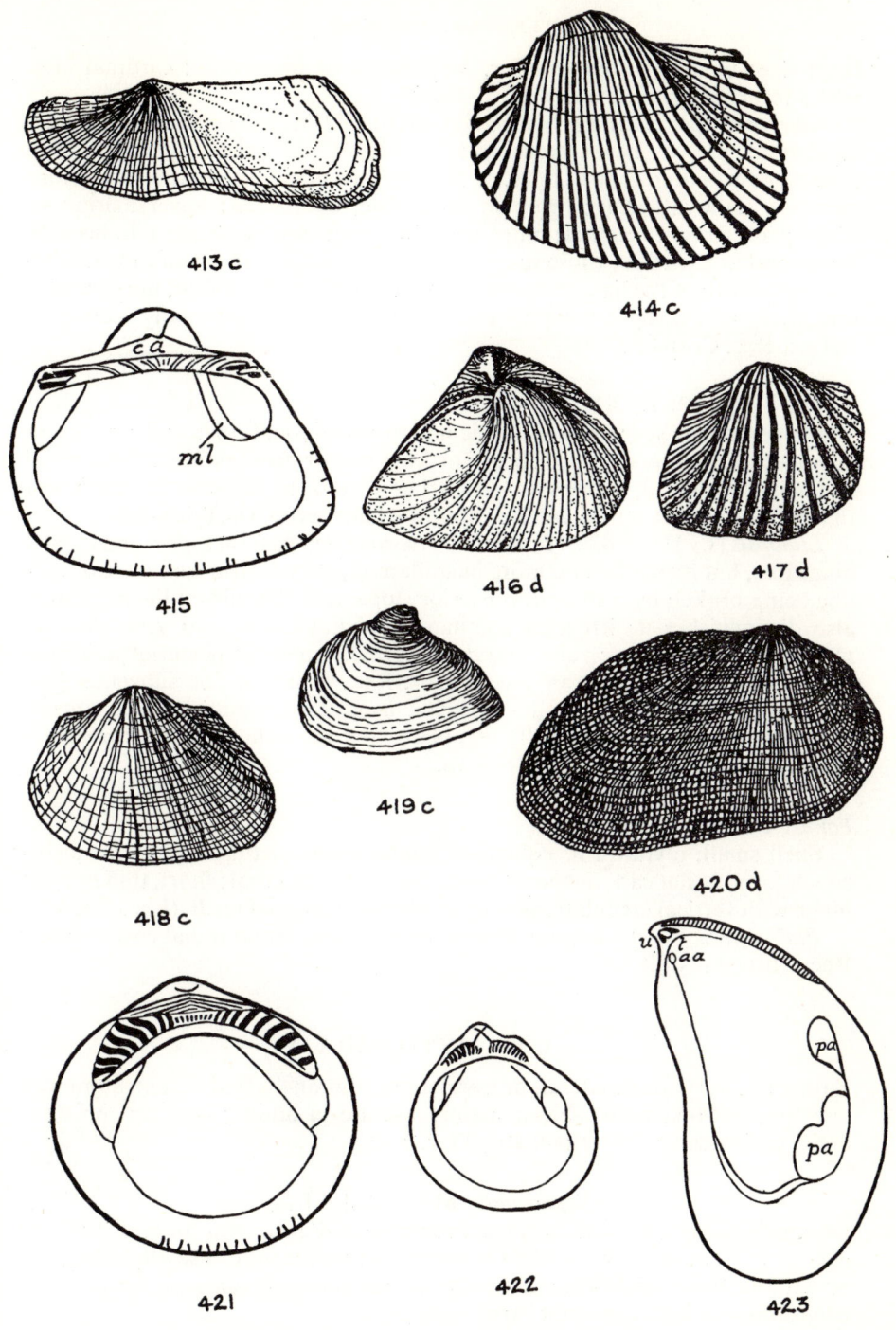

413 c

414 c

415

416 d

417 d

418 c

419 c

420 d

421

422

423

large size; ligament duplivincular, amphidetic or prosodetic; cardinal area with a few chevron grooves. (L.Cret.-Rec.)

Glycymeris [*Axinaea, Pectunculus*] (*s.s.*) (Fig. 421): shell solid, suborbicular, equilateral, closed; beaks erect, small, more or less opisthogyrate; surface smooth or with fine growth lines, or radially costate or striate, the latter more apparent on weathering; ligament area wide, chevroned; lower margin of hinge plate strongly curved; hinge teeth in a curved series tending to become horizontal at each end, those in the centre more or less truncated or obliterated by growth of the ligament area; adductor scars well marked, the posterior one often with a slight flange; valve margins generally internally scalloped. L.Tert.-Rec. Cosmop.

Family LIMOPSIDAE

Hinge line arched; shell suborbicular to subtrigonal, of small size, equilateral or inequilateral, closed; beaks orthogyrate or prosogyrate; subumbonal ligament area small, triangular, not striated, sunk in between the teeth and thus separating them into anterior and posterior series. (U.Trias.-Rec.)

Limopsis (*s.s.*) (Fig. 422): resembles *Glycymeris* in general aspect and curved hinge line, but is smaller and more inequilateral, the uniform curvature of outline being broken by a straight posterior slope, and the sunken ligament area also distinguishes it; irregular commarginal undulations and very obscure, narrow, radial furrows; valve margins internally smooth; posterior adductor scar larger than the anterior one. M.Jur.-Rec. Cosmop. The subgenus *Pectunculina* differs in having radial costellae as well as commarginal ornament, the valve margins are internally crenulate, and the anterior adductor scar is slightly buttressed. Cret.-Rec. Cosmop.

Family PHILOBRYIDAE

Shell small, mytiliform, equivalve; umbones projecting or with prodissoconch forming flat cap; ligament internal or partly external; short, thin byssus; hinge with vertical crenulations with or without marginal teeth. (Eoc.-Rec.)

Philobrya is Pleistocene and Recent only. *Cosa:* strong radial costae. Eoc.-Rec. Austral., N.Z.

Order MYTILOIDA

Equivalve, very inequilateral, anisomyarian filibranchs and eulamellibranchs with prismato-nacreous shells; mainly byssate in adults; ligament opisthodetic, parivincular; integripalliate. (Dev.-Rec.)

Superfamily MYTILACEA

Obliquely elongated, with beaks prosogyrate and at or near anterior end; no ears or byssal notch, but a slight byssal gape; ornament of growth lines, with or without fine radial lines; internally nacreous or subnacreous; hinge either edentulous, or with dysodont cardinal teeth

186

$$\frac{1}{2a, \ 2b}$$

or taxodont both before and behind the ligament in direct relation to the radial sculpture; ligament long, external, opisthodetic, supported by nymphs, sometimes becoming subinternal behind; anisomyarian, the posterior scar generally bilobed; periostracum strong. (Dev.-Rec.)

The superfamily is distinguished from the Pteriacea by the absence of ears, byssal notch and cardinal area and the opisthodetic position of the ligament. Owing to their superficial resemblance to the Mytilidae, the members of the family Dreissenidae were at one time also placed in the Mytilacea, but they are now placed in the order Veneroida.

Family MYTILIDAE

Anterior adductor scar small, varying in position from high up in the umbonal cavity to well below it, but never on a myophoric septum; interior nacreous to subnacreous; habitat marine, with tolerance of estuarine conditions in *Mytilus*; other characters as for the superfamily. (Dev.-Rec.)

Mytilus (*s.s.*) (Fig. 423): transversely elongated, pointed anteriorly, beaks quite terminal; anteroventral margin nearly straight, dorsal margin gently curved, posterior margin declivous and rounded; sculpture usually of growth lines only; ligament subexternal, very long; hinge teeth

$$\frac{1}{(2a, \ 2b)}$$

set on a very short hinge plate, often obsolete; anterior adductor scar high up in umbonal cavity, posterior scar separated into a smaller upper (retractor) and larger bilobed lower scar; pallial line not parallel to valve margin, being nearest to its anterior portion. Plio.-Rec. N.Eur., N.Amer., S.Amer., S. Austral., N.Z., S.Pac. (extra-tropical, littoral). *Modiolus* [*Volsella*] (*s.s.*) (Fig. 424): trapeziform, inflated; beaks distinctly behind the anterior end, the antero-ventral margin usually bulging forwards; no teeth or hinge plate; umbonal ridge well rounded; anterior adductor scar well developed below the umbonal cavity, posterior adductor scar not bilobed; surface smooth or occasionally with very weak radial costellae; periostracum usually hirsute. Dev.-Rec. Cosmop. (tropical and temperate). *Brachidontes* (*s.s.*) (Fig. 425): modioliform to submytiliform; fine radial sculpture (the costellae increasing in number by bifurcation and intercalation, especially in the anterior and posterior regions); valve margins internally crenulate except anteriorly; dysodont teeth present, in front of and behind ligament; anterior adductor scar present or absent; ligament fairly short, internal to subexternal. Jur.-Rec. Cosmop. (especially tropical). *Septifer* (*s.s.*) (Fig. 426): outline as in *Mytilus*, but surface with divaricate radial costellae as in *Brachidontes* which likewise produce a taxodont appearance on the hinge; umbonal slope sharply rounded; a triangular septum for the anterior adductor scar extends from the terminal umbo for about half the length of the ligament (one-third of the hinge line);

187

posterior adductor scar long, lobed; valve margins internally crenulate. Trias.-Rec. Cosmop. [The marine *Septifer* differs from the freshwater *Dreissena* in its nacreous interior and presence of radial (divaricate) ornament; it also appears to have, like other Mytilidae, a less evolved type of gill (filibranch).] ***Lithophaga*** (*Lithodomus*) (*s.s.*) (Fig. 427): shell closed, inflated-subcylindrical, elongate, somewhat tapering posteriorly; umbones obtusely rounded, at or near anterior end; hinge edentulous; ligament elongate, narrow, subexternal; anterior adductor scar rather ventrally placed, only a little smaller than the posterior one; surface smooth or with vague transverse ornament or radial costallae at the posterior end; valve margins internally smooth. ?Carb.-Rec. Cosmop. (a warm water form, boring in coral reefs etc.). ***Musculus*** [*Modiolaria*] (*s.s.*) (Fig. 428): modioliform, but beaks still farther from anterior end than in *Modiolus* (yet anterior to mid-line), moderately inflated, with broadly rounded umbonal keel, and anterior and posterior portions of the surface with radial costellae, the two portions separated by a triangular smooth area with fine growth lines only; hinge and adductor scars much as in *Brachidontes*; valve margins internally crenulate. Jur.-Rec. Cosmop. ***Crenella*** (Fig. 429): small to very small, obliquely oval, inflated, with small, incurved, prosogyrate beaks anterior to mid-line; fine radial ornament decussated by commarginals, often divaricate along mid-line of umbonal slope and usually bifurcating; hinge mytiloid, with small knobs anteriorly, behind which is a deep pit for the ligament which is wholly internal; valve margins internally finely crenulate, crenulations sometimes extending on to the hinge itself to produce a pseudotaxodont appearance. ?Cret.-Rec. Cosmop.

Superfamily PINNACEA

Usually equivalve; elongate in an oblique direction, triangular, with terminal beaks and posterior truncation; byssal gape narrow and long, posterior gape wide; ornament radial and commarginal in varied combination; outer shell layer usually thin, sometimes thick, prismatic; inner nacreous layer not extending for more than two-thirds the length from the beaks; anterior adductor scar relatively small, about its own length behind the beak and occupying nearly the full height of the shell; posterior adductor scar large, about half way between the two ends, and occupying the dorsal half of the height of the

---

Figs 424–432. BIVALVIA: MYTILACEA, PINNACEA AND PTERIACEA
Figs 424–429 after Soot-Ryen, in Moore; Fig. 430 after Cox and Hertlein, in Moore; Fig. 432 after Hertlein and Cox, in Moore; Fig. 431 original.
424d. *Modiolus modiolus* (Linné), Rec. Norway. T. × 0·5.
425d. *Brachidontes modiolus* (Linné), Rec. Florida. T. × 1.
426c. *Septifer bilocularis* (Linné), Rec. Indo-Pac. T. × 0·8.
427d. *Lithophaga lithophaga* (Linné), Rec. France. T. × 0·75.
428c. *Musculus discors* (Linné), Rec. Denmark. T. × 2.
429d. *Crenella decussata* (Montagu), Rec. Norway. T. × 5.
430d. *Pinna rudis* (Linné), Rec. Barbados. T. × 0·3.
431. *Pteria hirundo* (Linné), Rec. Medit. T. × 1.
432c. *Pinctada margaritifera* (Linné). Rec. Japan. T. × 0·3.

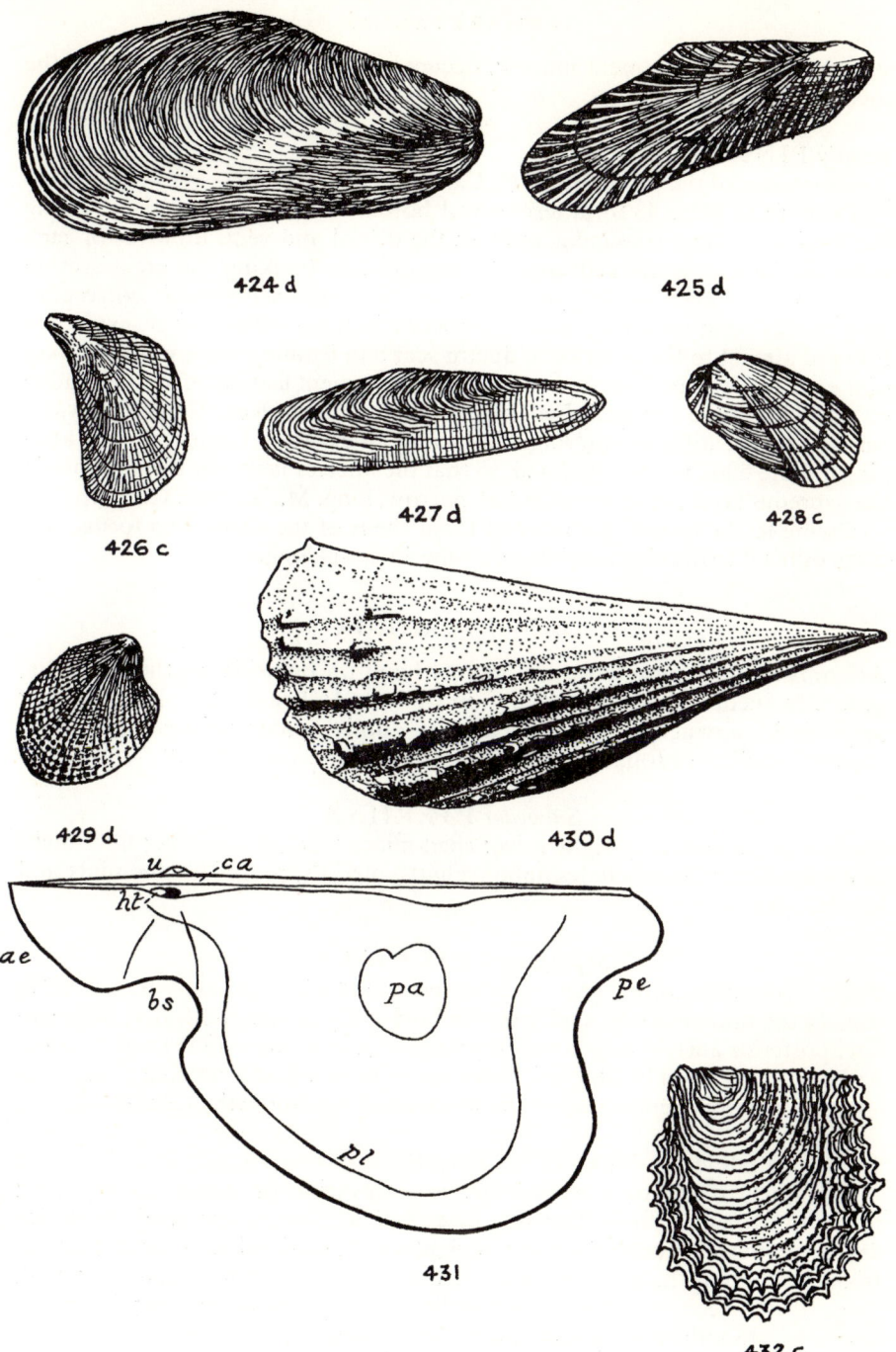

424 d

425 d

426 c

427 d

428 c

429 d

430 d

431

432 c

shell; edentulous; ligament narrow, occupying at least half the length of the shell. (L.Carb.-Rec.)

Family PINNIDAE

Characters of the superfamily. (L.Carb.-Rec.)

Pinna (*s.s.*) (Fig. 430): anteroventral border more or less straight, giving the shell an acutely triangular outline; the dorsal and ventral halves of each valve are bent at a marked angle to one another (making the cross section rhomboidal or even square), and may be differently sculptured; corresponding to this bend internally is a deep groove in the nacreous layer, extending forward almost to the anterior adductor scar and forming the lower boundary of the posterior scar; ligament subinternal; ornament usually of spinose radial costae with some commarginal undulations. L.Carb.-Rec. Cosmop. (warm water). *Atrina:* differs from *Pinna* in having an excavated anteroventral margin, and the flank is not angulated, so that the interior lacks the deep groove in the nacreous layer; ligament external, narrow, long. M.Jur.-Rec. Cosmop.

Owing to the fragile character of fossil shells of the above two forms, it is often difficult to decide to which genus specimens belong.

Order PTERIOIDA

Anisomyarian or monomyarian, mainly lying on one side on the sea floor; generally inequivalve and usually inequilateral; ligament opisthodetic or amphidetic, alivincular, multivincular or duplivincular; prismato-nacreous, crossed lamellar, or foliate internally; integripalliate. (Ord.-Rec.)

Suborder PTERIINA

Byssate or cemented by right valve; shell microstructure variable; filibranchs or eulamellibranchs; includes anbonychiids, pteriids, pectinids, anomiids and limids. (Ord.-Rec.)

Superfamily PTERIACEA

Beaks prosocline, often near or at anterior end of hinge margin; left valve usually the more convex; shell extended obliquely posteroventrally; biauriculate, posterior auricle large and often wing-like, anterior auricle small; permanently attached by a byssus; prismato-nacreous or crossed lamellar; hinge line straight, mainly edentulous; ligament external, opisthodetic. (Ord.-Rec.)

Family PTERIIDAE [AVICULIDAE]

Obliquely ovate to suborbicular; monomyarian or nearly so in adult; feebly to strongly inequivalve; very inequilateral, anterior ear small, posterior ear large; byssal notch anteriorly in right valve; cardinal area long, narrow, triangular, going back from beak; ligament simple, in an oblique hollow; hinge teeth weak or absent; pallial line usually discontinuous; internally nacreous; smooth forms predominate. (Trias.-Rec.)

Pteria [*Avicula*] (Fig. 431): nacreous, obliquely subovate, with a large pos-

190

terior ear and a smaller anterior ear, smooth or with radial rows of small spines; slightly inequivalve, right valve flatter than left valve; beak distance less than one-quarter; anterior ear bounded by a groove which leads to the broad byssal notch; umbones small, prosogyrate; hinge long, straight, with one small tooth in front of beak in each valve, that of the right valve the more anterior, a long, low lamella behind beak in right valve, and a corresponding socket in left valve; umbonal cavity broad and shallow; cardinal area long, narrow; ligament opisthodetic; byssal notch shallow in left valve, deep in right valve. Trias.-Rec. Cosmop. (warm seas). *Pinctada* [*Meleagrina*] (Fig. 432): this is the pearl 'oyster', and differs from *Pteria* in being less oblique and with thicker valves, in having a hinge line shorter than the height of the shell, a broader cardinal area, a narrower byssal notch and deeper groove leading to it, no hinge teeth and a more pit-like umbonal cavity; no definite posterior wing; obliquely subrectangular; right valve slightly more convex than left valve; anterior ear distinctly limited; radial costae and commarginal lamellae; adductor scar somewhat behind the middle. Mio.-Rec. Mainly Indo-Pac.

Family BAKEVELLIIDAE

Obliquely elongate and auriculate like *Pteria*, but with serial ligament pits; usually with some hinge teeth; anterior adductor much reduced or absent; smooth or with radial costae; internally nacreous. (Perm.-Eoc.)

Aviculoperna: obliquely trapeziform; muscle impression bilobed; posterior wing obtuse, anterior auricle good; one short, oblique anterior tooth in left valve, two in right valve; one short posterior tooth in each valve; left valve, or both, with radial costellae. Jur., Eoc. Eur. [This is the only member of the family that survived the Mesozoic.]

Family ISOGNOMONIDAE [PERNIDAE, MELINIDAE]

Subquadrate or submytiliform, with or without posterior wing, rarely with anterior auricle, usually higher than long; with or without byssal gape below the terminal beaks; subequivalve to very inequivalve, the left valve the more convex; hinge line straight, usually edentulous in adult; ligament external, multivincular; internally nacreous; surface smooth or scaly; monomyarian; pallial line usually broken into small pits. (U.Perm.-Rec.)

Isognomon [*Perna, Melina*] (*s.s.*) (Fig. 433): mytiliform-subquadrate, beaks at anterior end of hinge, pointed; anterior margin excavate, with a narrow byssal gape; posterior wing obtuse; ligamental grooves fairly numerous; surface lamellose. U.Trias.-Rec. Cosmop.

Family MALLEIDAE [VULSELLIDAE]

Ostreiform, usually non-auriculate, more or less gaping posteriorly but without byssus, not cemented, often higher than wide, subequivalve or inequivalve; monomyarian; edentulous; ligament area triangular, internal or external; shell internally nacreous. (Jur.-Rec.)

This family is a derivative of the Pteriidae, adapted to a nestling life. Living

191

embedded in sponges, they have lost byssus and ears. Having a triangular ligament area and being more or less irregular in shape and sculpture, they are liable to be taken for oysters, from which they are distinguished: (1) by the absence of a scar of fixation on either valve, (2) by the valves often gaping, which those of *Ostrea* never do, (3) by the shell being internally nacreous (the thin external layer is often lost), and (4) in many cases by their equivalve character.

Vulsella (Fig. 434): vertically elongate, linguiform, subequivalve, compressed, gaping fore and aft; beaks opisthogyrate; surface with commarginal lamellae; ligament like that of *Ostrea*; adductor scar subcentral, flush. U.Cret.-Rec. Cosmop. *Pseudoheligmus* (Fig. 435): gibbose and cordiform, usually longer than high, nearly equivalve, with scaly or radially costate surface; there is a depressed region posterior to the umbos on which long, narrow, and at times bifurcating fissures occur in the margin of the shell; adductor scar submedian, raised; external shell layer prismatic, internal layer cavernous. U.Cret.-M.Eoc. Eur., M.East., Africa. *Heligmopsis* (Fig. 436): vertically elongate-oval, not gaping, subequivalve; a few costae anteriorly and posteriorly or in hinder region only, forming interlocking projections on margin; muscle scar elongate, ventral in position, posteriorly raised. U.Cret.-U.Eoc. Eur., Egypt. *Euphenax* (Fig. 437): variable outline and inflation, more or less ovate, usually closed, subequivalve to inequivalve; shell structure peculiar: a thick, fragile outer layer of radially elongate, prismatic, thin-walled, hollow cells, weakly joined by thin concentric laminae to a more compact inner layer, which is itself chambered near the umbones, and may support a number of thin-walled chambers on its inner side; muscle scar not raised; surface with or without radial costellae. M.Eoc. Jamaica, W.Pakistan, M.East, Somalia.

Superfamily PECTINACEA

Usually inequivalve, suborbicular or with tendency to equilateral and vertically elongate form; with anterior and posterior ears; free, byssate or cemented, byssal notch usually below right anterior auricle; right valve below when at rest; sculpture usually radial, rarely divaricate or smooth; inner layer of shell nacreous or crossed-lamellar aragonite in Palaeozoic, foliate calcite in Mesozoic, often with thin outer prismatic layer; monomyarian; ligament area

---

Figs 433–440. BIVALVIA: PTERIACEA AND PECTINACEA

Fig. 435 after H. Douvillé; Figs 436–437 after Hertlein and Cox, in Moore; Fig. 440 after Hertlein, in Moore; Figs 433–434 and 438–439 original.

433. *Isognomon ephippium* (Linné), Rec. Carib. × 0·75.

434b, e. *Vulsella deperdita* Lamarck, U.Eoc. (Bartonian). Hampshire. b × 1·3, e × 1·1.

435e. *Pseudoheligmus nigeriensis* (R. B. Newton), Pal. Sudan. × 0·75.

436c. *Heligmopsis petrocoriensis* (Coquand), Coniacian. T. × 1.

437. *Euphenax jamaicensis* (Trechmann), M.Eoc. W.Pakistan. T. × 1·3. (Transverse section showing internal chambers).

438a, b. *Pecten maximus* (Linné), Rec. × 0·75.

439. *Chlamys islandicus* (Müller), Rec. T. × 0·75.

440c, d. *Minnivola isomeres* Iredale, Rec. Austral. T. × 1.

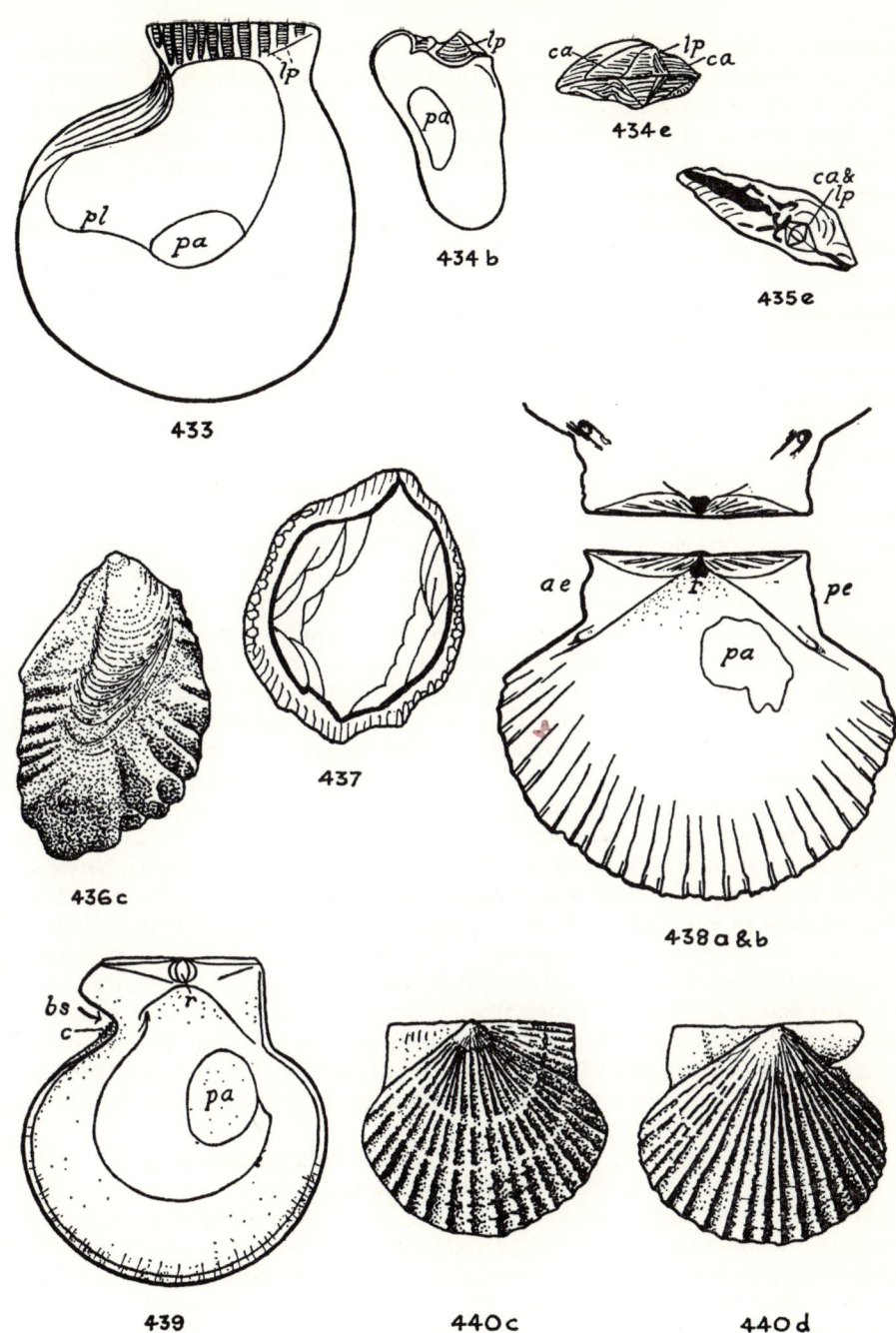

433

434 b

434 e

435 e

436 c

437

438 a & b

439

440 c

440 d

alivincular, becoming internal in most post-Palaeozoic forms; hinge isodont, i.e. with dental lamellae symmetrically placed before and behind the resilium (there being no true hinge plate, the dental lamellae were termed *cardinal crura* by Dall, similar lamellae on the margin of each ear being termed *auricular crura*). Some genera are fixed by byssus or cementation, others (e.g. *Pecten*,) are active swimmers. (Ord.-Rec.)

The ears, byssal notch and dental lamellae show affinity with the Pteriacea, from which the Pectinacea are separated by the non-nacreous interior (of post-Palaeozoic forms) and resilium. The anatomy of the nervous, respiratory and other systems shows that *Amusium* should be separated from the Pectinidae and *Plicatula* from the Spondylidae. There has been much discussion as to the families in which some genera should be placed; the following arrangement takes some anatomical characters into account.

Family PECTINIDAE

Free, or fixed by byssus, or cemented by right valve; more or less inequivalve and inequilateral, orbicular to oval; auriculate, anterior ear straight; umbones central; closed or gaping; often with radial costae and commarginal threads, or with commarginal lamellae, sometimes smooth; ligament obsolete, external in a narrow groove, internal in a resilium (amphidetic); hinge taxodont in young, with obsolete teeth in adults; usually with crura; adductor scar eccentric posteriorly; internally porcellanous. (Trias.-Rec.)

This family affords a striking example of change in habits: although belonging to the branch in which fixation is the chief control of structure, its members are actually the most active of all Bivalvia, swimming in a jerky manner (with intervals of rest on the sea bottom) by sudden closing of the valves so as to shoot out water, with violent recoil. This habit is accompanied by the development of numerous eyes along the mantle border.

So far as shell form is concerned, the main division may be drawn between forms with a well marked byssal notch and only slightly inequivalve shell, byssus-fixed in early life at least (*Chlamys*) and forms in which the notch is lost and the shell is very inequivalve (*Pecten*): the former covers the great majority of Tertiary species. Species of Pectinidae attained a much larger size in Miocene and Pliocene times than previously. More than one hundred generic and subgeneric names have been proposed for members of the family; since some of them seem to be based on relatively minor differences (such as details of the ornament), a number of these names will probably eventually be sunk into synonymy of others.

Pecten (*s.s.*) (Fig. 438): orbicular, equilateral, with median umbones; right valve convex, left valve flat or slightly concave, the inner margins coarsely scalloped; right valve with coarse, rather flat costae, sometimes with fine radials and commarginal threads as well; left valve costae finer and with wider intervals; ears almost equal, without any byssal notch; three or four cardinal crura before and behind resilium pit; auricular crura ending in tubercles; crura strongly wrinkled; right valve adductor scar indented below. U.Eoc.-Rec. Eur., Medit. [It is evident that some earlier forms previously placed in this group belong elsewhere; modern types of *Pecten* do not seem to occur earlier than the U.Eocene.] The subgenus *Flabellipecten* has more numerous costae which on the right valve are flatter; left valve umbonal area more convex. L.Mio.-Rec. Eur., Asia, N.Amer., C.Amer. *Chlamys* (*s.s.*) (Fig. 439): subequivalve, left valve only slightly the more inflated, moderately convex; suborbicular to subtrigonal, slightly higher than wide; equilateral but for the ears, of which the anterior are the larger; right anterior ear with byssal notch and *ctenolium* (a row of teeth guiding the byssal threads); radial costae numerous and close-set, often imbricate, grooved or spiny, sometimes with secondaries, corrugating the whole thickness of the valve; dental lamellae obsolete; adductor scar large, occupying nearly one-third the height of the shell; pallial line distant from valve margin, often obscure. Trias.-Rec. Cosmop. The subgenus *Aequipecten* is orbicular, about as long as high or even a little longer, the left valve is the more inflated, the muscle scar is smaller, and the auricular crura end in tubercles; the left valve auricles are subequal, and the costae are relatively fewer in number and have fine radial grooves on and off them; byssal notch smaller. Jur.-Rec. Cosmop. (but not E.Pac.). The subgenus *Argopecten* [*Plagioctenium*] differs from *Aequipecten* in having both valves inflated, longer auricles, fuller and wider umbones, in being more inequilateral, and in lacking radial grooves on the costae and interspaces (there being fine commarginal lines); the byssal notch is smaller. L.Mio.-Rec. Cosmop. (warm seas). The subgenus *Lyropecten* attains a large size, both valves are convex (the left one slightly more so), the costae are strong and the whole surface is over-run by fine longitudinal striae; the posterior ear is a little smaller than the anterior one, and the byssal notch is deep. L.Mio. (Vaqueros)-M.Plio. Carib., N.Amer. The subgenus *Nodipecten* is large, the valves are more or less equally convex, and the large corded costae often bear nodes or humps; the anterior ear is the larger, the byssal notch is moderately shallow, and the ctenolium is well defined. M.Mio.-Rec. Atl., Carib. The subgenus *Antipecten* is large and has relatively very large ears, the right valve is flat and the left valve convex, and the byssal notch is very deep; the right valve has coarse costae with subsidiary radial ornament, and the left valve costae are unequal and unequally spaced. Burd.-M.Mio. Eur. *Patinopecten* (*s.s.*): large, right valve gently convex, left valve nearly flat; the costae are flatter on the right valve, finer and sharper on the left valve; the ears are subequal and the byssal notch is not sharply defined; cardinal crura weak. U.Olig.-Rec. N.Pac., N.E.Pac. *Minnivola* (Fig. 440): a curious cross between *Pecten* and *Chlamys*; it has the form, ornament, convex right valve, and flat left valve of *Pecten*, but possesses a distinct byssal notch and ctenolium like *Chlamys*; costae on left valve finer than those on right valve. U.Mio.-Rec. Austral., Pac., M.East, W.Pakistan, S.Africa. *Hinnites* (Fig. 441): large, heavy, subequivalve, edentulous, with small ears and irregular, squamose radial ornament; valve margins internally smooth; right valve fixed by cementation in adult, with consequent irregularity of shape; a polyphyletic group. U.Eoc.-Rec. Cosmop. (warm seas).

Family AMUSIIDAE [ENTOLIIDAE]

Often fairly small and thin, smooth or with feeble ornament; often with internal radial lirae; hinge short or moderately long, rarely with transverse folds. (L.Carb.-Rec.)

Although members of this family (like *Pecten*) swim about by clapping their valves and are very active, the gills differ from those of the Pectinidae by lacking interlamellar junctions, the nervous system and shell structure differ, and there are few or no ocelli along the mantle margin; the group is evidently more primitive.

Amusium [*Amussium*] (Fig. 442): relatively large, smooth or almost smooth (there may be a few weak costae in the umbonal area); suborbicular, gaping anteriorly and posteriorly; valves not convex, right valve slightly the more inflated; inner surface with paired or unpaired radial lirae which do not reach the ventral margin; auricles subequal; byssal notch extremely small; no ctenolium; hinge with pair of cardinal crura; auricular crura present. L.Mio.-Rec. Cosmop. (warm seas). *Lentipecten* [*Pseudentolium*] (*s.s.*): like *Amusium*, but lacks the internal lirae, and has a somewhat deeper byssal notch. L.Eoc.-U.Plio. Eur., India, N.Z. *Amussiopecten:* like *Amusium* except that the ornament is more like that of *Pecten* but becoming obsolete in the later stages of growth. U.Olig.-U.Mio. Eur., Medit., M.East, Indonesia, E.Africa, N.Amer., S.Amer., Japan. *Propeamussium* (*s.s.*) (Fig. 443): small, valves thin and not very convex, laterally gaping; right valve with moderate byssal notch; left valve with radial striae, right valve more or less smooth; internally with a few lirae which stop a long way from the margin. L.Jur.-Rec. Cosmop. *Palliolum* (*s.s.*) (Fig. 444): small, equivalve, slightly inflated; surface with fine vermicular or reticulate marks or obsolete radial threads; cardinal margin sharply cross-striated; one pair of small cardinal crura; no internal lirae, and valve margins internally smooth; right valve anterior ear more strongly ornamented, with good byssal notch and ctenolium. ?Eoc.-Rec. Medit., Africa.

Family PLICATULIDAE

Shell more or less irregular and ostreiform, subequilateral, not very inflated, not attaining a very large size; attached by umbo of right valve; inequivalve to subequivalve, closed; hinge short, with short crura on each side of a deeply sunk resilium pit; with small or no ears, and no byssal sinus; ornament of coarse radial folds (simple or divaricate) or fine radial threads, commarginally lamellose; monomyarian, the single rather small adductor scar situated

---

Figs 441–446. BIVALVIA: PECTINACEA
Figs 441–444 after Hertlein, in Moore; Figs 445–446 original.
441c. *Hinnites crispa* (Brocchi), Plio. Italy. T. × 0·5.
442a, d. *Amusium pleuronectes* (Linné), Rec. Indo-Pac. T. × 1.
443d. *Propeamussium anconitanum* (Foresti), L.Mio. Italy. × 1.
444d. *Palliolum incomparabile* (Risso), Rec. Medit. T. × 2.
445. *Plicatula marginata* Say, U.Mio. (Yorktown). Maryland. × 0·9.
446a, b. *Spondylus* cf. *aurantius* Lamarck, Rec. Seychelles. × 0·8.

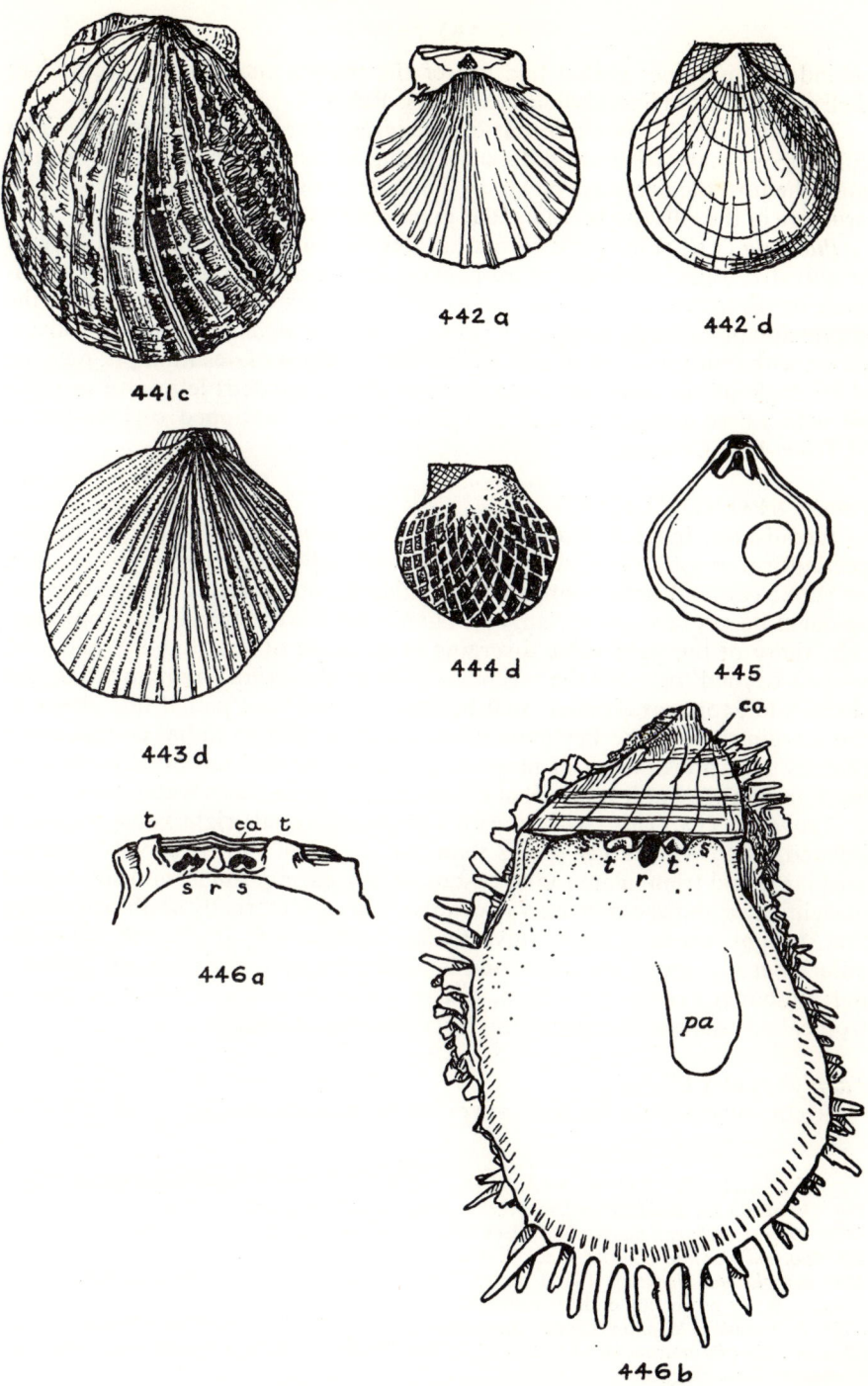

441c

442 a

442 d

443 d

444 d

445

446 a

446 b

behind the mid-line; pallial line nearer the margin than in the Pectinidae; shell structure as in Spondylidae. (M.Trias.-Rec.)

The anatomy differs from that of *Spondylus* more than that of *Spondylus* differs from that of *Pecten*, differences being in the characters of the mantle edge, the lips, the nervous system, in the absence of a foot, and in the simpler structure of the gills (no interlamellar junctions).

Plicatula (*s.s.*) (Fig. 445): somewhat deltoidal in outline, subequivalve to inequivalve, right valve the more convex; ornament of radial costae, squamose or divaricate, sometimes almost smooth; area of attachment takes the impression of the body to which it is fixed; cardinal areas small in both valves; valves with crura in front of and behind the deeply sunk resilium pit, meeting at an angle of less than 90°, both transversely crenulated; left valve resilium pit with raised margins; valve margins internally scalloped or crenulated. M.Trias.-Rec. Cosmop.

Family SPONDYLIDAE

Inequivalve, inflated, fixed by the right valve which is the larger; outline pectiniform or ostreiform, subequilateral, with indistinct ears; no byssal notch or gape; cardinal area triangular, amphidetic, much larger in right valve; ligament deeply sunk in a triangular ligament pit, with strong crura on either side, those of the right valve diverging at an angle of less than 90° and with sockets beyond them for the reception of left valve crura; monomyarian, the single adductor scar situated well behind the mid-line; pallial line closer to the margin than in the Pectinidae; ornament of spinose radial costellae and threads of various orders of magnitude (secondary and tertiary); inner shell layer aragonitic, outer layer calcitic, lamellar. Lives in warm water. (Jur.-Rec.)

Spondylus (*s.s.*) (Fig. 446): auricles small, subequal; right valve the more inflated, often with larger spines than the left valve and higher, its cardinal area large and triangular; radial costae usually strong; hinge with solid crura in right valve and crura in left valve; valve margins internally scalloped; hinge line straight; crura of right valve thicker and curved; in the left valve the anterior crus is more horizontal and lamellar, the posterior thick and pyramidal; umbonal cavities very deep, rest of cavity relatively shallow. A mainly tropical genus, flourishing on coral reefs. Jur.-Rec. Cosmop.

Family DIMYIDAE

Right valve a little the less convex, fixed; valves ostreiform, suborbicular,

---

Figs 447–451. BIVALVIA: PECTINACEA AND ANOMIACEA
Fig. 447 after Cox and Hertlein, in Moore; Figs 448–451 original.
447d. *Dimya deshayesiana* Rouault, Eoc. France. T. × 1·7.
448. *Placuna placenta* (Linné), Rec. Ceylon. T. × 0·3.
449. *Indoplacuna sindiensis* (Vredenburg), M.Mio. (Gaj.) Tyrah Valley, Kachh. × 0·9.
    x, x, freely projecting ends of resiliifer.
450b, d. *Anomia ephippium* (Linné), Rec. Britain. T. × 0·75. *bm*, byssal muscle scars.
451a, a`. *Carolia placunoides* Cantraine, U.Eoc. Egypt. × 0·75. a, byssal sinus open; a`, byssal sinus plugged.

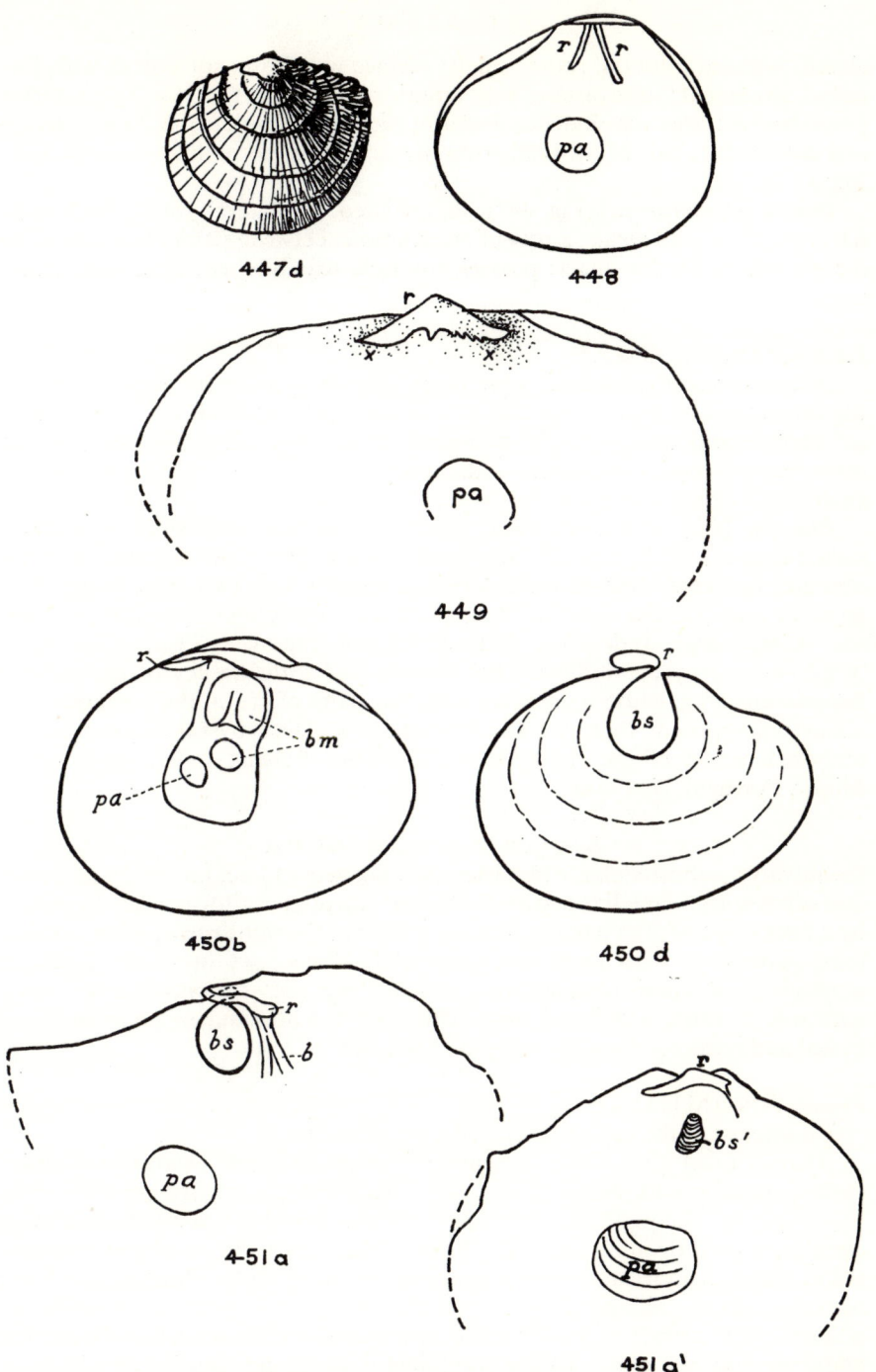

447d

448

449

450b

450 d

451a

451a'

199

closed, without auricles, rather small; surface lamellose, sometimes with fine radial ornament; dimyarian, but anterior adductor the smaller; interior porcellanous; area amphidetic, obscure; ligament obsolete, resilium internal and alivincular; pair of cardinal crura in each valve; no pallial sinus. (M.Jur.-Rec.)

Dimya [*Deuteromya*] (Fig. 447): thin-shelled, equilateral, rather flat; cardinal crura short and weak; a row of crenulations between pallial line and valve margin which are far apart; posterior muscle scar bilobed. Eoc.-Rec. Widespread.

Family PLACUNIDAE

Inequivalve, suborbicular, often large, internally nacreous; monomyarian, the single adductor scar large and subcentrally placed; adult free; no additional pedo-byssal muscle scars; non-auriculate; hinge short, edentulous, but often bearing crura for support of the internal resilium; often with fine vermicular radial ornament. (Eoc.-Rec.)

Placuna [*Placenta*] (*s.s.*) (Fig. 448): compressed, orbicular, sometimes saddle-shaped, with delicate ornament, translucid; cardinal area narrow, obscure; ligament internal and external, internally with two crura (meeting at an acute angle) in the right valve, the anterior crus being the shorter; not sessile. L.Mio.-Rec. Indo-Pac., India, Mekran, Indonesia, Eur. [The term 'window-pane oyster' indicates the flatness and translucency of the valves.] *Indoplacuna* (Fig. 449): right valve with two pairs of crura, the posterior pair almost superposed; there are the remains of a closed byssal foramen in the umbonal region of the right valve. (**73**). Mio. Burma, India, Egypt, Asia Minor, Portugal, N.Africa.

Superfamily ANOMIACEA

Inequivalve, suborbicular, often irregular, non-auriculate, internally nacreous and often with a lamellar texture; attached (throughout life or only in youth) by a horny byssal plug passing through a hole in the right valve; hinge edentulous, sometimes with large divergent resilial processes or crura; ligament amphidetic, internal, attached to crura or deep scars; monomyarian, often with one to three additional scars which are the attachment impressions of byssal and retractor muscles. (?Perm., Cret.-Rec.)

Family ANOMIIDAE

Characters of the superfamily. (?Perm., Cret.-Rec.)

Anomia (*s.s.*) (Fig. 450): subcircular, often irregular or distorted, thin-shelled, usually with radial striae, very inequivalve, left valve cup-like, right valve flat; by unequal growth in early life the byssal notch becomes a large circular or oval opening under the beak of the right valve, the byssus being a thick, short, cylindrical, horny plug for attachment; hinge short, edentulous; ligament short, immediately above the byssal foramen; monomyarian, adductor subcentral; in the left valve, two byssal muscle scars (one beside the adductor and one above it) are combined with the adductor scar within an oblong area. ?Perm., Cret.-Rec. Cosmop. *Placunanomia* (*s.s.*): fairly large, often with about three folds, and with fine radial ornament; byssal foramen of right valve usually closed in adult (shell not attached); right valve with a pair of crura above the byssal plug; the left valve has a byssal muscle scar in addition to the adductor scar, the muscle functioning as an additional adductor. Mio.-Rec. N.Pac., N.Amer., Carib., S.Amer. *Pododesmus* (*s.s.*): shell heavier than in *Anomia*, but it is likewise smooth or with fine radial threads; right valve attached by its whole surface as well as the byssus; byssal foramen closed by a small plug well within hinge margin; edentulous; left valve with one additional muscle scar. Mio.-Rec. N.Amer., S.Amer., Japan. The subgenus *Monia* has radial costae and the right valve perforation is very large; byssal scar large. ?Olig., Mio.-Rec. N.Amer., N.Z., England. *Carolia* (*s.s.*) (Fig. 451): flattened, thick-shelled, sometimes of large size, with fine radial ornament; left valve the more convex; byssal plug may be much reduced; right valve with a long, broad, obtusely bent resiliifer. M.Eoc.-U.Eoc. Egypt, N.Africa, Carib. [The ancestral stock of this remarkable form appears to be an undescribed species of *Anomia* from the Middle Eocene (Khirthar) of Sind, which differs from typical *Anomia* in having a flat left valve: it also has sculpture typical of *Carolia and Placuna* (commarginal, with fine radial lines between), but is otherwise a true *Anomia*. *Carolia* starts life like this species, but passes through a series of revolutionary stages: the foramen becomes relatively smaller and is finally closed; at the same time the ligament sinks and the process carrying it becomes longer and broader and obtusely bent, being converted into a resiliifer, with a corresponding triangular pit in the left valve.]

Superfamily LIMACEA

Inequilateral, nearly always equivalve, typically biauriculate, oval, obliquely oval-subtriangular to obliquely subquadrate (somewhat produced anteriorly), higher than long, with small beaks; smooth or with radial ornament, commarginal ornament rare; triangular ligament area between beak and hinge line, containing the triangular ligament pit (which is thus external, not as in *Pecten*); byssus present or absent, when present emerging by a slight anterior gape, not a notch; sometimes with posterior gape also; hinge edentulous or taxodont; monomyarian, adductor scar placed higher up and farther back than in Pectinidae. Those forms that have no byssus may be as active as *Pecten*. Shell with outer calcitic layer and middle and inner layers of non-nacreous aragonite. (L.Carb.-Rec.)

Family LIMIDAE
Characters of the superfamily. (L.Carb.-Rec.)

Lima [*Mantellum*] (Fig. 452): obliquely oval to oval-subtriangular, not greatly inflated; lunular region strongly depressed; strong, scaly radial costae; hinge line rather short, edentulous; anterior auricle the smaller; valve margins internally scalloped; narrow anterior and posterior gape. Jur.-Rec. Cosmop. *Limaria* [*Promantellum*] (*s.s.*): differs from *Lima* in lacking the depressed lunular area, in being more inflated and having a wider posterior gape, in being rather thin-shelled, and in having finer and only slightly roughened costae. Mio.-Rec. Cosmop. *Ctenoides:* differs from *Lima* in being more symmetrical, in having divaricate radial scaly costae, in lacking the depressed lunular area, in having distinct lateral teeth, and in the valve margins internally being smooth or only very weakly scalloped. U.Jur.-Rec. Cosmop. *Acesta* (*s.s.*): sometimes attains a very large size for members of the family, relatively thin-shelled, with moderate byssal gape, the ligament pit is oblique, the anterior auricle is very small, and the ornament consists of very fine and numerous costellae which may be absent on the flank of the shell. U.Jur.-Rec. Cosmop. *Limatula:* similar to *Lima*, but almost equilateral, strongly inflated, lacking the depressed lunular area, not gaping, and the radial costae are confined to the median part of the surface. Trias.-Rec. Cosmop. *Limea* (*s.s.*): resembles *Limaria* in form and ornament, but has the valve margins internally crenulate, and has a taxodont hinge, which with its central ligament pit is very similar to that of *Limopsis* (Limopsacea, p. 184). [The essential difference between the genera lies in the monomyarian condition of *Limea* and dimyarian of *Limopsis*; the form of the latter is also more rounded.] Mio.-Rec. Eur. [Jurassic forms previously placed in this genus are now referred to *Pseudolimea*.]

Suborder OSTREINA

Inequivalve; cemented by the left valve which is the larger; monomyarian; shell built up of lamellae of porcellanous calcite actually forming the outer surface, while prismatic calcite may line and even fill the spaces between the lamellae; hinge line short, edentulous; large triangular amphidetic ligament (mainly acting as a resilium) the insertion of which consists of a central triangular pit for fibrous ligament flanked by narrow triangular areas for lamellar ligament; ornament commarginal-lamellar, with or without radial ornament of every degree of fineness or coarseness. (U.Trias.-Rec.)

Like other cemented forms, oysters vary greatly in shape in adjustment to their surroundings, while the external ornament is obscured in the adolescent part of the shell which usually takes the imprint of the surface to which it is attached. The characters of the hinge and adductor scar are more constant, but even these may vary with environment within the limits of a species. The methods of discriminating genera are therefore not very satisfactory. Many supraspecific names have been proposed for species in the family Ostreidae (e.g. see Vialov, **72a**), and it seems very probable that a fair proportion will

---

Figs 452–458. BIVALVIA: LIMACEA AND OSTREACEA
Fig. 452 after Cox and Hertlein, in Moore; Figs 453–458 original.
452c. *Lima lima* (Linné), Rec. Philippines. T. × 0·5.
453b. *Ostrea edulis* Linné, L.Plio. Suffolk. T. × 0·6.
454b. *Ostrea latimarginata* Vredenburg, M.Mio (Gaj). Kachh. × 1·1.
455b. *Ostrea sculpturata* Conrad, U.Mio. (Yorktown). Maryland. × 0·75.
456b. *Crassostrea gryphoides* (Schlotheim), U.Mio. Touraine. × 0.5.
457b. *Crassostrea saccellus* (Dujardin), U.Mio. Touraine. × 1·1.
458b. *Crassostrea prismatica* (Gray), Neogene (Alexandria). Namaqualand, S.Africa. × 0·38.

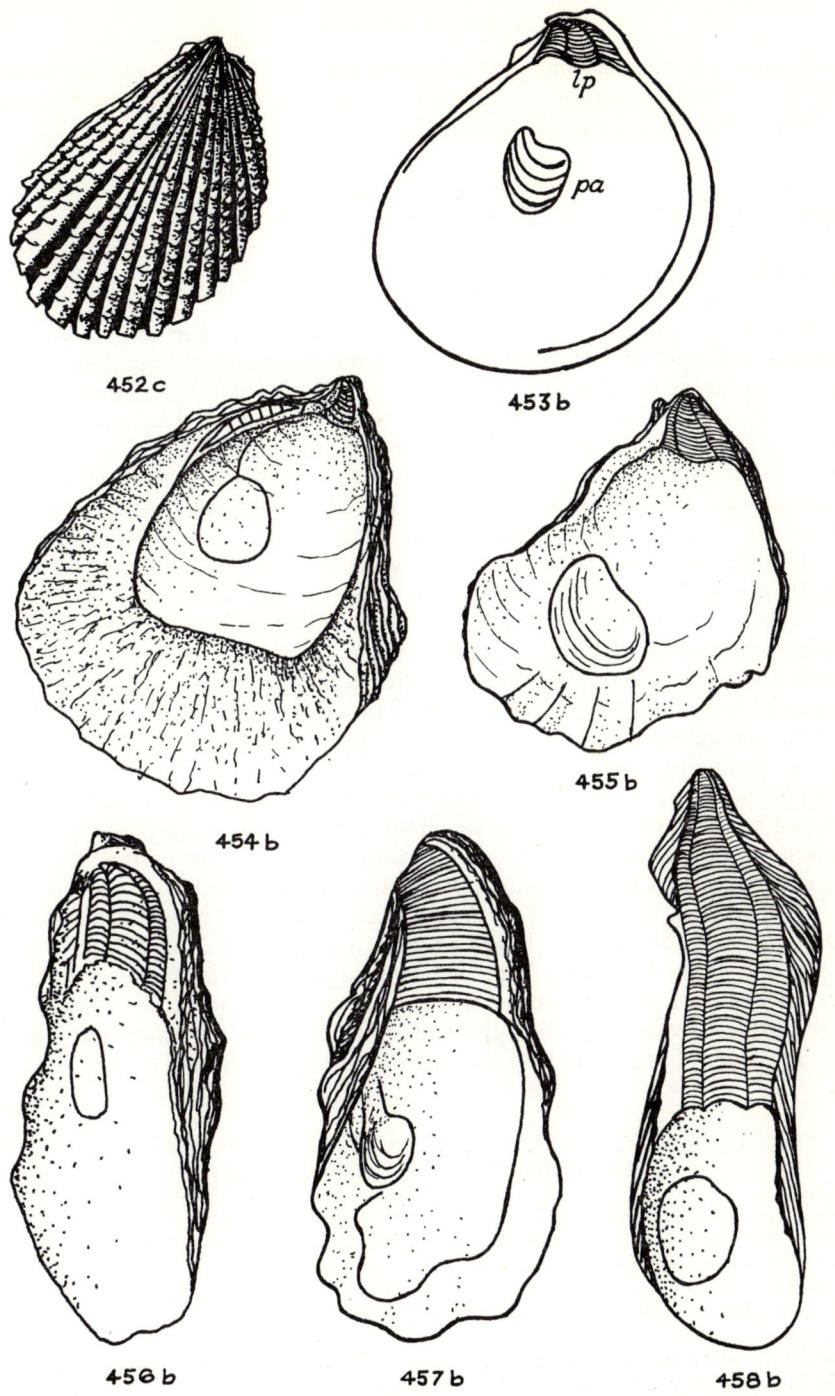

452 c

453 b

lp

pa

454 b

455 b

456 b

457 b

458 b

eventually fall into the synonymy of earlier names. Only a few of the more important Tertiary genera are referred to in dealing with the family Ostreidae below. Habitat marine and estuarine.

Superfamily OSTREACEA
Characters of the suborder. (U.Trias.-Rec.)

Family OSTREIDAE
Characters of the suborder. (U.Trias.-Rec.)

Ostrea (*s.s.*) (Figs 453–455): strongly inequivalve, the left valve being larger and deeper and with rather coarse folds extending to the margin; right valve smaller, nearly flat and without folds; inequilateral; ligament pit deep in left valve, shallow, flat or even slightly convex in right valve; adductor scar extending slightly in front of mid-line and about half-way between hinge and ventral margin; lateral margins near beak crenulated. Cret.-Rec. Cosmop.

Crassostrea (Figs 456–458): very thick-valved and greatly lengthened in the dorso-ventral direction; both valves lamellar, the deeper left valve with very feeble radial ornament, the flat right valve smooth; ligament area greatly lengthened in the same direction as the whole shell; ligament pit very deep in left valve, convex and protuberant in right valve; adductor scar semilunar, variable in position; probably a polyphyletic genus, not all its species being derived from the same ancestral stock; attains a large size. Cret.-Rec. Cosmop.

Lopha [*Alectryonia*] (*s.s.*) (Figs 459, 460): much less inequivalve as a rule than *Ostrea*, the valves differing more in shape than in size; coarse folds on both valves, producing strongly zigzag interlocking margins; these folds radiate from the umbo and may increase in number by irregular bifurcation; where crossed by the chief growth lines they rise into scaly projections which may be prolonged into tubular outgrowths, those of the left valve helping to attach the shell; ligament pit and muscle scar with much the same range of variation as in *Ostrea*. In species which fix themselves to mangrove roots or the stems of *Gorgonia*, the shell becomes long and narrow and the folds, instead of radiating from the umbo, diverge from a central ridge on the right valve, and in the left valve from a central groove which fits the supporting root or stem; the folds of the left valve may then give rise to projections which clasp the support on each side, giving to the whole shell a grotesque resemblance to a cater-

---

Figs 459–465. BIVALVIA: OSTREACEA, UNIONACEA, TRIGONIACEA AND LUCINACEA
Original.
459b. *Lopha clot-beyi* (Bellardi), U.Eoc. Egypt. × 1·1.
460b. *Lopha haitensis* (G. B. Sowerby), L.Mio. (Thomonde). Haiti. × 0·56.
461. *Pycnodonte vesicularis* (Lamarck), U.Cret. (*mucronata* zone). Norwich. T. × 0·85.
    co, corrugations.
462. *Unio pictorum* (Linné), Rec. Britain. T. × 0·56.
463. *Potomida littoralis* (Linné), Rec. × 0·56.
464d. *Eotrigonia subundulata* (Jenkins), M.Mio. (Balc.). Muddy Creek, Victoria. T. × 1·1.
465a, b, d, f. *Linga columbella* (Lamarck), L.Mio. Vienna B. a, b × 1·5; d, f × 1·1.

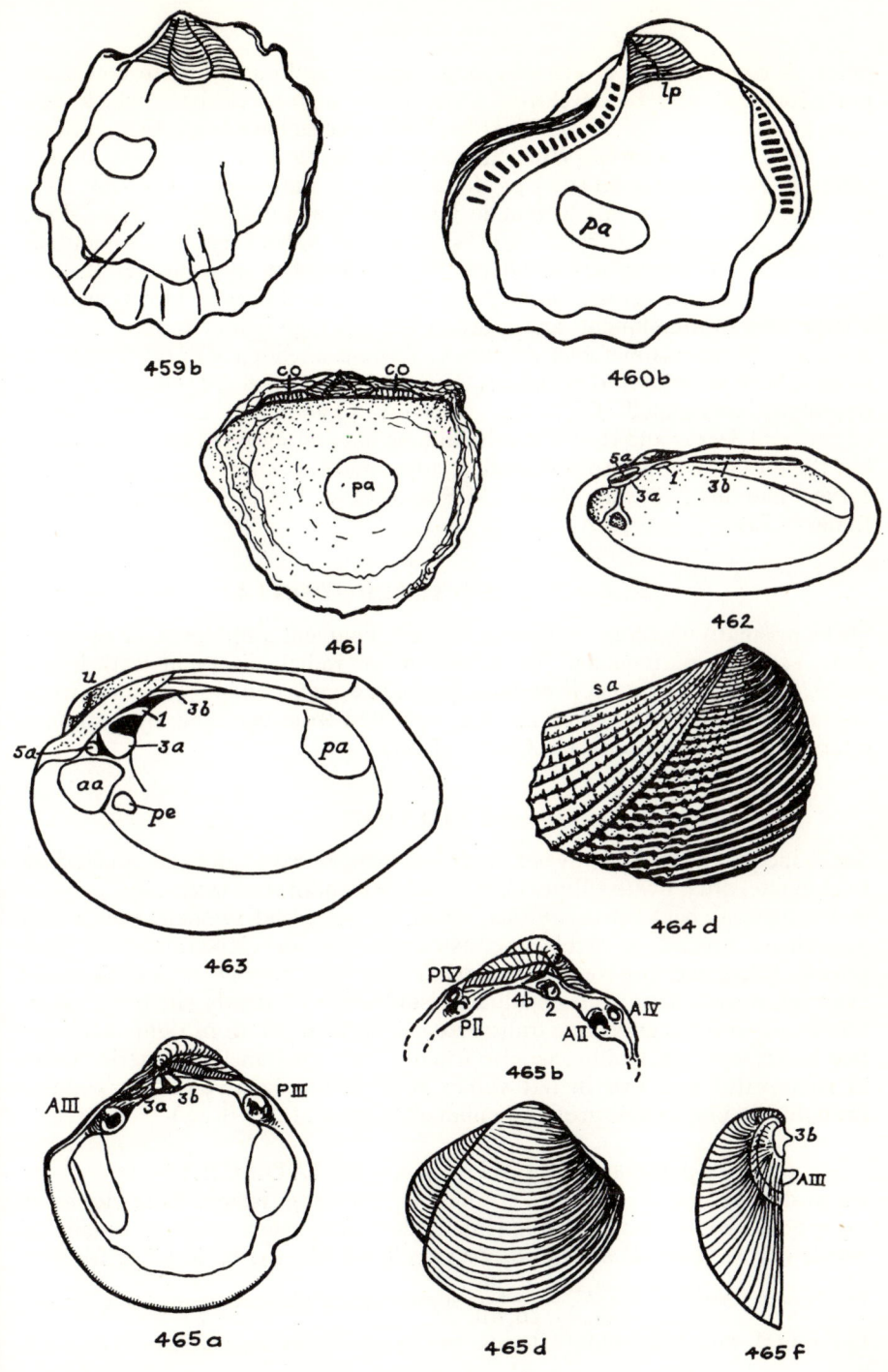

459b

460b
lp
pa

461
co *co*
pa

462
sa
3a *1* *3b*

463
u
5a
3b
1
3a
aa
pe
pa

464d
sa

465b
PIV
PII
4b *2*
AII
AIV

465a
AIII
3a *3b*
PIII

465d

465f
3b
AIII

pillar. A similar habit of growth, long, narrow, and curved (with concavity posterior) may also be developed without any stem to clasp, in which case there is a central ridge on both valves. Such species have been distinguished as a subgenus **Arctostrea**. When the shell is narrow and enlongated dorso-ventrally in either *Lopha* (*s.s.*) or *Arctostrea*, the muscle scar does not shift proportionately but tends to remain relatively near the hinge. U.Trias.-Rec. Cosmop. **Pycnodonte** [*Pycnodonta*] (Fig. 461): left valve very deep, sometimes approaching a hemispherical shape; right valve flat or concave. It thus resembles in shape some Jurassic 'Gryphaeas', but differs in having usually a large area of attachment, and a broad but not high ligament area. Surface smooth; left valve sometimes very thick. There is a tendency to the formation of ears, both anterior and posterior, the margins of which are often finely and irregularly corrugated. Cret.-Rec. Cosmop. **Cubitostrea**: the left valve is costate and deeper and the right valve almost flat and smooth like *Ostrea* (*s.s.*), but the costae diverge from a median ridge, and the valves are very strongly curved and bent like an elbow. M.Eoc-M.Olig. Eur., N.Amer., S.Amer., India, N.Z.

Subclass PALAEOHETERODONTA

Shells prismato-nacreous, equivalve, closed; ligament amphidetic or opistho-detic, external, parivincular; hinge with a few radial, divergent teeth below beaks, sometimes striated, in some Unionoida even becoming taxodont; when lateral teeth are present they are not separated from cardinals by a smooth interval as in Heterodonta. (M.Cambr.-Rec.)

Order UNIONOIDA

Shells inequilateral, equivalve or nearly so, compressed to globose, subcircular to elongate, often ovate, elliptical, trigonal or trapeziform, internally nacreous in Unionacea; beaks prosogyrous, small, or large and swollen; lunule and escutcheon absent or feeble; periostracum well developed; surface with growth lines, and (in some Unionacea) also with ribs or nodes; ligament external, opisthodetic; integripalliate; isomyarian or nearly so; hinge eden-tulous, or with (usually) one transverse subumbonal tooth in right valve and two in left valve, in Unionacea also usually with one lamellar posterior tooth in right valve and two in left valve; Anthracosiacea atypically taxodont; probably a polyphyletic group of eulamellibranchs. (Dev.-Rec.)

Superfamily UNIONACEA [NAIADACEA]

Equivalve (except fixed forms), strongly inequilateral, isomyarian (with same exception); beaks usually ornamented; dentition, when present, usually of rather rugose cardinal and posterior lamellar teeth; the basic hinge formula has been referred to as $\frac{5a, 3a, 1, 3b}{4a, 2a, 2b, 4b}$, some or all of the teeth being lamellar and horizontal, thus resembling laterals; but in reality the separation of cardinals and laterals has not taken place and all may be considered cardinals. (The lamellar teeth are often called 'pseudocardinals', an unnecessary qualification: 'pseudolaterals' would be rather better.) (?Perm., Trias.-Rec.)

The fresh-water mussels date at least from the Triassic. There are, according to Simpson's conservative estimate, 1,200 Recent species, half of them living in North America; the number has, of course, now increased. The variety of form and hinge details among these is probably greater than in any other superfamily.

Classification is difficult. Two relatively small and fairly well defined families are found mainly in the Southern hemisphere (Mutelidae, Etheriidae); the remainder are distributed through the Northern hemisphere mainly, and their separation into families is rather less satisfactory. Zoologists have classified them largely by gill structure, unknown in fossils, the only shell character to which they attach value being the ornament of the adolescent shell (beak sculpture). Owing to the tendency to erosion of the beaks of fresh-water mussels, this sculpture is easily destroyed, but it is occasionally well preserved in fossils.

In the following outline the classification followed is that of Thiele (37), Newell (67a), Vokes (72b) and Haas, in Moore (66a).

Family UNIONIDAE
Shape more or less elliptical to ovate, height one-half to two-thirds length, siphonal area usually not well defined; equivalve; ornament usually commarginal, sometimes coarse, rather irregular; traces of 'double-loop' sculpture on beaks, if unworn; umbonal cavity deep; anterior muscle scars, adductor and pedal scars deep, especially the former, the posterior shallow or even faint. (Trias.-Rec.)

Unio (*s.s.*) (Fig. 462): elliptical, beak distance one-quarter; teeth all fairly horizontal and lamellar, recorded as $\dfrac{5a,\ 3a,\ 1,\ 3b}{4a,\ 2a,\ 2b,\ 4b}$, 1 being almost vestigial, anterior cardinals short and sloping forward (2a, 4a in echelon; 3a, 5a side by side), the posterior very long and horizontal (2a, 3a, 4a corrugated and denticulate). Trias.-Rec. Eur., Asia, Africa. *Potomida* [*Psilunio, Migranaia*] (*s.s.*) (Fig. 463); differs greatly from *Unio* in the nature of the hinge; the changes in size and shape of the teeth seem correlated with greater thickness of shell and with forward curvature of the beak; the anterior teeth have swung round through 90°, as compared with *Unio*, so as to point obliquely backwards, at the same time losing their lamellar character and becoming thick and somewhat pyramidal, while the posterior teeth have become wedge-like rather than lamellar by thickening at the base; rounded, higher posteriorly; beaks a little anterior to mid-line, with numerous wavy ridges. Olig.-Rec. Eur., ?E.Asia. *Anodonta* (*s.s.*): elliptical with submedian beaks; moderate posterior ridge and wing; all hinge teeth are lost, the valves being connected by ligament only. [This degeneration is one to which other families are equally liable, and discrimination between toothless forms is difficult.] U.Cret.-Rec. Eur., N.Amer. *Lampsilis* (*s.s.*): thinner-valved than the Unioninae, with umbo more nearly median, and height about two-thirds length; surface nearly smooth, except siphonal area which has rough radial folds; this area is bent at an angle to the general surface, concave and obliquely truncated, and continued above level of hinge as a nymph, giving the posterior end an 'alate' appearance (like the 'ear' of a dysodont); teeth recorded as in *Unio*, but 1 is wanting, 3b shorter and increasing in height to posterior end where it is abruptly truncated, 2b and 4b also short. ?Trias., Olig.-Rec. N.Amer., C.Amer. **Lamellidens**: transversely elliptical, more pointed behind; beaks with ridges, rest of surface smooth; all teeth long. Eoc.-Rec. Indonesia, Burma. **Parreysia** (*s.s.*): rounded-rhomboid, with convex ventral margin, valves thick and inflated; beaks and flanks with chevron-shaped ridges; cardinal teeth heavy, striated; lamellar teeth short. Mio.-Rec. S.Asia. **Amblema** (*s.s.*): valves thick, transversely oval, with oblique radial costae which curve upward posteriorly; hinge heavy. L.Cret.-Rec. N.Amer., C.Amer. **Prisodon** [*Hyria*]: pteriiform, with growth lines and very fine radial ornament; hinge narrower medially; left valve with two thin cardinals which may split into denticles and two granular lamellar teeth; right valve with two or more similar cardinals and one lamellar tooth. Mio.-Rec. S.Amer.

Family MARGARITIFERIDAE

More or less elongate, anteriorly rounded, equivalve, compressed; beaks low, usually with feeble commarginal ornament only; left valve usually with three teeth, right valve usually with two teeth. (U.Cret.-Rec.)

Margaritifera [*Margaritana*] (*s.s.*): valves thick, oblong, rounded in front, somewhat pointed behind, gently concave ventrally; beak distance one-fifth; growth line sculpture only; siphonal area not clearly defined but slightly concave; teeth $\dfrac{\text{3a, 3b}}{\text{4a, 2a, 2b}}$, 3a and 4a acutely pyramidal, 2a jagged, 2b and 3b long but low and ill-defined; no posterior lamellar teeth. U.Cret.-Rec. Holarctic circum-polar.

Family MUTELIDAE

Shell resembles those of Unionidae, but, except when irregularly taxodont, often lacks hinge teeth; umbones smooth or with weak radial ornament (cf. Unionidae). (?Trias., Cret.-Rec.)

Mutela (*s.s.*): elongate-oblong (height one-third length), obliquely truncated posteriorly, with low, smooth beaks and low, smooth posterior ridge; hinge teeth faint or lacking. Cret.-Rec. Africa. **Iridina** is Recent only. Its subgenus **Pleiodon** is oval-subquadrate and has taxodont teeth. ?Cret., Pleist./U.Plio-Rec. ?Brazil, Africa.

Family ETHERIIDAE

Free in youth, attached and irregular in adult, nacreous, lamellar, edentulous. (Plio.-Rec.)

Etheria: inequivalve; ligament external, twisted, partly sunk in groove in left valve. Plio.-Rec. Africa, Madagascar.

The Unionacea as a whole show parallel radiations of structure. Thus there are imitations among them of the taxodont and heterodont types of hinge, and of the burrowing and fixed habits of life with their effects on the shell. *Solen*-like (p. 234) fresh-water mussels are known from E.Asia, Africa and S.America; oyster-like fixed forms from the two latter continents. These great variations have suggested to various zoologists from time to time that the group is polyphyletic, but its great antiquity and its retention of such primitive features as the nacreous interior do not support such a view.

Order TRIGONIOIDA

Nacreous, equivalve, trigonal, posteriorly more or less truncate and carinate, smooth or ornamented, mainly marine (one small group fresh-water); ligament external, opisthodetic; right valve usually with two teeth, left valve with three (median one stronger); no lateral teeth; anterior adductor scar usually, and posterior adductor scar sometimes, buttressed; usually integripalliate; filibranchs. (?M.Ord., Dev.-Rec.)

Superfamily TRIGONIACEA
Characters of the Order. (?M.Ord., Dev.-Rec.)

Family TRIGONIIDAE
Compared with the more primitive family Myophoriidae, the median tooth of the left valve is prominent and extremely broad, with a median groove, a strongly concave lower surface, and conspicuous transverse ridges on either side; the two main teeth of the right valve are widely divergent, not resting on a hinge plate; the usual interpretation of the dentition is $\frac{5a, 3a, 3b}{4a, 2b, 4b}$, but Odhner has suggested that it is $\frac{3a, 1, 3b+PI}{2a, 2b, PII}$; classification has to be based on the external characters of the shell, the general outline, the characters of the umbones, the prominence of the marginal carina, the width of the area, modifications in the convexity of the flank, and the ornament of both the flank and the area; escutcheon usually present. (M.Trias.-Rec.)

This family, so widespread and represented by so many genera in the Mesozoic, is represented by only two genera in Tertiary and Recent faunas.

Eotrigonia (Fig. 464): fairly small, oblong, moderately inequilateral; umbo slightly projecting; marginal and escutcheon carinae obtuse, the former marked by a rounded costa; area rather wide, with several delicately serrate radial costae; flank ornamented with slightly oblique, narrow costae sloping towards the marginal carina and anteriorly with a few serrate radial costellae; morphologically this group is very close to *Trigonia* (*s.s.*), but the shell is thinner, so that internal crenulations of the posterior margin correspond to the radial costae of the area. Pal.-U.Mio. Austral. (Type species: *Trigonia subundulata* Jenkins). *Neotrigonia*: trigonally ovate, very slightly inequilateral; umbones broadly rounded, depressed; flank, area and escutcheon (not

209

impressed) ornamented with tuberculate radial costae, two of which coincide with the obtuse angles which represent the marginal and escutcheon carinae. U.Mio.-Rec. Austral. (Type species: *Trigonia pectinata* Lamarck.)

Subclass HETERODONTA

Hinge teeth heterodont; complex crossed-lamellar or prismatic eulamellibranchs; hinge plate usually well built, carrying little-alternating and interlocking cardinal teeth, of which the right valve median cardinal is sometimes opisthocline, often with anterior and posterior laterals; ligament opisthodetic, usually external, but frequently more or less sunken and showing division into ligament proper and resilium; equivalve or inequivalve; isomyarian, integripalliate or sinupalliate; internally porcellanous; ornamentation varied. (M.Ord.-Rec.)

The Heterodonta may be classified according to the plan of the hinge teeth, as worked out by F. Bernard and Munier-Chalmas. Two types were recognized by them, according to whether the pivotal tooth is in the left valve (2, Lucinoid type) or the right (1, Corbiculoid type). The careful researches of H. Douvillé on Mesozoic heterodonts have shown, however, that certain Lucinoid hinges may take on a form which, in the absence of the evidence of ancestral history, may be taken, and has been taken, for corbiculoid; these groups are the Crassatellacea and the Carditacea.

Order VENEROIDA

Usually equivalve and isomyarian, active or nestling, rarely burrowing or sessile; cardinal teeth with or without anterior and posterior laterals, the latter behind the ligament. (M.Ord.-Rec.)

Suborder LUCININA

Includes the Superfamilies Lucinacea, Leptonacea, Gaimardiacea, Cyamiacea and Carditacea. (M.Ord.-Rec.)

Superfamily LUCINACEA

Outline mostly rounded to subtrigonal, with prosogyrate to straight beaks; hinge of lucinoid type, teeth sometimes all obsolete; equivalve; lunule usually larger on left valve, escutcheon weak; integripalliate; sculpture commarginal, sometimes combined with radial; often with anterior and posterior areas; in the families Lucinidae, Cyrenoididae, Thyasiridae and Ungulinidae the anterior adductor scar is long and narrow. (Silur.-Rec.)

Family LUCINIDAE

Orbicular to ovate or subtrigonal, varying from compressed to inflated; beaks fine, prosogyrate; lunule and escutcheon variable; ligament marginal or subinternal, but no resilium; siphonal area frequently marked off by a fold; sculpture as for superfamily; hinge varying from complete lucinoid type with full number of laterals, through various stages of reduction to an edentulous condition; valve margins internally sometimes finely crenulate; ventral extension of anterior adductor scar separate from pallial line. (Silur.-Rec.)

Lucina [*Phacoides*, *Dentilucina*] (*s.s.*) (Fig. 467): a little more transverse and less inflated than *Linga*, umbones less prominent, lunule larger and lanceolate, and valve margins internally more finely crenulate; two distinct areas, the posterior the larger; the spaced commarginal ornament is more lamellar posteriorly; hinge of right valve with AIII, 3a (obsolete), 3b (oblique) and PIII; hinge of left valve with AIV, AII, 2, 4b (oblique), PII and PIV. U.Cret.-Rec. Eur., Africa, Asia. *Linga* (*s.s.*) (Fig. 465): heavy, subcircular, globose; lunule small, cordiform, delimited by a groove; posterior area marked off by a strong groove; ligament external, but sunk; sculpture of commarginal lamellae; hinge of right valve with AIII, 3a, 3b (bifid) and PIII; hinge of left valve with AIV, AII, 2 (bifid), 4b, PII and PIV; valve margins internally crenulate. ?Pal., L.Eoc.-Rec. Eur., Africa, N.Amer., C.Amer. The subgenus *Bellucina* [*Cardiolucina*] (Fig. 466) is small, thick-valved, globose and ornamented with strong commarginal lamellae which have radial intercalary threads; lunule little sunk; umbones a little behind mid-line, the anterior end being slightly produced and the posterior end truncated and with a distinct area; right valve hinge with AIII, 3a (feeble), 3b (strong) and PIII; left valve hinge with AIV (feeble), AII, triangular 2 and 4b, PII and PIV (feeble); valve margins internally strongly crenulate. M.Eoc.-Rec. Cosmop. *Callucina* (*s.s.*): rounded, a little truncated posteriorly, with serrate commarginal ornament and obscure radial threads; areas feebly marked; valve margins internally finely crenulate; hinge weak, right valve with AIII, 3b (bifid) and PIII (feeble), left valve with trace of AIV, AII, 2, 4b and PII (feeble). Aptian-Rec. Eur., Africa, N.Amer., Asia. *Ctena* [*Jagonia*] (*s.s.*) (Fig. 468): small, suborbicular, anterior end somewhat produced, the beaks being pointed and behind the mid-line; no areas; valves thin but inflated; lunule small, deep; narrow radial costae and narrow commarginal beads; valve margins internally crenulate; right valve hinge with AIII, 3a, 3b (somewhat bifid) and PIII; left valve hinge with AIV, AII, 2, 4b, PII and PIV; the posterior laterals are stronger and the anterior and posterior laterals more equidistant than in *Codakia*. M.Eoc.-Rec. Cosmop. *Loripes* (*s.s.*) (Fig. 469): small, suborbicular, inflated, areas obsolete; feeble commarginal ornament; no escutcheon; valve margins internally smooth; ligament deeply sunk, long; right valve hinge with AIII (feeble), 3b (large, triangular) and PIII (weak); left valve hinge with AII (feeble), 2 (small), 4b and PII (very weak). L.Mio.(Aquit.)-Rec. Eur., Africa, S.Amer., Japan. *Codakia* (*s.s.*) (Fig. 470): fairly large, orbicular, flattened, without areas; regular reticulate ornament, the radial element as strong as, or predominant over, the commarginal; lunule short, deep; ligament deeply sunk, wide, not absolutely internal; valve margins internally crenulate; right valve hinge with AIII (prominent), 3a (small but distinct), 3b (obtusely bifid) and PIII (small); left valve hinge with AIV and AII (well developed), 2 (narrowly and obtusely bifid), 4b (thin) and small PII and PIV; valves rather thick; anterior adductor scar rather short; warm-water habitat. Pal.-Rec. Cosmop. *Miltha* (*s.s.*) (Fig. 471): more or less large, valves thick, compressed, usually wider than high, areas poorly defined; beaks small, pointed, a little anterior to mid-line; very fine, serrate commarginal striae; lunule small, depressed, asymmetrical; ligament area wide, sunk but not internal; no lateral teeth; right valve hinge with 3a (small) and 3b (narrowly triangular, more or less bifid); left valve hinge with 2 (strong, more or less bifid) and 4b (thinner); valve margins internally smooth. U.Cret.-Rec. Cosmop. *Megaxinus* (*s.s.*) (Fig. 472): suborbicular, slightly higher than wide, beaks a little anterior to mid-line; valves thick, inflated, ornamented with commarginal lamellae; lunule small, deep; nymph flattened, long, wide; edentulous; valve margins internally smooth; anterior adductor scar long, not parallel to the pallial line; areas vague, the posterior the better. Olig.-Rec. Eur., Carib., N.Amer., Red Sea. *Lucinoma* (Fig. 473): regularly rounded except for a fairly straight postero-dorsal margin; regular commarginal lamellae; no areas; right valve hinge with AI (very feeble), 3a and 3b (narrow, bifid); left valve hinge with AII (small), 2 (narrowly bifid) and 4b; lunule narrow, long, not sunk; ligament area long, narrow, sunk, straight; valve margins internally smooth. Olig.-Rec. Eur., Red Sea, N.Amer., C.Amer., N.Pac., N.Atl., W.Atl., Medit., Austral., N.Z., Japan. *Myrtea* [*Myrtaea*] (*s.s.*) (Fig. 474): ovate-triangular, subequilateral, somewhat inflated, with small, pointed beaks and obsolete areas; ornament of commarginal lamellae which become spinose on the postero-dorsal margin; lunule and escutcheon long and narrow, the lunule being larger on the left valve than on the right; ligament long, narrow, external; right valve hinge with AIII, 3b (simple) and PIII; left valve hinge with AII (weak), 2 (small), 4b and PII (weak); valve margins internally smooth. Mio.-Rec. Eur., Carib., N.Amer. (Chickasawhay), tropical and temperate seas. *Gibbolucina* (*s.s.*) (Fig. 475): small, subtrigonal, inflated, with well developed areas and irregular growth lines; beaks pointed; lunule long, wide, deeply sunk; escutcheon fairly distinct; nymph flat, wide, short; no lateral teeth; right valve hinge with 3a (small or lacking) and 3b (bifid); left valve hinge with 2 and 4b; valves internally callous and with fine radial furrows; valve margins internally smooth. L.Eoc.-Rec. Eur., Africa,

---

Figs 466–479. BIVALVIA: LUCINACEA

Figs 466, 469–476 and 478–479 after Chavan, in Moore; Figs 467–468 and 477 original.
466c. *Linga* (*Bellucina*) *semperiana* (Issel), Rec. T. × 1·4.
467d. *Lucina michelottii* Mayer, Burdigalian. Aquitaine. × 3.
468c. *Ctena squamosa* (Lamarck), Pleist. Greece. × 5.
469b. *Loripes lucinalis* (Lamarck), Rec. England. T. × 2·4.
470c. *Codakia orbicularis* (Linné), Rec. W.Indies. T. × 1.
471b. *Miltha childreni* (Gray), Rec. Brazil. T. × 1.
472b. *Megaxinus rostratus* (Pecchioli), Plio. Italy. T. × 1·6.
473b. *Lucinoma borealis* (Linné), Rec. Eur. × 1.
474d. *Myrtea spinifera* (Montagu), Rec. Medit. T. × 0·7.
475a, d. *Gibbolucina callosa* (Lamarck), M.Eoc. France. T. × 1.
476d. *Gibbolucina* (*Eomiltha*) *contorta* (Defrance), Pal. T. × 0·7.
477. *Pseudomiltha mutabilis* (Lamarck), M.Eoc. Paris B. × 0·56.
478. *Anodontia edentula* (Linné), Rec. Indian Ocean. T. × 1.
479d. *Divaricella angulifera* von Martens, Rec. Mauritius. T. × 1.

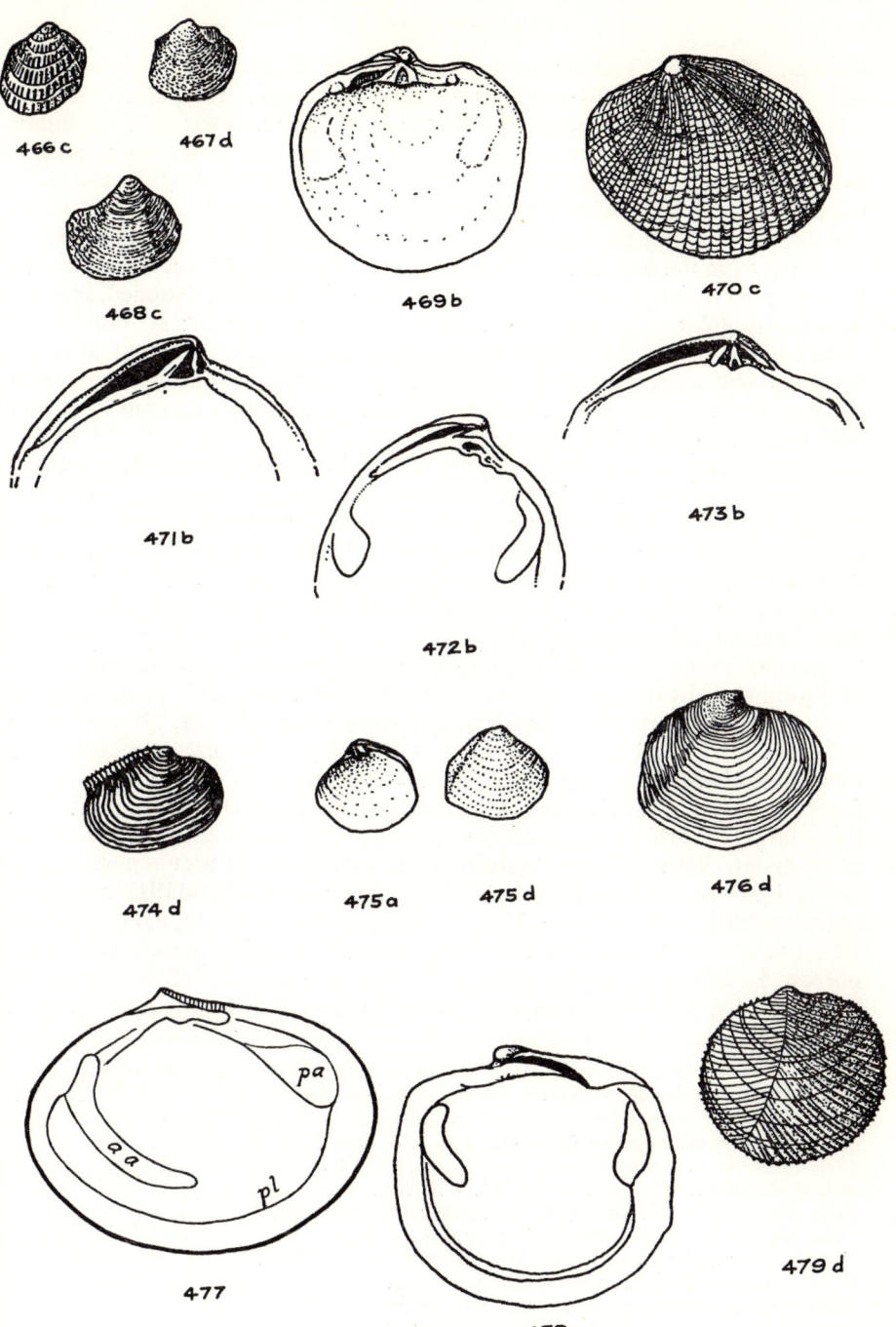

466 c   467 d

468 c

469 b

470 c

471 b

472 b

473 b

474 d

475 a   475 d

476 d

477

pa

a a

pl

478

479 d

213

Austral., ?Red Sea. The subgenus *Eomiltha* (Fig. 476) is larger, higher than wide, compressed, subquadrate; areas mere flexuosities, the postero-ventral end being somewhat produced; ornament of fine commarginal lamellae; lunule elongate, fairly narrow, not sunk; ligament sunk but not internal; nymph not long; no lateral teeth; hinge of right valve with 3a (thin and close to edge) and 3b (very widely bifid); hinge of left valve with 2 (slightly bifid) and 4b; valve margins internally smooth; anterior adductor scar very long and diverging from the pallial line. Cret.-Rec. Eur., Amer., E.Africa. *Pseudomiltha* (*s.s.*) (Fig. 477): large, suborbicular to transversely elliptical, flattened, smooth or with irregular commarginal striae; beaks small, straight; no lunule; no areas; hinge differs from that of *Eomiltha* in being edentulous; anterior adductor scar extremely long and curved parallel to the pallial line; valve margins internally smooth; interior of valves with traces of radial lines and with two buttresses. ?Jur., Pal.-Olig. Eur., M.East, Somalia, India, W.Pakistan, Afghanistan. *Anodontia* (*s.s.*) (Fig. 478): fairly large, subcircular, globose, slightly truncated posteriorly, without definite areas, smooth except for commarginal growth lines and vague radial lines; umbones incurved, beaks prosogyrate; lunule short, wide, poorly defined; ligament deeply sunk, not on a prominent nymph; valves thin; edentulous except for a trace of 3b; valve margins internally smooth; anterior adductor scar not very elongate. L.Eoc.-Rec. Cosmop. *Divaricella* (*s.s.*) (Fig. 479): suborbicular to oval, posterior end a little less strongly rounded than anterior end; umbones rounded; ornament of prominent divaricate costellae in chevrons; lunule small, protruding on right valve; valve margins serrate; right valve hinge with 3a, 3b (somewhat bifid) and PIII (weak); left valve hinge with AII (obsolete), 2, 4b (feeble) and PII (almost invisible); anterior adductor scar flexuous, its lower end diverging from the pallial line. Plio.-Rec. Asia, Austral., E.Africa. *Divalucina* (Fig. 480): fairly large, suborbicular, solid, little inflated; umbones small, erect; ornament of divaricate flat bands with narrow separating furrows; lunule narrow, not sunk; right valve hinge with AIII, 3a, 3b (widely triangular) and PIII (obsolete); left valve hinge with AIV (weak), AII, 2 (triangular, bifid), 4b, PII (weak) and PIV (weak); valve margins internally smooth; anterior muscle scar long, narrow, largely separate from the pallial line to which it is close and almost parallel. Olig.-Rec. N.Z., Austral., Japan, Ceylon. *Divalinga* (*s.s.*) (Fig. 481):

Figs 480–490. BIVALVIA: LUCINACEA AND LEPTONACEA
Figs 481–482, 484 and 487–490 after Chavan, in Moore; Figs 480, 483 and 485–486 original.
480c. *Divalucina cumingi* (Adams and Angas), Rec. T. × 0·75.
481b. *Divalinga quadrisulcata* (d'Orbigny), Rec. Florida. T. × 2.
482b. *Lucinella divaricata* (Linné), Rec. France. T. × 4.
483. *Diplodonta senegalensis* Reeve, Rec. W.Africa. × 1·2
484b. *Ungulina cuneata* (Spengler), Rec. Senegal. T. Enlarged.
485c. *Thyasira bisecta* Conrad, L.Olig. Oregon. × 0·56.
486a, b. *Fimbria fimbriata* (Linné), Rec. Philippines. T. × 0·75.
487b. *Erycina pellucida* Lamarck, M.Eoc. France. T. × 4·8.
488b. *Kellia suborbicularis* (Montagu), Rec. Spain. T. × 3.
489b. *Lepton squamosum* (Montagu), Rec. Ireland. T. × 3.
490b. *Montacuta substriata* (Montagu), Rec. Britain. T. × 6.

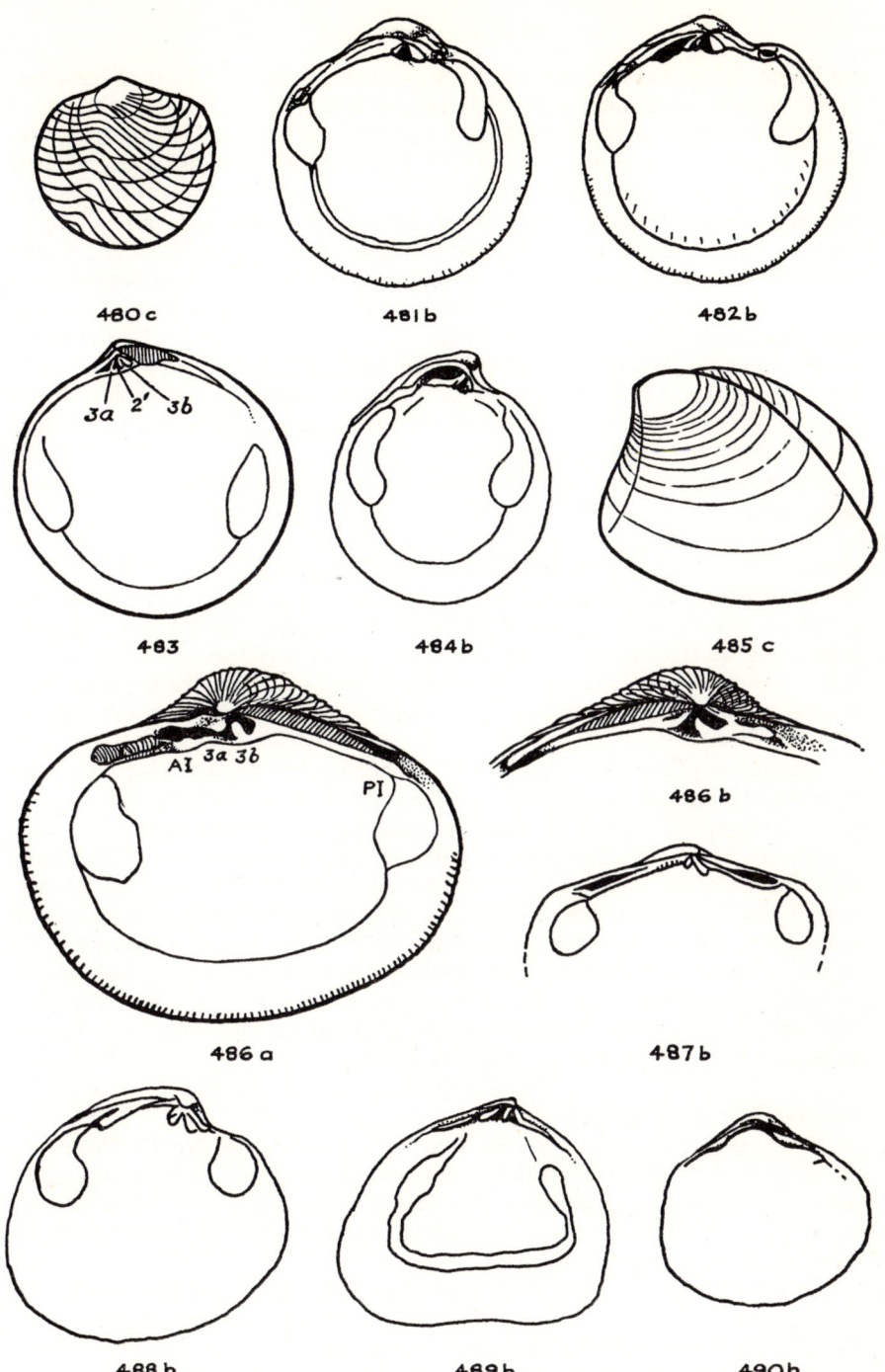

480 c

481 b

482 b

483

3a 2' 3b

484 b

485 c

486 a

AI 3a 3b

PI

486 b

487 b

488 b

489 b

490 b

orbicular, fairly convex, without definite areas; ornament as in *Divalucina*; lunule narrow, disymmetrical; right valve hinge with AIII, 3a (obsolete), 3b (triangular and bifid) and PIII; left valve hinge with AIV (small), AII (a small, prominent button), 2 (solid), 4b, PII and PIV; posterior lateral teeth distinctly farther from the cardinal teeth than the anterior ones; valve margins internally crenulate; anterior adductor scar not very long, its lower end only a little separated from the pallial line. Aquit.-Rec. Eur., N.Amer., Carib. *Lucinella* (*s.s.*) (Fig. 482): fairly small, suborbicular, with undulating ornament and fairly distinct posterior area; lunule wide, oval, little disymmetric; ligament much sunk; right valve hinge with AIII, 3a (very small), 3b (widely triangular) and PIII; left valve hinge with AIV, AII, 2, 4b (narrow and bifid), PII (obsolete) and PIV; valve margins internally crenulate; anterior adductor scar not very long, its lower end only slightly diverging from the pallial line. M.Mio.-Rec. Eur., Indian Ocean.

Other Eocene members of this family belong to less well known genera and subgenera.

Family UNGULINIDAE [DIPLODONTIDAE]

Suborbicular to subtrigonal, valves thin, equivalve, more or less smooth or punctate, without definite areas; umbones prosogyrate, rather low; ligament subinternal to semi-external; no lateral teeth; cardinals 2 and 3b nearly always bifid; adductor scars moderately elongate, not detached from pallial line, posterior scar usually larger; integripalliate; valve margins internally smooth. (U.Cret.-Rec.)

Diplodonta [*Taras*?] (*s.s.*) (Fig. 483): suborbicular, moderately inflated, inequilateral; smooth or with fine commarginal striae, with or without fine punctae; ligament somewhat sunk, but external and on a nymph; right valve hinge with 3a, 3b (bifid) and 5b (very weak); left valve hinge with 2 (bifid) and 4b; antero-dorsal margin of both valves internally with a long, fine groove; anterior adductor scars only moderately elongate, narrower than the posterior ones. Cret.-Rec. Cosmop. *Ungulina* (Fig. 484): differs from *Diplodonta* in the ligament being partly external but mostly internal, in lacking the long groove internally along the antero-dorsal margins of the valves, and in its nestling habit which leads to irregularity in shape; oval, higher than wide; muscle scars subequal. Mio., ?Rec. France. (Rare as fossils.)

Family THYASIRIDAE

Valves thin, rounded, subquadrate or triangular; equivalve, usually with a radial posterior flexure, smooth or with commarginal ornament; beaks small, prosogyrate; usually edentulous; ligament subexternal, on a groove, resilium sunk; valve margins internally smooth; pallial line often punctate; adductor scars lucinoid, not strong. (M.Trias.-Rec.)

Thyasira [*Axinus, Cryptodon*] (*s.s.*) (Fig. 485): small, outline rounded to subtrigonal, sometimes subquadrate; more or less inflated; lunule short, broad, fairly well defined; escutcheon long, narrow; posterior area sharply defined and folded so as to give a doubly sinuated posterior outline; smooth except for accentuated growth lines; hinge edentulous, but in the left valve there is a notch in the hinge plate below the umbo, and behind this the plate projects slightly (vestigial tooth?); ligament long and narrow, becoming internal by the overlapping of the edge of the escutcheon. Cret.-Rec. Eur., Pac., Austral. [Post-Palaeocene forms are mainly boreal; *T. bisecta* Conrad, Olig.-Rec., N.Pacific (Japan to California) is an important index of the varying temperature of this region, attaining a length of nearly three inches, and surviving now only in the Puget Sound.]

Family FIMBRIIDAE [CORBIDAE]

Transversely oval to suborbicular, with rounded umbones; valves thick; lunule and escutcheon distinct; radial costae, with or without overlying commarginal ornament; hinge plate heavy; usually two strong cardinal teeth and good lateral teeth in each valve; valve margins internally crenulate; ligament external; adductor muscle scars oval (the anterior ones not elongate as in the Lucinidae); teeth stronger than in the Lucinidae. (Trias.-Rec.)

Essentially a Mesozoic family, *Fimbria* is one of the only two genera surviving into Tertiary and Recent times.

Fimbria [*Corbis*] (Fig. 486): transversely oval, inflated, subequilateral; beaks small, prosogyrate; lunule small, lanceolate; ligament slightly sunk; ornament cancellate, the commarginal ridges more conspicuous and continuous than the radial and overlying them; almost integripalliate; right valve hinge with AI, 3a, 3b (grooved) and PI; left valve hinge with AII, 2, 4b and PII; anterior lateral teeth much closer to cardinal teeth than the posterior laterals; anterior adductor scar broader below. M.Jur.-Rec. Cosmop. (The prior name *Fimbria* Bohadsch, 1761 is invalidated by Opinion 125 of the I.C.Z.N.)

Superfamily LEPTONACEA [ERYCINACEA]

Usually small and with thin valves, equivalve; ligament internal, occasionally marginal; cardinal teeth usually tubercular, but laterals long; byssiferous; marine. (?Cret., Pal.-Rec.)

Possibly a polyphyletic assemblage of forms, mostly of 'nestling' habit, i.e. living in hollows in rocks or under the protection of other organisms (sponges, echinoderms, etc.). The hinge is degenerate and therefore does not clearly show their affinities; the majority seem to be degraded Lucinacea, but it is possible that some are members of other stocks that have undergone a parallel degeneration.

Family ERYCINIDAE

Generally inequilateral, the anterior side the longer, rather compressed; beaks prosogyrate; ligament internal, the resilium causing a notched hinge line, sometimes also with a marginal portion; lateral teeth present; cardinal teeth 1 and 2 usually well developed, 4b inconstant. (Pal.-Rec.)

Erycina (*s.s.*) (Fig. 487): oval-transverse, usually nearly equilateral, smooth or with feeble radial striae, valves thin; resilium oblique, wide open, ligament extended posteriorly; right valve hinge with AI, 3a (faint), 1 (trigonal) and PI; left valve hinge with AII, 2, 4b (small) and PII. Pal.-Mio. Eur.

Family KELLIIDAE

Ovate, orbicular, subtrigonal or subquadrate, inflation variable; beaks often fairly prominent, often submedian, sometimes anterior to mid-line; smooth and often glossy, occasionally punctate; external ligament small, internal resilium in a posteriorly directed, oblique groove; hinge with one or two cardinal teeth and one or two posterior lateral teeth in each valve, anterior laterals usually poorly developed. (Pal.-Rec.)

Kellia [*Kellya*] (Fig. 488): suborbicular, inflated, smooth and polished or with growth lines; valves thin; beaks small, incurved, prosogyrate; right valve hinge with AIII beyond 3a, 1 (strong, oblique), PI (distinct) and PIII (weak); left valve hinge with 2a (curved), 2b (conical, pointed) and PII (strong). Eoc.-Rec. Cosmop.

Family LEPTONIDAE

Subequilateral or with the small beaks behind the midline, usually compressed; no gape; resilium short, thick; cardinal teeth few; anterior lateral teeth well developed, not separated from anterior cardinals; posterior laterals present, lamellar; pallial line not close to ventral margin. (?Cret., Pal.-Rec.)

Lepton (*s.s.*) (Fig. 489): rounded-subquadrate, more or less equilateral; surface punctate; right valve hinge with AI (joined to 1), 3a (very oblique), 1 (strong), PI and PIII; left valve hinge with AII (joined to 2a), 2a (oblique), 4b (vertical) and PII; posterior laterals lamellar. ?Cret., Eoc.-Rec. Eur.

Family MONTACUTIDAE

Transversely ovate, suborbicular or subtriangular, more or less convex, with beaks submedian, in front of or behind mid-line; ligament internal; pseudocardinal teeth, and laterals which are usually double on right valve; shell sometimes covered by mantle. (Eoc.-Rec.)

Montacuta (Fig. 490): small, subquadrate, inflated, posterior end truncated; beaks prosogyrate, submedian; valves thin; commarginal threads and radial striae or wide-spaced costellae; two anterior and posterior laterals in right valve, one pair in left valve, with small resilium pit in between; integripalliate; valve margins internally smooth. Eoc.-Rec. Eur. *Mysella* (*s.s.*): elongate-subtriangular, with commarginal ornament, beaks anterior to mid-line; resilium without crests; lateral teeth good, and pseudocardinals 1 and 2 present. Mio.-Rec. Cosmop. The subgenus *Rochefortia* has a similar outline, but the resiliifer has crests forming a bifid tooth. Plio-Rec. Indian Ocean, Austral., N.Amer. *Tellimya:* elongate-elliptical, posteriorly somewhat truncated; beaks a little opisthogyrate, well behind mid-line; anterior laterals oblique, right valve one longer. Mio.-Rec. Eur., Amer.

Family GALEOMMATIDAE

Shell partly covered by mantle, more or less elongate transversely; valves

218

thin, gaping ventrally; small resilium and marginal ligament; more or less edentulous or with weak cardinal teeth which hook on to curved, obscure lateral teeth; integripalliate. (U.Eoc.-Rec.)

Galeomma is Recent only. *Spaniorinus* (Fig. 491): transversely suboval, a little shorter and somewhat truncated posteriorly, with submedian beaks; right valve with one tooth anterior to resilium and traces of small ones in front and behind; left valve with one oblique tooth. Eoc.-Plio. N.Amer., Eur.

Superfamily CYAMIACEA

Small, equivalve, valves fairly thick, common in southern seas; resilium close to nymph; upper hinge teeth long, hooked, laterals distinct; differs anatomically from the Leptonacea (e.g. possessing two posterior mantle openings and two gill plates). (Jur.-Rec.)

Family CYAMIIDAE [PERRIERINIDAE]

Suborbicular to subtrigonal; ligament internal; hinge with a few divergent cardinal teeth (some bifid) and distant lateral teeth (some more or less taxodont). (Mio.-Rec.)

Cyamium is Recent only. *Perrierina* (Fig. 492): subovate, posterior end longer; right valve hinge with 1, 3a–b and several oblique taxodont laterals; left valve hinge with AII (curved), 2a and 2b (adjoined), 4b and several oblique taxodont laterals. Plio.-Rec. N.Z., Antarctic.

Family SPORTELLIDAE

Suborbicular to transversely oval-subquadrate; resilium internal, ligament external and on a prominent nymph; lateral teeth not fully developed; one or two cardinal teeth in each valve (1, 2 and 4b usually developed); integripalliate; valve margins internally smooth. (Jur.-Rec.)

Sportella (Fig. 493): transversely oval, subequilateral, beaks slightly behind mid-line, smooth; resilium rudimentary; nymph long, flat; right valve hinge with AIII (short), 3a (small), 1 (strong, triangular) and 3b (well developed); left valve hinge with AII (short), 2a above 2b and 4b (small, divergent). Pal.-Mio. Eur. *Hindsiella:* transversely oval, beaks anterior to mid-line, with strong growth lines; right valve hinge with AIII (long) and 1 (rounded); left valve hinge with AII (long) and 2 (curved); ventral margin deeply incised corresponding to a radial fold from the umbones. Pal.-U.Eoc. Eur.

Family TURTONIIDAE

Oval-subtriangular, with beaks well anterior to mid-line; hinge plate narrow; right valve hinge with AIII and AI as extensions of 1 and AIII–3a by tubercular 3b; left valve hinge with AIV (thin), AII close to 2 (grooved) and 4b (minute); posterior laterals distant, very small. (Mio.-Rec.)

Turtonia: characters of the family (previously placed in the Veneridae). Mio.-Rec. Eur., Greenland, Alaska, Japan.

219

Family NEOLEPTONIDAE

Very small, obliquely oval to subquadrate, usually higher than wide, with beaks submedian or posterior to mid-line; ligament internal; hinge line with no or one cardinal tooth, but with long, usually curved laterals on each side. (Plio.-Rec.)

Neolepton: suborbicular, a little truncated posteriorly; beaks submedian; right valve hinge with AIII, AI, PI and PIII (beyond PI); left valve hinge with AIV, AII and 2 (at an angle) and PII. Plio.-Rec. Eur.

Superfamily CARDITACEA

Shell cordiform, trigonal, trapezoidal, occasionally mytiliform, usually with radial ornament and valve margins internally crenulate; lunule small; escutcheon weak; hinge of lucinoid type, the posterior cardinals elongate on account of the prosogyrate beaks; lateral teeth small if present; ligament usually external, sometimes internal; integripalliate; byssiferous; marine. (?Ord., Dev.-Rec.)

Family CARDITIDAE

Equivalve, valves thick; shell subtrigonal, trapeziform, ovate or mytiliform; usually with radial costae; valve margins internally crenulate; ligament external; hinge of *Astarte* type, modified by the strong forward throw of the umbo; 2 short, vertical or with a slight forward slope; 3b long and curving forward; 4b very long and sloping, underlying the long ligament almost horizontally; 3a very small or obsolete; lateral teeth in various stages of obsolescence; pedal scars as in the Crassatellacea. (Dev.-Rec.)

Cardita [*Arcinella auctt.*, *Mytilicardita*] (*s.s.*): transversely oval-sub-quadrate-modioliform, with concave ventral margin; radial costae beaded; right valve hinge with elongate 3b; left valve hinge with divergent cardinals; anterior laterals weak. Pal.-Rec. Eur., Africa, Asia, Austral. The subgenus *Jesonia* [*Mytilocardia*] is elongate, modioliform, with the beaks very near the anterior end; scaly costae; lunule sunk; right valve hinge with 3a (thin) and 3b

---

Figs 491–502. BIVALVIA: LEPTONACEA, CYAMIACEA, CARDITACEA AND CRASSATELLACEA

Figs 491–493, 495–496 and 499 after Chavan, in Moore; Figs 494, 497–498 and 500–502 original.

491. *Spaniorinus cossmanni* (Dall), Tert. U.S.A. T. × 3.
492. *Perrierina taxodonta* Bernard, Rec. Stewart I. T. × 10.
493. *Sportella dubia* (Deshayes), M.Eoc. France. T. × 1·5.
494. *Cardites antiquatus* (Linné), Rec. T. × 1·25.
495d. *Glans trapezia* (Linné), Rec. Medit. T. × 2·5.
496c. *Venericardia imbricata* (Gmelin), M.Eoc. Paris B. T. × 0·6.
497. *Venericardia* (*Venericor*) *planicostata* (Lamarck), M.Eoc. Hampshire. T. × 0·5.
498. *Pteromeris orbicularis* (J. de C. Sowerby), L.Plio. (Coralline Crag). Suffolk. × 1·25.
499c. *Condylocardia sanctipauli* Munier-Chalmas, Rec. St. Paul I. T. Enlarged.
500. *Crassatella ponderosa* (Gmelin), M.Eoc. Paris B. T. × 0·56.
501. *Eucrassatella turgidula* (Conrad), M.Mio. (Choptank). Maryland. × 0·56.
502. *Astarte basteroti* Lajonkaire, L.Plio. (Coralline Crag). Suffolk. × 1.

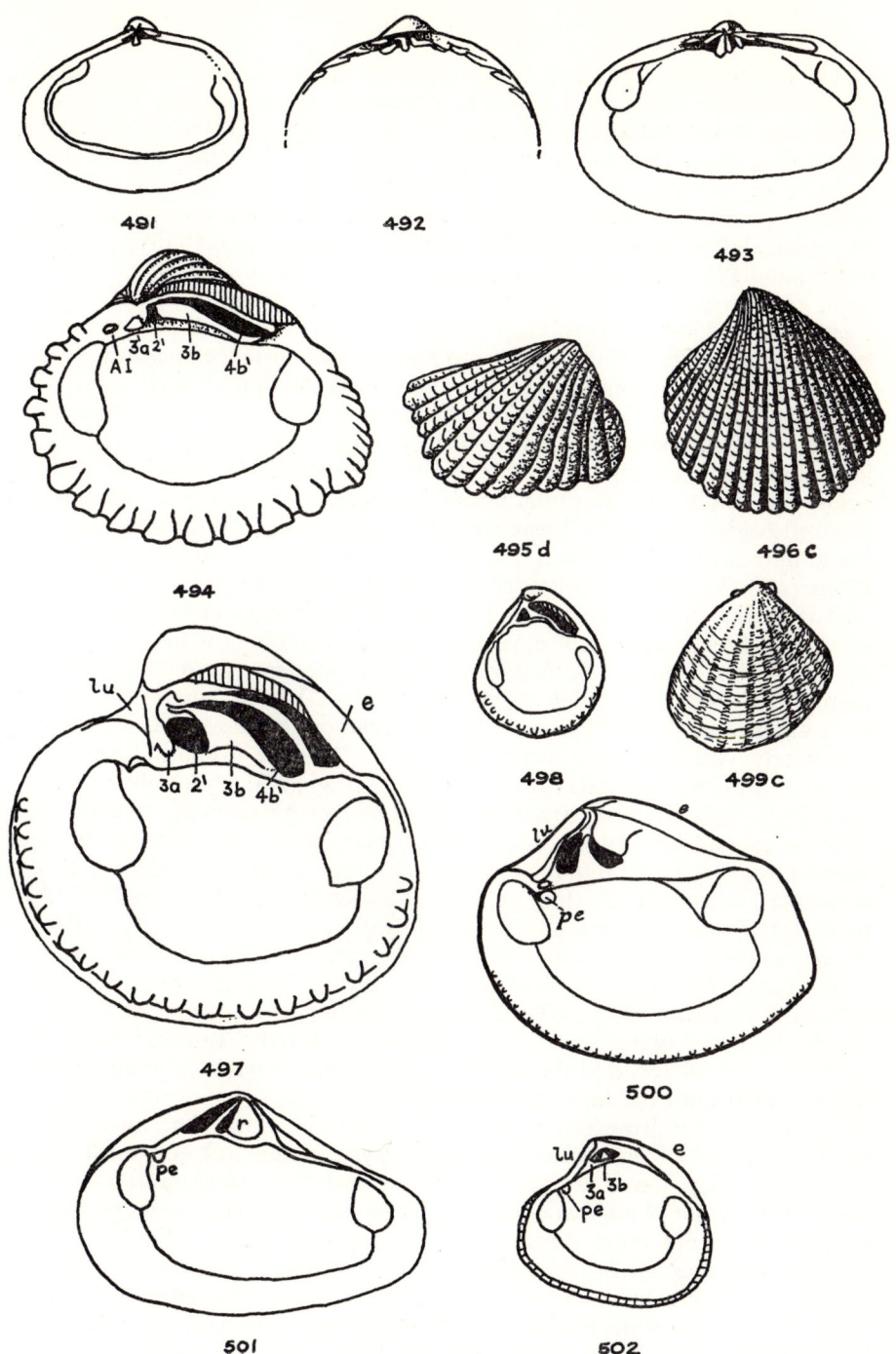

491

492

493

494

$3a\ 2'$ $3b$ $4b'$
$A\ I$

495 d

496 c

497

lu e
$3a\ 2'$ $3b$ $4b'$

498

499 c

500

lu e
pe

501

r
pe

502

lu e
$3a\ 3b$
pe

221

(thick, horizontal); left valve hinge with AII (pointed, low) 2 (triangular), 4b (horizontal, long, thin) and PII (present); upper surface of 3b and upper and lower surfaces of 4b striated; valve margins internally unequally fluted. Olig.-Rec. Cosmop. *Carditamera* (*s.s.*): also elongate, but the beak distance is one-quarter; outline transversely oblong-subquadrate; right valve hinge with AI (well developed), 3a (small), 3b (long, very oblique) and PI (small); left valve hinge with AII (well developed), 2 and 4b divergent, PII (small) and PIV (faint); costae rugose, broad and flat anteriorly, posteriorly high, keeled, more or less spinose. U.Eoc.-Rec. N.Amer., Eur., ?E.Africa. *Cardites* (Fig. 494): ovoid-subtrapezoidal, rounded in front, slightly truncated behind, moderately inflated; beaks prominent, beak distance one-fifth; lunule small, convex, often concealed by the beaks; escutcheon undefined; ornament of strong, beaded costae much wider than their interspaces in the anterior part of the shell, but becoming finer and less beaded posteriorly; valves thick, inner margins fluted; right valve hinge with AI (vestigial), 3a (small or lacking, below the lunule) and 3b (long, nearly horizontal, curved above, cuspate or grooved below): left valve hinge with AII (minute), 2 (short, pyramidal), 4b (long, narrow) and PII (minute). L.Eoc.-Rec. Eur., Asia, Austral., Indo-Pac. *Glans* (*s.s.*) (Fig. 495): posteriorly truncated, giving the valves a square or oblong outline, umbonal slope well marked off; umbones well anterior to mid-line; lateral teeth quite well developed; right valve hinge with AIII (weak), AI (small, but definite), 3a (rather small), 3b (elongate-triangular), PI (not strong) and PIII (weak); left valve hinge with AII (fairly strong), 2, 4b (at right angles to 2) and PII (weak); strong squamose costae especially medio-posteriorly; valve margins internally fluted. Pal.-Rec. Medit. Eur. Asia Africa, Carib., widespread. *Venericardia* (*s.s.*) (Fig. 496): ovate-subtriangular, cordiform, inflated, with thick valves; umbones prominent, prosogyrate, anterior to mid-line; lunule small, deep; escutcheon long, narrow; beaded radial costae; valve margins internally crenulate; right valve hinge with AI (obsolete), 3a (small), 3b (long) and PI (obsolete); left valve hinge with AII (obsolete), 2 (large, entire, somewhat elongate) and 4b (long, subparallel to 2); compared with *Cardita*, the hinge plate is straighter below and the cardinal teeth are less divergent. ?U.Cret., Pal.-Eoc. Eur., Africa, N.Amer., N.Z. The subgenus *Venericor* (Fig. 497) consists of an important group of forms in which the hinge plate is very high and carries massive cardinal teeth; the umbones are inflated, and the uniform plain costae, narrow in youth, gradually broaden until the interspaces disappear and the whole surface becomes smooth except for growth lines. Pal.-U.Eoc. Eur., N.Amer. [The variations of the type species *Venericardia* (*Venericor*) *planicosta* Lamarck and allied forms in time and space have been described by Stewart (**72**).] *Megacardita:* shows a somewhat similar change in costation, but is more elongate in shape and its cardinal teeth are more divergent (as in *Jesonia*). ?Eoc., Olig.-Rec. Eur., N.Africa, Austral., N.Z. *Pleuromeris:* small, cordiform, inflated, only slightly inequilateral, and the umbones are prominent; strong, beaded radial costae; valve margins internally crenulate; right valve hinge with AIII, AI, 3b and PI; left valve hinge with AII, 2, 4b, PII and PIV. ?Eoc., Mio.-Rec. N.Amer.,

Austral., ?Africa. **Pteromeris** (*s.s.*) (Fig. 498): small (usually less than 2 cm. in any measurement), compressed, rounded-subtrigonal in outline, with beaded radial costae; valve margins internally crenulate; right valve hinge with 3b (triangular) and PI; left valve hinge with 2 and 4b (obsolete). Aquit.-Rec. N. Amer., Eur., N.Z. (rather boreal).

Family CONDYLOCARDIIDAE

Always small, with distinct prodissoconch and internal ligament; ovate, suborbicular, subtrigonal to cordate, mostly higher than long; usually with costae (rarely smooth); cardinal teeth feeble, often with lateral teeth close to them. (Eoc.-Rec.)

Condylocardia (Fig. 499): obliquely oval-subtrigonal, with low, transversely striate costae; right valve hinge with AI (long), AIII–3a, 3b and 5b; left valve hinge with AII, 2a, 4b, 6b and PII. M.Eoc.-Rec. Carib., S.Atl., Eur., Pac., N.Z.

Suborder ASTARTEDONTINA

Comprises the Crassatellacea, Cardiacea, Chamacea, Tridacnacea, Mactracea, Solenacea and Tellinacea. (Ord.-Rec.)

Superfamily CRASSATELLACEA [ASTARTACEA]

Usually of small or medium size, solid and rounded-triangular to trapezoidal; sculpture nearly always commarginal, costate but often degenerating to striate or smooth; lunule and escutcheon distinct; beaks prosogyrate, fine and pointed; with well marked pedal muscle scars above the adductor scars; ligament external or internal; integripalliate or almost so; hinge lucinoid, but owing to the strong development of 3b and occasional presence of a small 5b assuming a corbiculoid appearance; hinge plate broad, the right valve usually with one median cardinal tooth and the left valve with two cardinal teeth enclosing its socket, as well as more or less definite indications of other cardinal teeth or lateral teeth. (Ord.-Rec.)

The crenulation of the valve margins, not being associated with external radial sculpture, is not a persistent character, being usually present but sometimes absent in the same species. This is presumably due to the periodic formation of the crenulations before a resting stage in growth (like varices in gastropods). It may be noted here that Cossmann and Peyrot used the notation 3a, 1, 3b (right valve) and 2a, 2b (left valve) for cardinal teeth, instead of the notation 3a, 3b, 5b and 2, 4b accepted here.

Family CRASSATELLIDAE

Outline varying from suboblong to trigoniiform or occasionally subcircular (astartiform), rounded in front, usually truncated and rostrate behind, often inequilateral; valves thick; beaks prosogyrate; ornament as in *Astarte* (*q.v.*), commarginal folds or threads often degenerating to smoothness; hinge with large resilium pit; cardinal teeth $\dfrac{3a, 3b, 5b}{2, 4b}$, 3a and 5b often obsolete; lateral

223

teeth variable, usually weak; some cardinal teeth often striated; adductor muscle scars deep. Habitat warm water (contrast to Astartidae). (Dev.-Rec.)

Crassatella [*Crassatellites*] (*s.s.*) (Fig. 500): outline subquadrate-sub-trigonal, with posterior angulation; ornament of commarginal lamellae or rugae; beaks low, small and close, prosogyrate; lunule and escutcheon very deeply sunk, with a fissure between them, immediately below the umbo, leading to the ligament pit; hinge plate very deep, triangular, flanked by lunule and escutcheon; resilium pit large, more or less triangular, extending from sub-umbonal fissure little more than half-way towards ventral margin of hinge plate, sometimes divided into anterior broad and posterior narrow portions by a slight ridge; right valve hinge with AI (short, very feeble), 3a (very feeble), 3b (large, wedge-shaped, vertically below the umbo, its upper end margining part of the resilium pit), 5b (very feeble) and PI (obscure); sockets for AII, 2 and 4b deep; left valve hinge with AII (short, less obvious than its socket), 2 and 4b about equal, with large socket between them, and PII (ill-defined); sockets for 3a and 5b are more obvious than the teeth that should fit them; 3b crenulate above and below, 2 and 4b crenulate underneath; in both valves the posterior part of the hinge plate is largely occupied by a smooth surface without hinge structures; umbonal cavity very deep, but more or less solidified; valve margins internally crenulate. U.Cret.-M.Mio. Eur., N.Amer., S.Amer., Carib., Asia. *Bathytormus:* transversely subtrigonal, the resilium pit extending to the ventral margin of the hinge plate, absorbing most of the smooth area and pushing forward the cardinals so that 4b is vertical and 3b slopes forward; also, PII is longer and more lamellar and the edge of the right valve escutcheon projects as a long PIII with well marked socket in the left valve; AI lamellar; valve margins internally crenulate. U.Cret.-Rec. N.Amer., Eur., Mexico, India, Japan, Ghana, S.Africa. *Eucrassatella* (*s.s.*) (Fig. 501): similar to *Bathytormus*, but the valve margins are internally smooth; ventral margin rounded. Pal.-Rec. Austral., N.Z., Eur. The subgenus *Hybolophus* is flattened and has opisthogyrate beaks, and is rostrate posteriorly; the anterior laterals are very short, and 3a is well separated from the edge of the lunule. Mio.-Rec. N.Amer., S.Amer. [Forms in the Miocene of Virginia and Maryland—e.g. *E.* (*H.*) *undulata* (Say)—attain a very large size.] *Crassinella* [*Pseuderiphyla*, *Gouldia auctt. non* type species]: small, subtrigonal, umbones flattened and the small beaks opisthogyrate; lunule narrow; escutcheon broad; ligament largely internal; right valve hinge with AI (long), 3a, 3b and PI; left valve hinge with 2, 4b and PII (long); valve margins internally smooth; commarginal ridges, sometimes also with fine radial lines. M.Eoc.-Rec. Carib., N.Amer., S.Amer., E.Pac. [There seems to be no name for the group including *Crassatella sulcata* (Solander) and *Crassinella concentrica* (Dujardin); while generally similar to *Crassinella* and the valve margins being smooth internally, there is a posterior area, and Cossmann and Peyrot (8) record the hinge characters to be two cardinal teeth in each valve, and an AI and a PII. L.Eoc.-Mio. Eur., India, Burma, Russia.] *Crassatina:* similar to *Crassinella*, but the valve margins are crenulate internally. Pal.-Rec. Eur., Africa, Japan.

224

Family ASTARTIDAE

Rounded-trigonal to subquadrate, inequilateral, subequivalve; valves thick; beaks pointed, prosogyrate; lunule distinct; escutcheon narrow, long; ligament external; lateral teeth inconstant; adductor muscle impressions well marked; integripalliate; with commarginal (rarely divaricate) ornament, at least in young stages. (?M.Ord., Dev.-Rec.)

Certain species (Fig. 503) exhibit an occasional inversion of hinge teeth, 2 and 4b being then found in the right valve, 3a and 3b in the left valve (cf. *Chama*). Throughout the Tertiary era the headquarters of this family were Boreal or Arctic, and the range of its several members extended to various distances southwards at different times.

Astarte (*s.s.*) (Figs 502, 503): rounded-trigonal, compressed, only slightly inequilateral, with commarginal striae or lamellae; valves thick; beaks pointed; lunule and escutcheon well marked and smooth, the lattter very narrow; right valve hinge with AIII (obsolete), AI (distinct), 3a (small), 3b (stout, vertical, trigonal), 5b (very small) and PIII (not distinct from valve margin); left valve hinge with AII (not distinct from valve margin), 2 and 4b (subequal, rather strong, diverging on either side of the socket for 3b), PII (small) and PIV (obsolete); hinge sometimes inverse; valve margins internally crenulate. M.Jur.-Rec. Cosmop. *Digitaria* [*Woodia*] (Fig. 504): small, rounded, inequilateral, ornamented with oblique curved lines not parallel to the valve margin; no lunule or escutcheon; hinge similar to that of *Astarte*, but tooth 3b bifid; valve margins internally smooth. Olig.-Rec. Eur., S.Africa.

Superfamily CARDIACEA

Size variable; equivalve; usually with radial ornament, either on siphonal region only or, more usually, over whole surface; umbones fairly central; ligament external; cardinal teeth and lateral teeth usually well developed, right valve without a median cardinal tooth, the cardinals usually conical, anterior laterals sometimes lacking; usually integripalliate; adductor muscle scars more or less equal. (U.Trias.-Rec.)

Family CARDIIDAE

Shell usually as high as, or higher than, long; no definite hinge plate; cardinal teeth peg-like, simple, projecting prominently, the line joining teeth of either valve at right angles to the line joining the sockets—a very characteristic pattern; lateral teeth lamellar; valves nearly always costate, their ventral margins internally crenulate or fluted; ligament external, parivincular, usually short; integripalliate. (U.Trias.-Rec.)

Cardium (*s.s.*) (Fig. 505): large, subquadrate, inflated; umbones median, orthogyrate or slightly prosogyrate, rather swollen; sharp radial costae giving rise to coarsely corrugated interlocking margins; ligament short, on prominent nymphs; complete hinge formula

$$\frac{\text{AIII, AI, 3a, 3b, PI}}{\text{AII, 2, 4b, PII, PIV}}'$$

225

with 4b high and almost vertically above 2; integripalliate; posterior gape. Mio.-Rec. Eur., W.Africa. The subgenus **Bucardium** [*Ringicardium*] (Fig. 506) is rounded and has strong, wide costae and lamellae posteriorly developing spines which project like bars over an extremely narrow posterior gape; hinge formula

$$\frac{\text{AIII, AI, 3a, 3b, PI, PIII}}{\text{AII, 2, 4b, PII, PIV}}.$$

L.Mio.-Rec. Eur., W.Africa. **Acanthocardia** (*s.s.*) (Fig. 507): obliquely rounded-oval, slightly truncated posteriorly where there is a very small gape; costae few and wide, with spinose upgrowths rising from a median groove; inflated; hinge similar to that of *Bucardium*, but without PIV; cardinal teeth arranged diagonally, 3a obliquely above 3b, 4b obliquely above 2. L.Mio. (Aquit.)-Rec. Eur., Medit. **Trachycardium** (*s.s.*) (Fig. 508): typically higher than broad (in the proportion 4:3); in adaptation to this shape the hinge is short and usually sharply bent, so that anterior and posterior lateral teeth are nearly at right angles, and the central part of the hinge plate does not bulge downwards as in *Cardium*, while the anterior lateral teeth are buttressed from within the umbonal cavity; almost equilateral; moderately wide, rather flattened costae with hoop-shaped lamellae, the scales being produced only along the posterior side of the central costae; cardinal teeth 3a and 4b are small. Mio.-Rec. Carib., C.Amer., N.Amer. (Suwannee), S.Amer. **Papyridea** (Fig. 509): transversely oval-oblong, rather compressed, thin-valved, with posterior and anterior gape; umbones low; beak distance about one-third; posterior gape serrated as in *Bucardium*; radial costae weak, with transverse lamellae or spines anteriorly and posteriorly; cardinal teeth small, tending to be obsolete, 2 and 3b present; three well developed lateral teeth in each valve. Mio.-Rec. N.Amer., Carib., E.Pac., Transatlantic. **Fragum** (*s.s.*) (Fig. 510): inflated, rhombic-subtriangular, higher than long, posteriorly truncated, siphonal area marked off by an angular or subangular ridge; posterior cardinal margin high; many flat costae with thin obscure beads, those on the siphonal area being coarser and ending in deep serrations as in *Bucardium*, but there is no gape; hinge similar to that of *Acanthocardia* and *Trachycardium*, the hinge plate medially projecting downwards, but the posterior lateral teeth

---

Figs 503–512. BIVALVIA: CRASSATELLACEA AND CARDIACEA
Figs 506, 508–509 and 512 after Keen, in Moore; Figs 503–505, 507 and 510–511 original.
503a, a′, b, b′. *Astarte solidula* Deshayes, U.Mio. Touraine. × 1·7. Normal and inverse individuals.
504c. *Digitaria digitaria* (Linné), L.Plio. Walton, Essex. × 3.
505a, b. *Cardium costatum* (Linné), Rec. W.Africa. T. × 0·6.
506c. *Cardium* (*Bucardium*) *ringens* Bruguière, Rec. W.Africa. × 1.
507a, b. *Acanthocardia echinatum* (Linné), Rec. Medit. × 0·75.
508c. *Trachycardium isocardia* (Linné), Rec. W.Indies. × 0·5.
509c. *Papyridea soleniformis* (Bruguière), Rec. W.Indies. T. × 1.
510. *Fragum unedo* (Linné), Rec. Indo-Pac. × 0·75.
511. *Nemocardium semistriatum* Deshayes, M.Eoc. Bracklesham. × 1·1.
512d. *Laevicardium oblongum* (Gmelin), Rec. Eur. T. × 0·5.

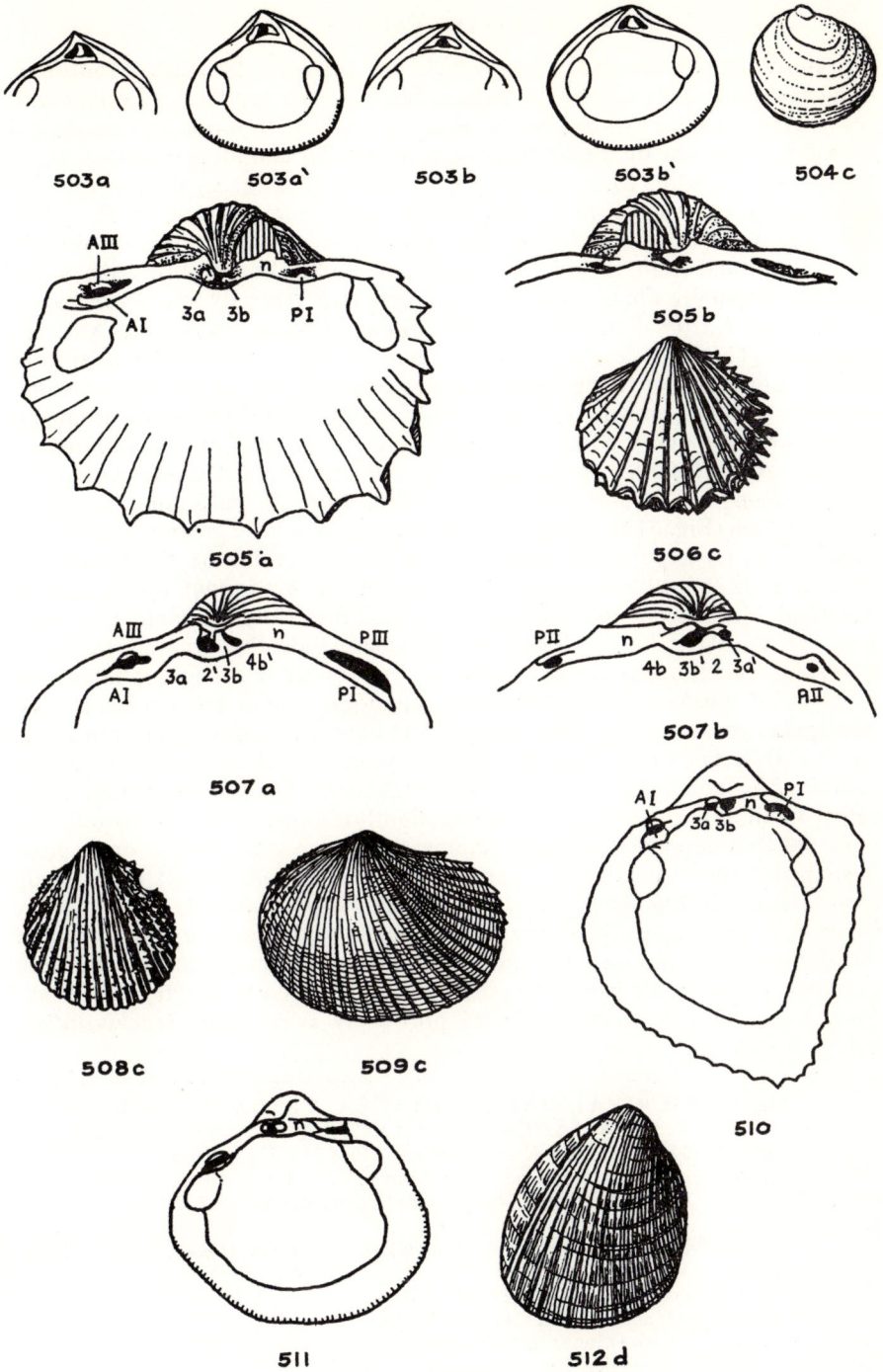

503a   503a'   503b   503b'   504c

AIII
AI   3a 3b   n   PI

505b

505a   506c

AIII   n   PIII
AI   3a 2'3b 4b'   PI

507a

PII   n
4b 3b' 2 3a'
AII

507b

AI   PI
n
3a 3b

510

508c   509c

511   512d

are closer to the cardinal teeth which are peg-like. Mio.-Rec. Indo-Pac. *Trigoniocardia* (*s.s.*): similar to *Fragum*, but less high and flank of valves with few, narrow costae bearing beads, their intervals with looped commarginal lamellae; the costae on the siphonal area are smaller and more crowded. Mio.-Rec. E.Pac., Carib. (San Sebastian), N.Amer. (Suwannee, Chickasawhay), S.Amer. *Nemocardium* (*s.s.*) (Fig. 511): ovate-subquadrate, with coarse spinose costae on the siphonal area, elsewhere with finer, almost smooth costae; cardinal teeth 3a and 4b are small, 3b and 2 short and peg-like; valve margins internally finely crenulate anteriorly and ventrally, but more coarse posteriorly. L.Cret.-Rec. Eur., Carib., Asia, N.Amer., S.Amer., Austral., Indo-Pac. [The Jurassic and Cretaceous genus *Protocardia* differs in having no radial costae anterior to the siphonal area.] *Discors:* ovate in shape, sometimes posteriorly truncated, with oblique lines of sculpture crossing radial ornament on the anterior and median areas and faint radial lines on the siphonal, a general radial structure appearing on decortication; hinge similar to that of *Loxocardium* and *Acanthocardia*. Eoc.-Mio. Eur., W.Pakistan, Indonesia. *Laevicardium* (Fig. 512): obliquely suboval, height a little greater than width, valves thin; no gape; faint indications of smooth costae on the flanks, the extreme ends smooth; hinge similar to that of *Trachycardium*, but 3a and 4b are small; lateral teeth upturned; valve margins internally crenulate. Eoc.-Rec. Eur., W.Atl., W.Pakistan, Indonesia, N.Amer., Indo-Pac. (mainly tropical). *Cerastoderma:* transversely oval-subquadrate, longer than high, inflated; flat or rounded costae with transverse crenulations, costae broader than their interspaces; beaks orthogyrate; hinge plate narrow; cardinal teeth not large, 3a and 4b being small. Mio.-Rec. Eur. [More boreal than *Acanthocardia;* includes the common cockle *C. edule* (Linné)]. *Loxocardium* (Fig. 513): rather small, suborbicular-subquadrate, slightly oblique, truncated posteriorly; siphonal area depressed; no gape; numerous costae with straight or inverted V-shaped transverse lamellae; cardinal teeth 2 and 4b small; more finely costate than *Acanthocardia*. Eoc.-Mio. Eur. *Plagiocardium* (*s.s.*): obliquely suboval, a little higher than long; hinge similar to that of *Acanthocardia*, but 2 and 3b smaller, and anterior laterals nearer the cardinals; small costae with granular or triangular peduncles, the costae connected by small ridges. Pal.-Mio. Eur., Indonesia. [Some forms previously referred to *Trachycardium* belong here.]

Figs 513–520. BIVALVIA: CARDIACEA, TRIDACNACEA, CHAMACEA AND MACTRACEA

Fig. 514 after Gillet; Fig. 513 after Keen, in Moore; Figs 515–520 original.
513c. *Loxocardium formosum* (Deshayes), Eoc. France. T. × 1·5.
514d. *Lymocardium fittoni* (d'Orbigny), Sarmatian. Cricov, Rumania. × 0·75.
515. *Prosodacna* sp., 'Dacian'. Rumania. × 0·75.
516a, d. *Oraphocardium oraphense* (Davidashvili), L.Rumanian (Duab Stage). Guria, Transcaucasia. × 1.
517. *Tridacna compressa* Reeve, Rec. Ceylon. × 0·56.
518a, b, c. *Chama squamosa* Solander, U.Eoc. Barton, Hampshire. × 1.
519a, b. *Arcinella arcinella* (Linné), L.Mio. Shoal R., Florida. T. × 0·75.
520. *Mactra antiquata* Spengler, Rec. Philippines. × 1·25.

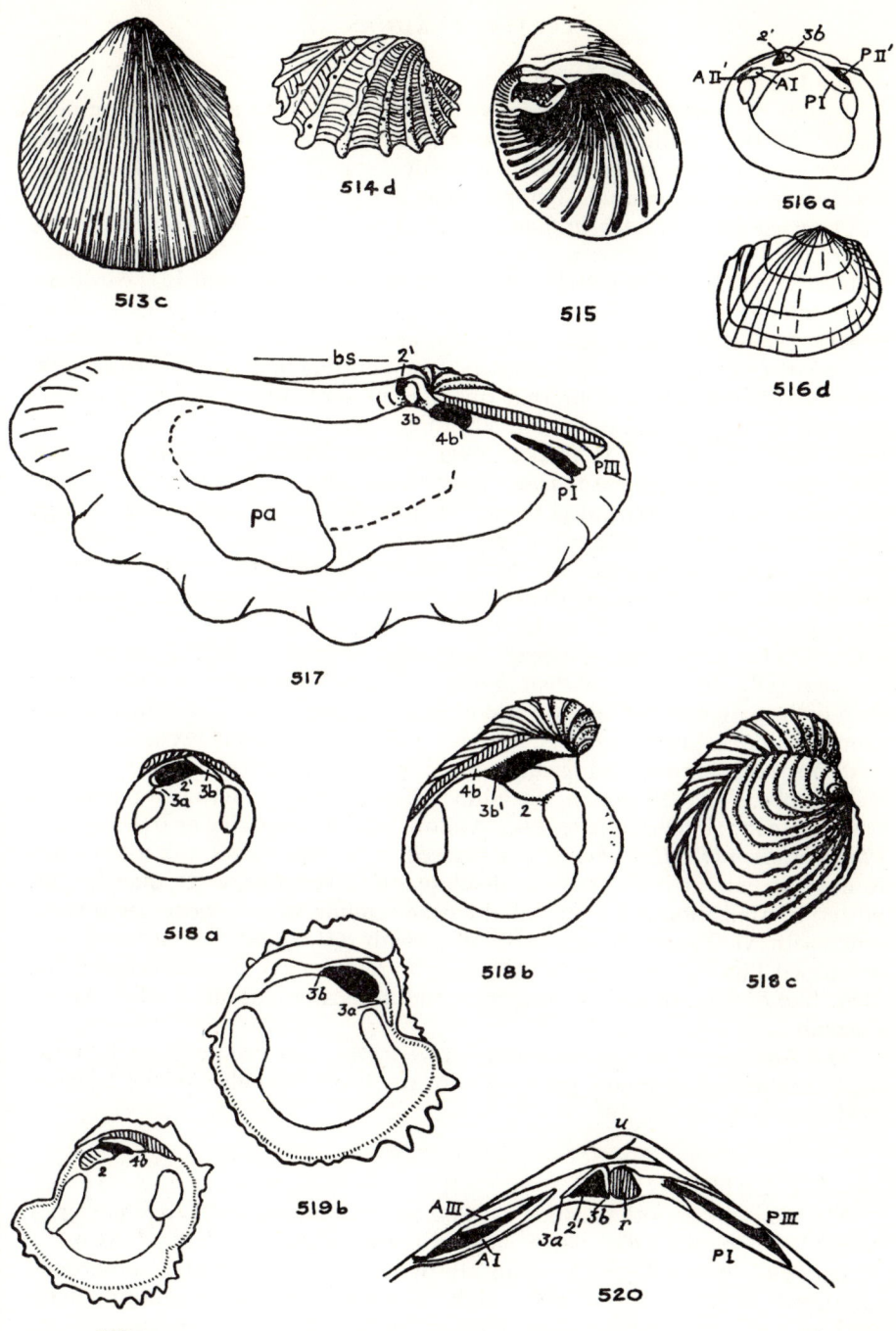

513 c

514 d

515

516 a

516 d

517

518 a

518 b

518 c

519 b

519 a

520

Family LYMNOCARDIIDAE

Costate or smooth, often dorsally subalate, some with a gape; hinge normally with two cardinals in each valve (sometimes obsolete), with variable, distant lateral teeth; integripalliate or with a small to medium pallial sinus; brackish water, very variable, and probably polyphyletic. (Mio.-Rec.)

Lymnocardium [*Limnocardium*] (*s.s.*) (Fig. 514): oblong-subquadrate, generally truncated; costae rather sparse, smooth or spiny; right valve hinge with AIII, AI, 3b, PI and PIII; left valve hinge with AII, 2 and PII; cardinals small, laterals far apart and well developed; with a small pallial sinus or none. U.Mio.-L.Plio. E.Eur., Near East. (estuarine or fresh-water). *Prosodacna* [*Psilodon*] (*s.s.*) (Fig. 515): obliquely oval-subtriangular, with broad, low costae or almost smooth; internal ribs present; beaks far forward, prosogyrate, spiral; no cardinal teeth; AI very thick, overhanging the deep adductor scar; AIII thin, low; AII thick; integripalliate. Plio. Aralo-Caspian, Eur. (estuarine or fresh-water). *Adacna:* transversely suboval, posterior half higher than anterior half, gaping at both ends; valves thin; broad, flat costae; a large pallial sinus; no teeth except for very faint vestiges of 2 and 3b. U.Plio.-Rec. Aralo-Caspain. (estuarine or fresh-water). *Monodacna* (*s.s.*): rounded-subquadrate, length greater than height, with broad costae and a posterior gape; hinge with 2 and 3b, no laterals or only traces of them; a slight pallial sinus. U.Plio.-Rec. Aralo-Caspian. (estuarine or fresh-water). *Didacna* (*s.s.*) is recent only. Its subgenus *Pontalmyra* is oval-subtriangular, a little longer than high, posterior end slightly and obliquely truncate, with prominent, subcentral umbones; ornament like that of *Cerastoderma*, costae becoming faint; posterior keel sharp in young stage, more or less obsolete in adult; no lateral teeth, only 2 and 3a present in adult; valve margins internally fluted; no posterior gape; integripalliate. Plio. Aralo-Caspian. (estuarine or fresh-water). *Phyllocardium* [*Phyllicardium*]: compressed, with thin valves and a tendency to develop ears at both ends; oval-subquadrate, length greater than height, posterior truncation oblique; radial costae tending to disappear; right valve hinge with AI, 3b and PI; left valve hinge with AII, 2 and PII; all teeth well developed. Plio. Aralo-Caspian. (estuarine or fresh-water). *Oraphocardium* (Fig. 516): like *Phyllocardium*, but posterior end vertically truncated. Plio. Aralo-Caspian.

The Adacnids and Lymnocardiids have in the past been regarded as distinct families, but are now grouped with the five subfamilies of the Lymnocardiidae.

Family LAHILLIIDAE

More or less smooth cardiids with no anterior lateral teeth; one posterior lateral tooth in each valve; valve margins internally smooth. (U.Cret.-Mio.)

Lahillia [*Amathusia*] (*s.s.*): ovate-subtrigonal, smooth or with commarginal ornament; beaks nearly median, high; cardinal teeth and posterior lateral teeth as in *Cardium*, but no anterior lateral teeth. U.Cret.-Mio. S.Amer., N.Z., Austral., Antarctic.

Superfamily TRIDACNACEA

An aberrant group which has adopted the byssal-fixation habit resulting in the abortion of the anterior adductor and forward shift of the posterior adductor to a central position; medium size to very large; equivalve, with smooth or spiny radial ornament and valve margins usually deeply fluted internally; valves thick; ligament external; hinge with two oblique cardinal teeth and one or more lateral teeth; integripalliate. (?U.Cret., Eoc.-Rec.)

Family TRIDACNIDAE

Characters of the Superfamily. (?U.Cret., Eoc.-Rec.)

Tridacna (Fig. 517): valves thick and solid; large, subtrigonal-suboval, nearly twice as long as high; beaks prosogyrate, nearly subcentral; lunule region excavated by a wide and long byssal opening; post-umbonal part of hinge straight, with fairly long ligament; surface with about five very coarse radial costae with scaly elevations, causing the valve margins to be very strongly fluted; right valve hinge with 3b (large), PI and PIII; left valve hinge with 2 (small), 4b (small) and PII; posterior adductor scar shifted in front of beak, very large, narrower medially; inner layer of shell reflected over margins of byssal opening, bearing oblique byssus-guiding ridges; unattached, but living among coral reefs. L.Mio.-Rec. Eur., Indonesia. [The relation of *Tridacna* to the normal bivalve shell may perhaps be better interpreted if we take as a reference line that separating the main part of the gills in front from the siphonal openings and posterior adductor behind. Such a line in Fig. 517 would run from the umbo above the adductor, dipping about 20° below the horizontal. Re-orientating the shell on this line as vertical, we see that there has been a great antero-posterior compression (especially in front of the reference line) and a great vertical extension (mainly behind the reference line). These gigantic shells, living on or boring in coral reefs, may attain a length of over four feet and a weight of abour four hundredweight. In non-boring species and the allied *Hippopus* (Mio.-Rec.), byssus is vestigial, its opening being lost but replaced by a concave area still keeping the byssus-guiding ridges on its margin. There is an old record, over a hundred years old, from the Miocene of Poland, but this has never been confirmed and may be regarded as erroneous.] *Avicularium* [*Lithocardium*]: trigonal, height nearly twice length, very inequilateral, umbones well forward, ventral margin pointed, surface divided into two unequal regions by a straight ridge; hinge teeth

$$\frac{\text{3b, PI, PIII}}{\text{2, 4b, PII}};$$

anterior adductor scar very small, the posterior large and at some distance from hinge; no byssal gape, but a notch in anterior margin may mark exit of byssus; costae almost smooth. M.Eoc.-L.Olig. Eur., Carib. *Byssocardium:* similar to *Avicularium*, but with large serrated byssal gape in front; anterior adductor scar doubtful; valve margins internally smooth. M.Eoc.-L.Mio. Eur.

Superfamily CHAMACEA

One valve fixed, even if only in young; more or less inequivalve, with commarginal or radial ornament or both; beaks prosogyrate; ligament parivincular; hinge plate more or less thick, with few teeth; muscle scars large, subequal; integripalliate. (U.Cret.(Danian)-Rec.)

Although included by Newell (67a) and Vokes (72b) in the otherwise exclusively pre-Tertiary Hippuritoida, the hinge and other characters seem to place this superfamily in the Veneroida where it had previously been believed to belong (and where it is placed by Moore, 66a).

Family CHAMIDAE

Characters of the Superfamily. (U.Cret.(Danian)-Rec.)

Chama (*s.s.*) (Fig. 518): valves thick, very inequivalve, left valve the more inflated; fixed by the left valve; outline suborbicular, beaks prosogyrate; ornament commarginal and irregularly lamellate with flattened spines in irregular rows; ligament external, more or less sunk in a groove; right valve hinge with 3a, 3b and PI; left valve hinge with 2 (very large, often grooved), 4b (long, curved, obscurely bifid) and PII (rudimentary). L.Eoc.-Rec. Cosmop. (warm seas). *Arcinella* [*Echinochama*] (Fig. 519): differs from *Chama* in being practically equivalve, and in having a deeply impressed but convex lunule; attached in youth only; spinose; hinge as in *Chama*, but all species appear to be inverse (the hinge being the only feature capable of showing inversion). L.Mio.-Rec. N.Amer., Carib., E.Pac., S.Amer., C.Amer., *Pseudochama* (*s.s.*): like *Chama*, but inverse and fixed by the right valve, nepionic shell with commarginal ornament only; no PII in adult. L.Mio.-Rec. Indonesia, E.Pac., S.W.Pac., C.Amer., N.Amer. (Tampa), Medit. [Although regarded by some as a synonym of *Chama*, the fact that its range is so much shorter seems to warrant its recognition.]

Superfamily MACTRACEA

Equivalve, triangular or oval or more or less elongate, closed or gaping; external ligament small or lacking, resilium present; tooth 2 bifid, like an inverted V; teeth 3a and 3b widely divergent; lateral teeth variable; sinupalliate except for the Anatinellidae and Cardiliidae; surface rarely ornamented except by simple growth lines, sometimes scaly. (U.Cret.-Rec.)

Family MACTRIDAE

Rounded-trigonal to oblong; often closed or with a slight gape; beaks prosogyrate; lunule and escutcheon not well defined; opisthodetic ligament very short; resilium triangular or ovoid in a deep pit, usually causing a downward protuberance of the hinge plate; 4a reduced to a thin lamina on front margin of resiliophore, or wanting; pallial sinus more or less deep; valve margins internally smooth; periostracum present. (U.Cret.-Rec.)

Mactra (*s.s.*) (Fig. 520) is Recent only. The subgenus *Barymactra* is large, orbicular-subtrigonal, with thick valves; hinge teeth set close together; pallial sinus short, descendent. Eoc.-Mio. Eur., S.W.Africa. [In the later Tertiary

European and American stocks (subgenera) evolved independently.] *Spisula* (*s.s.*) (Fig. 521) is Recent only; the ligament and resilium have no lamina separating them. The subgenus *Pseudoxyperas* is transversely oval, with striate lateral teeth, and tooth 4a present; the pallial sinus is long and oval. Burd.-Rec. Eur. [Local subgenera developing as in *Mactra*.] *Mulinia:* the ligament is sunk into the resilium pit and not externally visible; thicker-valved and more inflated than *Mactra* and *Spisula*; lateral teeth short, smooth; pallial sinus short, small, angular; estuarine. Rec. only. C.Amer., S.Amer., E.Africa. *Raeta* (*s.s.*) (Fig. 522): suborbicular-subtriangular, slightly rostrate posteriorly; valves thin and ornamented with strong commarginal undulations; posterior lateral teeth present. Eoc.-Rec. N.Amer., C.Amer., S.Amer., Eur. *Eastonia:* oval-subtriangular, with blunt umbones and bluntly rounded posterior end; ligament not separated from nymph; lateral teeth short, the anterior ones crowded against the cardinals; posteriorly gaping; fine radial ornament (except at extreme ends) as well as commarginal growth lines; pallial sinus deep. L.Mio.-Rec. Eur. Medit. *Lutraria* (*s.s.*) (Fig. 523): transversely suboval, beak distance one-third; ornament of prominent growth lines; hinge as in *Spisula*, but no lateral teeth; pallial sinus half the length of the shell; a permanent burrower, gaping at both ends. Olig.-Rec. Eur., Indo-Pac., N.Z. The genus was unknown at any time in North America where its place is taken by *Tresus* [*Schizothaerus*], which is ovate-truncate, with a more rounded resilium and very small laterals. U.Mio.-Rec. N.Amer., E.Asia.

Doubtfully placed in the Mactridae (or ?Isocardiidae) is *Blagraveia* (Fig. 524): trigonal-ovate, inequilateral, posteriorly truncated, moderately inflated; valves thin, closed, ornamented with commarginal undulations; escutcheon narrow, deep, bounded on each valve by a sharp ridge; lunule cordate, bounded on each valve by a narrow, incised groove which is represented on the inside of the shell by a thin, projecting lamina; ligament external, opisthodetic; 3a prominent, 3b narrow and divergent, 2a strong and bifid, 2b weak; no lateral teeth. L.Eoc.-M.Eoc. India, W.Pakistan, Somalia, Iran.

Family MESODESMATIDAE

Trigonal, cuneiform or oval, sometimes more elongate than is usual for mactrids, close; valves solid; beaks opisthogyrate; hinge similar to that of Mactridae, with lateral teeth nearly always developed; with or without a small pallial sinus. (Eoc.-Rec.)

Mesodesma (*s.s.*) is Recent only. *Mactropsis* (Fig. 525): ovate-subtriangular, with heavy hinge plate and distinct cardinals; laterals striate; ligament and resilium fused; small pallial sinus. Eoc. N.Amer.

Family CARDILIIDAE

Cordiform, height greater than length, equivalve, with prominent, proso-gyrate beaks; valves small, thin, usually with radial ornament; *Mactra*-like hinge, but no lateral teeth; ligament external, on a nymph; resilium internal, connecting projecting chondrophores; integripalliate; one genus edentulous. (?Eoc., Olig.-Rec.)

233

Cardilia (Fig. 526): inflated; two cardinals per valve; posterior adductor scar buttressed; radial costae at least in part. L.Mio.-Rec. Indo-Pac., Japan.

Superfamily SOLENACEA

Shell usually laterally compressed and more or less distinctly elongate, ends gaping; ligament external, attached to a ledge; hinge variable, usually without lateral teeth; pallial sinus mostly not very deep; burrowers. (L.Cret.-Rec.)

Family SOLENIDAE

Extremely inequilateral, with terminal or nearly terminal beaks; hinge with only one cardinal tooth in each valve. (L.Eoc.-Rec.)

Solen (Fig. 527): elongate-subrectangular, ends truncated and gaping; nymph narrow, long; umbones terminal; smooth; anterior adductor scar long, parallel to the dorsal margin; slight anterior clavicle; hinge

$$\frac{3a}{2a};$$

the ventral pallial line is withdrawn from the margin and doubled against its anterior section; the pallial sinus has shifted forward nearly to the posterior adductor scar. Eoc.-Rec. Cosmop. *Solena* (*s.s.*): differs from *Solen* in having the beak distance about one-sixth, and in having a short anterior adductor scar. Olig.-Rec. E.Asia, C.Amer., Eur. The subgenus *Eosolen* is similar to *Solena*, but has an external oblique groove running forward from the umbo. Eoc. Eur., N.Amer. The subgenus *Plectosolen* is narrower, posteriorly rounded, and has a curved dorsal margin. L.Eoc.-M.Eoc. Eur. N.Amer.

Family CULTELLIDAE

Transversely oblong, gaping at both ends, beaks not usually terminal; hinge of lucinoid type (one to three cardinal teeth); ligament external, short, on nymphs; sinupalliate. (L.Cret.-Rec.)

Cultellus (*s.s.*) (Fig. 531): oblong, with rounded and gaping ends; somewhat compressed; length more than three times the height; beak distance about one-quarter; right valve hinge with 3a (large, wedge-shaped) and 3b (oblique);

Figs 521–531. BIVALVIA: MACTRACEA, SOLENACEA
AND TELLINACEA

Figs 522–523, 525–529 and 531 after Keen, in Moore; Fig. 524 after Keen and Casey, in Moore; Figs 521 and 530 original.

521a, b. *Spisula solida* (Linné), Rec. Britain. T. × 1·1.
522c. *Raeta plicatella* (Lamarck), Rec. W.Indies. T. × 1.
523. *Lutraria lutraria* (Linné), Rec. Medit. T. × 1.
524d. *Blagraveia corrugata* Cox, M.Eoc. India. T. × 1.
525d. *Mactropsis aequorea* (Conrad), Eoc. Alabama. T. × 1·5.
526c. *Cardilia semisulcata* (Lamarck), Rec. Japan. T. × 1·3.
527c. *Solen vagina* Linné, Rec. Eur. T. × 0·4.
528c. *Siliqua radiata* (Linné), Rec. Indonesia. T. × 0·3.
529c. *Pharus legumen* (Linné), Rec. Medit. T. × 0·5.
530a, b. *Ensis ensis* (Linné), Rec. Britain. T. × 0·66.
531. *Cultellus lacteus* (Spengler), Rec. Indonesia. T. × 0·7.

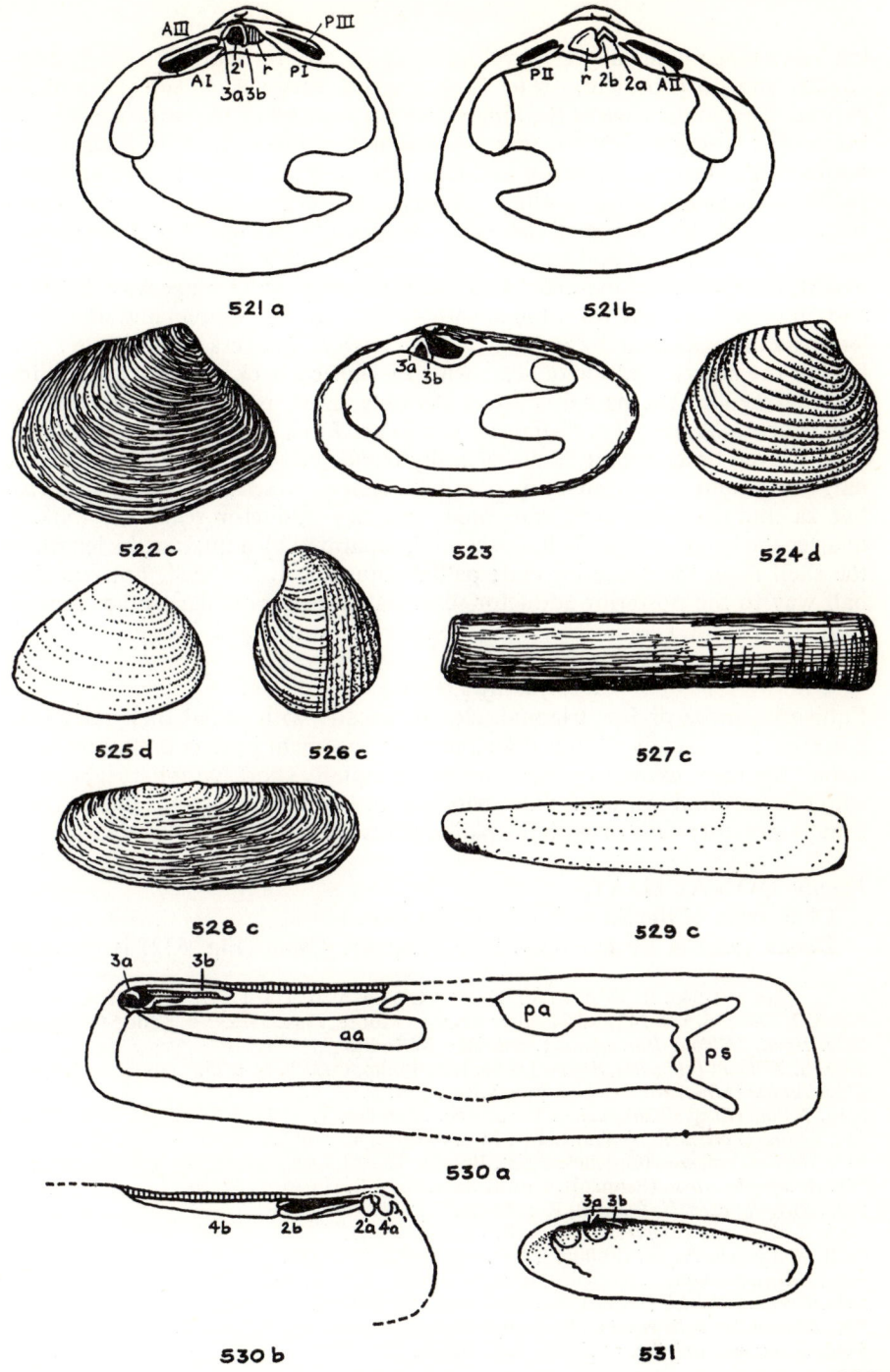

521 a

521 b

522 c

523

524 d

525 d

526 c

527 c

528 c

529 c

530 a

530 b

531

235

left valve hinge with 4a (well developed), 2a (nearly vertical), 2b (nearly horizontal) and 4b (obsolete)—thus 3a is gripped between 4a and 2a, and 3b between 2b and the edge of the hinge plate corresponding in position to 4b; no buttresses; anterior adductor scar oval-triangular, just in front of hinge teeth, with a ridge above it; posterior adductor scar similar, shifted a little forward; pallial sinus small, not extending in front of posterior adductor scar. L.Eoc.-Rec. Eur., W.Africa, Asia, Indo-Pac. *Siliqua* (*s.s.*) (Fig. 528): transversely oblong, with rounded and gaping ends; compressed; length about three times height; beak distance one-quarter to one-third; right valve hinge with 3a (thin) and 3b (obsolete); left valve hinge with 4a (present), 2 (simple) and 4b (obsolete); teeth supported by a long, nearly vertical buttress or 'clavicle', just behind the anterior adductor scar, which is shifted back slightly; pallial line rather irregular, the pallial sinus moderately short, rounded, large. L.Eoc.-Rec. Eur., N.Amer., Transatlantic, Indo-Pac. *Ensis* ('razor-shell', Fig. 530): length seven times height, ends vertically truncated and gaping, beaks terminal; dorsal and ventral margins parallel, slightly curved; hinge as in *Pharus*, but 2a and 2b completely separated; anterior adductor scar very long, a quarter the length of the shell, horizontal, separated by a quarter the length of the shell from the posterior end; pallial sinus short, rounded, not reaching half way to the posterior adductor scar; pallial line irregular, rather remote from the margin. L.Eoc.-Rec. Eur., N.Amer., Transatlantic, E.Pac.

Superfamily DONACACEA

Equivalve, more or less trigonal, closed, usually with radial shell structure; posterior part usually shorter than anterior; ornament both commarginal and radial; ligament extremely short, borne on equally short nymphs; right valve with two cardinal teeth, left valve with one or two cardinal teeth; lateral teeth usually present; with or without pallial sinus. (Trias.-Rec.)

Family DONACIDAE

Characters of the Superfamily. (Trias.-Rec.)
Donax (*s.s.*) is Recent only. The subgenus *Chion* (Fig. 532) is elongate

Figs 532–545. BIVALVIA: DONACACEA AND TELLINACEA
Figs 532, 534–536, 539–543 and 545 after Keen, in Moore; Figs 533, 537–538 and 544 original.
532c. *Donax* (*Chion*) *denticulatus* Linné, Rec. W.Indies. T. × 1.
533a, b. *Tellina* (*Tellinella*) *virgata* Linné, Rec. Philippines. T. × 0·75.
534c. *Tellina* (*Moerella*) *donacina* Linné, Rec. Eur. T. × 1.
535c. *Tellina* (*Eurytellina*) *punicea* Born, Rec. W.Indies. T. × 1.
536. *Tellina* (*Peronaea*) *planata* Linné, Mio. Austria. T. × 0·7.
537. *Macoma calcarea* (Gmelin), Pleist. Britain. T. × 1·1.
538. *Arcopagia crassa* (Pennant), Rec. Britain. T. × 1.
539c. *Strigilla carnaria* (Linné), Rec. Carib. T. × 1.
540b, d. *Oudardia compressa* (Brocchi), Plio. Eur. × 1.
541d. *Apolymetis meyeri* (Philippi), Rec. Indonesia. T. × 0·5.
542c. *Gastrana matadoa* (Gmelin), Rec. W. Africa. T. × 1.
543c. *Gari amethystus* (Wood), Rec. Indonesia. T. × 0·75.
544. *Macrosolen hollowaysi* (J. Sowerby), M.Eoc. Hampshire. T. × 0·75.
545d. *Solecurtus strigilatus* (Linné), Rec. Medit. T. × 0·35.

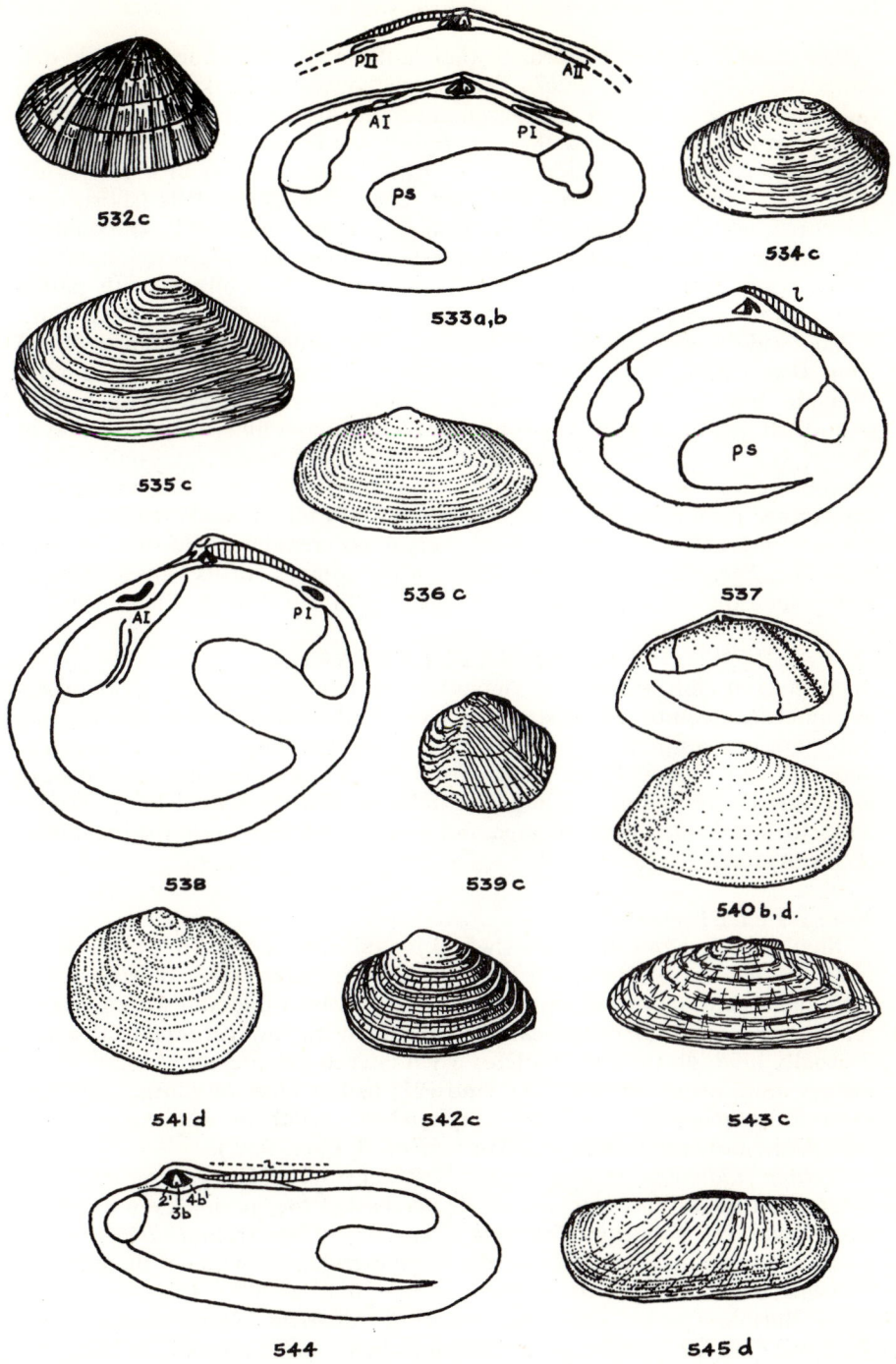

532 c

533 a,b

534 c

535 c

536 c

537

538

539 c

540 b, d.

541 d

542 c

543 c

544

545 d

wedge-shaped, thicker posteriorly than anteriorly, inequilateral, beak distance two-thirds; valves thick; beaks opisthogyrate; flanks with punctate radial sculpture as well as commarginal ornament; siphonal area marked off by a bend and bearing scaly threads; right valve hinge with AIII (long, conspicuous), AI (long, conspicuous), 3a (thin, nearly horizontal), 3b (stout, triangular, roughened, bifid), PI (short, conspicuous) and PIII (short, conspicuous); left valve hinge with AII (long), 2 (long, roughened), 4b (smaller) and PII (short, conspicuous); ligament short, on slightly projecting nymphs; pallial sinus deep, its apex broadly U-shaped, partly confluent with pallial line; valve margins internally denticulate. L.Eoc.-Rec. Carib., E.Asia, Eur., S.Pac., widespread (except in cold seas). The subgenus **Paradonax** differs from *Donax* in having a smaller and shallower chondrophore, in being more elongate and nearly smooth, and in lacking anterior lateral teeth. L.Mio. (Aquit.)-Rec. Eur., Carib. The subgenus **Hecuba** has a hinge similar to that of *Donax*, but the outline is subequilateral-trigonal, the beaks being only a little behind the mid-line, the siphonal area being marked off by a subspinose keel, the surface ornamented with commarginal and radial threads, and the valve margins internally are smooth or only very finely crenulate. Mio.-Rec. Burma, Indo-Pac. **Egerella:** differs from *Donax* in the thinness of its valves, in being nearly smooth, and in the absense of lateral teeth. Pal.-Eoc. N.Amer., Eur.

Superfamily TELLINACEA

Shell short or elongate, ovoid, trigonal or rarely oblong, usually more or less asymmetrical, compressed, usually with external ligament which may be sunk but is not in a resilium pit; with lucinoid hinge, but longer than in Lucinacea and with very narrow hinge plate, cardinals tending to be bifid; pallial sinus large, sometimes wholly or largely confluent with the pallial line; anterior adductor scar sometimes elongated, but never so markedly as in the Lucinidae. (U.Trias.-Rec.)

Family TELLINIDAE

Shell usually elongate, often slightly inequivalve (asymmetric at posterior end), the left valve typically a little the larger; beaks small, opisthogyrate; siphonal area more or less distinctly marked off by a fold and with a general tendency to rostration or truncation; hinge plate narrow, cardinal teeth small, 2 usually bifid, 4b thin or obsolete; when laterals are present, AI and PI are always more prominent than AII and PII; pallial sinus very large; ligament external, attached to a narrow nymph; shell smooth or with commarginal ornament, more rarely with radial ornament. (L.Cret.-Rec.)

Tellina [*Liotellina*] (*s.s.*) is evidently Recent only. [The generic name has often been loosely used as a waste-paper basket for species, even as old as Mesozoic.] The subgenus **Tellinella** (Fig. 533) differs from *Tellina* in being rostrate and carinate posteriorly (the slope carrying a few radial threads) and in being ornamented with distinct commarginal lamellae. Olig.-Rec. Indo-Pac., Burma, Carib., Eur. The subgenus **Moerella** [*Donacilla* Gray *non* Philippi] (Fig. 534) differs from *Tellina* in being more inflated, in being distinctly inequilateral, in the posterior end being rounded-truncate and shorter than the anterior end, and in the pallial sinus almost reaching the anterior adductor scar; there is very fine commarginal ornament; 3a is bifid and 2 very small; laterals strong, anterior ones closer to cardinals. L.Eoc.-Rec. Eur., N.Amer., Pac., widespread. The subgenus *Eurytellina* (Fig. 535) is compressed compared with *Moerella*, is more equilateral, is not rostrate, has a smooth, flattened siphonal area, and the flanks are ornamented with fine impressed commarginal lines; 4b small; left valve laterals the weaker; anterior laterals closer to cardinals. Mio.-Rec. Widespread. The subgenus *Peronaea* (Fig. 536) is oval and subequilateral, with left valve slightly the flatter, almost smooth, has no AII, AI is closer to the cardinals, and the pallial sinus is confluent almost to the anterior adductor scar. Olig.-Rec. Eur., Asia. *Macoma* (*s.s.*) (Fig. 537): similar to *Moerella* in general form but less inflated; hinge also similar, but lateral teeth are completely absent; pallial sinus fairly wide and horizontal, attaining the mid-line, not confluent with the pallial line. Mio.-Rec. N.Amer., Asia, M.East, N.Eur., Arctic (especially boreal and temperate seas). The subgenus *Psammacoma* comprises warmer water forms with blunter posterior truncation and pallial sinus coalescent with the pallial line for half its length. L.Mio.-Rec. Indo-Pac., Carib. S.Amer., W.Atl. *Arcopagia* (*s.s.*) (Figs 394, 538): more inflated, oval-rounded, anterior end somewhat produced; ornament of closely spaced commarginal lamellae; valves thick; right valve hinge with AI (strong, prominent), 3a (small, oblique), 3b (bifid) and PI (strong); left valve hinge with AII, 2 (strong, triangular, a little oblique, bifid), 4b (thin, oblique) and PII (rudimentary); pallial sinus large, oval, ascendant, not confluent with pallial line. ?Cret., Eoc.-Rec. Eur., Egypt. *Strigilla* (*s.s.*) (Fig. 539): form and hinge of *Arcopagia*; pallial sinus deep, adherent to the pallial line in both valves; ornament of rugae which divaricate once or twice. Mio-Rec. Eur., Carib. *Oudardia* (Fig. 540): small, transversely ovoid, posteriorly obliquely truncated, siphonal area limited by a fine ridge, beaks behind mid-line; sometimes with fine, oblique ornament; valves thin; right valve hinge with AI (close to cardinals), 3a and 3b (not bifid); left valve hinge with 2 (slightly bifid) and 4b; anterior adductor scar elongate, with a radial buttress ('clavicle') immediately behind it; pallial sinus partly confluent with pallial line and reaching forward close to the clavicle. ?Eoc., Olig.-Rec. Eur., N.Amer. *Apolymetis* [*Capsa* Lamarck *non* Bruguière, *Metis* H. and A. Adams *non* Philippi, *Polymetis* Salisbury *non* Walsingham] (Fig. 541): oval-subquadrate, with median beaks, convex, somewhat truncated behind, and with a sinuous fold bounding the siphonal area; weak commarginal ornament or smooth; right valve hinge with 3a (thick, short) and 3b (unequally bifid); left valve hinge with 2 (thick, triangular, bifid) and 4b (thin); pallial sinus large, subrhombic, reaching mid-length, not confluent with pallial line. U.Eoc.-Rec. Indonesia, Burma, Indo-Pac., Eur. *Gastrana* (Fig. 542): ovoid-trigonal, posteriorly rostrate and with a blunt fold, postero-ventral margin slightly concave; beaks a little anterior to mid-line; valves very thin; right valve hinge with 3a (thick) and 3b (markedly bifid); left valve hinge with 2 (markedly bifid) and 4b (thin); ornament of fine commarginal lamellae and feeble radial lines; pallial sinus large and gibbous, rather descendent, reaching mid-line, separated from pallial line by a long and narrow interval. L.Eoc.-Rec. Eur., W.Africa.

Family PSAMMOBIIDAE [GARIDAE]

Outline tending towards an oblong or oval-transverse shape, not very inequilateral; typically with a small gape at both ends; cardinal teeth telliniform; lateral teeth usually absent; ligament borne on well-marked nymphs; posterior adductor scar often in a more anterior and dorsal position than usual; pallial sinus deep. (U.Cret.-Rec.)

This family differs from the Tellinidae in its adaptation to a burrowing life.

Gari (*Psammotaea*] (*s.s.*) (Fig. 543): transversely oval, rounded in front, somewhat truncated behind; valves thin; siphonal area weakly set off; flanks with commarginal ornament, sometimes also with oblique striae; pallial sinus gibbous, deep, more or less ascendent. Mio.-Rec. Eur., N.Z., Asia. The subgenus **Psammobia** is similar to *Gari*, but the siphonal area is marked off by a strong ridge and there are two or three pedal muscle scars instead of one; right valve hinge with 3a (feebly bifid) and 3b (distinctly bifid); left valve hinge with 2 (bifid) and 4b (thin); sometimes traces of PI and PII. L.Mio.-Rec. Eur.

Macrosolen (Fig. 544): oblong, increasing in height behind, length usually at least three times greatest height; beak distance about one-fifth; closed; compressed; a weak furrow just below post-umbonal slope; posterior adductor scar shifted antero-dorsally; pallial sinus not extending to midline; hinge formula

$$\frac{3b}{2, 4b}.$$

L.Eoc.-L.Mio. Eur., Egypt, W.Pakistan, Arakan Coast.

Family SOLECURTIDAE

Elongate-subrectangular in outline, compressed, gaping at both ends; smooth or with oblique rugae; hinge weak, with two cardinal teeth in each valve; beaks near the mid-line; ligament external; pallial sinus usually deep. (L.Eoc.-Rec.)

Solecurtus [*Psammosolen, Macha, Solenocurtus*] (Fig. 545): subrectangular, with bluntly rounded, gaping ends; beaks only a little anterior to mid-line; ornament of commarginal growth lines and rather widely spaced oblique rugae that disappear on the anterior end; right valve hinge with 3a (long, conical, bent upward) and 3b (compressed, oblique); left valve hinge with 2 (short) and 4b (slender, very oblique); ligament long; pallial sinus deep and U-shaped, reaching the mid-line. Eoc.-Rec. Medit., Eur., Russia, Indonesia, N.Z., Indo-Pac., Carib., N.Amer. (widespread in warm seas). **Azorinus** [*Azor, Zozia*] (*s.s.*): similar to *Solecurtus*, but lacking the oblique rugae; pallial sinus broad, rounded. Plio.-Rec. Eur., Medit. **Tagelus** (*s.s.*): similar to *Solecurtus*, but more elongate and lacking the oblique rugae, and with short nymphs; left valve with only one cardinal tooth; hinge plate very slight; posterior adductor scar shifted antero-dorsally; pallial sinus extending forwards as far as, or beyond, the beak, partly confluent. Olig.-Rec. E.Atl., E.Pac., S.Amer., N.Amer., Carib., Medit., W.Africa. **Pharus** (Fig. 529): transversely oblong, anterior end rounded, posterior end slightly truncated, beak distance about two-fifths; about five times as long as high; right valve hinge with 3a and 3b (weak); left valve hinge with 4a, 2a, 2b and 4b; 2a and 2b not well separated; short, backwardly-sloping buttress; anterior adductor scar long, narrow, horizontal, occupying the second fifth of the length of each valve, reaching the buttress; posterior adductor scar as in *Siliqua*; pallial sinus obtusely pointed, scarcely reaching beyond the posterior adductor scar. L.Mio.-Rec. Eur., Medit.

Family SCROBICULARIIDAE
Orbicular or oval, usually flattish; resilium internal; cardinal teeth of tellinid type; no lateral teeth; pallial sinus deep; valve margins internally smooth; no posterior fold. (Pal.-Rec.)
Scrobicularia (Fig. 546): oval and subequilateral; commarginal growth lines only; hinge formula

$$\frac{3a, 3b}{2};$$

pallial sinus large, descending abruptly in front and resting directly on the pallial line; resiliophore triangular. Eoc.-Rec. Eur., Indonesia.

Family SEMELIDAE
Like the Scrobiculariidae, but usually fairly well ornamented, with external as well as internal ligament (the internal one in a small chondrophore), lateral teeth present, and a slight posterior fold. (Pal.-Rec.)
Semele [*Amphidesma*] (*s.s.*) (Fig. 547): form and pallial sinus of *Arcopagia*, but may be subcircular in outline; resiliophore rather long; both valves with two cardinal teeth and strong anterior and posterior laterals; ornament of commarginal lamellae frilled by radials; pallial sinus deep, ascendant, free of pallial line. Eoc.-Rec. Carib., N.Amer., S.Amer., Eur., W.Pakistan, most tropical seas. *Abra* [*Syndosmya*] (*s.s.*) (Fig. 548): small, oval, oval-transverse to trigonal, rather compressed, smooth; resiliophore well developed; right valve hinge with AI (distinct but slender), 3a, 3b (heavier than 3a) and PI (distinct but slender); left valve hinge with AII (very weak), 2a, 2b (very small or absent) and PII (very weak); pallial sinus deep, very wide, partly confluent with the pallial line. Pal.-Rec. Eur., N.Amer., Carib., M.East, Celtic, N.Atl. (abyssal), Indo-Pac., E.Pac.

Superfamily ARCTICACEA [CYPRINACEA, CYPRICARDIACEA]
Oval to elongate; beaks more or less prosogyrate and in front of mid-line; ligament external; hinge 'cyprinoid', with two or three cardinals in each valve and usually also distinct lateral teeth (the family Euloxidae has no laterals); pallial line usually entire; marine. (M.Dev.-Rec.)

Family ARCTICIDAE [CYPRINIDAE]

Suborbicular, ovate, trigonal or trapeziform, evenly inflated or with posterior carina; equivalve, inequilateral, closed; ornament of commarginal striae, or smooth; ligament on nymphs; valve margins usually internally smooth; beaks strongly prosogyrate, hinge rotated forward with them, making anterior laterals very short, cardinals sloping obliquely, posterior laterals relatively long; hinge formula:

$$\frac{\text{(AIII), AI, 3a, (1), 3b, (5b), (PI), (PIII),}}{\text{AII, (2a), 2b, 4b, (PII), (PIV),}}$$

teeth in parenthesis being irregular in their occurrence. (U.Trias.-Rec.)

Arctica [*Cyprina*] (Fig. 549): valves thick, suborbicular, rather inflated, with commarginal striae; lunule and escutcheon not defined; hinge plate heavy, well defined, its free margin double-curved; right valve hinge with AIII (small), AI (very small, conical, crenulated), 3a (narrowly triangular, sloping backwards), 1 (like 3a, only slightly divergent), 3b (long, horizontal, its crest sloping off in front and behind, bifid) and PI (far back, oblique, lamellar, long, striated); there is no socket for 2b, but a flat area of the hinge plate; left valve hinge with AII (pyramidal, jagged, crenulated), 2a (wedge-like, sloping back, close to AII), 2b (represented by a flat area not projecting beyond the median plane), 4b (thin and small) and PII (very slight); ligament long and thick, on fairly prominent nymphs; integripalliate. L.Cret.-Rec. Eur., Russia, ?India, Arctic, Boreal, Celtic, N.Amer.

Family TRAPEZIIDAE [LITHOPHAGELLIDAE]

Rather elongate, with beaks situated well forward; sometimes keeled posteriorly; smooth or with commarginal lamellae, sometimes also with fine radial ornament; hinge plate narrow, usually with two cardinals in each valve and one posterior and one small anterior lateral; usually integripalliate. (Cret.-Rec.)

Trapezium [*Libitina, Cypricardia*] (*s.s.*) (Fig. 550): valves thick, subovalsubtrapeziform, with subterminal beaks, posteriorly obliquely angular; ornament of commarginal lamellae with fine radial striae; ligament external; valve margins internally smooth; integripalliate; a nestler. Eoc.-Rec. Indo-Pac.,

---

Figs 546–554. BIVALVIA: TELLINACEA, ARCTICACEA
AND DREISSENACEA
Figs 547–548 and 550–551 after Keen, in Moore; Figs 547–548 and 550–551 original.
546. *Scrobicularia piperata* (Linné), Pleist. Sussex. × 1.
547b. *Semele proficua* (Pulteney), Rec. Carib. T. × 1.
548c. *Abra tenuis* (Montagu), Rec. England. T. × 1.
549a, b. *Arctica islandica* (Linné), × 0·75. *x'*, flat area corresponding to position of 2*b'*.
550c. *Trapezium oblongum* (Linné), Rec. Indonesia. T. × 1.
551c. *Euloxa latisulcata* (Conrad), Mio. Virginia. T. × 1.
552. *Dreissena polymorpha* Pallas, Rec. England. T. × 1·3.
553a, f. *Congeria subglobosa* (Partsch). T. a × 0·75, f × 0·5.
554c. *Congeria rhomboidea* M. Hoernes, Pontian. Rumania. × 0·75.

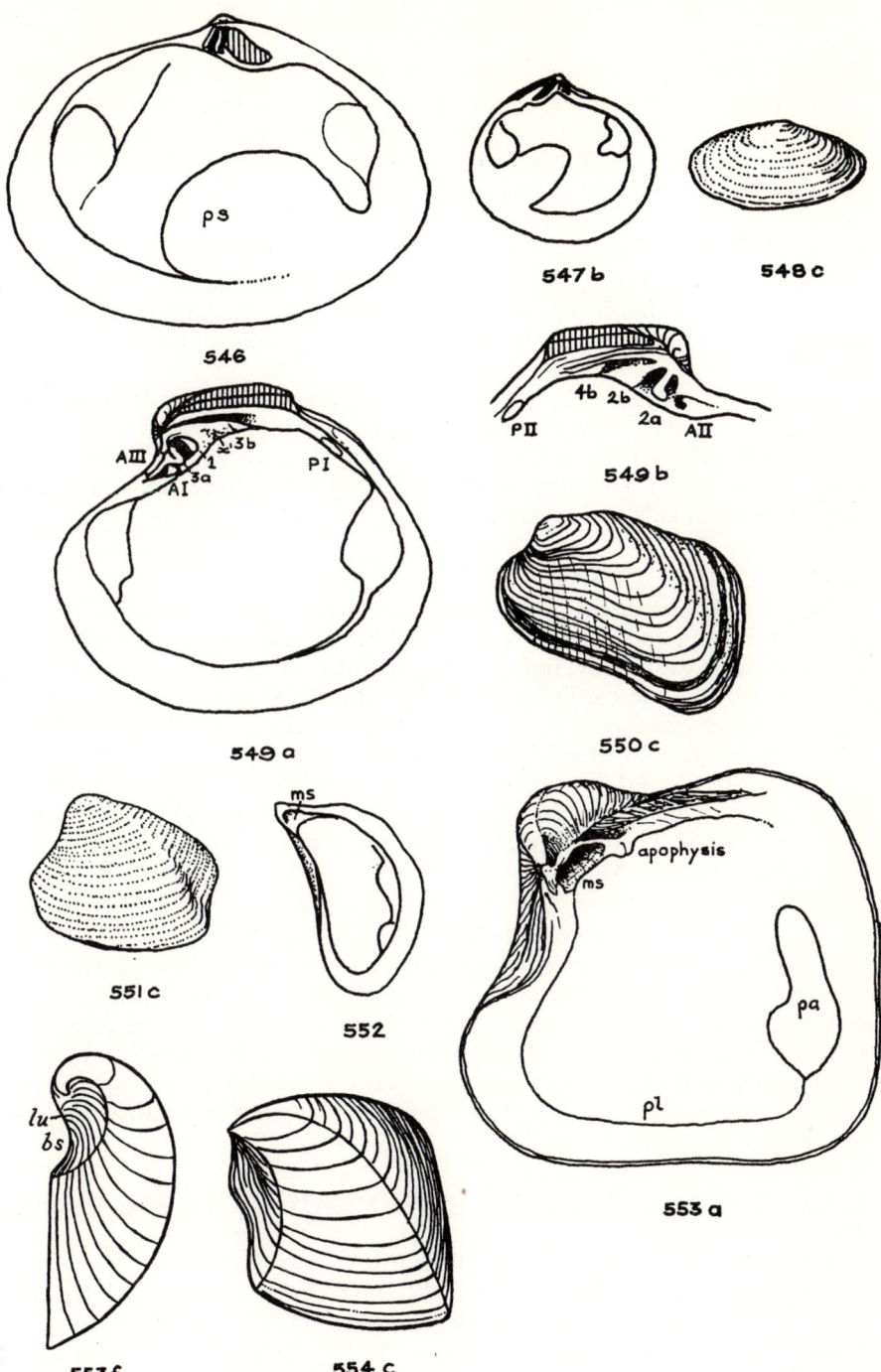

546

547 b

548 c

AIII
AI
1
3b
3a
PI

549 a

PII
4b 2b
2a AII

549 b

550 c

551 c

ms

552

ms
apophysis
pa
pl

553 a

lu-
bs

553 f

554 c

N.Amer., India, Eur. **Coralliophaga** (*s.s.*): similar to *Trapezium,* but elongate modioliform, valves thin, with fine radial ornament and lamellar commarginal ornament posteriorly, with a slight posterior gape, and with a small, wide pallial sinus; right valve hinge with 1, 3b and PI; left valve hinge with AII, 2, 4b and PII. Eoc.-Rec. Eur., Carib., Indonesia.

Family EULOXIDAE

Subtriangular-subtrapeziform, with or without a posterior keel having a depressed area behind it; ornament of commarginal undulations and threads or with radial costellae as well; no lateral teeth; pallial sinus practically nil. (Mio.)

Euloxa (Fig. 551): beaks prosogyrate, situated well forward; right valve hinge with 1 (large, triangular) and 3b (narrow, sloping backwards, with a median groove); left valve hinge with 2a (small, narrow), 2b (large, thick, wedge-shaped) and 4b (long, narrow, slightly curved); valve margins internally smooth. U.Mio. N.Amer.

Family KELLIELLIDAE

Very small, closed, equivalve, suborbicular to gently caudate, with beaks rather prominent and anterior to the mid-line; no gape; smooth or with commarginal striae or threads; ligament usually internal; anterior laterals long or close to and like the cardinals; posterior laterals inconstant; cardinal teeth variable, never more than two in each valve; lunule present; integripalliate; valve margins smooth internally. (Eoc.-Rec.)

Kelliella [*Kellyella*] (Fig. 558): inflated, suborbicular, beaks well anterior to mid-line; ligament feeble, external; right valve hinge with AI (close to cardinals), 1 and 3b (obsolete); left valve hinge with 2 (widely bifid) and 4b. Eoc.-Rec. Eur., Atl., Medit.

Superfamily DREISSENACEA

Mytiliform, modioliform or rhomboid, beaks nearly or wholly terminal; anterior surface usually flattened or concave, almost at right angles to the general surface; a slight byssal gape usually present; ligament marginal, subinternal; hinge edentulous, always with a septum in the anterior angle, on which is situated the anterior adductor scar (cf. *Septifer*); internally porcellanous to subnacreous (cf. Mytilidae); posterior adductor scar rather elongate; fresh-water. (Eoc.-Rec.)

Family DREISSENIDAE [DREISSENSIIDAE] (I.C.Z.N. Direction 41)

Characters of the Superfamily. (Eoc.-Rec.)

Dreissena [*Dreissensia*] (*s.s.*) (Fig. 552): mytiliform, with pointed, terminal beaks which are bent gently forwards; internal septum in umbonal region, but no apophysis as in *Congeria*; smooth; byssal depression on right valve; posterior adductor scar sub-bilobed; integripalliate. Eoc.-Rec. Eur., Africa.

Congeria (Figs 553, 554): very variable in form, primitively mytiliform, some-times developing ears, in other cases becoming rhomboidal or globose; smooth or with fine commarginal marking; myophoric septum thick, with an apophysis behind it; posterior adductor scar often '6'-shaped; integripalliate. L.Olig.-Plio. Eur., W.Asia. [Primitive species are difficult to separate from *Dreissena*; in the U.Miocene and L.Pliocene of E.Europe the genus attained its maximum of abundance, size and variety of form, but it rapidly became extinct after the Pontian; the supposed Recent African and American species are now referred to *Mytilopsis* (see below).] *Dreissenomya* (Fig. 555): beaks prosogyrate, subterminal; anterior outline a normal convex curve (atypical for the family); modioliform, a little expanded posteriorly; gaping behind, and with a pallial sinus (atypical for the family); septum obsolete. Plio. Aralo-Caspian. *Mytilopsis:* resembles the more primitive forms of *Congeria*; mytili-form, with antero-ventral marginal gape; commarginal growth lines; posterior adductor scar long, bilobed; apophysis directed downwards into the umbonal cavity, with two muscle scars as in *Congeria*. L.Mio.-Rec. S.Amer. (Mancora), W.Africa, Eur., Indonesia. [Species of this genus have been referred to *Congeria*, as have a small number of Recent species from Equatorial West Africa.]

Superfamily ISOCARDIACEA [GLOSSACEA]

Suborbicular to subtrigonal, with commarginal striae or costae, sometimes with a well defined siphonal area, inequilateral, usually equivalve; beaks strongly prosogyrate, sometimes spirally enrolled, so that the cardinal teeth are very oblique; hinge 'cyprinoid', with two or three cardinals in each valve, teeth 2a and 3a more or less obsolete, usually with good laterals; ligament external; usually integripalliate. (U.Trias.-Rec.)

Family ISOCARDIIDAE [GLOSSIDAE]

Hinge teeth dragged forward by the prosogyrate spiral beaks, the cardinals becoming horizontal instead of vertical; no definite lunule or escutcheon, al-though lunular area is depressed; ligament and resilium deeply sunk; two lamellar cardinals in each valve, laterals variable; integripalliate; valve margins internally smooth. (Pal.-Rec.)

The names Glossacea and Glossidae, dating from Poli, are not here accepted; Poli gave different names to the shell and the soft parts, and this is not strictly Linnéan.

Isocardia [*Glossus*] (*s.s.*) (Fig. 556): suborbicular-cordiform, globose, siphonal area not defined; smooth or with commarginal growth lines; all hinge teeth lamellar and horizontal; right valve hinge with 1 (short, thin), 3b (longer, bilobed) and PI (small); left valve hinge with 2b (thin, bilobed), 4b (long) and PII (strong, triangular); no escutcheon; lunular area depressed; nymph prominent, long. L.Olig.-Rec. Celtic, Eur. *Meiocardia* [*Miocardia*] (Fig. 557): inflated, subtrigonal, with the siphonal area bounded by a sharp keel forming an angular salient on the outline; ornament of rather strong commarginal furrows; hinge formula

245

$$\frac{\text{AI, 1, 3b, PI, PIII}}{\text{AII, 2b, 4b, PII}}$$.

Pal.-Rec. Eur., M.East, Burma, India, Indo-Pac.

Family VESICOMYIDAE [PLIOCARDIIDAE]
Suborbicular, ovate, to transversely elliptical; lunule usually well defined; hinge with up to three teeth; valve margins usually internally smooth; integripalliate or sinupalliate. (L.Mio.-Rec.)

Vesicomya (*s.s.*): suborbicular, beaks anterior to midline; smooth; lunule delimited by an incised line; hinge like that of *Isocardia*, but without lateral teeth. Mio.-Rec. Atl., Eur., Indonesia. *Pliocardia* (Fig. 559): posteriorly rostrate, the postero-ventral margin being gently emarginate; ornament of commarginal rugae; valves thick; beaks well anterior to mid-line; lunule as in *Vesicomya*; right valve hinge with AI (broad, short) and 3b (long, curved, thicker posteriorly); left valve hinge with 2 (widely bifid, posterior part heavier) and 4b (thin, long); no posterior laterals; valve margins internally with oblique grooves except at posterior end. Plio. Carib. [In the original description of the genus Woodring (**42**) considered all the hinge teeth to be anterior lateral teeth.]

Superfamily GAIMARDIACEA
Small, valves thin, equivalve, swollen, beaks anterior to mid-line; smooth, or with commarginal ornament or radial costae; ligament opisthodetic, external or sunk; hinge plate thin, usually with one cardinal tooth in left valve, and right valve with 1 beneath bifid 3; lateral teeth sometimes weak or absent; integripalliate; byssiferous. (Mio.-Rec.)

Family GAIMARDIIDAE
Characters of the Superfamily. (Mio.-Rec.)

Gaimardia is Pleistocene to Recent only. *Kidderia:* obliquely suboval, with broad umbones, smooth; ligament deeply sunk; no gape and sinuosity of the ventral margin as in *Gaimardia*. Mio.-Rec. N.Z., southern seas.

Figs 555–565. BIVALVIA: ARCTICACEA, DREISSENACEA, ISOCARDIACEA, CORBICULACEA AND VENERACEA
Figs 558–559 and 564 after Keen, in Moore; Figs 560–561 after Keen and Casey, in Moore; Figs 555–557, 563 and 565 original.
555. *Dreissenomya aperta* (Deshayes), Pontian. Rumania. $\times$ 0·64.
556. *Isocardia humana* (Linné), Rec. England. T. $\times$ 0·5.
557. *Meiocardia moltkiana* (Gmelin), Rec. Indo-Pac. T. $\times$ 0·39.
558b, c. *Kelliella miliaris* (Philippi), Rec. Norway. T. a $\times$ 6, c enlarged.
559d. *Pliocardia bowdeniana* (Dall), Plio. Jamaica. T. $\times$ 4.
560c. *Corbicula fluminalis* (Müller), Rec. Asia Minor. T. $\times$ 1.
561c. *Polymesoda* (*Geloina*) *coaxans* (Gmelin), Rec. Indonesia. T. $\times$ 0·25.
562a, d. *Pisidium amnicum* (Müller), Rec. Eur. T. $\times$ 3.
563a, b. *Venus verrucosa* Linné, Rec. Britain. T. $\times$ 0·75.
564c. *Ventricolaria rigida* (Dillwyn), Rec. Carib. T. $\times$ 0·5.
565. *Periglypta reticulata* (Linné), Rec. $\times$ 0·75.

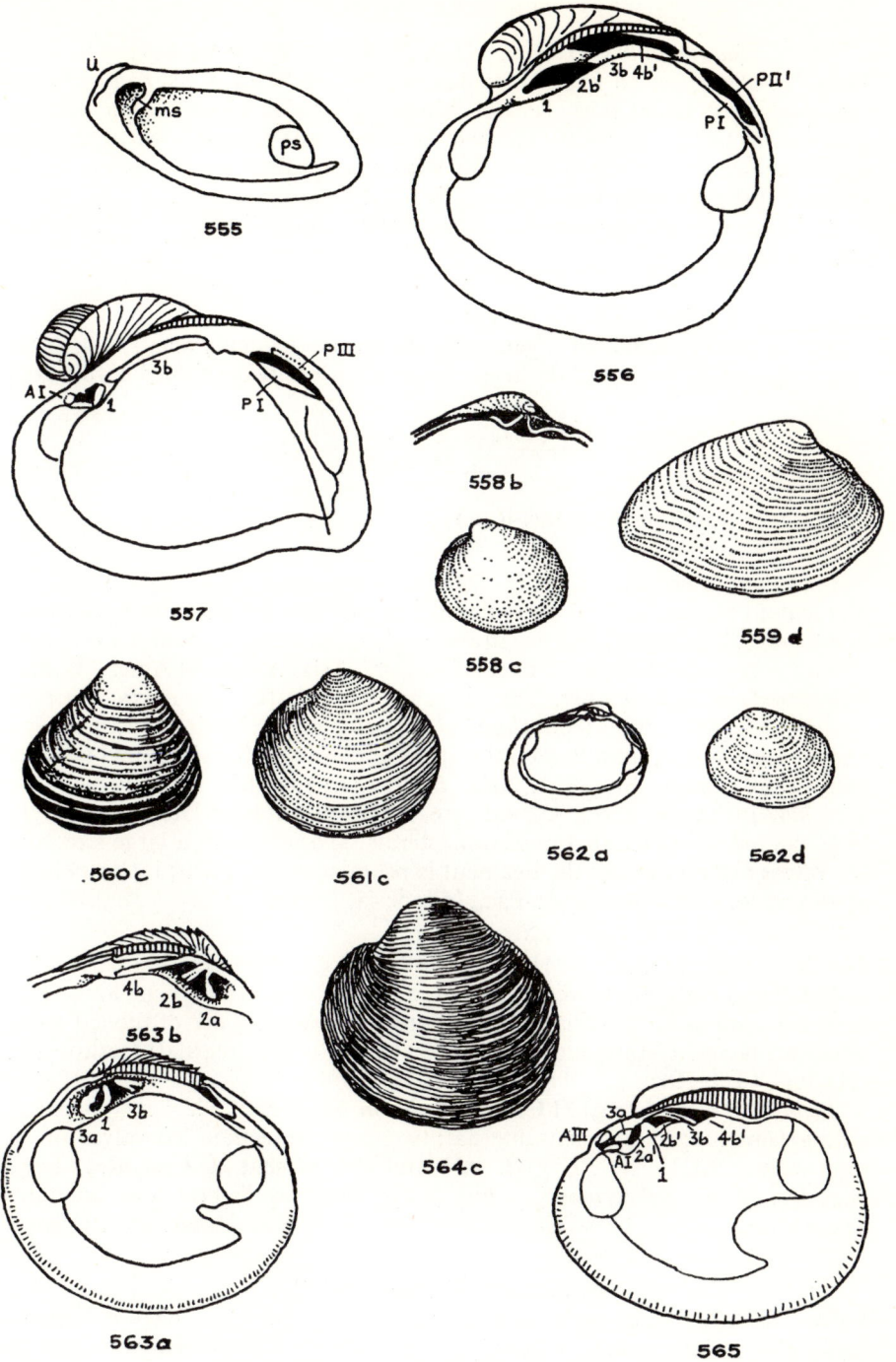

555

556

557

558 b

558 c

559 d

.560 c

561 c

562 a

562 d

563 b

563 a

564 c

565

247

Superfamily CORBICULACEA [CYRENACEA]

Rounded-triangular to oval, porcellanous, usually with weak commarginal ornament, with beaks fairly central; hinge corbiculoid when fully developed, with lateral teeth; ligament external, on prominent nymphs; valve margins internally smooth; integripalliate, or with small pallial sinus; mainly fresh-water, but occasionally estuarine, some fossil forms even marine. (Jur.-Rec.)

As in other fresh-water molluscs the periostracum is thick, yet the beaks are often eroded.

Family CORBICULIDAE [CYRENIDAE]

Valves thick; pallial sinus generally absent, occasionally small; ornament nearly always of commarginal striae; usually with fully developed corbiculoid hinge:

$$\frac{\text{AIII, AI, 3a, 1, 3b, PI, PIII}}{\text{AII, 2a, 2b, 4b, PII}};$$

laterals sometimes serrate. (Jur.-Rec.)

Corbicula [*Cyrena*] (*s.s.*) (Fig. 560): oval to subtrigonal; fully developed corbiculoid hinge; cardinal teeth all bifid or grooved; lateral teeth longer than in *Polymesoda* (*Geloina*) and usually strongly cross-striated; pallial line simple or with a very slight sinus. U.Cret.-Rec. Eur., Asia. Africa, N.Amer., W.Pac. *Polymesoda* is Recent only. The subgenus **Geloina** [*Cyrena auctt. non* type species] (Fig. 561) differs from *Corbicula* in being larger, having thicker valves, and having shorter, smooth lateral teeth; there is no pallial sinus, and teeth 3a and 4b are simple. Eoc.-Rec. Asia. *Batissa* (*s.s.*): differs from *Polymesoda* (*Geloina*) in having short, curved, cross-striated anterior laterals and very long, curved, cross-striated posterior laterals; species attain a large size, and the valves are very thick; the ligament is prominent and thick; integripalliate or almost so. U.Jur.-Rec. Indo-Pac., Indonesia, Burma.

Family PISIDIIDAE [SPHAERIIDAE] (I.C.Z.N. Declaration 27)

Small and with thin valves, ovate, subquadrate or subtriangular; hinge very narrow; cardinal teeth variable, usually two per valve, those of the left valve separate; lateral teeth elongate; ligament more or less internal; fresh-water. (?U.Jur., Cret.-Rec.)

Pisidium [*Pisum*] (*s.s.*) (I.C.Z.N. Opinion 335) (Fig. 562): inequilateral, anterior end longer; ligament internal; two cardinal teeth in left valve, one in right valve; AIII, AI, PI, PIII, AII and PII present. U.Cret.-Rec. Eur., N.Amer., Asia. *Sphaerium* [*Cyclas*] (*s.s.*): more equilateral and generally larger than *Pisidium*; beaks rounded. ?U.Jur., Cret.-Rec. Holarctic, Africa.

Superfamily VENERACEA

Valves usually thick; ornament predominantly commarginal, sometimes radial also, striate, costate or lamellate, occasionally spiny (chiefly on anterior boundary of siphonal area); beaks usually anterior to mid-line, prosogyrate; ligament external, opisthodetic, nymph not rising above dorsal margin; cardinal teeth 3a, 1, 3b, 2a, 2b and 4b usually present; lateral teeth variable, AII often rudimentary or absent; pallial sinus more or less developed, seldom entirely lacking. (L.Cret.-Rec.)

Family VENERIDAE

Equivalve, closed; lunule and escutcheon usually well marked; cardinal teeth typically corbiculoid, 1 and 2b usually thicker than 3a and 2a; anterior laterals well developed, obsolete or wanting; posterior laterals feeble or wanting; pallial sinus varying in size and shape; valve margins internally smooth or crenulate (**68, 66a**). (L.Cret.-Rec.)

This may be regarded as the dominant bivalve family of the Tertiary era, as scarcely any other shows such an increase in numbers and variety. It is now divided into eleven subfamilies, as follows.

Subfamily VENERINAE

Ornament often both radial and commarginal, sometimes commarginal only; anterior lateral tooth present, especially in left valve. (M.Eoc.-Rec.)

Venus [*Clausina, Ventricola*] (*s.s.*) (Fig. 563): outline rounded, with less curvature along postero-dorsal margin and slight posterior truncation; beak distance one-quarter to one-fifth; lunule medially convex, bounded by a groove; escutcheon ill-defined; sculpture of strong commarginal ridges, somewhat scaly, crenulated by discontinuous radial sculpture (the latter often more conspicuous on worn shells), with divaricate radial costae posteriorly; right valve hinge with AI (minute), 3a (lamellar, oblique, nearly parallel to lunule), 1 (thick, vertical, curved, trigonal) and 3b (long, straight, backwardly inclined, slightly bifid); left valve hinge with AII (minute), 2a (oblique, forwardly inclined), 2b (thick, grooved, backwardly inclined) and 4b (long, thin, nearly horizontal); pallial sinus short, pointed; valve margins internally crenulate. L.Mio. (Aquit.)-Rec. Eur., Medit., N.Africa. *Ventricoloidea:* ovate-subtriangular, with commarginal lamellae and fine threads in between, but no radial ornament; right valve hinge with AIII, AI, 3a, 1 (bifid) and 3b (bifid); left valve hinge with AII, 2a, 2b (bifid) and 4b; pallial sinus shallow, pointed; valve margins internally crenulate. Olig.-Rec. Eur., W.Pakistan, N.Amer., Indonesia. *Ventricolaria* (Fig. 564): suborbicular, globose; no radial ornament; lunule deeply sunken, smooth and flat. Olig.-Rec. N.Amer., Carib., S.Amer., Medit., E.Pac. *Periglypta* [*Cytherea* Bolten *non* Fabricius] (Fig. 565): ovate, inflated, beak distance one-sixth to one-seventh; reticulate ornament; ligament sunk; right valve hinge with AIII (very small), AI (very small), 3a (narrow, subparallel to lunular margin), 1 (bifid) and 3b (bifid); left valve hinge with AII (small), 2a, 2b (strongly bifid) and 4b (simple, narrow, long); pallial sinus short, blunt, squarish; valve margins internally crenulate. Olig.-Rec. Eur., M.East, W.Pakistan, India, Burma, Indo-Pac., E.Pac., Carib., N.Amer.

Subfamily CIRCINAE

Sculpture usually of more or less dichotomous radial costae; equivalve, often subequilateral; anterior lateral teeth present; practically no pallial sinus. (Pal.-Rec.)

Circe (*s.s.*) (Fig. 566): compressed, ovate-subtriangular to subquadrate, ornamented with commarginal lamellae; prodissoconch flattened, the beaks pointed, low, with more or less divaricate ornament; ligament deeply sunk; right valve hinge with AIII, AI, 3a (thin, short), 1 (perpendicular, subtriangular) and 3b (narrow, oblique, bifid); left valve hinge with AII (rather long), 2a (perpendicular), 2b (thick, bifid) and 4b (long, thin); pallial sinus very shallow; valve margins internally smooth. L.Olig.-Rec. Eur., Indo-Pac., Indonesia.

Subfamily SUNETTINAE

Ligament in a very deeply excavated escutcheon; shell smooth or with commarginal ornament; anterior laterals present, elongate; small but distinct pallial sinus. (Eoc.-Rec.)

Sunetta (*s.s.*) (Fig. 567): transversely oval-trigonal, compressed, beaks a little behind mid-line; ornamented with fine commarginal grooves; lunule long and narrow; escutcheon long and deeply re-entrant; right valve hinge with AIII (long), AI (long), 3a, 1 and 3b; left valve hinge with AII, 2a, 2b and 4b (short); pallial sinus semi-elliptical, moderately deep; valve margins internally crenulate. Mio.-Rec. Eur., India, Austral., Indo-Pac. *Meroena* (Fig. 568): similar to *Sunetta*, but more equilateral, 1 and 2b are bifid and 3b well grooved; valve margins internally smooth. Eoc. Eur. *Dosiniopsis* (Fig. 569): suborbicular, with fine commarginal ornament, rather compressed; right valve hinge with AIII, AI, 3a, 1 (bifid), 3b (bifid) and PI (long); left valve hinge with AII, 2a (bifid), 2b and 4b (bifid); lunule ill-defined (cf. *Dosinia*); pallial sinus short, equilateral-triangular, not ascendant (cf. *Dosinia*); valve margins internally smooth; AI and AII coarsely grooved. Pal.-U.Eoc. Eur., N.Amer. [Sometimes placed in the Pitarinae.]

Subfamily MERETRICINAE

Valves usually longer than high, with subdued ornament; hinge with

---

Figs 566–575. BIVALVIA: VENERACEA

Figs 566–567, 569–571 and 573 after Keen, in Moore; Figs 568, 572 and 574–575 original.
566d. *Circe scripta* (Linné), Rec. Pac. T. × 0·75.
567d. *Sunetta scripta* (Linné), Rec. Pac. T. × 0·75.
568. *Meroena semisulcata* (Lamarck), M.Eoc. Paris B. × 1·25.
569a, c. *Dosiniopsis meeki* Conrad, Eoc. Maryland. T. × 0·5.
570d. *Meretrix meretrix* (Linné), Rec. Indonesia. T. × 0·5.
571b, d. *Tivelina rustica* (Deshayes), Eoc. France. T. × 1·5.
572. *Tivela baini* R. B. Newton, Neogene (Alexandria). S.Africa. × 0·6.
573. *Grateloupia irregularis* (Basterot), Mio. T. × 1.
574. *Grateloupia* (*Cytheriopsis*) *hydana* (Conrad) [= *moulinsi* Lea], M.Eoc. (Claiborne). Alabama. × 1·25.
575. *Pitar perovatus* (Conrad), M.Eoc. (Claiborne). Alabama. × 1·4.

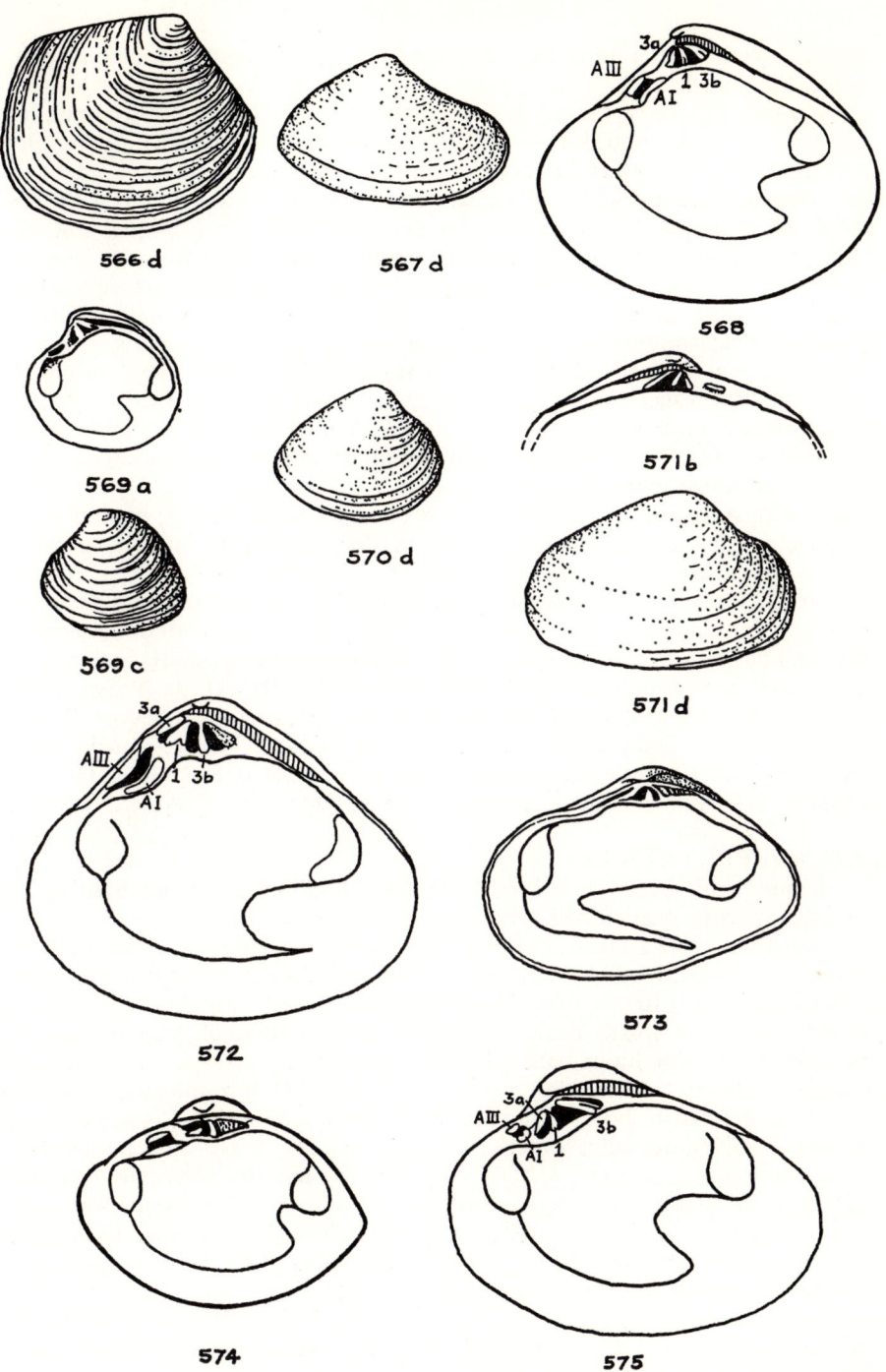

566 d

567 d

568

569 a

570 d

571 b

569 c

571 d

572

573

574

575

cardinal teeth tending to radiate, and anterior laterals usually double in right valve. (U.Cret.-Rec.)

Meretrix [*Cytherea* Lamarck *non* Fabricius] (Fig. 570): ovoid-trigonal, inequilateral; beaks anterior to mid-line, prosogyrate; smooth; no lunule; escutcheon ill-defined; right valve hinge with AIII, AI, 3a (thin), 1 (more or less perpendicular) and 3b (oblique, semi-bifid); left valve hinge with AII (trigonal, prominent), 2a (perpendicular, semi-bifid), 2b (oblique) and 4b (thin, prominent, horizontal); pallial sinus short, a gentle arch; valve margins internally smooth. U.Mio.-Rec. Indonesia. *Tivelina* (Fig. 571): not very large, oval-cuneiform, rather compressed; post-umbonal ridge with a shallow furrow parallel to it; smooth or with commarginal ornament; hinge like that of *Meretrix*, but all teeth simple and 2b thick; pallial sinus short, rounded. L.Eoc.-L.Mio. Eur. *Tivela* (*s.s.*) (Fig. 572): trigonal, height nearly equal to length, equilateral; lunule long, narrow, ill-defined; escutcheon short, well-defined; smooth or with fine commarginal ornament; nymph subdivided; right valve hinge with AIII, AI, 3a (thin), 1 (stout) and 3b (thin); left valve hinge with AII, 4a (small), 2a (thin), 2b (stout) and 4b (stout); behind 4b, a sloping rough area as in *Mercenaria*, but smaller; hinge plate short, high; pallial sinus elliptical, reaching mid-line. Mio.-Rec. N.Amer., C.Amer., W.Africa, Indo-Pac. *Grateloupia* (*s.s.*) (Fig. 573): differs from *Tivela* in its elongate shape, nearly $1\frac{1}{2}$ times as long as high, with consequent differences in the inclination of the various hinge structures; subtrigonal, posteriorly attenuated, subequilateral; right valve hinge with AIII, AI, 3a (large), 1, 3b and PI; left valve hinge with AII, 4a, 2, 4b and PII; rugosities behind 4b and 3b; pallial sinus fairly deep, at least reaching midline. L.Mio.-M.Mio. Eur. The subgenus *Cytheriopsis* [*Grateloupina*] (Fig. 574) is less elongate, subrhombic in outline, and has a short pallial sinus. Eoc.-Mio. N.Amer., S.Amer.

Subfamily PITARINAE

Inequilateral, beaks anterior to mid-line; cardinal teeth not tending to radiate; anterior laterals well developed. (L.Cret.-Rec.)

Pitar [*Pitaria*] (*s.s.*) (Fig. 575): ovate to ovate-subtriangular, moderately to strongly inflated, rounded behind; beaks prominent, prosogyrate, beak distance one-third to one-fifth; lunule cordate, limited by an incised line, not sunk; escutcheon long, feeble; ornament of commarginal growth lines or threads; right valve hinge with AIII, AI, 3a (vertical), 1 (close to 3a) and 3b (nearly horizontal, bifid); left valve hinge with AII (prominent), 2a (thin, straight), 2b (heavy, short, triangular, bevelled on top) and 4b (long, slender); 2a and 2b joined dorsally, sometimes also 3a and 3b; pallial sinus moderately deep, attaining mid-length, tongue-shaped; valve margins internally smooth. Pal.-Rec. W.Africa, Eur., Asia, Indo-Pac., N.Amer., widespread. The subgenus *Calpitaria* is similar to *Pitar*, but the pallial sinus is more U-shaped. Eoc. Eur., N.Amer., W.Pakistan. *Macrocallista* (Fig. 576): large, transversely elliptical, smooth and polished; beak distance about one-quarter to one-third; lunule narrowly cordiform, limited by an incised line; escutcheon undefined; right valve hinge with AIII (low), 3a, 1 (very thin, prominent, close to 3a) and 3b (thin, long, bifid); left valve hinge with AII (prominent), 2a (grooved), 2b (triangular, thick) and 4b (very thin); pallial sinus not quite reaching mid-line, somewhat rhomboidal; valve margins internally smooth. Eoc.-Rec. W.Atl., N.Amer., Indonesia. *Chionella* (*s.s.*) (Fig. 577): ovate-trigonal, not very elongate, smooth and polished; beak distance one-third; right valve hinge with AIII, AI, 3a, 1 and 3b (narrow, grooved); left valve hinge with AII, 2a, 2b (bevelled) and 4b; pallial sinus not reaching mid-line, U-shaped to subrhombic; valve margins internally smooth. Eoc.-Olig. Eur., W.Pakistan. The subgenus *Costacallista* [*Callista* Mörch *non* Leach] (Fig. 578) is similar to *Chionella*, but the surface is ornamented with commarginal obliquely incised lines separated by intervening flat ribbons, and 2a is also grooved; lunule well marked, narrow; escutcheon indistinct. Pal.-Rec. Indo-Pac., Eur., Medit., Egypt, N.Amer., Zanzibar, W.Pakistan, India, Burma, Indonesia, N.Z., widespread. [This has sometimes been regarded as a sub-genus of *Macrocallista*, but the much less elongate outline and more condensed hinge seem to group it with *Chionella*.] *Amiantis* (*s.s.*) (Fig. 579): large, ovate to ovate-trigonal, with swollen umbones and commarginal wrinkles; valves thick; right valve hinge with AIII, AI, 3a (small), 1 and 3b; left valve hinge with AII, 2a, 2b and 4b (very long); pallial sinus only moderately deep, pointed; valve margins internally smooth; escutcheon long, narrow, moderately deep; 2a–2b and 3a–3b are united above in arch-form, and there is a roughened area between the ligament and 4b, the rugosities extending over part of the surface of 4b and its socket. Mio.-Rec. N.Amer. (Vaqueros). *Pelecyora* [*Sinodia*] (*s.s.*) (Fig. 580): trigonal and high to orbicular, strongly inflated; umbones tumid, strongly prosogyrate; beak distance about one-third; lunule large, cordate, weak, limited by an incised line; escutcheon indistinct; ornament of simple growth lines or smooth, flat bands separated by narrow incised lines; right valve hinge with AIII, AI, 3a (short), 1 and 3b (bifid); left valve hinge with AII (feeble, situated low down), 2a, 2b (heavy) and 4b; pallial sinus not very short, ascendant, its tip subangular; valve margins internally smooth. M.Eoc.-Rec. Eur., N.Amer., Red Sea, Indonesia. The sub-genus *Cordiopsis* (Fig. 581) has less tumid umbones, the escutcheon has an obtuse angulation at its border, there is a flat space between 3b and the nymph, AII is nearer 2a, and the pallial sinus is a little shorter and more rounded. L.Eoc.-Mio. Eur., Egypt, Iran, Russia, W.Pakistan, India. *Saxidomus* (Fig. 582): transversely oval-subquadrate, large, with ornament of commarginal rugae; valves thick; no lunule or escutcheon; cardinals 1, 3b, 2b and 4b bifid; small AI and AII present; pallial sinus deep, tongue-shaped, horizontal; adductor muscle scars large, deeply impressed. L.Mio.(Aquit.)-Rec. N.Amer., Japan.

Subfamily DOSINIINAE

Lenticular, usually suborbicular (rarely subquadrate), usually with commarginal striae or threads (rarely smooth); left valve with an anterior lateral tooth. (U.Cret.-Rec.)

Dosinia (*s.s.*): usually orbicular and compressed, with well marked lunule,

253

and somewhat straighter post-umbonal margin; beaks prosogyrate; beak distance about one-quarter; sculpture finely commarginal; right valve hinge with AIII, AI, 3a (short), 1 (parallel to 3a) and 3b (narrowly bifid); left valve hinge with AII (close to 2a), 2a (thin, vertical), 2b (large, unequally bifid) and 4b (thin, curved); pallial sinus rather deep, ascendant, pointed; valve margins internally smooth. L.Olig.-Rec. Tropics, Pac., N.Z. [The genus (*s.l.*) seems to have originated in New Zealand.] The subgenus *Asa* (Fig. 583) differs from *Dosinia* (*s.s.*) in having a long, narrow escutcheon. Mio.-Rec. Eur., W.Africa.

Subfamily CYCLININAE

Resembling Dosiniinae in form, but without anterior lateral teeth or incised lunule. (L.Cret.-Rec.)

Cyclina (*s.s.*) (Fig. 584): suborbicular, not strongly inflated, with traces of radial ornament as well as growth lines; no lunule or escutcheon; right valve hinge with 3a, 1 and 3b (bifid); left valve hinge with 2a (bifid), 2b (bifid) and 4b (weak); pallial sinus deep, angular; valve margins internally crenulate. Olig.-Rec. Asia.

Subfamily GEMMINAE

Small, with marginal grooves and denticles but no lateral teeth; valve margins internally crenulate. (Eoc.-Rec.)

Gemma (Fig. 585): small, subequilateral, ovate-subtrigonal, with fine commarginal striae; lunule indefinite; right valve hinge with 3a (rudimentary), 1 and 3b; left valve hinge with 2a, 2b (conical, arcuate) and 4b; pallial sinus triangular, narrow, ascendant. Eoc.-Rec. N.Amer., W.Atl.

Subfamily CLEMENTIINAE

Inequilateral, without escutcheon; valves thin; hinge without lateral teeth; sculpture subdued or lacking; valve margins internally smooth. (U.Cret.-Rec.)

Clementia (*s.s.*) (Fig. 586) (74): ovate, very inequilateral, inflated, orna-

---

Figs 576–587. BIVALVIA: VENERACEA

Figs 576, 579–580, 582 and 584–587 after Keen, in Moore; Figs 577–578, 581 and 583 original.

576c. *Macrocallista nimbosa* (Lightfoot), Rec. Carib. T. × 0·25.
577. *Chionella laevigata* (Lamarck), M.Eoc. Paris B. × 1·1.
578a, b. *Chionella* (*Costacallista*) *erycina* (Lamarck), Rec. T. × 0·75.
579c. *Amiantis callosa* (Conrad), Rec. Calif. T. × 0·23.
580. *Pelecyora hatchetigbeensis* (Aldrich), Eoc. T. × 1.
581. *Pelecyora* (*Cordiopsis*) *suborbicularis* (Goldfuss), U.Eoc. (Headon). I. of Wight. T. × 0·75.
582c. *Saxidomus nuttalli* Conrad, Rec. Calif. T. × 0·5.
583. *Dosinia* (*Asa*) *lentiformis* (J. Sowerby), Pleist. Walton, Essex. × 0·75.
584c. *Cyclina sinensis* (Gmelin), Rec. China. T. × 0·5.
585d. *Gemma gemma* (Totten), Rec. W.Atl. T. × 7·5.
586c. *Clementia papyracea* (Gray), Rec. Pac. T. × 0·5.
587c. *Paphia rotundata* (Linné), Rec. W.Pac. T. × 0·5.

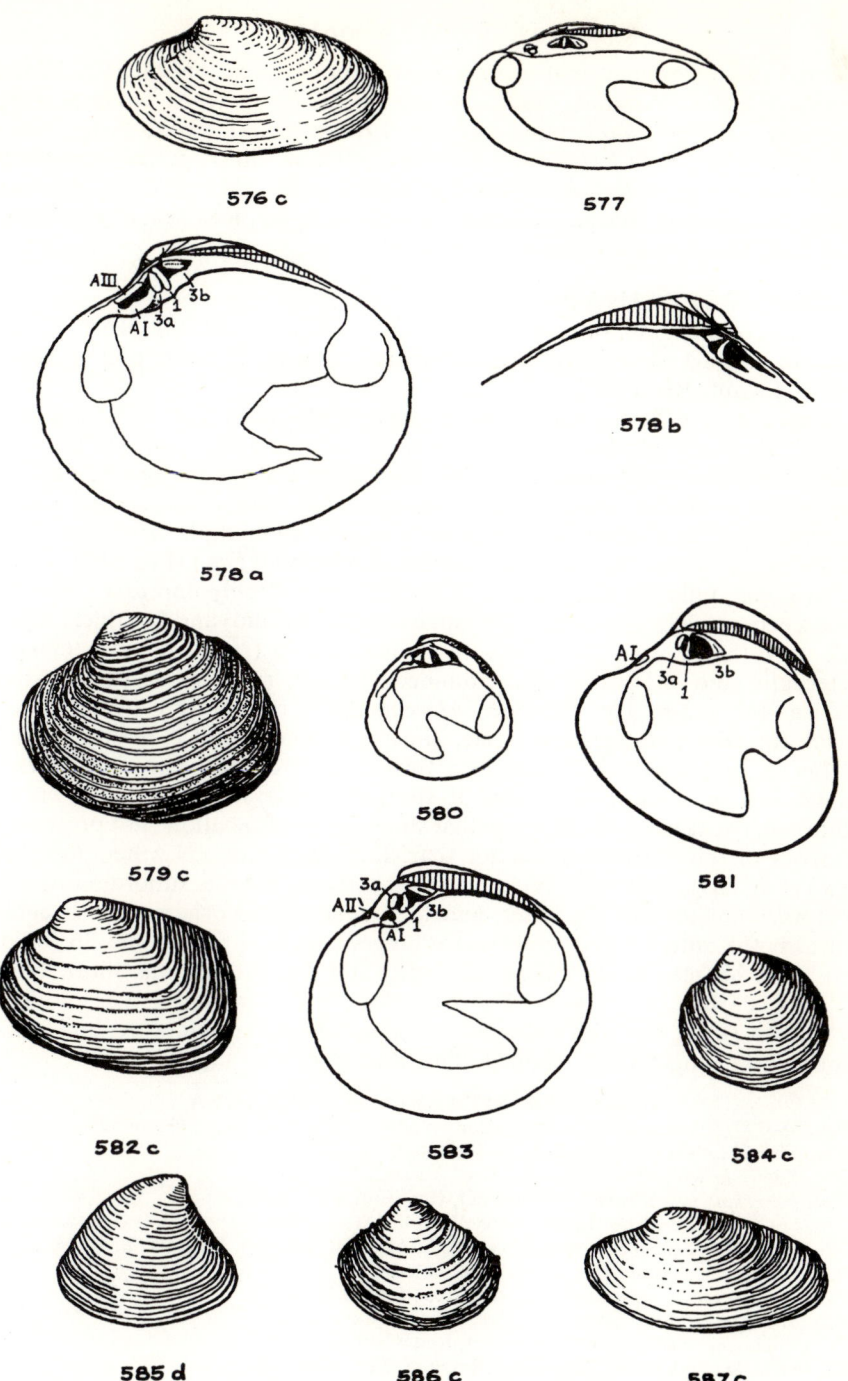

576 c

577

AIII
AI 3a 1 3b

578 a

578 b

579 c

580

AI
3a 3b
1

581

582 c

3a
AII
AI 1 3b

583

584 c

585 d

586 c

587 c

mented with commarginal coarse corrugations (seen on internal casts); ligament short, external, sunk; no lunule or escutcheon; cardinal teeth concentrated; right valve hinge with 3a, 1 (close to 3a) and 3b (long, bifid); left valve hinge with 2a, 2b and 4b (long, oblique, thin); pallial sinus long, wide, ascendant, tapering to an asymmetric apex. M.Eoc.-Rec. Austral., India, W.Pakistan, Burma, E.Africa, Indonesia, Egypt, Red Sea, Iran, Austria, C.Amer., Carib., widespread in warm seas.

Subfamily TAPETINAE

Ovate to elongate, often polished; valve margins internally nearly always smooth, at least posteriorly; hinge plate narrow, with 3a entire and no lateral teeth. (L.Cret.-Rec.)

Tapes (*s.s.*) is Pleistocene to Recent only. *Paphia* (*s.s.*) (Fig. 587): similar to *Tapes*, regularly elongate-oval, usually almost smooth, and pallial sinus short, subrectangular, ascendent. Plio.-Rec. Indo-Pac. The subgenus *Callistotapes* has a similar outline, but is ornamented with commarginal deeply incised lines like *Costacallista*, and 4b is nearer the nymph. Mio.-Rec. Eur., W.Pakistan, India, Burma, N.Z. *Marcia* [*Levimarcia*] (*s.s.*) (Fig. 588): ovate-subtrigonal, inflated, inequilateral, smooth; lunule feebly impressed, convex; escutcheon undefined; right valve hinge with 3a, 1 (bifid) and 3b (thick, bifid); left valve hinge with 2a (bifid), 2b (bifid) and 4b (rugose); valve margins internally smooth; pallial sinus rounded, distinct, not very deep. Mio.-Rec. Africa, Indonesia. The subgenus *Mercimonia* is transversely ovate to subquadrate, with commarginal striae, and the posterior end is blunter; lunule rather large, excavated, limited by a furrow; escutcheon narrow; right valve hinge with 3a, 1 (close to 3a) and 3b (narrow, bifid); left valve hinge with 2a (thin), 2b (weak) and 4b (thin); pallial sinus extremely shallow, not protruding in front of the posterior adductor scar. Eoc.-Mio. Eur., N.Amer. *Katelysia* (*s.s.*) (Fig. 589): externally resembles *Costacallista*, but has subordinate radial sculpture not seen in the latter, and the commarginal ornament is irregular towards the anterior end; right valve hinge with 3a, 1 (bifid) and 3b (bifid); left valve hinge with 2a (bifid), 2b (bifid) and 4b; pallial sinus short, blunt. ?Pal., Eoc.-Rec. Eur., Carib., N.Amer., Pac., N.Z. *Venerupis* (*s.s.*) (Fig. 590): similar to *Tapes*, but with radial as well as commarginal sculpture, and pallial sinus a little longer and U-shaped. Plio.-Rec. Eur.

Figs 588–598. BIVALVIA: VENERACEA

Figs 588–590, 593 and 595–597 after Keen, in Moore; Figs 591–592, 594 and 598 original.
588c. *Marcia opima* (Gmelin), Rec. Indonesia. T. × 0·75.
589c. *Katelysia scalarina* (Lamarck), Rec. Austral. T. × 1.
590d. *Venerupis saxatilis* (F. de Bellevue), Rec. Eur. T. × 1.
591a, c. *Chione cancellata* (Linné), Rec. W.Indies. T. × 1·35.
592. *Chione* (*Lirophora*) *latilirata* (Conrad), M.Mio. (St. Mary's). Virginia. T. × 1·2.
593c. *Clausinella fasciata* (Da Costa), Rec. Medit. T. × 1.
594a, b. *Mercenaria mercenaria* (Linné), Rec. Transatlantic. T. × 0·6.
595a, d. *Timoclea ovata* (Pennant), Rec. Medit. T. a × 2, d × 1·3.
596. *Cooperella subdiaphana* (Carpenter), Rec. Calif. T. × 1·3.
597. *Petricola* (*Rupellaria*) *lithophaga* (Retzius), Rec. Medit. T. × 1.
598. *Petricolaria pholadiformis* (Lamarck), Rec. Essex. T. × 1·25.

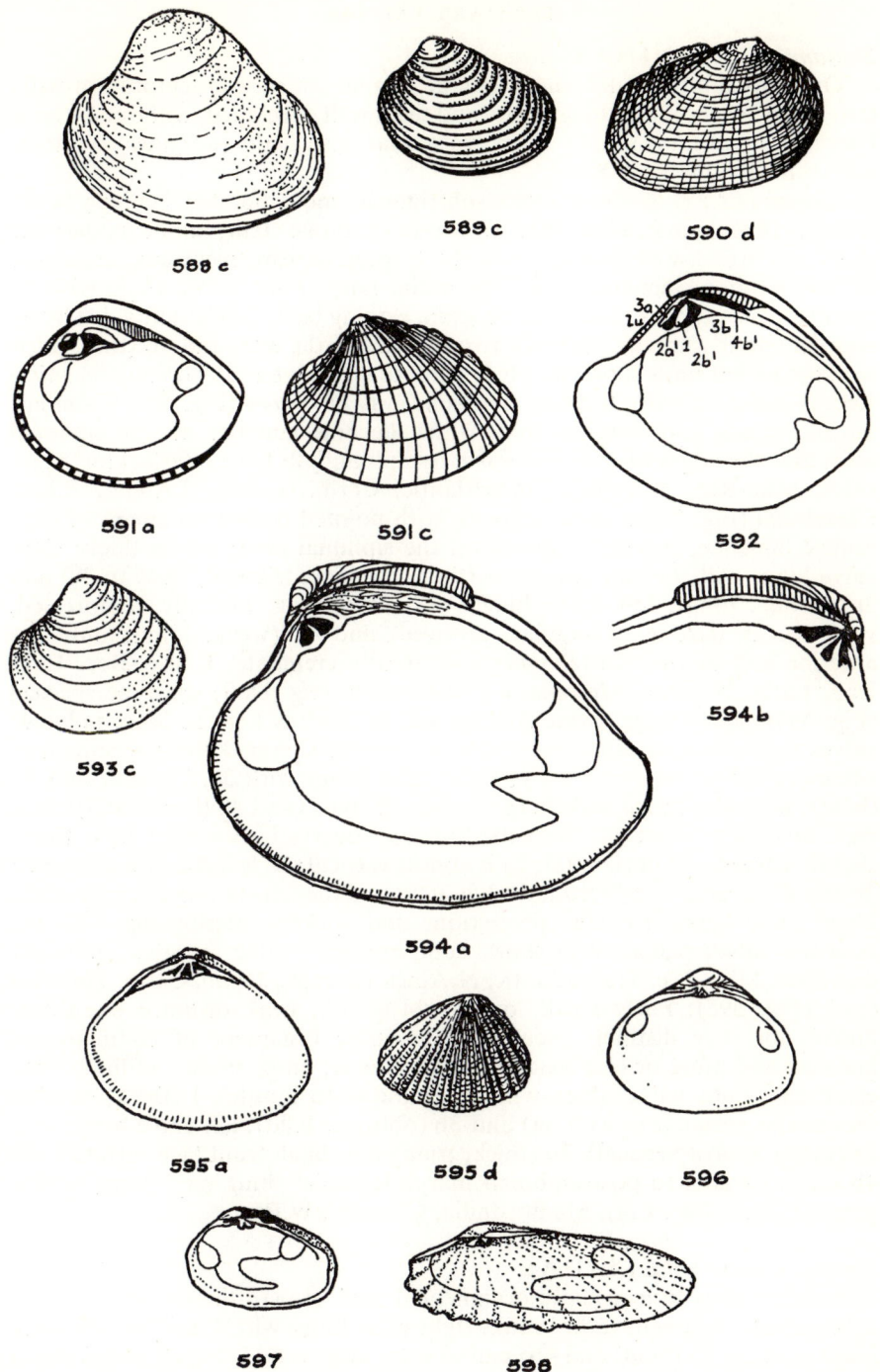

588 c

589 c

590 d

591 a

591 c

592

593 c

594 a

594 b

595 a

595 d

596

597

598

257

Subfamily CHIONINAE

Ovate to subtriangular, inequilateral; lunule impressed; sculpture usually (not always) cancellate; hinge plate and teeth well developed, without anterior laterals; pallial sinus usually short; valve margins usually (not always) internally crenulate. (M.Eoc.-Rec.)

Chione (*s.s.*) (Fig. 591): ovate-subtrigonal, more pointed behind; beaks strongly prosogyrate, beak distance one-third to one-sixth; lunule moderately short and broad, well defined; escutcheon long, narrow, reentrant; ornament of commarginal lamellae frilled by radial lines; right valve hinge with 3a (small), 1 (stout, pyramidal) and 3b (thin, sloping back, bifid); left valve hinge with 2a (stout, curved), 2b (with two notches at the end) and 4b (thin, horizontal); pallial sinus short, pointed; valve margins internally crenulate. Mio.-Rec. N.Amer., Carib., S.Amer. The subgenus *Lirophora* (Fig. 592) is similar, but has strong commarginal folds which become lamellar on the siphonal area, and the edges of the right valve nymph and 4b have interlocking rugosities. Mio.-Rec. N.Amer., Transatlantic, Carib. (Lares), S.Amer., E.Pac. *Clausinella* (Fig. 593): ovate-trigonal, with pointed beaks; ornament of concentric lamellae, especially strong on the siphonal area; valves thick; right valve hinge with 3a (thin, nearly vertical), 1 (thick, trigonal, close to 3a) and 3b (strong, long); left valve hinge with 2a (strong, laterally compressed, vertical), 2b (strong, triangular, grooved) and 4b (weak, long); only the anterior half of the ventral margin internally crenulate. L.Mio.-Rec. Eur., Iran, India, N.Amer. *Mercenaria* [*Venus auctt.* (e.g. Dall) *non* type species] (Fig. 594): fairly large, ovate-subtrigonal, at least as high as long, inflated; valves thick; beak distance one-tenth; no defined siphonal area; ornament of commarginal growth lines only; right valve hinge with 3a, 1 (bifid) and 3b (bifid); left valve hinge with 2a (grooved), 2b (bifid) and 4b (horizontal); 3a is thin, diverging about 30° from the lunule; 1 curves backwards; 3b is thick, slightly curved and horizontal; 2a is almost vertical; 2b is inclined backwards; 3b and 4b diverge widely from the ligament and the intervening flat part of the hinge plate bears irregular projections and sockets interlocking with the opposite valve; pallial sinus short, subtriangular; valve margins internally crenulate. Mio.-Rec. Transatlantic, N.Amer. (Byram), Japan, Carib. *Timoclea* (*s.s.*) (Fig. 595): fairly small, ovate-subtrigonal; beak distance one-third; lunule not very distinct; escutcheon smooth; ornament of commarginal lamellae and more or less beaded radial threads; hinge small, cardinals concentrated; right valve hinge with 3a (parallel to lunule), 1 (thick, curving forward to parallelism with 3a) and 3b (oblique, bifid); left valve hinge with 2a (thick, almost vertical), 2b (thick, triangular, bifid, front face vertical) and 4b (thin, parallel to post-umbonal margin); pallial sinus very short, subtrigonal. L.Mio.-Rec. Eur., Medit., India, C.Amer., W.Pac.

Family COOPERELLIDAE

Small; valves thin, usually ornamented with growth lines only; ovate to subquadrate; ligament rather sunk; right valve hinge with 1 and 3b; left valve hinge with 2a, 2b (bifid) and 4b; pallial sinus deep and U-shaped. (Mio.-Rec.)

Cooperella (*s.s.*) (Fig. 596): ovate, slightly inequilateral, a little longer than high; 2b deeply bifid and 3b bifid. Mio.-Rec. N.Amer., S.Amer., E.Pac.

Family PETRICOLIDAE
Shell oval to cylindrical, beak distance often about one-seventh; the cylindrical form and slight posterior gape are due to the animal's boring habit, and the shell is sometimes deformed; no lunule or escutcheon; usually two cardinal teeth in each valve, occasionally a third in the left valve; 2b bifid; no laterals; usually with radial ornament, but valve margins internally smooth; pallial sinus large, deep. (Eoc.-Rec.)

Petricola (*s.s.*) is Recent only. The subgenus *Rupellaria* (Fig. 597) is oval-oblong, very inequilateral, posteriorly attenuated and compressed; no definite lunule; radial ornament; right valve hinge with 1 and 3b; left valve hinge with 2a (small), 2b (bifid) and 4b; pallial sinus deep, rounded. Eoc.-Rec. Eur., N.Amer., Japan, Pac. *Petricolaria* (*s.s.*) (Fig. 598): elongate, oblong-cylindrical, pholadiform, with radial costules; right valve hinge with 1 and 3b (small); left valve hinge with 2b and 4b; pallial sinus long, elliptical, horizontal. Mio.-Rec. Eur., N.Amer. [The Recent species *Petricolaria pholadiformis* (Lamarck) has a striking external resemblance to the common *Pholas dactylus*, to be explained by similarity of habits and not as a case of protective mimicry, since a mollusc in a burrow needs no such protection.]

Family RZEHAKIIDAE [ONCOPHORIDAE]
Ovate, strongly inequilateral; right valve hinge with 1 and 3b; left valve hinge with 2a, 2b and sometimes a weak 4b; anterior adductor scar crescentic, margined behind by a ridge; pallial sinus very shallow but distinct. (U.Mio.)

Rzehakia [*Oncophora*]: beaks low, beak distance about one-quarter; smooth; the ridge by the adductor scar leaves a deep groove on the internal cast; brackish water. M.Mio.-U.Mio. Eur.

Order MYOIDA

Thin-shelled, with well developed siphons and united mantle margins; inequilateral; equivalve or inequivalve; isomyarian or anisomyarian; edentulous or with one cardinal tooth per valve; lunule and escutcheon weak or absent; shell complex crossed lamellar. Probably polyphyletic. (Carb.-Rec.)

Suborder MYINA
Ligament external, on nymphs, sometimes a resilium also; pallial sinus present. (Perm.-Rec.)

Superfamily MYACEA
Shell usually more or less small (except some Myidae); elongate to ovate, subequivalve, internally porcellanous; ligament with internal chondrophore, one valve asymmetrical on account of the apophysis which fits under the dorsal margin of the other valve; right valve hinge with one cardinal (3, by

analogy), or none; left valve hinge with a ridge behind the resiliifer or united with it; pallial sinus small or nil, sometimes large; valve margins internally smooth. (U.Jur.-Rec.)

The resilium is carried asymmetrically between resiliifers which vary within the superfamily from inclined to horizontal.

Family MYIDAE

More or less equivalve; posterior gape; ligament and resilium internal, opisthodetic; large chondrophore in left valve, small one in right valve; left valve resiliifer horizontal; chondrophore merges with dorsal margin posteriorly; transversely ovate-subtriangular to subquadrate, rather compressed, with commarginal ornament; no cardinal teeth; rather boreal in distribution; burrowers. (Pal.-Rec.)

Derivatives of the Corbulidae, which have resumed the burrowing habit of their remote Palaeozoic ancestors that the Corbulidae had abandoned. They have returned to an oblong, rather compressed shape with posterior gape, and an almost equivalve condition, but as usual in such cases have kept some heirloom of their intermediate ancestors such as could never have been acquired by an unbroken line of burrowing forms. This special feature, as Douvillé has demonstrated, is the asymmetric resilium-apparatus, with a horizontal resiliifer in the left valve and an inverted resilium pit in the right (Fig. 599).

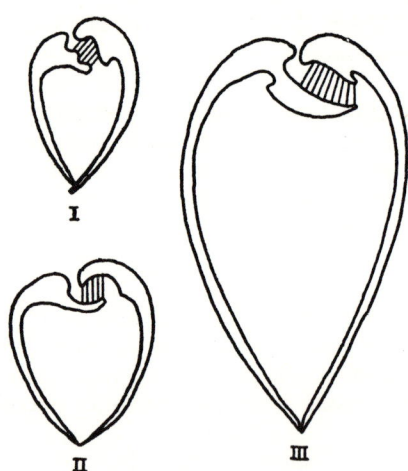

Fig. 599

Fig. 599. DIAGRAMMATIC SECTIONS OF THE SHELLS of I, *Corbula*, II, *Bicorbula* and III, *Mya*, showing the change in position of resilium and resiliifer and return to approximately equivalve form superficially. Right valve is on right in each case. Original.

Mya (*s.s.*) (Fig. 600): typical family characters; valves thick, with commarginal growth lines only; not very elongate, posteriorly truncate; valve margins internally smooth; a true burrower L.Mio.(Aquit.)-Rec. N.Amer., Transatlantic, Eur., Japan. *Tugonia* (*s.s.*): more perfectly equivalve than *Mya*, very globose, and has some reticulate ornament; posterior area set off by a constriction; chondrophores subequal; pallial sinus small. L.Mio.(Aquit.)-Rec. Eur., Indo-Pac. *Sphenia* (Fig. 601): small, inequilateral, subquadrate; left valve the smaller; vestiges of tooth 3 and its socket are retained; pallial sinus small to rather large; a nestler in cavities in rocks, but not a borer, its shape generally deformed to fit its cavity. Pal.-Rec. Eur., Transatlantic, N.Amer., E.Pac., Asia.

Family CORBULIDAE

Small to medium size, usually inequilateral, inequivalve to subinequivalve (left valve the smaller), valves often thick; hinge of left valve with a projecting spatulate resiliifer, fitting into a gap in the right valve (except in *Corbula*); right valve with strong tooth 3, behind which is the gap, with resiliifer lying on outer wall of umbonal cavity; external ligament also present; chondrophore does not merge into dorsal margin (cf. Myidae); pallial sinus small or absent. (U.Jur.-Rec.)

Corbula [*Aloidis*] (*s.s.*) (Figs 602, 599): ovate-subtriangular, often keeled posteriorly; valves thick; right valve a little the larger; beaks prominent, opisthogyrate; posteriorly rostrate; moderately strong commarginal ornament; right valve hinge with a strong anterior cardinal and an obscure posterior lateral; left valve hinge with a deep socket in front of the non-projecting chondrophore and an obscure lateral lamella behind the chondrophore. Cret.-Rec. W.Africa, Eur., Carib., C.Amer., N.Amer., W.Pakistan, India, Burma, Indonesia, Transatlantic, etc., widespread in shallow water deposits. *Varicorbula* [*Agina auctt. non* Turton] (Fig. 603): much more inequivalve than *Corbula*, the right valve being distinctly larger and more convex, its margins flanging those of the left valve which has much weaker commarginal ornament and sometimes also fine, scant radial threads; inequilateral but not rostrate; left valve with a chondrophore; pallial sinus practically nil. L.Eoc.-Rec. Eur., W.Pakistan, India, Burma, Indonesia, N.Amer., Indo-Pac. *Caryocorbula:* similar to *Corbula*, but with a projecting chondrophore in the left valve. Eoc.-Rec. N.Amer., S.Amer., E.Asia, N.Z. *Bicorbula* (Figs 604, 599): larger than *Corbula*, attaining a length of 44 mm., the smallest species (*B. ficus*) being 17 mm. long; the essential difference is in the resiliifer, which is practically horizontal instead of oblique (Fig. 599); there is also a slight notch or fissure in the right valve umbo, communicating with the resilium space; inequivalve; posteriorly rostrate, but posterior keel weak; left valve commarginal ornament much weaker than that of the right valve; pallial sinus broad and shallow. U.Cret. (Maastr.)-Eoc. Eur., Egypt, Somalia, Iran, W.Pakistan, India, Burma, Indonesia. *Lentidium* (*s.s.*) (Fig, 605): small, slightly inequivalve, compressed, *Tellina*-like; valves (atypically) thin; beaks small, beak distance two-thirds; right valve hinge with tooth 3 supported in some species by a buttress, followed by a resilium pit connected with the external ligament by an umbonal fissure (larger, relatively, than that of *Bicorbula*); left valve hinge with a deep socket for 3, and behind this is a bilobed spatulate resiliifer; adductor scars subequal; pallial sinus practically nil. Pal.-Rec. Eur., Medit. The subgenus **Corbulomya** is similar, but the adductor scars are unequal, and there is a shallow pallial sinus. Eoc.-Plio. Eur.

Family ERODONIDAE

Elongate-triangular, corbuliform, more or less smooth, inequivalve; left valve with a broad, projecting chondrophore as in *Mya*; no pallial sinus. (U.Eoc.-Rec.)

Erodona: U.Eoc.-Rec. N.Amer., C.Amer., S.Amer., Eur.

Family PLEURODESMATIDAE

Transversely subtrigonal-subquadrate, ventral margin almost straight or gently concave; umbonal ridge distinct; beaks anterior to mid-line, prosogyrate; smooth or with growth lines; ligament external; resilium internal, fairly large and long; hinge intermediate between that of *Corbula* and *Panopea*; right valve hinge with 3a (strong) and 3b (rudimentary); left valve hinge with 2 (salient); muscle scars and pallial line faint; no pallial sinus. (L.Mio.)

Pleurodesma: L.Mio. Eur.

Family RAETOMYIDAE

More inequivalve, more inflated, and less gaping than *Mya*; beaks high, beak distance three-fifths; valves thin, ovate; ornament of strong commarginal undulations; left valve chondrophore larger and projecting. (M.Eoc.-U.Eoc.)

Raetomya: left valve chondrophore with three lobes. M.Eoc.-U.Eoc. Egypt, Nigeria, S.W.Africa, N.Africa, Cameroons, Senegal. [This genus appears to be the direct ancestor of *Mya*; it is less completely adapted to a burrowing life.]

Family SPHENIOPSIDAE

Small, with the form of *Cuspidaria* (ovate-trigonal, posteriorly rostrate);

Figs 600–608. BIVALVIA: MYACEA, GASTROCHAENACEA
AND HIATELLACEA

Figs 601, 603 and 605–606 after Keen, in Moore; Figs 600, 602, 604 and 607–608 original.
600a, b. *Mya truncata* Linné, Rec. × 0·85.
601c. *Sphenia binghami* Turton, Rec. England. T. × 1.
602a. *Corbula carinata* Dujardin mut. *hoernesi* Benoist. × 1·5.
603c, d. *Varicorbula gibba* (Olivi), Rec. England. T. × 2.
604a, b. *Bicorbula* sp., [M.Eoc., Paris B.?]. × 1·4.
605d. *Lentidium mediterraneum* (Costa), Rec. Italy. T. × 3.
606c. *Gastrochaena cuneiformis* Spengler, Rec. Indo-Pac. T. × 1.
607. *Hiatella arctica* (Linné), Rec. T. × 1·33.
608. *Panopea americana* Conrad, M.Mio. (Choptank). Maryland. × 0·375.

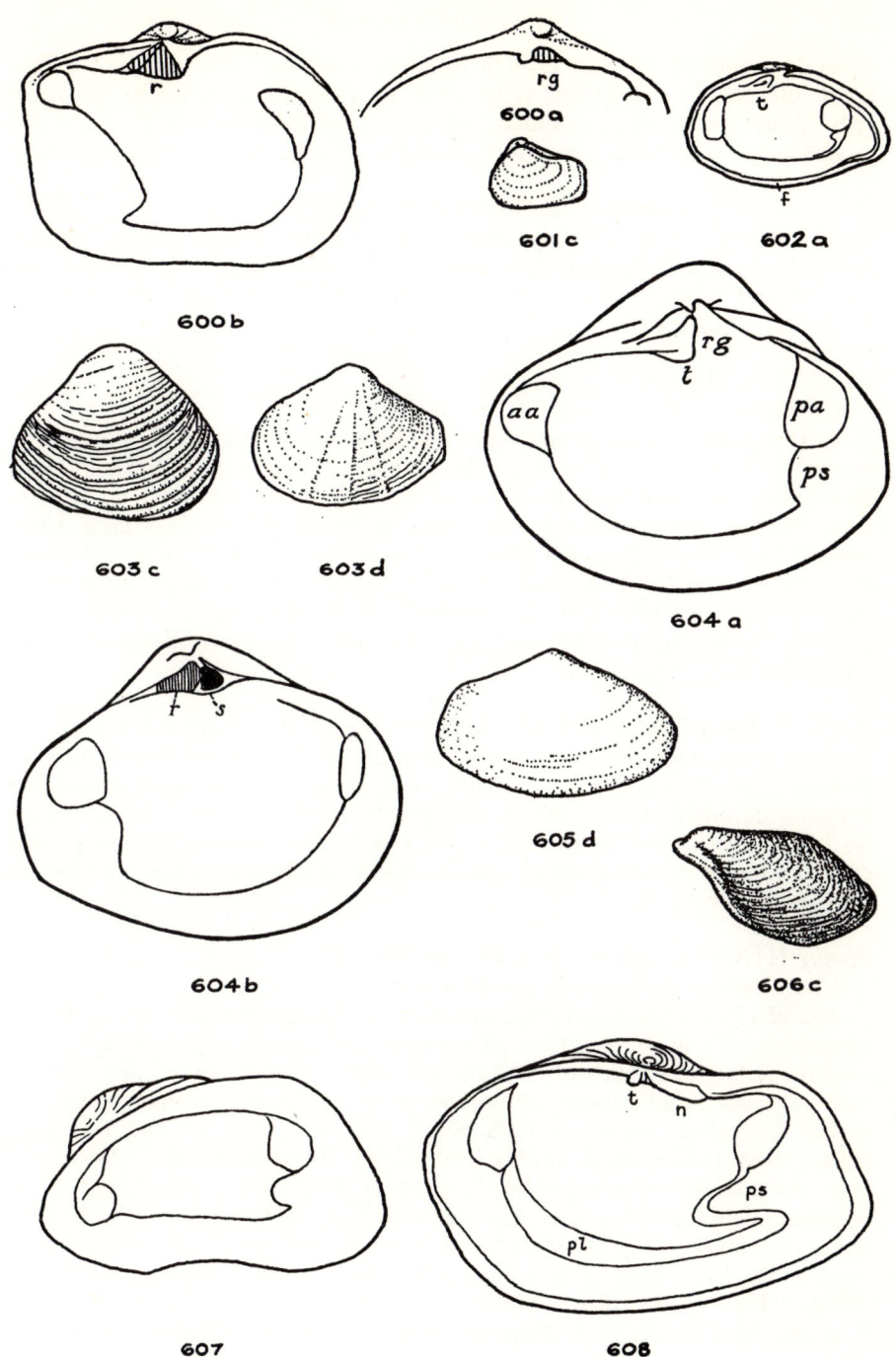

600b

600a

601c

602a

603c

603d

604a

604b

605d

606c

607

608

equivalve; smooth or with growth lines or commarginal undulations; ligament internal, in a deep pit; right valve with two laminar teeth; left valve edentulous; pallial sinus rounded, not very deep. (M.Eoc.-Rec.)

Spheniopsis: trigonal, posteriorly rostrate. M.Eoc.-Mio. Eur., N.Amer.

Superfamily GASTROCHAENACEA

More or less elongate, fairly small, inequilateral, widely gaping anteroventrally; ligament external; edentulous; pallial sinus deep; anterior adductor scar smaller than posterior one; shell equivalve, with beaks far forward; burrowers. (U.Jur.-Rec.)

Family GASTROCHAENIDAE

Characters of the Superfamily. (U.Jur.-Rec.)

Gastrochaena (*s.s.*) (Fig. 606): modioliform, higher posteriorly; valves thin, smooth or almost so; pallial sinus deep, somewhat angular; secretes in addition to the valves an adventitious tube, but only when necessary for protection. ?U.Jur., U.Cret.-Rec. Carib., Eur., N.Amer., Indonesia, Celtic, Lusitanian, Indo-Pac. **Eufistulana** [*Fistulana auctt. non* Müller] (*s.s.*): shell more elongate than in *Gastrochaena*, anteriorly keeled, costate, and with a denticulate margin; it always secretes a protective tube which in limestones is often preserved as a subcylindrical cast. Eoc.-Rec. Eur., Egypt, W.Pakistan, Somalia, Indo-Pac.

Superfamily HIATELLACEA [SAXICAVACEA]

Equivalve or nearly so, more or less inequilateral; ovate or oblong to subquadrate, usually gaping at both ends, often irregular, with rather coarse and irregular commarginal ornament; hinge with cardinals 3 or 2, or toothless; no lateral teeth; pallial line often not confluent, sinus variable; ligament external, borne on well marked nymphs. Common in cold waters; nestlers or burrowers. (Perm.-Rec.)

Family HIATELLIDAE [SAXICAVIDAE]

Characters of the Superfamily. (Perm.-Rec.)

Hiatella [*Saxicava*] (*s.s.*) (Figs 395, 607): small in size (up to $3\frac{1}{2}$ cm. long); irregularly oblong-subquadrate, posterior end a little the higher; inequivalve; very inequilateral (beak distance one-quarter); more or less gaping behind; cardinal teeth 3 and 2 present in young, adult edentulous; pallial sinus deep; surface irregularly rugose; a rock borer. Olig.-Rec. N.Atl., Eur., N.Amer., E.Pac., N.Z. (widespread, mainly cold water). **Panopea** [*Panope, Glycimeris*] (Fig. 608): large, more or less elongate, with thick valves, rounded in front, truncated behind, with ornament of commarginal striae, gaping at both ends; beaks near mid-line; one cardinal tooth in each valve, only feebly interlocking; nymphs large and high; pallial sinus rather deep, narrow and tongue-shaped. ?Trias., L.Cret.-Rec. Eur., Medit., N.Amer., S.Amer., Boreal, Austral., S.Africa, Magellanic, N.Z. **Panomya** (Fig. 609): differs from *Panopea* in its smaller size, and in being shorter in proportion to height, and in having a smaller pallial sinus, and especially in the breaking up of the pallial line into an irregular series of rounded scars; equivalve; left valve with two small cardinals, right valve with one; ornament of broad commarginal folds, with in addition a median radial depressed area. ?L.Cret., Plio.-Rec. Arctic, Boreal, N.Amer., Eur., Medit. *Cyrtodaria:* elongate, beak distance three-fifths; edentulous; pallial sinus very small; hinge plate callous. Pal.-Rec. Eur., Arctic.

Suborder PHOLADINA [ADESMACEA]

Equivalve, inequilateral, sometimes elongate, sometimes spherical, gaping; usually with radial ornament; no hinge plate; small chondrophore and internal ligament usually present; valves held together by muscles; hinge degenerate through boring habit, without teeth; hinge margin more or less reflected over umbonal region, carrying anterior adductor upon it; a curved myophoric apophysis projects freely into umbonal cavity; accessory shelly structures usually present in addition to the valves; mostly marine borers. (?Carb., Jur.-Rec.)

Superfamily PHOLADACEA

Characters of the Suborder. (?Carb., Jur.-Rec.)

Family PHOLADIDAE

More or less elongate, sometimes subspherical, sometimes with an umbonal-ventral sulcus; umbonal process reflected over umbones; externally more or less spiny anteriorly; dorsally with several accessory plates, variable in number; edentulous; pallial sinus deep; ligament obsolete; antero-ventral gape sometimes closed by callum. (?Carb., Jur.-Rec.)

Pholas (*s.s.*) (Fig. 610): very inequilateral (beak distance about one-quarter), elongate (height barely one-third length), more or less cylindrical in the middle, compressed towards both ends, more abruptly and with a blunt projection in front; surface marked by growth lines with radial rows of scale-like hollow elevations, crowded in front and dying away behind (the pattern indicates use as a file in boring, though signs of war and tear are not obvious); anterior adductor scar on front part of reflected surface; posterior adductor scar in normal position; pallial sinus very deep, reaching beyond mid-line; loose accessory plates along dorsal surface: a pair of large, thin, triangular *protoplaxes* in front and a pair of smaller, solid *mesoplaxes* behind (covering the reflected area), and a single long *metaplax* (covering the post-umbonal region); the reflected shell is divided into two layers, one plastered over the umbonal region, the other separated from it by about twelve vertical partitions (formed by finger-like projections of the mantle); valve margins generally deeply embayed antero-ventrally for protrusion of the foot; myophoric apophysis broad and grooved. Mio.-Rec. Eur., N.Africa. [*Pholas* bores in various rocks, by preference in firm clays or limestones, but micaceous crystalline rocks (schists etc.) may be attacked.] *Barnea* (*s.s.*): similar to *Pholas*, but the reflected area is undivided, the pedal embayment extremely narrow, and the apophysis is rod-like; there is only one lanceolate protoplax.

265

Mio.-Rec. Eur., Indo-Pac. *Zirfaea* [*Zirphaea*] (Fig. 611): similar to *Pholas*, with pedal embayment present, but the reflected area is undivided, the apophysis is rather broad but not grooved, and an external groove runs downwards and a little backwards from the umbo; there is one rudimentary protoplax and one rudimentary mesoplax. Mio.-Rec. Eur., N.Atl., N.Amer., N.Pac. *Xylophaga*: short, globose, showing an accentuation of the outline of *Zirfaea*, the pedal embayment taking the form of a right-angled re-entrant; apophysis short, curved; mesoplax divided; posterior gape. U.Cret.-Rec. Cosmop. [Unlike previous genera *Xylophaga* bores in wood, and in this respect, as well as in shell structure, might appear to relate it to *Teredo*, but it seems that we have to do with a case of parallel development.]

A group of specialized Pholadidae, never common as fossils, are now briefly mentioned; they are characterized by accessory shelly structures protecting the foot and siphons which the valves can no longer enclose. A median curved plate, the callum closes the pedal gape, while one or more cup-like structures (*siphonoplaxes*) protect the siphons. *Martesia* (*s.s.*): rather like *Zirfaea*, but with callum but no siphonoplaxes; one large oval mesoplax, one long metaplax, two narrow hypoplaxes, but no protoplax. ?Carb., Jur.-Rec. Eur., N.Amer., Carib., Transatlantic, Indo-Pac., E.Pac., temperate and tropical seas. *Pholadidea* (*s.s.*): rather like *Zirfaea*, but with callum and one siphonoplax; two protoplaxes, one rudimentary mesoplax and one rudimentary metaplax. L.Eoc.-Rec. Eur., E.Pac., N.Z. *Teredina*: *Teredo*-like, but with a callum; valves completely fused to the long tube; the two protoplaxes and two mesoplaxes are fused. U.Cret.-M.Mio. Eur. *Jouannetia* (*s.s.*): more globose, and has a large callum. U.Cret.-Rec. Eur., Indo-Pac., Carib., W.Africa, temperate and tropical seas.

Family TEREDINIDAE

Relatively small, free, equivalve, covering only anterior end of long animal, without accessory plates; shell auriculate, widely gaping at each end, with pallets; very globose, trilobed, with an umbonal-ventral ridge; edentulous; a long curved apophysis beneath umbones; chondrophore and small internal ligament present; borers. (?Cret., Pal-Rec.)

The classification of supraspecific categories is based on the anatomy and characters of the pallets.

Teredo (*s.s.*) (the ship-worm, Fig. 612): shows relationship to the Pholadi-

---

Figs 609–614. BIVALVIA: HIATELLACEA, PHOLADACEA,
PHOLADOMYACEA AND PANDORACEA

Fig. 610 after Turner, in Moore; Fig. 613 after Cox, in Moore; Figs 609, 611–612 and 614 original.

609. *Panomya norvegica* (Spengler), Rec. Boreal. T. × 0·56.
610a, c. *Pholas dactylus* Linné, Rec. Malta. T. × 0·5.
611a, c. *Zirfaea crispata* (Linné), Rec. T. × 0·67.
612a, c. *Teredo* sp., Rec. England. × 2·25.
613c. *Pholadomya ambigua* (J. Sowerby), L.Jur. England. × 3·5.
614a, b. *Pandora inaequivalvis* (Linné), Rec. T. × 1·25.

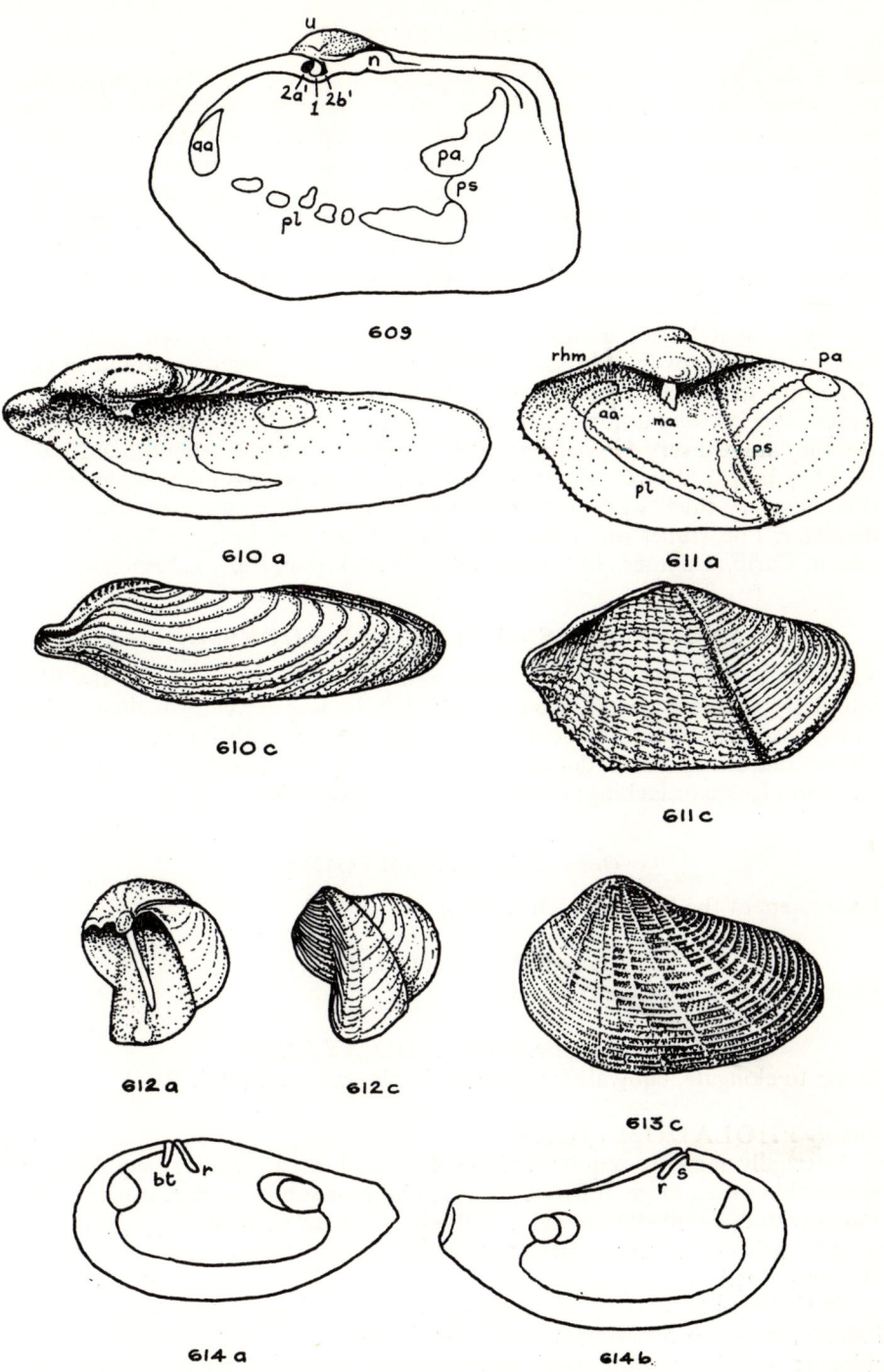

609

610 a

611 a

610 c

611 c

612 a

612 c

613 c

614 a

614 b

dae in the form of its valves and the presence of a long curved apophysis, but the valves scarcely make a pretence of enclosing the body and the long siphons secrete a calcareous tube, incompletely septate towards the distal end; the valve surfaces are divided into two main areas, both sculptured with fine commarginal ridges, but in the anterior area these are broken up into fine denticles; the changes in direction of the ridges, parallel to the changing contour, lead to further subdivision into five areas in all, at least in some species; the classification of Recent species is based upon the *pallets*—small, loose, trowel- or feather-like calcareous bodies, which have been found fossil but unfortunately not usually in association with the shell (**46**). Eoc.-Rec. Worldwide, except in cold seas. [Fossil remains of *Teredo* are found in the driftwood in which it bored, when such wood is preserved in sediments. Recent species were notorious for their ravages on the hulls of wooden ships, and they are still very destructive to submarine wooden piles, pier foundations, etc.] ***Kuphus*** [*Cuphus, Cyphus*): burrows in sand and mud on the sea floor; the tube has very thick walls and may attain a metre in length and 10 cms. in diameter. Eoc. (tubes only)-Rec. Indo-Pac., Iran, Red Sea, E.Africa, Madagascar, Carib., C.Amer., Indonesia, India, W.Pakistan, Eur., widespread.

Subclass ANOMALODESMATA [ANOMALODESMACEA]

Valves often thin, short to long, often nestling, equivalve to inequivalve and internally nacreous; nearly isomyarian; siphons well developed, thus pallial sinus usually present; ligament usually (not always) with internal chondrophore and lithodesma; dorsal margin without distinct hinge plate, and dentition feeble or lacking; usually burrowers. (Ord.-Rec.)

Order PHOLADOMYOIDA

Characters of the Subclass. (Ord.-Rec.)

Suborder PHOLADOMYINA

Characters of the Order. (Ord.-Rec.)

Superfamily PHOLADOMYACEA

Ovate to elongate, equivalve; ligament simple, external. (Ord.-Rec.)

Family PHOLADOMYIDAE

Internally nacreous; equivalve (except in one Jurassic genus); inequilateral; valves thin, often large, inflated, ovate, oblong or subtriangular, usually with radial costae; ligament external, opisthodetic, on strong nymphs; pallial sinus usually present; usually edentulous; usually gaping at posterior or both ends. (L.Carb.-Rec.)

Pholadomya (*s.s.*) (Fig. 613): elongate-oval, equivalve, very inequilateral, inflated, more so anteriorly; beaks far forward and only moderately prominent; anterior region short and rounded, posterior region drawn out, truncated, gaping behind; ornament of oblique radial costae (corrugations, not thickenings), siphonal area sometimes with difference of costation; hinge practically edentulous; ligament external, short, on strong nymphs; pallial sinus large; valves very thin and friable, ornament fully shown on internal cast; essentially a Mesozoic genus (acme in Jurassic), but still fairly common in some Tertiary formations. U.Trias.-Rec. Carib., Eur., Russia, Egypt, N. Amer., S. Amer., N.Z., N.Africa, India, W.Pakistan, Burma (unpublished record), deep Atl., widespread. [Restriction to abyssal depths seems a recent event, fossil species being found in shallow-water deposits.]

Superfamily PANDORACEA [ANATINACEA]

Sedentary or burrowing; nacreous, at least internally; valves thin, inequivalve, rather elongate or gaping; no heterodont dentition, but hinge margin reinforced by buttresses or denticles; ligament and resilium usually with lithodesma. (Trias.-Rec.)

Family PANDORIDAE

Fixed or free, compressed, internally nacreous, inequivalve, usually cuneiform, with dorsal edges of valves overlapping; hinge formed of lamellar ridges; resilium internal, opisthodetic, usually with a ventral lithodesma; pallial line simple, broken up into a series of spots; usually rostrate posteriorly but without gape. (Pal.-Rec.)

Pandora (*s.s.*) (Fig. 614): valves of medium thickness, inequivalve (right valve quite flat, left valve convex and overlapping it), very inequilateral (beak distance one-third), general outline subcrescentic, antero-dorsal margin convex, posteriorly rostrate and with well-marked escutcheon; hinge of right valve having no hinge plate, bearing two buttress-teeth fitting into gaps in the hinge plate of the left valve which may have an anterior lamina; resilium oblique (nearer vertical than horizontal), not carried on resiliifers but fitting into grooves on inner surface of valves; (these features give to the interior of each valve an illusory resemblance to *Placuna* (p. 200), increased by the nacreous translucence, but the chevron-like structures in the two cases are quite different); smooth or with growth lines; pallial line distant from margin, discontinuous; no pallial sinus; lithodesma obsolete. Olig.-Rec. Eur., N.Amer., W.Pac.

Family LATERNULIDAE [ANATINIDAE]

Valves thin, internally nacreous, more or less equivalve, gaping posteriorly; edentulous; longer than high, with low beaks; resilium on prominent spoon-like resiliifers, each supported by two diverging buttresses; no external ligament; sinupalliate. (U.Trias.-Rec.)

Laternula [*Anatina*] (*s.s.*) (Fig. 615): valves thin, nearly equivalve; height about half length, beak distance four-sevenths, anterior region rounded-oblong, posterior region attenuated, truncated and widely gaping; pallial sinus wide but not very deep; feeble commarginal ornament. U.Cret.-Rec. (rare in the Tertiary). Eur., Indo-Pac., N.Z., N.Amer.

269

Family CLEIDOTHAERIDAE

Chama-like, internally nacreous, with prosogyrate beaks; very inequivalve, right valve very deep and fixed, left valve flat, with one cardinal tooth; ligament internal, with lithodesma; no pallial sinus. (Mio.-Rec.)

Cleidothaerus: Mio.-Rec. Austral., N.Z.

Family PERIPLOMATIDAE

More or less inequivalve, the right valve more convex; internally nacreous; posterior end shorter, not gaping; edentulous; ligament internal, on a chondrophore, with or without lithodesma; pallial sinus wide, short. (U.Cret.-Rec.)

Periploma (*s.s.*) (Fig. 616): obliquely oval-quadrate, with opisthogyrate beaks, more or less smooth; beak distance two-thirds; chondrophore narrow, oblique, not fixed posteriorly, with small triangular lithodesma; posterior clavicle; anterior adductor muscle scar long and narrow, posterior one short; pallial line close to ventral margin, with short, sharply rounded sinus. U.Cret.-Rec. Eur., N.Amer., S.Amer., N.Z., E.Asia.

Family LYONSIIDAE

Internally nacreous, usually inequivalve (the left valve the larger); outline usually oblong; edentulous; resilium in a long, narrow resiliifer, reinforced by a lithodesma; slight pallial sinus; surface smooth, granulose or with fine radial lines. (Pal.-Rec.)

Lyonsia (*s.s.*) (Fig. 617): valves very thin, elongate-oblong, slightly produced behind, slightly inequivalve, slightly gaping; surface marked with fine, wavy radial lines, crossing growth lines; beaks slightly anterior to mid-line; resilium opisthodetic, in a groove extending from umbo towards posterior adductor scar, those of the two valves diverging backwards so that the large lithodesma (within the resilium) has an arrow-head shape; muscle scars (except the left posterior), pallial line and sinus all indistinct. Pal.-Rec. Boreal, Eur., Celtic, Arctic, N.Amer. (a cold water form).

Family MARGARITARIIDAE

Subelliptical-subcylindrical, transverse, with low beaks, posterior and

---

Figs 615–622. BIVALVIA: PANDORACEA, CLAVAGELLACEA AND POROMYACEA

Figs 616–617 and 620–621 after Keen, in Moore; Figs 615, 618–619 and 622 original.

615. *Laternula truncata* (Lamarck), Rec. Philippines. × 0·675.
616b, c. *Periploma margaritaceum* (Lamarck), Rec. W.Indies. T. × 1.
617b. *Lyonsia norvegica* (Gmelin), Rec. England. T. × 1.
618a, b. *Thracia corbuloides* Deshayes, Rec. Medit. × 0·9.
619c. *Penicillus* sp., Pleist. (Sicilian). Sicily. × 0·75. *wp*, 'watering-pot' apparatus.
620d. *Poromya granulata* (Nyst and Westendorp), Rec. Atl. T. × 3.
621c. *Verticordia cardiiformis* (Sowerby), Plio. England. T. × 2.
622b. *Cuspidaria* sp., Pleist. (Sicilian). Sicily. × 1·5.

Note: there are no Figs 623–632.

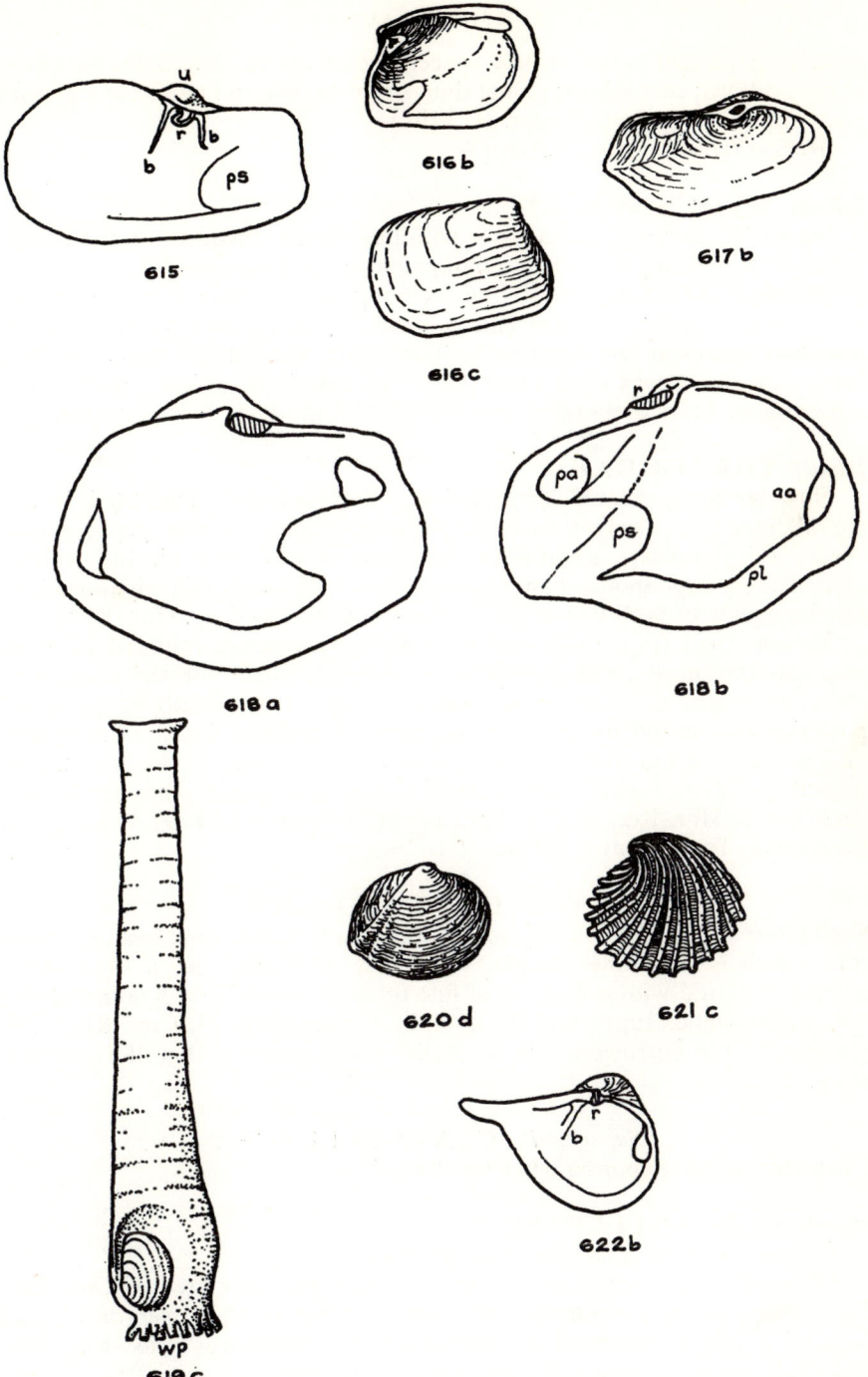

615

616 b

616 c

617 b

618 a

618 b

619 c

620 d

621 c

622 b

anterior gape and a few widely spaced radial costae; internally nacreous; nymph callous; edentulous; pallial sinus obsolete; anterior adductor scar oval, posterior one larger and longer. (U.Mio.)

Margaritaria: U.Mio. N.Amer.

Family MYOCHAMIDAE

Internally nacreous, edentulous, with lithodesma; inequivalve, the convex valve dorsally fitting the other; small pallial sinus. (Mio.-Rec.)

Myadora [*Myodora*] (*s.s.*): free, compressed, equilateral, ovate-trigonal with pointed beaks; anterior end rounded, posterior end truncated; commarginal ornament and a prismatic outer layer; the flat left valve fits into a furrow on the convex right valve the dorsal margin of which is turned in at right angles; lithodesma falciform. Mio.-Rec. Indo-Pac., N.Z., Austral.

Family THRACIIDAE

Non-nacreous; oblong to trapezoidal, often posteriorly truncated, inequilateral, inequivalve (right valve the larger); slight posterior gape; surface chalky or granulose; edentulous; ligament mostly internal, the resilium attached to a large spoon-shaped chondrophore directed obliquely posteriorly on hinge plate of both valves; pallial sinus moderately deep. (Jur.-Rec.)

Thracia (*s.s.*) (Fig. 618): outline irregularly oblong, rounded in front, attenuated, truncated and slightly gaping behind; valves not very thick, non-nacreous, height about three-quarters the length, only slightly inequilateral, posterior area set off by a low ridge, with commarginal striae; some species only slightly inequivalve; ligament external and resilium on low, unbuttressed resiliifers, usually with a small lithodesma; pallial sinus moderately deep, obtuse. Jur.-Rec. Medit., Eur., Egypt, S.Amer., N.Amer., N.Z., Celtic, Lusitanian, Transatlantic, E.Pac.

Suborder CLAVAGELLINA

Shell internally nacreous, fairly small, one or both valves sunk in a calcareous tube which is sometimes simple, sometimes perforated like a sieve at the anterior end and with a number of fine tubules; edentulous; ligament external, opisthodetic, supported by nymphs; sinupalliate. (U.Cret.-Rec.) (An aberrant line of burrowers, and a parallel development to that of *Teredina*—see p. 266.)

Superfamily CLAVAGELLACEA

Characters of the Suborder. (U.Cret.-Rec.)

Family CLAVAGELLIDAE

Characters of the Suborder (U.Cret.-Rec.)

Clavagella (*s.s.*): left valve fixed to tube, right valve free. U.Cret.-Rec. Eur., India, Austral. *Penicillus* [*Brechites* (not binomial), *Aspergillum*] (*s.s.*) (Fig. 619): both valves fixed to tube; anterior adductor scar weak. Mio.-Rec. Indo-Pac., India, E.Africa, S.Austral.

Order SEPTIBRANCHOIDA

Free, more or less equivalve, smooth or with small spines, sometimes with radial or commarginal ornament; ovate to subtrigonal or elongate, usually not gaping; chondrophore with a lithodesma; hinge with or without teeth; pallial sinus nil or small. (Cret.-Rec.)

This is the most highly specialized group of bivalvia, the only ones which have abandoned the microphagous for a carnivorous diet, and having the gill lamellae modified into a muscular partition across the mantle cavity; the shell gives no hint of such specialization.

Superfamily POROMYACEA [SEPTIBRANCHIA]
Characters of the Order. (Cret.-Rec.)

Family POROMYIDAE
Shell internally nacreous; usually rather small, rounded, cordate or cuneiform, with commarginal and sometimes also radial ornament; ligament external; hinge plate usually with a tooth in one valve. Usually deep-water forms. (Cret.-Rec.)

Poromya (*s.s.*) (Fig. 620): ovate-subtriangular, plump, subequilateral, ornamented with fine granules arranged in radial lines, post-umbonal slope marked by a ridge; strong cardinal tooth in right valve, socket and PII in left valve; no pallial sinus. Cret.-Rec. Atl., Medit., E.Pac., N.Amer. (deep water).

Family VERTICORDIIDAE
Shell internally nacreous, small, cordiform, inequilateral, usually with prominent, incurved beaks; no true lunule, but usually a deep, rounded depression in dorsal margin immediately in front of beaks; ligament internal, with a lithodesma; hinge normally with one cardinal tooth in each valve; no pallial sinus; usually costate and often prickly. (Pal.-Rec.)

Verticordia (*s.s.*) (Fig. 621): rather small, suborbicular, equivalve, inflated; 'lunular depression' deeper on left valve; right valve with one cardinal tooth, left valve edentulous except for a posterior lateral; ornament of narrow radial costae which extend beyond ventral margin as denticulations; valve margins internally fluted. Pal.-Rec. Eur., N.Amer., S.Amer., Transatlantic, Deep Atl., Carib., Indo-Pac., N.Z.

Family CUSPIDARIIDAE
Shell non-nacreous, small, corbuliform, posteriorly rostrate, with beaks near the middle or slightly anterior to mid-line; surface smooth or with strong commarginal or radial ornament; hinge edentulous or with teeth; resilium internal, with a lithodesma; with or without an internal posterior buttress; pallial sinus slight. (U.Cret.-Rec.)

Cuspidaria [*Neaera*] (*s.s.*) (Fig. 622): nearly equivalve (left valve more convex than right valve), very inequilateral, being rounded in front and ros-

273

trate behind (like some forms of *Corbula,* but much less inequivalve); chondrophore a narrow, spoon-shaped plate in each valve; no cardinal teeth; right valve with a posterior lateral tooth having a linear socket above it into which the margin of left valve fits; commarginal ornament; slight pallial sinus. U.Cret.-Rec. Medit., Eur., N.Amer., N.Z. (mainly deep-water to abyssal, cf. Vol. II, Fig. 1.)

SHORT GLOSSARY OF TECHNICAL TERMS APPLIED TO BIVALVIA

Actinodont. With numerous hinge teeth radiating downwards from the beaks.

Adductor muscle. A closing muscle, crossing from one valve to the other, and by its contraction drawing the valves together.

Adductor scar. Impression on interior of shell where adductor muscle was attached.

Alate. Having 'wings' or 'ears'.

Alivincular. Type of ligament located between cardinal areas of valves, with lamellar layer both anterior and posterior to fibrous layer.

Amphidetic. Extending both in front of and behind the beaks (applied to cardinal area or ligament).

Anisomyarian. Having one of the two adductor scars conspicuously smaller than the other.

Anterior lateral tooth. Lateral tooth situated in front of beaks.

Anterodorsal margin. That part of dorsal margin in front of beaks.

Apophysis. A projecting structure for attachment of a muscle.

Arcticoid. Type of heterodont dentition intermediate between lucinoid and corbiculoid types.

Auricle. Ear-like extension of dorsal region of shell, often separated from rest of shell by a notch.

Auricular crura. The most anterior and posterior of the hinge teeth of Pectinacea.

Auriculate. Alate, *q.v.*

Beak. Point where growth of shell started (on or above hinge); tip of the umbo.

Beak distance. As here used, the distance of the beaks from the anterior end (expressed as a fraction of the total length).

Buttress. A vertical or oblique ridge on the inner surface of a valve, supporting some part of the hinge.

Byssus. A secretion of conchiolin by the foot, serving to fix the animal to a rock or other firm body; it may take the form of a short cylinder, or (more usually) a bunch of long fine silky threads.

Callum. A secondary calcareous structure, present in some Pholadidae, forming an anterior extension of the shell and closing the pedal gape in the adult.

Cardinal area. The part of the surface of either valve lying between the beak and the hinge line and bent at an angle to the general surface; partly or wholly occupied by the ligament.

Cardinal crura. In Pectinacea, the hinge teeth lying immediately before and behind the resilium pit.

Cardinal teeth. In Heterodonta and Palaeoheterodonta, those hinge teeth which lie directly beneath the beak, in a more or less vertical or oblique, rarely horizontal, position.

Carina. A keel or sharp edge at the junction of two surfaces.

Carinated. Keeled, provided with a carina.

Caudate. With narrow tail-like extremity.

Chevron. A linear structure with an acute bend in the middle, V-shaped (cf. geniculate).

Chondrophore. Resiliifer, *q.v.*

Clavicle. Buttress, *q.v.*

Closed. Not gaping anywhere along the margins.

Commarginal. With direction part of surface of shell determined by former position of shell margin. (Replaces the previous usage of 'concentric', particularly with regard to ornament.)

Commissure. The line of junction of the two valves.

Compressed. Flattened in a transverse direction, so that the thickness (from right valve to left) is diminished.

Corbiculoid. Type of heterodont dentition with three cardinal teeth in each valve, the middle one of the right valve (1) being in a median position below the beak; there are usually three cardinal teeth in each valve: 3a, 1, 3b and 2a, 2b, 4b. (This term replaces the previously used term 'cyrenoid'.)

Cordate or *Cordiform*. Heart-shaped, usually as viewed from the anterior end.

Crenulate. Bearing a regular series of rather fine notches (e.g. valve margins crenulate).

Ctenolium. A series of tooth-like projections below the byssal notch of some Pectinidae, serving to guide the byssal fibres.

Dentition. The whole series of hinge teeth.

Dimyarian. Having two adductor scars (whether equal or unequal).

Divaricate ornament. Consisting of two series of lines which, while both approximately radial, diverge sharply.

Donaciform. Shaped like *Donax*, trigonal, wedge-like, with short posterior end and elongated, rounded anterior end.

Duplivincular. Type of ligament with lamellar component repeated as series of bands, each with its two edges inserted in narrow grooves in cardinal areas of valves (e.g. *Arca*).

Dysodont. With small, weak teeth near beaks (some Mytilacea).

Ear. See *auricle*.

Edentulous. Having no hinge teeth.

Equilateral. Having the anterior and posterior halves of each valve approximately symmetrical about a plane passing through the beaks.

Equivalve. Having the right and left valves mirror images of one another (the hinge excepted).

Escutcheon. A depressed external area on either side of the hinge line behind the beaks (equivalent to a posterior cardinal area).

Excurrent. Forming passage for current of water expelled from mantle cavity (applied to siphon or mantle opening).

Fluted. With series of narrow, parallel, rounded excavations.

Gaping. Having the valves so shaped that certain portions of their margins do not meet those of the opposite valve.

Geniculate. Knee-like, bent obtusely in the middle (compare *chevron*).

Globose. Tending towards a globular shape.

Heterodont. Having hinge teeth of two distinct types: *cardinals* under the beak, *laterals* before and/or behind.

Hinge. Collective term for structures of dorsal region which function during opening and closing of valves.

Hinge line. That part of the margin of a valve which is not a free edge, but is in permanent contact and continuity with the corresponding part of the other valve.

Hinge plate. The flat vertical internal surface along the hinge line, from which the hinge teeth arise.

Hinge teeth. Projections of shell substance from the hinge plate, projecting beyond the median plane into sockets in the hinge plate of the opposite valve.

Hypoplax. Elongate accessory plate extending along posterior end of ventral margin in some Pholadidae.

Incurrent. Forming passage for current of water drawn into mantle cavity (applied to siphon or mantle opening).

Inequilateral. Not divisible into symmetrical anterior and posterior halves along a plane passing through the beak.

Inequivalve. In which one valve is not a mirror image of the other.

Integripalliate. Having a pallial line parallel to the valve margin, not indented by a pallial sinus.

Inversion. The displacement of asymmetric parts so that what is usually right becomes left and vice versa. In equivalve shells the only structures capable of inversion are the hinge teeth; in inequivalve shells the whole shell may be inverted. (See Chamidae and Astartidae, pp. 232, 225.)

Isodont. With small number of symmetrically arranged hinge teeth (e.g. *Spondylus*).

Isomyarian. Having two adductors approximately equal in size.

Keel, keeled. Carina, carinate.

Lateral teeth. Hinge teeth more or less horizontally disposed and often lamellar, situated either (1) in front of the cardinal teeth (*anterior laterals*) or (2) behind the ligament (*posterior laterals*).

Lenticular. Shaped like a biconvex lens.

Ligament. A structure of thickened conchiolin lying in the hinge line and in physical continuity with the periostracum of both valves; by its elasticity it tends to keep the valves open in opposition to the closing action of the adductors.

Lithodesma. A calcification of the interior of the resilium.

Lucinoid. Type of heterodont dentition in which the pivotal tooth is in the left valve (2) and there are usually only two cardinal teeth in each valve: 3a, 3b and 2, 4b.

Lunule. A depressed external area on either side of the hinge line in front of the beaks (equivalent to an anterior cardinal area).

Mantle. The sheet of soft tissue which lines the inside of the shell and by which the shell has been secreted.

Mesoplax. Elongate accessory plate lying across umbonal region in some Pholadidae.

Metaplax. Long, narrow accessory plate covering gap between posterodorsal margins in some Pholadidae.

Monomyarian. Having only one adductor scar (originally the posterior, but more or less central in position) (e.g. Ostreidae).

Multivincular. The type of ligament which is set upon a number of depressions on the cardinal area (e.g. *Arca, Isognomon*).

Muscle scars. Areas of the internal surface of the shell, usually excavated but sometimes raised, to which the muscle fibres were attached.

Myophoric apophysis, lamella or *septum.* Any projection of the internal surface of the shell to which a muscle was attached; when rod-like it is an apophysis, when thin and platy a lamella or septum.

Nacreous. Type of shell structure consisting of thin leaves of aragonite parallel to the inner surface of the shell and exhibiting characteristic lustre (e.g. *Nucula*).

Nymph. A vertical shell lamina, behind the beak, to which the ligament is attached.

Opisthocline. Sloping (from lower end) in posterior direction (e.g. hinge teeth, body of shell).

Opisthodetic. Type of ligament (the most usual kind) lying entirely behind the beaks.

Opisthogyrate (umbo). Curved so as to look towards the posterior end (cf. orthogyrate, prosogyrate).

277

Orthogyrate (umbo). Curved so as to look towards the umbo of the other valve.

Ovoid. Egg-shaped (with reference to the outline, not necessarily to the solid form).

Pallet. Small calcareous structure present in Teredinidae, one of a pair closing end of boring when siphons are retracted.

Pallial. Pertaining to the mantle.

Pallial line. The line on the inner surface of the valves marking the inner margin of the thickened mantle lips.

Pallial sinus. A posterior embayment of the pallial line, the scar of the retractor muscles of the siphons.

Parivincular. An elongate type of ligament located posterior to the beaks.

Pedal. Pertaining to the foot.

Pedal gape. Opening between the valve margins for protrusion of the foot.

Pedal scars. Scars of attachment of the protractor and retractor muscles of the foot.

Periostracum. The external layer of the shell, composed of uncalcified conchiolin.

Polyphyletic. Derived from several distinct ancestral stocks (and therefore unjustifiably united in the same genus, family, etc.).

Porcellanous. Having the texture and lustre of porcelain (contrast with *nacreous*).

Posterior lateral tooth. Lateral tooth situated behind the ligament.

Posterodorsal margin. That part of the dorsal margin behind the beaks.

Prodissoconch. The earliest formed (nepionic) shell.

Prosocline. Sloping (from lower end) in anterior direction.

Prosogyrate (umbo). So curved as to look towards the anterior part of the shell (the most usual type of umbo).

Protoplax. Flat accessory plate, in one or two pieces, located at anterior end of dorsal margin in some Pholadidae.

Resiliifer. The portion of the hinge plate which bears the resilium (*q.v.*): it may be a simple pit (*Mactra, Crassatella*) or a projecting knob (*Laternula*) or plate (*Mya*) or pair of narrow ridges (*Placuna*).

Resilium. The portion of the original ligament which is sunk below the level of the valve margins and is under compression instead of tension; often termed 'internal ligament'.

Rostrate. Having the posterior region drawn out and marked off by a slight reversal of curvature.

Scalloped. With series of regular internal flutings corresponding to ends of external costae (term applied to shell margin).

Schizodont. Having very coarse divergent teeth (*Unio, Trigonia*): this does not constitute a natural division.

Sinupalliate. Having the pallial line embayed at its posterior end by the scar of the retractor muscles of the siphons.

Siphonal area. A posterodorsal triangular area, with its apex at the beak and its base extending over the incurrent and excurrent openings of the mantle; often more ornamented than the rest of the surface (*Protocardia, Trigonia, Pinna*). The escutcheon, when present, intervenes between the siphonal area and the posterior part of the hinge line.

Siphonoplax. Tubular secondary calcareous structure forming posterior extension of shell in some Pholadidae.

Siphons. Two tubular extensions (incurrent and excurrent) of the mantle lips at the posterior end of the shell.

Socket. Depression for reception of hinge tooth of opposite valve.

278

Taxodont. With numerous short hinge teeth, some or all transverse to the margin (*Nucula, Arca*).

Teleodont. With differentiated cardinal and lateral teeth and with elements also constituting a rather complicated hinge (e.g. *Venus*).

Truncate. Having the natural continuation of the curved outline replaced by a straight line or line of much less curvature (usually at the posterior end).

Umbo. The apical region of the very asymmetric cone of which each valve consists, more or less curved over towards the hinge line and including the earliest-formed shell (prodissoconch).

Valve. One of the two elements constituting the shell in bivalvia.

Valve margin. The free margin of each valve, as distinct from the hinge line.

Vestigial. Reduced to so small a size as to be useless and barely recognizable.

SOME FRENCH TECHNICAL TERMS

Charnière. Hinge.
Crochet. Beak.
Cuilleron. Resiliifer.
Écusson. Escutcheon.

Manteau. Mantle.
Plateau cardinal. Hinge plate.
Tronqué. Truncate.

SOME GERMAN TECHNICAL TERMS

Band. Ligament.
Hauptzahn. Cardinal tooth.
Klappe. Valve.
Schliessmuskel. Closing muscle, adductor.
Schloss. Hinge.
Schloss-platte. Hinge plate.
Schloss-rand. Hinge line.
Schloss-zahn. Hinge tooth.
Seitenzahn. Lateral tooth.
Wirbel. Umbo.
Zahngruben. Tooth sockets.

(For Bibliography see end of next chapter.)

Chapter IV

TERTIARY GASTROPODA

REFERENCE-LETTERS following figure numbers (where there is no letter, 'a' may be understood in the absence of other indication):
a, ordinary apertural view; a^1, part of same enlarged.
b, view of labral profile, or growth line.
c, apical view.
d, basal or umbilical view, or internal view of limpet type.
e, side view of limpet type, or side view other than a or b.
f, protoconch.
g, surface ornament.
h, operculum, external view; h^1, internal view.
i, view showing interior.
j, internal cast.

REFERENCE-LETTERS to parts of figures:

ac, anterior canal or notch.	lnb, lateral notch and band.
an, anal notch.	lt, labral tooth or teeth.
ap, aperture.	ltg, labral tooth and groove.
ap.sh., apertural shelf.	n, nacre.
ap.u., apical umbilicus.	p, perforation (excretory).
ca, callus.	pc, posterior canal.
cf, columellar folds.	pg, pulmonary groove.
co, columella or columellar lip.	pr, protoconch.
	r, ramp.
ct, columellar teeth.	s, shoulder.
dig, digitations.	sd, sutural dip.
d.sh., doubled shelf.	sf, siphonal fasciole.
ev, lip-eversion (spout).	sg, sutural groove.
f, protoconch.	sh, sutural shelf.
fb, attached foreign bodies.	shl, shoe-horn lamina.
fl, flange.	sp.g., spiral groove.
fos, fossula.	tr, truncation of columella.
hsm, horse-shoe muscle scar.	u, umbilicus.
l, labrum (outer lip).	uf, umbilical funicle.
ln, lateral (excretory) notch.	vx, varix.

Primitively, the gastropod shell may have been a simple open cone, like that of the limpet, *Patella* (Fig. 639), though all known limpet-like forms appear to be degenerations from the more usual type in which the shell is a tube, slowly increasing in diameter, wound round an axis to form a helicoid spiral. In these helicoid shells each turn of the spiral is called a *whorl*. The whorls are sometimes loose (*Tenagodus*, Fig. 682) or only just touching (*Clathrus*, Fig.

701); usually they are more or less tightly pressed together, the spiral line along which they are seen to meet, externally, being called the *suture*. Each whorl then conceals, in greater or less degree, the whorl preceding it in growth, and only the *last whorl* (which has sometimes been called the *body whorl*) exhibits its full shape. The visible parts of all the whorls except the last are collectively called the *spire*: this may form, according to the extent of the inclusion of one whorl by the next, anything from nine-tenths of the whole shell (*Zaria*, Fig. 678) down to a minute fraction, or may even be entirely concealed by the last whorl (*Simnia*, Fig. 732).

The nomenclature of the parts of the shell is confused by the varying orientation adopted in describing it. In France it was customary to figure a shell with the apex of the spire downwards, in most other countries in the reverse position. Accepted technical terms inconsistently implied now one and now the other position, as will presently appear. In the following account the terms 'upper' and 'lower' refer to the shells placed apex upwards: 'anterior' is approximately lower, 'posterior' approximately upper.

All whorls, including the last, have in common an *upper suture* and a *side* (*lateral area, flank*); all except the last have a *lower suture*, the last alone showing a *base* (concealed in other whorls) and an *aperture*. The side of a whorl may be *flat* (Fig. 669) or more or less *rounded* (Fig. 785), or *angular*, i.e. with one or two obtuse bends in its outline (Figs. 678, 810). The sutures may be *flush* with the general surface, or more or less *impressed* (Fig. 835), or *channelled* (sunk in a groove, Fig. 802). When the side is angulated, each angle may be raised into a *carina* (or *keel*, Fig. 755). When there is only one angle it is usually near the upper suture and is termed the *shoulder* (*s*, Fig. 810), the part of the side between it and the suture being called the *ramp* if sloping (*r*, Fig. 787) or the *shelf* if horizontal (*sh*, Fig. 790).

The base of the last whorl may be simply *rounded* (Fig. 647), or *flattened* (Fig. 727), or there may be a reversal of curvature to form a *neck* (Fig. 767). When the base is quite flat, the side-margin may be extended below it as a *flange* (*fl*, Fig. 706).

When the coiling is not very tight, the last whorl may surround an empty space, the *umbilicus* (more precisely, *basal umbilicus*, *u*, Fig. 681), which may be broad, narrow, or *rimate* (almost closed up). A true umbilicus is continued up through all the whorls to just below the apex; an opening which does not extend above the last whorl is a *false umbilicus*. Shells with an umbilicus are described as *umbilicate* (or *perforate*, which is a bad term); those without as *anomphalous* (or *imperforate*, which is a bad term). In these latter the inner walls of all the whorls coalesce into a solid pillar, the *columella* (Fig. 699).

In a few cases (*Haliotis* (*Padollus*), Fig. 638; *Sinum*, Fig. 742), the whorls are not part of a true tube, but of a split tube or gutter, the inner wall of each whorl being non-existent. In such cases there is neither umbilicus nor columella, or alternatively we may say that the umbilicus is merged in the general shell cavity. The shell may even depart still farther from its tubular character and resemble a rolled-up sheet of paper (*Terebellum*, Fig. 723; *Philine*, Fig. 832).

When the increase in diameter of the growing tube is proportional to its increase in length, very regular forms of shell are produced, the spire being generally *conical* (Figs 643, 727), with straight sides (or at least straight tangents to the sides). If the diameter increases rather more rapidly than the length, the sides become concave and the spire is *coeloconoid* (previously referred to as *extraconic*, which is a bad term, Figs 637, 741); if less rapidly, they become convex and the spire is termed *cyrtoconoid* (previously referred to as *conoidal*, a term with a broader meaning—Figs 642, 857) or (in the case of an acutely pointed shell) *subulate* (Fig. 868). Where this latter tendency is carried to an extreme and the diameter after a time remains constant (or even finally contracts) the *pupiform* or *cylindrical* shape results (Fig. 876). These terms may be applied to the whole shell in cases where the spire forms its main bulk; in other cases the particular shape of the base may have to be allowed for in framing a descriptive term for the whole.

The axis of coiling being placed vertically, the suture is not far from horizontal: the angle by which it diverges from horizontality is the *dip* of the suture or *sutural angle*. If this angle be very low, the shell requires many whorls to attain a given length, and is said to be *multispiral* (or *polygyral*, Fig. 876); with a high sutural angle the same length is attained in few whorls and the shell is *paucispiral* (or *oligogyral*, Figs 847, 860). Sometimes, immediately before reaching the aperture, the suture may not only become horizontal but even reverse its dip and become a *rising suture* (Figs 629*b*, 691*b*). In the genus *Distorsio* (Fig. 747) such a rising suture is found at intervals along the spire, producing the strange distortion of shape. When, in an anomphalous shell, the last whorl completely embraces and conceals the earlier one, the shell is *convolute* (some cypraeids); if the apex is umbilicate and the earlier whorls can be seen in the umbilicus, the shell is *involute* (*ap.u.*, Fig. 728*b*). Similar conditions combined with a very broad lower umbilicus result in a *discoidal* shell (Fig. 853).

The great majority of gastropod shells are *dextral* (coiled in a right-handed spiral), i.e., when the apex is upwards and the aperture faces the observer, the aperture is to the observer's right hand. Much more rarely, left-handed (*sinistral*) shells are found (Fig. 778); these may be individual 'sports' in a normally dextral species, or characterize one or more species of an otherwise dextral genus, or may be the rule for a whole genus or family. Discoidal shells may be interpreted as either dextral or sinistral: the anatomy of the animal will determine which is correct. If the process by which a turbinate shell gradually passes into a discoidal shell be continued beyond the discoidal stage, a downwardly directed spire will appear and the shell will apparently be sinistral though the animal will remain anatomically dextral: such shells are termed *ultradextral* (e.g. *Lanistes*, Fig. 661); the term *hyperstrophic* applies both to this condition and the vice versa condition.

If tangents be drawn along opposite sides of the spire, the angle at which they meet above the apex is called the *spiral angle*: this is usually more or less acute, but may be a right angle, or obtuse, or even (when there is an apical umbilicus) re-entrant.

282

When the base of the last whorl is flat or only slightly rounded and the spiral angle moderately acute (45° to 90°), the whole shell is *conical* or *trochiform* (Fig. 641) unless coeloconoid or cyrtoconoid. When the base is more or less hemispherical the shell becomes top-shaped or *turbinate* (Figs 646, 741); with an increased height of spire and very acute spiral angle this passes into the *turreted* or *elongate* form (Fig. 678); with a lengthening of the base into a semi-ellipsoid the turbinate form passes into the *ovoid* (Fig. 737), or with some elongation of the spire, *ovoido-conical* or *bucciniform* (Fig. 762). If the base be straight-sided, forming an inverted cone, the shell is *biconical* (Fig. 819); with increasing obtuseness of the spiral angle (Fig. 818) this passes into *obconical*. If a turreted form has the base drawn out into a long neck it becomes *fusiform* (Fig. 785).

The form of the aperture is of the greatest importance in the classification of gastropod shells. When damaged, as it often is in fossils, its form can be reconstructed to some extent from the growth lines, each of which has been temporarily (or momentarily) an apertural margin.

In limpet-shaped shells the apertural margin or *peristome* is simply the edge of the open cone and is usually elliptical or oval; in loosely coiled shells it is still only the edge of the cylindroidal tube and is usually circular (*Clathrus*, Fig. 701; *Tenagodus*, Fig. 682). In tightly coiled, and especially in anomphalous shells, a more complicated outline is usually found, the outline of the protuberant penultimate whorl breaking the continuity of the curvature and forming the *parietal* portion of the aperture (marked *ca* in Figs 848, 847a).

The peristome is usually divisible into an *outer lip* or *labrum* (the free edge of the outer surface of the last whorl) and an *inner* or *columellar lip*. At the anterior end the two lips may pass imperceptibly into one another or be defined by an abrupt change in curvature or by a notch. At the posterior end they may also pass into one another, in which case the peristome is said to be *continuous*, or they may meet abruptly, forming a *discontinuous* peristome. In the former case the inner wall of the last whorl separates from the outer wall of the penultimate whorl to form a free edge, like the labrum with which it is continuous; in the latter case it continues in contact with the outer surface as a layer of *callus* (i.e. the inner shell layer turned inside out and plastered over the external ornament, if any), sometimes having a well defined edge, in other cases thinning away indefinitely. This callus marks the extent to which the mantle lobe spreads out beyond the aperture. In some gastropods mantle lobes extend out on both sides of the aperture, lapping over more or less of the external surface of the shell, or even completely covering it (*Cypraea*, Fig. 728): in such cases a thick layer of callus (then called *enamel*) may partially or completely cover up the external sculpture (if any) and the whole shell shows a smooth and polished surface, often with colour ornament but no sign of growth lines. The outer lip may be regarded from two points of view: (1) looked at edgewise (view 'a' of many figures) with the aperture facing the observer, it shows either the *outline* of the ordinary whorl section with its basal portion unconcealed, or such modifications of it as are referred to in the next paragraph: (2) viewed from a direction at right angles to the first view (view

283

'b' of many figures), it is seen in *profile*. This *labral profile* (which determines the growth lines throughout the shell) may be straight (Fig. 319b), simply curved (Fig. 736b), or sinuous (Fig. 693b); the most usual type of sinuosity is that of a reversed S, which is termed *parasigmoidal*. It may be vertical (Fig. 835b) or oblique: in the latter case it is most usually inclined so that the upper end is ahead of the lower (*prosocline*, Fig. 643b) (previously termed *antecurrent*), more rarely in the opposite way (*oposthocline*, Figs 724b, 726b, 714b) (previously termed *retrocurrent*). Sometimes there is a marked difference of inclination in the basal and lateral portions of the labrum (Trochidae, *Xenophora*). When the profile is sinuous, the angle at which it meets the suture may often differ from the general inclination, and it is distinguished as *prosocline*, *normal* or *opisthocline* to the suture. The presence of a lateral notch or slit in the labrum may produce a V-shaped re-entrant angle on the profile (ln, Figs 820–824, 826). More rarely a mantle infold to form a labral tooth or spine may give rise to a salient 'V' and spiral groove (Figs 806, 761).

When shell growth is continuous the labrum does not differ in thickness from the shell along any other growth line, but when the animal indulges in alternate periods of growth and stability the labrum in stable periods is usual thickened and/or everted. Thickening may be equal along its whole length, or be concentrated at certain points on its inner face, giving rise to *teeth* (*lt*, Figs 747, 758), or at certain points on its free edge, producing *digitations* (Fig. 725), or a *wing* (Fig. 721). When growth is resumed, the animal may find it necessary to resorb some of these excrescences if they are in the way of the growing shell; but usually they will remain, in part or whole, forming a *varix* (*vx*, Fig. 718). Such varices may be recognized along the spire at regular (*Favartia*, Fig. 753), semi-regular (*Sassia*, Fig. 746) or irregular (*Rimella*, Fig. 715) intervals.

At the junction of the labrum and parietal region or inner lip there may be a groove or *gutter* (*exhalant channel*) —an appropriate term in the earlier French orientation of the shell—or a *notch*, or a *canal* (i.e. a split tube) running some way up the spire (*pc*, Figs 721, 718).

The columellar lip may be straight, sinuous, or *excavate* (concave in outline). It may bear teeth like those of the labrum (Figs 747, 758), or wrinkles (Fig. 744)—both confined to the immediate neighbourhood of the aperture— or *columellar folds* which continue spirally up the columella (Figs 699, 797). At its anterior end the columellar lip may continue imperceptively into the labrum, or may be abruptly *truncated* (*tr*, Fig. 874), or between it and the labrum there may be a slight eversion or *spout* (*ev*, Fig. 740), or a *notch* of varying depth (Fig. 686), or both lips may be drawn out into a *canal* (*ac*, Fig. 788): if there is no notch or canal the shell is *holostomatous*, in other cases, *siphonostomatous*. This *anterior canal* may be short or long, straight or curved or sharply bent (*reflected*, Fig. 697).

A notch or even a spout causes an indentation of the growth lines the cumulative effect of which appears as a definite band on the neck called the *basal* or *siphonal fasciole* (*sf*, Fig. 762); this may be flat or raised; it is only visible for a short distance until it disappears under the columellar callus.

284

Some gastropods have an *operculum* by which the aperture is closed when the animal's body is entirely withdrawn: this must not be thought of as a second valve, as it is not hinged to the shell and is secreted by the foot, not the mantle. The operculum may be either horny (not preserved in fossils) or calcareous; it may fit the aperture or be smaller than it. It may be *spiral*, growing regularly around the initial point, and then either *multispiral* or *paucispiral* (Fig. 736 *h,h*[1]); or it may grow unequally from the initial point, becoming either *imbricate* (broadly oval) or *unguiculate* (claw-like); the outline is usually smooth, but in some cases it bears blunt projections, and is termed *articulate*.

The *ornament* (or *sculpture*) of gastropod shells consists of variations in the thickness of the shell secreted at different times or different places. Variation in time produces *collabral* ornament, of which the growth line constitutes the 'motif'. Such ornament is *commarginal* in limpets (as in bivalvia) and more or less *vertical* (or *axial*) in ordinary gastropods. Variation in space, i.e. in the amount secreted at different points along the mantle border, leads to *transcurrent* ornament—*radial* in limpets, *spiral* in ordinary gastropods.

Increased activity in either case produces ornament in relief: according to its degree we can distinguish *threads* (slight and narrow), *ribbons* (low and broad), *costae* (height and breadth fairly equal), *lamellae* (high and thin). Local or temporary diminution of secretive activity may give rise to ornament in intaglio: *striae* (very fine), *grooves* (deeper and wider), or *punctae* (small pits).

When both vertical and spiral ornament are developed, a *reticulate* (or *cancellate*) ornament results. The vertical ornament may sometimes rise up in a scaly manner along the growth lines, giving a *muricate* surface (*Murex*). A somewhat similar appearance on the spiral threads may be termed *imbricate*.

At the points where vertical and spiral costae (or other grades of relief sculpture) cross one another there is often a concentration of secretive activity, resulting, in the case of the finest ornament, in a *granule*; coarser ornament gives a rounded *tubercle* or *knob* or a pointed *spine*.

Varices take their place in the ranks of vertical ornament, sometimes obviously of different style from the rest, in other cases merely as accentuated ornament of the ordinary vertical type.

The apex of the shell is formed by the *protoconch* or larval shell (*f* or *pc*, Figs 785, 787, 789, 792, 797), which consists of an initial embryonic shell, more or less globular, and 1–4 whorls usually smooth and polished, rarely with fine ornament. The passage from the larval to the adult shell is sometimes gradual, but much more frequently is abrupt, being marked by a varix, a sudden appearance of ornament, or a change in the direction and character of the coiling. This sudden change corresponds to a change of life in the animal, from a free-swimming to a bottom-crawling existence, and the number of larval whorls is roughly a measure of the length of the larval life. Protoconchs with only one whorl after the embryonic shell are *paucispiral*, those with two, three or more are *multispiral*.

In certain cases (*Architectonica*, Pyramidellidae) the protoconch appears

285

to be sinistral, while the rest of the shell (*teleoconch*) is dextral: such shells are termed *heterostrophic*. As there is no corresponding reversal of asymmetry in the animal itself, these apparently sinistral protoconchs are really ultradextral. This is obviously the case with *Architectonica*, where the protoconch is perfectly inverted in respect of the teleoconch, projecting downwards into the basal umbilicus. In the Pyramidellidae the axis of the protoconch forms an obtuse angle or a right angle with that of the teleoconch. Those shells in which there is no such apparent inversion are termed *homoeostrophic*. In these also the axis of the protoconch may be at an angle with that of the teleoconch (*deviated* protoconch). Often the protoconch as a whole is more globose than the teleoconch (Fig. 785).

In some cases the protoconch is wholly or partly horny: the horny part is then cut off by a septum and is lost in fossils, so that the apex comes to have an asymmetric pointed end (Fig. 792f). In some Volutidae which lay few and large eggs, the embryonic shell is enormous in proportion to the size of the teleoconch.

There has been much discussion as to the value of the protoconch in classification. The idea that every different type of protoconch represents a different ancestral form, and that therefore the primary divisions should be based upon the protoconch, may be dismissed at once: it is too obvious that adaptation to differences in larval life is one of the main controlling factors. Nevertheless, the dominance of the naticoid form among protoconchs does suggest some such form as an early ancestor to a large part of the class. In the classification of families and genera, the protoconch is one of the characters to be considered, though different systematists differ greatly in the degree of importance they attach to it. However, the characters of the protoconch in some families (e.g. the Turridae) do seem to be of considerable importance at the generic level.

The apex of a gastropod shell being usually unprotected is liable to mechanical injury; being also part of the shell that has been longest in existence it is the most exposed to the solvent action of the surrounding water. This action is greatest in the case of fresh waters, and though fresh-water gastropods have the protection of a periostracum that is thicker than in most marine shells (as is also the case with the bivalvia), they often show considerable *decollation*, i.e. destruction of the apex by solution (Fig. 683).

It is an attractive but futile subject for speculation how palaeontologists would have classified the Gastropoda, had they been an extinct class. Except for the broad distinction between holostomatous and siphonostomatous shells, which does not seriously conflict with zoological divisions, no shell characters—not even that most important feature, the growth line or labral profile—can furnish a sure diagnosis of any division higher than the family, or perhaps superfamily. The bases of zoological classification that have been used are:

(1) certain features of the nervous system, expressed in the terms *Streptoneura* and *Euthyneura*,

(2) features of the mantle chamber and respiratory system, expressed in

286

terms *Prosobranchia, Opisthobranchia* and *Pulmonata*, as well as in the ordinal names *Aspidobranchia* and *Pectinibranchia*,
(3)   the tooth-pattern of the chitinous radula or rasping tongue, expressed in the terms *Docoglossa, Rhipidoglossa*, etc.

As none of these characters are determinable in fossils, such terms as are used here will be used without definition, though the distinguishing shell characters will be indicated as far as they exist.

Class GASTROPODA

Mollusca with unchambered, single-apertured, nearly always univalve (extremely rarely 'bivalve') shells, with or without an operculum, or with shells in various stages of degeneration or without shells. (Cambr.-Rec.)

Subclass PROSOBRANCHIA [STREPTONEURA]

The typical shell-bearing gastropoda, mainly marine in habitat; the majority with shells coiled in a helicoid spiral, the more primitive being holostomatous, the more advanced, siphonostomatous; a few open-conical, a few planispiral; usually with horny or calcareous operculum. (Cambr.-Rec.)

Order ARCHAEOGASTROPODA

Shell spiral, limpet-shaped, or curved; usually with a nacreous layer. (Cambr.-Rec.)

Suborder PLEUROTOMARIINA

Shell often conispiral, sometimes patelliform, auriform or discoidal; usually with exhalant notch or slit except in patelliform genera with apical fissure; inner shell layer nacreous except in patelliform genera. (U.Cambr.-Rec.)

Superfamily FISSURELLACEA

Protoconch spiral; shell conical, porcellanous, with perforation, slit, notch or emargination for passage of exhalant current; muscle scar horse-shoe-shaped, open anteriorly. (Trias.-Rec.)

Family FISSURELLIDAE

Characters of the Superfamily; no operculum. (Trias.-Rec.)
Fissurella (*s.s.*) (keyhole limpet, Fig. 633): broadly conical, with apical hole only slightly anterior to the middle; radial costae; margin in one plane, more or less simple, not folded; hole internally surrounded by callus the posterior margin of which is truncated. Eoc.-Rec. Carib., N.Amer., S.Amer., Eur., Asia, N.Pac. **Lucapina:** like *Fissurella*, but inner margin thin and finely crenulate. Mio.-Rec. N.Amer., Carib. **Lucapinella:** like *Fissurella*, but apical hole large, its internal marginal callus not truncated posteriorly, apex sub-central, and posterior margin a little elevated; inner margin only crenulate

287

anteriorly and posteriorly. L.Mio.-Rec. N.Amer., Carib., S.Amer., E.Pac. *Diodora* (*s.s.*) (Fig. 634): like *Fissurella*, but apex bent over backwards, hole small, muscle scar with ends curved in, margin internally crenulate, and lateral margins may be a little raised. U.Cret.-Rec. Medit., cosmop. *Emarginula* (*s.s.*) (Fig. 635): conical, with spirally coiled protoconch situated well behind the middle, elliptical in plan; ornament of radial costae and subordinate commarginal ridges; well-marked parallel-sided slit in anterior margin, its scar forming a continuous band to the apex. M.Trias.-Rec. Eur., N.Amer., Carib., S.Amer., N.Z., warm seas. *Hemitoma* [*Subemarginula*] (*s.s.*) is Recent only. Its subgenus *Montfortia* has the apex central, blunt and feebly curved backwards; slit nil or short, no slit-band; numerous alternating larger and smaller costae and fine commarginal growth lines; anterior end with a costa a little to the right of the mid-line. Eoc.-Rec. Austral., Eur., N.Amer., S.Amer., S.Pac. *Puncturella* (*s.s.*): apex posterior to middle and slightly enrolled to the right; hole rounded, closely in front of apex; slit-band very short, internally forming a septum; externally with finely crenulated radial costae. Olig-Rec. Eur., Japan, northern seas (moderately deep water). *Rimula:* outline oval, apex near posterior end and inclined to the right; a long hole in the middle of the anterior side, at the end of a small slit-band leading to the apex; no septum; aperture entire; radial costae and commarginal threads. Cret.-Rec. Eur., Carib., N.Amer., S.Amer., Indo-Pac. *Scutus* [*Parmophorus*] (*s.s.*) (Fig. 636): oblong, depressed, apex posteriorly eccentric, curved backwards; ornament of commarginal growth lines only; no slit and no slit-band; anterior end slightly notched, posterior end rounded. Mio.-Rec. Austral., N.Z., Indo-Pac., Eur.

Superfamily PLEUROTOMARIACEA

Shell mostly conispiral, more rarely flattened and discoidal or auriform; internally nacreous; whorls with a slit-band and terminal slit or a row of holes

---

Figs. 633–646. GASTROPODA: FISSURELLIDAE, PLEUROTOMARIIDAE, HALIOTIDAE, PATELLIDAE, ACMAEIDAE, TROCHIDAE, ANGARIIDAE AND TURBINIDAE

Fig. 637 after H. Woodward; Figs 633, 639 and 643 after Wenz; Figs 634–636, 638, 640–642 and 644–646 original.

633c. *Fissurella nimbosa* (Linné), Rec. W.Indies. T. × 0·5.
634d, e. *Diodora graeca* (Linné), Plio. Medit. d × 1·4, e × 1·125.
635c. *Emarginula crassa* J. Sowerby, ?U.Plio. Suffolk. × 0·56.
636d. *Scutus australis* Blainville, Rec. N.Z. × 0·56.
637b. *Mikadotrochus beyrichii* (Hilger), Rec. Japan. × 0·55.
638d. *Haliotis* (*Padollus*) sp., Rec. Indo-Pac. × 0·56.
639c. *Patella caerulea* Linné, Rec. France. T. × 0·66.
640e. *Acmaea* (*Tectura*) *crenulata* Broderip, Rec. Seychelles. × 1.
641. *Tectus crenularis* (Lamarck), M.Eoc. Paris B. × 0·7.
642a, b. *Gibbula* (*Steromphala*) *cineraria* (Linné), Pleist. St. Andrew's, Scotland. × 1·5.
643a. *Calliostoma conulus* (Linné), Rec. Medit. T. × 1.
644a, b. *Margarites margaritula* (Mérian), M.Olig. Weinheim, Rhenish Hesse. × 2·25.
645a, b. *Angaria delphinus* (Linné), Rec. Indo-Pac. T. × 0·6.
646a, b. *Turbo* (*Marmarostoma*) sp., Plio. Zanzibar. × 0·9.

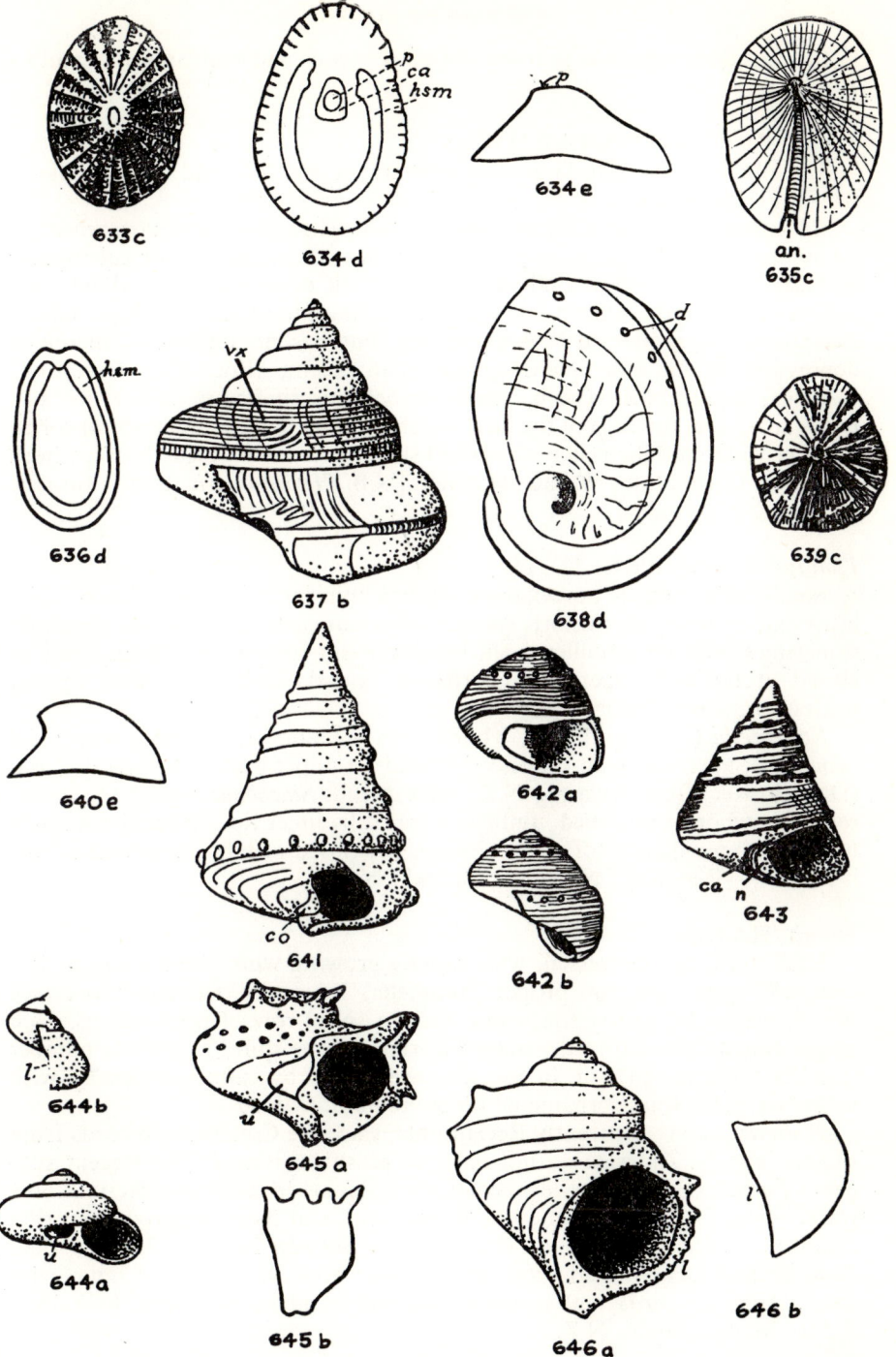

633 c

634 d

634 e

an.
635 c

636 d

637 b

638 d

639 c

640 e

641

642 a

642 b

643

644 b

644 a

645 a

645 b

646 a

646 b

in this position; opercula in living forms corneous and multispiral. (Cambr.-Rec.)

Family PLEUROTOMARIIDAE

Usually fairly large to large, trochiform; slit and slit-band near middle of whorl. (Trias.-Rec.)

Mikadotrochus (Fig. 637): trochiform, whorls gently convex; ornament of crenulate spiral threads; narrow umbilicus; last whorl rounded or subangular at base; slit-band below mid-whorl; labral slit extending back about one-eighth of a whorl; labrum strongly prosocline above it. Plio.-Rec. Japan. **Perotrochus** is similar to *Mikadotrochus*, but has no umbilicus, merely a depression on the base; width greater than height. ?Eoc., Olig.-Rec. Eur., S.Africa, Japan, N.Z., Carib. **Entemnotrochus** is similar to *Mikadotrochus*, but is angular at the edge of the base which is rather flat, has weak spiral cords, a rather wider umbilicus, and the labral slit extends back at least half a whorl. Eoc.-Rec. Carib., Eur., W.Pac., N.Amer. [N.B. **Pleurotomaria** itself is Jurassic and Cretaceous only.]

Family SCISSURELLIDAE

Small, thin-shelled, porcellanous with a thin inner nacreous layer; spire low-conic or turbinate; whorls smooth or ornamented, the last one large and sometimes with an umbilicus; slit-band and slit present, the latter open or closed at the end; operculum corneous, round, multispiral, with central nucleus. (U.Cret.(Danian)-Rec.)

Scissurella (*s.s.*): spire low, whorls with a blunt keel, usually with axial ornament; aperture oval, slit and slit-band on upper part of whorls. U.Cret. (Danian)-Rec. Eur., Austral., N.Z., widespread. **Sinezona:** similar to *Scissurella*, but whorls not keeled, umbilicus sealed by inner lip, slit-band not much more than one-quarter of a whorl long, slit closed at labrum so that a hole is left. Plio.-Rec. Eur., N.Amer., Austral., N.Z.

Family HALIOTIDAE

Shell auriform, depressed, with rapidly growing whorls and more or less eccentric spire, without proper columella, internally nacreous; aperture broad, occupying nearly the whole base; a spiral row of perforations, each originating as a notch in the outer lip, runs up the spire, a variable number (usually three to ten) still in use, the rest closed by callus; externally with rather irregular spiral ornament; no operculum. (Mio.-Rec.)

Haliotis (*s.s.*) is apparently Recent only, since the Cretaceous records from Europe and N. America do not seem to be substantiated. The Recent subgenus **Padollus** (Fig. 638) is less broad, the holes are in callous projections on a spiral costa, there is a spiral costa above them, and there are prominent thin collabral lamellae above that. The subgenus **Sulculus** is a little broader than *Padollus* and lacks the adapical spiral costa; area above row of holes with spiral striae or cords and irregular transverse ridges or nodes. Mio.-Rec. E.Atl., Medit., Japan, N.Z.

290

Suborder PATELLINA [DOCOGLOSSA] (Limpets)

Shell conical or cap-shaped, bilaterally symmetrical; no perforation, marginal notch or internal septum; muscle scar semicircular or horse-shoe-shaped, open anteriorly; inner layer sometimes iridescent but not nacreous; no operculum. Habitat littoral, clinging to rocks. (?M.Sil.-?L.Trias., M.Trias.-Rec.)

The anatomy of the limpets shows them to be less asymmetrical than any other true gastropods, and their shells might pass for bilaterally symmetrical and truly primitive, but for the spiral protoconch (usually worn away). As with other limpet-like shells, however, the open-conical form is evidently a degeneration from an ancestral coiled shell, though the degeneration must have set in at a much earlier stage in evolution than in any other case. Examples of limpet-like shells in other families and orders are the Fissurellidae, *Capulus* and *Hipponix* among prosobranchs, *Umbraculum* among opisthobranchs, *Siphonaria*, *Valenciennius* and *Ancylus* among pulmonates. Less close approximations to the limpet-like form are shown successively by *Cheilea* and *Crucibulum*, by *Crepidula*, and by *Calyptraea* and *Trochita*, some forms of *Theodoxus*, and *Concholepas*. All these show much clearer evidence of ancentral spiral coiling than do the true limpets, and in most of them the limpet-form is associated with a more or less sedentary life and microphagous habit.

Superfamily PATELLACEA

Characters of the Suborder; apex not turned backwards. (?M.Sil.-?L.Trias. M.Trias.-Rec.)

Family PATELLIDAE

Internally iridescent to porcellanous. (?Jur. Eoc.-Rec.)

Patella (*s.s.*) (Fig. 639): outline oval, usually a little narrower in front; apex nearly central, very slightly anterior to the middle; apical angle more or less obtuse, not curved, protoconch not seen; ornament of strong radial costae and commarginal growth lines; internally iridescent. ?U.Cret., Eoc.-Rec. Eur. *Helcion* (*s.s.*): outline oval; rather thin-shelled; capuliform, apex bent over far forward and almost above anterior margin; moderately solid scaly radial costae. ?Jur. Cret.-Rec. S.Africa Eur. Pac. *Nacella* (*s.s.*): similar to *Helcion*, apex fairly near to anterior margin; with low simple costae; shell pellucid. Eoc.-Rec. S.Amer., Antarctica, Eur.

Family ACMAEIDAE

Shell conical, porcellanous, often smaller than Patellidae; apex central or near anterior margin; externally with radial ornament or smooth; protoconch, conical, not spiral. (M.Trias.-Rec.)

Acmaea (*s.s.*) is Pleistocene to Recent. The subgenus *Tectura* (Fig. 640) is oval in outline, the apex is strongly bent over and in front of the middle, and the shell has feeble radial and commarginal ornament. Plio.-Rec. Eur., Medit., Atl., Indian Ocean. The subgenus *Parvacmea* is small, rather thin-shelled, the

291

apex is near the anterior end, and the shell externally is ornamented with fine radial threads crenulated by commarginal growth lines. Olig.-Rec. N.Z.

Family LEPETIDAE

Small, colourless, conical or capuliform, with apex in front of middle; smooth or with weak ornament. (Mio.-Rec.)

Lepeta (*s.s.*): apex only slightly anterior to middle, the tip of the muscle scar in front of it; externally with fine radial costae; the spiral protoconch is lost in the adult. Plio.-Rec. Eur., N.Atl., Arctic, N.Pac.

Superfamily COCCULINACEA

Like the Patellacea, but apex turned backwards. (Mio.-Rec.)

Family COCCULINIDAE

Small, conical to capuliform; apex posterior to middle; deep water habitat. (Mio.-Rec.)

Cocculina (*s.s.*): apex only slightly posterior to middle, the spiral protoconch usually lost; extremely fine radial and commarginal ornament. Mio.-Rec. Carib., N.Amer., Eur.

Family LEPETELLIDAE

Low to obliquely conical, smooth; apex median or behind middle, not spiral; aperture rounded or oval. (Mio.-Rec.)

Lepetella is Recent only. The subgenus *Tectisumen* has a high, nearly central apex, is nearly circular in outline, and the anterior and posterior margins are raised up. Mio.-Rec. N.Z.

Suborder TROCHINA

Usually conispiral, occasionally discoidal; labrum simple; inner shell layer (in some forms, complete shell) nacreous; operculum calcareous or corneous in Trochacea, calcareous and multispiral in some other forms, otherwise unknown. (L.Ord.-Rec.)

Superfamily TROCHACEA

Conical, turbiniform or subglobose; aperture entire; inner shell layer (and outer layer in some genera) nacreous; operculum calcareous or corneous, spiral. (Trias.-Rec.)

Although the typical *Turbo* and *Trochus* are easily distinguished by their respectively 'turbinate' and 'trochiform' shape, this distinction does not extend through the two families: there are trochiform Turbinidae and turbinate Trochidae. Among Recent forms the operculum, calcareous in Turbinidae, horny in Trochidae, forms a simple criterion; and, though it is not usually found with the shell in fossils, Cossmann has pointed out that the rigid calcareous operculum is correlated with a peristome lying in one plane, while the flexible horny operculum allows of a difference of plane for the basal and lateral portions of the labrum, and such a difference is shown by most (though not all) Trochidae.

292

In general also the aperture of Trochidae is more depressed than that of Turbinidae, and teeth are more frequently present on the peristome.

Family TROCHIDAE

Peristome discontinuous or nearly so, columellar lip and labrum not in the same plane; aperture rhomboidal to rounded; except in the Monodontinae, basal portion of labrum concave in profile, increasing in curvature towards the angle and continuing on the lateral area in a straight line or very slightly convex curve to meet the suture at an acute angle (about 30°); columellar lip vertical, often meeting labrum at an aprupt angle, near which it may bear a tooth or fold. (Trias.-Rec.)

Subfamily TROCHINAE

Conical, rarely turbiniform, ornamented with costae or nodes; with or without small umbilicus, base medially excavated; aperture quadrate, with markedly discordant lips; peristome discontinuous, labrum sharp and strongly prosocline; columellar lip straight, curved or toothed, usually at a distinct angle to base of shell. Three genera, all surviving into the Caenozoic. (U.Cret.-Rec.)

Trochus (*s.s.*): conical, with more or less flattened base, periphery angular; ornament of granular spiral cords; umbilical area with callous coating; columellar lip with one or more folds. Mio.-Rec. Indo-Pac., Africa, Austral. *Tectus* (*s.s.*) (Fig. 641): spire longer and more acute, bearing costae or several rows of tubercles, base almost smooth; no umbilicus; strong fold on columella, which ends abruptly in front of it. U.Cret.-Rec. Indonesia, Indo-Pac., Japan, India, Ceylon, W.Pakistan, E.Africa. *Clanculus* (*s.s.*): rather small, subturbinate, with crenulate spiral ornament; false umbilicus; teeth and folds on both lips, greatly constricting the aperture. U.Cret.(Maastr.)-Rec. Red Sea, Eur., Indo-Pac., W.Pakistan, Indonesia, Austral., warm seas (except Amer.).

Subfamily MONODONTINAE

Turbinate, high-conical or littoriniform, base convex and without umbilicus; usually with spiral ornament or smooth; umbilical area and base of columella buried in callus; peristome discontinuous; labrum very prosocline in most forms, basal and lateral portions in one plane; columellar lip usually with one or more teeth. (?Trias., M.Jur.-Rec.)

Monodonta (*s.s.*): thick-shelled, globose-turbinate, with weakly crenulate spiral ornament; columella with one or two anterior teeth and labrum internally denticulate. Olig.-Rec. Indonesia, Eur., Austral., Indo-Pac. *Cantharidus* (*s.s.*): spire high-conical, so that height of shell is greater than width; smooth or with weak spiral furrows; aperture less than half the height of the shell; columella with a median fold but no distinct tooth. Mio.-Rec. Austral., N.Z. *Thalotia* (*s.s.*): high-conical like *Cantharidus*, but with finely crenulated spiral threads, columella with a solid tooth, and labrum and floor internally folded.

293

Mio.-Rec. Austral., Indonesia, Indo-Pac. **Bankivia** (*s.s.*) is Pleistocene to Recent only. Its subgenus **Leiopyrga** is also high-conical and has the labrum and floor internally folded; ornament of crenulated spiral threads; columella curved, little twisted, not truncated anteriorly; unusual in having a small umbilicus. Mio.-Rec. S.Pac., Austral. **Turcica** [*Ptychostylis*]: spire high, with flattened sides and deeply sunken sutures; ornament of crenulated spiral threads; columella with one or two large teeth on a strong fold. Mio.-Rec. Austral., N.Amer., Japan. **Tegula** (*s.s.*): trochiform, with flattened base and false umbilicus; whorls with crenulated to beaded spiral threads; columella with a strong anterior tooth. Mio.-Rec. Pac.-S.Amer., N.Amer., Pac. The subgenus **Chlorostoma** tends to be cyrtoconoid, is smooth or has beads arranged in oblique rows on its upper surface, and the columella has two teeth anteriorly. Mio.-Rec. N.Amer., S.Amer., Japan, Pac. The subgenus **Omphalius** [*Neomphalius*] is similar to *Chlorostoma*, but is nearly smooth and has a deep false umbilicus. Mio.-Rec. Japan, N.Amer., C.Amer., Pac. The genus **Diloma** is Recent only. Its subgenus **Oxystele** is rounded-conical, smooth or with spiral striae, with an irregularly thickened callous area on the base; labrum strongly prosocline; columella without fold. Mio.-Rec. S.Africa, W.Africa, Eur., Japan.

Subfamily GIBBULINAE

Turbinate, usually with umbilicus; usually with spiral ornament, rarely smooth; peristome usually discontinuous; labrum strongly prosocline; columellar lip usually smooth, arising from umbilical margin, more rarely with a weak tooth. (U.Jur.-Rec.)

Gibbula (*s.s.*): whorls stepped, with deep sutures; ornament of fine spiral furrows and threads and a row of nodes posteriorly; last whorl subangular at base; rather scaly oblique growth lines. U.Cret.-Rec. Medit., Eur., Atl., S.Amer. The subgenus **Steromphala** (Fig. 642) has the apex and periphery more rounded; there are fine spiral striae, but no nodes. Mio.-Rec. Eur. The subgenus **Colliculus** is small, has strong spiral cords all over, a very narrow umbilicus, a less strongly prosocline labrum, and a slight fold on the columella. Eoc.-Rec. Medit., Eur., Atl., Indonesia, Japan, N.Z., Africa. **Phorculus:** small, low-turbinate, spire cyrtoconoid; coarse and fine spiral ornament; umbilicus fairly wide and deep; columella with a feeble anterior tooth. Eoc.-Mio. Eur., N.Amer., S.Amer.

Subfamily CALLIOSTOMATINAE

Trochiform or turbinate, usually with flattened base; anomphalous or with small umbilicus; aperture oblique, rounded quadrate; peristome discontinuous; columellar lip straight, oblique, meeting labrum obtusely; labrum strongly prosocline; columellar lip meeting parietal region at an abrupt angle, smooth or with anterior denticle. (L.Cret.-Rec.)

Calliostoma (*s.s.*) (Fig. 643): anomphalous; crenulated spiral threads on early whorls, later whorls nearly smooth; conical, base rather flattened; distinct columellar tubercle. Mio.-Rec. Medit., Atl. **Astele** (*s.s.*): like *Calliostoma*, but with an umbilicus bounded by a crenulated cord, and spiral threads beaded throughout. Cret.-Rec. Cosmop.

Subfamily MARGARITINAE

Small, thin-shelled, internally iridescent; conical-turbinate to sublenticular; umbilicus usually more or less wide; peristome discontinuous; both lips thin; columellar lip usually simple; labrum only gently prosocline. (Trias.-Rec.)

Margarites [*Eumargarita*] (*s.s.*) (Fig. 644): turbinate, smooth or nearly so, with fairly wide umbilicus. U.Cret.(Danian)-Rec. Northern seas, Eur., Iceland, N.Amer. [Shows some resemblance to certain land snails but is internally iridescent and the labral profile is more oblique.] The subgenus *Periaulax* is trochiform, carinate at the edge of the base and the umbilicus has a beaded thread around its margin. U.Cret.-Plio. Eur., N.Amer., Africa, Burma, Indonesia.

Subfamily SOLARIELLINAE

Conical-turbinate, with umbilicus; aperture nearly circular. (U.Cret.-Rec.)

Solariella (*s.s.*): rather low-turbinate, with sunken sutures and spiral cords, the adapical ones being a little beaded; umbilicus limited by a beaded spiral cord. Cret.-Rec. Cosmop.

Subfamily UMBONIINAE

Depressed-turbinate or low-trochiform; umbilicus partly or wholly filled by a callous pad. (U.Cret.-Rec.)

Umbonium (*s.s.*): broadly turbinate-lenticular, with linear sutures, smooth or with spiral striae; callous pad completely filling umbilicus. Plio.-Rec. Indonesia, Indo-Pac., Japan. *Ethalia* (*s.s.*): broadly turbinate, smooth, umbilicus rarely completely filled by callus. Mio.-Rec. Pac.

Family ANGARIIDAE [DELPHINULIDAE]

Turbinate, discoidal in youth, umbilicate; whorls with spiral rows of spines, nodes or crenulations; aperture mostly circular, internally nacreous; peristome continuous; labral profile straight, nearly vertical or gently oblique; operculum as in Trochidae. (Trias.-Rec.)

Regarded by some as a subfamily of the Trochidae.

Angaria (*s.s.*) [*Delphinula*] (Fig. 645): low-turbinate; spiral rows of nodes or spines, some of the latter being curved or branched; umbilicus wide, usually limited by a row of nodes. M.Jur.-Rec. Indonesia, Eur., Africa, W.Pakistan, Austral., Indo-Pac.

Family STOMATIIDAE

Shells with few whorls, often auriform and with a large aperture so that the inside of the last whorl is wholly visible; anomphalous; operculum usually lacking. (Trias.-Rec.)

Stomatia is Recent only. *Roya:* limpet-shaped, outline oval, apex at about four-fifths of length; muscle scar horse-shoe-shaped. Mio.-Rec. N.Z., Indo-Pac.

Family TURBINIDAE

Solid, globose, turbinate, usually well ornamented, rarely smooth; aperture circular or subcircular, internally nacreous; peristome entire, in one plane; columella smooth, curved; operculum calcareous, spiral, with central or eccentric nucleus. (M.Trias.-Rec.)

Subfamily TURBININAE

Turbinate, usually large; aperture circular; operculum heavy, subcircular, externally convex. (U.Cret.-Rec.)

Turbo (*s.s.*): turbinate, smooth, last whorl large and rounded; anomphalous; columella moderately widened and callous, anteriorly with a feeble auriform expansion; operculum internally flat. Olig.-Rec. Philippines, Indo-Pac., Indonesia, Red Sea. The subgenus *Marmarostoma* [*Senectus*] (Fig. 646) differs from *Turbo* in being ornamented with scaly spiral cords and having subangular whorls, and in having a small umbilicus and low siphonal fasciole; the operculum is externally asymmetrically convex. Olig.-Rec. Indo-Pac., Indonesia, W.Pakistan, E.Africa, Red Sea, Eur., N.Amer., Carib. The subgenus *Sarmaticus* differs from *Turbo* in being broader and in having a relatively larger aperture; there are weak spiral threads and a blunt shoulder with weak nodes; the columellar callus is considerably broader. U.Cret.-Rec. S.Africa, Egypt, Eur. *Tectariopsis:* conical-turbinate, with flattened whorls ornamented with rows of small crenulations; last whorl usually bicarinate, posteriorly with a shelving shoulder; base with crenulated spiral threads and axial folds; anomphalous; columellar lip slightly expanded anteriorly and with a tubercle; the flaring labrum and floor have a row of crenulations inside with lirae on their inner side. Eoc. Eur.

Subfamily ASTRAEINAE

Trochoid, with more or less carinate and spinose periphery; base flattened; operculum oval. (M.Trias.-Rec.)

Astraea (*s.s.*): trochiform; whorls with crenulated spiral threads and a spinose periphery throughout, the flattened spines overlapping the sutures; base concave, with wide, deep umbilicus. Mio.-Rec. N.Z., Austral. The subgenus *Astralium* is similar, but has a less flattened base and is anomphalous. Eoc.-Rec. Indian Ocean, Indonesia, Eur., Carib., ?S.Amer., N.Z. The subgenus *Pachypoma* is similar to *Astralium*, but is not spinose. Mio.-Rec. Carib. *Bolma* (*s.s.*) (Fig. 647): turbinate; early whorls with axial folds and an anterior spinose keel, but first two whorls discoidal; last whorl with a nodose shoulder, two weak spiral keels, and crenulated spiral threads; sutures deeply inset; early stage with umbilicus, but in adult this is covered by a broadly spreading callus; aperture oblique, circular; operculum with spiral costae. Mio.-Rec. Medit., Eur., Azores, Asia Minor, Indo-Pac.

Subfamily HOMALOPOMATINAE

Small, with spiral ornament well developed, axial ornament absent or weak. (Pal.-Rec.)

Homalopoma (*s.s.*): rounded-turbinate, with spiral ornament; anomphalous; operculum with both horny and shelly layers, multispiral. Pal.-Rec. Medit., Eur., N.Africa, N.Amer., S.Amer. **Cirsochilus** (*s.s.*): low-turbinate, with spiral carinae but only fine spiral ornament on base; umbilicus deep, moderately wide, with angular margin; columellar lip thin parietally, with an anterior auricle; labrum flaring, with a varix behind the edge. M.Jur.-Plio.-?Rec. Eur., Indonesia, Carib., S.Amer., Indo-Pac.

Family LIOTIIDAE

Fairly small, rounded-turbinate to discoidal; axial ornament usually well developed, spiral ornament sometimes well developed; usually, but not always, with an umbilicus; aperture circular, labrum often strongly thickened; operculum internally horny, externally calcareous or with calcareous granules; apertural area nacreous. (Trias.-Rec.)

This group has been regarded by some as a subfamily of the Turbinidae, but their small size and general characters seem to merit it being regarded as a separate family.

Liotia: low-turbinate, with strong cancellate ornament; well developed umbilicus limited by a crenulated spiral cord; aperture circular, peristome continuous, and labrum thickened. Mio.-Rec. S.Amer., Carib., N.Amer., Indonesia. **Liotina** (*s.s.*): like *Liotia*, but with very strong, sharp, frilled axial costae and spiral threads; aperture small and circular, with very strong varix, apertural callus frilled outside. Eoc.-Rec. Eur., N.Africa, W.Pakistan, Indo-Pac., Japan, Austral. **Pareuchelus:** turbinate, with spiral cords latticed by axial ornament; umbilicus small: labrum flaring; aperture circular, with an anterior auricle. Eoc.-Plio. Eur., Indonesia.

Family COLLONIIDAE

Small, turbinate to subdiscoidal, not nacreous; spire low; smooth or with predominantly spiral ornament; umbilicus present; peristome continuous; operculum calcareous, paucispiral. (U.Cret.(Danian)-Plio.)

Although regarded by some as a subfamily of the Turbinidae, the non-nacreous shell and generally small size seem to warrant the group being regarded as a family.

Collonia (*s.s.*): low-turbinate, with convex whorls and very fine spiral ornament; labrum varicose; umbilicus limited by a crenulated cord. Pal.-Plio. Eur., N.Amer., Austral.

Family CYCLOSTREMATIDAE

Very small, low to rounded-turbinate, not nacreous, usually with an umbilicus; smooth or ornamented. (U.Cret.-Rec.)

Cyclostrema: spire very low to flat; strong axial costae; aperture circular, oblique; wide umbilicus. Pal-Rec. Carib., N.Amer., Eur., Indo-Pac., tropical

297

seas. *Circulus:* very small, discoidal with very low spire, usually with spiral threads; umbilicus limited by a crenulated cord; aperture circular, peristome discontinuous. Mio.-Rec. Medit., Carib., Atl.

Family SKENEIDAE

Small, not nacreous; turbinate to discoidal; smooth or with weak ornament, more rarely with costae; with or without an umbilicus; peristome usually continuous. (U.Jur.-Rec.)

Skenea is Pleistocene and Recent only. *Norrisella:* minute, low-turbinate, smooth; columellar lip a little widened over the umbilicus. Eoc.-Mio. Eur., Indonesia.

Family PHASIANELLIDAE

Elongate-oval-conical to rounded, smooth or with weak spiral ornament, not nacreous; aperture oval; peristome discontinuous; anomphalous or with a small umbilicus; operculum calcareous, with eccentric nucleus. (Pal.-Rec.)

Phasianella (Fig. 648): elongate-oval-conical, whorls not numerous, smooth, moderately convex; aperture ovate, with posterior gutter; anomphalous; labrum thin, straight, slightly prosocline; columellar lip with narrow but thick band of callus bearing one slight ridge. Mio.-Rec. Austral., N.Africa, Indonesia, Indo-Pac. *Tricolia* (*s.s.*): like *Phasianella*, but short-oval-conical and with more rounded aperture; a slight umbilicus may be present. Pal.-Rec. Carib., Eur., N.Amer.

Family VELAINELLIDAE

Internally nacreous; very aciculate, cylindrical-turreted, smooth; loosely polygyrate, but with no columellar wall. (Eoc.)

Velainella, a unique genus. Eoc. Eur.

Suborder NERITOPSINA

Usually coiled and ovate or globular, rarely capuliform or patelliform; spire small, with few whorls; outer shell layer calcitic, inner layer aragonitic; sur-

---

Figs 647–657. GASTROPODA: TURBINIDAE, PHASIANELLIDAE, NERITIDAE, PSEUDOMELANIIDAE, CYCLOPHORIDAE AND VIVIPARIDAE

Figs 647–657 original.
647a, b. *Bolma baccata* (Defrance), U.Mio. Touraine. × 2.
648a, b. *Phasianella australis* (Gmelin), Rec. Austral. T. × 0·6.
649a, c. *Nerita* (*Amphinerita*) *polita* Linné, Rec. Seychelles. × 1.
650a, c. *Theodoxus concavus* (J. de C. Sowerby), L.Olig. (U.Headon). I. of Wight. × 1·5.
651a, e. *Velates perversus* (Gmelin), L.Eoc. Paris B. T. × 1·5.
652. *Velates perversus* (Gmelin), M.Eoc. (L. Khirthar). Baluchistan. T. Internal cast. × 0·75.
653a, b. *Bayania lactea* (Lamarck), M.Eoc. Paris B. × 1·63.
654a, b. *Cyclophorus* (*Litostylus*) *involvulus* (Müller), Rec. India. × 0·75.
655. *Ferussina tricarinata* (M. Braun), U.Olig. Mainz B. × 2.
656a, b. *Viviparus suevicus* Wenz, M.Mio. Nr. Ulm, Bavaria. × 0·95.
657. *Viviparus bifarcinatus* (Bielz), Plio. (L.Rumanian). Campina, Rumania. × 1·1.

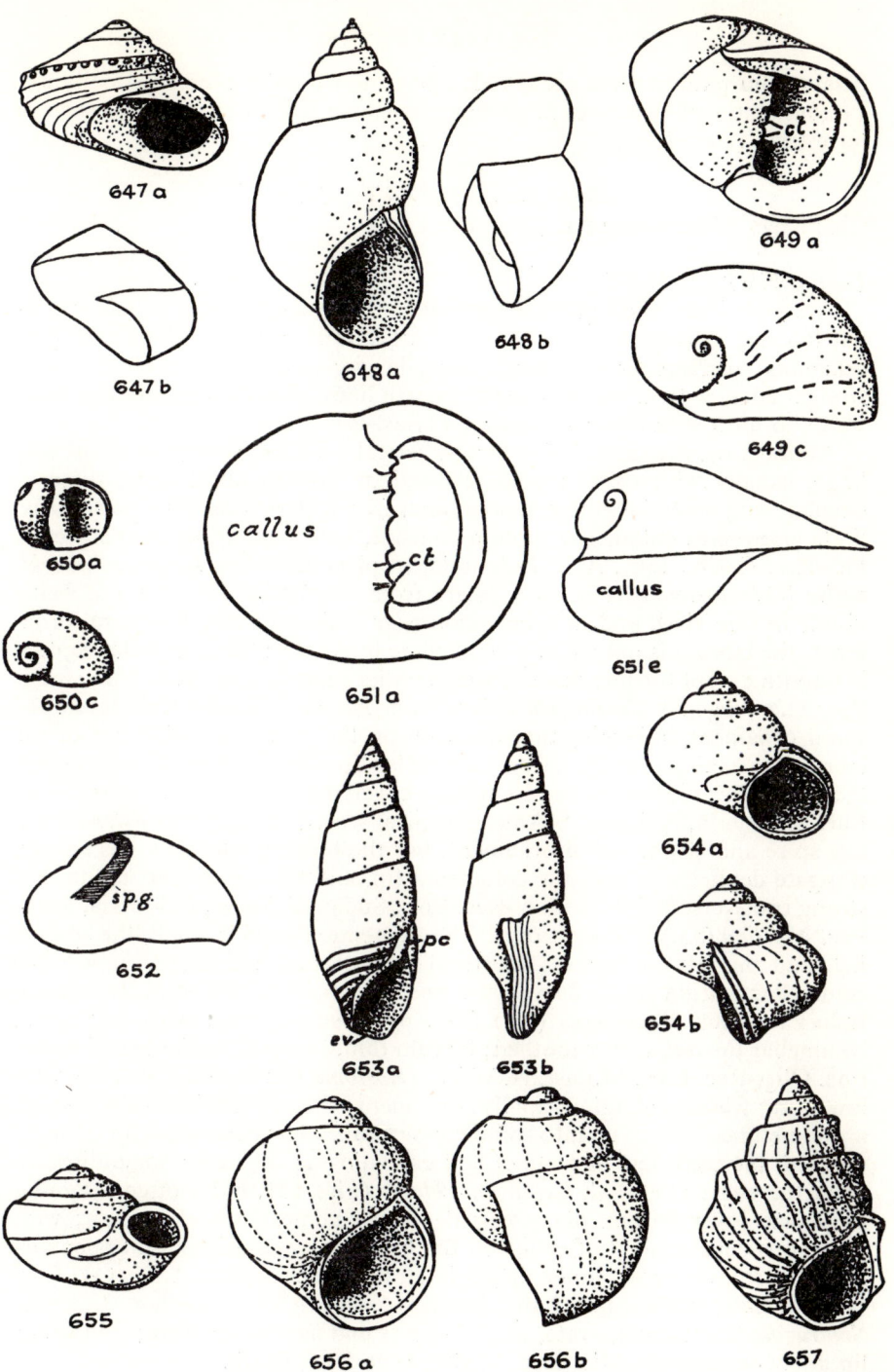

647 a

647 b

648 a

648 b

649 a

649 c

650 a

650 c

651 a

callus

ct

651 e

callus

652

s.p.g.

653 a

ev

pc

653 b

654 a

654 b

655

656 a

656 b

657

ficial colour pattern often preserved; operculum usually present, usually calcareous, and with processes projecting from inner side. Marine, fresh-water or terrestrial. (Dev.-Rec.)

Superfamily NERITACEA
Characters of the Suborder. (Dev.-Rec.)

Family NERITIDAE

Anomphalous; shell globose-turbiniform with small spire, capuliform or patelliform; usually thick-shelled, with inner walls resorbed; internally porcellanous; aperture semicircular; columellar region flattened, shelf-like, with straight edge, with or without teeth; growth lines often curving sharply backwards to meet suture at right angles. (Trias.-Rec.)

Nerita (*s.s.*): rounded-oval to hemispherical, with low spire; last whorl very large, usually with spiral threads crossed by growth lines; aperture semicircular, with posterior gutter; labrum internally with two denticles posteriorly, lirate elsewhere; columellar septum concave, with two strong median teeth. Pal.-Rec. Carib., Eur., Asia, Indonesia, tropical seas. The subgenus *Amphinerita* [*Odontostoma*] (Fig. 649) differs from *Nerita* in the spire not being visible in side view, and in being smooth or with only very fine spiral ornament; the labrum is only lightly denticulate internally; the columellar septum is smooth except for fine denticulations on its edge. U.Cret.-Rec. Eur., Indo-Pac. The subgenus *Theliostyla* differs from *Nerita* in having well developed spiral ornament, in lacking the two teeth on the posterior inside end of the labrum, in having the edge of the columellar lip with small teeth only, and in the columellar septum having its surface covered with granules. U.Cret.-Rec. Eur., Indonesia, N.Amer., S.Amer., tropical seas. The subgenus *Ritena* has a low spire and strong spiral ornament, but the labrum internally has strong, elongate denticles as does the columellar lip, and the columellar septum has strong transverse wrinkles. Eoc.-Rec. Eur., Sudan, S.Amer., Indonesia, Indo-Pac. *Neritina* (*s.s.*) is Recent only. The subgenus *Dostia* is small, the spire is flat, and the labrum is thin; apertural callus broad and shield-shaped, the columellar margin gently curved and finely denticulate. Eoc.-Rec. Indonesia, Indo-Pac., Eur. *Theodoxus* (*s.s.*) (Fig. 650): small, oval, with low spire; columellar lip very feebly toothed; labrum thin, smooth; fluviatile and estuarine. Olig.-Rec. Eur., M.East, N.Africa. *Otostoma* [*Desmieria*]: globose, with low spire; whorls enlarging rapidly; ornament of weak collabral ridges which are sometimes broken up into spiral rows of tubercles; columellar lip strongly toothed, the teeth few in number and extending as ridges for some distance from the edge. Cret.-Pal. Cosmop. *Velates* (Figs 651, 652): often attaining large size (over 10 cm. diameter); depressed conical, ovoid in plan; with enormous development of callus on the base and encroaching on upper surface; aperture narrow, inner lip fairly straight, denticulate. U.Cret.-U.Eoc. Eur., M.East, W.Pakistan, India, Burma, Asia, S.Africa, N.Amer., C.Amer. *Smaragdia* (*s.s.*): small, ovate, with low spire and narrow aperture; columellar lip finely denticulate. Mio.-Rec. Medit., Indo-Pac., Carib.

300

Family NERITOPSIDAE

Globose, with short spire and large last whorl; ornament of beads or nodes; anomphalous; inner lip broad and smooth; inner walls not resorbed; operculum not spiral. (Carb.-Rec.)

Neritopsis (*s.s.*): ornament of beaded spiral threads and sometimes weak collabral threads; labrum interally weakly folded; aperture circular; inner lip without tubercle. Trias.-Rec. Cosmop.

Family HELICINIDAE

Land snails with few whorls, conical, globose or lenticular; inner walls resorbed; columellar lip spreading over umbilicus, sometimes forming a callous pad. (U.Cret.-Rec.)

Three separate stocks of Gastropods have taken to a land life: (1) the Pulmonata, devoid of an operculum, springing probably from Opistho-branchs and forming the great bulk of the terrestrial molluscs, (2) the Cyclophoridae and their allies, Pectinibranchs allied probably to the Littorinidae and at one time known as 'operculate pulmonates', (3) the Helicinidae and a few relatives, which are Aspidobranchs. Their shells have much similarity to those of some true Pulmonates, but the Helicinidae are distinguished by a semicircular or semi-elliptical operculum (except for the Proserpininae, which have none); the aperture is simple and semicircular with a slight anterior notch, except for the Prosperpininae in which it is constricted by a number of spiral lamellae. The group is abundant in North, Central and South America, the Caribbean and the Pacific.

Helicina (*s.s.*): rounded-conical; labrum reflected; umbilical area with a callous pad bearing a pit; columella thickened below. ?Neogene-Rec. Carib., C.Amer., S.Amer. *Dimorphoptychia:* lenticular, with low-conical spire, and periphery with a blunt keel; practically anomphalous; aperture with three parallel parietal folds and a basal fold. U.Cret.-Pal. Eur. [The occurrence of this form in the Palaeocene of the Paris Basin raises an interesting problem of geographical distribution.]

Family PHENACOLEPADIDAE

Low-conical to capuliform, with apex bent backwards and behind the middle. (Eoc.-Rec.)

Phenacolepas is Recent only. *Plesiothyreus:* low-capuliform, with fine radial striae; apex over posterior margin. Eoc.-Rec. Eur., China.

The following Superfamily is of doubtful subordinal position in the Archaeogastropoda:

Superfamily AMBERLEYACEA

Usually littoriniform or turbinate, last whorl occasionally expanded and discoidal; sometimes sinistral; aperture subcircular or subangular at base of columella; spiral ornament usually dominant, sometimes noded or cancelled by collabral ornament; where known, shell structure nacreous. (Trias.-Olig.)

301

Family AMBERLEYIDAE

Dextral, turbinate to low-turriculate, with a distinct nacreous layer; usually with good spiral and collabral ornament; usually anomphalous; peristome usually discontinuous. (Trias.-Olig.)

Eucyclus: high-conical to turbinate, anomphalous; whorls imbricate, with simple and prickly spiral keels and lamellar growth lines; columella fairly straight and vertical, meeting the floor at an angle; labrum posteriorly prosocline. Trias.-Olig. Eur., S.Amer. [This is the only genus in the Superfamily that survived the Mesozoic.]

Order CAENOGASTROPODA [PECTINIBRANCHIA]

Shell without distinctive ordinal characters, but never nacreous, either holostomatous or siphonostomatous; marine, fresh-water and terrestrial. (Ord.-Rec.)

The following family is of uncertain Superfamily affinities:

Family PSEUDOMELANIIDAE

Elongate-conical to turreted, with numerous whorls which are smooth or carry only weak ornament; protoconch smooth; aperture oval, holostomatous or nearly so, usually angulated posteriorly; labrum thin, straight or gently curved, never strongly oblique. (Perm.-Mio.)

Bayania (Fig. 653): subulate, sutures bordered by an extremely narrow shelf; sculpture of fine spiral furrows, crossed by rather coarse costae in the earlier whorls, dying down on later whorls which may become smooth; aperture obliquely ovate, with deep and narrow posterior gutter and slight anterior spout giving rise to a siphonal fasciole so near the inner lip as to be inconspicuous. U.Cret.(Maastr.)-Mio. Eur., N.Amer.

Order MESOGASTROPODA

Shell not nacreous, of very variable form, nearly always spiral, rarely cap-shaped; younger, more highly developed forms often with an anterior siphonal channel; operculum usually horny, less often calcified. (Ord.-Rec.)

Superfamily CYCLOPHORACEA

Form variable, often turbinate; land and fresh-water. (Carb.-Rec.)

Family CYCLOPHORIDAE

Usually dextral and turbinate, more rarely lenticular or turreted, smooth or ornamented; aperture rounded, sometimes with a notch or breathing tube at the suture; peristome simple or thickened; operculum more or less calcareous or horny, usually multispiral. (L.Carb.-Rec.)

These land shells are distinguished from the true Pulmonata by the presence of an operculum. Their habitat is rather tropical.

Cyclophorus (*s.s.*) is Pleistocene and Recent only, and its subgenus *Lito-*

stylus (Fig. 654) is Recent only. **Leptopoma** (*s.s.*): turbinate, with smooth, convex whorls and a narrow umbilicus; parietal callus thin; labrum flaring, but not thickened. Mio.-Rec. Indo-Pac., Eur. The subgenus **Trocholeptopoma** is thicker-shelled, has a little spiral ornament, and the last whorl is angular at the edge of the base. Pal.-Rec. Eur., Indo-Pac. **Leptopomoides:** like *Leptopoma*, but sometimes with fine spiral threads, and the aperture more or less notched posteriorly. Olig.-Rec. Eur., Ceylon. **Palaeocyclophorus:** rounded-turbinate with only gently convex whorls; last whorl angular at the edge of the base, but the angulation dies out before reaching the aperture; base gently convex; a wide umbilicus limited by a blunt keel; peristome widened, feebly reflected. U.Cret.-U.Eoc. Eur. **Protocallia:** small, oval, with rounded apex, paucispiral, smoothed over with callus; aperture rounded-oval, both lips and floor widened and expanded. Olig. Eur. **Cardiostoma:** rather acutely conical, whorls with a median blunt keel; oblique radial costellae; base flattened, and umbilicus very narrow to lacking; aperture obliquely cordiform, horizontal, with a tooth at the end of the columella; peristome widely expanded. U.Eoc. Eur. **Ferussina** [*Strophostoma*] (*s.s.*) (Fig. 655): a remarkable extinct group in which the last whorl is upwardly deviated, so that the aperture faces obliquely or even directly upwards; low-heliciform; small umbilicus; peristome widened. Eoc.-U.Olig. Eur. **Craspedopoma** [*Bolania*—*nom. nud.*] (*s.s.*): more or less turbinate, practically anomphalous but with rimate base; last whorl slightly produced; aperture contracted, with continuous circular peristome. Pal.-Rec. E.Atl. islands, Eur.

Family VIVIPARIDAE [PALUDINIDAE]

More or less high-turbinate; whorls growing rapidly, usually convex, rarely keeled, usually smooth, more rarely with spiral threads or rows of nodes; aperture oval, angulated posteriorly; peristome continuous, simple; anomphalous or with small umbilicus; operculum horny, thin. Fresh-water. (?Carb., ?Trias., Jur.-Rec.)

Viviparus [*Paludina*] (*s.s.*) (Figs 656, 657, 658, 659): globose-turbinate, usually with smooth rounded whorls and obtuse apex; anomphalous or with narrow umbilicus; growth lines straight, prosocline; aperture subcircular, bluntly pointed behind, not notched in front; peristome continuous, in one plane. ?Carb., Jur.-Rec. Eur., Asia, N.Amer., S.Amer. [Certain forms from the Pliocene of the Near East develop two or three simple or noded keels and resemble the Recent genus *Tulotoma* which, however, has the aperture gently channelled anteriorly.]

Family AMPULLARIIDAE [PILIDAE]

Sometimes very large, globose, rarely discoidal; usually dextral, sometimes (*Lanistes*) ultradextral; umbilicus usually present; whorls convex, smooth or weakly ornamented; aperture large, oval, narrower posteriorly than in the Viviparidae. Amphibious. (?Carb., ?U.Cret., Eoc.-Rec.)

Ampullarius (*s.s.*) (Fig. 660): large, dextral, with smooth, convex whorls and distinct, low spire; aperture a little widened but not thickened; labral profile vertical; umbilicus moderately wide. ?Cret., ?Mio., Plio.-Rec. S.Amer.,

N.Amer. **Pila** [*Ampullaria*] (*s.s.*): like *Ampullarius*, but shell more constricted and narrower anteriorly, and columellar callus less well developed. Mio.-Rec. Africa, Asia, Indo-Pac. **Lanistes** (*s.s.*) (Fig. 661): globose, ultradextral; last whorl rounded or gently carinate; umbilicus fairly wide, limited by a keel U.Eoc.-Rec. Africa, Asia Minor, Madagascar. **Pseudoceratodes:** ultradextral, inflated discoidal, with kidney-shaped aperture; links *Lanistes* and *Pila*. U.Eoc.-Olig. Africa, W.Pakistan.

Superfamily VALVATACEA

Small, thin-shelled, turbinate to discoidal with few rounded whorls and open umbilicus; usually smooth, rarely with costae or spiral threads; aperture circular, vertical; peristome thin, continuous. Fresh-water. (?Carb., Jur.-Rec.)

Family VALVATIDAE

Characters of the Superfamily. (?Carb., Jur.-Rec.)

Valvata (*s.s.*): planorbiform to low-turbinate, loosely coiled, smooth, widely umbilicate; aperture circular, oblique, parietal callus thin. U.Jur.-Rec. Eur., Asia, N.Central Africa, N.Amer., ?S.Amer., Holarctic.

Superfamily LITTORINACEA

High-turbinate to rounded, smooth or with spiral and/or axial ornament; aperture ovate, in one plane, with continuous peristome; columellar lip callous, often flattened; operculum horny or calcified. (Carb.-Rec.)

Family LITTORINIDAE

Rounded to high-turbinate, usually anomphalous, with rounded or oval, holostomatous aperture; smooth or with spiral, rarely also axial, ornament; labrum simple, prosocline; columella thick; non-nacreous. (U.Cret.-Rec.)

Littorina (*s.s.*): nearly spherical, with very low spire, smooth; aperture circular, bluntly angular posteriorly; columellar lip concave. Plio.-Rec. Eur., N.Amer., northern seas. The subgenus **Algaroda** (Fig. 662) is more turbinate,

---

Figs 658–669. GASTROPODA: VIVIPARIDAE, AMPULLARIIDAE, LITTORINIDAE, POMATIASIDAE, HYDROBIIDAE, BITHYNIIDAE, RISSOIDAE AND RISSOINIDAE
Figs 660–661 after Wenz; Figs 658–659 and 662–669 original.
658. *Viviparus dazmanianus* (Brusina), Plio. (L.Rumanian). S.Russia. × 1.
659. *Viviparus fuchs-sadleri* (Neumayr), Plio. (L.Rumanian). S.Russia. × 1.
660. *Ampullarius urceus* (Müller), Rec. S.Amer. T. × 0·33.
661. *Lanistes carinatus* (Olivier), Rec. N.Africa. T. × 0·5.
662a, b. *Littorina* (*Algaroda*) *littorea* (Linné), Rec. Britain. T. × 1·1.
663. *Pomatias bisulcatum* (Zieten), L.Mio. Frankfurt. × 2.
664a, b. *Dissostoma mumia* (Lamarck), M.Eoc. Paris B. T. × 1.
665a, b. *Lithoglyphus naticoides* Férussac, Pleist. Hungary. × 3.
666a, b. *Bithynia tentaculata* (Linné), Pleist. Essex. × 2·25.
667. *Rissoa* (*Rissostomia*) *membranacea* (Adams), Rec. T. × 4·5.
668a, b. *Alvania curta* (Dujardin), U.Mio. Touraine. × 8.
669a, b. *Keilostoma turricula* (Bruguière), M.Eoc. Paris B. T. × 1.

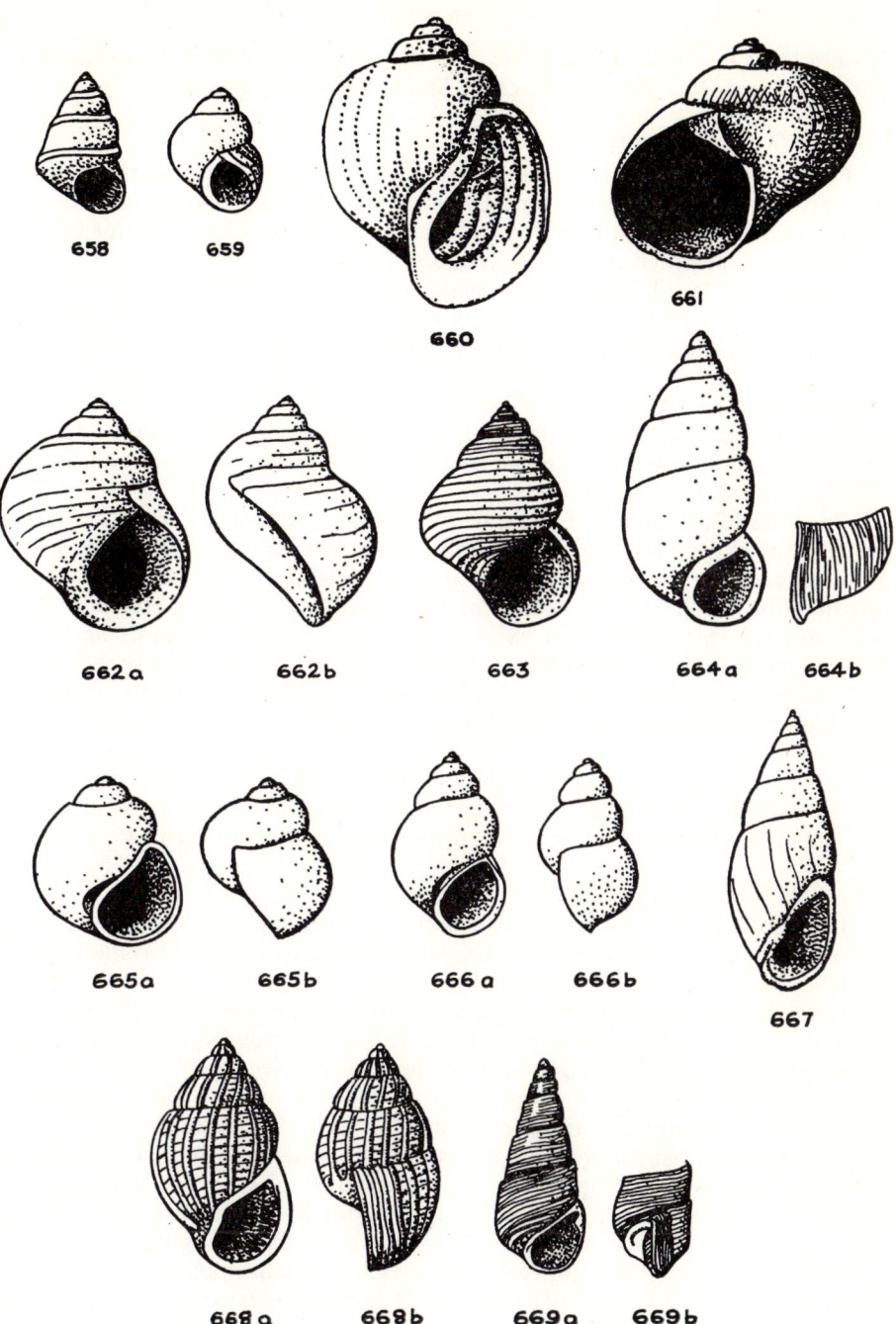

658   659

660

661

662 a   662 b   663   664 a   664 b

665 a   665 b   666 a   666 b

667

668 a   668 b   669 a   669 b

305

is smooth or has weak spiral striae, and the apertural callus is discontinuous parietally. Eoc.-Rec. Eur., N.Amer., S.Amer., northern seas.

Family LACUNIDAE

Forms resembling Littorinidae but smaller and thinner and usually having a small umbilicus bounded externally (opposite the columellar margin) by an angle; aperture as in Littorinidae but with more or less of an anterior sinus; smooth or ornamented as in Littorinidae. (M.Trias.-Rec.)

Lacuna (*s.s.*): globose, with small spire and rapidly growing whorls bearing very fine spiral striae; aperture large, slightly produced anteriorly but not notched. ?Plio., Pleist.-Rec. Eur., N.Amer., northern seas.

Family POMATIASIDAE

Low- to high-turbinate land shells with convex whorls; smooth or with spiral ornament; aperture rounded to oval, in one plane; peristome simple or a little reflected. (U.Cret.-Rec.)

Pomatias [*Cyclostoma*] (Fig. 663): oval-conical, with fairly convex whorls bearing spiral ornament; narrow umbilicus; aperture circular or subcircular; peristome simple; labral profile straight, very slightly prosocline. Olig.-Rec. Medit., Eur., Asia Minor. *Tudorella:* like *Pomatias*, but more elongate and with distinct spiral ornament. Mio.-Rec. Eur., N.Africa. *Dissostoma* (Fig. 664): slender-conical to subulate, with spiral ornament and growth lines on later whorls, a narrow umbilicus, and an obliquely oval aperture the margins of which are somewhat thickened. Pal.-L.Olig. Eur.

Superfamily RISSOACEA

Usually small, oval-conical to turreted; whorls more or less convex, smooth or ornamented; aperture oval or round, usually unchannelled. (Perm.-Rec.)

Family HYDROBIIDAE

High-conical to low-turbinate, smooth or ornamented, with or without umbilicus; peristome continuous, sometimes thickened; labral profile nearly vertical. Mostly fresh- or brackish-water. (Perm.-Rec.)

Hydrobia [*Paludestrina*] (*s.s.*): conical to elongate oval-conical, smooth, with narrow umbilicus; aperture obliquely oval, somewhat angulated posteriorly; labral profile straight and vertical. Perm.-Rec. Eur., Asia, Africa, N.Amer. *Prososthenia:* base contracted, giving it a pupiform shape; rather coarse axial ribbing; labrum thickened. Plio. Eur. *Amnicola:* rounded-oval, smooth, with small umbilicus; peristome continuous, thin. U.Jur.-Rec. N.Amer. *Lithoglyphus* (Fig. 665): turbinate, with short spire and apical angle about 90°, smooth; peristome very oblique (about 40°), labral profile straight; aperture approaching semicircular, but rounded in front, angulated behind; columellar lip nearly straight, with much callus; small umbilicus. Plio.-Rec. Eur., Asia. Doubtfully placed in the family is *Potamaclis:* very long-turreted, with posterior angulation to the oval aperture; peristome continuous; anomphalous. ?Pal., Olig. Eur.

306

Family TRUNCATELLIDAE

Usually elongate and more or less cylindrical, sometimes oval-conical, apex often truncated in adult; smooth or ornamented; aperture oval. Fresh-water, terrestrial, or on sea-shores. (Pal.-Rec.)

Truncatella (*s.s.*): cylindrical, smooth or weakly costate, last whorl nearly one-half the height, anomphalous; aperture obliquely oval, somewhat angular posteriorly. Pal.-Rec. Eur., Asia, N.Africa, Austral. *Pyrgula* (*s.s.*): turreted, whorls with simple or noded median keel; aperture oval, somewhat angulated posteriorly, its long axis nearly vertical; a second, more anterior keel on last whorl; anomphalous. Plio.-Rec. Eur., Asia Minor.

Family STENOTHYRIDAE

Small, inflated-oval to elongate-oval, smooth; a few feebly convex whorls, the last one bent down; anomphalous; aperture rounded-oval, somewhat constricted. (Pal.-Rec.)

Stenothyra [*Nematura*]: usually rounded-oval with rather cyrtoconoid spire and large last whorl; aperture rather small, obliquely oval. Plio-Rec. Asia, Austral. *Stenothyrella:* more elongate-oval that *Stenothyra*; parietal callus feebly appressed. Pal.-Plio. Eur.

Family BITHYNIIDAE [BULIMIDAE] (I.C.Z.N. Opinion 475)

Globose to oval-conical, with moderately convex, smooth or rarely ornamented whorls; usually anomphalous; aperture oval or rounded; peristome continuous; operculum usually strongly calcified. (Eoc.-Rec.)

Bithynia [*Bulimus, Bithinia*] (*s.s.*) (Fig. 666): oval-conical with moderately high spire; whorls convex, smooth; last whorl large, usually with an umbilical slit; aperture rounded-oval, posteriorly angulated; labral profile only slightly prosocline. Eoc.-Rec. Eur., Asia, N.Africa. The subgenus *Tylopoma* differs from *Bithynia* in the whorls being costate and somewhat shouldered posteriorly, and in lacking an umbilicus. Plio. E.Eur., Asia.

Family MICROMELANIIDAE

Turreted-conical to oval-conical; whorls numerous, smooth or ornamented; aperture vertical or nearly so, oval; peristome continuous; usually anomphalous. (Cret.-Rec.)

Micromelania (*s.s.*): slender-turreted, whorls medially angular, with a few spiral keels or threads carrying fine prickles where crossed by collabral costellae; anomphalous. Plio.-Rec. Eur., Asia. *Stalioa* (*s.s.*): oval-conical, with moderately prominent spire and much the same shape as *Bithynia*; ornament of fine spiral striae; small umbilicus; aperture oval, bluntly angular posteriorly; labrum vertical, externally varicose. U.Cret.(Danian)-Plio. Eur. *Nystia:* cylindroidal, with eroded apex, smooth or with fine collabral ornament; narrow umbilicus; aperture obliquely oval, with thickened peristome. Pal.-Mio. Eur.

Family RISSOIDAE

Small, turbinate to elongate, with rounded whorls; sculpture a cancellate combination of axial and spiral ridges, or of either series alone, rarely smooth; aperture oval, usually pointed above; peristome continuous, usually thickened or expanded; labrum vertical or prosocline, never opisthocline, straight; aperture not notched anteriorly. (Jur.-Rec.)

Rissoa (*s.s.*): elongate-oval-conical, with collabral costae and spiral striae, costae fading out on base of last whorl; labrum varicose, internally with a posterior denticle. Olig.-Rec. Eur. The subgenus *Rissostomia* (Fig. 667) is like *Rissoa*, but has no spiral ornament and collabral costae on the early whorls only, the later whorls being smooth. Plio.-Rec. Eur. *Alvania* (*s.s.*) (Fig. 668): oval-conical, with solid collabral costae latticed by weaker spiral threads; labrum varicose, internally lirate. Pal.-Rec. Cosmop.

Family RISSOINIDAE

Like the Rissoidae, but aperture channelled anteriorly and labrum nearly always opisthocline. (Jur.-Rec.)

This group is included by some authors in the Rissoidae, but the anteriorly channelled aperture and opisthocline labrum seem to warrant its recognition as a family.

Rissoina (*s.s.*): elongate-oval-conical, with convex whorls bearing solid, curved collabral costae with fine spiral striae in the intervals; labrum with solid varix; aperture obliquely oval, a little channelled and produced anteriorly. Pal.-Rec. Cosmop. *Keilostoma* [*Paryphostoma*] (Fig. 669): a fairly large genus (up to at least 25 mm. in height); turreted, with flat-sided whorls ornamented with fine spiral grooves, sutures deeply incised; aperture asymmetric, pointed behind, with arcuate labrum and obtusely bent inner lip; peristome continuous, thickened; aperture slightly emarginate anteriorly; labrum with a strong varix, posteriorly opisthocline. U.Cret.-Olig. Eur., Asia, N.Amer.

Family CAECIDAE

Very small, with spiral protoconch followed by a feebly curved tube; smooth or ornamented; internally often with transverse lamellae. (Pal.-Rec.) (87c)

Caecum (*s.s.*): strong annular ornament; peristome circular, simple, somewhat thickened; last septum conical. Olig.-Rec. Eur., Carib., N.Amer., Austral., N.Z., warm seas.

Family SYNCERIDAE

More or less high oval-conical, with rather flattened or convex whorls; aperture rounded-oval, usually angulated posteriorly; peristome simple. (Mio-Rec.)

Syncera [*Assiminea*] (*s.s.*): whorls fairly convex, smooth; anomphalous; aperture rounded; peristome continuous. Mio.-Rec. Eur., N.Amer., coasts of warm seas.

Family ACMEIDAE

Cylindrical, with blunt apex and feebly convex whorls, smooth or with collabral costellae; anomphalous; aperture oval; peristome thickened and labrum sometimes varicose. (Eoc.-Rec.)

Acme (*s.s.*): collabral ornament; aperture posteriorly angulated; slight swelling on neck; labrum varicose. Olig.-Rec. Eur.

Family TORNIDAE [ADEORBIDAE]

Rather small, low-turbinate to discoidal; whorls few, smooth or ornamented; aperture oblique, rounded to oval; labrum sharp; usually with a wide umbilicus. (Jur.-Rec.)

Tornus [*Adeorbis*]: low-turbinate, with rather flattened base and a few spiral threads; last whorl large, the umbilicus limited by a keel; peristome continuous. Pal.-Rec. Cosmop. *Teinostoma* (*s.s.*): small, smooth or with extremely fine spiral striae; umbilicus covered by a broad, flat callus. Jur.-Rec. Cosmop. [For many years this genus was placed in the Skeneidae.)

Family OMALOGYRIDAE

Very small and discoidal; whorls rounded, smooth or ornamented; aperture round. (Eoc.-Rec.)

Omalogyra: smooth. Eoc.-Rec. Cosmop.

Superfamily CERITHIACEA

Usually more or less turreted, less commonly conical or discoidal; whorls smooth or (more usually) with spiral ornament which may be noded, sometimes also with collabral ornament; aperture often anteriorly channelled; labrum often prominent, simple or thickened; columella simple or with folds, often twisted anteriorly. (Trias.-Rec.)

Family TURRITELLIDAE

Usually slender, turreted and anomphalous; whorls numerous, with spiral ornament and curved growth lines; aperture not large, rounded or subquadrate, rarely channelled anteriorly; columella smooth, concave; labrum thin, arcuate, prosocline at the suture. (Cret.-Rec.)

[In Marwick's generic revision of the Turritellidae (**87b**) the most reliable criteria for classifying members of the family were suggested to be (1) the trace of the labrum, (2) the ontogeny of the primary spirals (the first four being lettered A, B, C and D from the top downwards), and (3) the protoconch.]

Subfamily TURRITELLINAE

Labrum parasigmoidal or with a double sinus; columella curved or slightly twisted. (Cret.-Rec.)

Turritella (*s.s.*) (Fig. 670): turreted-conical, ornamented with spiral threads; lateral and basal portions of labrum form one wide shallow sinus sweeping back from the adapical suture and crossing the base in a straight line; on neanic whorls the sinus is narrower and situated higher up; primary spirals appear in the unusual order B–A–C and a secondary starts between A and B before C begins. Cret.-Rec. Cosmop. *Archimediella* (*s.s.*) (Fig. 671): labral curve very asymmetric and shallow; primaries B and C form two main keels; neanic whorls generally tricarinate, but sometimes, by retardation of A and C, they are unicarinate, although C soon strengthens. Cret.(?*s.l.*)-Rec. Widespread. In the subgenus *Torculoidella* (Fig. 672) primaries B and D dominate; labral curve moderately deep and more symmetric. Mio.-Rec. Eur., C.Amer., S.Africa. *Haustator* (*s.s.*) (Fig. 673): neanic whorls tricostate, primaries starting in the order C–B–A; labral profile with a fairly deep and fairly oblique lateral sinus, somewhat asymmetric. Cret.-Rec. Cosmop. The genus *Torquesia* (*s.s.*) is Cretaceous. Its subgenus *Ispharina* (Fig. 674) has the labral sinus deep and situated rather high up; peribasal primary spiral beaded. Eoc. Eur., Africa, Asia, N.Amer.

Subfamily PROTOMINAE
Labral profile with a wide, shallow sinus; aperture anteriorly with a deep notch; siphonal fasciole present. (Olig.-Rec.)
Protoma (*s.s.*) (Figs 675, 676): turreted-conical, with flat-sided whorls and spiral ornament; primary spirals starting in the order B–C–A. Olig.-Rec. Widespread.

Subfamily PAREORINAE
Aperture effuse over columella which has a spiral ridge; labrum parasigmoidal, convex on the base. (Cret.-Rec.)
Pareora: small, with multispiral protoconch and spiral threads; aperture oval; labrum posteriorly broadly and deeply notched, anteriorly prominent

---

Figs 670–685. GASTROPODA: TURRITELLIDAE, MATHILDIDAE, ARCHITECTONICIDAE, VERMETIDAE, THIARIDAE AND MELANOPSIDAE

Figs 670–674 after Guillaume (modified); Fig. 679 after Wenz; Figs 675–678 and 680–685 original.
670b. *Turritella terebralis* series.
671b. *Archimediella turris* series.
672b. *Archimediella* (*Torculoidella*) *subangulata* series.
673b. *Haustator imbricataria* series.
674b. *Torquesia* (*Ispharina*) *hybrida* series.
675b. *Protoma obeliscus* Grateloup.
676b. *Protoma quadrifoliata* Basterot.
677a, b. *Sigmesalia multisulcata* (Lamarck), M.Eoc. Paris B. a × 1·9.
678. *Zaria duplicata* (Linné), Rec. E.Indies. T. × 0·56.
679. *Mathilda quadricarinata* (Brocchi), U.Plio. Sicily. T. × 1.
680. *Gegania sulcata* (Pilkington), U.Eoc. (Barton). Hampshire. × 2·25.
681a, b, d. *Architectonica* (*Stellaxis*) *bicingulata* (R. B. Newton), U.Eoc. Ameki, S.Nigeria. × 2.
682e. *Tenagodus anguinus* (Linné), U.Mio. Touraine. T. × 1·9.
683a, b. *Thiara amarula* (Linné), Rec. Madagascar. T. × 0·67.
684. *Melanoides inquinata* (Defrance), U.Pal. London B. × 0·75.
685a, b. *Melanopsis fusiformis* J. Sowerby, U.Eoc. (Headon). Hampshire. × 2·25.

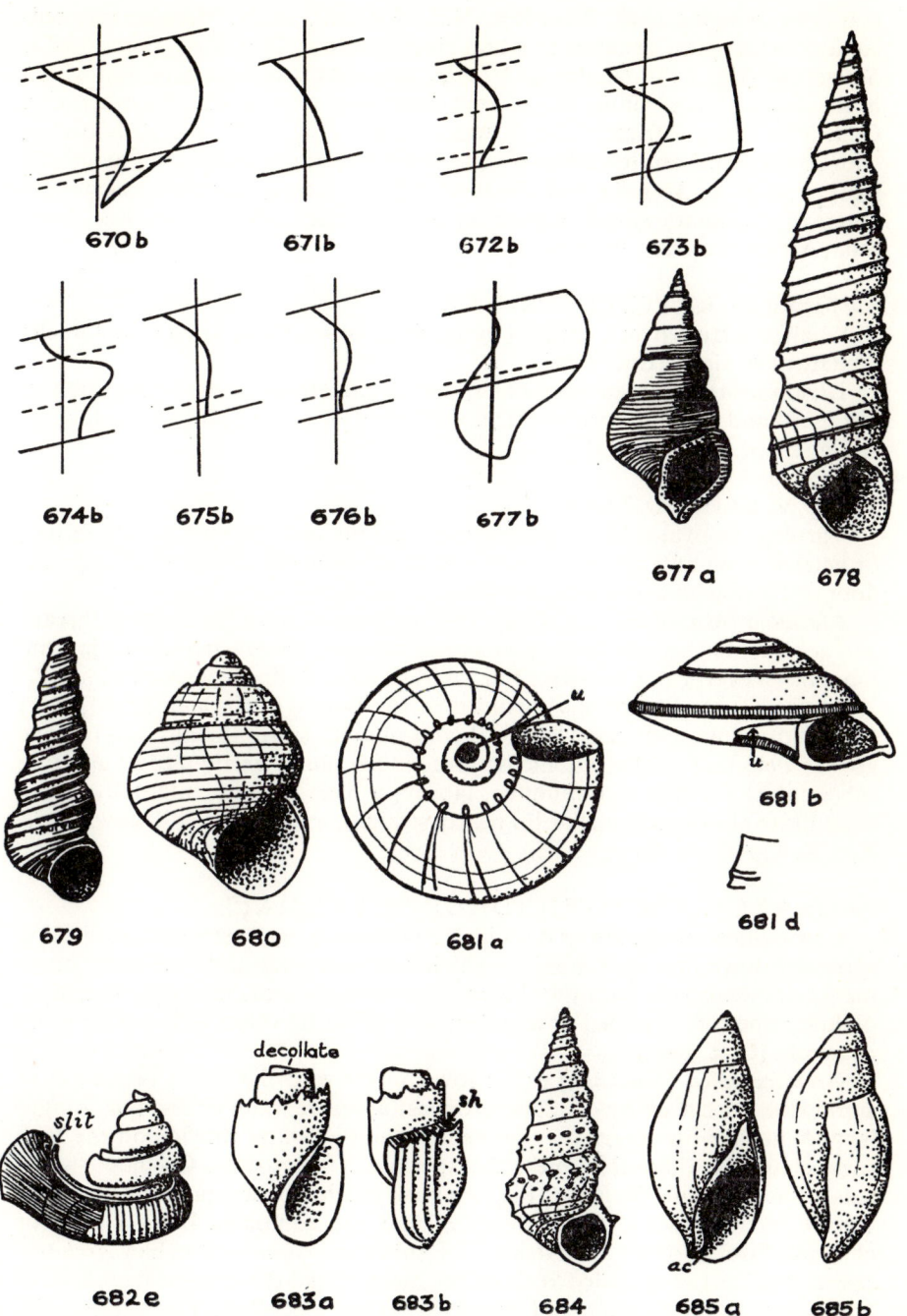

670 b   671 b   672 b   673 b

674 b   675 b   676 b   677 b   677 a   678

679   680   681 a   681 b   681 d

682 e   683 a   683 b   684   685 a   685 b

and later swinging back. Mio.-Rec. N.Z., Austral., ?Italy. *Mesalia:* turreted-conical, with spiral ornament; labral sinus shallow. Cret.-Rec. Eur., Africa. *Sigmesalia* (Fig. 677): similar to *Mesalia*, but usually less turreted and with a much deeper labral sinus, the labrum becoming very prominent anteriorly. Cret.-Mio. Eur., Africa, Asia, N.Amer. *Zaria* (Fig. 678): turreted-conical, with spiral ornament including several spiral keels; spiral ridge on columella weaker than in *Mesalia* and *Pareora*; labrum deeply notched, prominent anteriorly; primary spirals appearing in the order C–B–A. Cret.-Rec. Eur., Asia, Africa, Austral., Indo-Pac.

Subfamily TURRITELLOPSINAE

Small; aperture ovate, entire; labrum thin, more or less straight; columella arcuate or straight. (Eoc.-Rec.)

Glyptozaria: ornament cancellate, with crenulations at the intersection of the spiral and collabral ornament; no collabral ornament on base; columella straight. Eoc.-Rec. Austral.

Family MATHILDIDAE

Turreted to oval-conical, ornamented; protoconch oblique, partly retuse; or heterostrophic; labrum thin, not notched; columella smooth, little curved; floor often angulated or feebly notched. (Trias.-Rec.)

Mathilda [*Mathildia*] (*s.s.*) (Fig. 679): turreted-conical, with spiral threads and accentuated growth lines; protoconch oblique; narrow umbilicus; labrum more or less vertical. Cret.-Rec. Eur., Africa, Asia, Indo-Pac., Austral., N.Z., N.Amer., Carib. Doubtfully included in the family is *Gegania* [*Tuba*] (*s.s.*) (Fig. 680): globose-turbinate to oval-conical, with spiral threads and oblique growth lines and a false umbilicus; protoconch blunt, partly retuse; aperture subcircular; peristome continuous; labrum a little oblique, internally somewhat lirate above; columellar lip nearly straight, a little everted below. Cret.-Rec. Eur., Atl., N.Amer., N.Z.

Family ARCHITECTONICIDAE [SOLARIIDAE]

Low-conical, turbinate or discoidal; protoconch heterostrophic, projecting vertically down into the umbilicus which is usually wide and with spiral ornament; whorls usually with spiral ornament; labrum straight, nearly vertical to oblique; aperture rounded or angular; columellar lip often with one or more channels. (Cret.-Rec.)

Architectonica [*Solarium*] (*s.s.*): low-conical with flat base; ornament of spiral ribbons cut up by collabral grooves; aperture more or less quadrangular; columellar lip anteriorly channelled; the two outer threads on the base form channels where they meet the aperture. U.Cret.-Rec. Cosmop. (warm seas). The subgenus *Stellaxis* (Fig. 681) has much less ornament on the spire whorls (merely two anterior threads), and the columella has no anterior channel. Eoc. N.Amer., W.Africa. The subgenus *Nipteraxis* is like *Architectonica*, but has two rounded keels on the margin, and the ornament consists of crenulated spiral threads and collabral striae. Eoc.-Plio. Eur., Indonesia,

N.Amer. **Torinia** (*s.s.*): turbinate, with somewhat narrower umbilicus and more oval aperture; ornament of crenulated spiral threads. Eoc.-Rec. Eur., Indonesia, Japan, Austral., Indo-Pac. The subgenus **Climacopoma** is low-conical with a flat, smooth base and a wide umbilicus; spire whorls slightly concave, smooth except for one anterior crenulated spiral thread; spire slightly cyrtoconoid. Cret.-Eoc. Eur., Africa, Indonesia, N.Amer.

Family VERMETIDAE

More or less irregularly coiled, at least in later whorls, tube-like, ornament variable; aperture rounded or with a long, narrow slit; some forms with an internal arrangement of one to three septa. (?Dev., M.Trias.-Rec.)

Evidently related to the Turritellidae, these are aberrant forms in which, owing to a sessile habit, the shell often largely loses its original turreted form and becomes a loose spiral or altogether irregular. Fossil forms are often difficult to distinguish from tubicolous annelids (*Serpula* etc.), but microscopic thin sections of the shell show three layers instead of the two of *Serpula*, which in contrast has a porcellanous inner layer appearing brown in thin sections; other differences are the presence in the vermetids of a protoconch, an operculum (horny), and internal septa in some forms, as well as a deep slit in the aperture of others. Mainly tropical and subtropical.

Vermetus: irregularly spirally wound, with longitudinal ornament and growth lines; each lateral wall with an internal lira, and columellar lip with another low one; adherent. Plio.-Rec. W.Africa, Eur., Japan, warm seas. **Petaloconchus:** spirally coiled and then becoming free; cancellated ornament of spiral threads crossed by scaly growth lines often giving rise to spirally arranged crenulations; later whorls with two internal lirae on the columellar wall. Mio.-Rec. N.Amer., Carib., C.Amer., Eur., Atl., Medit. **Serpulorbis** [*Lemintina—nom. nud.*]: whorls inter-adherent or free, irregularly wound, with three or more external small crenulated keels and crenulated longitudinal threads; internally with occasional transverse plates which are concave outwards, and some transverse lamellae on columellar side. ?U.Cret., Eoc.-Rec. Cosmop. **Vermicularia** (*s.s.*): early whorls coiled like a *Turritella* but later whorls becoming disjunct although still spirally arranged; whorls carinate and with spiral threads and arcuate growth lines; no internal plates or lamellae. Pal.-Rec. Indo-Pac., Austral., Eur., N.Amer., Carib. **Tenagodus** [*Tenagodes, Siliquaria*] (*s.s.*) (Fig. 682): loosely spiral in youth, irregular later, with spiral lines broken up, where crossed by growth lines, into rows of fine tubercles or spines; tube with a longitudinal slit throughout, analogous to that of *Pleurotomaria* etc., but remaining permanently open for the whole length of the shell, though it may be constricted at intervals to form a row of elliptical perforations. M.Trias.-Rec. Indo-Pac., Austral., India, W.Pakistan, Eur., Africa, N.Amer., C.Amer.

Family THIARIDAE [MELANIIDAE]

Usually more or less turreted or rounded, often truncated above, smooth or

313

ornamented; aperture essentially in one plane, anteriorly notched or channelled, but without long canal; anomphalous; peristome sometimes discontinuous; fluviatile to brackish-water. (U.Cret.-Rec.)

Subfamily THIARINAE

Aperture elongate-oval, posteriorly angulated, anteriorly rounded, notched or with a short channel; peristome simple. (U.Cret.-Rec.)

Thiara [*Melania*] (*s.s.*) (Fig. 683): oval-turreted, with a row of backwardly directed spines separated from the suture by a narrow shelf, otherwise smooth or with fine spiral ornament; generally decollated; aperture oval, posteriorly angulated, anteriorly feebly notched; columellar lip fairly straight, somewhat callous; labrum straight in profile with slight parasigmoidal curve. ?Pal., Mio.-Rec. Indo-Pac., Asia, Africa, Eur. *Melanoides* [*Eumelania*] (*s.s.*) (Fig. 684): turreted-conical, with feebly convex whorls bearing spiral ornament and gently concave collabral costellae, sometimes with crenulations or a row of small spines; aperture oval, not notched; columella concave, callus thin parietally; labrum somewhat prominent anteriorly. Pal.-Rec. Africa, Eur., Asia, Indonesia, M.East. The subgenus *Tarebia* is less turreted, has a very straight growth line, more definitely shouldered whorls and more distinctly cancellated ornament. Pal.-Rec. Indo-Pac., Eur., Africa, Asia. *Hemisinus* (*s.s.*): turreted-conical, subulate, smooth; labral profile nearly straight and nearly vertical; aperture anteriorly with a slight notch. U.Cret.-Rec. S.Amer., C.Amer., Carib., Eur.

Subfamily PALUDOMINAE

Globose to oval-conical, usually anomphalous; smooth or ornamented; aperture fairly large, oval, at most feebly channelled anteriorly; columellar lip usually callous. (U.Cret.-Rec.)

Paludomus (*s.s.*): globose, littoriniform, with weak spiral ornament. Pal.-Rec. Indo-Pac., Asia, Eur. *Pyrgulifera* (*s.s.*): like *Thiara*, but shorter and stouter and with weak spiral ornament in addition to the spinose costae, the ramp more distinct; aperture anteriorly channelled; slight umbilical slit. U.Cret.-Eoc. N.Amer., Eur., Asia. *Cornetia:* turbinate, with a few spiral keels bearing small spines where crossed by collabral costae; aperture large, rounded, posteriorly somewhat angular; peristome solid. Pal. Eur. *Stomatopsis* (*s.s.*): elongate-conical, with collabral costae; anomphalous; aperture internally rounded, much widened and flaring, with much columellar callus; labrum gently prosocline, prominent anteriorly. U.Cret.(Danian)-Pal. Eur.

Family MELANOPSIDAE

Subulate to turreted, smooth or ornamented; aperture oval, pyriform or auriform, always with an anterior notch, the consequent deviation of the growth lines giving rise to a well-marked siphonal fasciole (cf. *Protoma* and the siphonostomes); columellar lip callous, bent anteriorly and truncated at junction with sinus; habitat fluviatile. (Cret.-Rec.)

314

Subfamily MELANOPSINAE

Labrum little sinuous. (Cret.-Rec.)

Melanopsis (*s.s.*) (Fig. 685): subulate, smooth, with rather short spire; aperture with anterior sinus, giving rise to siphonal fasciole; labrum nearly straight, thin; columellar lip smooth, callous; columella strongly twisted. U.Cret.-Rec. Eur., Medit., N.Africa, Asia, Indo-Pac. The subgenus *Lyrcaea* (Fig. 686) is more bucciniform, with rather imbricate whorls and feebly twisted columella. Eoc.-Rec. Medit., Eur., N.Africa, Asia. *Faunus* (*s.s.*): turreted-conical, terebriform, with numerous flat whorls which are smooth or at most with a few weak spiral striae and growth lines; labrum posteriorly notched and floor notched deeply. U.Cret.(Danian)-Rec. Indo-Pac., Asia, ?Eur.

Subfamily PIRENINAE [MELANATRIINAE]

Turreted to globose-oval-conical, smooth or ornamented, anomphalous; aperture usually anteriorly angular or feebly channelled, without distinct notch. (U.Cret.-Rec.)

Pirena [*Melanatria*] (*s.s.*) (Fig. 687): turreted-conical, with spiral threads and collabral costae which on later whorls become spinose, thus forming a ramp; aperture notched anteriorly and posteriorly; labrum prominent. Pal.-Rec. Madagascar, W.Pakistan, Eur., ?Japan, ?S.Amer. *Brotia* (*s.s.*): low-turreted-conical, with a row of spines just above the suture; aperture anteriorly angular. ?Cret., Plio.-Rec. Burma, Indonesia, Indo-Pac.

Family PLEUROCERIDAE

More or less turreted, smooth or ornamented; aperture anteriorly sinuous, angular or feebly channelled, not notched; labrum thin. (U.Cret.-Rec.) [Except for one Recent genus from E.Asia, all these fluviatile forms are from America.]

Pleurocera (*s.s.*), apart from one doubtful record from the Upper Cretaceous of S.America, is Pleistocene to Recent only. *Goniobasis* (*s.s.*): oval-conical to turreted-conical; smooth or ornamented; last whorl fairly convex; anomphalous; aperture oval, narrow posteriorly, feebly angular and slightly notched anteriorly; columellar lip not twisted. U.Cret.-Rec. N.Amer.

Family PLANAXIDAE

Oval-conical, anomphalous, smooth or with spiral ornament; aperture oval, posteriorly angulated, anteriorly with a narrow notch; columellar lip callous, anteriorly flattened; labrum sinuous, nearly always lirate internally. (Eoc.-Rec.)

Planaxis (*s.s.*): littorinoid, ornamented with spiral furrows. Eoc.-Rec. Indo-Pac., Indonesia, Eur., Carib.

Family MODULIDAE

Turbinate, with large last whorl; whorls convex, with spiral and collabral

315

ornament; aperture obliquely rounded; columella concave, usually with a tooth. (U.Cret.-Rec.)

Modulus: spire not very high; whorls convex, with spiral threads and rather flat collabral costae; narrow umbilicus; columella very concave, anteriorly strongly twisted and with a solid tooth. Olig.-Rec. Carib., N.Amer., Eur., warm seas.

Family POTAMIDIDAE

More or less turreted-conical, with numerous, usually ornamented whorls; aperture oval, with a short anterior channel; operculum rounded; brackish-water. (U.Cret.-Rec.)

Subfamily POTAMIDINAE

Usually with crenulated spiral threads, with or without collabral costae. (U.Cret.-Rec.)

Potamides (*s.s.*): whorls flat or convex, with spiral rows of crenulations and curved growth lines which may become lamellar on the last whorl; sometimes a median crenulated carina; aperture small and rounded, anteriorly notched or feebly channelled; anomphalous; columella short, concave; labrum concave posteriorly, prominent anteriorly. Eoc.-Plio. Eur., W.Pakistan, Indonesia, Carib., S.Amer. The subgenus *Ptychopotamides* has three or more spiral rows of crenulations of which the uppermost one is a little stronger than the others, the aperture is channelled posteriorly, the columellar callus is somewhat widened, and the columella carries a blunt fold. Pal.-Plio. Eur., Medit., ?N.Amer. The subgenus *Potamidopsis* (Fig. 688) is turreted and slightly coeloconoid and has a distinct, denticulate keel close to the anterior suture in addition to spiral rows of crenulations above; the aperture is rounded-quadrate and much widened; the labrum is greatly expanded, but there are no varices except close behind the aperture. Eoc. Eur. *Pirenella:* small forms with rounded aperture, very slight canal, almost straight labrum,

---

Figs 686–698. GASTROPODA: MELANOPSIDAE, POTAMIDIDAE, DIASTOMIDAE AND CERITHIIDAE

Figs 686–698 original.

686. *Melanopsis* (*Lyrcaea*) *aquensis* Grateloup, L.Mio. Aquitaine. × 0·9.
687a, b. *Pirena spinosa* (Lamarck), Rec. Madagascar. × 0·6.
688a, b. *Potamides* (*Potamidopsis*) *tricarinatus* (Lamarck), U.Eoc. Paris B. T. × 1·1.
689a, b. *Tympanotonos* (*Eotympanotonus*) aff. *funatus* (Mantell), U.Pal. Mt. Bernon, Marne. × 1.
690b. *Cerithidea ventricosa* (J. Sowerby), U.Eoc. (Headon). I. of Wight. × 2·6.
691a, b. *Pyrazus ebeninus* (Bruguière), Rec. T. × 0·75.
692a, b. *Terebralia palustris* (Bruguière), Rec. T. × 0·75.
693b. *Batillaria concava* (J. Sowerby), U.Eoc. (Headon). I. of Wight. × 1·1.
694. *Diastoma costellatum* (Lamarck), M.Eoc. Paris B. T. × 0·75.
695a, b. *Cerithium nodulosum* Bruguière, Rec. T. × 0·5.
696a, b. *Thericium* sp., Rec. × 1·5.
697a, e. *Clava* sp., Rec. × 0·75.
698. *Bittium reticulatum* (Da Costa), Rec. T. × 5.

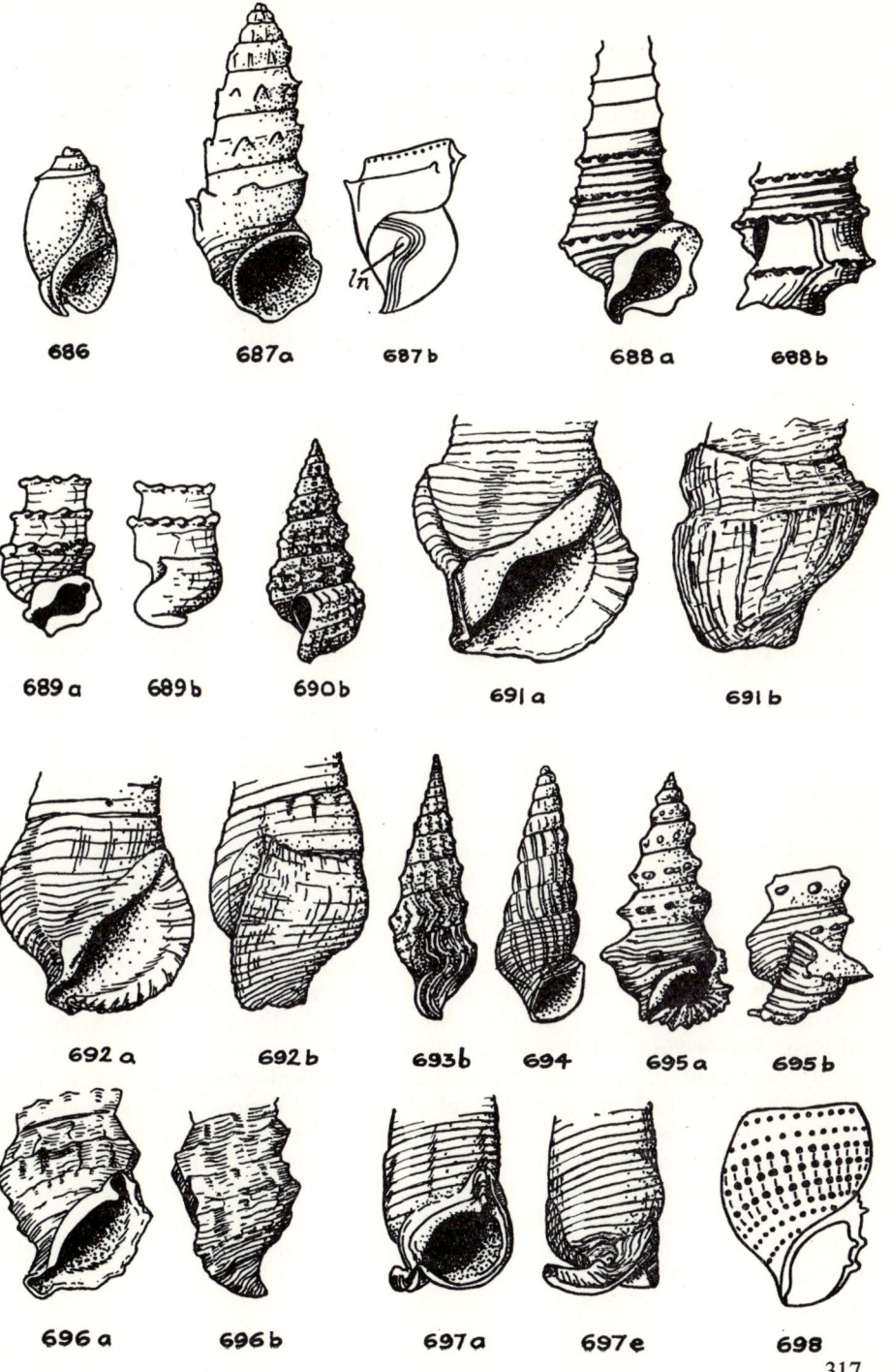

686   687a   687b   688a   688b

689a   689b   690b   691a   691b

692a   692b   693b   694   695a   695b

696a   696b   697a   697e   698

and ornament of spiral rows of close-set granules. U.Cret.-Rec. Medit., Eur., E.Africa, Asia, Indo-Pac. *Tympanotonos* (*s.s.*): turreted-conical; aperture sub-quadrangular, labrum with hood-like anterior projection, columellar lip straight; ornament changes with growth, and may be quite variable, consisting of crenulated spiral threads with or without development of nodes and spines; where later whorls develop spines, these are on the middle spiral thread. U.Cret.-Rec. Africa, Eur., Asia, Indonesia, Indo-Pac., N.Amer., S.Amer. In the subgenus *Eotympanotonus* (Fig. 689) it is the uppermost spiral thread that develops spines. Eoc. Eur. *Vicarya:* attaining a large size; whorls at first rather flat, with crenulated spiral threads or spiral furrows, later often developing a row of nodes or spines; columella very callous and much widened, with an internal spiral fold; labrum with a deep lateral notch, producing a spiral band like that of *Pleurotomaria*, below which the growth lines run very obliquely forward. L.Eoc.-Mio. W.Pakistan, Burma, Indonesia, Philippines, Japan. [N.B. The age of the type species, *Vicarya verneuili* (d'Archiac and Haime), is Miocene, not Danian as sometimes quoted.] *Morgania* [*Irania*]: elongate-pupiform, apex coeloconoid, later stages cyrtoconoid, whorls flat, smooth or with spiral striae; aperture rounded-quadrate, anteriorly deeply notched; labrum deeply notched laterally, but not so deeply as in *Vicarya*. U.Cret. (Danian)-Eoc. India, W.Pakistan, Iran, ?Carib., ?Peru. *Cerithidea* (*s.s.*) (Fig. 690): turreted-conical, with collabral and spiral ornament sometimes crenulated at their intersections; last whorl with a varix in addition to that of the labrum; aperture anteriorly snout-like, unchannelled. U.Cret.-Rec. Mada-gascar, Eur., Asia, Japan, Indonesia, N.Amer., Indo-Pac. *Telescopium:* large, turreted-conical, with numerous low, flat-sided whorls; ornament of spiral furrows, intervals smooth or crenulated; base flattened, anomphalous; aperture rounded-quadrate, anteriorly truncated and deeply notched; columella with two solid folds; columellar lip thin; labrum posteriorly con-cave. U.Cret.-Rec. India, W.Pakistan, Indonesia, E.Africa, Eur., S.Amer. *Pyrazus* (*s.s.*) (Fig. 691): labrum expanded like that of *Potamidopsis*, but giving rise to very regular varices which, occurring vertically above one another in successive whorls, give to the whole shell the appearance of a five- or six-sided pyramid; spiral threads as well; aperture scarcely channelled posteriorly, anteriorly with a short, oblique canal; columellar lip callous; labrum little sinuous. U.Cret.-Rec. Austral., N.Z., Indonesia, Japan, Asia, M.East. *Terebralia* (*s.s.*) (Fig. 692): turreted, often rather swollen or pupi-form; ornament of spiral ridges broken by vertical grooves into rows of oblong knobs; labial and other irregularly disposed varices; columella short, with an oblique fold; columellar lip callous; parietal fold; anteriorly with a short, truncated canal; labrum reflected, internally denticulate, posteriorly concave, anteriorly prominent. U.Cret.-Rec. Asia, Indo-Pac., Austral., M.East, Eur. The subgenus *Pyrazisinus* differs from *Terebralia* in lacking a columellar fold and in having collabral ornament consisting of coarse vertical costae or nodules which fade out near the last whorl. Olig.-Plio. N.Amer., S.Amer.

318

Subfamily BATILLARIINAE

Turreted; spiral ornament and sometimes also weak collabral ornament; aperture with or without a distinct canal. (U.Cret.-Rec.)

Batillaria (*s.s.*) (Fig. 693): turreted-conical, with crenulated spiral threads and weak, curved collabral costae; aperture pyriform, notched anteriorly, with continuous peristome and callous columellar lip; labrum notched laterally. U.Cret.-Rec. Austral., Japan, Indo-Pac., W.Pakistan, Iran, Eur.

Family DIASTOMIDAE

Turreted shells, with spiral and often also collabral ornament; labral profile slightly oblique and nearly straight; aperture small, oval; a very slight, spout-like anterior eversion of the peristome. (?Trias,. U.Cret.-Rec.)

Diastoma (Fig. 694): sculpture of fine spiral lines with imbricate appearance, crossed by rather coarse, slightly oblique costae and coarser varices at irregular intervals, the last at about 180° from the adult aperture; spiral lines only, alternately coarse and fine, on the base; last quarter-whorl becoming detached from spire; aperture oval, with posterior channel and slight anterior spout, giving rise to an extremely narrow siphonal fasciole; labrum very slightly sinuous, slightly prosocline, not carrying a varix. U.Cret.-Plio.-?Rec. Eur., Egypt, India, Austral., N.Amer., S.Amer. [Distinguished from *Bayania* (Pseudomelaniidae), with a very similar aperture, by the slight obliquity of the growth lines and persistence of collabral ornament.] **Sandbergeria:** oval-conical, with finely crenulated reticulate ornament but no varices; otherwise much as in *Diastoma*. U.Cret.-Mio. Eur., India.

Family LITIOPIDAE

Small, oval-conical to turreted-conical, usually with weak spiral and collabral ornament or smooth; aperture oval, with continuous peristome; columella straight, anteriorly often truncated or feebly channelled. (?Jur., U.Cret.-Rec.)

Litiopa: buccinoid, smooth or with fine spiral striae, anomphalous; protoconch with collabral costellae. Eoc.-Rec. Eur., N.Africa, Japan, Austral., warm seas.

Family CERITHIIDAE

Usually turreted, sometimes large, usually with spiral and collabral ornament; well-marked anterior canal; operculum oval; habitat marine. (U.Cret.-Rec.)

Subfamily CERITHIINAE

Usually turreted-conical and with strong ornament; aperture with a distinct canal or notch; operculum oval, paucispiral, with marginal nucleus. (U.Cret.-Rec.)

Cerithium (Fig. 695) is Pleistocene to Recent only (Carib., Red Sea, Indo-Pac.), but the type species is illustrated since many other forms, now considered not to be congeneric with it, have been called '*Cerithium*'; the labrum

319

is only slightly concave on the lateral area, nearly vertical, and it projects at its anterior end across the canal which is rather short; there is a posterior gutter bounded by a well-marked parietal fold; spiral threads and nodose collabral costae; antilabial varix; labrum internally denticulate. *Gourmya:* oval-conical, with rather constricted last whorl and shorter spire than *Cerithium*; ornament of spiral threads, with nodes also in some forms and also an antilabial gibbosity; aperture rounded-oval; parietal fold and oblique columellar fold present; canal short, truncated; labrum internally weakly lirate. Eoc.-Rec. Indo-Pac., E.Africa, Eur. *Tiaracerithium* (*s.s.*): not very large, turreted-conical; whorls imbricate, posteriorly with a row of nodes with spiral threads anterior to them; last whorl with antilabial varix. U.Cret.-Rec. Eur., Indo-Pac. *Ptychocerithium:* turreted-conical; ornament of collabral costae and spiral threads with crenulations at their intersections; last whorl with antilabial varix, its base with strong spiral threads; aperture small, with narrow posterior channel and short, truncated, weakly bent back anterior canal; labrum not very sinuous, externally with a varix, internally folded. ?U.Cret., Pal.-Rec Eur., Iran, India, Ceylon, W.Pakistan, Indonesia, Pac., ?N.Amer. *Thericium* [*Vulgocerithium*] (*s.s.*) (Fig. 696): turreted-conical, tending to be cyrtoconoid; whorls moderately convex, with spiral threads and rows of crenulations, sometimes medially subangular where collabral costae form small spines; last whorl with antilabial varix; aperture oval, posteriorly channelled and with a parietal fold; anterior canal short and deep; labrum not prominent. U.Cret.-Rec. Eur., N.Africa, ?W.Pakistan, Indonesia, Austral., Carib., N.Amer., warm seas. *Clava* [*Rhinoclavis*] (*s.s.*) (Fig. 697): subulate, with rather solid collabral costae and spiral threads, sometimes becoming smooth on later whorls; aperture ovoid, with a very oblique axis; columella uniformly excavate, with a fold which is barely visible even on a perfect specimen; anterior canal strongly recurved and projecting horizontally away from the aperture; posterior gutter forming a slight canal; labral profile nearly straight and vertical; parietal fold. Pal.-Rec. Indo-Pac., Asia, Medit., N.Africa, Eur., N.Amer., C.Amer. The subgenus *Pseudovertagus* has the general form of *Clava,* but usually has a row of nodes or short spines just above the middle of the whorls, but is otherwise fairly smooth; columella sometimes with a swelling, but without a fold. Pal.-Rec. Indo-Pac., Eur. *Bellatara* [*Bellardia*]: large, elongate-oval-conical; whorls flat-sided and smooth, except for heavy knobs developed on the last two whorls; antilabial gibbosity; parietal fold and narrow gutter; short, rather backwardly bent canal; columellar lip indistinct; labrum anteriorly prominent, posteriorly curved back. Eoc.-Olig. (unpublished). Eur., W.Pakistan.

Subfamily BITTIINAE

Usually rather small, with canal reduced to a sublateral notch; operculum like that of Cerithiinae, but with more numerous whorls; a smooth or lirate protoconch; the presence or absence of varices, and the nature of the ornament are the principle characters used for supraspecific classification. Habitat marine. (Pal.-Rec.)

Bittium (*s.s.*) (Fig. 698): turreted-conical; protoconch with two spiral lirae; teleoconch with spiral threads and collabral costellae bearing crenulations at their intersections; last whorl with an antilabial varix, base without collabral ornament; aperture small, oval; labrum internally laciniate, straight; peristome discontinuous. Pal.-Rec. Eur., Medit., Atl., Carib., N.Amer., ?Japan.

Subfamily CAMPANILINAE

Turreted-conical, with numerous whorls; last whorl with long, twisted neck and short canal. (U. Cret.-Rec.)

Campanile (*s.s.*) (Fig. 699): turreted, usually attaining a very great size (up to over 50 cm. high); early whorls smooth or with crenulated spiral threads, later whorls developing a row of nodes below the suture; last whorl bluntly angular, its base moderately convex and smooth; aperture large, oval, posteriorly a little detached and channelled, anteriorly with a moderately long and slightly backwardly bent canal; columellar lip callous, with a small median fold and an oblique anterior fold along the canal; labrum posteriorly notched, anteriorly very prominent. U.Cret.-Rec. Eur., Iran, W.Pakistan, Indonesia, Austral., N.Amer., S.Amer. **Serratocerithium** (Fig. 700): slender turreted-conical; early whorls with crenulated spiral threads, later also with a spinose keel below the suture; aperture obliquely oval, slightly channelled posteriorly and with a weak parietal fold; anteriorly with an oblique, broad, short canal which is feebly bent backwards; columella with one oblique fold; labrum concave, symmetrical on the lateral area, not projected across the canal; no varices. Eoc. Eur.

Family CERITHIOPSIDAE

Small, turreted, often with cancellate or spiral ornament; aperture rounded-quadrate, with a short, notched anterior canal; labrum fairly straight, not prosocline; operculum rounded-oval, paucispiral. Rather fragile, cold-water, marine shells. (L.Cret.-Rec.)

Cerithiopsis (*s.s.*): slender-turreted, with smooth protoconch; teleoconch with spiral rows of axially arranged crenulations; base of last whorl with one or two smooth spiral threads above a short neck; columella not twisted; aperture anteriorly with a short, deeply notched, somewhat backwardly bent canal. Cret.-Rec. Eur., Indo-Pac., Japan, Austral., N.Amer., S.Amer. **Cerithiella** [*Lovenella, Newtonia, Newtoniella*] (*s.s.*): whorls crenulated by collabral and spiral ornament; last whorl bluntly angular anteriorly, base smooth; columella very concave, anteriorly with a fold. Cret.-Rec. Eur., Austral. Carib., N.Amer., widespread.

Family TRIFORIDAE

Small, subulate, often sinistral; protoconch smooth or ornamented; teleoconch usually with crenulated spiral threads; aperture circular, constricted, the anterior and sometimes also the posterior canal forming a closed tube. Habitat marine. (U.Cret.-Rec.)

Triforis (*s.s.*): dextral, with smooth protoconch; peristome detached;

anterior and posterior canals both forming tubes. U.Eoc. Eur. *Triphora* (*s.s.*): sinistral, with latticed protoconch; anterior canal forming a tube, bent back; aperture posteriorly channelled. U.Cret.-Rec. Austral., N.Z., Indo-Pac., Japan, Eur., N.Amer., Carib., widespread.

Superfamily SCALACEA

Elongate-conical to turreted or rounded (Janthinidae); aperture rounded or somewhat angular anteriorly. (Jur.-Rec.)

Family SCALIDAE [SCALARIIDAE, EPITONIIDAE]

Turreted, whorls only just in contact; aperture and whorl section circular (rarely with flattened sides); peristome continuous; labral profile straight, often oblique and prosocline; suture usually very deep; labrum raised into a more or less prominent circular lip; costae often united to one on the adjacent whorl, thus forming a continuous obliquely-vertical ornament; surface between costae smooth or ornamented with spiral ridges or grooves; varices often present; protoconch smooth, obtuse. (Jur.-Rec.)

Subfamily SCALINAE

Lamellar costae; peristome continuous; loosely coiled; small siphonal fasciole and anterior auricle sometimes present. (U.Cret.-Rec.)

Scala [*Epitonium, Scalaria*] is Recent only; it is rather turbinate (height only about twice the breadth), the whorls are barely in contact, the intervals between the ribs are smooth, and there is a distinct umbilicus. The subgenus *Crisposcala* differs from *Scala* in having a smaller umbilicus, fine spiral striae in the intervals between the costae, and the costae have small spines just below the suture. Eoc., ?Rec. Eur., N.Africa, W.Pakistan, India, Austral., N.Amer. *Clathrus* (*s.s.*) (Fig. 701): more slender than *Scala*, costae continuous from whorl to whorl and with more or less smooth intervals; peristome solid; aperture with an anterior auricle practically covering a small siphonal fasciole. Eoc.-Rec. Eur., N.Amer., Carib., ?S.Amer., N.Z., warm seas. *Boreoscala:* similar to *Clathrus*, but the intervals between the costae are spirally grooved. Eoc.-Rec. Newfoundland, Iceland, Eur., Japan, northern seas. *Cirsotrema*

Figs 699–709. GASTROPODA: CERITHIIDAE, SCALIDAE, HIPPONICIDAE AND CALYPTRAEIDAE
Figs 699–700 after Cossmann; Figs 701–709 original.
699a, b, i. *Campanile giganteum* (Lamarck), M.Eoc. Paris B. T. a, b × 0·1, i × 0·2.
700b. *Serratocerithium serratum* (Bruguière), M.Eoc. Paris B. T. × 0·75.
701. *Clathrus clathrus* (Linné), Rec. T. × 1·65.
702e. *Hipponix cornucopiae* (Lamarck), M.Eoc. Paris B. T. × 1·1
703d, e. *Cheilea martiniana* (Reeve), Rec. × 1·1.
704d. *Calyptraea chinensis* (Linné), Rec. Britain. T. × 1·5.
705e. *Calyptraea chinensis* (Linné), L.Plio. Suffolk. T. × 1·9.
706e. *Trochita aperta* (Solander), U.Eoc. (Headon). Hampshire. × 1·16.
707d. *Trochita* sp., Rec. × 1.
708d, e. *Sigapatella calyptraeiformis* (Lamarck), Rec. N.Z. × 0·75.
709d. *Crucibulum spinosum* G. B. Sowerby, Pleist. Calif. × 1.1

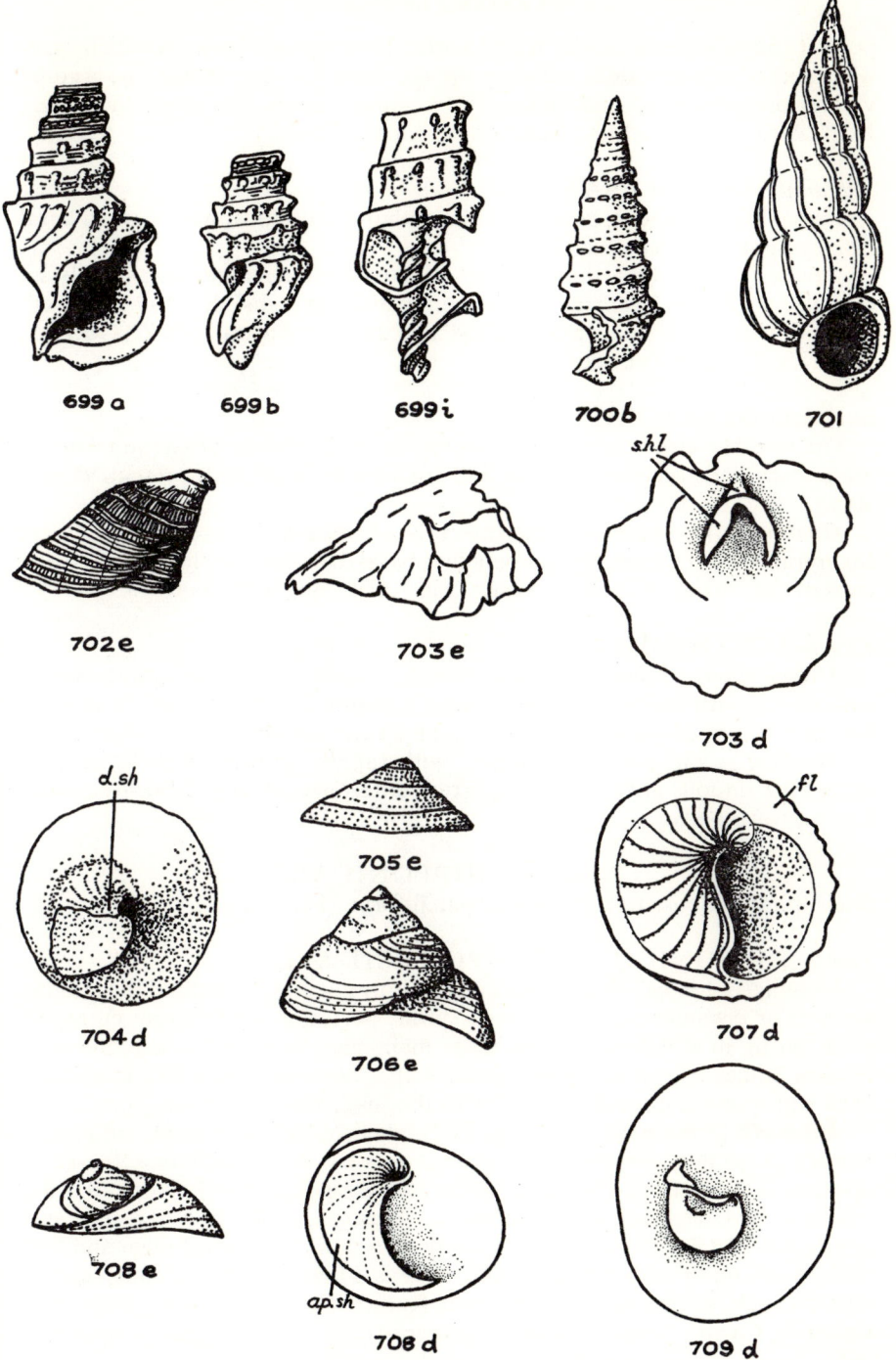

699 a   699 b   699 i   700 b   701

702 e   703 e

shl

703 d

d.sh

704 d

705 e

706 e

fl

707 d

708 e

ap.sh

708 d

709 d

(*s.s.*): turreted-conical; solid, frilled costae the intervals between which carry well-marked spiral ridges running up the backs of the costae; occasional varices present; basal disk limited by a spiral thread. Eoc.-Rec. Indo-Pac., N.Z., Africa, Medit., Eur., N.Amer., C.Amer., Carib.

Subfamily ACRILLINAE
Cancellate ornament; peristome thin, subdiscontinuous; well-marked basal disk present. (U.Cret.-Rec.)
Acrilla (*s.s.*): rather thin-shelled, turreted-conical; numerous closely-set costellae with spiral threads in their intervals; basal disk limited by a well-marked thread. U.Cret.-Rec. Indo-Pac., Japan, Austral., N.Z., M.East, N.Africa, Eur., W.Pakistan.

Subfamily OPALIINAE
Opaline texture; usually with strong costae; surface often punctate or granulose; peristome continuous and thickened; basal disk present. (Cret.-Rec.)
Opalia (*s.s.*): solid costae the intervals between which are smooth except for spiral rows of extremely fine punctae; basal disk solid, limited by the lower ends of the costae. Mio.-Rec. Austral., N.Amer., northern seas.

Family JANTHINIDAE
Thin-shelled, globose to oval-conical, usually with low spire; whorls smooth or with feeble collabral striae; aperture large, oval, often angular between end of columella and floor; labrum curved, sharp. (Mio.-Rec.)
Janthina (*s.s.*): globose-turbinate, with rapidly growing whorls; smooth except for sinuous growth lines; aperture anteriorly angular. Mio.-Rec. Eur., Carib., warm seas.

Superfamily HIPPONICACEA
Shell spiral, globose to turbinate or patelliform. (Perm.-Rec.)

Family HIPPONICIDAE [HIPPONYCIDAE]
Irregularly conical, limpet-like or capuliform, with apex behind the middle; smooth or ornamented; protoconch spiral; posterior shell margin embayed, so as to lie in a different plane from main part of margin; attachment to foreign bodies is by a calcareous plate with a horse-shoe scar like that in the shell (this plate is not often found with the fossil shell). (U.Cret.-Rec.)
Hipponix [*Hipponyx*] (*s.s.*) (Fig. 702): shell obliquely conical, with backwardly projecting curved apex; sometimes very thick, with bevelled edge; aperture oval to circular; surface rough, with irregular growth lines and radial costellae. U.Cret.-Mio. Eur., N.Africa, ?Austral., N.Z., N.Amer. The subgenus *Rothpletzia* [*Neomonopleura*] tends to a more cylindrical form and lives under reef conditions; the calcareous plate is a low asymmetric cone, and the whole simulates one of the Cretaceous Rudists (Bivalvia). ?Eoc., Olig.-Mio. Canaries, Carib., ?Italy. *Cheilea* [*Mitrularia*] (Fig. 703): irregularly conical or

324

cyrtoconoid, apex a little eccentric towards the posterior; internally like *Crucibulum*, but in place of the small open cone there is a plate curved into a half-cone widely open in front (hence 'shoe-horn limpet'); irregular com-marginal and radial ornament. Eoc.-Rec. Indo-Pac., N.Z., Japan, Eur., N.Amer., C.Amer., Carib.

Family FOSSARIDAE

Globose to high-turbinate, often with spiral ornament, sometimes with collabral ornament as well; last whorl large, usually umbilicate; aperture oblique, in one plane, with usually straight columella; labrum simple, usually prosocline. (Perm.-Rec.)

Fossarus: globose-turbinate, with low spire; whorls convex, often with a posterior ramp, with a few small spiral keels with spiral threads between them; small umbilicus; aperture large, suboval, oblique. ?U.Cret., Eoc.-Rec. Medit., Eur., N.Africa, Austral., Carib., S.Amer., warm seas.

Family VANIKOROIDAE

Subglobular, with very low spire; last whorl very large, with spiral threads and fairly solid, oblique collabral costae; aperture very large, obliquely rounded; umbilicus. (?U.Jur., U.Cret.-Rec.)

Vanikoro (the only genus): ?U.Jur., U.Cret.-Rec. Indo-Pac., Austral., Indonesia, Japan, India, Eur., N.Amer., Carib.

Superfamily CALYPTRAEACEA

Shell spiral or capuliform; operculum horny and with terminal nucleus in spiral forms, lacking in others. (Cret.-Rec.)

Family CALYPTRAEIDAE

The shell is more or less limpet-like in form, but the cavity is not freely open below, being divided up by a lamina or shelf which may be spiral, flat or conical; shell thin; surface smooth, striate or costulate; lateral surface project-ing as a flange around the flat base; labral profile very oblique to suture; apex central to terminal, twisted or not. (Cret.-Rec.)

Calyptraea [*Galerus*] (Figs 704, 705): shell conical or coeloconoid, with central apex, and showing no external spiral suture; smooth except for growth lines; the central convexity of the labrum is pronounced and asymmetric, there being scarcely any concave portion at the outer end, while the change to the columellar concavity is abrupt, and the free margin of the latter is doubled back to meet the shelf, forming a flattened tube; the whole shelf is oblique and the aperture narrow (cf. *Trochita*); the floor is usually convex. U.Cret.-Rec. Eur., Iceland, Indo-Pac., Austral., Japan, India, N.Amer., Carib., S.Amer., warm seas. *Trochita* [*Trochatella*] (Figs 706, 707): cyrtoconoid, with rounded, central apex; whorls separated externally by a depressed spiral suture; base concave, the margin forming a flange; aperture confined to base, depressed, subpyriform; columellar lip short, vertical, uniformly excavate; basal portion of labrum forming a flat spiral shelf which is feebly convex in the middle,

325

concave towards either end, one end swinging sharply round to meet the basal flange tangentially, the other joining the columella by a smooth curve but also bearing a slight free edge diverging slightly from the columella; surface ornamented by radial or oblique costellae, which may be broken up into rows of asperities. Pal.-Rec. S.Amer., N.Amer., Eur. *Sigapatella* (Fig. 708): similar to *Trochita*, but apex eccentric and more prominent, with concave basal labral profile; there is a narrow eccentric umbilicus, but no radial ornament. U.Cret.-Rec. N.Z., Austral., Pac. *Crucibulum* (*s.s.*) (cup-and-saucer limpet, Fig. 709): patelliform, subconical, ornamented with radial striae or costellae; margin more or less scalloped; shelf replaced by a small cone, open downwards and attached by a radial lamina. Mio.-Rec. N.Amer., S.Amer., Carib., Eur. In the subgenus *Bicatillus* the inner cone has almost disappeared, only its lamina of attachment remaining. Mio.-Rec. Indo-Pac., Indonesia, Eur. *Crepidula* (*s.s.*) (slipper-limpet, Fig. 710): oval-oblong, with apex close to the end remote from aperture and curved to one side; protoconch spiral; surface smooth or weakly costate; marginal flange as in previous genera; shelf a flat plane, no longer spiral, its free edge (basal profile of labrum) practically straight. U.Cret.-Rec. Medit., Eur., Austral., Japan, N.Amer., S.Amer., warm seas. [Recent species are remarkable for the formation of 'chains', series of individuals so grouped that the excretory parts of their mantle chambers are close together: each newcomer to the chain is a male, but changes through hermaphrodite to female as others join. Such chains are found fossil (U.S.G.S., Prof. Paper 59, pls. ix, x).] In the subgenus *Dispotaea* the apex is not quite marginal, and the septal lamella descends from the apex, being fixed along both sides, the left side strongly curved. L.Mio.-Rec. N.Amer., Carib., S.Amer. The subgenus *Siphopatella* is oval in outline, with a spiral, marginal apex; rather flat; septum medially compressed into a funnel, concave on the left. Plio-Rec. Indo-Pac., Indonesia, India. In the subgenus *Crepipatella* (Fig. 711) there is no radial ornament, and the profile of the apertural shelf is medially strongly convex. Plio.-Rec. Magellanic, N.Amer., S.Amer.

Family CAPULIDAE

Cap-shaped, conical; apex more or less spiral and post-centrally placed; surface smooth or sculptured; aperture rounded or oval; usually without any trace of a septum; internally with horse-shoe-shaped muscle scar. (U.Cret.-Rec.)

Capulus (*s.s.*) (Fig. 712): fine radial sculpture between coarse commarginal

---

Figs 710–714. GASTROPODA: CALYPTRAEIDAE, CAPULIDAE AND STROMBIDAE

Fig. 714 after Hoernes; Figs 710–713 original.

710d, e, k. *Crepidula fornicata* (Linné), Rec. Britain. T. d, e × 0·75; k is a chain of eight individuals, the oldest numbered I, the youngest VIII, × 0·9.

711d. *Crepidula* (*Crepipatella*) *dilatata* Lamarck, Rec. Peru. T. × 0·6.

712d, e. *Capulus hungaricus* (Linné), Rec. Devon. T. × 0·75.

713a, b. *Strombus pugilis* Linné, Rec. W.Indies. T. × 0·75.

714b. *Pereiraea gervaisi* (Vézian), U.Mio. Carniola. T. × 0·5.

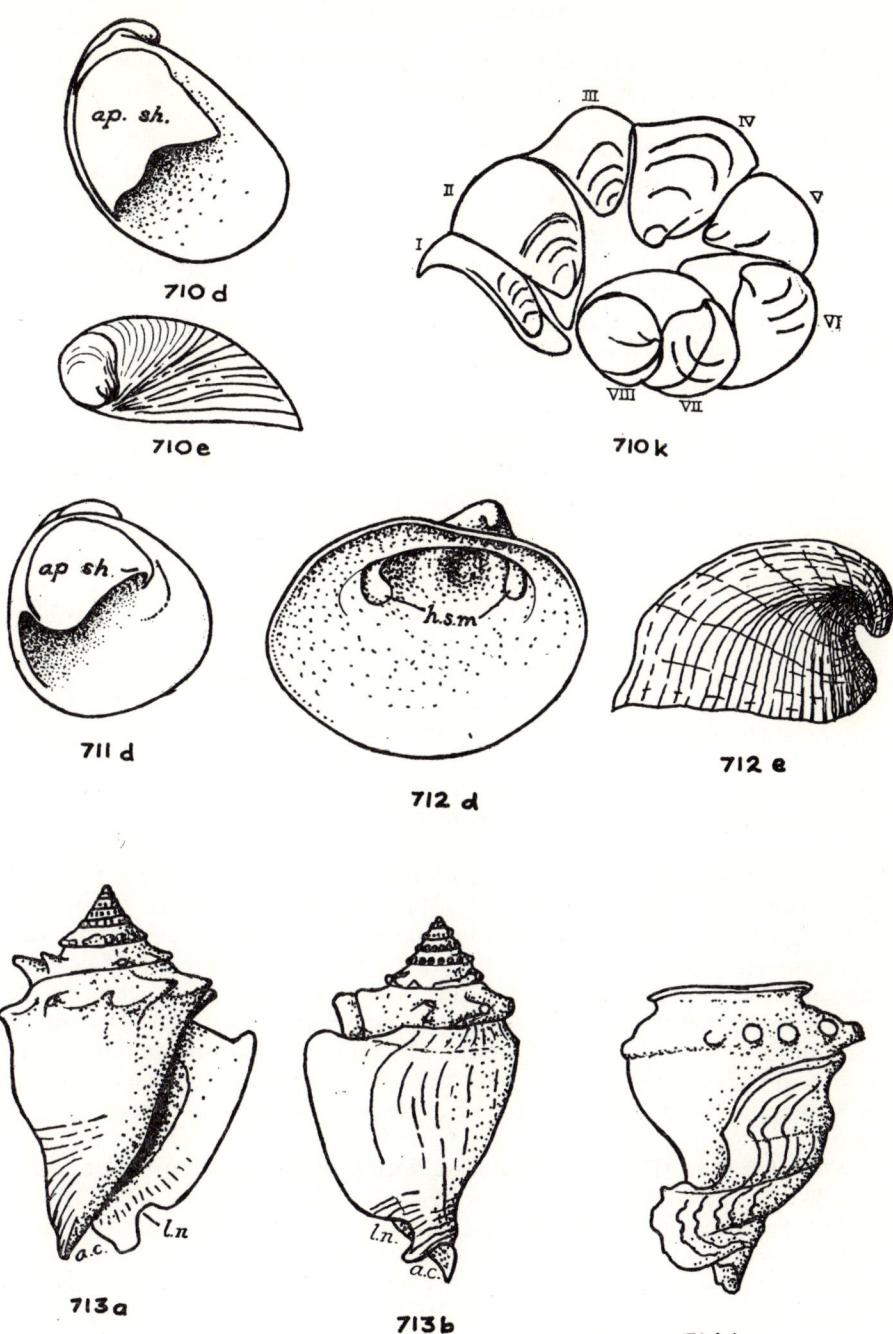

710 d

710 e

710 k

711 d

712 d

712 e

713 a

713 b

714 b

growth lines; mus.le scar widely open in front and remote from anterior shell margin. U.Cret.-Rec. Eur., Medit., India, ?Indonesia, Atl., N.Amer., Carib.

Family TRICHOTROPIDIDAE

Shell spiral, turbinate to turreted, rarely discoidal, usually umbilicate; columella pointed anteriorly, there often being an incipient canal; labrum prosocline; operculum horny, with terminal nucleus. (Cret.-Rec.)

Trichotropis (*s.s.*): turbinate, with large last whorl; moderately narrow umbilicus; whorls developing two spiral keels with oblique growth line striae in between; aperture anteriorly somewhat channelled; labrum posteriorly strongly prosocline. Mio.-Rec. Japan, N.Amer., S.Amer., Iceland, northern and southern seas.

Superfamily STROMBACEA

Low-conical to turreted; columella sometimes produced and aperture anteriorly channelled; labrum often reflected, widened and digitate. (?U.Trias., Jur.-Rec.)

Family STROMBIDAE

Rather thick-shelled, sometimes very large, usually with more or less high spire; smooth or ornamented; aperture high and narrow, columella anteriorly usually forming a rostrum, the floor to the right of the rostrum often distinctly notched; labrum simple or widened and alate, sometimes with projections; columellar lip callous; labrum usually normal to suture, sometimes definitely opisthocline (*Terebellopsis*), rarely slightly prosocline; active jumpers, confined to warm seas with maximum in the tropics. (Cret.-Rec.)

Both the Strombidae and Aporrhaidae are more or less fusiform, owing to the drawing out of the columellar region into a grooved rostrum which, as pointed out by Cossmann, differs from the anterior canal of other fusiform siphonostomes in being independent of the siphonal notch, which (if present) is near the base of the rostrum. The labrum is more or less thickened and may be expanded into a wing, often digitate; and there is a frequent tendency for the posterior canal to extend up the spire. The extension of the posterior canal up the spire, rarely altogether absent, though often slight, develops into a continuous wrapping-round of the spire by an upward extension of the later whorls in *Calyptraphorus*, *Orthaulax* and *Seraphs*. Varices occur sporadically in Strombidae, but their presence or absence is not always even of specific value.

Strombus (*s.s.*) (Fig. 713): biconical, with a moderately short spire; early whorls with spiral threads and a row of small nodes, last whorl with a row of nodes or spines but no spiral threads; aperture long and rather narrow, more or less parallel-sided; labrum greatly thickened, more or less expanded, not digitate, with well marked notches at both ends; growth lines straight and vertical above the shoulder, oblique and arcuate below it; columellar lip almost straight, ending in a pointed rostrum bent dorsalwards; columellar callus more widely spread posteriorly. Mio.-Rec. Carib., N.Amer. The subgenus *Aliger* [*Monodactylus*] has a rather longer, gently coeloconoid spire and a finger-like upgrowth from the posterior end of the labrum. Mio.-Rec. Red Sea, Eur., Indonesia, Carib., Indo-Pac. The subgenus *Dilatilabrum* has a short, gently coeloconoid spire, a dentate keel on the last whorl, and a very broad labral wing which is not notched anteriorly and only a little produced posteriorly. Eoc. Eur. The subgenus *Eustrombus* is large, with blunt spines which on the last whorl are on an angulation; aperture long and fairly broad, posteriorly widened, anteriorly with a truncated rostrum and a notch; labrum thick, widened, broadly notched posteriorly, gently notched anteriorly, often folded internally; columellar callus thin, widely spread. Eoc.-Rec. Carib., cosmop. *Canarium* (*s.s.*): biconical with moderately high spire; spiral threads and nodes which are more upturned than in *Strombus*; aperture long, anteriorly with a distinct rostrum and moderate notch; columellar lip straight, with narrow callus; labrum not much widened, anteriorly with a broad notch, internally folded. Olig.-Rec. Indo-Pac., Red Sea, Africa, Eur. The subgenus *Labiostrombus* [*Gallinula*] has costae on the early whorls, but only spiral furrows on the later whorls, the last whorl being only bluntly angular; aperture long, with a short, bent anterior rostrum, a broad, deep anterior notch, posteriorly with a narrow channel produced up on to the penultimate whorl to which it is fused; labrum thickened, internally smooth, with broad anterior notch. Plio.-Rec. Indo-Pac., Indonesia, N.Amer. *Pereiraea* (Fig. 714): biconical, with coeloconoid spire; whorls anteriorly with a keel bearing pointed nodes which on the last whorl become hollow spines; last whorl with four strong spiral threads and curved growth lines; aperture pyriform, posteriorly with a channel going back along the suture, anteriorly with a distinct, pointed rostrum and a broad notch; labrum alate, with anterior hood, posteriorly opisthocline; columellar callus well spread over ventral surface. ?Eoc.-Mio. Eur., Africa. *Lambis* [*Pterocera*] (*s.s.*): oval, with a few spiral threads and nodes; aperture narrow, with an anterior rostrum which is curved to the right; labrum expanded, with six other curved digitations having a closed channel, the uppermost one rising right up the spire and surpassing it. Eoc.-Rec. Indo-Pac., Indonesia, Africa, Eur., W.Pakistan, Asia Minor, Tibet. *Rimella* (*s.s.*) (Fig. 715): not large, fusiform; whorls rather convex, with axial costae and occasional varices; neck of last whorl with spiral threads; aperture rather narrowly oval, anteriorly with a short, pointed rostrum and a broad notch, posteriorly with a narrow posterior channel (formed by the fusion of the outer and inner lips) extending up to near the apex, curving at its end but not re-descending; labrum without digitations, orthocline, internally smooth, widened and somewhat reflected. ?Cret., Pal.-Olig. Eur., Africa, Madagascar, W.Pakistan, Tibet, ?N.Amer., Indonesia, S.Amer. *Orthaulax* (*s.s.*) (Figs 716, 717) (**78**): pyriform, oval-conical or fusiform, smooth, sometimes with a dorsal swelling; the small spire is just visible through a layer of enamel secreted by the mantle; this callus overlaps the spire continuously, and the earlier whorls become solidified, so that about half the shell is a solid mass, features that may have been adaptations to withstand the battering action of waves on coral reefs etc.; aperture narrow, posteriorly with a narrow channel, anteriorly with

329

a short, backwardly curved rostrum and a deep notch on the floor beside it; labrum thin, practically straight. Mio. Carib., N.Amer., C.Amer. The subgenus *Veatchia* is similar to *Orthaulax*, but the spire is wholly covered by the last whorl. Pal. Trinidad. *Dientomochilus* (*s.s.*): inflated-fusiform; whorls convex, with axial costae and spiral threads but no varices except for a labral varix; aperture narrow, the posterior channel curving backwards up over the penultimate whorl, anteriorly with a short beak curved to the right; labrum somewhat widened, straight, anteriorly notched, internally denticulate; small notch also in floor of aperture. Eoc. Eur. The subgenus *Varicospira* has the same general form and varices as *Rimella*, but the posterior channel is like that of *Dientomochilus*; there are incised spiral lines between the costae. U.Cret.-Rec. Indo-Pac., Indonesia, India, W.Pakistan, Africa, Eur., N.Amer. *Ectinochilus* (*s.s.*) (Fig. 718): differs from *Rimella* in having a second (labral) sinus like *Strombus*; the costae are weaker on the apertural side of the last whorl. Eoc.-Olig. Eur. [American Upper Cretaceous and Eocene forms belong to three different subgenera; the posterior canal may double back completely from near the apex.] *Tibia* [*Rostellaria*] (*s.s.*) (Fig. 719): fusiform, multispiral, spire sometimes slightly coeloconoid; axial costae (and sometimes also spiral threads) in early whorls, later smooth or with a few weak spiral furrows; rostrum narrow and pointed, straight; columella excavate, with much callus; labrum digitate anteriorly, with a deep bay between digitations and rostrum, and forming posteriorly a wing extending more or less up the spire with the posterior canal and columellar callus, usually curving over on the penultimate whorl; growth line parasigmoidal, normal or slightly prosocline at the suture. Eoc.-Rec. Indo-Pac., Eur., Africa, Asia. *Cyrtulotibia* (Fig. 720): similar to *Tibia*, but spire and rostrum shorter, rostrum bent somewhat to the right, and often developing a strong shoulder on the last whorl; posterior channel slit-like, curving back along suture; labrum gently opisthocline, with only one short anterior spine. U.Eoc. Nigeria. *Hippochrenes* (Fig. 721): fusoid, smooth, with short rostrum which may be bent to the right; labrum without digitations, with a much expanded wing extending up to (or nearly to) the apex. U.Cret.-Olig. Eur., Arabia, W.Pakistan, Tibet, Russia. *Calyptraphorus* (*s.s.*) (Fig. 722): fusiform, early whorls with axial costae which, however, are in the adult masked by enamel which spreads over the whole shell and may bear coarse

---

Figs 715–723. GASTROPODA: STROMBIDAE
Fig. 717 after Cooke; Fig. 719 after Wenz; Figs 715, 716, 718 and 720–723 original.
715. *Rimella fissurella* (Linné), M.Eoc. Paris B. T. × 1·1.
716. *Orthaulax gabbi* Dall, M.Mio (Chipola). Alum Bluff, Florida. × 1·25. Young specimen with spire incompletely enveloped.
717x, y. *Orthaulax aguilladensis* Maury, Mio. × Thomonde, Haiti, × 0·56; y external and internal casts associated, St. Croix, × 0·375.
718. *Ectinochilus canalis* (Lamarck), M.Eoc. Paris B. T. × 2·5.
719. *Tibia fusus* (Linné), Rec. China. T. × 0·25.
720a, b. *Cyrtulotibia unidigitata* (R. B. Newton), U.Eoc. Ameki, S.Nigeria. T. × 1·1.
721. *Hippochrenes fissura* (Coquand and Brongniart), M.Eoc. Paris B. × 0·75.
722a, e. *Calyptraphorus trinodiferus* Conrad, L.Eoc. Bell's Landing, Alabama. × 1.
723. *Terebellum terebellum* (Linné), Rec. Indo-Pac. T. × 0·75.

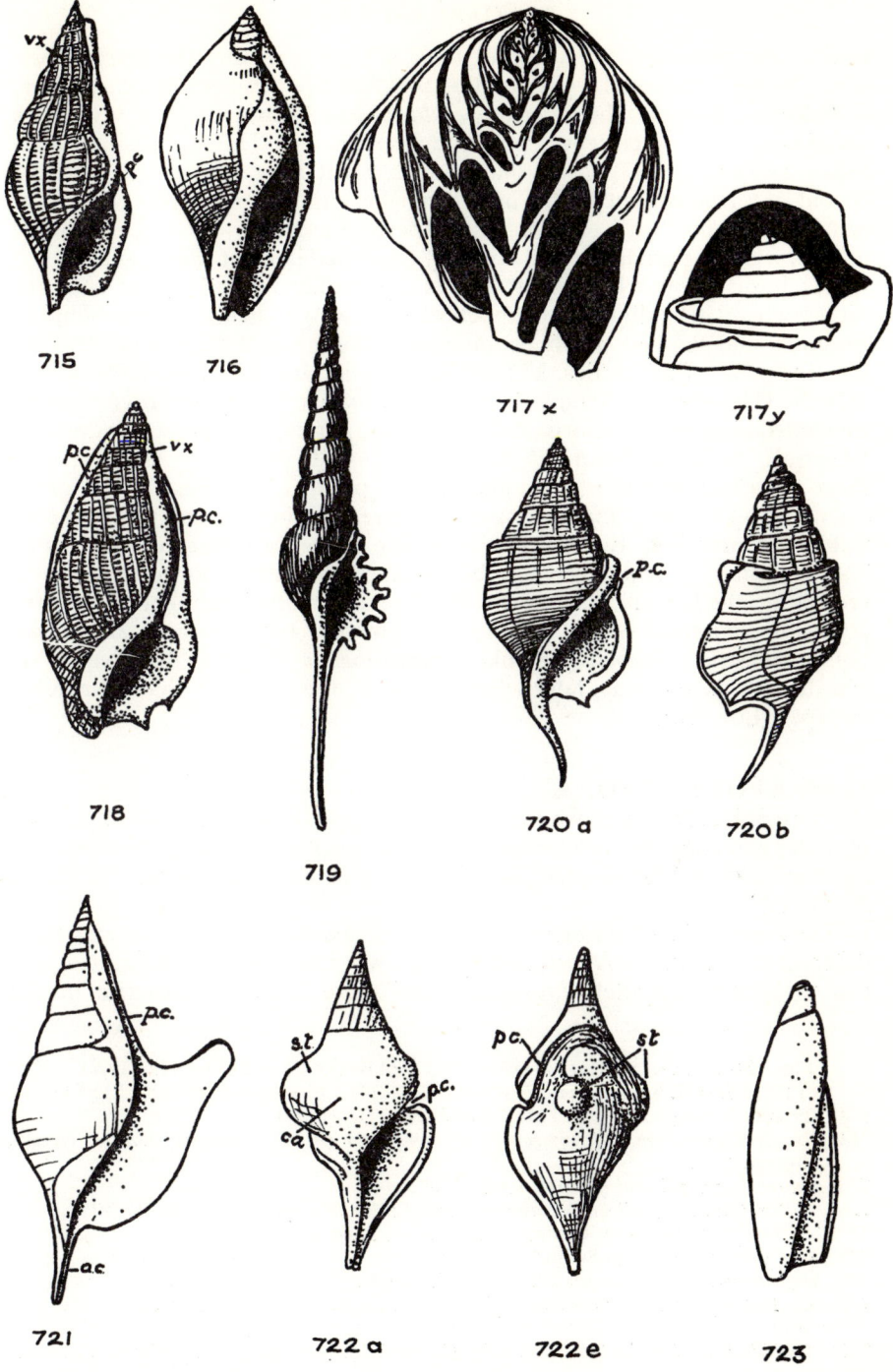

715

716

717 x

717 y

718

719

720 a

720 b

721

722 a

722 e

723

knobs; the posterior canal curves back without reaching the apex; labral wing feebly thickened. U.Cret.-Eoc. N.Amer., S.Amer., Eur., Africa, W.Pakistan, Burma. The subgenus *Aulacodiscus* is covered by a very thick calcareous layer which leaves only the aperture free. U.Eoc. S.Amer. *Terebellum* (*s.s.*) (Fig. 723): cigar-shaped with blunt apex, oligogyral, smooth; devoid of a true columella, the inner lip being simply the parietal wall of the penultimate whorl with a slight callus; aperture narrow, very acutely pointed behind, widening gradually and abruptly truncated in front where its outline is spiral; both lips nearly straight, the labral profile vertical except near the suture where it is slightly opisthocline. The whole shell is of the simplest structure, and can very nearly be imitated by rolling up a sheet of paper. Pal.-Rec. Indo-Pac., Eur., Africa, M.East, W.Pakistan, India, Indonesia, N.Amer., ?Carib. The subgenus *Seraphs* is like *Terebellum*, but the spire is completely enveloped by the later whorls. Pal.-Olig. Eur., Africa, W.Pakistan, Indonesia, ?Austral., N.Amer. *Semiterebellum* (*s.s.*): fusoid, practically smooth, with short rostrum; aperture long, posteriorly pointed, the posterior channel extending up the spire and turning over before reaching the apex; labrum only feebly widened. Eoc. Eur., W.Pakistan. The subgenus *Africoterebellum* (Fig. 724) is similar to *Semiterebellum*, but is more lanceolate, the rostrum is less projecting, the labrum is more opisthocline and more lobate below, and the posterior channel does not ascend the spire but turns over at the suture. U.Eoc. Nigeria. *Terebellopsis:* like *Semiterebellum*, but the first five or six whorls with axial costae, later whorls smooth; the posterior channel reaches the apex. Eoc. Eur.

Family APORRHAIDAE

More or less fusiform; whorls convex, with spiral ornament including rows of nodes and sometimes also axial costae; columella produced into a rostrum; aperture narrow, labrum widened and with one or more digitations. (?U.Trias., Jur.-Rec.)

The Aporrhaidae and Strombidae seem to be parallel developments from different stocks. In the Aporrhaidae the growth line swings round at its upper end to become definitely prosocline, and there is no true anterior notch;

Figs 724–731. GASTROPODA: STROMBIDAE, APORRHAIDAE, STRUTHIOLARIIDAE, XENOPHORIDAE AND CYPRAEIDAE
Fig. 729 after Vredenburg; Figs 724–728 and 730–731 original.
724a, b. *Semiterebellum* (*Africoterebellum*) *elongatum* (R. B. Newton), U.Eoc. Ameki, S.Nigeria. × 1.
725. *Aporrhais pes-pelicani* (Linné), Rec. × 0·75.
726a, b. *Struthiolaria* (*Pelicaria*) *coronata* Tate, L.Plio. Victoria, Australia. × 0·75.
727d, e. *Xenophora cumulans* (Brongniart), M.Eoc. Paris B. × 1.
728a, b. *Cypraea pantherina* Solander, Rec. × 0·75.
729. *Megalocypraea ranikotensis* Schilder, Pal. Jhirak. × c.0·25.
730a, c. *Austrocypraea contusa* (M'Coy), M.Mio. (Balcombian). Muddy Creek, Victoria. × 0·75.
731c. *Umbilia leptorhyncha* (M'Coy), Olig. (Janjukian). Neumerella, Gippsland, Victoria.

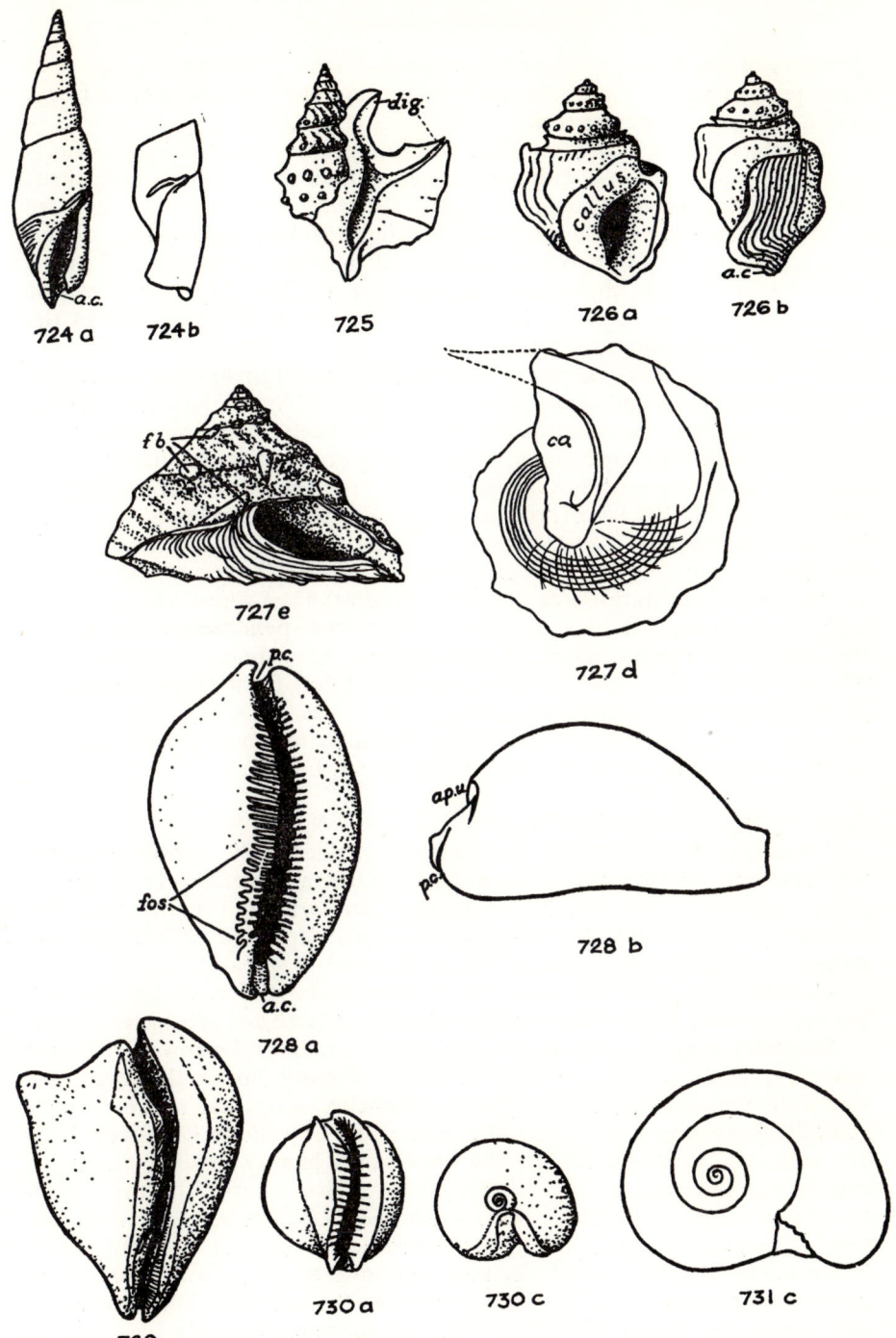

724 a

724 b

725

726 a

726 b

727 e

727 d

728 a

728 b

729

730 a

730 c

731 c

members of the family are crawling animals, mainly Mesozoic in age, only three genera and five subgenera surviving through the Tertiary; Recent forms live in cold and temperate seas.

Aporrhais [*Chenopus*] (*s.s.*) (Fig. 725): more or less fusiform; whorls convex or angular, at first with axial costellae, then with two (and sometimes a weaker third) rows of nodes on keels, also with spiral threads; aperture narrow, the anterior rostrum bent back; wing with two major lateral digitations (corresponding to two keels or rows of tubercles on the whorls) and a third feeble anterior digitation; the anterior rostrum and the posterior digitations (adherent to the spire) are formed half by the wing and half by the columellar callus, separated by a groove, too narrow to be a true canal; there are therefore four subequal processes in all; columellar lip and parietal region strongly callous. U.Cret.-Rec. Eur., Asia, Africa, N.Amer., S.Amer., Medit., Atl. The subgenus *Araeodactylus* has three spiral keels on the spire whorls but no costae, nodes or tubercles, and there is only one (horizontal) labral digitation. Pal. Eur. The subgenus *Maussenetia* is ornamented with spiral threads and a keel on the last whorl; the labrum is expanded into a broad wing with five feeble, triangular digitations and a longer posterior digitation which is free from the spire. U.Cret.(Maastr.)-Pal. Eur., Carib. *Drepanocheilus:* curved axial costellae and spiral threads, fading out on the last whorl which has two keels; rostrum short; labrum not extended up on to penultimate whorl, with one upwardly curved, short digitation only. Cret.-Eoc. N.Amer., Eur., Africa, M.East, Asia. The subgenus *Arrhoges* differs from *Drepanocheilus* in the axial and spiral ornament persisting on to the last whorl which is not carinate, and in the labrum extending posteriorly up on to the penultimate whorl. U.Cret.-Rec. N.Amer., Eur., Madagascar, E.Pac.

Family STRUTHIOLARIIDAE

A small family of somewhat turbinate shells, allied to the Aporrhaidae, but with rostrum, anterior sinus and wing all reduced, sometimes to vanishing point; aperture oval, usually angulated anteriorly and posteriorly, the columella being uniformly excavate; columellar lip widened and callous; growth lines slightly prosocline at the suture. (Pal.-Rec.)

The headquarters of this family are in New Zealand, Tertiary and Recent, but with representatives in Australia, S.America (Tert.) and Kerguelen (Rec.).

Struthiolaria (*s.s.*): ornament of fine spiral threads and a row of shoulder tubercles; height of aperture equal to height of spire; labrum thickened and with a slight hood; aperture posteriorly channelled, anteriorly with a distinct, short canal forming a feeble fasciole; columellar lip callous. Olig.-Rec. N.Z. The subgenus *Struthiolarella* has the whorls only bluntly angular, early with axial costae, later with nodes and spiral furrows; last whorl with strong spiral threads anteriorly. Pal.-Olig.-?Mio. Argentine, N.Z. The subgenus *Pelicaria* (Fig. 726) has a flattened, planorbid-like protoconch (not a prominent one as in *Struthiolaria*), and has a marked posterior notch on later whorls; the labrum is prominent anteriorly; there is much callus on the last whorl, the ramp being converted into a gutter. Mio.-Rec. N.Z. The subgenus *Tylospira* has a higher spire, a short, less broadly oval aperture, and the columellar callus is wider. Mio.-Rec. N.Z., Austral.

Family XENOPHORIDAE

Conical-trochiform, more or less sharply keeled, upper side often with agglutinated stones or shells; aperture very oblique, with discontinuous peristome and simple labrum; floor broadly notched. (?U.Trias., ?Jur., Cret.-Rec.)

Xenophora (Fig. 727): general build rather similar to that of *Trochus*, but with a flanged base, so that the aperture is invisible in side-view; the basal profile of the labrum is sickle-shaped or gamma-shaped (like an inverted L): the outer end of this labral curve is strongly projected, and continues on the upper (lateral) surface in an extremely oblique direction to meet the suture at a very acute angle, the whole growth line thus showing *Trochus*-characters in a very exaggerated form; anomphalous or with a narrow umbilicus; ornament on base formed by growth ridges crossed by spiral grooves which cut them up into squarish blocks; on the upper surface the very oblique growth ridges are crossed by more prominent irregular wrinkles, not spiral but about at right angles to the growth lines; a common feature is the incorporation into the shell of foreign bodies (pebbles, small shells, etc.); base feebly concave; floor deeply notched. ?U.Trias., Cret.-Rec. Carib., Eur., Africa, W.Pakistan, Ceylon, Indonesia, Japan, N.Z., N.Amer., warm seas. *Tugurium* (*s.s.*): like *Xenophora*, but without agglutinated foreign bodies; keel wavy or with lappets; wide umbilicus; base concave by keel, convex by umbilicus. ?Eoc., Olig.-Rec. Indo-Pac., Indonesia, India, W.Pakistan, Africa, Eur. The subgenus *Trochotugurium* is like *Tuguium*, but has agglutinated foreign bodies near the sutures and keel, and the umbilicus is narrower; the floor is deeply notched. U.Cret.-Rec. Eur., Africa, N.Amer., S.Amer., warm seas. The subgenus *Haliphoebus* is like *Tugurium* in lacking agglutinated foreign bodies, but the floor is deeply notched and the keel usually has regular digitate projections. Eoc.-Rec. Indo-Pac., Eur.

Superfamily CYPRAEACEA

Shell ovoid, strombiform or elongate, with apertural surface usually flattened; spire usually more or less completely hidden, apex sometimes in an umbilicus; surface smooth or ornamented with costae or knobs, partly or wholly covered by a deposit of callus (enamel) formed by extended mantle lobes; aperture narrow, usually notched at both ends; labrum usually crenulated; columellar lip frequently crenulated also; towards the anterior end the columella is usually flattened and slightly excavated, forming a *fossula* (*fos*, Fig. 728), bounded in front by a well-defined oblique ridge and continued backwards by a narrow groove (*columellar furrow*), the crenulations usually crossing both fossula and groove. In the Lamellariidae the shell is thin, reduced and auriform, there being no canal, fossula or teeth. (U.Jur.-Rec.)

The classification of the Cypraeacea is not easy; much work has been done on the group by Schilder (**91, 92**).

335

Family CYPRAEIDAE

Ends more rounded than sides; back and ventral side usually both covered with enamel; spire often covered over, but not involute; teeth sharp; fossula thickened, usually concave, ribbed or with inner marginal teeth. Recent forms live with corals and sponges. (U.Jur.-Rec.)

Subfamily CYPRAEINAE

The group with the most thickly enamelled surface and best developed fossula; shape usually oval, narrower in front (sometimes cylindroidal), not rostrate, occasionally with thickened margins; spire pointed to flat; ventral side flat; no dorsal furrow; columella ribbed or with a furrow. (L.Eoc.-Rec.)

Cypraea (*s.s.*) (Fig. 728): oval to pyriform, more or less globose, with spire only slightly projecting and buried in enamel; aperture moderately wide, curved, a little to the right of the mid-line, deeply notched at both ends; labrum and columella subparallel, both crenulated; anterior part of labrum, opposite fossula, more sloping than remainder and with longer teeth; columellar furrow and fossula transversely ribbed. The young cowry differs greatly in form from the adult, being more like a *Ficus*. M.Mio.-Rec. Indo-Pac., Indonesia, Africa.

Subfamily CYPRAEORBINAE

Oval to pyriform, inflated; back usually smooth, occasionally with swellings; spire usually hidden by callus; both ends sometimes with canals more or less extended into rostra, their ends appearing notched in dorsal view; fossula concave, smooth; columella without furrow; teeth sometimes weakly developed. In internal moulds the birostrate form is usually lost. Some members of the subfamily may attain a large size, up to about a third of a metre in length. (U.Jur.-Rec.)

Cypraeorbis: pyriform, with rather flat ventral side; spire a little projecting; aperture narrow, with short teeth. ?U.Cret., M.Eoc.-L.Mio. N.Amer., Carib., Eur., Asia. *Bernaya* (*s.s.*): rather similar to *Cypraeorbis*, but less pyriform, ventral side not so flat, spire short and covered over, aperture wider anteriorly, and columellar teeth longer. U.Jur.-L.Olig. Eur., W.Africa, W.Pakistan, E. Asia. *Gisortia*: pyriform, with flat spire and carinate shoulder; lips posteriorly rostrate and bent back; fossula vague; apertural teeth obsolete. Pal.-U.Eoc. Eur., Egypt, Iran, Afghanistan, W.Pakistan, India, Formosa, Carib. (**95**). *Megalocypraea* (Fig. 729): obconical, spire flattened and covered over, anterior end pointed; back with a transverse swelling posteriorly and a few knobs anterior to it; aperture strongly curved posteriorly; fossula broad, concave, smooth; columellar lip rarely with teeth; labrum with a few teeth anteriorly. U.Cret.(Maastr.)-U.Eoc. Eur., N.Amer., S.Amer., W.Pakistan, Tibet. *Vicetia:* like *Megalocypraea*, but more cylindrical, ventral side flatter, aperture straighter, and back with two transverse swellings producing a saddle-like form; fossula present; labrum with teeth; columellar lip occasionally with a few anterior teeth. L.Eoc.-U.Eoc. Eur., N.Africa, Iran, W.Pakistan, India.

336

Subfamily NARIINAE [EROSARIINAE]

Oval, spherical or pyriform, ventrally flattened and often with radiating ridges; back smooth or with granules and sometimes with a longitudinal furrow (especially at the ends); aperture narrow, symmetrically placed; fossula usually narrow and ribbed; no columellar furrow; margin of upper surface more or less swollen, with tendency to formation of pits between it and the inflated centre; teeth (especially at ends) usually long. (?U.Cret., L.Eoc.-Rec.)

Naria is Recent only. *Erosaria* (*s.s.*): oval-pyriform, slightly birostrate; apertural teeth (especially the outer ones) long, tending to extend over the under surface and traceable into the partitions between the pits of the upper surface; aperture wider anteriorly; terminal columellar tooth usually bifid, bending forward to make the sharp boundary of the anterior canal; fossula shallow, ribbed. M.Mio.-Rec. Indo-Pac., Indonesia, E.Africa. *Monetaria* is Pleistocene to Recent only. The subgenus *Ornamentaria* is oval (not angular) in outline, the sides are thick and callous but without punctae, the inner teeth are anteriorly radial, and the inner edge of the fossula is wholly smooth. L.Mio.-Rec. Indo-Pac., Indonesia, N.Africa. *Austrocypraea* (Fig. 730): globose-pyriform, back smooth or nearly so, sides rounded, the marginal keel being more on the ventral side which is convex; fossula ribbed. M.Olig.-M.Mio. Austral.

Subfamily CYPRAEOVULINAE

Pyriform to cylindrical, sometimes fairly large; sides (especially the labrum) often margined; spire usually flat and umbilicate; outer teeth long; fossula usually narrow, slight, ribbed; columellar furrow ribbed or with inner marginal teeth. (?Eoc., L.Olig.-Rec.)

Cypraeovula is Recent only. *Umbilia* (*s.s.*) (Fig. 731): pyriform, large, with curved and bent apex, ornamented with extremely fine crenulations; ventral side flattened; aperture curved posteriorly, a little wider anteriorly, a little to the right of the mid-line; outer teeth longer; terminal columellar tooth consisting of three radial folds; fossula narrow, smooth; no columellar furrow. ?Eoc., M.Olig.-Rec. Austral., ?W.Pakistan. [Previously classified near *Gisortia*, until Schilder indicated its relationship to *Cypraeovula*]. *Erronea* (*s.s.*): oval-subpyriform, with umbilicate spire; sides and ends compressed, ventral side fairly flat; aperture straight, only curved posteriorly; labrum feebly margined; fossula shallow, ribbed; columellar furrow feeble; inner marginal teeth of fossula weak. U.Plio.-Rec. Indo-Pac., Java, N.Africa. The subgenus *Adusta* is more pyriform, and the inner marginal teeth of the fossula are distinct. ?Eoc., L.Olig.-Rec. Indo-Pac., Indonesia.

Family AMPHIPERATIDAE

Not large, back with thin or no enamel; spire usually involute; the functional posterior outlet is cut into the curved end of the labrum, the usual posterior canal being represented by a faint depression only; fossula usually vague

337

and smooth; columellar furrow vague and columellar teeth often absent. (Cret.-Rec.)

Subfamily AMPHIPERATINAE

Pyriform to fusiform, back sometimes with a transverse keel; spire involute; aperture wider anteriorly; labrum margined; usually with no columellar teeth, only teeth on the labrum; fossula feeble, smooth. (L.Eoc.-Rec.)

Amphiperas is Recent only. *Simnia* (*s.s.*) (Fig. 732): fusiform, with pointed ends; back, at least at ends, with spiral ornament; no teeth; labrum sharp-edged; fossula very weak; columella twisted anteriorly. M.Mio.-Rec. Medit., Eur. The subgenus *Calpurna* [*Neosimnia*] differs from *Simnia* in the columella not being twisted anteriorly, and in the back often having a transverse keel; there are only faint spiral grooves at the ends. L.Eoc.-Rec. Medit., Eur., Carib., N.Amer., S.Amer., Indo-Pac.

Subfamily PEDICULARIINAE

No enamel layer; costate or latticed, back without furrow; spire project-ing in youth, later involute; fossula smooth or with inner marginal teeth. (U.Cret.(Maastr.)-Rec.)

Pedicularia (*s.s.*): small, spire deeply umbilicate to involute; no teeth; fos-sula vague, only rarely with teeth on the inner side; bowl-shaped, back with spiral costae; aperture wide. Plio.-Rec. Medit., E.Atl., Indo-Pac. *Cypraedia* (*s.s.*): inflated-pyriform, with cancellate ornament; spire involute, projecting in young; aperture narrow, posteriorly curved; labrum little margined, with long teeth; fossula very narrow, smooth, only ribbed externally; terminal columellar tooth consisting of two to four transverse ribs. U.Cret.(Maastr.)-M.Olig. Eur., W.Pakistan, Burma, Indonesia, Austral., Carib., N.Amer.

Subfamily SULCOCYPRAEINAE

Pyriform to oval, surface sometimes with ornament so fine as to look apparently smooth; spire involute; aperture very narrow, usually curved pos-teriorly; both lips with teeth; fossula smooth, rarely with inner marginal teeth; no columellar furrow. (Cret.-Rec.)

Eocypraea (*s.s.*) (Fig. 733): inflated-pyriform; aperture curved posteriorly,

---

Figs 732–741. GASTROPODA: CYPRAEIDAE, ERATOIDAE AND NATICIDAE

Figs 732–741 original.
732. *Simnia patula* (Pennant), Rec. Britain. × 1·5.
733. *Eocypraea hiantula* Cossmann, M.Eoc. Paris B. × 1·3.
734. *Erato laevis* (Donovan), U.Mio. Touraine. × 2·6.
735. *Trivia avellana* (J. Sowerby), Pleist. (Red Crag). Suffolk. × 1·5.
736a, b, h, h'. *Naticarius canrena* (Linné), Rec. T. × 0·75.
737a, b. *Polinices mamilla* (Linné), Rec. T. × 0·75.
738a, b. *Neverita olla* (M. de Serres), U.Mio. Touraine. × 1·1.
739a, b. *Euspira catena* (Da Costa), Pleist. Suffolk. × 1.
740a, b. *Globularia sigaretina* (Deshayes), M.Eoc. Paris B. × 0·75.
741a, b. *Crommium willemeti* (Deshayes), M.Eoc. Paris B. T. × 0·75.

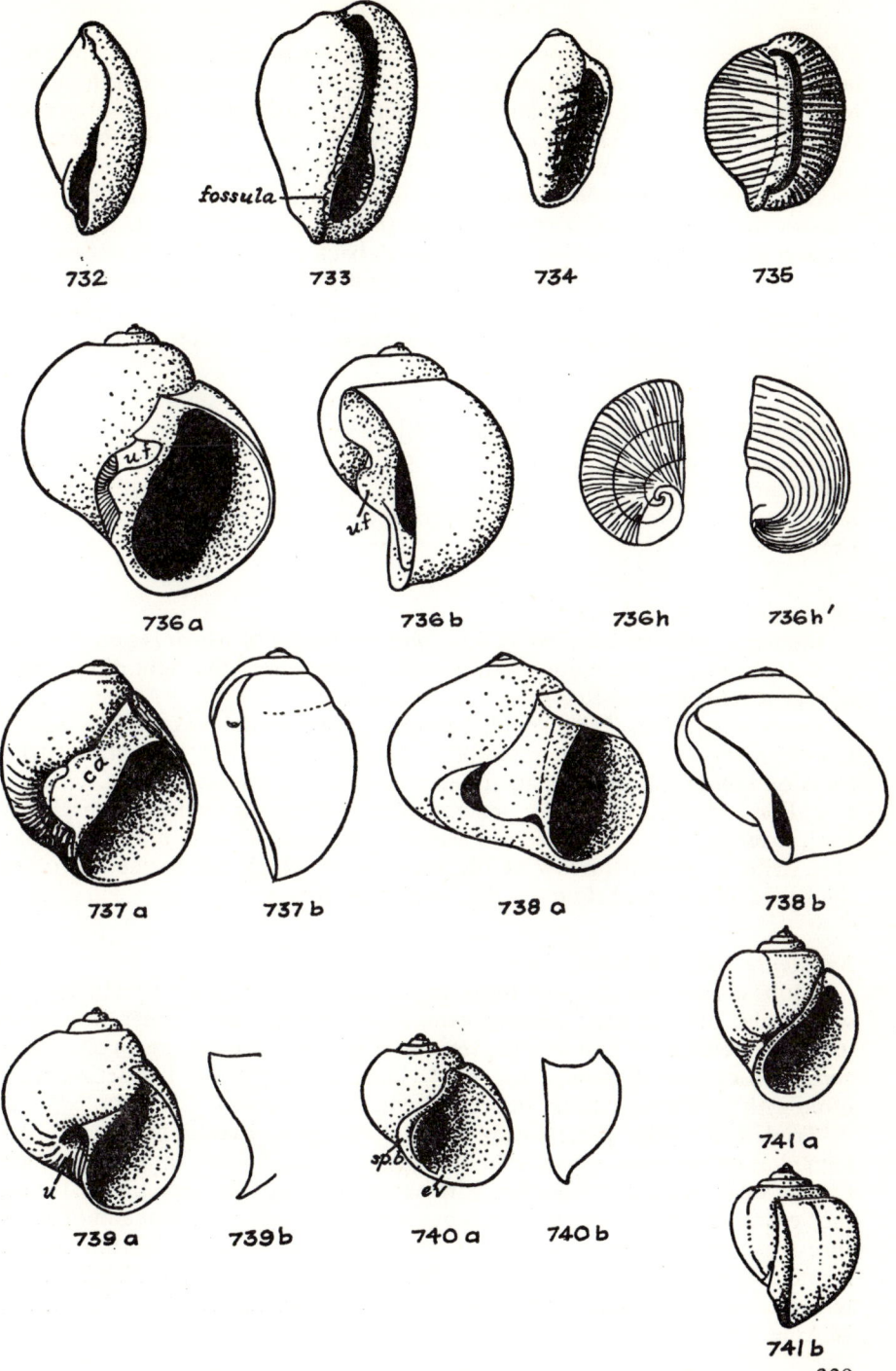

732

733

fossula

734

735

736 a

736 b

736 h

736 h'

737 a

737 b

738 a

738 b

739 a

739 b

740 a

740 b

741 a

741 b

a little wider anteriorly; posterior end of labrum bending sharply inwards; labrum feebly margined; fossula fairly broad, its inner edge smooth; outer teeth long, inner teeth posteriorly short; terminal columellar tooth of one or two ridges. Cret.-U.Mio. Eur., Africa, W.Pakistan, Indonesia, N.Amer., Carib., S.Amer., ?N.Z. *Eotrivia:* rather small, inflated; spire and labrum somewhat projecting posteriorly; with spiral ornament except for a vertical back furrow which is a little removed from the mid-line. M.Eoc.-U.Eoc. Eur. *Transovula:* elongate, fusiform, with smooth back except for spiral lines at the ends; aperture wider anteriorly, with distinct posterior canal; labrum margined; teeth short and delicate; terminal columellar tooth consisting of one or two short ridges; two long, strong columellar folds. L.Eoc.-U.Eoc. Eur., N.Amer.

Family ERATOIDAE [TRIVIIDAE]
Usually small, biconical to subspherical, with enamel layer, smooth or noded or costate; labrum often margined; aperture narrow, with reflected labrum; both lips denticulate; posterior canal feeble or lacking; fossula present. (U.Cret.(Danian)-Rec.)

Subfamily ERATOINAE
Biconical, with broadly projecting spire the height of which is not attained by the labrum; aperture not axial, posteriorly ending to the right of the midline. (U.Cret.(Danian)-Rec.)
Erato (*s.s.*) (Fig. 734): small, biconical, with prominent, callous spire; labrum margined; in shape resembling *Marginella* (p. 383) but without true folds on columella, only crenulations and smooth fossula; smooth. U.Eoc.-Rec. Eur., Medit. *Eratotrivia:* general form of *Erato*; labrum not margined; teeth extended as costae covering ventral side, back, fossula and columella; back with a vertical furrow at the margins of which the costae are often thickened into warts. L.Eoc.-L.Olig. Eur., N.Amer.

Subfamily TRIVIINAE
Globose-subspherical, with deep spire usually covered over; aperture axial, with wide, deep canals; labrum extending slightly above inner lip; teeth usually extended as costae over ventral side, back, fossula and columella; vertical back furrow usually present, dividing the costae into two sets; costae often with small crenulations between them. (U.Eoc.-Rec.)
Trivia (*s.s.*) (Fig. 735): subspherical, with spire covered over; teeth extended as costae over ventral side, back, fossula and columellar furrow; back furrow weak or absent; aperture narrow; crenulations between the costae on the ventral side and the lateral sides but not on the back; anterior end of aperture truncated, posterior notch very feeble; posterior end of labrum curved round almost at right angles to axis; fossula shallow, not separated from anterior canal by a ridge. U.Eoc.-Rec. Eur., N.Africa, Medit., Atl., Carib., N.Amer., N.Z.

340

Family LAMELLARIIDAE

Usually thin-shelled, auriform, with rapidly growing whorls and small spire, smooth or with spiral striae; anomphalous; aperture very large, with sharp labrum; no canal, fossula or teeth. Systematic position based on the anatomy. (?Pal.-Rec.)

Lamellaria (*s.s.*): last whorl very large, semi-oval; aperture very large, obliquely oval, with continuous peristome. ?Pal.-Rec. Eur., N.Amer., widespread.

Superfamily ATLANTACEA [HETEROPODA]

Shell feebly calcified and symmetrical or absent. A parallel development to the pteropods (p. 399), which are also adapted to a pelagic life in warm seas. (Perm.-Rec.)

Family ATLANTIDAE

Dextral, planispiral, with apex visible or covered over by later whorls; last whorl sometimes with a keel or flange. (Perm., Cret.-Rec.)

Atlanta: small initial portion conical, the remainder discoidal and with whorls in contact; last whorl keeled and with a flange; aperture transversely oval, with a slit at the keel. U.Cret.-Rec. Atl., Eur., Austral., N.Amer., C.Amer., Carib., warm seas. *Eoatlanta:* small, discoidal, smooth, without keel; whorls rounded in section, last whorl disjunct. Eoc. Eur.

Family CARINARIIDAE

Shell open-conical, with posterior spiral apex. (Eoc.-Rec.)

Carinaria: capuliform, usually somewhat laterally compressed; last whorl anteriorly carinate, smooth or with accentuated growth lines. Eoc.-Rec. Medit., Eur., Carib., warm seas.

Superfamily NATICACEA

Shell varying from auriform to turbinate to globose, with obtuse protoconch; spire distinct but usually fairly low; usually smooth, occasionally with spiral ornament rather weakly developed; last whorl large; umbilicus open or closed (often by a pad of callus), often with an umbilical cord (*funicle*); aperture semi-circular or oval, holostomatous; peristome in an oblique plane, primarily discontinuous, but becoming continuous when there is a large deposit of callus on the columellar border; columella straight or excavate; labrum not thickened, fairly straight and oblique, prosocline but often becoming less oblique on approaching the suture; operculum horny or calcified. (Trias.-Rec.)

Family NATICIDAE

Characters of the Superfamily. (Trias.-Rec.)

Subfamily NATICINAE

Usually globose, with a low spire; mostly smooth and umbilicate, umbilicus with a funicle separated from parietal callus by a sulcus; operculum with an external calcareous layer, usually thick. (Cret.-Rec.)

Natica (*s.s.*): globose, with low spire; smooth and umbilicate, the umbilicus with an extremely obsolete median funicle; parietal callus extending laterally in a tongue over the upper end of the umbilicus; aperture semicircular; labrum straight or nearly so, little oblique; operculum externally with two marginal costae. Mio.-Rec. Indo-Pac. The subgenus *Cochlis* is like *Natica*, but the umbilicus has a distinct median funicle and the posterior tongue of parietal callus is very small or absent. Pal.-Rec. Indo-Pac., cosmop. *Naticarius* (*s.s.*) [*Natica* Lamarck *non* Scopoli] (Fig. 736): like *Cochlis*, but the simple funicle ending in the anterior part of the umbilicus; operculum externally costate all over. Eoc.-Rec. Carib., Indo-Pac., Austral., Indonesia, India, N.Amer., C.Amer., warm seas.

Subfamily POLINICINAE
Sometimes attaining a fairly large size, oval-conical to rounded; last whorl large, with usually distinctly oblique aperture; umbilicus open or closed, the parietal callus sometimes being very thick, the funicle weak or absent; operculum horny. (U.Cret.-Rec.)

Polinices (*s.s.*) [*Mammilla, Mamillaria*] (Fig. 737): ovoid, higher than broad, spiral angle only slightly obtuse, whorls rather flat-sided, spire small; labrum oblique, sinuous, almost normal to the suture; ventral surface somewhat flattened; middle and posterior callus-extensions united into a thick mass partly or completely filling umbilicus and posterior angle of aperture, but not grooved transversely. U.Cret.(Maastr.)-Rec. Indo-Pac., cosmop. *Neverita* (*s.s.*) (Fig. 738): depressed, only two-thirds as high as broad, with low spire and flattened ventral surface; parietal callus forming a large plug over the umbilical region; floor not callous. L.Eoc.-Rec. Medit., cosmop. *Cepatia:* globose, with obtusely-conical, straight-sided spire and rounded base; aperture little oblique, semicircular; umbilicus filled with a tongue-like callous plug. Pal.-Olig. Eur., N.Africa, W.Pakistan, N.Amer. *Euspira* (*s.s.*) [*Lunatia*] (Fig. 739): small to large, globose to oval-subpoliniciform; aperture semicircular; an even spread of callus over the parietal region, reduced in width at the top of the columella leaving an umbilicus which practically always has no funicle. Cret.-Rec. Eur., cosmop.

Subfamily AMPULLOSPIRINAE
Globose to oval-conical, mostly thick-shelled, mostly smooth; umbilicus open (without funicle) or closed, sometimes with an anterior limiting polished band, the peristome being anteriorly everted; inner lip usually callous, the callus sometimes being very thick. (Trias.-Rec.)

Ampullospira: ovate-globose to turbinate, with distinct spire, whorls often with a well marked (often concave) ramp below the suture; parietal callus similar to that of *Euspira*, but not very thick; umbilicus minute or nil; aperture semicircular, faintly everted anteriorly; labrum not very oblique. [Essentially Mesozoic, comprising many of the Jurassic 'Naticas', with some Tertiary survivors]. Jur.-Mio. Eur., Africa, W.Pakistan, E.Asia, N.Amer. *Ampullella* [*Ampullina auctt. non* Bowdich (*nom. nud.*)]: thick-shelled and rather globose; whorls stepped, with spiral rows of fine punctae; umbilicus small, limited externally by a well defined limb; aperture narrowly semilunar; columellar and parietal callus quite thick. Jur.-Mio. Eur., Africa, Asia, Indo-Pac., Japan, N.Amer., Carib. *Globularia* (*s.s.*) [*Cernina*] (Fig. 740): usually fairly large to large, thick-shelled and globose, with a small spire; umbilicus nearly always covered over by the thick callus of the inner lip; anterior limb limited by a ridge; smooth; aperture semicircular. Jur.-Rec. Eur., E.Africa, W.Pakistan, Indo-Pac. The subgenus *Ampullinopsis* [*Megatylotus*] attains a large size, the whorls are stepped and have very faint spiral ornament, and the parietal region and columellar lip are covered by a wide and thick spread of callus. L.Eoc.-Mio. N.Amer., Eur., N.Africa, W.Pakistan, Burma, Indonesia, ?Japan. *Euspirocrommium:* high-spired, narrowly oval-conical, with convex whorls and deep sutures; aperture simple; anomphalous. Cret.-Mio. Eur., N.Africa, Egypt, M.East, W.Pakistan, India, Ceylon, Indonesia, Austral., N.Amer., Carib. *Pachycrommium:* oval-conical, spire broader than in *Euspirocrommium*, whorls slightly stepped; aperture semicircular; callus on inner lip quite well developed, with a distinct anterior limb; anomphalous. Eoc.-Mio. Carib., Eur., W.Pakistan, Tibet, Indonesia. *Crommium* (Fig. 741): globose, with short, distinct, coeloconoid spire; last whorl shouldered, ornamented with fine spiral lines; small umbilicus without marginal band; aperture narrowly semicircular, with prominent anterior eversion of the peristome. U.Cret.-Mio. Eur., Iran, E.Africa, W.Pakistan, Indonesia, N.Amer., S.Amer.

Subfamily SININAE [SIGARETINAE]
Usually thin-shelled; auriform to globose-poliniciform, with large last whorl; usually with spiral ornament. (Cret.-Rec.)
  Sinum [*Sigaretus*] (Fig. 742): shaped like *Haliotis*, but smaller and without perforations or inner nacreous layer; depressed, auriform; aperture large and gibbous; labrum thin, oblique, gently convex, prosocline; no proper columella, and no limb or funicle; anomphalous; ornament of spiral lines and growth lines. ?U.Cret., Eoc.-Rec. Africa, cosmop. *Sigaretotrema:* like *Sinum*, but less auriform and with an umbilicus. Pal.-Rec. Eur., W.Pakistan, India, Indonesia, N.Z., Austral., N.Amer., Carib.

Superfamily TONNACEA [DOLIACEA]
Often very large, with spire moderately high to low; last whorl fairly large to very large and more or less inflated; aperture anteriorly with a short channel or longer canal; varices sometimes present. (Cret.-Rec.)

Family TONNIDAE [DOLIIDAE]
Sometimes attaining a large size; rather thin-shelled, inflated oval, with fairly short spire; whorls convex, with spiral ornament, without varices except occasionally for a thickened labrum; last whorl very large, the aperture wide

and anteriorly notched, the columella somewhat produced anteriorly. (?Maastr., Eoc.-Rec.)

Tonna [*Dolium*]: large, thin-shelled, globose, with rather low spire; ornament of broad spiral ribbons and fine spiral threads crossed only by fine growth lines; labrum straight, prosocline; last whorl large, with wide aperture which is deeply notched in front, but not forming a canal; siphonal fasciole prominent, traceable on the columella; small, narrow umbilicus. Plio.-Rec. Medit., Eur., W.Africa, W.Pakistan, Indonesia, Japan, Carib., warm seas.

Eudolium: similar to *Tonna*, but spire a little higher, no siphonal fasciole, no umbilicus, callus on the inner lip thin and less extensive, the columella showing a few feeble folds, and labrum externally margined and internally denticulate. Olig.-Rec. Medit., Eur., W.Pakistan, Indonesia, Indo-Pac. ***Malea:*** globose, with low spire, differing from *Tonna* in lacking an umbilicus, having a narrower aperture, the labrum being thickened and internally with long denticles, the columella being twisted and with a few strong folds, and the parietal region also having a few strong folds; the genus therefore shows a superficial resemblance to some members of the Cassididae. Eoc.-Rec. S.Amer., Eur., Carib., C.Amer., N.Amer., Indo-Pac.

Family CASSIDIDAE (helmet-shells)

Globose to elongate-oval, with short spire and depressed-conical protoconch; last whorl large and often inflated; the surface may be smooth, more or less cancellated, or tuberculate; aperture varying from auriform to very narrow (cowrie-like); anterior canal short, more or less recurved; posterior canal variable; labrum nearly vertical, slightly prosocline, usually thickened and sometimes toothed; columella with a large spread of callus, more or less rugose; varices may be present. (U.Cret.-Rec.) (**90, 104**)

Cassis [*Cassidea*] (*s.s.*) (Fig. 743): large to very large, with short spire; shell with faint cancellate ornament, at least one row of coarse tubercles at the shoulder, and varices; aperture narrow, oblique, with coarse transverse ridges along both lips; anterior canal short, deep, recurved at right angles, forming a strong siphonal fasciole which disappears under the columellar callus and is bounded above by a groove leading to a small false umbilicus; no posterior channel; the columellar callus ends in a free edge, detached from the last whorl or united with the penultimate varix. Pal.-Rec. Cosmop.

Figs. 742–750. GASTROPODA: NATICIDAE, CASSIDIDAE, CYMATIIDAE BURSIDAE AND FICIDAE

Figs 742–750 original.
742a, b. *Sinum clathratum* (Gmelin), M.Eoc. Paris B. × 2·25.
743. *Cassis exigua* Tenison Woods, M.Mio. (Balcombian). Muddy Creek, Victoria. × 1·5.
744. *Galeodea nodosa* (Solander), M.Eoc. Paris B. × 0·75.
745a, b. *Sconsia striata* (Lamarck) subsp. *sublaevigata* (Guppy), Mio. Columbia. × 0·75.
746a, b. *Sassia arguta* (Solander), U.Eoc. (Barton). Hampshire. × 1·36.
747. *Distorsio* sp., Rec. × 0·75.
748. *Gyrineum* (*Aspa*) *marginatum* (Gmelin), U.Plio./Pleist. Asti, Italy. × 0·75.
749. *Ficus ficus* (Linné), Rec. Seychelles. T. × 0·75.
750. *Ficopsis cowlitzensis* (Weaver), M.Eoc. (Cowlitz). Nr. Olequah, Washington, × 1·1.

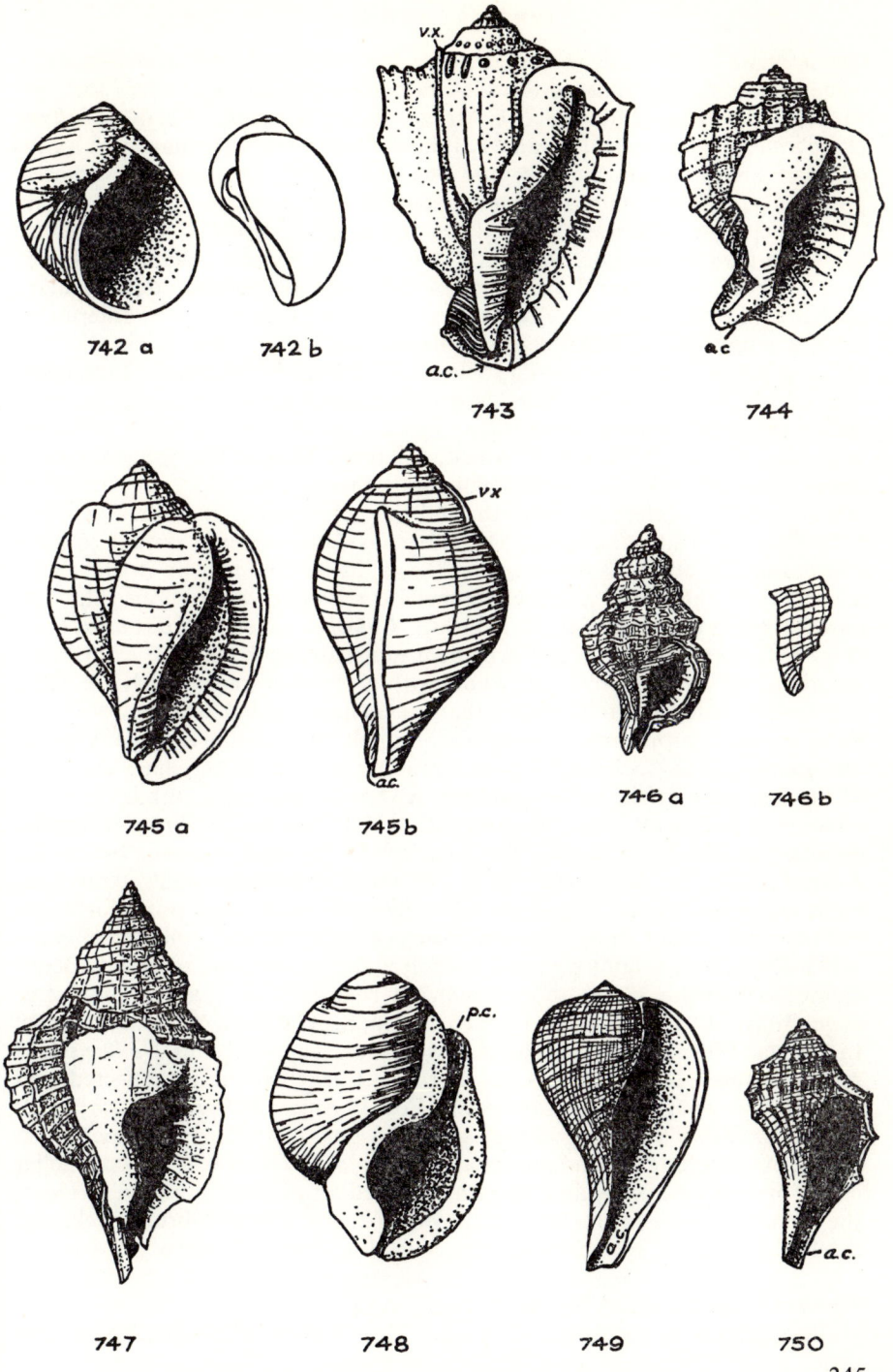

742 a **742 b**

743

744

745 a **745 b**

746 a **746 b**

747 **748** **749** **750**

Cypraecassis (*s.s.*): oval, the aperture very narrow and with a well-marked posterior canal, thus making the shell rather cowrie-like; spire very low; no varices. Mio.-Rec. Indo-Pac., E.Africa, Eur., N.Amer., Carib. **Phalium** [*Bezoardica*]: oval-conical, relatively thin-shelled; whorls angular, with a row of small spines on the shoulder and one or more varices; aperture pointed posteriorly, with a short, strongly recurved and notched anterior canal; labrum varicose, internally denticulate, the anterior denticles forming short spines on the outline. Eoc.-Rec. Indo-Pac., Japan, N.Amer. **Semicassis** (*s.s.*): ovate, whorls with spiral ornament which may be beaded, not angular; aperture (like that of *Phalium*) wider than in *Cassis*, without spines anteriorly; usually with only a labral varix; anterior columellar fold thick. U.Cret. (Maastr.)-Rec. Japan, Indo-Pac., India, Ceylon, W.Pakistan, Indonesia, Austral., N.Africa, Medit., Eur., N.Amer. The subgenus **Tylocassis** has cancellate ornament, and the columellar rugae are broken up into small granules. Mio.-Rec. Carib., N.Amer., C.Amer., S.Amer., Medit. The subgenus **Casmaria** is smooth or has a row of small nodes, there are no varices except at the labrum which is internally smooth, and the columella is smooth. Eoc.-Rec. Indo-Pac., Austral., ?N.Amer. The subgenus **Eocasmaria** is similar to *Casmaria*, but the labrum is not varicose, the callus on the inner lip is wider posteriorly than anteriorly, and the columella carries one very oblique fold with an excavation above it. Pal. W.Pakistan. **Galeodea** [*Morio, Cassidaria*] (*s.s.*) (Fig. 744): oval-conical, spiral ornament much stronger than the axial, early whorls usually shouldered, last whorl with a series of keels which are often tubercular; anterior canal longer than in *Semicassis*, but less recurved and therefore more projecting, not notched; though the siphonal fasciole is not so well-marked as in *Cassis* and has no bounding groove, yet the sharp elevation of the free edge of the callus produces a very distinct false umbilicus; columella with few or no rugae; usually with labial varix only; labrum internally denticulate. Pal.-Rec. Medit., Eur., Africa, W.Pakistan, Burma, Indonesia, Austral., N.Z., Atl. **Sconsia** (*s.s.*) (Fig. 745): oval, with rather weak spiral ornament and a few varices; aperture narrowly oval, constricted posteriorly, anteriorly with a very short, scarcely recurved or projecting, unnotched canal which forms only a faint siphonal fasciole; labrum varicose, internally denticulate; columellar callus thin, somewhat rugose, with slightly raised free edge in front. U.Eoc.-Rec. Carib., N.Amer., C.Amer., S.Amer., Japan, Indonesia, Burma, W.Pakistan, N.Africa, Eur. The subgenus **Doliocassis** is relatively small, has no varices, the labrum is thickened but not reflected, and the anterior canal is deeply notched. M.Eoc. N.Amer. **Morum** [*Oniscia*] (*s.s.*): biconical, with a very low, obtuse spire and long last whorl; the aperture is drawn out to a long slit and the canal reduced to a mere notch; the columellar callus has no free edge, and bears a scattered set of short horizontal ridges; the ornament consists of a few spiral rows of coarse knobs; labrum thickened, externally reflected, internally dentate. Mio.-Rec. Carib., N.Amer., C.Amer.

Family CYMATIIDAE [TRITONIDAE]

Bucciniform to fusiform, of variable size; whorls with varices; aperture oval, anteriorly with a moderately long canal; labrum thickened and usually dentate internally; parietal region often with a posterior tooth limiting a gutter; columella usually with folds. (U.Cret.-Rec.)

While some genera atypically have no varices, some have three varices to two whorls, and others have two varices to each whorl (rising in lines up the two sides of the shell). The varices differ from those of the Bursidae in never being channelled posteriorly (in the Bursidae the labial varix, at least, is channelled posteriorly). The ornament is often reticulate in early whorls, the vertical costae tending to become coarser and ill-defined in later whorls: about ten primary spiral ribbons, of which three are visible on the spire whorls, but the varying dip of the suture may locally expose others: the second is the strongest, and the posterior one has a tendency to break up into two or more. Labrum straight and almost vertical in profile, with about seven teeth, corresponding to the interspaces between the primary spirals. Columella uniformly excavate or with slight parietal protuberance, twisted into the canal in front, covered by a callus which is often wrinkled.

The shells of this family are distinguished from those of the Muricidae (p. 349) by their non-muricate surface, the usually different spacing of the varices, often wrinkled columella and open anterior canal. They appear in the Cretaceous and increase in importance through the Tertiary. Of warm water habitat and attractive appearance they received early attention from naturalists, and the abundance of generic names given has resulted in much confusion of nomenclature (**103**).

Cymatium (*s.s.*): fusiform, large; last whorl with three strong varices, triangular; anterior canal moderately long, straight; columella with a median fold. Eoc.-Rec. Carib., Formosa, warm seas. The subgenus *Lampusia* differs from *Cymatium* in having a bend in the columella at the origin of the canal; a parietal fold sets off a posterior gutter, and the callus of the inner lip carries numerous folds and wrinkles. Mio.-Rec. Indo-Pac., Indonesia, W.Africa, Eur., warm seas. *Sassia* (Fig. 746): of medium size, moderately elongate, with subangular to rounded whorls; ornament reticulate to decussate, with small tubercles at the intersections; varices irregular (about one per whorl); aperture oval, with sharply defined, short, oblique and slightly twisted canal; columella uniformly excavate; parietal fold limiting a well defined posterior gutter; callus of inner lip with wrinkles better developed anteriorly. U.Cret.(Danian)-Rec. Eur., Madagascar, W.Pakistan, Austral., N.Amer., warm seas. *Ranella:* fusiform, with cancellate ornament bearing crenulations at the intersections and additional finer spiral threads; two varices per whorl, forming lines up the two sides of the shell; aperture rounded-oval, anteriorly with a fairly long canal which is inclined to the left; columellar lip with wrinkles; posteriorly, a parietal tubercle separates off a gutter. U.Cret.(Danian)-Rec. Medit., Eur., W.Pakistan, India, Burma, Japan, warm seas. *Distorsio* [*Persona*] (*s.s.*) (Fig. 747): bucciniform to fusiform; irregular, owing to great variations in dip of suture, the direction of growth changing after each resting-stage (marked by a varix); ornament cancellate, with crenulations at the intersections; aperture greatly constricted by the columellar folds and labral teeth; parietal callus

347

widely spread. Olig.-Rec. Indonesia, Burma, India, W.Pakistan, Indo-Pac., Carib., S.Amer., Eur., warm seas. *Charonia* [*Tritonium, Triton*] (*s.s.*): sometimes very large, buccinoid with rather high spire; whorls subangular, with broad, crenulated spiral threads; varices irregularly placed; apertural characters much like those of *Lampusia*, but the canal is quite short. U.Cret.-Rec. Indo-Pac., N.Z., Madagascar, Eur., warm seas. *Nassaria* [*Hindsia*] (*s.s.*): inflated oval-fusiform; whorls convex, with occasional varices; ornament of axial costae and spiral threads of more than one order; aperture oval, anteriorly with a narrow, moderately long, open, oblique canal which is slightly notched, thus forming a faint siphonal fasciole; inner lip with small folds, posteriorly with a denticle separating off a small gutter. Olig.-Rec. Indo-Pac., Indonesia, Formosa, Burma, India, W.Pakistan.

Family BURSIDAE

Fusiform to turbinate, spire more or less prominent; ornament spiral and axial, often crenulated or spinose; varices present, often continuous up the two sides of the shell; aperture much as in the Cymatiidae but channelled at both ends, the posterior channel not being a gutter. (Eoc.-Rec.)

Bursa (*s.s.*): whorls angular, with solid tubercles and two opposing rows of channelled varices forming spines; labrum varicose, internally denticulate; aperture anteriorly with a short, narrow, curved canal; columellar lip wrinkled. Eoc.-Rec. Indo-Pac., Japan, Indonesia, Eur., N.Amer. The subgenus *Apollon* [*Lampasopsis*] is less spinose, and the anterior canal is short, straight and a little oblique. Mio.-Rec. Indo-Pac., Burma, W.Pakistan, Eur. *Gyrineum* [*Bufonaria*] (*s.s.*): similar to *Bursa*, but whorls ornamented with crenulated spiral threads and a few rows of small spines, the columella is straighter, as is the anterior canal, and the varices are even more strongly spinose. Mio.-Rec. Indo-Pac., Formosa, Indonesia, India, W.Pakistan. N.Amer. The subgenus *Aspa* (Fig. 748) has a distinctly shorter spire and no spines; the ornament is degraded almost to smoothness, and there is only one row of small nodes near the suture; the anterior canal is a little shorter. Mio.-Rec. Medit., Eur., Africa.

Family FICIDAE [PYRULIDAE]

Rather thin-shelled, inverted pyriform to subfusiform; with one exception the ornament is a lattice of spiral and axial threads, and there may also be spiral rows of nodes or small spines; last whorl large; aperture elongate-pyriform, anteriorly with a distinct, sometimes quite long, fairly wide, un-notched canal; columellar callus usually very weakly developed. (U.Cret.-Rec.)

Ficus [*Pyrula*] (Fig. 749): spire usually very low; anterior canal moderately broad, moderately long, passing gradually into the aperture; ornament consisting of a lattice of collabral and spiral threads, sometimes of equal strength, sometimes with the collabral element somewhat finer; no spiral keels, nodes or spines. Eoc.-Rec. Indo-Pac., Austral., N.Z., Formosa, Indonesia, Burma, W.Pakistan, India, Eur., Carib., N.Amer., C.Amer., S.Amer. *Priscoficus:*

like *Ficus*, but spire slightly higher, anterior canal narrower, and last whorl with three carinae of which one, two or all carry coarse tubercles or blunt spines. Pal.-L.Eoc. Eur., N.Amer. **Ficopsis** (*s.s.*) (Fig. 750): similar to *Priscoficus*, but whorls with a distinct posterior ramp; last whorl simply latticed, with one or two keels which may be smooth or carry nodes. Eoc. N.Amer., Eur. The subgenus **Fulguroficus** is similar to *Ficopsis*, but has four or five carinae on the last whorl, all carrying blunt spines or nodes. Mio. Eur.

Superfamily MURICACEA

Shell usually with strong ornament and a long anterior canal; columellar lip without strong folds. (Cret.-Rec.)

Family MURICIDAE

More or less fusiform shells in which the growth lines have lamellar or spiny edges giving rise to a roughened, scaly ('muricate') surface, apart from definite ornament; aperture rounded or oval, with a more or less long, open or closed canal, not notched at the end; usually no siphonal fasciole recognizable; labral profile very nearly straight and vertical, though showing traces of a buccinid curve; varices generally present, sometimes spinose; protoconch papillose. (U.Cret.-Rec.)

Subfamily MURICINAE

Anterior canal short to very long; operculum usually with terminal nucleus. (Pal.-Rec.)

Murex (*s.s.*): spire moderately prominent; whorls with three spinose varices, the spines extending on to the flanks of the anterior canal; whorls subangular, with costae and spiral threads between the varices; aperture oval; columellar lip wrinkled anteriorly, sharply bent at the start of the long, nearly straight, closed anterior canal; labrum nearly straight, a little prosocline posteriorly. Mio.-Rec. Indo-Pac., Indonesia, N.Amer., C.Amer., S.Amer. The genus **Hexaplex** is Recent only. Its subgenus **Phyllonotus** [*Muricanthus*] is moderately high-spired, each whorl with three to four narrow, scaly varices; one or more axial costae with spiral ornament between the varices; aperture rounded, with a small posterior channel, anteriorly with a moderately short canal curved to the left; columellar lip externally detached; labrum varicose, internally folded. Mio.-Rec. Carib., N.Amer., C.Amer., S.Amer., Eur. [There is another group of species, exemplified by *Murex radix* Gmelin, which has been referred to *Muricanthus* by some conchologists, but which seems to belong to another subgenus of *Hexaplex* for which no name is available; this group has a stouter last whorl, a lower spire, an umbilicus, and there are seven to eight varices per whorl without intercalary costae. Eoc.-Rec. C.Amer., N.Amer., Eur., N.Africa, W.Pakistan, Indonesia, warm seas.] **Chicoreus** (*s.s.*): fusiform to inflated-fusiform; whorls convex, angular, depressed towards the suture; varices three to a whorl, but not absolutely continuous from whorl to whorl, bearing numerous tubular, frilled spines; between the varices are usually two nodular vertical costae; aperture rounded oval; anterior canal moderately long, its upper end oblique, but its lower end curved to the right. Olig.-Rec. Indo-Pac., Indonesia, Japan, Eur., Carib., N.Amer., warm seas. *Pterynotus* (*s.s.*): fusiform, with moderately high spire; three narrow, sharp, alate varices per whorl, forming a triangular pyramid; ornament of spiral furrows which continue on to the varices; aperture small, oval, with a fairly long, straight or gently curved, narrow anterior canal on to the flanks of which the varices continue; columellar lip moderately wide, slightly detached anteriorly; labrum straight, varicose, internally denticulate. Eoc.-Rec. China Sea, Indonesia, Austral., N.Z., Eur., warm seas. The subgenus *Pterochelus* [*Alipurpura*] (Fig. 751) differs from *Pterynotus* in having a long channelled projection at the posterior end of each varix, a small subspinose costa between each pair of varices, and a straight anterior canal. Eoc.-Rec. Pac., Austral., N.Z., Eur. The subgenus *Pteropurpura* differs from *Pterochelus* in having a gently curved anterior canal, and in the varices having smaller and unchannelled projections. Eoc.-Rec. Warm seas, Eur.

Subfamily TRITONALIINAE [OCENEBRINAE]

Operculum with lateral nucleus. (Eoc.-Rec.)

Tritonalia [*Ocenebra, Ocinebra*] (*s.s.*) (Fig. 715A): general form and ornament much as in *Phyllonotus*, but the canal is closed or nearly so and there is a ramp below the suture; compared with *Murex* the inner lip of the aperture is less strongly curved than the labrum; three to four varices per whorl. Olig.-Rec. Medit., N.Amer., Japan, Indonesia, ?S.W.Africa. *Eupleura:* shape and aperture as in *Muricopsis* (see p. 352), but with two opposing varices per whorl, cancellate ornament (the vertical element sometimes a little the stronger), and smooth columella. Eoc.-Rec. N.Amer., C.Amer., Eur. *Hadriania:* fusiform, whorls angular, with numerous costa-like varices which rapidly become obsolete on the ramp, and numerous spiral furrows; anterior canal inclined to the left and bent dorsalwards; varices dying out on base of last whorl. Eoc.-Rec. Medit., Eur., ?Austral. *Urosalpinx:* inflated-fusiform to buccinoid, with cancellate ornament (the axial element often stronger) but no varices; aperture oval, with a moderately short, oblique, narrow, open an-

---

Figs 751–761. GASTROPODA: MURICIDAE AND PURPURIDAE
Figs 751A and 756 after Wenz; Figs 751, 752–755 and 757–761 original.
751. *Pterynotus* (*Pterochelus*) *asper* (Solander), U.Eoc. Barton, Hampshire. × 1.
751A. *Tritonalia erinacea* (Linné), Rec. Medit. T. × 0·66.
752. *Trophonopsis muricatus* (Montagu), U.Plio./Pleist. Oakley, Essex. × 0·3.
753. *Favartia aquitanica* (Grateloup), U.Mio. Touraine. × 1.
754. *Typhis pungens* (Solander), M.Eoc. Paris B. × 1·5.
755. *Ecphora quadricostata* (Say), M.Mio. Virginia. × 0·66.
756. *Purpura fucus* (Gmelin), Rec. E.Atl. T. × 0·66.
757a, b. *Cymia charlesworthi* (Edwards), Olig. Hamstead, I. of Wight. × 1·9.
758. *Drupa arachnoides* (Lamarck), Rec. Indo-Pac. × 1·1.
759a, e. *Concholepas peruviana* Lamarck, Rec. Peru. T. × 0·95.
760. *Nucella tetragona* (J. de C. Sowerby), ?Pleist. Walton, Essex. × 1·1.
761a, b. *Nucella* (*Acanthina*) *crassilabrum* (Lamarck), Rec. Peru. × 0·9.

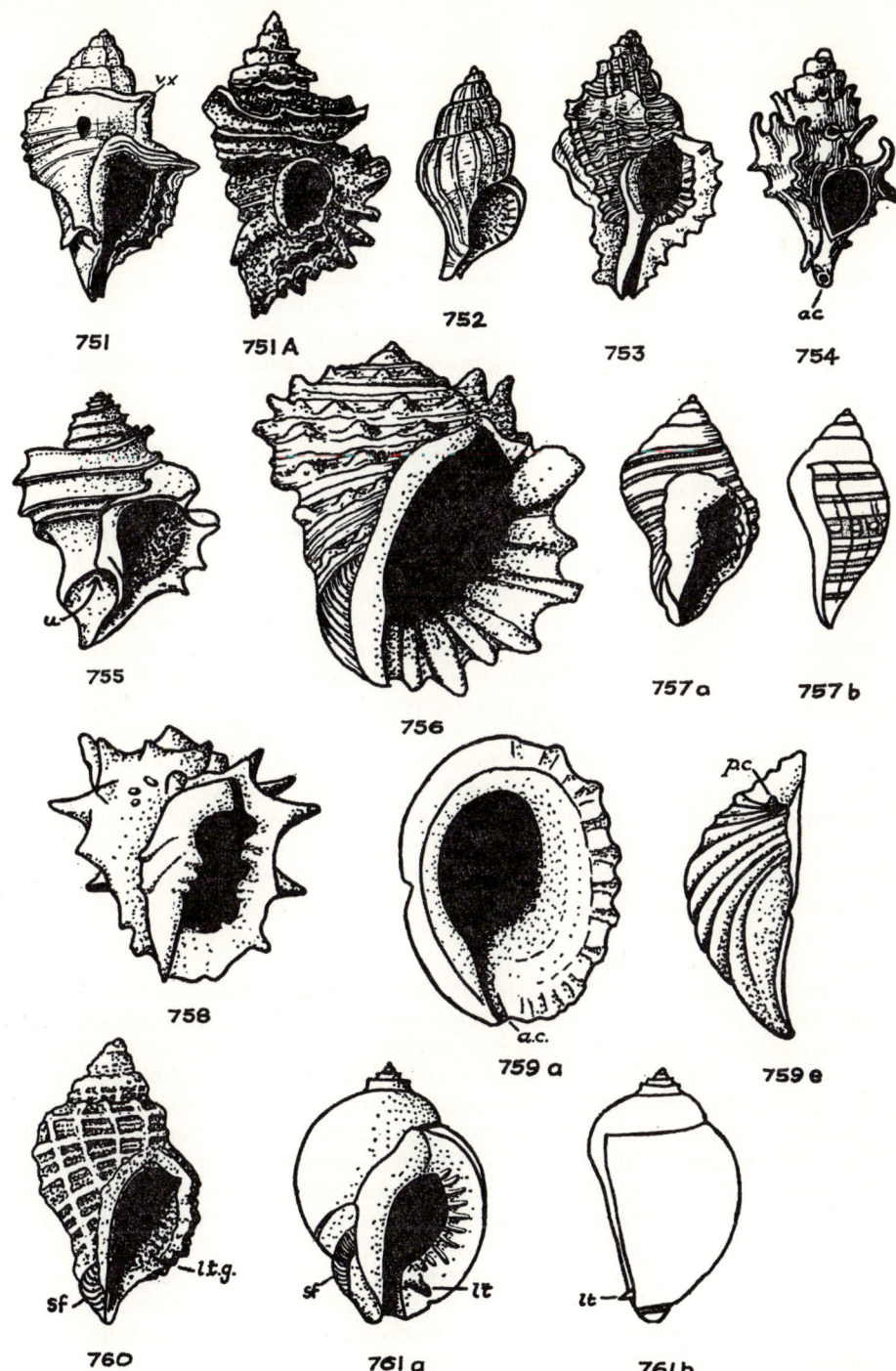

751 751 A 752 753 754

755 757 a 757 b

756

758 759 a 759 e

760 761 a 761 b

351

terior canal; low siphonal fasciole and umbilical depression present; labrum internally with solid denticles. Eoc.-Rec. N.Amer., S.Amer., Iceland, Eur., Atl., Indo-Pac., Indonesia, Austral.

Subfamily TROPHONINAE

Operculum with sublateral nucleus. (Eoc.-Rec.)

Trophon (*s.s.*): inflated-fusiform; whorls with a posterior ramp, numerous sharp varix-like costae which are weakly spinose at the shoulder, and finer spiral ornament between the costae; aperture broadly oval, with a moderately long, oblique anterior canal; siphonal fasciole and narrow, deep umbilicus present; columellar lip smooth; labrum internally sometimes feebly folded and with a small tooth at the start of the canal, externally varicose. Plio.-Rec. S.Amer., Magellanic, Antarctic, ?N.Z., ?Japan. *Trophonopsis* (*s.s.*) (Fig. 752): more fusiform than *Trophon*, and without siphonal fasciole or umbilicus; whorls convex or with a gentle posterior ramp; labrum internally denticulate. Eoc.-Rec. Medit., N.Amer., northern seas. The subgenus *Boreotrophon* differs from *Trophonopsis* in having weaker spiral ornament, and the labrum is smooth internally. ?Mio., Plio.-Rec. Northern seas, Iceland, N.Amer. *Favartia* (Fig. 753): short-fusiform; whorls convex, imbricate, with four to six rather rounded varices and spiral threads forming weak spines on the varices; aperture oval, anteriorly with a gently oblique, moderately long canal which is closed in adults; labrum with a strong, frilled varix, internally dentate; columellar lip smooth; siphonal fasciole present, but no true umbilicus. Eoc.-Rec. Indo-Pac., Austral., Indonesia, Medit., Eur., Atl., N.Amer. *Muricopsis*: general form somewhat like *Phyllonotus*, but more fusiform, columellar lip less concave, anterior canal not bent to the left, whorls shouldered, and columella with two obsolete folds; scaly siphonal fasciole and very narrow umbilicus also present; whorls with six to eight varices which are sometimes spinose; labrum with varix, internally with strong denticles. Pal.-Rec. Carib., N.Amer., Eur., Atl., Medit., W.Pakistan, India, Indonesia, Austral.

Subfamily TYPHINAE

Usually rather small, muriciform; hollow tubes at the shoulders of the whorls, alternating with or coalesced to varices; aperture small, ovate; anterior canal usually (but not always) closed, tubular, often twisted. (Pal.-Rec.) (**87a**)

Typhis (*s.s.*) (Fig. 754): four varices with hollow spines to each whorl, with a single tubular perforation of the shell between any two varices; small spines developed where a few spiral threads cross the varices; anterior canal closed, twisted, with small, spine-like siphonal fasciole adjacent to it. Pal.-Plio. Eur., N.Amer., C.Amer., Carib., N.Z. *Siphonochelus* [*Cyphonochelus*] (*s.s.*): four varices and tubes per whorl, the varices rather weak, smooth, the rather long tubes welded to the posterior face of each; anterior canal closed. Eoc.-Rec. S.Africa, Japan, Austral., Eur. *Laevityphis* (*s.s.*): four varices per whorl, with a small posterior spine; no spiral ornament; four tubes per whorl, situated between two varices and a little nearer the succeeding one; anterior canal closed. Eoc.-Rec. Eur., N.Amer., S.Amer.

Subfamily RAPANINAE

Buccinoid to fusiform, with a fairly short spire; aperture wide, with a fairly broad, open anterior canal; operculum with nucleus on lower side of outer margin. (U.Cret.-Rec.)

Rapana: large, inflated-fusiform to biconical with short spire; ornament of spiral keels and sharp axial costae which form spines on the shoulder; aperture broadly oval, with a short, backwardly curved anterior canal; prominent scaly siphonal fasciole and distinct umbilicus, against which the columellar callus becomes detached; labrum internally smooth. Mio.-Rec. Indo-Pac., Japan, Austral., Indonesia, W.Pakistan, Iran. *Chorus:* pyriform, with convex whorls bearing only feeble spiral ornament; aperture oval, with a moderately long, straight, oblique, notched anterior canal; siphonal fasciole low; labrum internally with a small spine at start of canal. Olig.-Rec. S.Amer. *Ecphora* [*Stenomphalus*] (Fig. 755): somewhat resembles broad forms of *Murex* in shape and aperture; differs from *Rapana* in being ornamented with four strong spiral keels only and in having a moderately long, oblique anterior canal. U.Cret.-Mio. N.Amer., Eur., ?Austral.

Subfamily COLUMBARIINAE

Fusiform, with a very long, straight anterior canal; whorls with a spinose shoulder; operculum pyriform, with terminal nucleus. (?U.Cret.(Danian), Eoc.-Rec.)

Columbarium: no strong spiral ornament or costae; shoulder spinose. Eoc.-Rec. Austral., Indo-Pac.

Family PURPURIDAE [THAIDIDAE]

Thick-shelled, buccinoid, often rather globose; frequently with nodose or spinose ornament in addition to spiral ornament; columellar lip straight or gently excavated, often flattened; labral profile straight, often prosocline, the only deviations being due either to a posterior gutter (*Purpura*, *Stramonita*) or to an infold spine (*Acanthina*); anterior canal very short or reduced to a notch; siphonal fasciole usually strong; no varices or murication. (U.Cret.-Rec.)

Purpura [*Thais*] (*s.s.*) (Fig. 756): spire short, last whorl large, with spiral rows of solid nodes; aperture wide, with a posterior channel and a very short anterior canal; columella straight, usually bent or twisted at the anterior end, flattened and covered with a broad callus; siphonal fasciole present. Olig.-Rec. Cape Verde Is., N.Amer., Japan, warm seas. The subgenus *Stramonita* is more oval-conical on account of its rather higher spire; ornament of numerous spiral threads and a few rows of nodes; aperture posteriorly channelled, there being a strong parietal fold; columellar lip with a few wrinkles; labrum internally laciniate; anterior canal very short, notched; siphonal fasciole well developed, separated from columella by an umbilical depression. Mio.-Rec. Medit., Eur., W.Pakistan, Indonesia, Austral., Carib., N.Amer., warm seas. *Cymia* [*Cuma*] (Fig. 757): biconical, with spiral ornament and broad, solid

353

costae or nodes; whorls concavo-convex or shouldered, the shoulder sometimes with short spines; aperture narrow, posteriorly deeply channelled, anteriorly with a short, deeply notched canal; columellar lip with a median fold and anterior wrinkles; labrum somewhat prosocline, internally strongly lirate; siphonal fasciole well developed, separated from columellar callus by an umbilical depression. Olig.-Rec. S.Amer., Carib., N.Amer., Eur., W.Pakistan, India, Indonesia, warm seas. *Drupa [Ricinula] (s.s.)* (Fig. 758): although with the form of *Purpura*, the aperture has an irregular and contracted outline on account of the presence of strong labral denticles and columellar teeth; the ornament is also more spinose. Plio.-Rec. Red Sea, Indo-Pac., E.Africa, ?N.Amer. The subgenus *Sistrum* is oval-biconical on account of its somewhat higher spire; ornament of noded spiral bands or only trellissed; aperture less constricted and labral teeth not grouped although unequal; siphonal fasciole well developed, and base sometimes somewhat excavated Eoc.-Rec. Indo-Pac., Indonesia, Eur. *Nassa [Jopas, Iopas] (s.s.)*: differs from *Purpura* in having a parietal area in the aperture, instead of a uniformly excavated inner lip; elongate-oval, with small conical spire, closely spaced spiral ornament and traces of axial ornament; aperture large, with a posterior channel limited by a parietal fold and a posterior labral denticle, and a short, broad, notched anterior canal; columellar lip slightly swollen medially; labrum internally finely folded, with an anterior denticle; siphonal fasciole well developed. Mio.-Rec. Indo-Pac., Eur., N.Amer. *Concholepas* (Fig. 759): the spire is very small, quite flat, and the last whorl almost completely embraces the earlier *Purpura*-like shell so that it approaches the shape of a limpet; ornament of scaly spiral threads; aperture very large and wide, anteriorly with a feeble channel instead of a canal; umbilicus limited externally by a siphonal fasciole almost indistinguishable from the spiral ornament; columella rather excavated. Eoc.-Rec. S.Amer., Austral. *Nucella [Polytropalicus] (s.s.)* (Fig. 760): like *Nassa*, but aperture broader, without a posterior channel limited by two denticles, and the labrum internally without the anterior denticle. Mio.-Rec. Eur., N.Amer., northern seas, Burma, Indonesia, southern seas. The subgenus *Acanthina [Monoceros]* (Fig. 761) is oval-conical, more inflated than *Nucella*; whorls convex or angular, smooth or with feeble spiral ornament; last whorl large, inflated-oval; aperture broadly oval, with a small, feebly notched anterior canal; labrum with a tooth opposite the beginning of the canal; siphonal fasciole, but no umbilicus; labral spine formed by an infold of the mantle, as it leaves an external spiral groove as a scar of its origin (as in *Pseudoliva*). Mio.-Rec. S.Amer., N.Amer., Iceland, Eur., Indonesia.

Family MAGILIDAE [CORALLIOPHILIDAE]

Muriciform to purpuriform, occasionally almost limpet-like, rarely with the last whorl becoming disjunct; spiral ornament always, axial ornament sometimes present; usually with a moderately developed anterior canal; an aberrant group of Muricacea spending a specialized life on or in madreporids. (U.Cret.-Rec.)

Magilus: rounded-oval, with the last whorl becoming disjunct and forming a straight or irregularly curved tube which has a small keel continuing from the columella; spiral ornament and growth lines. Eoc.-Rec. Indo-Pac., Red Sea, Medit., Madagascar. *Coralliophila* (*s.s.*): rounded-oval, with short, conical spire; whorls with spiral threads and sometimes weak axial costae; narrow siphonal fasciole; aperture broadly oval, with a short, open anterior canal which is little notched; columella little concave, flattened; labrum internally not dentate. Olig.-Rec. Cosmop.

Superfamily BUCCINACEA

Oval-conical to fusiform, often with spiral or axial ornament; columella usually without folds; anterior canal short to long. (Cret.-Rec.)

Family BUCCINIDAE

Oval-conical to oval-fusiform, with fairly large last whorl; whorls more or less convex, smooth or ornamented; aperture wide, anteriorly deeply notched or with a moderately long canal; siphonal fasciole often present; columella usually concave, mostly smooth. (U.Cret.-Rec.)

Subfamily BUCCININAE

Notch moderately deep and wide; siphonal fasciole thick; labral profile generally curved, prosocline above the shoulder, vertical on the base until curvature is reversed towards the siphonal fasciole. Mainly cold-water forms. (U.Cret.-Rec.)

Buccinum (*s.s.*) (whelk, Fig. 762): ovoid-conical (bucciniform) with convex whorls; last whorl large and swollen, not excavated except at junction with the short neck; ornament of fine spiral lines, crossed in early whorls by thick, ill-defined, vertical costae, not parallel to the growth lines, in later whorls by irregular growth lines only; aperture widely oval, with wide and moderately deep notch inflected dorsally as a very short and wide canal, continued by a broad siphonal fasciole rising steeply to pass under the columellar callus; labral profile vertical below the shoulder, sharply prosocline to the suture; columellar lip of irregular profile, not strongly excavated, without folds; anomphalous; labrum internally smooth. U.Olig.-Rec. Eur., Japan, N.Amer., northern seas. *Liomesus* [*Buccinopsis*]: like *Buccinum*, but with a shorter spire, less convex whorls, a grooved suture, no vertical costae, columella more smoothly excavated, and anterior canal more definitely marked off from the aperture; early whorls, at least, with weak spiral ornament, last whorl often smooth; labral profile nearly straight. Mio.-Rec. Eur., Iceland, N.Amer., northern seas. *Bendeia:* like *Liomesus,* but spire shorter and anterior canal longer and more inflected, ornament of two incised spiral lines near the posterior suture (apart from spiral threads on the base of the last whorl), the aperture is narrower, columella with a fold at the start of the canal, and base of last whorl more excavated. U.Eoc. Nigeria. *Agasoma:* inflated-fusiform with rather low spire and blunt apex; whorls gently concave, smooth except for a keel below the suture and a second keel on the last whorl limiting its base; aperture subrhombic, deeply channelled posteriorly, anteriorly with a moderately long, feebly notched canal which is curved to the right; labrum bisinuous; base of last whorl well excavated, with a fairly long neck; there is an umbilical slit, and the columella is bent at the start of the canal. Olig.-Mio. N.Amer.

Subfamily COMINELLINAE

Anterior notch deep; siphonal fasciole thick, margined by a distinct, fine keel and a depression. (U.Cret.-Rec.)

Cominella (*s.s.*) (Fig. 763): bucciniform, with conical spire, convex whorls and fine spiral and weak, curved collabral costae; siphonal fasciole thick, limited by a prominent keel; aperture broadly oval, posterior end angular, with a deep and narrow gutter, anteriorly with a short, broad, deeply notched canal; labral profile only slightly oblique, becoming more prosocline as it crosses the gutter; columella with an anterior fold. Pal.-Rec. N.Z., Austral., Indonesia, W.Pakistan, Africa, Eur., ?N.Amer., ?S.Amer. The subgenus *Ptychosalpinx* has no gutter, the anterior canal is longer, the labrum is smooth internally, and the columella is vaguely folded. Eoc.-Mio. N.Amer., C.Amer., Eur. *Lacinia:* fairly large, thick-shelled, globose, with very short coeloconoid spire; early whorls with collabral costae, later whorls smooth or with very faint spiral ornament; last whorl very large and inflated, with a solid siphonal fasciole and a false umbilicus; aperture fairly small, oval, posteriorly channelled, anteriorly with a very short, notched canal; labrum rather sinuous, internally smooth; inner lip strongly callous. Eoc.-Olig. N.Amer., Burma. *Laccinum:* like *Lacinia*, but last whorl more cylindrical, no umbilical depression, aperture less ample, and anterior canal better defined. U.Eoc. Nigeria. [The type species was originally placed in the genus *Athleta*.] *Searlesia* (Fig. 764): rather fusiform, superficially resembling a *Siphonalia*; ornament of spiral threads crossed by well-defined axial costae which may become obsolete on the last whorl; aperture rather narrowly oval, feebly channelled posteriorly, with well defined anterior canal continuing its long axis; labrum thickened, internally lirate, its profile with very gentle curvature; last whorl with a short neck and a fairly solid siphonal fasciole. Olig.-Rec. Eur., Iceland, N.Amer., Japan, Korea.

Figs 762–771. GASTROPODA: BUCCINIDAE, PYRENIDAE AND NEPTUNEIDAE

Figs 762–771 original.

762a, e. *Buccinum undatum* Linné, Rec. T. × 0·75.

763. *Cominella lineolata* (Lamarck), Rec. × 1·1.

764. *Searlesia costifera* (S. V. Wood), U.Plio./Pleist. Suffolk. × 0·75.

765. *Phos senticosus* (Linné), Rec. T. × 1·2.

766. *Pollia labiata* (J. de C. Sowerby), U.Eoc. Rordon, Hampshire. × 1·17.

767. *Laevibuccinum lineatum* Heilprin, L.Eoc. Wood's Bluff, Alabama. × 1·1.

768. *Babylonia areolata* (Lamarck), Rec. Indo-Pac. × 0·75.

769. *Pterygia major* (Sowerby), Rec. India. × 1·1.

770. *Mitrella sulcata* (J. Sowerby), Pleist. Walton, Essex. × 1·1.

771a, b. *Neptunea despecta* (Linné), Pleist. Butley, Suffolk. × 0·5.

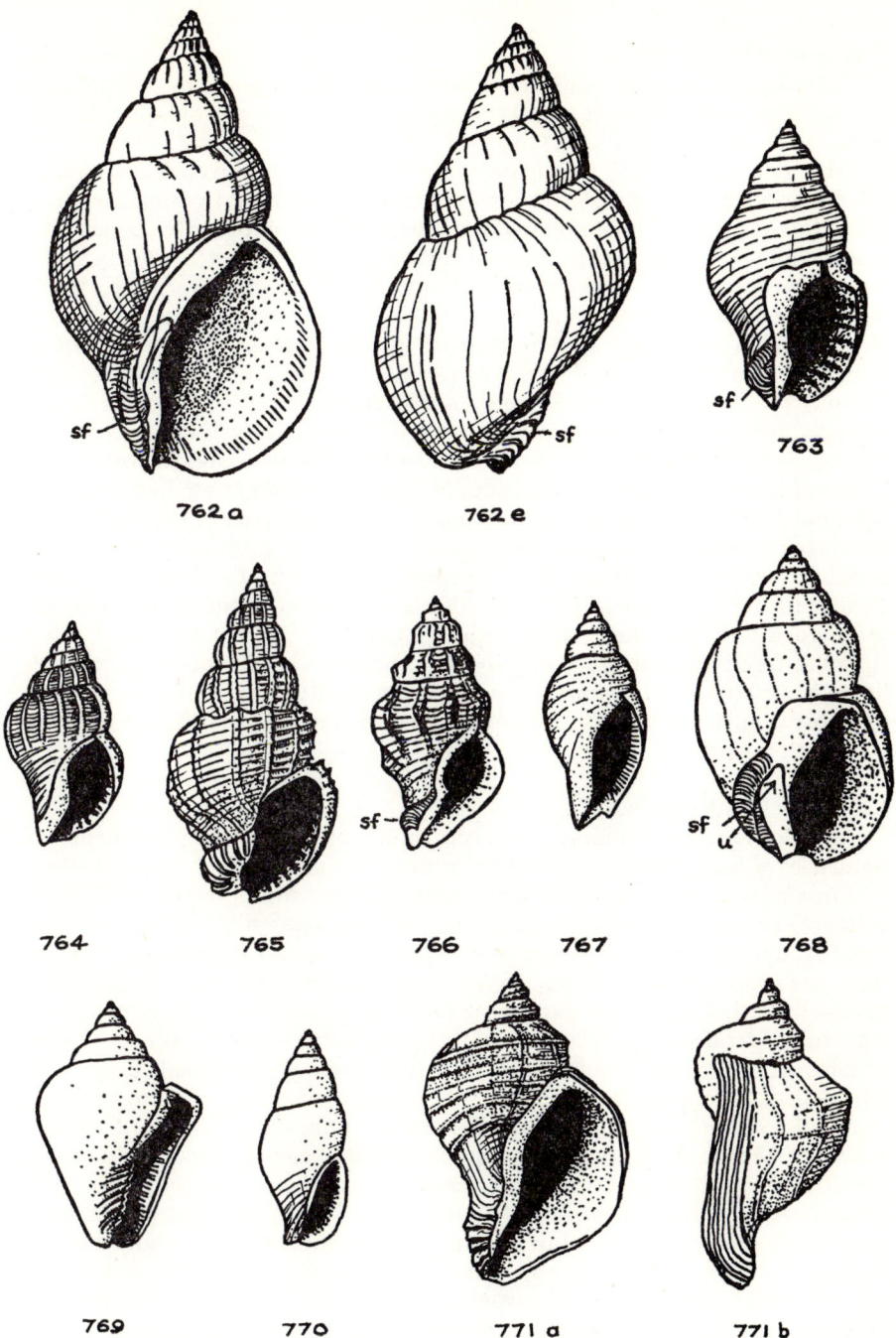

762 a

762 e

763

764

765

766

767

768

769

770

771 a

771 b

357

Subfamily PHOTINAE

Anterior notch rather shallow; siphonal fasciole margined by a depression but no keel; labral profile nearly straight, never strongly oblique. Mostly warm-water forms. (U.Cret.(Danian)-Rec.)

Phos (*s.s.*) (Fig. 765): elongate-oval-conical, more turreted than many other buccinids; whorls convex or angular, with collabral costae and finer spiral threads, the ornament being more delicate than in other buccinids; last whorl with short neck and prominent siphonal fasciole; aperture fairly small, ovate-subrhomboidal, with short anterior canal; labrum varicose, feebly prosocline, internally lirate; anterior part of columella twisted into a prominent fold bounded above by a narrow spiral groove; callus of inner lip often wrinkled. Eoc.-Rec. Indo-Pac., Austral., Japan, Indonesia, Burma, Eur., N.Amer., C.Amer., Carib., S.Amer. *Buccitriton:* like *Phos*, but whorls less strongly convex, sometimes even almost flat-sided, spire lower, occasional varices present, aperture more narrowly oval, anterior canal shorter, and siphonal fasciole less prominent. Eoc. N.Amer. *Tritiaria* (*s.s.*): like *Phos*, but siphonal fasciole low, aperture oval and not so constricted at the start of the canal, parietal fold present, anterior canal wider, and protoconch with some collabral costellae. U.Eoc.-Olig. N.Amer., S.Amer. *Terebrifusus:* rather slender-fusiform, with gently convex whorls; ornament of collabral costae and spiral furrows; last whorl fairly high, oval, with a distinct neck and a low siphonal fasciole; aperture narrowly oval, posteriorly channelled, anteriorly little constricted, with a short, broad anterior canal; labrum straight, smooth or feebly laciniate internally; columellar lip with numerous small folds. M.Eoc. N.Amer., S.Amer.

Subfamily PISANIINAE

Anterior notch only moderately deep; siphonal fasciole not prominent, not limited by a keel. Abundant in warm waters, mainly tropical. (U.Cret.-Rec.)

Pisania (*s.s.*): oval-fusiform, with feebly convex whorls; early whorls with collabral costae and spiral threads, later whorls with spiral threads only or smooth; aperture narrowly oval, with a posterior gutter limited by a parietal fold, anteriorly with a vague, short canal; labrum internally denticulate; columella with a fold at the start of the canal. Eoc.-Rec. Carib., N.Amer., Eur., ?Austral. *Metula* (*s.s.*): fusiform, with gently convex whorls and cancellate ornament; siphonal fasciole weak; aperture narrowly elliptical, posteriorly channelled, anteriorly with a broad, short, notched canal; labrum varicose, internally denticulate; no columellar teeth. Olig.-Rec. Carib., C.Amer., S.Amer., Medit., W.Pakistan, Indonesia. *Celatoconus* is like *Metula*, but is considerably stouter, and the labrum has a tooth at the start of the canal. Eoc.-Mio. N.Amer., Eur. *Cantharus* (*s.s.*): very inflated-fusiform, with convex whorls, strong collabral costae and finer spiral threads; last whorl with an excavate base, a short neck and a moderately developed siphonal fasciole; aperture oval, with a narrow posterior gutter, and a fairly short, notched anterior canal; labrum thickened, internally lirate, profile posteriorly prosocline; columellar lip gently concave, with many small folds. Eoc.-Rec. Indo-

358

Pac., W.Pakistan, India, Burma, Formosa, Japan, Austral., Indonesia, Eur. *Pollia* [*Tritonidea*] (Fig. 766): similar to *Cantharus,* but a little less inflated, costae strong but rather flat, labrum straight and prosocline, inner lip with one or two posterior wrinkles, two columellar wrinkles and several other small folds; anterior canal in continuation of axis of aperture. ?U.Cret., Pal.-Rec. Indo-Pac., Austral., Indonesia, Burma, Asia, E.Africa, Eur., N.Amer., S.Amer. *Suessonia:* oval-fusiform, with axial costae and spiral threads; no siphonal fasciole; aperture narrowly oval, posteriorly feebly channelled, anteriorly with a short, oblique, feebly notched canal; labrum sinuous, feebly varicose, internally finely denticulate; one columellar fold. Eoc.-Plio. Eur., ?N.Amer. *Janiopsis:* more fusiform than *Pollia,* with strong collabral costae and spiral threads of more than one order; siphonal fasciole very weak; aperture oval, posteriorly with a gutter limited by a parietal fold, anteriorly with a moderately short, little notched canal; columella with one or two folds; labrum posteriorly prosocline, internally coarsely denticulate. Eoc.-Plio. Eur., W.Africa, N.Amer.

Subfamily PISANIANURINAE [ANOCHETINAE]

Anterior canal short, wide, truncated; no siphonal fasciole; labral profile usually parasigmoidal, posteriorly prosocline. A small, warm-water group, mainly American. (U.Cret.-Rec.)

Pisanianura: inflated-oval-fusiform, with convex or angular whorls; collabral costae (forming nodes on shoulder when the whorls are angular) and spiral threads; last whorl with excavate base and short neck; aperture broadly oval, with short, truncated, anterior canal; labrum thick, internally smooth; columellar lip thin, smooth; protoconch with fine spiral striae. Eoc.-Rec. Eur., Atl. *Laevibuccinum* (Fig. 767): elongate-oval-fusiform; whorls nearly flat-sided, with canaliculate sutures; base gently excavate; sculpture of fine spiral grooves, stronger below the suture, crossed by fine growth lines producing a punctate appearance; aperture more than half the total height, narrowly oval, pointed behind, with short, broad canal in front; columella smoothly curved; labrum internally denticulate, its profile parasigmoidal. Eoc. N.Amer.

Subfamily BABYLONIINAE [LATRUNCULINAE]

With deep anterior notch and usually a well marked, flat siphonal fasciole; labral profile more or less sinuous, oblique; with or without umbilicus. (Olig.-Rec.)

Babylonia [*Dipsaccus, Eburna, Latrunculus*] (*s.s.*) (Fig. 768): oval-conical, smooth; whorl outline only slightly curved, but with a depressed shelf along suture giving the spire a stepped appearance; aperture oval, with deeply notched posterior gutter corresponding to outer part of sutural shelf and limited by a parietal fold; anterior notch very deep, with raised edges forming a rudimentary canal; columella excavated in a uniform curve, only slightly flattened and bent at anterior notch; siphonal fasciole very flat, margined externally by a very slight keel with depression beyond, and strongly curved so that between it and the columella is a wide area with coarse growth lines leading down to the umbilicus; labrum oblique, more or less sinuous, posteriorly prosocline, internally smooth. Olig.-Rec. Indo-Pac., Japan, Formosa, Indonesia, Burma, India, W.Pakistan, W.Africa, Eur.

Family PYRENIDAE [COLUMBELLIDAE]
Form biconical to oval or turreted; aperture generally oval-subrhomboidal, anteriorly with either a simple notch or a straight canal of varying length; labrum usually nearly straight and vertical, but slightly prosocline on suture, usually thick and crenulated within; columella scarcely excavated, twisted in front, with or without folds and generally with finer wrinkles; whorls smooth, with spiral ornament or also with collabral costae; last whorl more or less high; anomphalous; never attaining a large size, and including many smooth forms. (Eoc.-Rec.)
Pyrene [*Conidea*] (*s.s.*): oval to oval-conical, smooth except for spiral threads on the base of the last whorl; spire with low whorls, usually somewhat cyrtoconoid; whorls almost flat-sided, with distinct sutures; aperture long and narrow, posteriorly channelled, anteriorly with a very short, notched canal; labrum medially more or less thickened, internally denticulate; columella straight, feebly denticulate. Mio.-Rec. Indo-Pac., Eur., N.Amer., warm seas. **Pterygia** [*Columbella*] (*s.s.*) (Fig. 769): biconical, smooth or with close-set, weakly beaded spiral threads; aperture long, narrow, feebly sinuous, practically parallel-sided, an inward bulge of the labrum opposing the excavation of the columella; strong posterior gutter tending to a canal; no true anterior canal, but a twisted notch giving rise to a short, weak siphonal fasciole; columella with two folds and a few denticles above. Mio.-Rec. W.Africa, Indonesia, Eur., N.Amer., C.Amer., Carib., widespread. The subgenus **Alia** differs from *Pterygia* in being subfusiform and in the whorls being somewhat stepped. Mio.-Rec. N.Amer., Indonesia, Eur., warm seas. **Mitrella** (*s.s.*) (Fig. 770): oval-fusiform, with flat-sided whorls, smooth except for spiral threads on the base of the last whorl; labrum internally denticulate, its profile posteriorly gently excavate; aperture suboval, anteriorly without distinct canal, gently notched; columella smooth except for a few weak wrinkles. Mio.-Rec. Medit., Eur., E.Africa, Indonesia, Formosa, Japan, N.Amer., Carib., widespread. **Atilia:** oval-fusiform, with moderately convex, somewhat imbricate whorls, smooth or with collabral costae; aperture narrowly oval-subrhombic, posteriorly channelled, anteriorly with a short, truncated canal; labrum sinuous, rather thickened, internally denticulate; columella smooth. Mio.-Rec. Indo-Pac., Japan, Eur. **Strombina** (*s.s.*): inflated-fusiform, with fairly high spire; ornament of collabral costae which change to *Strombus*-like nodes on the last whorl which has a solid neck; aperture long, narrow, posteriorly with a shallow, extended channel, anteriorly with a short, notched canal; labrum thickened, internally denticulate; inner lip callous. Eoc.-Rec. Pac., N.Amer., C.Amer., Carib., Eur. **Anachis** (*s.s.*): inflated-fusiform, with moderately high, pointed spire and fairly well developed neck; whorls with a distinct shoulder, collabral costae and spiral threads; aperture narrowly oval, posteriorly feebly

360

channelled, anteriorly without distinct canal, merely notched; labrum thickened, internally denticulate; columella straight, folded. ?Eoc., Mio.-Rec. Indo-Pac., Austral., India, Japan, N.Amer., S.Amer., Eur. *Astyris* (*s.s.*): ovalconical, with gently convex whorls, usually smooth except for spiral striae on the base of the last whorl; neck short; aperture short, oval, posteriorly feebly channelled, anteriorly with a very short, vague, deeply notched canal; labrum straight, more or less thickened, internally denticulate; columellar lip narrow, smooth. L.Eoc.-Rec. N.Atl., Iceland, N.Amer., northern seas (Arctic, Celtic, Transatlantic).

Family NEPTUNEIDAE [CHRYSODOMIDAE]

Shell thick, varying between fusiform and bucciniform; protoconch papillate, oblique; whorls convex, ornamented with spiral lines, with or without collabral nodular costae; aperture oval, with or without a posterior gutter, with anterior canal sharply inflected to the left and backwards, not distinctly notched; siphonal fasciole usually weak; labral profile much as in *Buccinum* (basal portion vertical or slightly inclined, lateral portion straight and strongly prosocline); columella smooth, posteriorly excavated, anteriorly twisted. (U.Cret.-Rec.)

In the absence of the protoconch it is not always easy to distinguish some Neptuneidae from Buccinidae or Fusinidae.

Neptunea [*Chrysodomus*] (*s.s.*) (Fig. 771): bucciniform tending to fusiform; whorls rounded, with spiral ornament; last whorl with well-rounded shoulder on which the growth lines become prosocline; aperture oval, with short, inflected anterior canal; columella smooth, gently excavated, with a bend at the origin of the canal; rather weak siphonal fasciole and narrow umbilical depression present. U.Eoc.-Rec. Northern seas, Eur., Iceland, N.Amer., Japan. *Sipho* (*s.s.*): fusiform, more slender than *Neptunea*, with very gently convex whorls bearing weak spiral ornament; aperture narrowly oval, with a moderately long anterior canal; no siphonal fasciole. Olig.-Rec. Eur., Iceland, N.Amer., northern seas. *Parvisipho* (*s.s.*): similar to *Sipho*, but smaller, with very short, abruptly truncated anterior canal, ornamented with fine spiral furrows and occasionally extremely fine collabral costellae; labrum sometimes internally very finely crenulated. Pal.-Plio., ?Pleist. Eur., Japan, N.Amer. *Siphonalia* (Fig. 772): shape rather like that of *Neptunea*, but the anterior canal is strongly twisted and collabral costae or nodules are added to the spiral ornament, the whorls sometimes being bluntly angular; distinct siphonal fasciole, but no umbilical depression; labrum internally sometimes finely furrowed. ?U.Cret., Pal.-Rec. Japan, Formosa, Indo-Pac., Austral., Indonesia, Burma, ?S.Africa, Eur., N.Amer., ?S.Amer. *Coptochetus:* fusiform, with convex whorls, collabral costae and finer spiral furrows; aperture pyriform, with a distinct, moderately short, only slightly curved, truncated anterior canal; labrum straight, thickened, internally folded. Eoc.-Olig. Eur., S.Amer., Austral. *Cyrtochetus:* more bucciniform, with convex whorls and spiral ornament; aperture oval, anteriorly with a short but strongly twisted and notched canal; labrum straight, internally smooth or folded; siphonal fasciole

361

feeble. Eoc.-Mio. Eur., N.Amer., S.Amer. *Loxotaphrus:* form similar to that of *Cyrtochetus*, but whorls angular and latticed by collabral costellae and spiral threads; labrum varicose, internally laciniated; siphonal fasciole distinct, separated from inner lip by an umbilical slit. M.Mio.(Balc.) (*teste* Dr N. H. Ludbrook). Austral. *Buccinulum* (*s.s.*): fusiform, with conical spire; protoconch of two whorls, the first one and a half smooth, the last half whorl with collabral costellae; ornament of spiral threads; aperture oval, posteriorly channelled and limited by a parietal fold, anteriorly with a moderately long, slightly inclined canal; labrum parasigmoidal in profile, internally thickened and folded; columella with one fold at the start of the canal; siphonal fasciole feeble. Olig.-Rec. N.Z. The subgenus *Euthria* has collabral costae at least in the early whorls and may or may not have spiral threads; the anterior canal is a little longer. Eoc.-Rec. Medit., Eur., Africa. *Acamptochetus:* mitriform, with fairly high spire and spiral threads, occasionally also with curved collabral costellae; neck moderately long, without distinct siphonal fasciole; aperture narrowly oval, posteriorly weakly channelled, anteriorly with a fairly short, vaguely defined canal; labrum weakly varicose, internally folded; inner lip smooth. Mio.-Plio. Eur.

Family VOLEMIDAE [MELONGENIDAE, FULGURIDAE, BUSYCONIDAE]

Pyriform to fusiform, whorls usually shouldered; aperture moderately wide, with a moderately long or long anterior canal; columella usually smooth, sometimes with a few folds. (Cret.-Rec.)

Volema (*s.s.*): thick-shelled, pyriform-biconical; feeble spiral ornament, and last whorl may have a noded shoulder; aperture narrowly oval, with a well marked posterior gutter, anteriorly with a broad, short canal; siphonal fasciole present; columellar lip only gently excavated; labrum usually smooth internally, its profile becoming prosocline at the suture. Plio.-Rec. Red Sea, Indo-Pac., Indonesia, E.Africa. The subgenus *Melongena* [*Galeodes*] (Fig. 773) has collabral costae and spiral threads on the early whorls; the last whorl may have a double row of shoulder tubercles or they may die out; base of last whorl with a row of small spines above the siphonal fasciole; aperture a little

---

Figs 772–780. GASTROPODA: NEPTUNEIDAE, VOLEMIDAE AND NASSARIIDAE

Fig. 777 after H. Douvillé; Fig. 780 after Wenz; Figs 772–776 and 778–779 original.

772. *Siphonalia scalarina* (Lamarck), M.Eoc. Paris B. × 1·5.

773a, e. *Volema* (*Melongena*) *melongena* (Linné), Rec. × 0·75.

774. *Cornulina armigera* (Conrad), L. Eoc. Grigg's Landing, Alabama. T. × 0·75.

775. *Bruclarkia barkeriana* (Cooper), M.Mio. (Temblor), California. × 0·75. (A synthetograph).

776. *Pugilina subcarinata* (Lamarck), U.Eoc. Paris B. × 1·1.

777j. *Heligmotenia molli* H. Douvillé, M.Eoc. Tamaské, Sudan. T. × 0·45.

778. *Busycon perversum* (Linné), Rec. Transatlantic. × 0·5.

779a, b. *Sycostoma* sp. between *bulbus* (Solander) and *bulbiforme* (Lamarck), M.Eoc. Paris B. × 0·9.

780. *Nassarius mutabilis* (Linné), Rec. Medit. T. × 1.

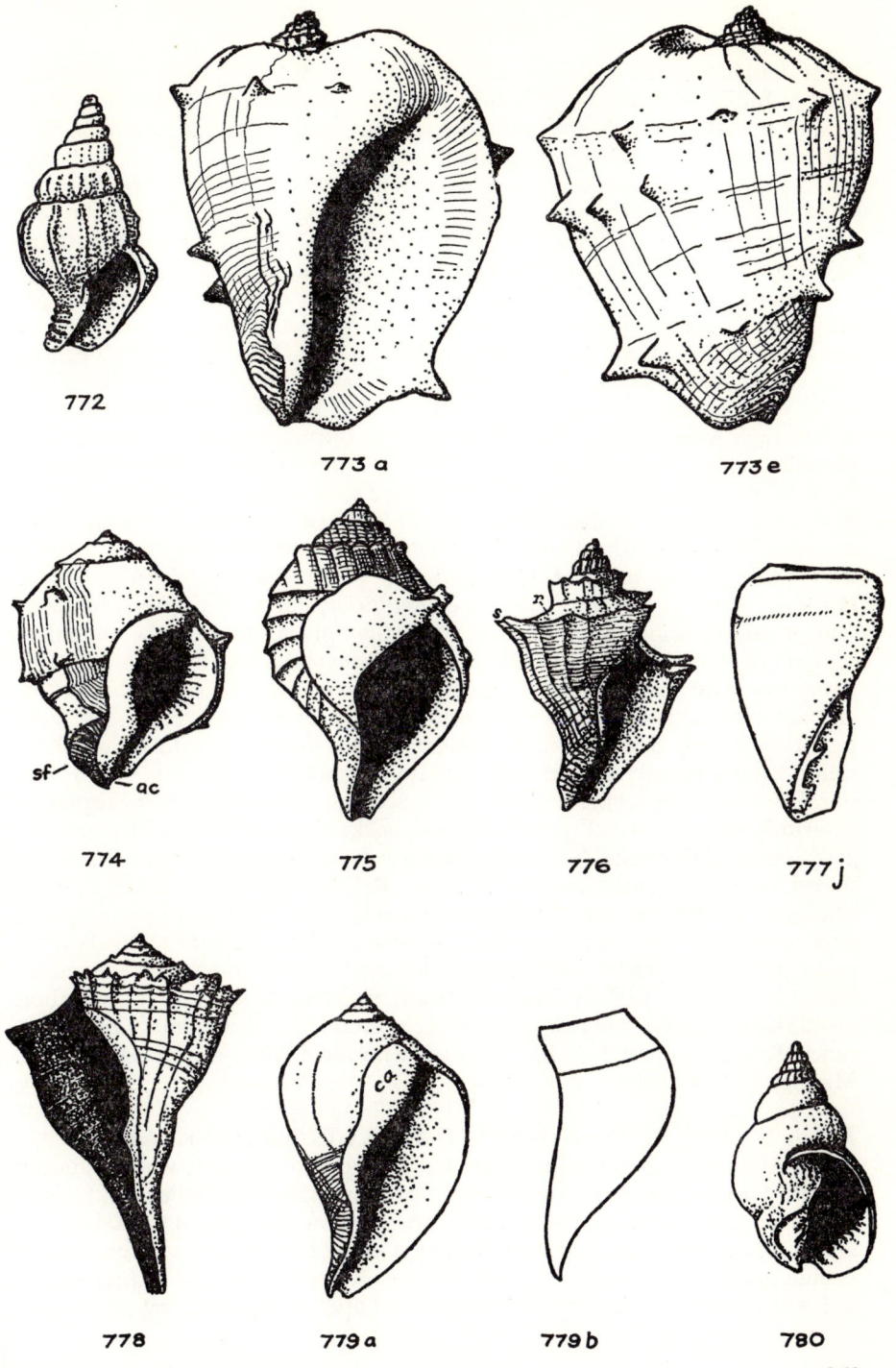

772

773 a

773 e

774

sf

ac

775

776

777 j

778

779 a

779 b

780

wider than in *Volema*. Eoc.-Rec. Carib., N.Amer., S.Amer., Indo-Pac., Japan, Burma, India, W.Pakistan, N.Africa, Eur. The subgenus *Cornulina* (Fig. 774) is more globose, and has a low, conical spire with an angle of about 90°; aperture symmetrically elliptical, anterior canal short and inclined; siphonal fasciole very broad and prominent; labrum internally laciniate, its profile very sinuous above the shoulder; there is a second row of tubercles on the last whorl a little above the level of the start of the canal; below this lower row are two spiral grooves formed by infolds of the mantle (cf. *Pseudoliva* and *Acanthina*). Eoc. N.Amer., S.Amer., W.Africa, Eur. *Bruclarkia* (Fig. 775): similar to *Cornulina*, but with a rather longer canal and three or four rows of small tubercles on the last whorl. U.Eoc./L.Olig.-Mio. N.Amer. *Pugilina* (Fig. 776): like *Volema*, but all whorls with an angular, spinose shoulder; canal well-defined, and aperture a little wider. Pal.-Rec. W.Africa, Carib., N.Amer., S.Amer., Eur., W.Pakistan, India, Burma, Indonesia, warm seas. *Hemifusus:* like *Pugilina*, but with a longer anterior canal and no siphonal fasciole. Pal.-Rec. Indo-Pac., Formosa, Japan, N.Amer., Indonesia, Burma, W.Pakistan, Egypt. *Heligmotenia* (Fig. 777): large, pyriform, with flat spire and shouldered last whorl which also has a keel; aperture narrow; columella with three strong folds. U.Cret. (Maastr.)-Eoc. Sudan, W.Africa, Asia. *Busycon* [*Fulgur*] (*s.s.*) (Fig. 778): like *Hemifusus*, but with a solid columellar fold, a slightly wider aperture, and a usually spinose keel on the shoulder; sometimes sinistral. Olig.-Rec. N.Amer. *Heligmotoma:* like *Busycon*, but with a very low spire and no spines, smooth, aperture with a broad posterior gutter, and anterior canal nearly as wide as the aperture; no columellar fold. Eoc. Egypt, N.Africa, S.W.Africa. *Levifusus:* inflated-fusiform, with rather short spire; whorls usually shouldered, dentate or nodose at the angle, with fine spiral threads; last whorl usually with a second noded keel, base excavated, anteriorly with a distinct, gently inclined neck; aperture oval, with a moderately long anterior canal; columella rather twisted, smooth; labrum internally folded, its profile gently sinuous. ?U.Cret., Pal.-Rec. N.Amer., S.Amer., Asia, Madagascar, Africa, Eur. *Sycostoma* [*Sycum*] (Fig. 779): pyriform, smooth, with somewhat coeloconoid spire; last whorl large; aperture elongate-oval, posteriorly with a feeble gutter, anteriorly with an indistinctly set off, rather short, broad and slightly inclined canal; siphonal fasciole present; labral profile little sinuous, slightly opisthocline posteriorly; labrum internally sometimes folded; columellar lip callous, without folds. U.Cret.-Olig. Eur., Madagascar, N.Amer.

Family NASSARIIDAE [NASSIDAE, ALECTRIONIDAE]

Usually of small to medium size, bucciniform to turreted; whorls smooth or with spiral or collabral ornament or both; aperture rounded or oval, with a very deep and oblique notch or short canal bent at an angle to the axis of the aperture; columella usually truncated by a very obliquely twisted fold; labral profile practically straight, slightly oblique. Recent forms are further distinguished from Buccinidae by their radula and operculum. (?U.Cret., Pal.-Rec.)

Subfamily NASSARIINAE [NASSINAE]

Aperture contracted; columellar fold strongly twisted. (?U.Cret., Pal.-Rec.)

Nassarius [*Nassa*] (*s.s.*) (Fig. 780): bucciniform, with moderately high spire; whorls convex, early ones with collabral costae, later ones smooth or with very fine incised spiral lines; last whorl oval, anteriorly with a short neck bearing a siphonal fasciole; aperture oval, posteriorly with a narrow gutter, anteriorly truncated and deeply notched; labrum moderately oblique, internally finely denticulate, its profile posteriorly prosocline; columella concave, ending in a fold forming margin of notch; columellar callus fairly widely spread posteriorly. Mio.-Rec. Medit., Eur., W.Africa, India, N.Z., Indo-Pac. *Demoulia* [*Desmoulea*]: very globose, all whorls ornamented with spiral furrows; aperture subcrescentic owing to large convex parietal portion; columella straight, with strongly twisted terminal fold; neck very short and bounded by a deep groove; labrum internally lirate; parietal region and columella with lirae. Mio.-Rec. Eur., W.Africa, Indo-Pac. *Arcularia* (*s.s.*): inflated-oval-conical, with low spire, smooth; no neck; aperture small, rounded-oval, posteriorly with a channelled gutter, anteriorly narrowly and deeply notched; labrum broadly swollen, internally smooth; columella limited anteriorly by a fairly strong fold; columellar callus very widely spread over ventral side and joined to the labral callus. Mio.-Rec. Medit., Eur., Indo-Pac., ?N.Amer. *Cyclope* [*Cyclops, Cyclonassa*] (*s.s.*): neritiform, smooth, with very low spire; last whorl large; aperture very oblique, rounded-rhombic, with posterior channel, anteriorly truncated and notched; labrum very oblique, internally smooth; columellar callus spreading over a large part of the shell; columella smooth. Plio.-Rec. Medit., Eur. *Phrontis* (Fig. 781): oval-conical, with relatively low spire; whorls convex, shouldered, with solid collabral costae; base of last whorl with spiral threads and a siphonal fasciole limited by a furrow; aperture oval, posteriorly with a channel-like gutter limited by a parietal fold, anteriorly truncated and deeply notched; labrum varicose, internally lirate, its profile prosocline; columellar lip smooth or finely wrinkled, its anterior end with a fold. Mio.-Rec. Indo-Pac., Japan, Indonesia, Africa, Eur., N.Amer. *Hinia* [*Hima*] (*s.s.*): oval-conical, with nearly flat-sided whorls bearing strong collabral costae and spiral furrows; siphonal fasciole limited by a furrow; columella with an anterior fold with a few small folds above it; labrum thickened, internally dentate, its profile prosocline; aperture oval, posteriorly with a gutter, anteriorly notched; columellar callus more widely spread posteriorly. Eoc.-Rec. Eur., Iceland, W.Pakistan, India, Indonesia, Indo-Pac., Formosa, Japan, Austral., N.Z., C.Amer., widespread. *Alectrion* (*s.s.*): elongate-oval-conical, with convex whorls and deep sutures; at least the early whorls with collabral costae frilled by spiral threads; last whorl oval, without distinct neck but with a siphonal fasciole; aperture broadly oval, not constricted anteriorly, posteriorly with a deep and narrow gutter, anteriorly deeply notched; labrum not varicose, its edge toothed, internally lirate; columella concave, with an anterior fold and wrinkles above it; parietal fold present. Eoc.-Rec. Indo-Pac., Japan, N.Z., Indonesia, N.Amer., S.Amer.

365

Cyllene (*s.s.*): oval-conical, with fairly low spire; whorls not very convex, with a fine sutural swelling posteriorly, ornamented with collabral costae (sometimes forming a nodose shoulder) and spiral threads both of which may become obsolete on the last whorl; aperture oval, with a posterior gutter and a very short, deeply notched anterior canal; labrum solid, internally lirate, its profile slightly opisthocline posteriorly; columellar lip callous and with weak wrinkles; siphonal fasciole present, with a furrow and a keel above it. [Some conchologists have placed this genus in the subfamily Cominellinae of the family Buccinidae.] Mio.-Rec. W.Africa, Indo-Pac., Indonesia, Burma, India, Eur.

Subfamily BULLIINAE [DORSANINAE]
Columella little twisted; aperture with a wide anterior notch. (Pal.-Rec.)
Bullia (*s.s.*): ovate to elongate, though never truly turreted, the last whorl usually being greater than half the height; whorls more or less convex, smooth (rarely with a few collabral costae in the early stages), the sutures more or less buried in enamel; aperture oval, posteriorly with a gutter, anteriorly broadly notched; labral profile feebly arcuate, the labrum itself thin and internally smooth; columella gently concave, smooth, slightly bent to the left anteriorly; strong siphonal fasciole margined by a keel; columellar lip callous, especially posteriorly. Eoc.-Rec. S.Africa, N.Amer. *Brachysphingus:* inflated-oval, with very short spire, completely smooth or with vague collabral costae on the last whorl; aperture with deep anterior notch giving rise to a broad siphonal band, posteriorly with a narrow gutter; labrum internally smooth; labral profile rather sinuous. Pal.-U.Eoc. N.Amer., Eur. *Dorsanum* (*s.s.*) (Fig. 782): elongate-oval-conical; whorls smooth, or with spiral striae, or with a row of crenulations at the suture and collabral costae (which may give rise to short spines posteriorly) below; last whorl with spiral furrows on the base and a very short neck; aperture broadly oval, posteriorly with a gutter, anteriorly broadly notched and without canal or with only a very short one; labrum rather thin, usually smooth internally; anterior end of columella with a twisted fold. Evidently related to *Bullia*. Mio.-Rec. W.Africa, Eur., S.Africa, Indonesia, Indian Ocean, S.Amer. Doubtfully included in the subfamily is *Molopophorus* (Figs 783, 784): oval, with whorls flattened and compressed posteriorly,

---

Figs. 781–791. GASTROPODA: NASSARIIDAE, FUSINIDAE AND VOLUTIDAE
Figs 781–791 original.
781. *Phrontis arcularia* (Linné), Rec. × 1.
782. *Dorsanum baccatum* (Basterot), L.Mio. Pont Pourquey, Gironde, France. × 1.
783. *Molopophorus anglonana* (Anderson), L.Mio. (Astoria). Oregon. × 1.
784b. *Molopophorus lincolnensis* Weaver, L.Olig. Chahalis V., Washington. × 1·9.
785. *Fusinus porrectus* (Solander), U.Eoc. Barton, Hampshire. × 1·25.
786a, b. *Streptochetus heptagonus* (Lamarck), M.Eoc. Paris B. × 1.
787. *Pleuroploca rugata* (Tate), M.Mio. (Balcombian). Muddy Creek, Victoria. × 1·25.
788. *Euthriofusus burdigalensis* (Grateloup), L.Mio. Gironde, France. T. × 0·75.
789. *Clavilithes* (*Clavellofusus*) *conjunctus* (Deshayes), M.Eoc. Paris B. × 0·75.
790f. *Clavilithes* (*Clavellofusus*) *tuberculosus* (Deshayes), M.Eoc. × 1·9.
791a, e. *Voluta musica* Linné, Rec. W.Africa. T. × 0·9.

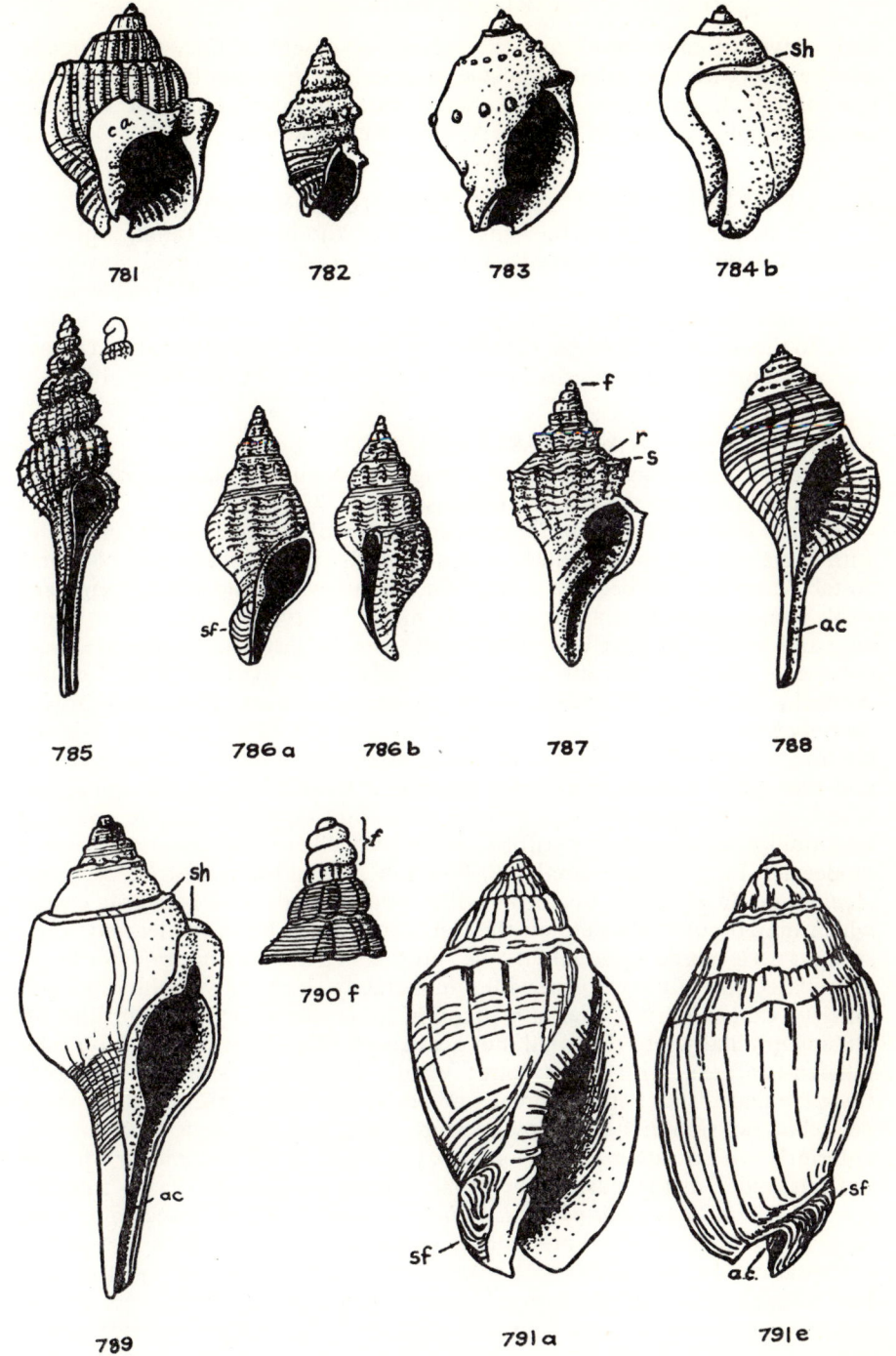

781

782

783

784 b

785

786 a

786 b

787

788

789

790 f

791 a

791 e

367

and a layer of callus bordering the suture; ornament of fine spiral threads usually also with fine collabral costae or costellae; siphonal fasciole present; aperture oval, anteriorly with a fairly broad notch; columella smooth; labrum simple. M.Eoc.-M.Mio. N.Amer., ?Eur., ?Japan.

Family FUSINIDAE [FUSIDAE, FASCIOLARIIDAE]

Shell fusiform through the lengthening of the anterior canal, which is straight (or only slightly curved) in continuation of the shell axis, and never deeply notched at the end; columella smooth or with folds; labral profile sometimes with forward bend above shoulder but usually recovering verticality close to the suture; no varices; usually with spiral threads and collabral costae, rarely smooth; labrum internally smooth or folded. (U.Cret.-Rec.) **(102)**

Subfamily FUSININAE [FUSINAE]

Anterior canal usually very long, straight or slightly twisted; nearly always without columellar folds or siphonal fasciole. (U.Cret.-Rec.)

Fusinus [*Fusus*] (*s.s.*) (Fig. 785): protoconch globose; shell rather elongate, with high, conical spire; whorls rounded, with spiral threads and collabral costae which tend to become weaker on the last whorl and which form very weak spines when the last few whorls medially become subangular; canal long and straight or nearly so, well set off from the last whorl; aperture oval; columella smooth; labrum sharp, internally lirate. U.Cret.-Rec. Ceylon, Austral., Indonesia, Formosa, India, Burma, N.Africa, Eur., widespread. **Streptochetus** (*s.s.*) (Fig. 786): fusiform, but shorter and stouter than *Fusinus*, tending more towards a buccinoid shape; ornament of noded collabral costae and fine spiral threads, the costae forming a twisted line up the spire; last whorl with a moderately long, oblique neck and a siphonal fasciole revealing an umbilical slit; aperture pyriform, posteriorly channelled, anteriorly with a moderately long, oblique canal. Pal.-Plio. Eur., W.Pakistan, Austral., N.Amer. **Aquilofusus** (*s.s.*): short-fusiform, with spiral threads and collabral costae which may become obsolete on the last whorl which has a moderately long, straight neck but no siphonal fasciole; protoconch at first smooth, later finely cancellate; aperture pyriform, with a moderately short, only slightly inclined anterior canal; labrum internally smooth, its profile sinuous; columellar lip passing gradually into the canal, not geniculate. Olig.-U.Mio. Eur. **Lirofusus:** inflated-fusiform, with short spire; whorls imbricate, with fine collabral costellae and distinct spiral threads or keels; last whorl with a rather short neck but no siphonal fasciole; aperture oval, not channelled posteriorly, anteriorly with a rather short, somewhat inclined canal; columellar lip bent but not folded at the start of the canal; labrum straight, internally smooth. Eoc. N.Amer. **Buccinofusus** [*Troschelia*]: inflated-fusiform, buccinoid, with moderately high spire; whorls convex, with spiral threads, with or without collabral costae; last whorl with a moderately long, somewhat twisted neck but no siphonal fasciole; labrum internally sometimes folded, its profile sinuous; apertue pyriform, with a moderately long, oblique anterior canal;

columella bent but not folded at the start of the canal. U.Cret.-Rec. N.Amer., S.Amer., Eur., W.Africa, Indonesia, Atl., Arctic.

Subfamily FASCIOLARIINAE

Anterior canal more or less inflected; columella nearly always folded; siphonal fasciole present or absent. (U.Cret.-Rec.)

Fasciolaria (*s.s.*): inflated-fusiform, often quite large, smooth (at least in later whorls); whorls convex, the last one with a twisted neck but no proper siphonal fasciole; aperture rather narrowly oval, posteriorly with a gutter, anteriorly with a moderately short, inclined canal; labrum internally lirate, its profile a little opisthocline at the suture; columella with one fold and one or two weaker ones above it; inner lip with vague callus. U.Cret.-Rec. Carib., N.Amer., C.Amer., S.Amer., ?Madagascar, ?Formosa. *Pleuroploca* (*s.s.*) (Fig. 787): fusiform; whorls angulated and with a ramp, ornamented with spiral threads and collabral costae which are spinose at the shoulder and obsolete on the ramp; last whorl with a fairly long neck and a siphonal fasciole; aperture narrowly pyriform, posteriorly with a gutter limited by a parietal fold, anteriorly with a fairly long, twisted, inclined canal; labrum internally lirate, its profile fairly straight, prosocline at the suture; columella with three folds; inner lip narrow, fairly callous. Eoc.-Rec. Indo-Pac., Austral., India, Eur., N.Amer. *Latirus* [*Lathyrus*] (*s.s.*): similar to *Pleuroploca*, but collabral costae solid and blunt, columellar folds less oblique, labral profile straight and vertical, and there is an umbilicus or umbilical slit. ?U.Cret., Eoc.-Rec. Austral., Indo-Pac., Japan, Indonesia, Burma, W.Pakistan, S.Africa, Eur., N.Amer., S.Amer. *Exilifusus* (*s.s.*): rather slender-fusiform, with fairly high spire; whorls convex, with rather fine collabral costae and spiral threads; last whorl with a fairly long, straight neck but no siphonal fasciole; aperture broadly oval, anteriorly with a distinct, long, narrow, straight canal; columella with two folds. Eoc. N.Amer., Eur. The subgenus *Dolicholatirus* differs from *Exilifusus* in having broader collabral costae, and the labrum is internally lirate. Pal.-Rec. Eur., N.Amer. The subgenus *Pseudolatirus* is like *Dolicholatirus*, but the whorls are angulated and with a smooth ramp, the spiral ornament is stronger, the collabral costae form low, small spines on the shoulder, and the labral profile becomes prosocline at the suture. Pal.-Plio. Eur., W.Pakistan, Indonesia, N.Amer. *Euthriofusus* (*s.s.*) (Fig. 788): inflated-fusiform, with rather short spire; early whorls angulated, with a long, steep ramp (at about 45°) above the shoulder, ornamented with subequal spiral threads and, below the shoulder, collabral costae; on later whorls the shoulder and collabral costae degenerate and finally disappear; last whorl with a long, straight neck but no siphonal fasciole; aperture pyriform, posteriorly with a gutter limited by a parietal fold, anteriorly with a long, perfectly straight canal (not quite as long as in *Fusinus*): labrum gently parasigmoidal in profile, internally with short lirae; columella subfolded at the start of the canal. Eoc.-Plio. Eur., W.Pakistan, India, Burma, Indonesia, N.Amer., S.Amer. *Clavilithes* [*Clavella*, *Rhopalites*]: fusiform, with relatively short spire, sometimes fairly large; protoconch low, of about two smooth whorls; early whorls with

369

nodes and spiral threads, but the nodes die out on later whorls; last whorl with a fairly long, straight neck and a slight umbilical depression, but no siphonal fasciole, and posteriorly (near the aperture) tending to form a small shelf; aperture pyriform, with a fairly long, straight anterior canal, posteriorly with a long, deep gutter; labrum internally smooth; columellar lip more callous posteriorly; columella on early whorls with two folds, but these later become obsolete. Eoc. Eur., N.Amer. The subgenus *Clavellofusus* (Figs 789, 790) is similar to *Clavilithes*, but the protoconch is cylindrical and consists of about three whorls, the last whorl lacks even the spiral ornament, the neck is more slender and lacks the umbilical depression, and the columella on the last whorl may have a small swelling or one or two low folds. Pal.-Plio. Eur., Egypt, W.Pakistan, India, Burma, Indonesia, Indo-Pac., Austral., N.Amer., S.Amer. The subgenus *Cosmolithes* is like *Clavellofusus*, but the last whorl is a little less inflated, both spiral and collabral ornament tend to persist on to the last whorl, and there are two quite strong columellar folds. Eoc. Eur. *Thersitea:* thick-shelled, inflated-fusiform, sometimes quite large and with irregular swellings; spire rather short, pointed; whorls smooth, more or less imbricate, sometimes carinate; aperture narrowly oval, posteriorly channelled, anteriorly rather twisted and with a short, straight canal; labrum convex in profile, becoming opisthocline posteriorly; columella concave, its callus widely spread. Eoc. N.Africa, Carib.

Superfamily VOLUTACEA

Oval to fusiform; many smooth forms, but others with collabral and spiral ornament; aperture anteriorly with a fairly long canal or only simply notched; columella usually with folds. (Cret.-Rec.)

Family VOLUTIDAE

Shell oval, biconical or fusiform, often of large size; protoconch smooth, varying greatly in size in different genera; aperture elongate, generally rather narrow, anteriorly more or less notched; labral profile approximately straight and vertical for the greater part of its course; columella not much excavated, callous, usually with a variable number of folds; siphonal fasciole very variable; many smooth forms, some wholly or partly with collabral costae and spiral threads, some with a spinose shoulder; in most forms the last whorl is relatively rather large. (U.Cret.-Rec.) **(89a)**

Subfamily VOLUTINAE

Ovate or strombiform, spire not very high; protoconch turbinate; mainly smooth, or with collabral costae which may form nodes or spines on the shoulder; columellar folds usually strong, passing parietally into lirations; anterior notch deep, forming a strong siphonal fasciole; labrum with a posterior sinus. (Eoc.-Rec.)

Voluta itself (Fig. 791) is Recent only. *Pseudaulicina:* like *Voluta*, but the four columellar folds are equal; thick-shelled, biconical; protoconch of four smooth whorls; teleoconch whorls convex, more or less angular, with spiral threads and collabral costae which form small spines on the shoulder when present; aperture rather narrow, posteriorly with a deep gutter, anteriorly wider and with a deep notch; labrum internally smooth; columella callous, with four slightly oblique, equal folds, with or without smaller folds above them, and ending in a twisted beak bounding the anterior notch; strong siphonal fasciole. Eoc. Eur., ?N.Amer.

Subfamily VOLUTILITHINAE

Fusiform, with shouldered whorls and fairly high spire; protoconch large, the apical whorl pointed; ornament of strong collabral costae which become spinose at the shoulder, with or without spiral threads; columella with one strong fold, sometimes with a few weak folds above it; anterior notch deep, forming a strong siphonal fasciole. (U.Cret.-Eoc.)

Volutilithes [*Eopsephaea*] (Fig. 792): spire higher than in *Voluta*; columella with a solid anterior fold and three or four feeble ones deep inside above it; spines on shoulder small, and spiral ornament almost invisible. U.Cret.-Eoc. Eur., N.Africa, Austral., W.Pakistan, Burma, N.Amer.

Subfamily ATHLETINAE

At first subfusiform with high spire and cancellate ornament, sometimes becoming *Cassis*-like or strombiform in the adult; parietal callus sometimes spreading on to spire; protoconch not large; last whorl with a rounded or angulated shoulder sometimes bearing nodes or spines; may become partly or wholly smooth in the adult; anterior canal straight, the notch shallow; no strong siphonal fasciole; one or more columellar folds. (U.Cret.-Rec.)

Athleta (*s.s.*): inflated-oval-conical, with short, coeloconoid spire; last whorl large, without distinct neck, ornamented with fine spiral furrows and a row of small spines along the shoulder; aperture narrow, with a deep gutter posteriorly, anteriorly fairly deeply notched; labrum externally thickened, internally denticulate; columellar lip only slightly concave, with three strong folds and one or two smaller ones above. Pal.-Mio. Eur., W.Pakistan, N.Amer. *Volutocorbis* (Fig. 793): less inflated and more fusiform than *Athleta*, with scabrous reticulate ornament; one strong columellar fold and three or four weaker ones above; labrum more of less reflected, internally denticulate. U.Cret.-Rec. N.Amer., S.Amer., Eur., Africa, W.Pakistan, Burma. *Volutospina* (Fig. 794): differs from *Volutocorbis* in lacking cancellate ornament, and in the callus of the inner lip being less widely spread; ornament consisting of sharp costae forming a crown of spines on the shoulder, and spiral threads; aperture posteriorly with two channels; labrum internally smooth. U.Cret.-Rec. Eur., Africa, Madagascar, India, W.Pakistan, Burma, Austral., N.Amer., S.Amer. *Neoathleta:* like *Volutospina*, but protoconch with four whorls, shell less pointed anteriorly, and second row of tubercles above those on the shoulder. Pal.-Mio. Eur., W.Pakistan, Burma.

Burnett Smith has traced the evolution of *Volutocorbis limopsis* of the Palaeocene of Alabama into a species now assigned to *Volutospina* (*V. petrosa*), which passes through a *Volutocorbis* stage with reticulate ornament

371

and acquires an angular shoulder with spines, while the vertical costae lose their distinctness and the spirals die away. This species persists with little change through Lower and Middle to Upper Eocene, but gives off several side-shoots (e.g. *V. tuomeyi* in Lower and *V. sayana* in Upper Eocene), which acquire new characters regarded as senile (phylogerontic), such as thickening of the shell, irregularity of growth lines, tendency of shoulder-spines to flatten horizontally and unite into a keel, and extension of callus upwards from posterior angle of aperture to envelop more and more of the spire (cf. the contemporary *Calyptraphorus* among Strombidae). Senile forms of this kind have received the generic name *Athleta* (*s.s.*), but they are evidently independent stocks since they show the senile characters in very different proportions. It also seems probable that the numerous species called *Volutospina* may have evolved from different *Volutocorbis* ancestors.

Subfamily LYRIINAE

Not large, thick-shelled, suboval with distinct sutures; usually costate, sometimes becoming smooth in adult; two to three columellar folds, parietal region above smooth or lirate; labrum straight, internally smooth or dentate; anterior notch deep, forming a short siphonal fasciole. (U.Cret.-Rec.)

Lyria (*s.s.*) (Fig. 795): elongate-oval, costate; protoconch small; a few spiral furrows on base of last whorl; aperture fairly narrow, with a posterior gutter; inner lip with three columellar folds and numerous wrinkles above; labrum thickened, internally smooth. U.Cret.-Rec. Austral., N.Z., Indo-Pac., Japan, Indonesia, Burma, India, W.Pakistan, Africa, Madagascar, Eur., N.Amer., S.Amer. *Mitreola* (Fig. 796): oval-mitriform, stouter than *Mitraria*; whorls slightly concavo-convex, convex part with spiral ornament and sometimes also a few feeble elongate nodes; four columellar folds; siphonal fasciole weak; labrum thickened, internally smooth except for one denticle just above the middle of the aperture. Pal.-Mio. Eur., ?Austral.

Subfamily FULGORARIINAE

Oval to fusiform, last whorl narrow or much inflated; protoconch small to

---

Figs 792–801. GASTROPODA: VOLUTIDAE, STEPSIDURIDAE AND OLIVIDAE

Fig. 799 after H. Douvillé; Figs 792–798 and 800–801 original.

792. *Volutilithes torulosus* (Deshayes), M.Eoc. Paris B. × 0·75.
793. *Volutocorbis scabriculus* (Solander), U.Eoc. Barton, Hampshire. × 1.
794. *Volutospina spinosa* (Lamarck), M.Eoc. Paris B. × 0·75.
795. *Lyria decora* Beyrich, L.Olig. Lattorf, Germany. × 0·9.
796. *Mitreola labratula* (Lamarck), M.Eoc. Paris B. × 1·5.
797. *Scaphella lamberti* (J. Sowerby), Pleist. (Red Crag). Suffolk. × 0·45.
798. *Caricella pyruloides* (Conrad), M.Eoc. Claiborne, Alabama. × 1.
799c, j, j'. *Eovasum soudanense* H. Douvillé, M.Eoc. Tamaské, Sudan. × 0·5. j, j', two views of internal cast.
800a, b. *Strepsidura* (*Strepsiduropsis*) *multispirata* R. B. Newton, U.Eoc. Ameki, S.Nigeria. × 1·5.
801a, e. *Oliva* sp., Rec. × 0·6.

372

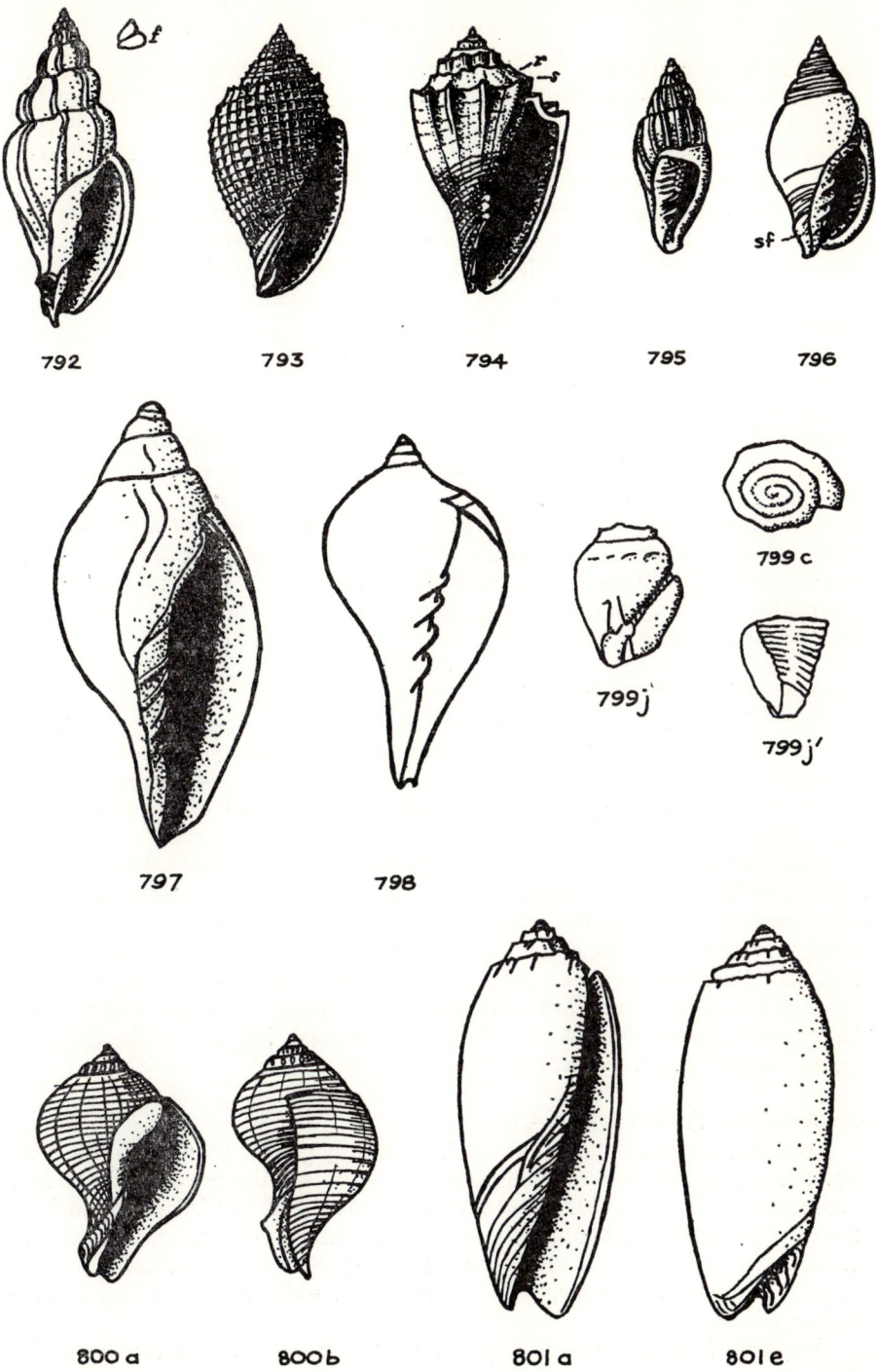

792   793   794   795   796

797   798   799 j   799 c   799 j'

800 a   800 b   801 a   801 e

373

bulbous, asymmetrical or with inclined axis, of few whorls; smooth or ornamented; anterior canal distinct; notch shallow, not forming a siphonal fasciole; labrum usually smooth internally. (Eoc.-Rec.)

Fulgoraria itself is Recent only. *Pterospira:* large, inflated-oval-fusiform;
protoconch large, horny, of one and a half whorls; whorls convex, with collabral costae and spiral threads, the costae becoming obsolete on the last
whorl which is large and inflated; aperture broad, widened and subalate posteriorly; labrum thin, internally smooth, posteriorly reaching the level of the
penultimate whorl; no anterior notch; three strong columellar folds. Eoc.-Rec.
Austral.

Subfamily CYMBIINAE

Sometimes large, last whorl usually broadly suboval; protoconch a callous,
truncated plug or turbinate-pupiform; anterior notch deep and forming a
strong siphonal fasciole; smooth or with short spines. (U.Cret.(Maastr.)-
Rec.)

Cymbium [*Yetus*]: protoconch a callous, truncated plug; last whorl smooth,
embracing all earlier whorls, even sometimes the protoconch; aperture with a
broad posterior gutter, anteriorly with a broad, deep notch; labrum thin,
internally smooth, vertical in profile. Eoc.-Rec. Indo-Pac., Indonesia,
W.Pakistan, N.Africa, Eur.

Subfamily ALCITHOINAE

Oval-biconical to subfusiform, smooth or with collabral costae sometimes
reduced to spines on the shoulder; protoconch variable; last whorl without a
marked basal contraction; anterior canal short, straight; anterior notch deep,
usually forming a strong siphonal fasciole; columella with two or more folds.
(Eoc.-Rec.)

Alcithoe: oval-fusiform, with moderately large protoconch; whorls gently
convex or shouldered, smooth or with collabral costae; columella with four
folds not visible from the exterior; labrum internally smooth, its profile
straight except for a small opisthocline posterior portion. Eoc.-Rec. N.Z.,
Austral., Pac. *Pachymelon:* short-fusiform, with large protoconch; whorls
more or less convex, with low collabral costae often becoming obsolete on the
last whorl; siphonal fasciole low; labrum thickened, internally smooth; columella with five to six distinct folds. Mio.-Rec. N.Z.

Subfamily SCAPHELLINAE

Oval-fusiform, with height of aperture usually greater than height of spire;
protoconch either with a membranous first part followed by a secondary
protoconch of irregular form and with a truncated apex, or fully calcified from
the start and turbinate; columella smooth or with two or more folds; anterior
canal short or moderately short, the notch shallow or deep; base of last whorl
usually not contracted; early whorls sometimes with collabral costae or cancellate. (U.Cret.-Rec.)

Scaphella [*Maculopeplum*] (Fig. 797): large, smooth except sometimes for the first two teleoconch whorls, thick-shelled; protoconch large, ending in an oblique point; elongate-oval-fusiform, with fairly short spire; four strong, equally spaced, oblique columellar folds; anterior canal moderately long. U.Cret.-Rec. Carib., N.Amer., Eur., Burma. *Caricella* (Fig. 798): differs from *Scaphella* in its shorter and coeloconoid spire, its more swollen upper part of last whorl, its longer and more distinct anterior canal, and its more prominent four columellar folds; protoconch scaphelloid; outline pyriform; first one or two teleoconch whorls with spiral ornament. U.Cret.-Eoc. N.Amer., Carib., ?N.Africa, India.

Subfamily CALLIOTECTINAE

Fusiform with more or less slender spire which is higher than the height of the aperture; nearly smooth or with sometimes curved costellae or costae and sometimes also spiral threads; anterior canal moderately long; columella usually straight, smooth or with weak folds which cannot be seen from the exterior. (Plio.-Rec.)

Calliotectum: oval-fusiform, with moderately convex whorls bearing closely-spaced collabral costae which become obsolete on the last whorl; small neck but no siphonal fasciole; aperture elongate-oval, without distinct canal, not notched; columella smooth. Plio.-Rec. S.Amer.

Subfamily ODONTOCYMBIOLINAE [ADELOMELONINAE]

Suboval, with a short anterior canal; columella folds few, slender, the anterior one stronger; smooth or with weak collabral costae and spiral striae. (Olig.-Rec.)

Miomelon [*Proscaphella*]: oval-fusiform, with convex whorls bearing a narrow depressed zone near the posterior suture, ornamented with weak collabral costae and spiral furrows; aperture small, moderately wide, anteriorly feebly notched; siphonal fasciole feeble; columella with three oblique folds; protoconch of several smooth whorls, with pointed apex. ?Olig.-Rec. S.Amer.

Subfamily VOLUTODERMINAE

Elongate-fusiform to broadly-fusiform with short spire; last whorl large, its base not contracted and passing gradually into the long, straight neck; anterior canal long, straight, deeply notched but not always forming a siphonal fasciole; columella straight, with one or more folds; ornament usually strongly cancellate or *Ficus*-like, in some cases the spirals forming strong cords in the adult. (Cret.-Rec.)

Carota: protoconch rather large, tilted; fine spiral ornament, and coarse nodes on the shoulder; last whorl elongate, anteriorly gently curved; labral profile with a deep notch at the shoulder; two or three columellar folds. U.Cret.-Rec. N.Amer.

Subfamily VOLUTOMITRINAE

Small, *Mitra*-like. (Plio.-Rec.)

Volutomitra is Recent only. **Microvoluta:** four equally strong columellar folds; aperture not notched anteriorly. Plio.-Rec. N.Z., Austral.

The subfamily relationships of the following three volutid genera are doubtful:

Indovoluta: inverted-conical with a flat or low spire, smooth; protoconch fairly large, of a few smooth whorls; no siphonal fasciole; aperture high, narrow and parallel-sided; columellar lip with five to eleven folds; labrum externally varicose, its profile with a deep U-shaped sinus above the rounded shoulder. Pal.-U.Eoc. W.Pakistan, India, Burma, Tibet. [Members of this genus were previously included in the Cretaceous *Gosavia* from which, however, they differ in having a varicose labrum and in lacking the scabrous reticulate ornament.] *Leptoscapha:* elongate-oval, with *Lyria*-like protoconch; weak costae on the early part of the teleoconch rapidly die out, and the greater part of the shell is ornamented with fine spiral striae only; columella with four folds, the uppermost the weakest; siphonal fasciole present; aperture narrowly oval, with posterior channel and deep anterior notch; labrum straight, varicose, internally smooth. Eoc. Eur., Austral. *Eovasum* (Fig. 799): thick-shelled, inverted conical with low coeloconoid spire, the shoulder bearing short spines and the concave ramp with sinuous spiral grooves; aperture high, narrow and parallel-sided; siphonal fasciole present; columellar lip with three to ten folds, the anterior ones oblique, the posterior ones nearly horizontal; labral profile parasigmoidal. ?U.Cret.(Maastr.), Eoc. Africa, W.Pakistan, S.Amer.

Family STREPSIDURIDAE [STREPTURIDAE]

Swollen, pyriform; anterior canal more or less arcuate, deeply notched, more distinct than in Buccinidae; keeled siphonal fasciole sometimes present; columella with folds; labrum internally smooth or denticulate. (U.Cret.-Olig., ?Mio.)

Strepsidura (s.s.): rounded-pyriform, with short, conical spire; ornament of fine spiral threads and collabral costae which may become obsolete on the last whorl; last whorl inflated, its base gently excavated; keeled siphonal fasciole present; columella with two folds; labrum internally somewhat thickened but smooth, its profile quite straight and vertical. Eoc.-Olig. Eur., W.Pakistan, Burma, Indonesia. The subgenus *Strepsiduropsis* (Fig. 800) differs from *Strepsidura* in having only one columellar fold, but has a convex pad above it on which are grouped six smaller folds; the anterior part of the shell is more constricted, the anterior canal is a little longer and more twisted, and the labrum is weakly crenulated internally. U.Eoc. Nigeria. *Peruficus:* differs from *Strepsidura* in having reticulate ornament, a straight neck and anterior canal, and a strong varix about one-third of the way back on the last whorl. Eoc. S.Amer.

Family OLIVIDAE

Shells tending to be long and narrow, but never truly fusiform since the anterior end is truncated and the spiral angle may be acute or obtuse; much or all of the surface coated with enamel; spire short; aperture long and narrow, posteriorly often with a channel sunk in callus, deeply notched in front; labrum internally nearly always smooth, its profile straight and vertical for most of its

376

course, sometimes feebly opisthocline posteriorly; columella straight or excavated, twisted and often bearing folds in front, smooth or wrinkled behind; siphonal fasciole broad and flat, often divided; members of the subfamily Pseudolivinae tend to be somewhat more globose. (U.Cret.-Rec.)

The family characters are unmistakable, and the three main subdivisions are clear.

Subfamily OLIVINAE

More or less cylindrical with low spire and smooth whorls; the posterior gutter notches the labral profile and is continued externally into a sutural groove. (U.Cret.-Rec.)

Oliva (*s.s.*) (Fig. 801): usually rather slender oval-cylindrical, with very long and narrow aperture (three-quarters to nine-tenths total height of shell); columellar lip straight to gently convex, with numerous low folds which are transverse behind but in front become oblique and broad and extend outwards on a callus which partly covers the siphonal fasciole. Eoc.-Rec. Carib., N.Amer., S.Amer., Japan, Indonesia, Burma, W.Pakistan. *Olivella* (*s.s.*): internal partitions of the spire resorbed; columella bent inward where it delimits the canal; columella with one strong fold and a few weaker anterior folds; parietal folds vague. U.Cret.-Rec. S.Amer., Carib., N.Amer., Eur., W.Pakistan, Burma, Indonesia, Japan. The subgenus *Callianax* (Fig. 802) is stouter and has a blunter spire. Eoc.-Rec. N.Amer., Carib., S.Amer., Eur. *Olivancillaria* (*s.s.*) (Fig. 803): pyriform, with short spire and fairly wide aperture; columella excavated and smooth except towards the anterior end; siphonal fasciole very broad and not raised. Olig.-Rec. S.Amer., India, Burma. The subgenus *Agaronia* is subulate in form, with aperture three-fifths to two-thirds the height of the shell, and there is no enamel layer on the spire. Eoc.-Rec. W.Africa, Eur.

Subfamily ANCILLINAE

Sutures more or less completely buried in callus; anterior part of last whorl also with a callous layer. (U.Cret.-Rec.) (*76a*)

Ancilla (*s.s.*): elongate-oval-cylindrical with slightly cyrtoconoid spire; aperture high, with very vague anterior notch; no parietal fold; columella feebly twisted and with fine folds; unvarnished zone of last whorl almost reduced to nil, practically reduced to a furrow; labrum with a small anterior denticle at the end of a furrow. U.Olig.-Rec. Burma, India, Indonesia, Indo-Pac. The subgenus *Sparella* is more olivoid, the labrum has no denticle and there is no furrow leading to it, and the columella is twisted and has stronger folds. Rec. Indo-Pac. [Eogene forms which have been referred to *Sparella* have a labial denticle and a large unvarnished zone on the last whorl and belong elsewhere—possibly in part to *Amalda* (see below).] *Ancillus* (*s.s.*) (Fig. 804): oval-conical to fusiform, with higher spire than *Ancilla*; labrum without denticle; last whorl with large unvarnished band; aperture with deep anterior notch; columellar folds short, unequal. U.Cret.-Rec. Eur., Africa, N.Amer., Austral. *Amalda* [*Sandella*] (*s.s.*): swollen, olivoid, conofusoid; last whorl with unvarnished band; anterior notch deep; labrum with a denticle at the end of a furrow; columellar folds rather fine, unequal; columella little twisted; enamel on spire thin. U.Cret.-Rec. C.Amer., N.Amer., Eur., W.Pakistan, India, Burma, Indonesia, Austral., Asia. The subgenus *Alocospira* (Fig. 805) is more fusoid than *Amalda*, the enamel extends in a tongue right up the spire, there are frequently weak spiral striae, and the protoconch is styliform. Eoc.-Rec. Austral., N.Z., Indonesia, India, W.Pakistan, Eur., N.Amer. The subgenus *Baryspira* is inflated biconical, the thin enamel spreads all round the apex, the columella is twisted, and the columellar folds are strong and less numerous. Mio.-Rec. Austral., N.Z., Indonesia, Indo-Pac. [This remarkable form is reminiscent of *Orthaulax, Calyptraphorus* and the gerontic *Athleta*.] The subgenus *Spinaspira* is more biconical, has a low spire, and the last whorl is narrower. Olig.-Plio. Eur., N.Z. *Olivula* (*s.s.*): form narrow-olivoid, anterior notch moderate, and enamel very thin; whorls not stepped; columellar folds in one group; fine spiral furrows and growth lines. Eoc. N.Amer. The subgenus *Ancillarina* is like *Olivula*, but the anterior notch is feeble and the columellar folds are divided into two groups; no spiral ornament. Eoc.-Mio. Eur. [This subgenus may be a synonym of *Tortoliva* which is, however, founded on one little known form from the M.Eocene of N.America.]

Subfamily PSEUDOLIVINAE

Oval to rounded-pyriform, with low spire; last whorl usually with a deep spiral furrow forming a denticle where it meets the labrum; columella rather concave, the callus of the inner lip being wider posteriorly. (U.Cret.-Rec.)

Pseudoliva (*s.s.*) (Fig. 806): rounded-oval to inflated-pyriform, with short, conical spire (generally similar to *Babylonia*, but more rounded); last whorl large; smooth or with accentuated growth lines; usually anomphalous; aperture broadly oval, deeply channelled posteriorly, with a deep anterior notch; columella bent to the right anteriorly; labial denticle present; siphonal fasciole

---

Figs 802–817. GASTROPODA: OLIVIDAE, MITRIDAE, TURBINELLIDAE, HARPIDAE, MARGINELLIDAE AND CANCELLARIIDAE
Fig. 811 a synthetograph, partly after Stewart; Figs 802–810 and 812–817 original.
802. *Olivella* (*Callianax*) *branderi* (J. Sowerby), U.Eoc. Barton, Hampshire. × 1.
803. *Olivancillaria braziliensis* Lamarck, Rec. T. × 0·75.
804. *Ancillus buccinoides* (Lamarck), U.Eoc. (Headon). Hampshire. T. × 0·75.
805. *Amalda* (*Alocospira*) *papillata* (Tate), Plio. (Kalimann). Grange Burn, Victoria. T. × 0·6.
806. *Pseudoliva fissurata* (Deshayes), U.Pal. Woolwich, nr. London. × 1·1.
807. *Mitraria elongata* (Lamarck), M.Eoc. Paris B. × *c*.0·75.
808. *Volvaria bulloides* Lamarck, M.Eoc. Paris B. T. × 1·9.
809. *Turbinella pyrum* (Linné), Rec. Ceylon. T. × 0·5.
810. *Tudicla rusticula* (Basterot), M.Mio. Vienna B. × 0·75.
811e. *Tudicla* (*Pseudoperissolax*) *blakei* (Conrad), M.Eoc. (Tejon). California. T. × 1·1.
812. *Cryptochorda stromboides* (Hermann), M.Eoc. Paris B. × 1.
813. *Marginella glabella* (Linné), Rec. Mauritius. T. × 0·8.
814. *Cancellaria* (*Bivetopsia*) *subcancellata* d'Orbigny, Mio. Vienna B. × 1.
815a, b. *Trigonostoma* (*Ventrilia*) *acutangula* (Faujas), L.Mio. Aquitaine. × 0·75.
816. *Unitas costulatus* (Lamarck), M.Eoc. Paris B. T. × 1·5.
817a, b. *Bonellitia pyrgota* (Edwards), U.Eoc. Brockenhurst, Hampshire. × 1·5.

378

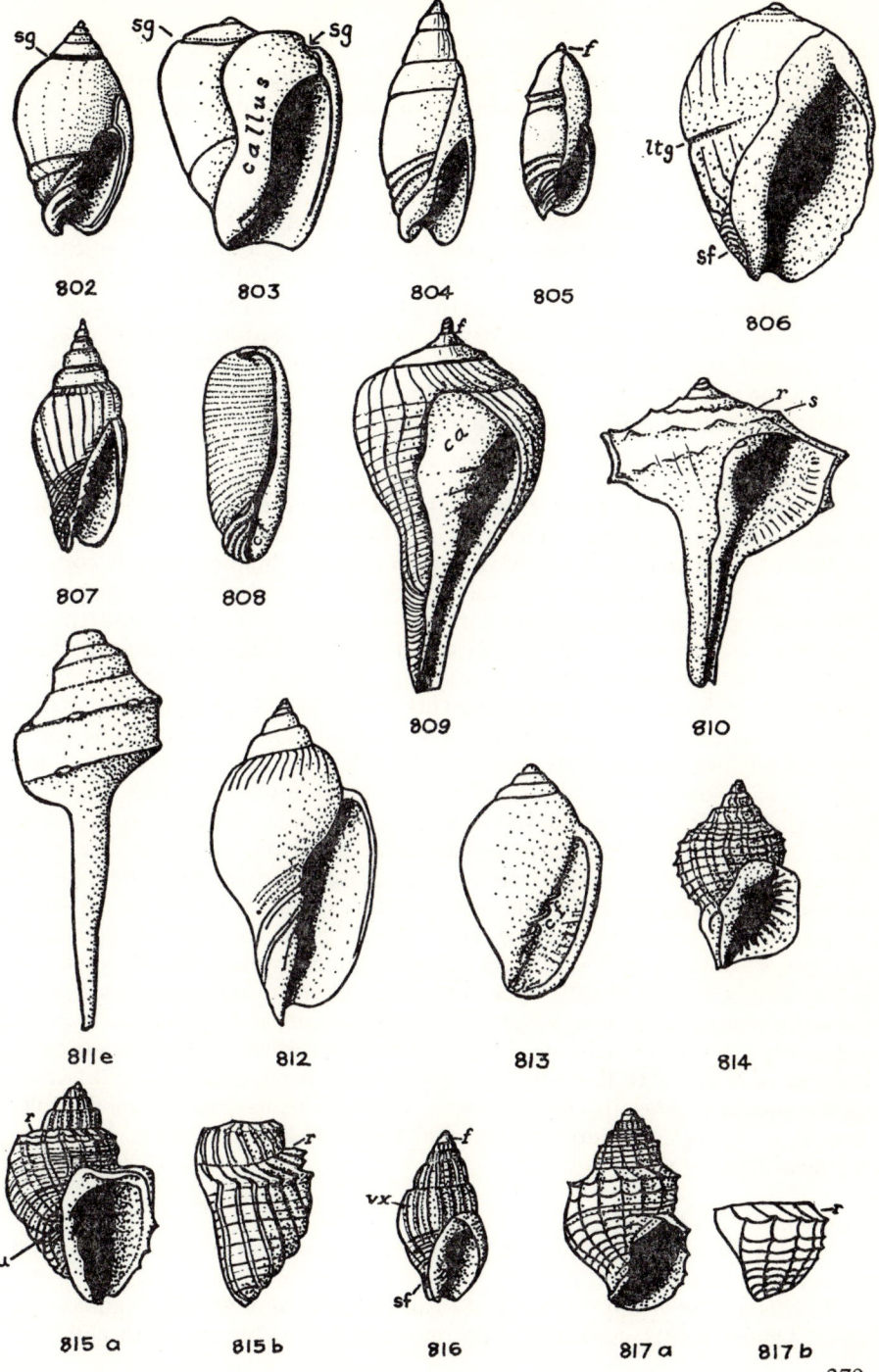

802   803   804   805   806

807   808   809   810

811e   812   813   814

815 a   815 b   816   817 a   817 b

developed. U.Cret.-Rec. W.Africa, S.Africa, W.Pakistan, E.Asia, Eur., N.Amer., S.Amer. The subgenus *Buccinorbis* is similar to *Pseudoliva*, but has a distinct umbilicus and is ornamented with fine spiral furrows. U.Cret.-Eoc. N.Amer., W.Africa, W.Pakistan, Austral.

Family MITRIDAE

More or less conoid, fusiform or biconical, with rather low to moderately high spire (except for *Volvaria*, which is involute); smooth or ornamented; often with a siphonal fasciole; aperture narrow, pointed behind, notched in front; columella straight or nearly so, with several folds increasing in strength backwards, and with well defined callus; labral profile approximately straight and vertical. (U.Cret.-Rec.) [Very rare in the Cretaceous, attaining its acme in Miocene and later times; the usually longer spire and reversed gradation in strength of the columellar folds distinguish this family from the Volutidae.]

Subfamily MITRINAE

Whorls usually smooth or with spiral ornament, rarely with collabral ornament; anterior canal vague, short. (U.Cret.(Maastr.)-Rec.)

Mitra [*Scabricola*] (*s.s.*): fusiform, with fairly high spire; whorls convex, with collabral costae and fine spiral threads; last whorl large, elongate-oval, with a siphonal fasciole; aperture narrow, posteriorly pointed, anteriorly with a short, broad canal which is broadly and deeply notched; columella with four or five folds; labrum internally denticulate. Plio.-Rec. Indo-Pac., Formosa, Indonesia, N.Amer., warm seas. The subgenus *Tiara* [*Cancilla*] is like *Mitra*, but the ornament consists of strong spiral threads with numerous fine collabral threads in their intervals; the base of the last whorl is less excavated, the labrum is smooth internally, and the siphonal fasciole is flat. Pal.-Rec. Austral., Indo-Pac., Formosa, Indonesia, Burma, India, W.Pakistan, E.Africa, Eur., N.Amer., C.Amer., S.Amer. *Fusimitra:* like *Tiara*, but with fine spiral ornament only, the last whorl with a rather more excavated base and a longer anterior canal, and the columella with only three folds. U.Cret.(Danian)-Eoc. N.Amer., Eur. *Mitraria* (*s.s.*) (Fig. 807): thick-shelled, fusiform but truncated in front, smooth or with very weak spiral ornament which may, however, be rather strong on the slightly excavated base; aperture narrow, truncated in front by a deep and wide notch; labrum thin, internally smooth, its edge anteriorly sometimes laciniate, its profile nearly vertical; columella straight, pointed in front, with five equidistant folds, the front one very weak; columellar border callous, its edge distinctly separated from the siphonal fasciole. Eoc.-Rec. Indo-Pac., Indonesia, Burma, India, W.Pakistan, Eur., N.Amer., S.Amer., widespread.

Subfamily VEXILLINAE

Usually costate, rarely smooth or with rows of nodes; anterior canal more or less bent. (U.Cret.-Rec.)

Vexillum [*Turricula*] (*s.s.*): oval-fusiform with moderately high-conical spire; whorls gently convex, usually more or less imbricate, with collabral costae (sometimes subspinose at a shoulder angle) and sometimes also with spiral threads; last whorl with a solid siphonal fasciole; aperture narrow, with a posterior gutter limited by a small parietal fold, anteriorly deeply notched and with a truncated canal; labrum internally denticulate, its profile straight; columella with four folds. U.Cret.-Rec. Indo-Pac., Philippines, Indonesia, Burma, S.Africa, Eur. *Conomitra:* small, oval or biconical, both ends equally constricted; spire short; smooth or with collabral costellae and sometimes also spiral furrows; no siphonal fasciole; aperture narrow, posteriorly channelled, anteriorly with a very vague, truncated, unnotched canal; labrum internally denticulate, its profile straight; columella with four only slightly oblique folds. Pal.-Plio. N.Amer., C.Amer., S.Amer., ?Indonesia, N.Africa, Eur.

Subfamily CYLINDROMITRINAE

Elongate-oval, with low or no spire; usually with spiral ornament; aperture long, narrow, anteriorly notched; labrum internally smooth, its profile more or less sinuous; columella straight, with two to ten folds. (Eoc.-Rec.)

Cylindromitra (*s.s.*): oliviform; last whorl large, with distinct siphonal fasciole; aperture anteriorly fairly deeply notched; fine spiral ornament; columella with at least nine folds. Plio.-Rec. Indo-Pac., Japan. *Volvaria* (*s.s.*) (Fig. 808): cylindrical, with involute, sunken spire; ornament of spiral furrows crenulated by growth lines; no distinct siphonal fasciole; aperture very narrow, a little wider anteriorly where it is truncated and only faintly notched; labrum internally smooth, its profile straight; columella with four narrow, oblique folds. Eoc.-Mio. Eur., Burma, Indonesia, N.Amer.

Family TURBINELLIDAE [XANCIDAE, VASIDAE]

Usually thick-shelled and more or less pyriform, with moderately to fairly high spire, smooth or with spiral and collabral ornament, sometimes noded; last whorl large, with or without siphonal fasciole; aperture usually fairly widely oval, with a moderately long or very long, usually unnotched anterior canal; labrum internally smooth or folded, its profile straight or sinuous; columella smooth or with one to five folds. (U.Cret.-Rec.)

Subfamily TURBINELLINAE [XANCINAE]

Columella with two to five folds; labral profile sometimes slightly prosocline posteriorly. (U.Cret.-Rec.)

Turbinella [*Xancus*] (*s.s.*) (I.C.Z.N. Opinion 489) (Fig. 809) (the 'chank' of Ceylon): varying from fusiform to pyriform, according to the elevation or depression of the spire; when the spire is depressed, the apex forms a cylindrical protuberance capped by a rounded protoconch; ornament of spiral furrows, growth lines and feeble nodes which become obsolete on the last whorl; last whorl inflated, with a fairly long neck bearing spiral threads, and with a low siphonal fasciole adjacent to a narrow umbilical slit; aperture oval-subcrescentic, with a long, very slightly notched anterior canal; inner lip convex in the parietal region, slightly excavate between that and the straight canal, and bearing three to five folds; labrum internally smooth, its profile

381

slightly prosocline at the suture. Olig.-Rec. Indo-Pac., Indonesia, India, W.Pakistan, Eur., Carib., C.Amer., N.Amer., S.Amer. *Vasum* (*s.s.*): shell biconical, very thick and spiny and with spiral threads; aperture rather narrow, with a fairly short, broad canal; anterior notch giving rise to a strong siphonal fasciole adjacent to a narrow umbilicus; labrum fluted, its profile straight; columella with three to five fairly strong folds. Eoc.-Rec. Indo-Pac., Egypt, Eur., N.Amer., S.Amer.

Subfamily TUDICLINAE [TUDICULINAE]

With one or two oblique columellar folds. (U.Cret.-Rec.)

Tudicla (*s.s.*) (Fig. 810): shape like the depressed species of *Turbinella*, and with the same protuberant apex, but with the anterior canal much more sharply defined, and with much more excavated columella bearing one blunt fold (scarcely more than the geniculation between aperture and canal); one or two rows of tubercles, of which the upper may pass into a keel; anterior canal long, narrow, unnotched; no siphonal fasciole; labral profile straight, posteriorly prosocline; labrum internally lirate. U.Cret.-Rec. Indo-Pac., Austral., India, Africa, Eur., N.Amer. The subgenus *Pseudoperissolax* (Fig. 811) has a moderately high spire, the last whorl has two noded, blunt carinae, and the neck is very long and slightly curved. Pal.-Eoc. N.Amer., Japan.

Subfamily PTYCHATRACTINAE

Canal slightly curved; no siphonal fasciole; columella usually with folds. (U.Cret.-Rec.)

This subfamily has also previously been placed either in the Buccinidae or the Fusinidae.

Ptychatractus: rather short-fusiform with conical spire; whorls convex, with deep sutures, ornamented with spiral threads; aperture narrow-pyriform, with short, straight, rather oblique canal; columella with two small, weak folds; labrum internally folded, its profile straight, prosocline posteriorly. Pal.-Rec. N.Amer., Eur.

Family HARPIDAE

Shell bucciniform to strombiform, with rather short spire, smooth or with widely spaced collabral costae; aperture wide, with deep anterior notch; labrum thickened, its profile practically straight and vertical; columella feebly concave, a little oblique to the shell axis, without folds, its callus spreading as a layer of enamel on to the ventral side of the shell, with a slight convexity where the siphonal fasciole underlies the callus, and ending in a point slightly inclined towards the labrum. (?U.Cret., Pal.-Rec.)

Harpa (*s.s.*): rounded-oval, with short, more or less imbricate spire; whorls with strong, lamellar, widely-spaced collabral costae; columella not twisted; columellar callus spreading far over the ventral surface. Eoc.-Rec. Indo-Pac., Eur., N.Amer., S.Amer. The subgenus *Eocithara* is like *Harpa*, but the callus of the inner lip does not spread so widely over the ventral surface, its well defined external margin being straight and vertical and anteriorly revealing a small umbilicus. Pal.-Mio. Eur., W.Pakistan, Burma, N.Amer., ?Carib. *Cryptochorda* [*Harpopsis*] (*s.s.*) (Fig. 812): fusiform, spire not very high, smooth; columella curved several times, without folds, anteriorly reaching further down than the labrum; columellar callus thin, spreading over ventral surface and the spire. Pal.-U.Eoc. Eur., N.Amer.

Family MARGINELLIDAE

Usually small; oval or biconical, sometimes cowrie-like; spire short or hidden; surface polished, sutures covered by enamel; aperture very narrow, very slightly notched in front; labral profile straight and nearly vertical, labrum thickened externally and often crenulated within; columella not excavated, usually with four or more folds decreasing in strength backwards; nearly always smooth, only very rarely with collabral or spiral ornament. (U.Cret. (Maastr.)-Rec.)

Mainly in tropical and subtropical seas. Those forms which resemble *Cypraea* may be distinguished by having folds, not mere teeth, on the columella.

Marginella (*s.s.*) (Fig. 813): oval to oval-conical; last whorl anteriorly constricted; aperture posteriorly channelled, anteriorly with an extremely short canal which is truncated and feebly notched, there being no siphonal fasciole; columella with four folds, the anterior one nearly vertical; labrum thickened, internally denticulate. Eoc.-Rec. W.Africa, Eur., Burma, Indonesia, Japan, Austral., N.Amer., C.Amer., Carib., S.Amer. *Cryptospira* (*s.s.*): oval-pyriform, with low spire; siphonal fasciole with raised edge; aperture not notched posteriorly; columella with five to six solid folds and a few finer folds in the parietal region; labrum thickened, internally smooth or finely denticulate. Mio.-Rec. Indo-Pac., Indonesia, Burma, India, E.Africa. *Gibberula* (*s.s.*): rather thin-shelled, pyriform to oval, with very low spire; aperture posteriorly narrowly channelled, anteriorly deeply notched; columella with four or more small folds, the three anterior ones best developed; labrum somewhat oblique in profile, not thickened, internally smooth or denticulate. U.Cret.(Maastr.)-Rec. W.Africa, Eur., N.Africa, India, Indonesia, Japan, N.Amer., warm seas. *Persicula* (*s.s.*): spire almost or completely hidden, the shell becoming cowrie-like; aperture high and narrow, posteriorly channelled, curved over and reaching the apex; labrum more or less oblique in profile, externally thickened, internally smooth or feebly denticulate; columellar callus anteriorly not swollen; columella with numerous transverse folds. Eoc.-Rec. W.Africa, Eur., N.Amer., Carib., S.Amer., warm seas.

Family CANCELLARIIDAE

Shell ovoid or bucciniform to almost turreted; ornament usually cancellate, often with tubercles or spines at the intersections of the collabral and spiral elements; base little excavated; umbilicate or anomphalous; aperture oval or rounded-triangular, with posterior gutter often remote from suture; labral profile straight, slightly oblique; labrum thickened, often denticulate internally; columella more or less straight, with two or three oblique folds, rarely smooth, ending pointedly in front, where there may be a notch or short canal. (U.Cret.-Rec.)

Of the four families Cancellariidae, Terebridae, Turridae and Conidae, the Cancellariidae is apparently the most primitive; yet the family does not appear before the Upper Cretaceous, and its sudden expansion in the Miocene is almost as striking as that of the Nassariidae or Pyrenidae.

Subfamily CANCELLARIINAE

Aperture with an anterior canal or notch; columella inflected to the left. (Pal.-Rec.)

Cancellaria [*Bivetia*] (*s.s.*): oval-conical, with the whorls latticed by spiral threads and collabral costae; small umbilicus adjacent to the solid siphonal fasciole; aperture wide, oval, anteriorly with a short, broad, deeply notched canal; columella with three folds; labrum internally denticulate. Mio.-Rec. Carib., C.Amer., N.Amer., S.Amer., Burma, Indo-Pac. The subgenus **Bivetopsia** [*Bivetopsis*] (Fig. 814) has thick collabral costae and finer spiral threads, the aperture is wider, the anterior notch is inflected dorsally, and the umbilicus is wider. Mio.-Rec. Carib., Pac. The subgenus **Merica** is elongate-oval-conical and finely latticed (rarely smooth); sutures deep and whorls usually with a ramp; aperture broadly oval, anteriorly with a short, feebly notched canal; labrum fluted, internally smooth or finely folded; siphonal fasciole distinct, low. Eoc.-Rec. Indo-Pac., N.Z., Indonesia, Burma, India, Eur., N.Amer., widespread. **Sveltia** (*s.s.*): turreted, with fairly high, conical spire; whorls more or less angular, with a steep ramp, ornamented with fine spiral furrows and collabral costae forming small spines on the shoulder; no distinct siphonal fasciole; columella with three folds, the anterior one very weak; labrum oblique in profile, flaring, internally thickened and folded. Eoc.-Plio. Eur., N.Amer. The subgenus **Sveltella** is not shouldered, has no spines, only two columellar folds, and a very small umbilicus. Pal.-Rec. Eur., N.Amer., Austral., warm seas. **Aphera** (*s.s.*): oval-conical, with latticed whorls; last whorl anomphalous or nearly so, without distinct siphonal fasciole; labrum thick, posteriorly prosocline in profile, internally denticulate; columella with two folds and sometimes a weak third anterior fold. Mio.-Rec. Pac., Carib., ?N.Z., Eur. The subgenus **Massyla** is like *Aphera*, but has weak collabral ornament, only two columellar folds, and the labral profile is more or less vertical. Mio.-Rec. Pac., Eur.

Subfamily TRIGONOSTOMINAE

No anterior canal or notch; columella inflected to the right. (Eoc.-Rec.)

Trigonostoma (*s.s.*): fairly large, elongate-oval-conical, with rather high spire and stepped, angular whorls the flanks of which bear collabral costae and spiral threads; last whorl with a wide umbilicus and a siphonal fasciole margined by a keel; aperture subtriangular, the points of the triangle being (1) the posterior gutter, well away from the suture, (2) the anterior beak, marking position of siphon, but not notched, (3) an obtuse bend in the columellar margin, which runs in one direction almost horizontally to the gutter while in the other it curves gently to meet the labrum at the beak; columella with three folds; labrum flaring, internally denticulate. Eoc.-Rec. Indo-Pac., Japan, Formosa, N.Z., Indonesia, Burma, India, W.Pakistan, Eur., N.Amer., Carib. The subgenus *Ventrilia* (Fig. 815) is more globose, the umbilicus is smaller, and there are only two weak, oblique columellar folds. Eoc.-Rec. Eur., Austral., Pac., N.Amer. The subgenus *Ovilia* has a low spire and very weak collabral ornament, thus resembling a *Tonna* but with wide umbilicus and two columellar folds. Mio.-Rec. Eur., Pac.

Subfamily ADMETINAE

Distinguished by: (1) a usually nearly vertical labral profile, (2) a more depressed protoconch, (3) the notching of the anterior beak, though without formation of a siphonal fasciole, (4) the disposition of the anterior columellar fold, which forms the margin of the beak and notch, (5) absence of umbilicus and of columellar callus (or slightness of the latter), and (6) thinner shell. (U.-Cret.-Rec.)

Admete (*s.s.*): small, bucciniform; cancellated, but collabral ornament tending to disappear on base; columellar folds dying away in adult; shell thin; cold-water form. Plio.-Rec. N.Amer., Eur., Iceland, northern seas, southern seas. *Unitas* [*Uxia*] (Fig. 816): elongate-ovoid, with irregular varices in addition to cancellate ornament, deep sutures and narrow oval aperture; labrum varicose, internally denticulate. U.Cret.-Mio. Eur., N.Africa, W. Pakistan, Austral., N.Z., N.Amer., C.Amer. *Bonellitia* (*s.s.*) (Fig. 817): shell less thin, with shouldered whorls; beak wider and less distinct from rest of aperture; columellar folds persistent, the anterior stronger than the other two; labrum thickened, internally denticulate. U.Cret.-Plio. Eur., Africa, Asia, Indonesia, Austral., N.Z., N.Amer., S.Amer.

Superfamily CONACEA

Fusiform, turreted-conical or inverted-conical, with very high to quite flat spire; aperture long and narrow to fairly short; columella without true spiral folds, occasionally with a few small, short folds. (Cret.-Rec.)

Family CONIDAE

Usually inverted-conical with low or flat spire and narrow whorls; last whorl very large, anteriorly regularly constricted; aperture high and narrow; columella straight, smooth. (?U.Cret., Pal.-Rec.)

Probably due to the dangers of rocky coasts and coral reefs, the outer shell becomes greatly thickened until the earliest whorls may be completely solidified, while the deeper-seated parts of the shell become resorbed until they are paper-thin. This internal character is of value in distinguishing true Cones from certain coniform Opisthobranchs and Pyrenidae.

Conus (*s.s.*): inverted-conical, with low, coeloconoid spire; whorls with broad, wavy nodes on the shoulder angle; last whorl smooth or with spiral

striae, with a broad siphonal fasciole; aperture high and parallel-sided, anteriorly broadly notched; labral profile posteriorly opisthocline, with a deep anal
notch above the shoulder. Plio.-Rec. Indo-Pac., Japan, Indonesia, Africa.
The subgenus *Chelyconus* (Fig. 818) has a higher spire, a rounded, smooth
shoulder, and the anterior half of the last whorl has weak spiral furrows. Eoc.-
Rec. Atl., Carib., Eur., Burma, Indonesia, Austral., N.Amer., C.Amer., warm
seas. The subgenus *Stephanoconus* (Fig. 819) has smooth or noded whorls and
the whole of the last whorl may carry spiral furrows. ?U.Cret., Eoc.-Rec.
Carib., Eur., Burma, Indonesia, Austral., warm seas. The subgenus *Conilithes*
is more slender, has a prominent, conical spire, the whorls have a finely noded
shoulder angle, and the labral profile is strongly opisthocline posteriorly. Eoc.-
Rec. Eur., Indonesia, N.Z., Austral., N.Amer. The subgenus *Lithoconus* has a
quite flat to low-coeloconoid spire, the shoulder is rather sharp, the last whorl
may have spiral striae anteriorly, and the labral profile is only feebly opisthocline posteriorly. ?U.Cret., Pal.-Rec. Indo-Pac., Indonesia, N.Z., Burma,
India, W.Pakistan, Africa, Eur., N.Amer., C.Amer., Carib., S.Amer., warm
seas. The subgenus *Leptoconus* is biconical, with imbricate whorls and a sharp
shoulder, and the labral profile is strongly opisthocline posteriorly, the anal
sinus being deep. ?U.Cret., Eoc.-Rec. Indo-Pac., Indonesia, Austral., Burma,
India, W.Pakistan, Eur., N.Amer., C.Amer., Carib., S.Amer., warm seas.
Hemiconus: biconical, strombiform; a row of tubercles half way between upper
and lower sutures of each whorl and on the shoulder of the last whorl which
has spiral furrows as well; labral profile somewhat opisthocline posteriorly;
aperture very parallel-sided, about two-thirds the height of the shell, not
notched anteriorly. Eoc.-Rec. Eur., Asia, Indonesia, Austral., Pac. The genus
Gastridium is Recent only. Its subgenus *Cleobula* is pyriform, the last whorl
has a rounded shoulder and some spiral ornament, and the labral profile is
little sinuous. Mio.-Rec. Indo-Pac., Indonesia, Burma, India, W.Pakistan,
Eur., ?N.Amer., C.Amer., Carib., S.Amer., warm seas. The subgenus *Den-*

Figs 818–832. GASTROPODA: CONIDAE, TURRIDAE, TEREBRIDAE,
ACTEONIDAE, RINGICULIDAE, SCAPHANDRIDAE AND
PHILINIDAE

Figs 820–821 and 824 after Wenz; Figs 818–819, 822–823 and 825–832 original.
818a, b, c. *Conus (Chelyconus) ponderosus* Brocchi, M.Mio. Vienna B. × 1.
819a, b. *Conus (Stephanoconus) calvimontensis* Deshayes, U.Eoc. Auvers, Paris B. × 1.
820. *Turris babylonia* (Linné), Rec. Indo-Pac. T. × 0·5.
821. *Crassispira bottae* (Valenciennes), Rec. Mexico. T. × 0·5.
822. *Turricula (Surcula) rostrata* (Solander), U.Eoc. Barton, Hampshire. × 0·75.
823a, b. *Bathytoma turbidum* (Solander), U.Eoc. Barton, Hampshire. × 1·25.
824. *Raphitoma histrix* (Jan), Rec. Medit. T. × 2.
825a, b. *Cryptoconus priscus* (Solander), M.Eoc. Paris B. × 1.
826a, b. *Genota ramosa* (Basterot), L.Mio. Aquitaine. × 1.
827. *Conorbis dormitor* (Solander), U.Eoc. Barton, Hampshire. T. × 1·5.
828a, e. *Subula* sp., Rec. × 0·36.
829. *Tornatellaea simulata* (Solander), U.Eoc. Barton, Hampshire. × 1·3.
830a, e. *Ringicula (Ringiculella) buccinea*(?) (Brocchi), M.Mio. Gironde. × 1·9.
831. *Scaphander edwardsi* (J. de C. Sowerby), M.Eoc. Hampshire. × 0·56.
832. *Philine aperta* (Linné), Rec. T. × 1·30.

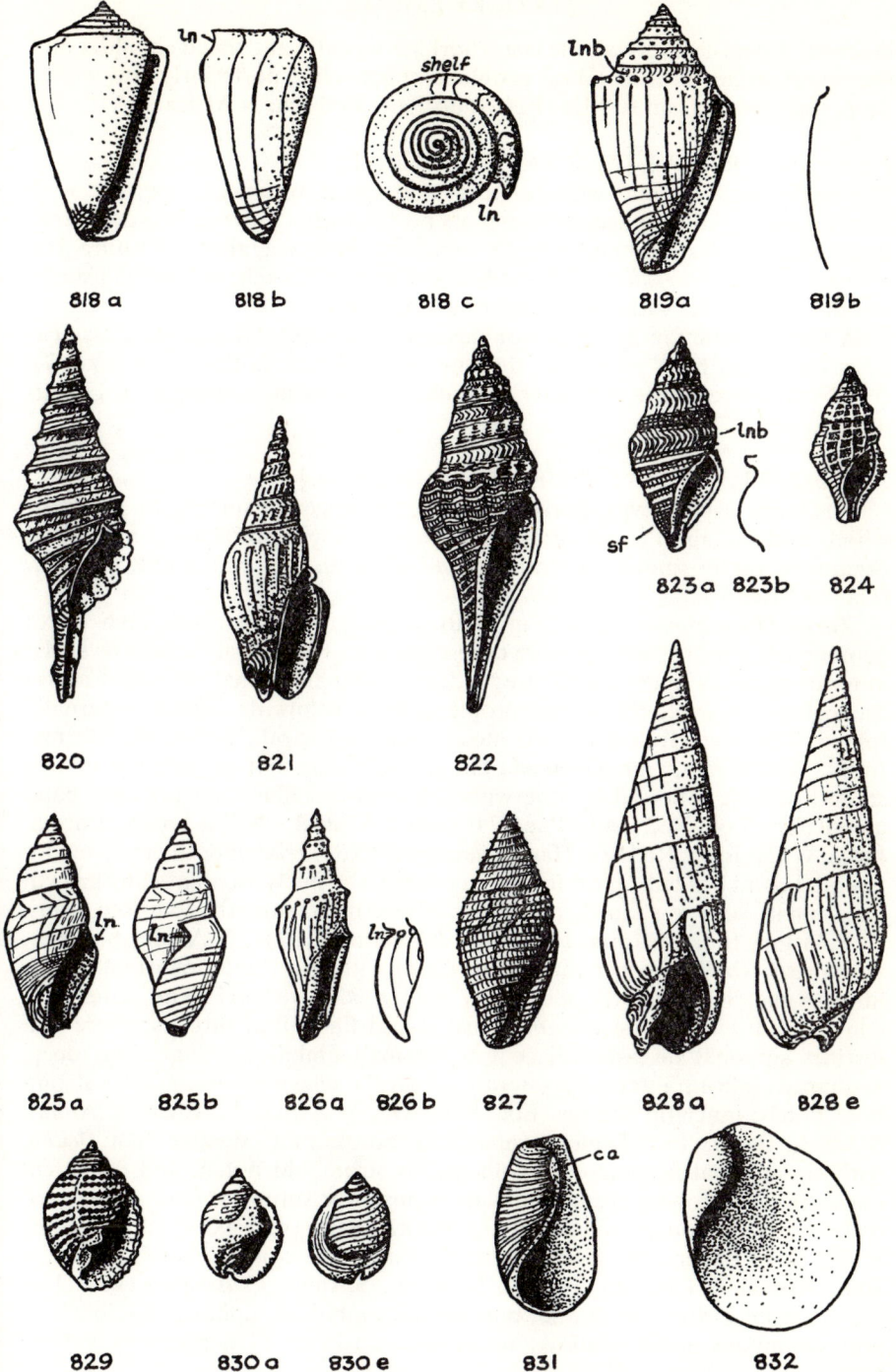

818 a   818 b   818 c   819a   819b

820   821   822   823a 823b   824

825a   825b   826a 826b   827   828a   828e

829   830a 830e   831   832

387

droconus is oval-cylindrical, the last whorl has a gently angulated shoulder and fine spiral striae, and the labral profile is opisthocline posteriorly and with a deep, rounded anal sinus. Mio.-Rec. Indo-Pac., Red Sea, N.Africa.

Family TURRIDAE [PLEUROTOMIDAE]

Usually fusiform and with collabral and spiral ornament; aperture with anterior canal; labrum usually sharp, its profile with a more or less deep anal notch near the posterior end, usually becoming vertical against the suture, but arching into a *Sigmesalia*-like hood below the lateral notch; columella usually smooth, sometimes with a few small folds. (U.Cret.-Rec.)

A very prolific family, very troublesome to classify. Of various classifications that have been proposed, the one adopted here is that of Wenz (**97**). By obsolescence of the characteristic notch some forms converge on Fusinidae.

Subfamily TURRINAE [PLEUROTOMINAE]

Fusiform, usually with fairly high spire; protoconch smooth or partly or wholly with collabral costellae; teleoconch whorls with spiral and/or collabral ornament, rarely smooth; labral profile with a distinct anal sinus. (U.Cret. (Danian)-Rec.)

Turris [*Pleurotoma*] (*s.s.*) (Fig. 820): acutely fusiform, with high spire; protoconch smooth; whorls more or less angular, with spiral threads or keels; anterior canal as long as aperture; a deep, rectangular anal sinus just above the peripheral keel, the labral profile convex below it; labrum internally folded. Eoc.-Rec. Indo-Pac., Formosa, Japan, Austral., Indonesia, Burma, India, W.Pakistan, Eur. *Gemmula* (*s.s.*): like *Turris*, but protoconch of two smooth whorls followed by one with collabral costellae, and the anal band crenulated. Eoc.-Rec. Indo-Pac., N.Amer., Austral., N.Z., Japan, Burma, India, W.Pakistan, Red Sea. *Hemipleurotoma:* like *Turris* (smooth protoconch), but anal band beaded, spiral threads tending to be finely crenulated by growth lines, anal sinus V-shaped, and labrum internally smooth. Pal.-Rec. Eur., W.Pakistan, India, Burma, Indonesia, ?Austral., E.Asia, N.Amer., S.Amer., warm seas. *Drillia:* rather inflated fusiform with short anterior canal bent a little to the right; whorls posteriorly with a concave anal band, anteriorly convex and with strong collabral costae and fine spiral threads; there is a distinct siphonal fasciole adjacent to a small umbilicus; anal sinus deep, U-shaped; labrum internally smooth, its profile also with an anterior stromboid notch; inner lip callous. Eoc.-Rec. Atl., W.Africa, N.Amer., C.Amer., S.Amer., Carib., Eur. *Eopleurotoma* (*s.s.*): a little more slender than *Drillia*, without siphonal fasciole or umbilicus, no stromboid notch, and ornament consisting of a row of crenulations on a posterior sutural thread and curved collabral costae on the flanks of the whorls, each costa being more or less noded at its upper end; U-shaped anal sinus in the concave band. Pal.-Olig. Eur., Africa, W.Pakistan, Burma, N.Amer., S.Amer. *Crassispira* (*s.s.*) (Fig. 821): similar in form to *Drillia*, but without umbilicus, siphonal fasciole low, and stromboid notch shallow; ornament consisting of a posterior sutural

388

thread and rather narrow collabral costae with finer spiral ornament in their intervals on the flanks; anterior canal short, wide, poorly differentiated; anal sinus as in *Drillia*. Eoc.-Rec. C.Amer., Carib., N.Amer., S.Amer., Eur., E.Africa, India, Burma, Indonesia, warm seas.

Subfamily CLAVATULINAE

Fusiform, usually with a fairly high spire and distinct neck; anal sinus only a little removed from the posterior suture; spiral ornament often weak and sometimes absent; protoconch usually smooth. (U.Cret.-Rec.)

Clavatula (*s.s.*): oval-fusiform to biconical; sutures situated between two strong spiral threads; anal sinus in a concave band; last whorl with collabral costellae and more or less crenulated spiral ornament, with short neck and low siphonal fasciole; aperture pyriform, with a fairly short, truncated and weakly notched anterior canal and with a callous parietal knob limiting a gutter; labrum internally smooth. Eoc.-Rec. S.Africa, Atl., N.Africa, Eur., W.Pakistan, Indonesia, Formosa, Japan, C.Asia, N.Amer. *Turricula* (*s.s.*): form of *Turris*, but more distinctly stepped, often of large size (80 mm. or more); fusiform, with high spire; whorls concavo-convex, the U-shaped anal sinus lying in the upper concave portion; ornament of a spiral thread just below the suture and weak collabral costae and spiral threads on the convex part of the whorls, rarely smooth; aperture with a long, straight, well differentiated anterior canal; labrum internally smooth. U.Cret.(Maastr.)-Rec. Indo-Pac., Indonesia, Formosa, Japan, Austral., S.W.Africa, Eur., N.Amer., C.Amer., S.Amer. The subgenus *Surcula* (Fig. 822) is more strongly ornamented, the elongate nodes or collabral costae on the convex portion of the whorls being more distinct; the anterior canal is sometimes a little bent, and there may be an obsolete siphonal fasciole. U.Cret.-Rec. Indo-Pac., W.Pakistan, India, Burma, Indonesia, Japan, C.Asia, W.Africa, Eur., N.Amer., C.Amer., Carib., S.Amer. The subgenus *Pleurofusia* is like *Turricula*, but the whorls have a ramp, and the nodes or collabral costae are crossed by strong spiral threads or lirae. Eoc.-Olig. N.Amer., S.Amer., Eur., W.Pakistan, Burma. *Surculites* (*s.s.*): fusiform-biconical, with angular whorls and a broad, flat anal sinus on the whole ramp; ornament of spiral threads and growth lines only; last whorl with one to three angulations; anterior canal not long, broad; labrum internally smooth. Eoc.-Mio. N.Amer., Eur.

Subfamily BRACHYTOMINAE

Fusiform, usually with fairly high spire; whorls with spiral and often also collabral ornament; distinct anal sinus present. (U.Cret.-Rec.)

Brachytoma: fusiform, with high, conical spire; whorls angular, with collabral costae and fine spiral threads; last whorl with a moderately long neck; aperture rather narrow, with a moderately long, feebly notched anterior canal; labrum somewhat thickened, its profile anteriorly with a stromboid notch, posteriorly with a fairly deep, slightly oblique anal sinus; small parietal callous knob present; rather strombiform in general appearance. Eoc.-Rec.

389

Pac., Japan, Formosa, Burma, India, W.Pakistan, warm seas. *Exilia* [?*Mitrae-fusus*] is doubtfully placed here; slender-fusiform, with high-conical spire and gently excavated base; whorls moderately convex, ornamented with collabral costae and fine spiral striae; neck long; aperture long and narrow, with a long, straight anterior canal; columella smooth or with a few weak, oblique folds internally; anal sinus very shallow. U.Cret.-Eoc.-?Mio. N.Amer., ?Eur. *Bathytoma* (Fig. 823): oval-fusiform to biconical; whorls angular, more or less concave posteriorly; ornament of variously combined collabral and spiral ridges, giving a spirally beaded appearance sometimes passing into smoothness; aperture narrowly subpyriform, nearly parallel-sided, angular posteriorly, with a moderately short, rather curved anterior canal; siphonal fasciole distinct; a geniculation of the columella almost forms a fold; anal sinus deeply U-shaped, on the angulation of the whorl; labrum internally more or less crenulated, its profile very prominent anteriorly. Eoc.-Rec. Eur., W.Pakistan, Burma, Indonesia, Indo-Pac., Austral., N.Z., N.Amer.

Subfamily CYTHARINAE

Usually rather small, oval to fusiform shells without operculum; protoconch smooth or ornamented; teleoconch whorls with spiral and usually also collabral ornament; anal sinus more or less distinct. (U.Cret.-Rec.)

Cythara: oval-fusiform, spire not very high; whorls more or less angular posteriorly, ornamented with collabral costae and fine, often finely crenulated spiral threads; aperture high and narrow (about three-fifths the height of the shell), parallel-sided, with a short, somewhat notched anterior canal; labrum varicose, internally lirate, its profile posteriorly with a rather flat anal sinus; columella with numerous small folds; protoconch smooth. Mio.-Rec. Indo-Pac., Austral., Japan, N.Amer., C.Amer. *Mangelia:* small, fusiform to elongate-bucciniform, with collabral costae and fine spiral lines in their intervals; columella smooth; anterior canal of moderate length, curved, not sharply set off from the aperture; neck moderately long; anal sinus rather broad and shallow, the labrum often with an internal swelling beneath it. Eoc.-Rec. Medit., Eur., ?India, ?Burma, Indonesia, Formosa, Japan, N.Amer., S.Amer., warm seas. *Borsonia* (*s.s.*): fusiform, with smooth protoconch; whorls medially more or less bluntly angular, with elongate nodes or short collabral costae and fine spiral threads, usually smooth above the angle; neck short; aperture narrowly pyriform, with a short anterior canal; labrum thin, internally smooth, its profile posteriorly with a moderate rounded anal sinus; columella with one fold. ?U.Cret., Pal-Rec. Eur., N.Africa, Indonesia, Austral., N.Z., N.Amer., C.Amer., S.Amer., warm seas. *Mitrolumna* (*s.s.*): oval-conical, with smooth protoconch and short spire; whorls gently convex, with spiral threads and often also with fine collabral costellae which may become obsolete on the last whorl; aperture very narrow, with only a vague, short anterior canal; labrum varicose, internally denticulate, its profile straight and without anal sinus; columella with two small folds (cf. *Conomitra*, which has four columellar folds). Eoc.-Rec. Medit., Eur., Indo-Pac., ?N.Amer. *Bela* (*s.s.*): buccinoid-fusiform, with fairly high spire and smooth protoconch;

whorls rather convex, with collabral costae and fine spiral threads; neck short; aperture oval, with very short anterior canal; labral profile sinuous, posteriorly with a feeble anal sinus; columella smooth; one of the least typical genera of the subfamily. Eoc.-Rec. Eur., Iceland, Burma, Indonesia, Japan, Austral., N.Amer., most seas. *Raphitoma* (*s.s.*) (Fig. 824): fusiform, with moderately high spire and strongly latticed whorls; protoconch at first with spiral ornament, then finely latticed; aperture oval, with a short, narrow anterior canal; labrum internally finely denticulate, its profile gently curved and with a distinct anal sinus posteriorly; columella smooth. Pal.-Rec. Medit., Eur., Iceland, ?Burma, Indonesia, Japan, ?Austral., ?N.Amer., warm seas.

Subfamily CRYPTOCONINAE

Fusiform to biconical; spiral ornament with or without collabral costellae; inner walls preserved except in *Cryptoconus* where there may be some resorption; no operculum. (U.Cret.-Rec.). [This subfamily has sometimes been attached to the Conidae, but its members differ from that family in the lack of an operculum.]

Cryptoconus (Fig. 825): biconical, with rather flat-sided whorls bearing spiral ornament only; aperture narrowly oval, nearly parallel-sided, with a vague, very short anterior canal; labrum internally smooth, its profile very convex below, prosocline at the upper suture, with a broad, distinct anal sinus on the shoulder; columella nearly straight, smooth (lacking the geniculation of *Bathytoma*). ?U.Cret., Pal.-Olig. Eur., W.Pakistan, N.Amer., Carib. *Genota* [*Genotia*] (*s.s.*): (Fig. 826) fusiform; whorls with a more or less crenulated angulation anterior to which there are collabral costae or costellae and finer spiral threads; aperture at least half the total height, narrow, nearly parallel-sided, with short, broad anterior canal which is truncated and notched, giving rise to a distinct siphonal fasciole on the neck; anal sinus rounded, on the sutural ramp but close to the shoulder; labrum internally smooth or occasionally denticulate. Pal.-Rec. W.Africa, N.Africa, Eur., W.Pakistan, Burma, Korea, Japan. *Conorbis* (Fig. 827): more perfectly biconical than *Cryptoconus* (the base being straighter-sided) with rather longer and more parallel-sided aperture and straighter columella; last whorl with a rounded shoulder above which is the broad, rounded anal sinus; labrum internally smooth; spiral ornament only; inner whorls partially resorbed. ?U.Cret., Eoc.-Olig. Eur., W.Pakistan, Indonesia, N.Amer., ?S.Amer.

Family TEREBRIDAE

Long, narrow, turreted shells with acute apex; aperture ovate to subrhomboidal, pointed behind, deeply notched in front; labrum thin; columella straight or excavate, abruptly bent at the origin of the short, wide anterior canal, and often carrying one or two small folds; siphonal fasciole conspicuous, margined by a keel; labral profile oblique and gently excavated for most of its length but changing curvature near the suture, the level of change being often marked by a spiral groove, the area between which and the suture forms a 'sutural band'; ornament of collabral costellae with or without spiral

391

threads, striae or rows of punctae, later whorls sometimes becoming smooth. A warm-water group. (?U.Cret., Eoc.-Rec.)

Terebra (*s.s.*): very slender; the distinct spiral furrow and spiral striae of the early whorls become obsolete on later whorls; columella with one weak fold. Eoc.-Rec. Indo-Pac., Austral., Formosa, Japan, Indonesia, W.Pakistan, Eur., Carib., N.Amer., C.Amer., S.Amer., warm seas. *Hastula* (*s.s.*): ornament of collabral costae only; no sutural band; one weak columellar fold; aperture somewhat constricted anteriorly. Eoc.-Rec. Indo-Pac., Austral., Japan, Burma, Eur., N.Amer., warm seas. *Duplicaria* [*Diplomeriza*]: shell with collabral costellae but no spiral ornament apart from the spiral band which is divided for most or all of its growth by a median groove; columella with two small folds. Olig.-Rec. Indo-Pac., W.Pakistan, India, Burma. *Strioterebrum* (*s.s.*): protoconch smooth; ornament of collabral costellae and spiral threads and a crenulated sutural band; one or more columellar folds. ?U.Cret., Eoc.-Rec. Eur., W.Pakistan, India, Burma, N.Amer., C.Amer., Carib., S.Amer., warm seas. *Subula* (*s.s.*) (Fig. 828): no spiral ornament except for the sutural band, which extends to the aperture; collabral costellae in the early stages only, fading out to growth lines in later stages; one weak columellar fold. Mio.-Rec. Indo-Pac., Burma, Eur., N.Amer.

Subclass EUTHYNEURA

Shell mostly spiral, occasionally much reduced, often covered by the mantle, sometimes completely lacking; spire prominent, more or less flat, or sunken; operculum lacking in a few genera; the visceral nerve cords do not usually cross, and are only secondarily orthoneurous (e.g. *Acteon*). (Carb.-Rec.)

Order CEPHALASPIDEA

Shell usually external, although occasionally more or less covered by the mantle; operculum present only in the Acteonidae. (Carb.-Rec.)

Superfamily ACTEONACEA

Characters of the Order. (Carb.-Rec.)

The Order Cephalaspidea together with the Orders Anaspidea, Thecosomata, Gymnosomata, Acochlidiacea, Sacoglossa, Notaspidea and Nudibranchia constitute the group known as Opisthobranchia.

Family ACTEONIDAE [ACTAEONIDAE]

Shell more or less oval, globose to subulate; smooth or ornamented with fine, punctate spiral grooves; spire generally short, angle more or less acute, sometimes sunken and involute; protoconch heterostrophic; aperture more or less auriform, wide and rounded in front, contracted and angular behind; labral profile nearly vertical and nearly straight; columella straight or twisted, often bearing very prominent spiral folds; horny, paucispiral operculum present. (Carb.-Rec.)

Subfamily ACTEONINAE [TORNATELLINAE]

Shell usually oval to oval-conical with a fairly short spire; usually ornamented with fine spiral furrows or spiral rows of pits; labrum sharp, internally sometimes weakly thickened; columella usually with one to three folds, sometimes with none. (Jur.-Rec.)

Acteon [*Actaeon, Tornatella*] (*s.s.*): oval to oval-conical, with conical spire, moderately convex whorls and deep sutures; ornamented with fine spiral grooves showing a punctate appearance under the lens; one columellar fold. U.Cret.-Rec. Medit., Eur., Africa, Asia, Amer., Austral., N.Z., widespread in warm seas. *Pupa* [*Solidula*]: like *Acteon*, but with two columellar folds, the anterior of which is stronger and bifid. Eoc.-Rec. Pac., Eur., India, Japan, N.Z. *Tornatellaea* (*s.s.*) (Fig. 829): like *Acteon*, but with two strong columellar folds, a wide and shallow notch at the anterior end of the aperture, and the labrum internally thickened and denticulate. Jur.-Olig., N.Amer., Eur., N.Africa, N.Z.

Subfamily LIOCARENINAE

Similar to *Acteon* in general form, but aperture narrow and rather oblique to the axis and with thickened margins; no columellar folds; resembles *Melampus*, but the inner whorls are not resorbed. (U.Cret.-Eoc.)

Liocarenus: ornament of very weak spiral striae only. U.Cret.-Eoc. Eur.

Subfamily ACTEONININAE [CYLINDROBULLININAE]

Oval-conical with stepped spire, or inverted-conical; usually smooth, occasionally with very fine spiral striae; aperture long, narrow, columella thickened but without folds. (Carb.-Pal.)

Acteonina is Carboniferous to Jurassic only. *Trochactaeonina* is Jurassic only. Its subgenus *Douvilleia* is oval, and the whorls have a ramp and a carinate shoulder; there is a trace of an umbilicus. Pal. Eur.

Family RINGICULIDAE

Small, inflated-oval to globose, with short spire, usually with punctate spiral striae, rarely smooth; aperture constricted by callus, folds and teeth; labrum always varicose, sometimes internally denticulate; columella short, with one or two folds, anteriorly truncated; floor usually notched; parietal region often with one or more teeth; no operculum. (Cret.-Rec.)

The only opisthobranch family with an anterior notch to the aperture. The family differs from the Acteonidae in the greater contraction of the aperture, the oblique profile of the labrum (which is thickened both outwardly and inwardly), and the absence of an operculum.

Ringicula (*s.s.*): oval-conical, with low to moderately high spire; aperture auriform, contracted behind, notched in front; columella with two very strong spiral folds; parietal region callous, with a tooth; labrum greatly thickened, especially at mid-length, internally denticulate, its profile straight and slightly opisthocline; ornament of fine spiral striae. U.Cret.-Rec. Eur., Africa, Asia, Indonesia, Austral., N.Z., Formosa, Japan, warm seas. The subgenus

Ringiculella (Fig. 830) has the labrum internally smooth, the spire is shorter, and the whole form is more globose; it may sometimes lack the spiral ornament. Eoc.-Rec. Eur., India, Japan, Carib., warm seas. *Gilbertina:* like *Ringicula*, but very globose, with obtuse spiral angle, the aperture not notched anteriorly, and the labrum has two teeth internally, the profile being the reverse of that of *Ringicula*, being prosocline on the suture. Pal.-Mio. Eur., N.Amer.

Family SCAPHANDRIDAE

Cylindrical, oval-conical or oval, smooth or with spiral grooves or rows of punctae; spire sunk or covered over; columella with or without folds. (U.Cret.-Rec.)

Scaphander (*s.s.*) (Fig. 831): oval-conical, distinctly narrower behind; involute, spire sunken, whorls not in contact; no columella (whole interior of spire visible from anterior end); sculpture of fine punctate spiral lines; aperture retort-shaped, very wide in front, channelled behind; apex without umbilicus. U.Cret.-Rec. Eur., W.Pakistan, India, Indonesia, Austral., N.Z., N.Amer., Carib., widespread. *Roxania:* oval, with deeply sunk, umbilicus-like spire; ornament of punctate spiral grooves; aperture high, narrower posteriorly; columella concave, truncated anteriorly; no parietal callus; small umbilical slit present; labral profile straight, vertical. U.Cret.-Rec. Eur., Medit., Indo-Pac., Indonesia, N.Z., Japan, Atl., N.Amer. *Cylichna* [*Bullinella*] (*s.s.*): elongate-cylindrical to subcylindrical, with spire deeply sunk and forming a narrow umbilicus; last whorl with fine spiral striae, occasionally partly smooth; aperture equals height of shell, a little wider anteriorly, notched anteriorly and posteriorly; columella with one weak fold. U.Cret.-Rec. Eur., Iceland, Greenland, Africa, Indonesia, Formosa, Japan, Austral., Antarctic, N.Amer., widespread. The subgenus *Cylichnella* differs from *Cylichna* in being more globose, the aperture rising little higher than the spire and not being notched posteriorly, and the columella being callous and carrying two folds the anterior one of which is short. U.Cret.(Danian)-Rec. Carib., N.Amer., Eur., Indonesia, Pac.

Family PHILINIDAE

Thin-shelled, oval, with incompletely enrolled spire and concave columella; columellar lip thin; aperture very wide open; last whorl large, evolute, often much widened; compared with *Bulla*, the shell is completely covered by the mantle, and being no longer of use for protection has either become very thin and imperfect, or has acquired peculiar shapes in adaptation to muscular attachments. (U.Cret.-Rec.)

Philine (*s.s.*) (Fig. 832): resembles *Scaphander*, but is much more loosely coiled and the aperture becomes very wide; last whorl smooth or with spiral striae. Eoc.-Rec. Medit., Eur., N.Z., Japan, N.Amer., S.Amer., widespread. The subgenus *Megistostoma* differs from *Philine* in being more rounded and in the labrum being flattened out behind and prolonged into a triangular projection. U.Cret.-Plio. N.Amer., Eur., ?Syria.

394

Family DIAPHANIDAE

Small, thin-shelled, inflated-oval to pyriform; spire very small or sunk; columella thin, with or without a fold. (Jur.-Rec.)

Diaphana (*s.s.*): aperture pyriform, narrower posteriorly; no columellar fold, columella not truncated anteriorly; ornament of fine spiral striae. Jur.-Rec. N.Atl., Eur., N.Amer.

Family BULLIDAE

Inflated-oval, with spire sunk in an umbilicus; last whorl smooth or with fine spiral striae; aperture equals height of shell, posteriorly narrow, anteriorly widened; labrum thin, its profile almost vertical; columella short, concave, smooth or with one or two folds. (Jur.-Rec.)

Bulla (*s.s.*) (Fig. 833): globose, only the last whorl visible; spire sunk in an apical umbilicus, involute; labrum surpassing apex; smooth or with spiral striae; columella smooth; compared with *Scaphander*, the whorls are in contact, and the aperture is less dilated in front. Jur.-Rec. Indo-Pac., W.Pakistan, Africa, Austral., Eur., Greenland, N.Amer., S.Amer., Carib., warm seas.

Family ATYIDAE

Thin-shelled, oval, usually with sunken spire; last whorl very large, usually with fine spiral ornament which may fade out medially; aperture lunate, sometimes higher than the shell; columella smooth or with a fold. (?U.Cret., Pal.-Rec.)

Subfamily ATYINAE

Body can withdraw wholly into shell. (?U.Cret., Pal.-Rec.)

Atys (*s.s.*) is Recent only. The subgenus *Aliculastrum* is elongate-oval (less globose than *Atys*), the aperture is little higher than the whole shell, and the columella has only a weak fold; there is a very narrow umbilicus, and the last whorl has fine spiral furrows at the ends. Pal.-Rec. Indo-Pac., India, N.Z., N.Amer., Carib., warm seas. *Haminaea* (*s.s.*): oval-cylindrical, *Bulla*-like, with very fine spiral striae or quite smooth; spire umbilicus very narrow or closed, inner whorls not visible; columella short, concave, smooth; thinner-shelled than *Bulla*, and has different anatomy. Eoc.-Rec. Medit., Eur., N.Amer., Carib., S.Amer., tropical seas.

Family ACTEOCINIDAE [TORNATINIDAE, RETUSIDAE]

Cylindrical, pyriform or oval-fusiform, posteriorly often truncated; spire very low, sunken or involute; last whorl smooth or with fine spiral striae, only rarely with collabral ornament; aperture long, narrow, wider anteriorly; columella usually smooth, sometimes with a weak fold; labrum nearly always opisthocline at the suture. (Jur.-Rec.)

Acteocina [*Tornatina*]: small, cylindrical or elongate-oval, with short spire (angle nearly 90°) and rounded anterior end; sutures of spire deeply grooved; last whorl smooth or with fine spiral striae; labrum posteriorly opisthocline and deeply notched; columellar lip thickened and reflected, with one strong

395

fold. Jur.-Rec. N.Amer., Carib., C.Amer., S.Amer., Japan, Eur., warm seas. *Retusa* (*s.s.*): like *Acteocina*, but tends to be pyriform-cylindrical; last whorl smooth or with fine spiral striae anteriorly, occasionally with accentuated growth lines posteriorly; columellar lip reflected and thickened. Jur.-Rec. Eur., Asia, N.Z., N.Amer., widespread. *Cylichnina* (Fig. 834): small, oval-cylindrical with rounded ends; spire sunk in a narrow umbilicus, not visible; last whorl anteriorly with fine spiral striae; aperture narrow, posteriorly surpassing spire; labrum medially somewhat constricted, its profile posteriorly opisthocline; columella short, concave, with a fold. Pal.-Rec. Eur., Medit., Atl., Indonesia, N.Amer., Pac., N.Z. *Rhizorus* [*Volvulella*]: small, elongate-oval-conical to fusiform, posteriorly pointed, anteriorly constricted, with involute spire; last whorl smooth or with spiral striae anteriorly and posteriorly; aperture equals height of shell, little widened anteriorly; labral profile straight, almost vertical; columella short, twisted, usually with a fold. Eoc.-Rec. Eur., N.Amer., C.Amer., Carib., Japan, N.Z., warm seas.

Order ENTOMOTAENIATA

Oval to slender-turreted; protoconch (where known) often heterostrophic. (Palaeozoic-Rec.)

Superfamily PYRAMIDELLACEA [PLOTIACEA]

Usually more or less turreted, holostomatous; protoconch heterostrophic, deviated or homoeostrophic; whorls smooth or ornamented. (Palaeozoic-Rec.)

Although previously placed between the Nerineacea and the Hipponicacea, this group has now been shown to belong to the Opisthobranchia.

Family PYRAMIDELLIDAE [PLOTIIDAE]

Small, oval-conical to turreted-conical; protoconch heterostrophic, inclined at an angle to the adult shell axis; whorls smooth or ornamented;

Figs 833–845. GASTROPODA: BULLIDAE, ACTEOCINIDAE, PYRAMIDELLIDAE, MELANELLIDAE, CAVOLINIDAE, APLYSIIDAE, UMBRACULIDAE, AND ELLOBIIDAE

Fig. 836 after Wenz; Figs 833–835 and 837–845 original.

833a, c. *Bulla ampulla* Linné, Rec. T. × 0·56.
834. *Cylichnina elliptica* (J. de C. Sowerby), U.Eoc. (U.Bracklesham). Hampshire. × 3·4.
835a, b. *Pyramidella* sp., Rec. × 1·65.
836. *Turbonilla lactea* (Linné), Rec. N.Africa. × 3.
837. *Melanella polita* (Linné), L.Plio. Suffolk. × 2·25.
838a, b, e. *Cavolina telemus* (Linné), Rec. Atlantic. × 1·5.
839d, e. *Vaginella depressa* Daudin, L.Mio. Aquitaine. × 3·4.
840. *Dolabella* sp., Rec. Ceylon. × 0·75.
841c, e. *Umbraculum* sp., Rec. × 0·375.
842. *Ellobium auris-judae* (Linné), Rec. India. × 0·75.
843. *Melampus* sp., Rec. Seychelles. × 1·75.
844. *Pythia scarabaeus* (Linné), Rec. Ceylon. × 0·75.
845. *Ovatella* (*Myosotella*) *pyramidalis* (J. de C. Sowerby), Pleist. Oakley, Essex. × 1·1.

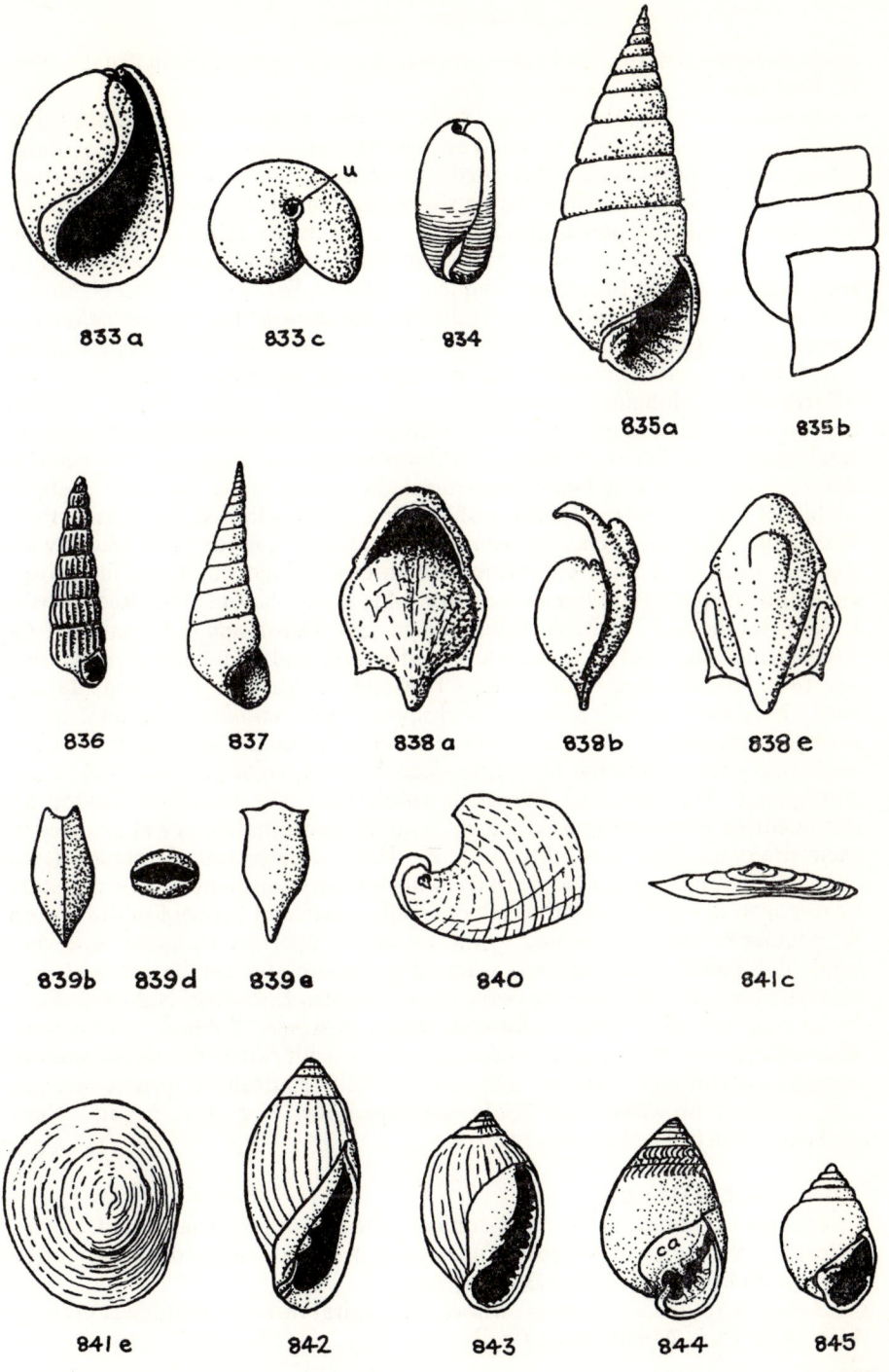

833 a   833 c   834

835 a   835 b

836   837   838 a   838 b   838 e

839 b   839 d   839 e   840   841 c

841 e   842   843   844   845

aperture holostomatous and columella usually with folds; labrum usually thin. (U.Cret.-Rec.)

Pyramidella [*Plotia*] (*s.s.*) (Fig. 835): usually somewhat larger than other members of the subfamily; turreted or elongate-turbinate, slightly cyrtoconoid or subulate; whorls almost flat-sided, smooth, with incised sutures; aperture auriform, holostomatous but with a slight eversion of the inner lip margining a minute umbilicus; peristome discontinuous; labrum thin, internally lirate, its profile vertical, almost straight; columella with three spiral folds of which the posterior is most prominent and least oblique. Pal.-Rec. Carib., N.Amer., S.Amer., Eur., India, Indonesia, Indo-Pac. *Otopleura:* like *Pyramidella*, but whorls imbricate, ornamented with collabral costae with spiral striae in their intervals, and umbilicus practically nil. Mio.-Rec. Indo-Pac., Indonesia, Eur. *Tiberia* (*s.s.*): elongate-oval-conical, with smooth, gently convex whorls; labrum internally smooth, its profile straight, vertical; two columellar folds; small umbilicus. Plio.-Rec. Japan, Eur., warm seas. The subgenus *Cossmannica* differs from *Tiberia* in being high-turreted-conical, in having only a slight umbilical slit, and in the labral profile being opisthocline at the suture. Pal.-Rec. Eur., N.Amer., Pac. *Syrnola* (*s.s.*): slender, cylindrical-conical, with smooth, nearly flat whorls; labrum thin, internally smooth, its profile prosocline and straight; only one columellar fold; anomphalous. Pal.-Rec. Japan, C.Asia, India, N.Z., Eur., N.Amer., warm seas. *Odostomia* (*s.s.*) differs from *Syrnola* in being shorter, subulate or turbinate, and in having a more oval aperture. U.Cret.-Rec. Eur., Iraq., E.Africa, Indonesia, Formosa, Japan, N.Z., N.Amer., Carib., S.Amer., widespread. *Chrysallida* (*s.s.*): very small, oval-conical, with more or less finely crenulated reticulate ornament; aperture oval; one weak columellar fold. Mio.-Rec. Pac., N.Amer., Carib., N.Z., Iraq, widespread. The subgenus *Pyrgulina* differs from *Chrysallida* in having an ornament of gently sinuous collabral costellae with incised spiral grooves in their intervals. Mio.-Rec. Japan, N.Z., Eur., widespread. *Turbonilla* (*s.s.*) (Fig. 836): elongate-turreted, with flat whorls bearing straight collabral costae; protoconch at right angles to shell axis, polygyrate and helicoid in the Group A, paucispiral and planorboid in the Group B; aperture rhombic; one very weak columellar fold; labrum internally smooth, its profile straight and vertical; base of last whorl smooth; anomphalous. Eoc.-Rec. N.Africa, Eur., Indonesia, Formosa, Japan, Austral., N.Z., N.Amer., S.Amer., warm seas. *Eulimella* (*s.s.*): small, slender-turreted-conical, with rather flat-sided, smooth whorls; aperture subrhombic; labrum internally smooth, its profile straight and vertical; no columellar folds; anomphalous. Eoc.-Rec. Medit., Eur., N.Africa, Austral., N.Z., Atl., Carib.

Family ACLIDIDAE

Small, oval-conical to turreted-conical, with smooth or ornamented, convex whorls; protoconch homoeostrophic; aperture oval, holostomatous; often with an umbilicus. (U.Cret.(Danian)-Rec.)

Aclis (*s.s.*): turreted-conical; whorls with spiral threads; umbilicus present. Mio.-Rec. Eur., Medit., Atl., Carib.

398

Family MELANELLIDAE [EULIMIDAE, STROMBIFORMIDAE]

Fairly slender turreted-conical, occasionally oval-conical, often smooth; protoconch homoeostrophic but usually at an angle to the main axis; aperture oval, holostomatous. (U.Cret.(Maastr.)-Rec.)

Mostly nestlers or parasites on Echinoderms, thus occupying a somewhat similar place in marine ecology to that of Leptonacea among bivalvia.

Melanella [*Eulima*] (*s.s.*) (Fig. 837): subulate, with curved axis and flat-sided, smooth whorls; labrum arcuate, its profile oblique; anomphalous. Eoc.-Rec. Indo-Pac., Japan, N.Z., Indonesia, India, N.Africa, Eur., N.Amer., Carib., S.Amer., warm seas. *Niso* (*s.s.*): like *Melanella*, but with a straight axis and a fairly large and deep umbilicus. U.Cret.(Danian)-Rec. Eur., India, Indonesia, Austral., N.Z., Japan, N.Amer., Carib., warm seas.

Family STILIFERIDAE

Subspherical to elongate-oval, with styliform apex. Always parasitic, particularly on Echinoderms. (Eoc.-Rec.)

Stilifer (*s.s.*) is Recent only. *Mucronalia:* elongate-oval to subulate, with straight spire and rather flat, smooth whorls; aperture elongate-oval; columella smooth, gently concave; peristome discontinuous; labral profile straight and vertical; anomphalous. Eoc.-Rec. Indo-Pac., Eur.

Order THECOSOMATA [PTEROPODA]

Shell spiral or bilaterally symmetrical; pelagic, free-swimming forms, derived from Opisthobranch stocks; the shell-bearing pteropods—the only ones which concern the palaeontologist—seem to be descended from Bullidae, but have undergone great modifications in adaptation to the swimming habit. (?U.Cret., Pal-Rec.)

Superfamily SPIRATELLACEA [EUTHECOSOMATA]

Shell always present, spiral and sinistral or bilaterally symmetrical. (?U.Cret., Pal.-Rec.)

Family SPIRATELLIDAE [LIMACINIDAE]

Shell spirally coiled, turbinate to discoidal, sinistral, with or without an umbilicus. (Eoc.-Rec.)

Spiratella [*Limacina, Spirialis, Valvatina*]: shell small, thin, translucent, smooth or with growth lines, turbinate with rounded whorls to nearly discoidal; whorls closely wound; aperture large and columella projecting; umbilicus present. Eoc.-Rec. Polar seas, N.Z., Austral., Africa, Eur., N.Amer., Carib., widespread.

Family CAVOLINIDAE

Shell thin, smooth or with fine growth lines, showing no sign of spiral coiling or of asymmetry; shape conical, urn-like or globular-labiate. (?U.Cret., Pal.-Rec.)

Cavolina [*Hyalaea*] (*s.s.*) (Fig. 838): globose; ventral surface inflated, dorsal surface flattened and produced anteriorly into a parabolic hood in front of the slit-like aperture, and posteriorly into a short spine; there are also smaller postero-lateral spines; dorsal side with some longitudinal folds. Mio.-Rec. Medit., Eur., Indonesia, N.Z., Fiji, Japan, N.Amer., C.Amer., Carib., warm seas. *Creseis:* shell a simple, very acute, straight or feebly curved cone, circular in section; smooth or with fine striae. Eoc.-Rec. Atl., Eur., N.Amer., N.Z., warm seas. *Clio* [*Cleodora, Balantium*]: triangular-pyramidal, with corners drawn out, dorsally keeled, apex acuminate. ?U.Cret., Pal.-Rec. Atl., Eur., Syria, Carib., widespread. *Vaginella* (Fig. 839): shaped like a flattened urn, being longer and narrower than *Cavolina*, dorsal and ventral surfaces almost alike, without hood and with only the posterior spine; lateral keels rounded, aperture somewhat widened. ?U.Cret., Pal.-Mio. Eur., Syria, Austral., N.Z., N.Amer., C.Amer.

Order SACOGLOSSA

Based on the anatomy. Shell, when present, external, thin, spiral and with much widened aperture. (Eoc.-Rec.)

Superfamily JULIACEA [TAMANOVALVACEA]

A most peculiar group of small forms with reduced shells; they secrete a second lid-like 'valve' and hence were for many years classified as Bivalvia. (Eoc.-Rec.)

Family JULIIDAE

Modioliform in outline. (Eoc.-Rec.)
Julia: the supposed 'umbo' is terminal, with a notch beside it; shell wider anteriorly. Olig.-Rec. Indo-Pac., Eur., N.Amer., Carib.

Order APLYSIACEA

Separated on anatomical grounds. (?Jur.-Rec.)

Family APLYSIIDAE

Shell thin, internal, leaf-shaped, rather flat, spire merely a swelling from columellar lip. (?Mio., Plio.-Rec.)

Subfamily DOLABELLINAE

Shell enrolled, spire covered by the thickened columellar lip; labrum posteriorly notched. (?Mio., Plio.-Rec.)
Dolabella (Fig. 840): smooth, like *Megistostoma* but more strongly enrolled, flat; coiled portion of shell much reduced and labral extension enlarged to the shape of a battle-axe. ?Mio., Rec. Indo-Pac., Red Sea, N.Amer.

Family AKERIDAE

Shell external, thin, oval to oval-cylindrical, anomphalous; spire truncated, visible, scarcely sunken; aperture constricted posteriorly, wider anteriorly; labral profile strongly arcuate, with wide embayment in front and very deep pointed notch behind; this, combined with the absence of a columella (as in *Scaphander*), makes the last half whorl seem detached from the rest of the shell; sutures channelled; protoconch heterostrophic. (?Jur.-Rec.)

Akera: characters of the Family. ?Jur.-Rec. Eur., N.Africa, M.East, W.Pakistan, N.Amer., S.Amer., widespread.

Order NUDIBRANCHIA (s.l.)

Shells only present in a few primitive groups (e.g. Umbraculidae). (Eoc.-Rec.)

Suborder NOTASPIDEA

Shell external, patelliform or capuliform, or internal and auriform. (Eoc.-Rec.)

Family UMBRACULIDAE [UMBRELLIDAE]

Shell external, patelliform or capuliform, rounded or oval in outline. (Eoc.-Rec.)

Umbraculum [*Umbrella*] (*s.s.*) (Fig. 841): depressed open-conical, with elliptical or oblong outline and subcentral apex and sinistral protoconch; distinguished from *Patella* by absence of definite radial costae (though obscure radial folds may be present) and by the muscle scars forming an unbroken ring (not a horse-shoe). Eoc.-Rec. Indo-Pac., Austral., Eur., warm seas.

Order BASOMMATOPHORA

Shell always present, spiral, patelliform or capuliform, with entire aperture; except for *Otina*, the body can be wholly retracted into the shell; an operculum is present only in the Amphibolids; there is a single pair of tentacles with the eyes at the base. (Carb.-Rec.)

The Basommatophora together with the Order Stylommatophora constitute the group known as Pulmonata. The Pulmonata are probably to be regarded as Opisthobranchs that began to adapt themselves to a fresh-water or terrestrial habit so far back in geological time that their divergence is considerable. Later repetitions of the process by various stocks of Prosobranchs have only led to the evolution of new families (Helicinidae, Cyclophoridae, Viviparidae, etc.). The most primitive families of Pulmonata have, however, returned to a more or less marine life without having ever travelled far from the shore (Siphonariidae, Ellobiidae). These, with the fresh-water pulmonates, constitute the Order Basommatophora.

Superfamily ELLOBIACEA

Shell spiral, only rarely sinistral; aperture usually constricted by teeth. (?Carb., Jur.-Rec.)

401

Family ELLOBIIDAE [AURICULIDAE]

Appearance of shell like that of Acteonidae, but generally without spiral ornament; columellar folds usually well marked; labrum thin, or thickened and finely denticulate, or with a few prominent teeth which, in combination with the columellar folds, greatly restrict the aperture; labral profile straight, vertical or slightly oblique; inner whorls (except in *Pedipes*) always more or less resorbed. (?Carb., Jur.-Rec.)

Recent forms largely tropical, frequenting mangrove-swamps, damp forests close to the sea, salt-marshes, river-mouths, intertidal shores, or even (in a few cases) clear, shallow, sea water. Fewer genera in the temperate zone, always close to the sea.

Subfamily ELLOBIINAE

Robust, elongate-oval to oval-conical, with moderately high spire; one columellar fold and one to four parietal folds; varices occasionally present; labrum more or less thickened, internally smooth. (?U.Jur., Pal.-Rec.)

Ellobium [*Auricula*] (*s.s.*) (Fig. 842): oblong-oval with short spire and much flattened whorls; spire whorls and upper part of last whorl with weak crenulated spiral threads; aperture elongate-auriform, pointed behind; one columellar fold and one to two parietal folds just above it; labrum thickened and expanded, occasionally leaving one antilabial varix; there is a minute umbilical slit. Some species attain a large size (8·5 cm. high). ?Jur., Eoc.-Rec. Philippines, Indo-Pac., Eur.

Subfamily CARYCHIINAE

Small, elongate-oval to oval-conical, rarely sinistral; smooth or with collabral costae; aperture small, oval; columella short, with one fold; one or more parietal and palatal folds. (Jur.-Rec.)

Carychium: elongate-oval, dextral; whorls convex, more or less smooth; peristome thickened and widened; one columellar fold, one parietal fold and one to two palatal folds; small umbilical slit. Jur.-Rec. Eur., Asia, Indonesia, Philippines, N.Amer., C.Amer., Carib.

Subfamily MELAMPODINAE

Oval to pyriform, usually with low spire and rather flattened whorls; aperture high and narrow; columella with one fold; one to five parietal folds. (?Jur., Cret.-Rec.)

Melampus (*s.s.*) (Fig. 843): inverted-oval-conical, smooth; whorls narrow; aperture very narrow, subcrescentic; labrum thin, internally with a number of parallel lirae a little distance from the edge, its profile straight; columella with one strong fold; one to five parietal folds. ?Jur., Cret.-Rec. N.Amer., Eur., coasts of warm seas. *Tralia* (*s.s.*): like *Melampus*, but the one columellar fold oblique; of the two parietal folds, the upper one is weaker and lies opposite a projection of the labrum inside which is a lira. M.Olig.-Rec. Carib., N.Amer., W.Africa, Indo-Pac., Eur.

Subfamily PEDIPEDINAE

Fairly small, rounded-conical to oval, with low spire and feebly convex whorls; last whorl large, more convex; aperture fairly wide, usually with two columellar folds and a palatal fold; smooth or with spiral ornament. (L.Eoc.-Rec.)

Pedipes itself is Recent only. *Marinula* (*s.s.*): elongate-oval-conical, smooth; two weak, oblique columellar folds, and a very strong parietal fold; the labrum may have one internal lira. M.Eoc.-Rec. Indo-Pac., Eur. *Laemodonta* [*Plecotrema*] (*s.s.*): small, oval-conical, with conical spire; whorls with spiral furrows; aperture not very large, oval; labrum thickened, internally with two or three teeth; one columellar fold; two parietal folds, the lower one bifid. Olig.-Rec. Pac., Eur.

Subfamily PYTHIINAE

Oval-conical to cylindrical, rarely sinistral, with more or less high-conical spire and fairly large last whorl; aperture small; labrum internally smooth or denticulate; one columellar fold and nought to five parietal folds. (Pal.-Rec.)

Pythia [*Scarabus*] (*s.s.*) (Fig. 844): regularly oval-conical, smooth; low varices, 180° apart, each followed by a sudden diminution (or even reversal) of sutural angle, give a compressed form (as though crushed) to the otherwise globose shell; aperture fairly narrow, oblique, crescentic; usually a narrow false umbilicus; one oblique columellar fold and two parietal folds; labrum internally with lirae alternating with the teeth on the inner lip. Eoc.-Rec. Indo-Pac., Eur. *Ovatella* (*s.s.*): elongate-oval-conical, smooth; anomphalous; labrum sharp, somewhat widened, internally more or less lirate; one columellar fold and two to five parietal folds. Plio.-Rec. Eur., Medit. The subgenus *Myosotella* [*Alexia*] (Fig. 845) has one lira inside the labrum; the aperture is shorter than in *Ellobium* and there are no varices. Pal.-Rec. Eur., Medit., E.Atl.

Subfamily CASSIDULINAE

Oval, with spiral ornament or smooth; last whorl large, more or less inflated; labrum sharp, internally with or without a lira; one columellar fold; one or two parietal folds. (U.Cret.-Rec.)

Cassidula (*s.s.*): *Cassis*-like, with spiral striae and narrow umbilicus; one oblique columellar fold; one parietal fold and usually a second higher up; labrum internally thickened and with an internal denticle fairly high up; more inflated than *Ellobium*. Mio.-Rec. Indo-Pac., E.Africa, Eur.

Superfamily AMPHIBOLACEA

Globose, dextral, with low spire and convex whorls; usually with an umbilicus. (Plio.-Rec.)

Family AMPHIBOLIDAE

Umbilicate. (Plio.-Rec.)

Amphibola: whorls more or less imbricate, smooth. Plio.-Rec. N.Z., Austral., Indo-Pac.

Superfamily SIPHONARIACEA

Patelliform or capuliform; externally with a ridge from the middle to the right anterior side, corresponding to an internal groove for a respiratory tube; muscle scar asymmetrically horse-shoe-shaped. (Jur.-Rec.)

Family SIPHONARIIDAE

Outline more or less elliptical or oval; apex median or slightly posterior; usually with radial costae. (Jur.-Rec.)

Siphonaria (*s.s.*) (Fig. 846): shell open-conical, with elliptical or sub-polygonal base; apex central or slightly posterior; surface with radial costae of varying strength, producing a scalloped margin; distinguished from *Patella* by asymmetry of outline or ribbing produced by the respiratory groove. U.Cret.-Rec. Indo-Pac., Austral., Siberia, Eur., S.Amer., Carib.

Family TRIMUSCULIDAE [GADINIIDAE]

Small, capuliform; apex posterior to middle, bent over, blunt; radial costellae and growth lines form a lattice; ends of horse-shoe-shaped muscle scar club-shaped. (?Pal., Olig.-Rec.)

Trimusculus [*Gadinia*]: characters of the Family. ?Pal., Olig.-Rec. Africa, Indo-Pac., Austral., Medit., Eur.

Family ACROREIIDAE

Smooth, with irregularly elliptical outline and pulmonary channel running

---

Figs 846–861. GASTROPODA: SIPHONARIIDAE, LYMNAEIDAE, PHYSIDAE, PLANORBIDAE, ACROLOXIDAE, PUPILLIDAE, VERTIGINIDAE, ORCULIDAE, CHONDRINIDAE, ENIDAE, SUCCINEIDAE AND ENDODONTIDAE

Figs 851, 854, 857 and 860 after Cossmann and Pissarro; Fig. 855 after Sandberger; Fig. 856 after Cox; Figs 846–850, 852–853, 858–859 original.

846c, d. *Siphonaria laciniosa* (Linné), Rec. × 1·1.
847a, b. *Lymnaea stagnalis* Linné, Rec. England. T. × 0·75.
848. *Radix auricularia* (Linné), Rec. England. T. × 1·1.
849d. *Valenciennius annulatus* Rousseau, Pontian. Rumania. T. × 0·375.
850. *Physa* sp., Rec. × 1·1.
851b. *Aplexa gigantea* (Michelin), Pal. Paris B. × 0·75.
852a, c, d. *Australorbis euomphalus* (J. Sowerby), U.Eoc. (Headon). I. of Wight. × 1·1.
853a, c. *Planorbarius corneus* (Linné), Rec. England. T. × 1·1.
854e, *Pseudancylastrum arenarium* (Cossmann), Pal. Paris B. × 4·2.
855a, e. *Pupilla impressa* (Sandberger), U.Olig. Mainz B. × 3·6.
856. *Vertigo vectensis* Cox, L.Olig. (Bembridge Lmst.). I. of Wight. × 11.
857. *Orcula plateaui* (Cossmann), Pal. Paris B. × 3·75.
858. *Abida secale* (Draparnaud), Rec. England. T. × 3·75.
859. *Zebrina detrita* (Müller), Rec. Hungary. × 1·1.
860a, b. *Succinea brevispira* Deshayes, U.Eoc. Paris B. × 3.
861a, b, d. *Anguispira alternata* (Say), Rec. U.S.A. T. × 1·1.

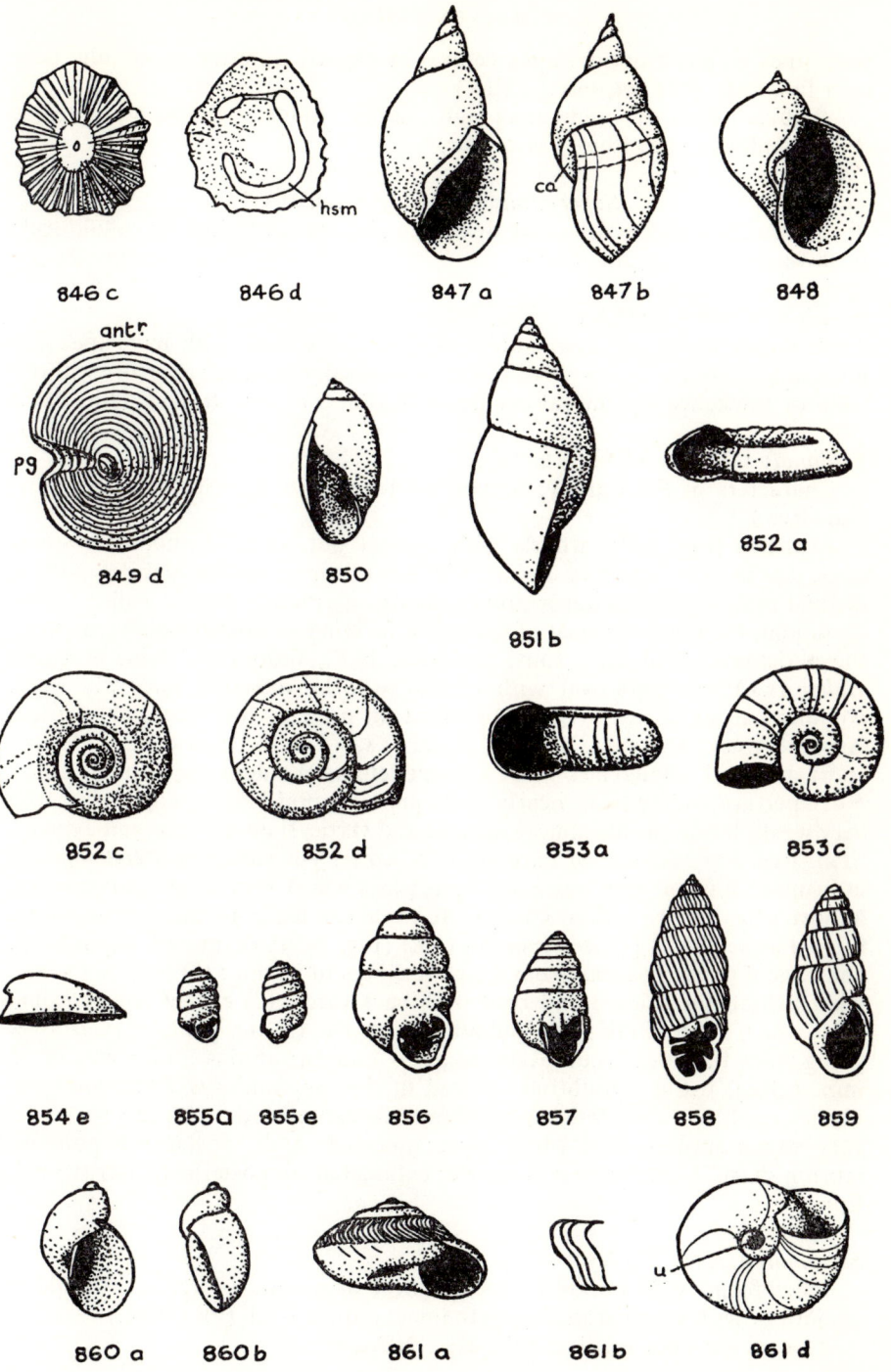

846 c   846 d   847 a   847 b   848

antr

pg

849 d   850   851 b   852 a

852 c   852 d   853 a   853 c

854 e   855 a   855 e   856   857   858   859

860 a   860 b   861 a   861 b   861 d

405

from apex to left anterior; apex central or slightly anterior to middle, bent over forwards. (U.Cret.-Eoc., ?Mio.)

Acroreia [*Acroria*]: moderately high; outline elongate-elliptical, pointed at the left anterior end. Eoc., ?Mio. Eur.

Superfamily LYMNAEACEA

Shell spiral, turreted to discoidal, dextral or sinistral, occasionally patelliform; fresh-water. (?Carb., Jur.-Rec.)

Family LYMNAEIDAE

Shell thin, ovoid and more or less elongate (oligogyral, with acute apex and inflated last whorl), rarely patelliform; occasionally sinistral; aperture without teeth or thickened lip; labral profile vertical and nearly straight. (Jur.-Rec.)

Subfamily LYMNAEINAE

Characters of the Family; limpet-like forms usually with enrolled apex. (Jur.-Rec.)

Lymnaea (Fig. 847): whorls with distinct but rounded shoulder; spiral angle about 30°; aperture about half total height; columellar lip with a distinct fold passing forwards into the peristome; labral profile slightly para-sigmoidal, the lower reversal of curvature forming an anterior embayment of the peristome. ?Pal.-Rec. Eur., N.Africa, N.Z., Indonesia, Asia, N.Amer. *Galba* (*s.s.*): elongate-oval with convex whorls; aperture narrowly oval; columella straight, not twisted; umbilical slit present. Jur.-Rec. Eur., Africa, Asia, N.Amer., C.Amer., S.Amer. *Radix* (*s.s.*) (Fig. 848): globose, with very short spire and inflated last whorl without definite shoulder; spiral angle about 50°; aperture widely oval, nearly threequarters total height; columellar fold very weak; labral profile almost straight and vertical; no anterior embayment. ?Pal.-Rec. Eur., Africa, M.East, Asia, N.Amer. The subgenus *Velutinopsis* is even more inflated and has a flat spire; last whorl very large and aperture almost subcircular. L.Plio. Crimea. It evidently leads to the extraordinary form *Valenciennius* [*Valenciennesia*] (*s.s.*) (Fig. 849): depressed capuliform, convergent on *Siphonaria*, from which it differs in having: (1) the apex some-what behind the middle and bent over backwards, (2) coarse commarginal folds which have a kink at the pulmonary groove, (3) the pulmonary grooves going from the apex to the right *posterior* margin; attains a diameter of 75 mm. L.Plio. E.Eur. Doubtfully placed in this Subfamily is *Pitharella* (pre-viously confused with *Douvilleia*): elongate-oval, smooth, with low spire and very vague sutures; last whorl large; aperture high, posteriorly pointed; labrum sharp, feebly curved in profile; columellar lip posteriorly narrow. Pal. Eur.

Subfamily LANCINAE

Patelliform, with no spire; apex central or a little posterior to middle; smooth or with commarginal growth lines; outline oval. (Plio.-Rec.)

Lanx (*s.s.*): apex median. Plio.-Rec. N.Amer.

406

Family CHILINIDAE

Elongate-oval, dextral, smooth, with moderately low, blunt spire; last whorl large, inflated-oval; aperture auriform; peristome continuous; labrum simple; columella with one or two folds. (L.Plio.-Rec.)

Chilina: characters of the Family. L.Plio.-Rec. S.Amer.

Family PHYSIDAE

Sinistral, ovate, smooth, thin shells; last whorl large, anomphalous; aperture oval, posteriorly pointed; columella not folded, twisted. The general aspect is not unlike a sinistral *Lymnaea*, but the build is narrower, as is the aperture. (?Carb., Jur.-Rec.)

Physa (*s.s.*) (Fig. 850): inflated-oval, smooth, with short, acute spire; aperture oval, posteriorly pointed, about three-quarters the total height; columella twisted; labral profile not very oblique, almost uniformly convex. ?Carb., Jur.-Rec. Eur., cosmop. *Aplexa* [*Aplecta*] (*s.s.*) (Fig. 851): like *Physa*, but spire longer, aperture less than half the total height, and labral profile oblique and nearly straight. Jur.-Rec. Eur., N.Amer., Carib., C.Amer., Asia.

Family PLANORBIDAE

Shell discoidal or with prominent spire; always sinistral (but easily interpreted as dextral owing to the discoidal shape: in following the descriptions below, the sinistral character must be borne in mind); aperture subcircular to subelliptical, modified by the protuberance of the parietal region (which, owing to the discoidal shape, forms the whole inner lip); peristome oblique, farther in advance below than above; labral profile concave, tending to V-shaped in some carinate forms, or becoming parasigmoidal by reversal of curvature towards upper suture (in dextral forms such reversal would make it sigmoidal). (Jur.-Rec.)

Subfamily PLANORBINAE

Not very large, usually discoidal with flat or somewhat sunken spire, occasionally oval to oval-cylindrical; whorls rounded or with one or more keels; base more or less deeply umbilicate. (Jur.-Rec.)

Planorbis: rather flatly discoidal, with upper and lower sides gently sunken; whorls growing slowly, with one carina; aperture obliquely oval, angular at the keel. U.Olig.-Rec. Eur., N.Africa, M.East, Madagascar, Asia, N.Z. *Gyraulus* (*s.s.*): usually discoidal, with flat or gently sunken upper side and a shallow umbilicus on the lower side; whorls growing fast, with spiral striae; aperture wider than high; labral profile parasigmoidal. Jur.-Rec. Eur., N.Africa, Asia, Indonesia, ?Austral., N.Amer. [The species *G. trochiformis* (Stahl) in the U.Miocene of Steinheim-am-Albuch, Württemburg, shows an enormous range of form, from discoidal to turbinate, and from smooth to strongly carinate; the discussion of its variations has given rise to a voluminous literature.] *Segmentina:* flat and lenticular, smooth, with more or less concave and deeply and narrowly umbilicate upper side and slightly convex but

407

medially sunken lower side; aperture very oblique, rounded or rounded-triangular; last whorl internally with (usually) three lirae. Olig.-Rec. Eur., Africa, Madagascar, Asia Minor, Asia, ?N.Amer. *Hippeutis:* shell 'thin' (i.e. height of last whorl less than one-quarter diameter of disk), more or less carinated but otherwise smooth; aperture very oblique (30° or more), oval to cordate; labral profile a simple curve or with slight reversal on base; no internal lirae. Pal.-Rec. Africa, Eur., Asia. *Australorbis* [*Planorbina*] (Fig. 852): moderately 'thin' shells (i.e. height of last whorl about one-quarter diameter of disk); whorls rather angulated; basal umbilicus extremely shallow, apical umbilicus deep or shallow but always defined by a ridge corresponding to upper angulation of whorl; aperture asymmetrical in outline; labral profile convex and strongly prosocline on upper and under faces, slightly concave and vertical between. U.Cret.-Rec. Carib., N.Amer., S.Amer., Eur., Asia. *Planorbarius* [*Coretus*] (Fig. 853): 'thick' shells (i.e. height of last whorl exceeds one-third the greatest diameter of the disk); whorls rounded; early whorls with spiral ornament, later with growth lines only; aperture not very oblique (20° to 25° from vertical), subcircular-lunate; labral profile a simple curve. U.Eoc.-Rec. Eur., N.Africa, Asia.

Subfamily BULININAE
Elongate-oval to oval-cylindrical, rarely discoidal; spire usually fairly low; last whorl and aperture large. (Jur.-Rec.)

Bulinus [*Bullinus, Isidora*] (*s.s.*): oval, with fairly low spire and blunt apex; whorls convex, with fine collabral costellae; last whorl large, with narrow umbilicus; aperture oval, pointed posteriorly; difficult to distinguish from some members of the Physidae. ?U.Plio., Pleist.-Rec. Africa, Eur., M.East.

Family NEOPLANORBIDAE
Small, usually dextral; spire flat, of very few whorls, or shell uncoiling and *Ancylus*-like. (Mio.-Rec.)

Neoplanorbis is Recent only. *Amphigyra:* capuliform and *Ancylus*-like (see below), of one and a half whorls only; spiral striae in later stages; aperture broadly oval, sometimes with a small septum. Mio.-Rec. N.Amer.

Family FERRISSIIDAE
Patelliform or capuliform, with apex more or less removed to the right posterior; smooth or with radial ornament; aperture widely elliptical, with or without a septum. (Eoc.-Rec.)

Ferrissia (*s.s.*): patelliform, with apex low and more or less posterior to middle; surface with radial striae all over or apically only. Eoc.-Rec. N.Amer., Eur., Africa, Asia, Austral.

Family ANCYLIDAE ('Fresh-water limpets')
Small, thin-shelled, patelliform to capuliform, fairly high; apex bent over, near posterior margin, slightly bent to the right; surface with radial striae; aperture broadly oval. (M.Olig.-Rec.)

Ancylus (*s.s.*): apex not spiral. M.Olig.-Rec. Eur., N.Africa, M.East.

Family ACROLOXIDAE

Small, patelliform to capuliform, with apex removed to left posterior. (U.Cret.-Rec.)

Acroloxus [*Velletia*]: patelliform, smooth, rather low, with apex behind and to left of middle; outline narrowly oval. U.Plio.-Rec. Eur. *Pseudancylastrum* (Fig. 854): capuliform, smooth or with radial striae; outline broadly oval. U.Cret.-Rec. Siberia, Eur.

Order STYLOMMATOPHORA (typical land snails)

Shell usually present and spiral, occasionally wholly or partly covered by the mantle. (?U.Cret., Pal.-Rec.)

These shells fall roughly into two shape categories—*heliciform* and *elongate*. Though neither is coextensive with a natural group it is convenient to give a description of the former type, to save repetitions in the systematic accounts.

The *heliciform* shell is unlike that of any marine gastropod, though it is most nearly approached among the Naticidae. It is depressed-turbinate to globose, the spire being low and obtuse, the last whorl rounded in outline and occupying about two-thirds the total height or more. The sutural angle is very low, but in an adult shell it usually increases markedly in the final quarter whorl. The base is rounded and may have a true umbilicus or a false umbilicus or umbilical fissure or be quite anomphalous. The aperture is typically outlined by two distinct curves, like the moon in a partial eclipse (*lunate*), the peristome forming an incomplete ellipse broken into by a smaller reverse curve formed by the outline of the penultimate whorl (*parietal* region). The incomplete ellipse may be an almost symmetrical horse-shoe, or very asymmetrical. The junction of the two curves may be abrupt at both ends (*Helicella*, Fig. 882; *Pleurodonte*, Fig. 886), or the inner end of the peristome may be bent into a vertical columella (*Grandipatula*, Fig. 863), or there may be a gradual change of curvature so that the aperture becomes more or less auriform (*Cepaea*, Fig. 880). Sometimes the whorls are more or less flat-sided with a keel immediately above the suture, the aperture becoming subrhomboidal (*Archaegopis*, Fig. 864). The labral profile is always more or less straight, oblique and prosocline on the suture.

Among elongated shells we very rarely find any that are truly turreted. Much commoner are the *bulimoid* and *pupiform* shapes. The former (from the much abused generic name *Bulimus*, which is really a synonym of *Bithynia*) is very broadly subulate or ovoid, the greatest width being at about half the height or lower down, the visible height of each whorl about fifty per cent greater than that of the previous whorl, and the aperture more or less auriform. The pupiform shell is cylindrical with a dome-like apex. Fusiform shells are occasionally produced, not as in *Fusinus*, but by a contraction of diameter in the latest whorls.

Many Stylommatophora restrict the aperture by forming *teeth*: these are referred to, according to position, as *columellar*, *parietal*, *palatal* (on vertical

409

part of labrum), and *basal* (anterior part of labrum). Pilsbry termed the two first *lamellae*, the two last *plicae* (without reference to shape).

The Stylommatophora are classified according to certain anatomical characters of the very complex hermaphrodite reproductive organs. No shell characters distinguish the main divisions, here given without diagnosis. Any attempt to classify on shell characters would lead to the association of forms which are homoeomorphic but unrelated, with consequent danger of mis-interpretation of palaeontological history.

Compared with the Basommatophora, there are two pairs of tentacles with the eyes borne on the tips of the posterior pair. The origin of the Stylom-matophora is obscure; it seems likely that they were evolved directly from primitive marine forms, not through the intermediary of fresh-water forms.

Suborder ORTHURETHRA

According to Pilsbry this primitive group is highly developed in the Pacific Islands, comprising parallel forms to all the higher types (Sigmurethra) except the degenerate slugs. In the great continental regions it is represented chiefly by diminutive forms, mainly pupiform, with a few heliciform and bulimoid genera. (?U.Cret.(Danian), Pal.-Rec.)

Superfamily PUPILLACEA [VERTIGINACEA]

Small, usually elongate-oval to turreted; aperture often with teeth. (Pal.-Rec.)

Family PUPILLIDAE

Elongate-oval to cylindrical; aperture with distinct angular and parietal lamellae and a strong columellar fold as well as palatal folds; these may some-times be reduced or supplemented by additional folds; apex usually depressed and increase in diameter ceasing after three or four whorls; surface usually ornamented with fine rugae or costellae of Endodontid type, straight and oblique; aperture twisted into a vertical plane tangential to the spire, so that the peristome is not parallel to the growth lines (this is a modification of the terminal deviation so usual in heliciform shells); peristome slightly everted, producing a minute false umbilicus. (Pal.-Rec.)

The apertural teeth are nearly always present, but are sometimes so far within that the shell must be cut or broken to verify them. When they are numerous, it is possible to name each individual tooth and recognize which are present in species presenting a lesser number. The three dominant teeth are usually a *columellar* and a *parietal* about in the centre of their portions of the apertural margin, and a *lower palatal* about the middle of the complete curve of the labrum: these three occupy the points of an equilateral triangle; other important teeth are the *angular* on the parietal margin near its junction with the labrum, an *upper palatal* half way up from the lower palatal, and a *basal* on the labrum close to the columella.

The family shows striking homoeomorphy (except in size) with certain neo-tropical families of Sigmurethra (Cerionidae, Urocoptidae), but is mainly holarctic.

410

Subfamily PUPILLINAE

Aperture relatively feebly dentate; no teeth in youth; strong tendency to reduction of teeth. (Pal.-Rec.)

Pupilla [*Pupa auctt.*] (*s.s.*) (Fig. 855): typically pupiform, with fine raised growth lines; peristome usually reflected or expanded; no angular fold; teeth reduced to three or four in number. U.Olig.-Rec. Eur., N.Africa, N.Asia, N.Amer.

Subfamily LAURIINAE

Teeth usually strong and well developed. (U.Olig.-Rec.)

Lauria (*s.s.*): only angular, parietal and columellar folds developed. U.Plio.-Rec. Eur., Canaries, Madeira, N.Africa, Transcaucasia.

Family VERTIGINIDAE

Small, oval to oval-cylindrical, with hollow axis which is, however, often closed at the end; aperture usually with two palatal folds, a parietal fold and a columellar fold, but the teeth are sometimes reduced. (Pal.-Rec.)

Subfamily VERTIGININAE

Aperture usually with six teeth (two parietal, one columellar, one basal and two palatal), but teeth sometimes lacking. (Pal.-Rec.)

Vertigo (*s.s.*) (Fig. 856): minute shells, very short and broad, usually dextral, rarely sinistral, usually with six teeth well developed and projecting well into the aperture; basal, angular and upper palatal folds may be reduced. Pal.-Rec. Eur., N.Africa, Asia, Japan, N.Amer., Carib., C.Amer.

Subfamily TRUNCATELLININAE

Usually cylindrical, with aperture toothless or nearly so. (Eoc.-Rec.)

Truncatellina: cylindrical, with blunt apex and convex whorls; narrow umbilicus; growth line rugae; aperture with one parietal fold and sometimes also one columellar fold and one lower palatal fold. U.Olig.-Rec. Madeira, Canaries, Cape Verde Is., Eur., N.Asia, C.Asia, Japan, Africa.

Subfamily NESOPUPINAE

Form, aperture and denticulation as in *Vertigo*; whorls convex, usually with distinct growth line costellae. (U.Olig.-Rec.)

Nesopupa (*s.s.*) is Recent only. Its subgenus **Indopupa** is oval, with convex whorls bearing growth line costellae and pits; angular, parietal, columellar and two palatal lamellae present. U.Olig.-Rec. Burma, Ceylon, Borneo, Philippines, India, M.East, Eur.

Family COCHLICOPIDAE

Elongate-oval to oval-cylindrical, with relatively high last whorl, smooth; aperture semi-oval, posteriorly pointed, with or without denticles. (Pal.-Rec.)

Cochlicopa: bulimoid, with last whorl slightly more than half the total height; anomphalous; peristome internally thickened, without teeth; columella anteriorly feebly twisted; labrum and floor sinuous. Pal.-Rec. Eur., Madeira, Azores, Iceland, Asia, N.Africa, N.Amer. *Azeca:* ovoid, with last whorl not more than half the total height; labrum posteriorly notched, with a marginal tooth and usually an additional internal palatal tooth; columella with a vertical lamella and a basal tooth; parietal region with a tooth and a lamella. Pal.-Rec. Eur.

Family PYRAMIDULIDAE

Small, thin-shelled, turbinate, with blunt apex and convex whorls bearing fine collabral growth line threads; last whorl more or less descendent; aperture rounded, without teeth; umbilicus open. (Eoc.-Rec.)

Pyramidula: characters of the Family. Eoc.-Rec. Eur., Japan.

Family ORCULIDAE

Cylindrical, with slowly growing whorls; aperture rounded; peristome more or less widened, with a long, strong parietal lamella, one or two columellar folds, sometimes a weak angular lamella, and occasionally also a palatal fold. (Pal.-Rec.)

Orcula (*s.s.*) (Fig. 857): cylindrical to cylindrical-conical, with umbilical slit; whorls feebly convex, later ones with growth lines or fine collabral costellae; one long parietal lamella, angular lamella small or absent, two deep-set columellar folds, and sometimes a lower palatal fold; peristome widened. Pal.-Rec. Eur., M.East, Caucasus.

Family CHONDRINIDAE

Low-conical to cylindrical; aperture rounded, usually with one angular fold, one or two parietal folds, one or two columellar lamellae, and two or more palatal folds. (M.Eoc.-Rec.)

Subfamily CHONDRININAE

Usually cylindrical with blunt apex and numerous whorls; columellar 'perforation' mostly closed; aperture usually with two columellar lamellae and various other lamellae and folds. (M.Eoc.-Rec.)

Chondrina (*s.s.*): cylindrical-conical, with five to ten convex whorls bearing growth line threads; umbilical slit present; aperture with six or more teeth. ?L.Mio., Pleist.-Rec. Eur., N.Africa. *Abida* (Fig. 858): elongate-pupiform; whorls with oblique growth line threads; aperture with at least six principal teeth and often several others, the palatal ones extending some distance in; teeth stronger than in *Chondrina*. M.Eoc.-Rec. Eur.

Subfamily GASTROCOPTINAE

Oval-conical to oval-cylindrical, with convex whorls bearing oblique growth line threads; peristome more or less widened; angular lamella and parietal lamella close together or fused; normally also with columellar lamella and palatal folds. (M.Olig.-Rec.)

412

Gastrocopta (*s.s.*): very small, short, with narrow umbilicus; aperture with columellar fold and two or more palatal folds, and sometimes also a basal fold; angular and parietal lamellae fused into a curved, simple lamella. Plio.-Rec. Azores, Cape Verde Is., Carib., N.Amer., C.Amer., Africa, Ceylon, Philippines, Hawaii. In the subgenus *Albinula* the angular and parietal lamellae are curved toward the right apertural angle, united anteriorly and with a bifid tip; columellar lamella, weak basal fold, two palatal folds and usually also a suprapalatal fold present. M.Olig.-Rec. N.Amer., C.Amer., Eur.

Family VALLONIIDAE

Heliciform, spire low to prominent, umbilicate; whorls with growth lines or fine collabral costellae; aperture rounded, with or without folds. (Pal.-Rec.)

Subfamily VALLONIINAE

Very small translucent shells, of about three and a half whorls; depressed heliciform; with rather small but well defined umbilicus; aperture nearly circular, very slightly cut into in the parietal region, peristome conspicuously thickened. (Pal.-Rec.)

Vallonia: characters of the Subfamily. Pal.-Rec. Eur., N.Africa, M.East, Asia, N.Amer.

Subfamily ACANTHINULINAE

Spire low to prominent; whorls convex, often with collabral costellae; aperture edentulous; peristome sharp, simple. (Pal.-Rec.)

Acanthinula: globose-turbinate, with narrow umbilicus; whorls convex or bluntly angular, with fine prosocline costellae; columellar lip fairly wide. Pal.-Rec. Eur., Canaries, Azores, Africa, Asia Minor, Transcaucasia.

Subfamily STROBILOPSINAE

Low-heliciform to turbinate, with convex or bluntly angular whorls bearing growth lines or costellae; aperture small, oblique, suboval; peristome thickened and reflected; two or three parietal lamellae. (M.Eoc.-Rec.)

Strobilops (*s.s.*): turbinate-heliciform, with cyrtoconoid spire and collabral costellae; two parietal lamellae and several other internal folds. M.Eoc.-Rec. N.Amer., C.Amer., S.Amer., Eur.

Family PLEURODISCIDAE

Low-heliciform, with open umbilicus, convex whorls and fine collabral costellae; aperture broadly lunate, without folds or lamellae, little oblique; peristome simple. (U.Olig.-Rec.)

Pleurodiscus: characters of the Family. U.Olig.-Rec. Eur., N.Africa, Asia Minor, M.East.

Family ENIDAE

Oval-conical to oval-cylindrical, usually dextral, sometimes with a narrow

413

umbilicus; aperture oval, higher than wide, occasionally dentate; peristome more or less everted. (?Pal., U.Eoc.-Rec.)

Subfamily ENINAE
Oval-conical to cylindrical-conical, with blunt apex; whorls gently convex, with growth line rugae; aperture edentulous. (U.Eoc.-Rec.)

Ena [*Bulinus*] (*s.s.*): elongate-oval-conical, with growth line rugae and fine spiral striae; labrum sharp, slightly flaring. U.Mio.-Rec. Eur., Asia Minor, Transcaucasia, M.East. *Zebrina* (*s.s.*) (Fig. 859): elongate-oval, bulimoid in form, dextral; peristome simple, without palatal fold. L.Plio.-Rec. Eur., M.East. *Napaeus* (*s.s.*): oval to oval-conical, with an umbilical slit; growth line rugae often wrinkled; peristome flaring. U.Eoc.-Rec. Canaries, Madeira, Azores, Eur.

Subfamily CHONDRULINAE
Elongate-oval or oval-conical, with one exception dextral; peristome callous, usually dentate. (L.Plio.-Rec.)

Chondrula (*s.s.*): elongate-oval, dextral; aperture with one parietal lamella, one columellar lamella, one palatal tooth, and often a small angular tooth. M.Plio.-Rec. Eur., Transcaucasia, M.East.

Subfamily JAMINIINAE
Much like the Chondrulinae but often sinistral. (U.Plio.-Rec.)

Jaminia (*s.s.*) is Pleistocene to Recent. The subgenus *Multidentula* is inflated-oval, dextral or sinistral; aperture with one angular tooth, one columellar lamella, one basal tooth and two or three palatal teeth. U.Plio.-Rec. Asia Minor, Eur., M.East.

Subfamily PACHNODINAE [CERASTUINAE]
Oval-conical, rather thin-shelled, with rather rapidly growing whorls; last whorl rather inflated; aperture edentulous; columella rather straight. (?Pal., U.Eoc., L.Mio.-Rec.)

Pachnodus is Recent only. Doubtfully included in the Subfamily is *Procerastus:* whorls moderately convex, with distinct growth line costellae; last whorl inflated-oval, with an umbilicus; peristome reflected. Pal.-U.Eoc. Eur.

Suborder HETERURETHRA
Shell thin and *Lymnaea*-like (with or without spire) or rudimentary. (Pal.-Rec.)

Superfamily SUCCINEACEA
Lymnaea-like, with large last whorl and aperture. (Pal.-Rec.)

Family SUCCINEIDAE
Characters of the Superfamily. (Pal.-Rec.)

Subfamily SUCCINEINAE
Characterized by the anatomy. (Pal.-Rec.)

Succinea (*s.s.*) (Fig. 860): shell thin, ovate, paucispiral, with short spire and large last whorl; aperture oval, more than half the total height, with thin labrum. Habitat marshy. Pal.-Rec. Eur., N.Africa, Asia, Austral., N.Amer.

Suborder SIGMURETHRA
Characterized by the anatomy. (?Palaeozoic, Cret.-Rec.)

Superfamily ENDODONTACEA
Shell usually rather low and widely umbilicate, more rarely approaching trochoid or turbinate; peristome simple, sharp; aperture occasionally with teeth or lirae. (Pal.-Rec.)

Family ENDODONTIDAE
Heliciform shells with thin, simple labrum; growth lines raised into low costellae with rounded tops, a steep face away from the aperture, and a 45° slope towards it; costellae tending to be more or less irregular; labral profile sinuous, parasigmoidal, strongly oblique and prosocline, but curving rapidly to meet the suture at right angles; last whorl rounded or keeled; aperture occasionally dentate. (Pal.-Rec.)

Subfamily PUNCTINAE
Small, conical to very low, usually with distinct growth line costellae, umbilicate; last whorl rounded or angular; aperture with or without a columellar lamella. (U.Olig.-Rec.)

Punctum (*s.s.*): low-heliciform, with rounded last whorl; growth line costellae and often also very fine spiral striae; first one and a half whorls with extremely fine spiral striae; aperture rounded-lunate, without basal lamella. U.Olig.-Rec. N.Amer., Eur., Africa, Asia.

Subfamily DISCINAE
Low-heliciform to almost discoidal, with convex or carinate whorls; protoconch extremely finely cancellate; later whorls with collabral costellae or costae which become sigmoidal on the base; umbilicus wide; aperture lunate, edentulous. (Pal.-Rec.)

Discus [*Patula*] (*s.s.*): very low-heliciform, with wide umbilicus; last whorl usually rounded. Pal.-Rec. Eur., N.Amer., Asia. **Anguispira** (*s.s.*) (Fig. 861): like *Discus*, but not so low, whorls often mainly subangular, and aperture wider than umbilicus. ?L.Eoc.-Rec. N.Amer.

Family ARIONIDAE
Shell small in comparison to body, sometimes partly or wholly covered by the mantle. (U.Mio.-Rec.)

415

Subfamily BINNEYINAE

Shell spiral or merely a bent plate, free or partly covered by the mantle. (U.Mio.-Rec.)

Craterarion: shell a thick, elongate-oval plate, convex above, concave below. U.Mio. N.Amer.

Superfamily ZONITACEA

Shell thin, heliciform to discoidal, usually with a wide umbilicus, occasionally reduced or lacking. (?U.Cret., Pal.-Rec.)

Family ZONITIDAE

Whorls more or less convex, usually smooth or weakly ornamented; aperture lunate, only exceptionally with small teeth; labral profile almost perfectly straight, oblique; terminal deviation of sutural dip very slight or non-existent. Distinguished from Helicidae by peristome being neither thickened nor reflected. (?U.Cret., Pal.-Rec.)

Subfamily ZONITINAE

Heliciform to discoidal, usually with a well developed umbilicus. (?U.Cret., Pal.-Rec.)

Zonites (*s.s.*) (Fig. 862): low-heliciform, with low, convex spire, polygyral, with deep and moderately wide umbilicus; growth lines rather coarse and granulated by spiral striae, weak or absent on base; floor sinuous. Olig.-Rec. Eur., Asia Minor. *Grandipatula* (*s.s.*) (Fig. 863): globose-hemispherical, with short, obtuse spire; last whorl nearly nine-tenths the total height; umbilicus wide and deep, subcarinate; aperture about as high as wide, obliquely oval, about one-quarter truncated by parietal region; growth line gently parasigmoidal; fine growth line costellae and widely spaced spiral striae, the

---

Figs 862–875. GASTROPODA: ZONITIDAE, VITRINIDAE, LIMACIDAE, SUBULINIDAE, MEGASPIRIDAE, CLAUSILIIDAE, FILHOLIIDAE, OLEACINIDAE AND TESTACELLIDAE

Figs 863, 869 and 870 after Cossmann and Pissarro; Figs 864, 868, 871 and 872 after Sandberger; Figs 867 and 875 after Pictet; Figs 862, 865, 866, 873 and 874 original.

862a, d. *Zonites algirus* (Linné), Rec. S.France. T. × 0·75.
863a, d. *Grandipatula hemisphaerica* (Michaud), Pal. Paris B. T. × 0·75.
864a, c, d. *Archaegopis discus* (Thomae), U.Olig. Mainz B. T. × 1·25.
865a, d. *Vitrea cristallina* (Müller), Rec. S.W.Ireland. T. × 6·75.
866. *Vitrina pellucida* (Müller), Rec. England. T. × 3·75.
867c. *Limax larteti* Dupuy, M.Mio. Sansan, Aquitaine. × 2·25.
868. *Scalaxis rillyensis* (de Boissy), Pal. Paris B. T. × 3·75.
869a, b. *Distoechia parisiensis* (Deshayes), Pal. Paris B. T. × 2·25.
870. *Palaeostoa exarata* (Michaud), Pal. Paris B. × 1.
871. *Canalicia articulata* (Sandberger), U.Olig. Mainz B. T. × 2·6.
872. *Laminifera rhombostoma* (Boettger), U.Olig. Mainz B. T. × 2·6.
873. *Filholia* sp., L.Olig. (Bembridge Lmst.). I. of Wight. × 0·5.
874. *Palaeoglandina costellata* (J. Sowerby), L.Olig. (Bembridge Lmst.). × 0·5.
875c, d. *Testacella larteti* (Dupuy), M.Mio. Sansan, Aquitaine. × 2·6.

416

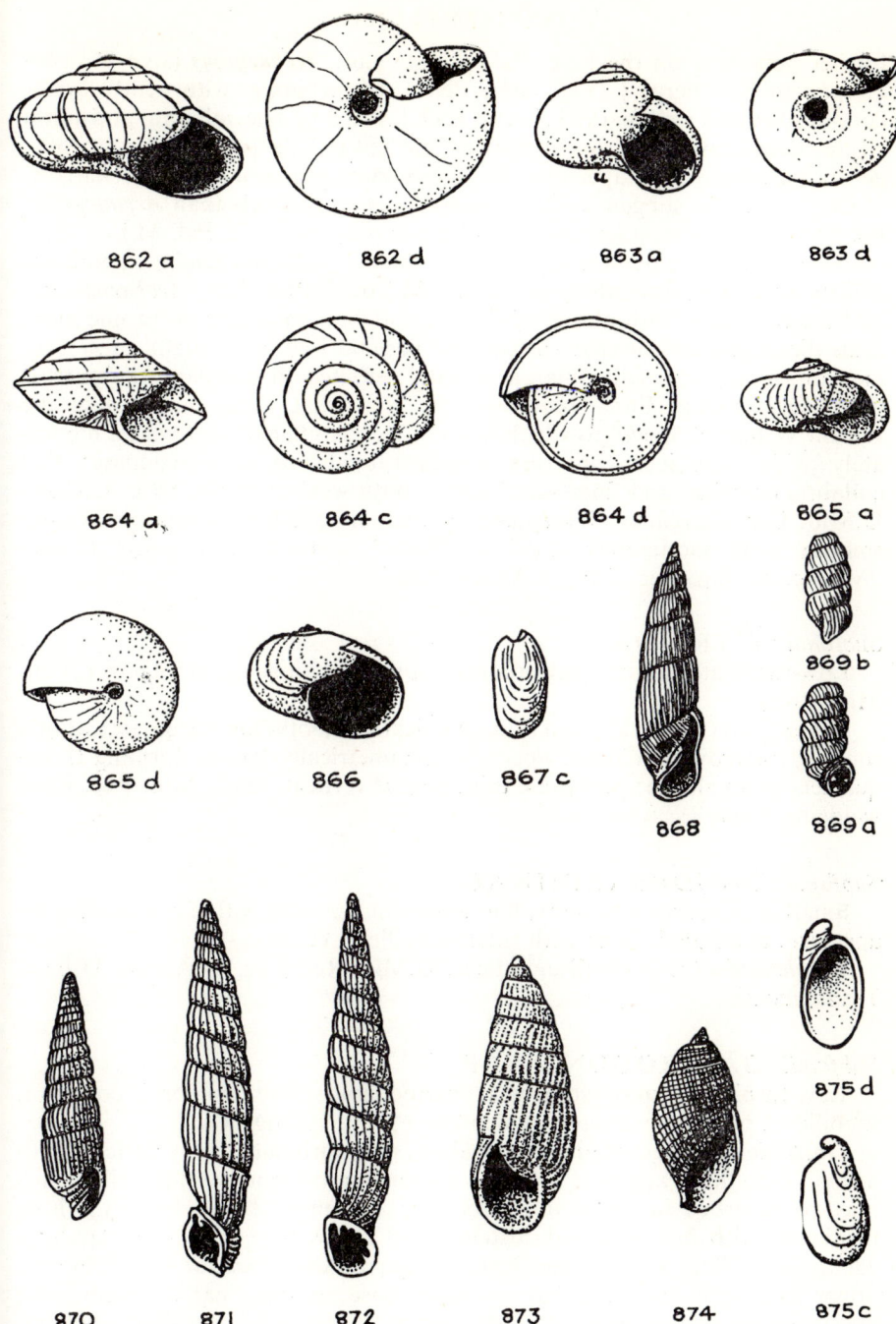

862 a   862 d   863 a   863 d

864 a   864 c   864 d   865 a

869 b

865 d   866   867 c   868   869 a

875 d

870   871   872   873   874   875 c

417

former persisting on the base. Pal.-M.Eoc. Eur. *Archaegopis* (*s.s.*) [*Trocho-morpha* of Sandberger] (Fig. 864): depressed-turbinate, with gently cyrto-conoid spire and well marked sutural keel, latticed by fine collabral and spiral striae; deep, rather narrow umbilicus; no sutural deviation; labral profile gently parasigmoidal; aperture subrhomboidal; protoconch rather narrow. U.Olig. Eur. The subgenus *Phacozonites* has fewer whorls than *Archaegopis*, the protoconch is broader, and the spire is less cyrtoconoid. Pal.-M.Eoc. Eur. *Archaeozonites:* less depressed and with a thicker shell than *Zonites*; umbilicus a little narrower, and no spiral striae. M.Eoc.-L.Plio. Eur. *Archaeoplecta:* globose-turbinate, with obtuse apex; fine collabral costellae; peristome more than three-quarters circular, broken by parietal region of slight curvature; margins acute, columellar margin slightly reflected to overhang the narrow umbilicus. U.Olig. Eur. *Omphalosagda:* somewhat dome-shaped like the Recent W.Indian *Sagda* (to which it is not closely related), but with a moder-ately wide umbilicus; aperture elliptical-lunate to circular-lunate; fine collabral costellae and finer spiral striae, both weaker on the base. U.Olig.-U.Mio. Eur. *Oxychilus* [*Hyalinia*] (*s.s.*): subdiscoidal, very low, practically smooth, with moderately small umbilicus; aperture transversely lunate. ?M.Eoc., L.Olig.-Rec. Eur., N.Africa, Asia.

Subfamily VITREINAE

Low-turbinate to discoidal; anomphalous or with small umblicus. (?Pal., Mio.-Rec.)

Vitrea (*s.s.*) (Fig. 865): shell thin, practically smooth, low-turbinate; whorls rounded; narrow umbilicus; aperture asymmetrically lunate, forming three-quarters of an ellipse; peristome thin, almost vertical. ?Pal., Mio.-Rec. Eur., Madeira, Canaries, Asia.

Subfamily DAUDEBARDIINAE

Small in proportion to body, flat, auriform, smooth, with large, horizontal aperture; anomphalous or with small umbilicus. (U.Mio.-Rec.)

Daudebardia (*s.s.*): small umbilicus. U.Mio.-Rec. Eur., N.Africa, M.East, Transcaucasia.

Subfamily GASTRODONTINAE

Low-turbinate, spire somewhat cyrtoconoid; fine collabral costellae; umbilicus; peristome smooth or dentate. (?Pal., U.Olig.-Rec.)

Gastrodonta is Recent only. *Ventridens* (*s.s.*): turbinate, finely latticed, with a small umbilicus; aperture with one columellar and one basal lamella which may be resorbed in the adult. ?Pal.-Rec. N.Amer. *Zonitoides* (*s.s.*): low-turbinate, with fine collabral costellae and fairly wide umbilicus; aperture edentulous. ?Pal., U.Olig.-Rec. Eur., N.Amer., Asia. *Janulus:* low-turbinate, upper side with fine collabral costellae, base smooth; narrow umbilicus; aperture with several rows of three to four palatal teeth. U.Olig.-Rec. Madeira, Canaries, Eur.

418

Family VITRINIDAE

Shell thin and transparent, of very few whorls, rapidly enlarging; anomphalous or with narrow umbilicus. (Pal.-Rec.)

Vitrina (*s.s.*) (Fig. 866): low-rounded, anomphalous; aperture very large, elliptical-lunate but with vertical columellar lip; peristome thin, labrum not very oblique, slightly arcuate and bending abruptly normal to suture. U.Olig.-Rec. Eur., Greenland, N.Asia, N.Amer. *Provitrina:* like *Vitrina*, but with a more depressed spire and a flattened base. Pal. Eur.

Family PARMACELLIDAE

Shell with a short spiral beginning or not spiral. (?U.Eoc., L.Plio.-Rec.)

Parmacella (*s.s.*): short spiral early stage; outline oval. ?U.Eoc., L.Plio.-Rec. Iran, Eur., N.Africa, Canaries, M.East.

Family MILACIDAE

Shell an oval, gently convex plate, with inconspicuous apex on the long axis. (?U.Eoc., L.Olig.-Rec.)

Milax: apex median, terminal. ?U.Eoc., L.Olig.-Rec. Eur., N.Africa, M.East.

Family LIMACIDAE

The familiar herbivorous slugs, related to Zonitidae but with shell reduced to a thin, flat or gently convex plate covering the lung chamber and concealed under the mantle; apex terminal, lateral; it has no spiral structure and is not unlike a single valve of *Lingula*. (U.Olig.-Rec.)

Limax (*s.s.*) (Fig. 867): like *Milax*, but thinner-shelled and apex asymmetrical. U.Olig.-Rec. Eur., N.Africa, M.East.

Superfamily ARIOPHANTACEA

Low-conical to low-turbinate; in some groups the shell is reduced and covered by the mantle. (?U.Cret., Pal.-Rec.)

Family EUCONULIDAE

Small, with slowly growing whorls; last whorl usually rounded, anomphalous or with small umbilicus. (?U.Cret., Pal.-Rec.)

Euconulus (*s.s.*): globose-turbinate with conical spire and deep sutures; fine growth lines; anomphalous; aperture lunate; peristome simple, sharp. ?U.Cret., Pal-Rec. Eur., Bermuda, N.Amer., widespread.

Superfamily ACHATINACEA

Elongate-oval to fusiform, sometimes very large; aperture oval or pyriform; columella usually truncated anteriorly. (U.Cret.(Danian)-Rec.)

Several Tertiary fossils previously placed in the genus *Achatina* are now referred to genera in other families, and several genera previously placed in the family Achatinidae are now placed in the family Subulinidae. As now conceived, the family Achatinidae did not appear before the Pleistocene.

419

Family FERRUSSACIIDAE
Small, usually smooth, elongate-oval to cylindrical shells often with blunt apex and usually anomphalous; aperture ovoid to auriculate, simple or with a few teeth. (U.Eoc.-Rec.)

Ferrussacia (*s.s.*): elongate-oval to oval-cylindrical, with cyrtoconoid spire less than half the height of the shell; aperture pyriform; columella with one oblique fold; labrum medially prominent, internally thickened. L.Mio.-Rec. N.Africa, Eur., Canaries, Madeira, Mauritius. *Coilostele:* thin-shelled, slender, more or less cylindrical; smooth or later whorls with collabral costellae; aperture small, narrowly oval, oblique; columella with one fold; inner whorls resorbed. U.Eoc.-Rec. India, N.Africa, M.East, Eur.

Family SUBULINIDAE [RUMINIDAE]
Elongate-oval to cylindrical, with numerous rather flat-sided whorls, smooth or with collabral costellae; aperture ovate, usually edentulous; columella sometimes truncated anteriorly. (Pal.-Rec.)

Subfamily SUBULININAE
Slender; columella usually truncated anteriorly. (U.Eoc.-Rec.)

Subulina is Recent only. *Opeas* (*s.s.*): small, elongate, slightly subulate, with obtuse apex and impressed sutures; aperture small, oval; columella not twisted or truncated, with straight free edge bounding a small false umbilicus; labrum thin, its profile oblique, slightly sigmoidal (being almost straight for most of its course, with strongly opisthocline curve to suture and reverse curve at opposite end); fine collabral costellae and sometimes also very fine spiral striae. U.Eoc.-Rec. Jamaica, E.Africa, Eur., China, tropical and subtropical Old and New Worlds.

Subfamily RUMININAE
Slender, turreted-conical to cylindrical-conical; aperture small, oval; columella not truncated anteriorly; columellar lip callous and reflected. (M.Mio.-Rec.)

Rumina: multispiral, cylindrical-conical but decollate far below apex (leaving only four or five whorls), the new tip closed by a septum; somewhat subulate, sutures slightly impressed; columella not twisted or truncated, callus lifted to the free margin of a minute false umbilicus; oblique collabral costellae which are stronger posteriorly. M.Mio.-Rec. Eur., N.Africa.

Subfamily OBELISCINAE
Usually elongate-oval-conical to high-turreted; aperture oval, posteriorly pointed; columella vertical, little truncated in adult. (Pal.-Rec.)

All are American forms except for one Recent South African genus; unknown between Palaeocene and Recent.

Obeliscus is Recent only. *Pseudocolumna:* sinistral, subcylindrical; whorls later becoming flat-sided. Pal. N.Amer.

420

Subfamily SCALAXINAE

Dextral or sinistral, slender-turreted-conical; collabral costellae; anomphalous; aperture small, somewhat oblique, narrowly pyriform; columella with a fold along the axis. (Pal.)

Scalaxis (Fig. 868): characters of the Subfamily. Pal. Eur.

Subfamily CYLINDRELLININAE

Cylindrical-conical, with blunt apex and collabral costellae, but smooth base limited by a keel; umbilicus; aperture elliptical; columella concave, with a fold. (Pal.-M.Eoc.)

Cylindrellina: cylindrical-conical, subulate; aperture large, with columellar fold and often also a palatal fold. Pal.-M.Eoc. Eur. *Distoechia* (Fig. 869): ovalcylindrical, apex usually broken off; labral profile vertical; aperture contracted, detached from last whorl, labrum doubled; one columellar fold, one basal fold and one palatal fold. Pal. Eur.

Family MEGASPIRIDAE

Shell cylindroidal, long and slender, slowly tapering to a rounded apex; with or without an umbilicus, axis hollow; aperture usually fairly small, rounded below, angular above; peristome thin; with spiral lamellae on columella and often inside labrum and on parietal region. (U.Cret.(Danian)-Rec.)

Megaspira is Recent only. *Palaeostoa* [*Eomegaspira*] (Fig. 870): collabral costellae; anomphalous or with an umbilical slit; peristome discontinuous; three columellar folds, parietal region with one strong lamella and one weak one on either side of it, and several palatal folds (the latter absent in *Megaspira*). U.Cret.(Danian)-L.Olig. Eur.

Superfamily CLAUSILIACEA

Always sinistral; slender, turreted-fusiform, with relatively small aperture; peristome continuous and more or less detached; usually with internal lamellae and a 'closing-plate' (*clausilium*). (U.Cret.(Danian)-Rec.)

Family CLAUSILIIDAE

Numerous slowly growing whorls, apex more or less acuminate; greatest diameter at penultimate whorl; last whorl contracting towards the aperture and with slight upward deviation at the end; ornament of oblique straight growth lines, more pronounced on the last whorl, especially on the base; aperture in a vertical plane not quite tangential to the spire, oval to subrhomboidal, with lamellae running some distance inwards; clausilium situated internally about the middle of the last whorl. (U.Cret.(Danian)-Rec.)

Unknown before the Upper Cretaceous (Danian), the earliest recorded species are Danian, Palaeocene and Lower Eocene, Paris Basin. These (*Proalbinaria* and its subgenus *Neniopsis*, *Palaeophaedusa* and *Oospiroides*) are members of a relatively primitive group, now almost entirely Oriental, though not quite so primitive as certain later genera in which the special family

421

features (clausilium, laminae) are wanting or rudimentary (*Balea*, Pleistocene to Recent, Europe, Azores; *Triptychia*, Upper Oligocene to Pliocene, Central and Southern Europe; *Eualopia*, Lower Miocene, Germany; *Alopia*, Recent, Carpathians). As fossils they are almost unknown outside Europe. At the present day there are nine subfamilies of which the most important are: (1) Clausiliinae (e.g. *Canalicia*, Fig. 871), the most highly developed group, almost exclusively European, but with stragglers in North Africa (Atlantic islands) and the Middle East, (2) Phaedusinae, including the Danian and early Eocene forms, now oriental, extending to Japan, but not beyond Wallace's Line, (3) Cochlodininae, Europe and North Africa, not extending back earlier than Lower Miocene, (4) Neniinae (e.g. *Nenia*), a neotropical group (except for three possible representatives in eastern Asia), mainly Peru to Colombia, but with an outlying species in Puerto Rico, with only one fossil representative in the Neogene, (5) Alopiinae, Europe, Middle East and North Africa, with no fossil representatives, (6) Fusulinae, Europe, Caucasus, Transcaucasia, Middle East, Arabia, East Africa, South Africa, unknown below the Pleistocene except for one possible representative in the Upper Eocene, and (7) Laminiferinae (e.g. *Laminifera*, Fig. 872), known only from Europe and traceable back to the Lower Eocene) (**75, 77**).

Subfamily CLAUSILIINAE

Upper lamella and spiral lamella fused; lower lamella present; subcolumellar lamella often deep-set and not visible from exterior; sutural folds feeble or lacking; principal fold often very short; with or without palatal folds and lunella; clausilium usually bent and strongly S-shaped. (?M.Eoc., U.Olig.-Rec.)

Clausilia (*s.s.*) is Pleistocene to Recent only. *Canalicia* (Fig. 871): whorls numerous, flattened, with curved collabral costellae; last whorl anteriorly constricted and with a neck fold; aperture anteriorly with a channel; labrum prominent, feebly thickened; principal fold long, parallel to suture; no palatal folds; lunella lacking or feeble. ?M.Eoc., U.Olig.-M.Mio. Eur.

Subfamily PHAEDUSINAE

Atypically with a few dextral forms; aperture relatively large; peristome more or less widened and reflected, but not so strongly as in the Neniinae; spiral lamella often fused with upper lamella; usually with numerous palatal folds; lunella present or absent. (U.Cret.(Danian)-Rec.)

Phaedusa (*s.s.*) is Recent only. *Proalbinaria* (*s.s.*) is Danian only; its subgenus *Neniopsis* is high-conical, with weaker collabral costellae; lower lamella prominent but not marginal; upper lamella usually not fused to spiral lamella. Pal.-?L.Olig. Eur. *Oospiroides:* inflated-oval-fusiform, with collabral costellae; last whorl anteriorly very constricted and with a distinct neck swelling; aperture subrhombic; upper lamella solid, marginal; lower lamella not marginal, simple or bifurcating; subcolumellar lamella externally bifid; one principal fold; two to four palatal folds; no lunella. Pal.-U.Eoc. Eur. *Palaeophaedusa:* fusiform, with oblique collabral costellae; aperture produced forward, subrhombic; upper lamella solid; lower lamella solid, marginal; sub-columellar lamella not visible from above; principal fold solid; lunella present. Pal. Eur. *Eualopia:* fairly large, inflated-fusiform, smooth or with extremely fine collabral costellae; peristome strongly flaring; upper lamella not marginal, spiral lamella feeble and fused to it or lacking; lower lamella not marginal; subcolumellar lamella weak; principal fold fairly long; one strong and up to two weaker palatal folds; no lunella; clausilium probably lacking. L.Mio. Eur.

Subfamily TRIPTYCHIINAE

Inflated-fusiform, with distinct collabral costellae becoming weaker on last whorl; aperture distinctly pointed posteriorly; peristome weakly reflected; upper lamella fused to spiral lamella; lower lamella and subcolumellar lamella prominent; principal fold, lunella and clausilium lacking. (U.Olig.-Plio.)

Triptychia (*s.s.*): apex not decollated; no sutural fold. U.Olig.-Plio. Eur.

Subfamily COCHLODININAE

Fusiform to club-shaped; upper lamella and spiral lamella separate; principal fold fairly long; palatal folds short; no lunella. (L.Mio.-Rec.)

Cochlodina: fusiform, with smooth or nearly smooth whorls; aperture oval to pyriform; lower lamella anteriorly truncated; subcolumellar lamella; two strong palatal folds and sometimes one or two others; clausilium plate anteriorly with two lappets. Plio.-Rec. Eur., N.Africa.

Subfamily NENIINAE

Fusiform, with last whorl strongly constricted and brought forward; aperture large, somewhat oblique; peristome much widened; subcolumellar lamella not visible from exterior. (?Neogene, Rec.)

Nenia is Recent only. Doubtfully included in the subfamily is *Cirrobasis:* not decollated, cylindrical-fusiform; last whorl detached; aperture oblique, oval; folds unknown. Neogene. S.Amer.

Subfamily LAMINIFERINAE

Fusiform; last whorl constricted, brought forward and detached as in the Neniinae; aperture pyriform; upper lamella fused to spiral lamella; principal fold present; palatal folds lacking or rudimentary; with or without lunella. (L.Eoc.-Rec.)

Laminifera (*s.s.*) (Fig. 872): whorls with collabral costellae; aperture obliquely drawn out, not channelled anteriorly; no neck keel; lower lamella close to upper lamella; subcolumella lamella often bifid; no palatal folds; lunella curved; clausilium plate pointed. L.Eoc.-Plio. Eur.

Doubtfully included in the Clausiliacea is the following:

Family FILHOLIIDAE

Sinistral, fairly large, pupiform or bulimoid shells; whorls nearly flat-sided, with slight subsutural depression and very fine collabral costellae; spire slightly cyrtoconoid; last whorl relatively high, anomphalous or with a slight umbilical slit; aperture pointed behind, rounded in front, usually with one columellar fold; labral profile straight, oblique, normal to the suture; parietal region of peristome more or less detached. (M.Eoc.-L.Olig.)

Filholia (Fig. 873): characters of the Family. M.Eoc.-L.Olig. Eur.

Superfamily OLEACINACEA

Oval to turreted-fusiform and capable of enclosing the whole body, or feebly spiral and capuliform and only covering posterior of body. (Pal.-Rec.)

Family OLEACINIDAE

Shell ovate-oblong to turreted-fusiform, thin, with polished surface; apex obtuse, whorls few (about seven), last whorl often at least two-thirds the total height; aperture more or less narrow, tending to retort-shaped; columella truncated, giving the appearance of an anterior notch, though the lip outline is not notched; labrum thin, profile sinous, parasigmoidal, convex for anterior half or more, then reversed to form a shallow bay, and again recurving to meet suture normally or retrally. (Pal.-Rec.)

At the present day this family is mainly American, but it has many Tertiary representatives in Europe, with one surviving genus *Poiretia* in Eastern Mediterranean lands. At one time it was customary to refer the European fossils to American genera such as *Oleacina* [*Glandina*] (Recent only), *Euglandina* (Pliocene to Recent) or *Varicella* (Recent only); but Pilsbry has pointed out that no generic distinctions can be made on shell characters alone, and that the geographical distribution of the American genera shows them to have evolved locally in relation to the breaking up of the Central American land area (the three genera quoted have their respective centres in Cuba, Mexico and Jamaica) so that it is unlikely that any of them ever reached Europe. European fossil forms are now referred to European genera.

Oleacina is Recent only. *Euglandina* (*s.s.*): elongate-oval, with fine collabral costellae and sometimes with spiral furrows also; last whorl large; aperture pyriform, posteriorly pointed; columella anteriorly truncated. Plio.-Rec. N.Amer., C.Amer., S.Amer. *Palaeoglandina* (Fig. 874): elongate-oval, somewhat inflated; ornament of fine collabral costellae (sometimes doubled) and spiral furrows, costellae sometimes forming a thickened band at the upper suture. Pal.-Plio. Eur. *Poiretia:* like *Palaeoglandina*, but more slender and without spiral ornament. M.Mio.-Rec. Eur., N.Africa, Caucasus.

Family TESTACELLIDAE

Shell feebly spiral and capuliform, only covering posterior of body. Carnivorous slugs. (M.Eoc.-Rec.)

Testacella (*s.s.*) (Fig. 875): shell and aperture of oval-elliptical outline; apex small, pointed, posterior. U.Olig.-Rec. Eur., N.Africa, Canaries, Madeira.

Superfamily ACAVACEA
Sometimes quite large, low to high-oval, with large protoconch. (Pal.-Rec.)

Family ACAVIDAE
Characters of the Superfamily. (Pal.-Rec.)

Strophocheilus (*s.s.*): elongate-oval, with back and front somewhat flattened; whorls with extremely fine, wavy spiral ornament; aperture vertical, elongate-oval, posteriorly pointed; peristome continuous; labrum reflected. Pal.-Rec. S.Amer. The subgenus **Megalobulimus** [*Bulimus auctt.*] is usually very large, like *Strophocheilus*, but more broadly oval, is smooth or has fine collabral costellae, and the labrum is less reflected. Pal.-Rec. S.Amer., Carib.

Superfamily BULIMULACEA
Usually oval-conical to subcylindrical; peristome more or less reflected, internally smooth or constricted by folds or teeth. (Cret.-Rec.)

Family BULIMULIDAE
Oval to oval-conical, sometimes lenticular, heliciform or cylindrical; anomphalous or umbilicate; aperture occasionally with a columellar fold or palatal tooth. (Mio.-Rec.)

Bulimulus (*s.s.*): more or less elongate-oval, fairly small, with slightly cyrtoconoid spire, practically smooth; aperture oval, with thin parietal callus. Mio.-Rec. S.Amer., C.Amer., Carib.

Family ODONTOSTOMIDAE
Low- to high-oval-conical, rarely lenticular, usually with a narrow umbilicus; aperture usually constricted by teeth and lamellae. (L.Mio.-Rec.)

Odontostomus is Recent only. **Hyperaulax** (*s.s.*): oval-conical, with fairly large last whorl, usually with fine collabral costellae and fine spiral furrows; umbilicus distinct; aperture oval, posteriorly pointed, with only one (parietal) tooth; peristome reflected and thickened. L.Mio.-Rec. S.Amer., N.Amer.

Family CERIONIDAE
Oval or oval-cylindrical, with cyrtoconoid spire, usually with collabral costellae; last whorl not high, with a narrow umbilicus; aperture oval, vertical, edentulous or with folds or teeth; peristome widened or reflected. (L.Mio.-Rec.)

More or less homoeomorphous with the Pupillidae, though generally of larger size.

Cerion (*s.s.*) (Fig. 876) is Pleistocene to Recent only. **Eostrophia:** whorls fairly low; like *Cerion*, but aperture edentulous. L.Mio. N.Amer.

Family UROCOPTIDAE
With one (discoidal) exception the shell is fusiform or turreted-conical to cylindrical, usually with collabral ornament; early whorls usually decollated; last whorl low, occasionally becoming detached; aperture small, more or less

425

rounded; peristome more or less widened, usually continuous. (Pal.-Rec.) Usually longer and more acuminate than typical Cerionidae.

Subfamily UROCOPTINAE

Usually fusiform to cylindrical, with early whorls decollated; columellar axis massive. (L.Mio.-Rec.)

Urocoptis is Recent only. *Cochlodinella:* cylindrical-fusiform, with decollated early whorls; fine collabral costellae; last whorl somewhat detached; peristome reflected. L.Mio.-Rec. Carib., N.Amer.

Subfamily EUCALODIINAE

Usually quite cylindrical and with hollow axis. (Pal.-Rec.)

Eucalodium (*s.s.*) is Recent only. *Holospira* (*s.s.*): cylindrical, not decollated, early whorls cyrtoconoid; smooth or with fine collabral costellae; aperture rounded-oval to pyriform; umbilical slit; last whorl somewhat detached; columella with four lamellae in penultimate whorl. Pal.-Rec. N.Amer., C.Amer.

The following two families are of doubtful affinities:

Family ANADROMIDAE

Oval-conical to oval, like members of the Bulimulidae; last half of last whorl often distinctly irregular. (Cret.-Eoc.)

Anadromus is Upper Cretaceous (Danian) only. *Vidaliella:* oval-conical to inflated-oval, ornamented with growth lines; last whorl rather inflated, a little more than half the height of the shell; aperture pyriform, very pointed posteriorly; parietal callus thick; labrum simple; umbilical slit. U.Cret.(Danian)-Pal. Eur. *Romanella:* elongate-oval-conical, with fine, oblique growth line costellae; last whorl a little less than half the height of the shell; aperture rounded-pyriform, posteriorly pointed; peristome thick, reflected; narrow umbilicus. M.Eoc. Eur., N.Africa.

Figs 876–887. GASTROPODA: CERIONIDAE, STREPTAXIDAE, POLYGYRIDAE, HELICIDAE AND CAMAENIDAE

Fig. 881 after Oppenheim; Figs 877 and 884 after Cossmann and Pissarro; Figs 878 and 883 after Zilch; Figs 876 879, 880, 882 and 885–887 original.

876. *Cerion uva* (Linné), Rec. Guadeloupe. T. × 1·1.
877a, b. *Rillya rillyensis* (de Boissy), Pal. Paris B. T. × 0·75.
878a, d. *Polygyra septemvolva* (Say), Rec. Florida. T. × 2.
879a, b. *Helix pomatia* Linné, Rec. S.England. T. × 0·5.
880a, b. *Cepaea lecointreae* (Collot), U.Mio. Touraine. × 1·1.
881a, d. *Dentellocaracolus damnatus* (Brongniart), U.Eoc. Roncà, N.Italy. T. × 0·75.
882a, d. *Helicella itala* (Linné), Rec. T. × 1·5.
883a, d. *Canariella hispidula* (Lamarck), Rec. Canaries. T. × 2.
884a, b. *Loganiopharynx rarus* (de Boissy), L.Eoc. Mt. Bernon, Marne. T. × 1·9.
885a, d. *Leptaxis undata* Lowe, Rec. × 1. (Immature specimen, not showing thickened peristome).
886. *Pleurodonte* sp., Rec. × 0·75.
887a, d. *Oreohelix strigosa* Gould, Rec. Utah. T. × 0·75.

426

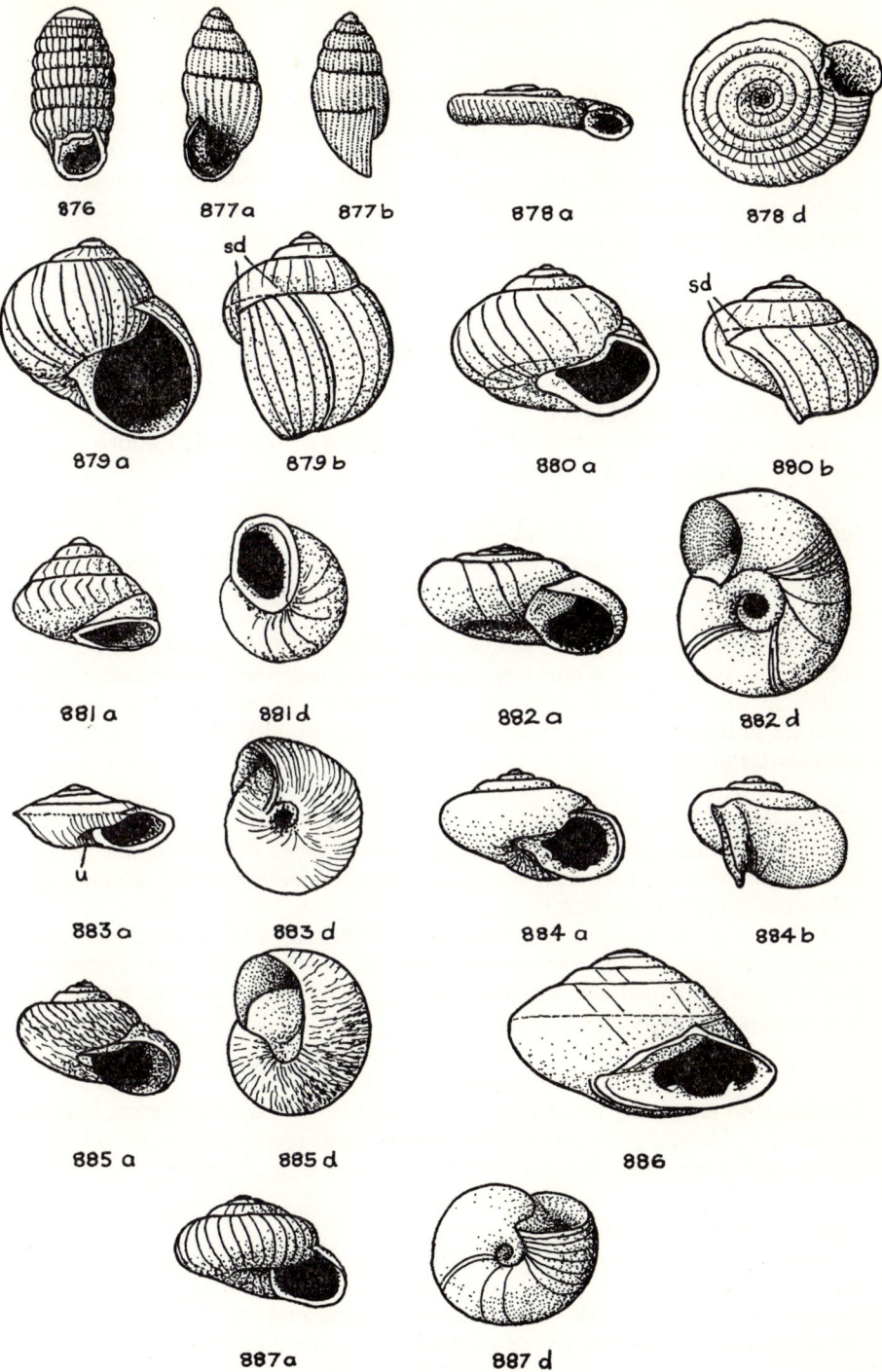

876

877a   877b

878a

878d

879a

879b

sd

880a

sd

880b

881a

881d

882a

882d

883a

u

883d

884a

884b

885a

885d

886

887a

887d

427

Family GRANGERELLIDAE

Cyrtoconoid to inflated-conical; last whorl large, its last part obliquely set up the spire; aperture rounded-triangular; labrum and floor reflected and thickened; one palatal tooth and one basal tooth. (Pal.)

Grangerella: rounded-conical, with fine, oblique growth line costellae; last whorl anteriorly flattened, its last part reaching apex of shell; aperture broadly triangular, oblique; labrum reaching apex. Pal. N.Amer.

Superfamily STREPTAXACEA

Discoidal, heliciform or turreted, often more or less irregularly wound; aperture simple or with small teeth. (U.Cret.-Rec.)

Family STREPTAXIDAE

Low-turbinate, elongate-oval, oval-cylindrical or turreted; smooth or with collabral costellae. (U.Cret.-Rec.)

Subfamily STREPTAXINAE

Low- to high-heliciform, smooth or with fine collabral costellae above; last whorl more or less bent down; aperture oval, edentulous or with teeth; peristome thickened and reflected. (U.Cret.-Rec.)

Streptaxis is Recent only. *Rillya* (Fig. 877): sinistral, pupiform, with fine collabral costellae; last whorl large, anteriorly constricted, sometimes with an umbilical slit, ventrally more or less flattened, descendent at aperture; aperture obliquely oval, vertical; columella occasionally with a fold. ?U.Cret. (Danian), Pal.-M.Eoc. Eur.

Subfamily ENNEINAE

Oval-cylindriform, pupiform to cylindrical-conical, usually regularly built; smooth or with collabral costellae; aperture often dentate or with a columellar fold. (U.Cret.(Danian)-Rec.)

Ennea is Recent only. *Gibbulinella:* cylindrical, with cyrtoconoid early whorls and fine, oblique growth line costellae; last whorl anteriorly rounded and with an umbilical slit; aperture suboval, edentulous; parietal callus thin. U.Cret.(Danian)-Rec. Eur., Canaries.

Superfamily POLYGYRACEA

Discoidal to lenticular or heliciform, sometimes with an umbilicus; aperture rounded to lunate, sometimes with lamellae and teeth; peristome simple or flaring. (?U.Cret., Pal.-Rec.)

Family POLYGYRIDAE

Low to discoidal, usually umbilicate; whorls numerous, growing slowly and regularly; aperture lunate to rounded-triangular, often with a parietal tooth and sometimes with other teeth; peristome more or less thickened and reflected; growth lines gently parasigmoidal; terminal deviation of sutural dip very short but well marked; often with Endodontid-like growth line costellae. (?U.Cret., Pal.-Rec.)

This purely American (and predominantly North American) family was included by Pilsbry in the Helicidae (*s.l.*) as its most primitive division (Protogona) along with some South African forms. The subfamilies are based on anatomical characters.

Subfamily POLYGYRINAE
(?U.Cret., Pal.-Rec.)

Polygyra (Fig. 878): low, subdiscoidal, with fine collabral costellae and very wide umbilicus; aperture oblique, kidney-shaped; peristome continuous, with thin parietal callus; one parietal tooth, but no palatal teeth. Plio.-Rec. N.Amer., Carib.

Subfamily TRIODOPSINAE
(L.Mio.-Rec.)

Triodopsis is Recent only. *Vespericola:* low-subspherical, anomphalous or with narrow umbilicus; last whorl with rounded periphery; weak growth line costellae; aperture broadly lunate, occasionally with a parietal tooth; ends of peristome fairly distant. L.Mio.-Rec. N.Amer.

Subfamily THYSANOPHORINAE
(Pal.-Rec.)

Thysanophora (*s.s.*): discoidal to low-rounded-conical, with narrow umbilicus; sutures deep; aperture lunate-oval; peristome simple; whorls with very oblique costellae which are more oblique than the growth lines. Pal.-Rec. C.Amer., Carib., N.Amer.

Superfamily HELICACEA
Globose with distinct spire to fairly low or oval, rarely discoidal, slender-conical or cylindrical; last whorl and aperture usually fairly large; peristome usually more or less widened, reflected and thickened. (?Dev., U.Cret.-Rec.)

Family HELICIDAE
Sometimes fairly large; usually globose or low-rounded, rarely discoidal or lenticular or turreted; whorls convex, only rarely angular; ornament usually of collabral growth line costellae, stronger ornament rare; anomphalous or with an umbilicus; aperture rounded to lunate. (L.Eoc.-Rec.)

This corresponds to Pilsbry's division Belogona Siphonadenia, character-ized by the maximum complexity of the hermaphrodite reproductive system. It is exclusively Palaearctic, with headquarters in Europe. The peristome and growth lines are, in general, less oblique than in other Stylommatophora, and the parasigmoidal profile tends to almost straightness.

429

Subfamily HELICINAE

Often fairly large, usually rounded to more or less low; whorls usually convex, only rarely angular; growth line costellae and sometimes other very weak ornament; last whorl usually rounded and anomphalous; aperture more or less oblique, broadly lunate to suboval; peristome continuous, thickened, often more or less reflected. (L.Olig.-Rec.)

Helix (*s.s.*) (Fig. 879): fairly large, globose-turbinate; spiral angle about 90°; growth lines usually fine, making surface almost smooth, in some cases rather coarse; umbilicus rimate, nearly or completely concealed by reflection of columellar lip; aperture little oblique, slightly higher than wide; inner margin of columellar lip passing either smoothly or somewhat abruptly, by reversal of curvature, into the parietal region, which is covered by a thin layer of callus and meets the labrum at about 90°; labral profile almost straight. L.Plio.(Pontian)-Rec. Eur., N.Africa, M.East. *Cepaea* [*Tachea*] (*s.s.*) (Fig. 880): like *Helix*, but apex more obtuse and aperture slightly wider than high and irregularly lunate, the basal portion of the peristome being almost rectilinear and passing into the inner lip by a very abrupt curve; anomphalous. M.Olig.-Rec. Eur.

Subfamily SPHINCTEROCHILINAE

Conical to more or less low-rounded, occasionally keeled; whorls smooth, crenulated or wrinkled; with or without umbilicus; aperture relatively small; peristome continuous. (U.Eoc.-Rec.)

Sphincterochila is Recent only. Its subgenus *Albea* [*Leucochroa* Martens *non* Beck] is globose, the sutures are slightly sunk, and the surface nearly smooth, chalky white; peristome a symmetrical three-quarter ellipse, columellar end thickened and reflected so as to reduce the very narrow umbilicus to a crescent, or it may be completely filled with callus; whorls usually rounded. Olig.-Rec. Eur., N.Africa, M.East. *Dentellocaracolus* (*s.s.*) (Fig. 881): globose to rounded-conical, more or less low or high above; whorls feebly convex, later somewhat swollen at the sutures, with transversely wrinkled costellae; last whorl slightly carinate; aperture horizontal. U.Eoc. Eur. The subgenus *Prothelidomus* is low-rounded and has unequal growth line costellae which may break up into warts; last whorl rounded or bluntly angular. U.Eoc. Eur.

Subfamily HELICELLINAE

Not very large, usually flat to rounded, rarely turreted; whorls convex, rounded or angular, with growth lines and rarely spiral ornament as well; aperture rounded-lunate, angular in keeled forms; peristome simple, sharp, not dentate. (U.Mio.-Rec.)

Helicella (*s.s.*) (Fig. 882): depressed, almost discoidal; whorls nearly elliptical in section; umbilicus wide and deep, without defined margin; peristome five-sixths of an ellipse, symmetrical; labrum sharp. ?L.Plio. (Pontian)-Rec. Eur. *Leucochroa* (*s.s.*): low to discoidal; whorls flattened above, with more or less distinct growth line costellae; last whorl sharply keeled and with a wide umbilicus; peristome simple, labral profile straight. Plio.-Rec. Eur., N.Africa.

Subfamily GEOMITRINAE

Discoidal, lenticular or conical, with or without an umbilicus; aperture rounded or semicircular; peristome more or less widened, internally thickened. (?U.Eoc., Olig.-Rec.)

Geomitra is Pleistocene to Recent only. *Ochthephila* is Recent only; its subgenus *Helicomela* is rounded and anomphalous, and has finely crenulated growth lines. ?U.Eoc., Olig.-Rec. Madeira, Eur. [The U.Oligocene zone-fossil *O.* (*H.*) *ramondi* (Brongniart) of central and western Europe belongs here.]

Subfamily HYGROMIINAE

Low-rounded to lenticular, with convex whorls and usually a rather narrow umbilicus; last whorl rounded or angular; aperture truncated oval; peristome simple, internally more or less margined. (Olig.-Rec.)

Hygromia: rounded-conical to lenticular, with gently convex whorls and fine growth lines; last whorl carinate and with a very narrow umbilicus; aperture oblique; labrum simple, sharp. Plio.-Rec. Eur. *Monachoides:* low-turbinate, with finely crenulated and scaly surface; last whorl rounded, with a moderate umbilicus; peristome internally solid. U.Olig.-Rec. Eur., N.Africa.

Subfamily HELICODONTINAE

Low to discoidal, with rather numerous whorls bearing distinct growth lines and often also crenulations; last whorl rounded to carinate, usually with an umbilicus; aperture rounded-triangular, rhombic or lunate; peristome reflected and margined. (L.Eoc.-Rec.)

Helicodonta [*Trigonostoma, Gonostoma*]: discoidal, thick, with many whorls and fairly wide umbilicus; peristome triangular (almost trifoliate); surface papillate. L.Olig.-Rec. Eur. *Canariella* (Fig. 883): lenticular, with small umbilicus and fine oblique costellae accompanied by spiral striae and spiral rows of small granules; last whorl keeled, more or less descendent at aperture which is oblique and transversely elliptical. ?M.Eoc., U.Olig.-Rec. Eur., Canaries. Doubtfully included in the subfamily is *Loganiopharynx* (Fig. 884): low-turbinate, with abrupt downward deviation of last whorl close to the aperture; the small umbilicus has a distinct swelling around it; aperture ear-shaped. L.Eoc.-M.Eoc. Eur.

Subfamily LEPTAXINAE

Rounded to low-turbinate, usually relatively strongly ornamented; last whorl anomphalous, sometimes carinate; aperture oblique, transversely oval; last whorl descendent near aperture; peristome simple or widened. (Olig.-Rec.)

Leptaxis (Fig. 885): depressed-globose to turbinate; surface marked with coarse wrinkles in the general growth line direction, but often very irregular (wrinkled); aperture elliptical or somewhat asymmetric owing to protrusion of parietal region; columellar lip more or less vertical, callous; peristome often thickened. L.Olig.-Rec. Madeira, Cape Verde Is., Azores, Eur.

431

Subfamily CAMPYLAEINAE

Usually low-turbinate to lenticular; whorls smooth or with growth line costellae and papillae; last whorl usually rounded, rarely carinate, with or without an umbilicus; aperture oblique, oval or lunate; peristome widened. (?M.Eoc., U.Eoc.-Rec.)

Campylaea (*s.s.*): lenticular, with low spire; whorls with distinct growth lines, with or without papillae; last whorl rounded, descendent at aperture; umbilicus moderately wide; aperture oblique, rounded-lunate; peristome bent back, the ends widely separated. Plio.-Rec. Eur., N.Africa. *Galactochilus:* thick-shelled, low-turbinate, with cyrtoconoid spire; smooth or with weak growth lines; last whorl descendent at aperture, with a narrow umbilicus limited by a blunt keel, its base convex; aperture oblique, lunate; peristome reflected and thickened; columellar callus partly spreading over umbilicus. U.Olig.-Plio. Eur. *Klikia* (*s.s.*): low-rounded, with dome-shaped or cyrto-conoid spire and fairly high last whorl; whorls with distinct growth lines and papillae arranged in oblique lines; umbilicus fairly small to nil; peristome thickened and strongly reflected; aperture oblique and broadly lunate. ?M.Eoc., Olig.-Plio. Eur.

Family CAMAENIDAE [PLEURODONTIDAE]

Low-rounded to discoidal, rarely higher; last whorl large, with or without an umbilicus; aperture occasionally with teeth or folds; peristome more or less widened, thickened and reflected. (U.Cret.-Rec.)

This is the division Epiphallogona (of Pilsbry) of the Helicidae (*s.l.*), now mainly distributed in the oriental region and northern Australia, as well as on the lands around the Caribbean.

Subfamily CAMAENINAE

Usually turbinate to lenticular or discoidal, predominantly dextral, rarely sinistral; last whorl sometimes descendent at aperture, with or without an umbilicus. (Eoc.-Rec.)

Camaena is Recent only. *Pleurodonte* [*Lucernella*] (*s.s.*) (Fig. 886): depressed-turbinate, with rather flat apex; spire almost flat-sided, gently cyrtoconoid; last whorl rounded or carinate, with rounded base; small umbilicus in youth, covered by callus of everted columellar lip in adult; aperture three-quarters elliptical, labrum thickened and everted, usually with two or more teeth (basal and lower palatal); parietal region sometimes also dentate; labral profile very oblique, growth lines of later whorls nearly straight except for the sutural swing, in some species tending to break up into irregular rows of small knobs. Plio.-Rec. Carib.

Subfamily AMMONITELLINAE

Narrowly wound, discoidal, with wide umbilicus; periphery of last whorl rounded; aperture oblique, rounded to lunate; peristome simple. (?L.Eoc., L.Mio.-Rec.)

Ammonitella: discoidal, concave above and below; aperture narrowly lunate; peristome internally somewhat thickened, edentulous. L.Mio.-Rec. N.Amer.

Subfamily OREOHELICINAE

Low to conical; surface with growth lines or rough; periphery in youth carinate or angular. (U.Cret.-Rec.)

Oreohelix (*s.s.*) (Fig. 887): depressed-turbinate, with spire sometimes slightly cyrtoconoid; whorls convex, with fine collabral costellae and sometimes also more or less crenulated spiral threads; last whorl with umbilicus of medium size; aperture subcircular-lunate. U.Cret.-Rec. N.Amer., C.Amer.

Family BRADYBAENIDAE [FRUTICICOLIDAE, EULOTIDAE]

Rounded to lenticular, occasionally elongate-oval-conical to turreted-conical, with or without an umbilicus; growth lines oblique and nearly straight except for the rapid curve to meet the suture at right angles. (Pal.-Rec.)

This is Pilsbry's Belogona Euadenia, with reproductive system almost reaching the height of elaboration of the true Helicidae.

Subfamily BRADYBAENINAE [FRUTICICOLINAE, EULOTINAE]

Low-rounded to lenticular, rarely turreted-conical, usually with an umbilicus; peristome usually reflected, ends discontinuous. (Pal.-Rec.)

Bradybaena [*Fruticicola, Eulota*] (*s.s.*): depressed-globose with fairly low cyrtoconoid spire, smooth, with a small umbilicus; aperture very broadly crescentic; labrum thin; peristome widened and internally more or less thickened. Plio.-Rec. Indonesia, Asia, Eur.

Family HELMINTHOGLYPTIDAE

Globose, conical, or lenticular, with or without an umbilicus; aperture not dentate, occasionally with internal folds; peristome usually widened, rarely reflected. Subfamilies based on anatomical characters. (U.Cret.-Rec.)

Subfamily HELMINTHOGLYPTINAE

Spherical to lenticular; labrum simple or widened. (U.Cret.-Rec.)

Helminthoglypta (*s.s.*): rounded to rather low, with conical spire; irregular growth lines; last whorl rounded, with or without a small umbilicus; peristome narrowly widened. Eoc.-Rec. N.Amer.

Subfamily CEPOLINAE

Usually spherical-conical; labrum simple. (?Plio., Pleist.-Rec.)

Cepolis (*s.s.*) is Recent only. The subgenus *Hemitrochus* is rounded-conical or rather low; whorls smooth or with growth line costellae; last whorl little descendent at aperture, with a narrow umbilicus or anomphalous; aperture lunate; peristome sometimes feebly widened, internally thickened. ?Plio., Pleist.-Rec. N.Amer., Carib.

433

GLOSSARY OF TECHNICAL TERMS
APPLIED TO GASTROPODA

Anal fasciole. Band on whorls formed by sinus, notch or slit of labrum close to adapical suture and anal opening.

Anomphalous. Without an umbilicus.

Anterior. In crawling forms, that part of the shell lying farthest from the apex; in high-spired conispiral shells, equivalent to abapical.

Aperture. Opening at last-formed margin of shell for protrusion of some soft parts.

Articulate operculum. With projecting processes (p. 285).

Auriform. Having a shape resembling a human ear.

Axial. In the direction of the axis about which the shell is coiled.

Base. That part of the surface of a shell that faces away from the apex.

Biconical. Resembling two cones (usually of unequal height) placed base to base.

Bucciniform. Shaped like a whelk shell, ovoido-conical.

Callus. Any additional deposit of the material forming the inner layer of the shell, especially that deposited beyond the aperture on its columellar border.

Canal. A narrow, tubular or semi-tubular extension of the aperture.

Cancellate. A combination of spiral and collabral ornament.

Carina. A prominent spiral ridge or keel.

Coeloconoid. Resembling a cone with concave sides (previously termed *extraconic*).

Collabral. Parallel to the growing edge of the shell (previously termed *concrescent*).

Columella. The solid or hollow pillar formed by coalescence of the inner walls of all the whorls of a shell.

Columellar fold. Spirally wound ridge on columella.

Columellar lip. Adaxial part of inner lip consisting of visible terminal part of columella.

Conical. As applied to helicoid shells, denotes a form generally resembling a cone with as near an approach to a flat base as the spiral growth permits.

Continuous peristome. Having a definite free edge throughout its whole course.

Convolute. With the last whorl completely embracing and concealing earlier ones, and without umbilicus.

Costa. Collabrally disposed round-topped elevation of moderate width and prominence (also termed *rib*).

Costate. Having costae.

Costella. Like costa, but smaller (also termed *riblet*).

Costellate. Having costellae.

Crenate. Scalloped or edged with rounded protuberances and notches alternately.

Cyrtoconoid. Like a cone with convex (instead of flat) sides (previously loosely termed *conoidal*).

Decussate. Having ornament consisting of two sets of obliquely disposed linear ridges that cross to form a series of Xs.

Deviated protoconch. One with its axis not in line with that of the rest of the shell.

Dextral. With genitalia on the right side of the head-foot mass or pallial cavity; includes ordinary dextrally wound shells and hyperstrophic shells.

Digitations. Long and narrow finger-like projections.

Discoidal. Approximating to planispiral, the spiral angle being 180° or even reentrant.

Discontinuous peristome. Having a definite free edge for the greater part of its course, the remainder being formed indefinitely by the surface of the last whorl.

Disjunct. Condition of whorls when not in contact.

Emarginate. With margin of labrum notched or excavated.

Enamel. Thick and polished callus.

Entire aperture. One with margin uninterrupted by siphonal canal or other emargination.

Exhalant channel. Channel at junction of outer and parietal lips, occupied by mantle fold by which exhalant current leaves mantle cavity (previously termed *gutter*— the *gouttière postérieure* of French writers).

False umbilicus. Depression at base of shell affecting only the last whorl.

Fasciole. A band in which the growth lines have a markedly different course from that seen on either side of it.

Folds. Ridges on an otherwise smooth surface; particularly, long spiral ridges on the columella.

Fossula. A hollow depression near the anterior end of the columellar lip of some cypraeids.

Funicle. A solid spiral cord within the umbilicus (e.g. *Naticarius*).

Fusiform. Spindle-shaped, thickest in the middle and tapering towards both ends.

Gibbous. Very convex, tumid.

Granule. A small rounded elevation on the shell surface.

Growth line. Any line marking what was at one moment the growing edge of the shell.

Heliciform. Shaped more or less like the shell of *Helix*.

Heterostrophic. Having the protoconch coiled in the opposite direction to the teleoconch.

Holostomatous. Having no notch or canal at the anterior end of the aperture.

Homoeostrophic. Having the protoconch coiled in the same direction as the teleoconch.

Hyperstrophic. Anatomically dextral, but shell falsely sinistral, being actually ultra-dextral; or vice versa.

Immersed. Condition of initial whorls when sunk and concealed by later whorls.

Impressed suture. One lying at the bottom of a groove.

Inner lip. See *Columellar lip*.

Involute. With last whorl enveloping earlier ones, but early whorls more or less visible in umbilici.

Keel. See *Carina*.

Labral profile. Appearance of labrum as seen when the aperture faces sideways from the observer.

Labrum. The outer lip, or that part of the apertural margin opposite to the columella.

Lamella. A general term for a thin plate, used especially for the teeth on the columellar and parietal regions (in certain land shells).

Last whorl. Last-formed complete volution of a coiled shell.

Lira. Fine linear elevation on shell surface or within outer lip.

Lirate. Bearing lirae.

Lunate. Shaped like a broad crescent, or moon in partial eclipse.

Mammillated. Having a rounded protuberance in the centre of a broader rounded area.

Microphagous. Feeding upon microscopic food.

Multispiral. Consisting of many whorls (shell) or spiral turns (operculum) in proportion to its size.

Muricate. Having the surface raised into scaly ridges and points.

435

Neck. The lowest (most anterior) part of siphonostomatous shells, marked off from the base by a reversed curvature in the outline.

Notch. An indentation of the peristome, breaking the uniform curvature of its outline.

Nucleus. Earliest formed part of protoconch or operculum.

Obconical. In the form of a cone with the base upwards.

Oligogyral. See *Paucispiral.*

Operculum. A calcareous or corneous plate closing the aperture of the shell, carried on the posterior part of the foot.

Opisthocline growth line. Directed towards the upper suture at an angle which is acute towards the aperture (previously termed *retrocurrent*).

Orthocline. At right angles to growth direction of helicone (often used for growth lines).

Outer lip. See *Labrum.*

Outline of labrum. Its appearance as seen when the aperture faces the observer (cf. *Profile*).

Ovate. Having the shape of an egg.

Palatal. Belonging to the outer lip (labrum) (referring commonly to folds and lamellae).

Parasigmoidal. Curved like a reversed S.

Parietal fold. Spirally wound ridge (projecting into shell interior) on parietal region.

Parietal region. That part of the outline of the aperture determined by the convexity of the penultimate whorl.

Patelliform. Limpet-shaped, forming a simple depressed cone.

Paucispiral. Having few whorls in proportion to its size.

Peristome. The whole margin of the aperture.

Phaneromphalous. With completely open umbilicus.

Planispiral. Coiled in a spiral which remains symmetrical to one plane.

Polygyral. See *Multispiral.*

Prosocline growth line. Directed towards the upper suture at an angle which is obtuse towards the aperture (cf. *Opisthocline*). (Previously termed *Antecurrent*.)

Protoconch. The larval shell consisting of the nucleus and one or more whorls, more or less demarcated from the teleoconch.

Pseudumbilicus. See *False umbilicus.*

Punctae. Small pit-like depressions of the surface of the shell.

Pupiform. Shaped like an insect pupa, cylindrical with rounded ends.

Pyriform. Pear-shaped.

Ramp. A sloping area below the suture (cf. *Shelf*).

Reticulate. Composed of crossing vertical and spiral ridges of equal strength.

Rib. See *Costa.*

Ribbon. A low broad spiral elevation.

Riblet. See *Costella.*

Rimate. Having the form of a narrow cleft.

Rostrate. Having a projecting rostrum or beak.

Scar. The mark which shows where a muscle has been implanted on the shell.

Sessile. Fixed to some solid surface.

Shelf. A horizontal area below the suture and following it round the shell.

Shoulder. Angulation of whorl forming edge of ramp or shelf.

Sigmoidal. Shaped like an S.

436

Sinistral. With genitalia on left side of head-foot mass or pallial cavity; shell a mirror image of dextral.

Sinus. Re-entrant curve of apertural margin or growth lines.

Siphonal canal. Tubular or canal-like extension of anterior part of aperture for enclosure of inhalant siphon.

Siphonal fasciole. Band of sharply curved growth lines near anterior end of columella marking successive positions of siphonal notch.

Siphonal notch. Sinus of aperture near end of columella for protrusion of inhalant siphon.

Siphonostomatous. Having a notch or canal at the anterior end of the aperture.

Spine. A pointed upgrowth of the shell surface.

Spiral angle. The angle contained between tangents to opposite sides of the spire.

Spout. A slight overturning of part of the apertural margin which does not disturb the curve of the outline (cf. *Notch*).

Stria. A fine linear depression on the surface of the shell.

Strombiform. Biconical with expanded labrum.

Styliform. Parallel-sided except at sharp-pointed apex (e.g. the protoconch of *Stilifer*).

Subulate. Awl-shaped, tapering with convex outline.

Sutural slope. Angle between suture and the horizontal.

Suture. The spiral line of external contact of successive whorls.

Teleoconch. The entire shell exclusive of the protoconch.

Thread. Fine linear surface elevation.

Tooth. An inward projection (other than a continuous spiral ridge) from some part of the peristome.

Transcurrent. Disposed at right angles to the growth lines.

Trochiform. Conical with a flat base.

Truncated. Ending abruptly as though cut off short.

Tubercle. A rounded eminence on the shell surface.

Turbiniform. Top-shaped, conical with a rounded base.

Turriculate. Elongate with acute spiral angle.

Ultradextral. With shell apparently sinistral but soft parts arranged dextrally.

Ultrasinistral. With shell apparently dextral but soft parts arranged sinistrally.

Umbilicate. Having an umbilicus; coiled round a hollow axis.

Umbilicus. Cavity or depression formed round shell axis between faces of adaxial walls of whorls where these do not coalesce to form a solid columella.

Unguiculate operculum. With a claw-like projection.

Varix. The remains of a thickened labrum, abandoned by the forward growth of the shell.

Whorl. One turn of a spiral shell through 360°.

Wing. More or less flattened expansion of labrum.

SELECT BIBLIOGRAPHY OF MOLLUSCA

The references given in this bibliography are mainly of two kinds (1) standard general works of various dates, (2) lately published papers dealing with special groups.

In addition to the papers cited below, many of those which will be found in the bibliographies at the end of each chapter of Vol. II deal with Tertiary Mollusca. Full lists of the literature are published every year in the *Zoological Record* (Zool. Soc. London).

I. MOLLUSCA generally.

1. ADAMS, H. and A. 1853–58. *The genera of recent Mollusca arranged according to their organization*, **1**, pp. 1–484, and **2**, pp. 1–661, (text); **3** (plates), (John van Voorst: London). [Generic Index at beginning of **1**; Cephalopoda and Proso-branchia in **1**; Opisthobranchia, Pulmonata and Bivalvia (Lamellibranchia) in **2**; classification quite out of date; few direct references to fossils.]
2. ALBRECHT, J. C. H. and VALK, W. 1943. 'Oligocäne Invertebraten von Süd-Limburg', *Meded. geol. Sticht.*, (C), **4**, (1), no. 3, 1–163, pl. 1–27.
3. ANDERSON, F. M. 1929. 'Marine Miocene and related Deposits of North Colombia', *Proc. Calif. Acad. Sci.*, (4), **18**, no. 4, 73–213, pl. 8–23.
4. BEETS, C. 1941. *Eine jungmiocäne molluskenfauna von der Halbinsel Mangkalihat, Ost-Borneo*, 1–219, pl. 1–9 (N. V. Boek- en Kunstdrukkerij v/h. Mouton & Co.: 's-Gravenhage).
5. BÖGGILD, O. B. 1930. 'The Shell Structure of Mollusks', *D. Kg. Danske Vidensk. Selsk. Skrifter, Nat. Math. Afd.*, (9), **2**, 233–325, 15 pls.
6. CHENU, J. C. 1859. *Manuel de Conchyliologie et de Palaeontologie Conchyliologique*, **1** [Cephalopoda, Gastropoda, etc.; **2**, Lamellibranchia and Index. Figures a number of species to most genera. Nomenclature Lamarckian but with good lists of synonyms] (V. Masson & Cie: Paris).
7. COOK, A. H. 1895. 'Mollusca', in *Cambridge Natural History*, **3** (Macmillan & Co.: London).
8. COSSMANN, M. and PEYROT, A. 1900–33. 'Conchologie Néogènique de l'Aquitaine', *Actes Soc. linn. Bordeaux*, 6 vols, 124 pls. [**1** and **2**, Lamellibranchia; **3–6**, Gastropoda; Plates separate, the text of **1** and **2**, uniform with the plates, the remainder 8vo text and 4to plates.]
9. COSSMANN, M. and PISSARRO, G. 1904–13. *Iconographie Complète des coquilles fossiles de l'Éocène des environs de Paris*, 2 vols (Soc. géol. Fr.: Paris). [Photographic figures only, of several thousand species.]
10. COX, L. R. 1927. Neogene and Quaternary Mollusca from the Zanzibar Protectorate, *Rept. Pal. Zanzibar Protect.*, 13–102, pl. 3–19.
11. COX, L. R. 1930. 'Miocene Mollusca: Pliocene Mollusca', *Monogr. geol. Dep. Hunter. Mus.*, IV, Reports on Geological Collections from the coastlands of Kenya Colony, 103–130, pl. 12–15.
12. COX, L. R. 1931. 'A Contribution to the Molluscan Fauna of the Laki and Basal Khirthar Groups of the Indian Eocene', *Trans. R. Soc. Edinb.*, **57**, (1), no. 2, 25–92, pl. 1–4.
13. COX, L. R. 1936. 'Fossil Mollusca from Southern Persia (Iran) and Bahrein Island, *Mem. geol. Surv. India Palaeont. indica*, N.S., **22**, Mem. no. 2, 1–69, pl. 1–8.

14. COX, L. R. 1948. 'Neogene Mollusca from the Dent Peninsula, British North Borneo', *Schweiz. palaeont. Abh.*, **66**, 1–70, pl. 1–6.
15. COX, L. R. 1952. 'Cretaceous and Eocene Fossils from the Gold Coast', *Bull. geol. Surv. Gold Cst.*, no. 17, 1–68, pl. 1–5.
16. DALL, W. H. 1890–1903. 'Contributions to the Tertiary Fauna of Florida . . .', Parts I–VI, *Trans. Wagner free Inst. Sci. Philad.* [Deals with morphology, classification, distribution, etc., of Tertiary Mollusca generally, far beyond the limits indicated by the title.]
17. DEY, A. K. 1961 (1962). 'The Miocene Mollusca from Quilon, Kerala (India), *Mem. geol. Surv. India Palaeont. indica*, N.S., **36**, 1–129, pl. 1–9.
18. EAMES, F. E. 1957. 'Eocene Mollusca from Nigeria: a Revision', *Bull. Br. Mus. nat. Hist., Geol.*, **3**, no. 2, 23–70, pl. 5–10.
19. FISCHER, P. 1880–87. *Manuel de Conchyliologie et de Paléontologie Conchyliologique*, 1–1369, 24 pls. (A. Lahure: Paris). [A very valuable work of reference, though out of date in classification. The plates are those of S. P. Woodward's manual (**44**).]
20. FLEMING, C. A. 1966. 'Marwick's Illustrations of New Zealand Shells, with a Checklist of New Zealand Cenozoic Mollusca, *Bull. N.Z. Dep. scient. ind. Res.*, **173**, 1–456, figs 1–1753.
21. FUCHS, T. 1879. 'Über die von Dr. E. Tietze aus Persien mitgebrachten Tertiärversteinerungen', *Denkschr. Akad. Wiss. Wien*, **41**, 99–108, pl. 1–6.
22. GLIBERT, M. 1957. 'Pélécypodes et Gastropodes du Rupélien Supérieur et du Chattien de la Belgique', *Mém. Inst. r. Sci. nat. Belg.*, no. 137, 1–98, pl. 1–6.
23. GÖRGES, J. 1952. 'Die Lamellibranchiaten und Gastropoden des Oberoligocänen Meeressands von Kassel', *Abh. hess. Landesamt Bodenforsch.*, **4**, 1–134, pl. 1–3.
24. GRANT, U. S. and GALE, H. R. 1931. 'Catalogue of the Marine Pliocene and Pleistocene Mollusca of California . . .', *Mem. S. Diego Soc. nat. Hist.*, **1**, 1–1036, pl. 1–32. [Deals extensively with molluscan morphology and classification beyond the limits indicated by the title, especially in respect of Pectinidae and Turridae.]
25. HANLEY, S. 1855. *Ipsa Linnaei Conchylia*, 1–556, 5 pls, (Williams & Norgate: London). [Of great value in determining what Linnaeus's genera and species actually are.]
26. HARRIS, G. D. and PALMER, K. V. M. 1946–47. 'The Mollusca of the Jackson Eocene of the Mississippi embayment (Sabine River to the Alabama River)', *Bull. Am. Paleont.*, **30**, no. 117: pt 1, 1–206, pl. 1–25; pt 2, 207–563, pl. 26–64.
27. JUNG, P. 1965. 'Miocene Mollusca from the Paraguana Peninsula, Venezuela', *Bull. Am. Paleont.*, **49**, no. 223, 389–652, pl. 50–79.
28. OLSSON, A. A. 1964. *Neogene Mollusks from Northwestern Ecuador*, 1–256, pl. 1–38 (Pal. Res. Inst.: Ithaca, N.Y.).
29. OYAMA, K., MIZUNO, A. and SAKAMOTO, T. 1960. *Illustrated Handbook of Japanese Paleogene Molluscs*, 1–244, pl. 1–71 (Dai-Nippon Printing Co., Ltd: Geol. Surv. Japan).
30. PELSENEER, P. 1906 'Mollusca', in Lankester, E.R., *Treatise on Zoology*, pt V (A. & C. Black: London).
31. RICHARDS, H. G. and PALMER, K. V. M. 1953. 'Eocene Mollusks from Citrus and Levy Counties, Florida', *Geol. Bull. Fla.*, no. 53, 1–67, pl. 1–13.
32. ROSSI, C. 1940. 'Fossili miocenici del sottosuolo della Gefara Tripolina (Libia)', *Annali Mus. libico Stor. nat.*, **2**, 211–249, pl. 20.

33. ROSSI, C. 1942. 'Molluschi paleogenici della Sirtica', *Annali Mus. libico Stor. nat.*, **3**, 109–193, pl. 8–11.

34. SANDBERGER, C. L. F. 1870–75. *Die Land und Süsswasser Conchylien der Vorwelt*, 2 vols (text, pp. 1–1000, and atlas, 36 pls) (C. W. Kreidels' Verlag: Wiesbaden). [Figures and describes a very large number of land and fresh-water shells, mainly Tertiary. Nomenclature must be corrected by reference to Wenz (**96**) so far as Gastropods are concerned.]

35. SORGENFREI, T. 1958. 'Molluscan Assemblages from the Marine Middle Miocene of South Jutland and their Environments', *Geol. Surv. Denmark*, (2), no. 79, (1–2), 1–503, pl. 1–76.

36. STCHEPINSKY, V. 1939. 'Faune miocene du vilayet de Sivas (Turquie)', *Maden Tetkik Arama Enstit. Yayinl.*, C, Monogr. no. 1, 1–63, pl. 1–10.

37. THIELE, J. 1929–31. *Handbuch der Systematischen Weichtierkunde*, **1**, 1–778 (G. Fischer: Jena). [Valuable for ideas of classification, though treating exclusively Recent Mollusca.]

38. TRYON, G. W. and PILSBRY, H. A. 1879–96. *Manual of Conchology*, **1–17** (G. W. Tryon: Philadelphia). [A long-continued publication, begun by Tryon, continued by Pilsbry, whose contributions to the anatomy and classification of Pulmonata are of first importance.]

39. VREDENBURG, E. W. 1925–28. 'Description of Mollusca from the post-Eocene Tertiary Formations of North-Western India', *Mem. geol. Surv. India*, **50**, 1–506, pl. 1–33. [Taxonomy needs modernizing.]

40. WEIR, J. 1925. 'Brachiopoda, Lamellibranchiata, Gastropoda and Belemnites: Kainozoic Lamellibranchiata', *Monogr. geol. Dep. Hunter. Mus.*, I, The Collection of Fossils and Rocks from Somaliland, 96–100, pl. 11–14.

41. WEIR, J. 1938. 'Additions to the Neogene Molluscan Faunas of Kenya', *Monogr. geol. Dep. Hunter. Mus.*, On a second collection of Fossils and Rocks from Kenya, 61–81, pl. 5–7.

42. WOODRING, W. P. 1925–28. *Miocene Mollusks from Bowden, Jamaica: Vol. I. Pelecypods and Scaphopods; Vol. II. Gastropods and Discussion of Results* (Carnegie Inst.: Washington). [A model monograph.]

43. WOODWARD, B. B. 1913. *The Life of the Mollusca*, 1–158, 32 pls (Methuen & Co. Ltd: London). [Popular account of molluscan ecology.]

44. WOODWARD, S. P. 1851–56. *Manual of the Mollusca*, 1–486, 25 pls (John Weale: London). [A useful work of reference in spite of its age.]

45. WRIGLEY, A. 1925. 'Notes on English Eocene and Oligocene Mollusca . . .', *Proc. malac. Soc. Lond.*, **16**, 232–248. [See also **102–104** below].

46. WRIGLEY, A. 1929. 'Notes on English Boring Mollusca, with descriptions of new species', *Proc. Geol. Ass.*, **40**, 376–383.

II. BIVALVIA

47. BERNARD, F. 1895–97. 'Note sur le développement et la morphologie de la coquille chez les Lamellibranches', *Bull. Soc. géol. Fr.*, (3), **23**, 104–154; **24**, 54–82, 412–449; **25**, 559–566,

48. CHAVAN, A. 1951. 'Essai critique de Classification des *Divaricella*', *Bull. Inst. r. Sci. nat. Belg.*, **27**, no. 18, 1–27.

49. COX, L. R. 1931. 'New Lamellibranch genera from the Tethyan Eocene', *Proc. malac. Soc. Lond.*, **19**, 177–187, pl. 20–21.

49a. COX, L. R. 1960. 'Thoughts on the Classification of the Bivalvia', *Proc. malac. Soc. Lond.*, **34**, (2), 60–68.

50. DAVIES, A. M. 1933. 'The Bases of Classification of the Lamellibranchia', *Proc. malac. Soc. Lond.*, **20**, 322–326.

51. DEPÉRET, C. and ROMAN, F. 1902–12. 'Monographie des Pectinidés néogènes de l'Europe et des Régions voisines', *Mém. Soc. géol. Fr.*, Mém. no. 26, 1–168, pl. 1–23.

52. DESIO, A. 1934. 'Lamellibranchi paleogenici della Sirtica e del Fezzan orientale', *Reale Accad. d'Ital.*, 1–48, pl. 6–13.

53. DOUVILLÉ, H. 1907. 'Les Lamellibranches cavicoles ou Desmodontes', *Bull. Soc. géol. Fr.*, (4), **7**, 96–114, pl. 2.

54. DOUVILLÉ, H. 1912. 'Classification des Lamellibranches', *Bull. Soc. géol. Fr.*, (4), **12**, 419–467.

55. EAMES, F. E. 1951. 'A Contribution to the Study of the Eocene in Western Pakistan and western India: B. The description of the Lamellibranchia from standard sections in the Rakhi Nala and Zinda Pir areas of the western Punjab and in the Kohat District', *Phil. Trans. R. Soc.*, (B), **235**, no. 627, 311–482, pl. 9–17.

56. EAMES, F. E. 1967. 'Notes on some *Anadara*', *Proc. malac. Soc. Lond.*, **37**, 303–308.

57. EAMES, F. E. and COX, L. R. 1956. 'Some Tertiary Pectinacea from East Africa, Persia, and the Mediterranean Region', *Proc. malac. Soc. Lond.*, **32**, (1–2), 1–68, pl. 1–20.

58. FRENEIX, S. and GORODISKI, A. 1963. 'Bivalves éocènes du Sénégal: Première Partie. Nuclacea, Arcacea, Mytilacea, Pectinacea, Anomiacea, Ostreacea', *Mém. Bur. Rech. géol. minièr.*, no. 17, 1–123, pl. 1–13.

59. GILLET, S. 1930. *Variation des Cardiidés dans le Bassin Dacique* (Bibliothèque Inst. français de Hautes Études en Roumanie: Paris).

60. GÖRGES, J. 1951. 'Die oberoligozänen Pectiniden des Doberges bei Bünde und ihre stratigraphische Bedeutung', *Palaeont. Z.*, **24**, (1–2), 9–22, pl. 1–3.

61. HEERING, J. 1942. 'Die Oligocänen Taxodonten Bivalven aus dem Peelgebiete (die Niederlande)', *Meded. geol. Sticht.*, (C), **4**, (1), no. 2, 1–42, pl. 1–4.

62. HEERING, J. 1944. 'Die Oberoligocänen Bivalven (mit ausnahme der Taxodonten) aus dem Peelgebiete (die Niederlande)', *Meded. geol. Sticht.*, (C), **4**, (1), no. 4, 1–48, pl. 1–10.

63. HEERING, J. 1950. 'Miocene Pelecypoda of the Netherlands (Peel-region)', *Meded. geol. Sticht.*, (C), **4**, (1), no. 10, 1–51, pl. 1–8.

64. LUDBROOK, N. H. 1955. 'The Molluscan Fauna of the Pliocene Strata underlying the Adelaide Plains: Part II. Pelecypoda', *Trans. R. Soc. S. Aust.*, **78**, 18–87, pl. 1–6.

65. MACNEIL, F. S. 1967. 'Cenozoic Pectinids of Alaska, Iceland, and Other Northern Regions', *Prof. Pap. U.S. geol. Surv.*, **553**, 1–57, pl. 1–25.

66. MERKLIN, R. L. and NEVESSKAJA, L. A. 1955. 'Identification of lamellibranchs from the Miocene of Turkmenia and western Kazakhstan', *Trudý paleont. Inst.*, **59**, 1–115, pl. 1–32.

66a. MOORE, R. C. (ed.). 1969. *Treatise on Invertebrate Paleontology: Part N. Mollusca 6*, 1–2, N1–N952 (University of Kansas Press).

67. NEUMAYR, M. 1884. 'Zur Morphologie des Bivalvenschlosses', *Sber. Akad. Wiss. Wien, math.-nat. Kl.*, **88**, 385–418, pl. 1–2. [A later, posthumous paper, 1891, 'Beiträge zur einer morphologischen Eintheilung der Bivalven', *Denkschr.*

441

Akad. Wiss. Wien, **58**, 701–801, deals mainly with pre-Tertiary bivalves.]

67a. NEWELL, N. D. 1965. 'Classification of the Bivalvia', *Am. Mus. Novitates*, no. 2206, 1–25.

68. PALMER, K. VAN W. 1927–29. 'Veneridae of Eastern America, Cenozoic and Recent', *Palaeontogr. am.*, **1**, no. 5, 209–428, pl. 32–76.

69. ROGER, J. 1944. 'Révision des Pectinidés de l'Oligocène du Domaine nordique', *Mém. Soc. géol. Fr.*, N.S., **23**, Fasc. 1, Mém. no. 50, 1–57, pl. 1–2.

70. ROGER, J. 1949. 'Le Genre *Chlamys* dans les Formations néogènes de l'Europe', *Mém. Soc. géol. Fr.*, N.S., **17**, Fasc. 7, Mém. no. 40, 1–294, pl. 1 (6)–28(33).

71. SCHENCK, H. G. 1934. 'Classification of Nuculid Pelecypods', *Bull. Mus. r. Hist. nat. Belg.*, **10**, no. 20, 1–78, pl. 3–5.

72. STEWART, R. B. 1930. 'Gabb's California Cretaceous and Tertiary Type Lamellibranchs', *Spec. Publs Acad. nat. Sci. Philad.*, no. 3, 1–314, pl. 1–17. [Really a second part to (**93**) *infra*. Clears up many points of nomenclature and classification.]

72a. VIALOV, O. S. 1936. 'Sur la classification des Huitres', *C.R.* (*Doklady*) *Ac. Sci. URSS*, **4**, (13), no. 1 (105), 17–20.

72b. VOKES, H. E. 1967. 'Genera of the Bivalvia: a systematic and bibliographic catalogue', *Bull. Am. Paleont.*, **51**, no. 232, 111–394.

73. VREDENBURG, E. W. 1924. 'On some fossil forms of *Placuna*', *Rec. geol. Surv. India*, **55**, 110–118, pl. 14–18.

74. WOODRING, W. P. 1926. 'American Tertiary Mollusca of the genus *Clementia*', *Prof. Pap. U.S. geol. Surv.*, **147**–C, 23–42, pl. 14–17.

III. GASTROPODA

75. BOETTGER, O. 1877. 'Clausilienstudien', *Palaeontographica*, **24**, Suppl. iii, Lief 6–7, 1–122, pl. 1–4.

76. BOUSSAC, J. 1912. 'Essai sur l'évolution des Cérithidés dans le Mésonummultique du Bassin de Paris', *Annls Hébert*, **6**, 1–90.

76a. CHAVAN, A. 1965. 'Essai de reclassification des Olividae Ancillinae (Gastropodes)', *Bull. Soc. géol. Fr.*, (7), **7**, 102–109.

77. COOKE, A. H. 1915. 'The genus *Clausilia*: a study of its geographical distribution . . .', *Proc. malac. Soc. Lond.*, **11**, 249–269.

78. COOKE, C. W. 1921. '*Orthaulax*: a Tertiary guide-fossil', *Prof. Pap. U.S. geol. Surv.*, **129**, 23–31, pl. 2–5.

79. COSSMANN, M. 1895–1925. *Essais de Paléoconchologie Comparée*, **1–13** (Les Presses Universitaires de France: Paris). [Deals in great detail, indexing most recorded species, with all fossil Gastropoda, excepting Pulmonata and Patellidae, Fissurellidae, Calyptraeidae and a few other Prosobranch families, the work being unfinished through death. Nomenclature Lamarckian. Stratigraphical statements must be treated critically. Every genus, subgenus and section recognized by the author is illustrated on the photographic plates. Generic Index at end of **13**. Specific index to each volume, except that there is a single index to the first four volumes at the end of **4**.]

80. EAMES, F. E. 1952. 'A Contribution to the study of the Eocene in Western Pakistan and western India: C. The Description of the Scaphopoda and Gastropoda from standard sections in the Rakhi Nala and Zinda Pir areas of the western Punjab and in the Kohat District', *Phil. Trans. R. Soc.*, (B), **236**, no. 631, 1–168, pl. 1–6.

81. GARDNER, J. 1944. 'The Molluscan Fauna of the Alum Bluff Group of Florida: Part VII. Stenoglossa (in part)', *Prof. Pap. U.S. geol. Surv.*, **142–G**, 437–486, pl. 49–51.

82. GARDNER, J. 1948. 'Mollusca from the Miocene and Lower Pliocene of Virginia and North Carolina: Part II. Scaphopoda and Gastropoda', *Prof. Pap. U.S. geol. Surv.*, **199–B**, 179–310, pl. 24–38.

83. GLIBERT, M. 1949. 'Gastropodes du Miocène Moyen du Bassin de la Loire: Première Partie', *Mém. Inst. r. Sci. nat. Belg.*, (2), Fasc. 30, 1–240, pl. 1–12.

84. GLIBERT, M. 1954. 'Pleurotomes du Miocène de la Belgique et du Bassin de la Loire', *Mém. Inst. r. Sci. nat. Belg.*, no. 129, 1–75, pl. 1–7.

85. GUILLAUME, L. 1924. 'Essai sur la classification des Turritelles. . . .', *Bull. Soc. géol. Fr.*, (4), **24**, 281–311.

86. HOERNES, R. 1895. '*Perairaïa Gervaisii* Vez. von Ivandol bei St. Bartelmae in Unterkrain', *Annls naturh. Mus. Wien*, **10**, 1–16, pl. 1–2.

87. INGRAM, W. M. 1947. 'Fossil and Recent Cypraeidae of the western regions of the Americas', *Bull. Am. Paleont.*, **31**, no. 120, 1–75, pl. 1–3.

87a. KEEN, A. M. 1944. 'Catalogue and Revision of the Gastropod subfamily Typhinae, *J. Paleont.*, **18**, (1), 50–72.

87b. MARWICK, J. 1957. 'Generic Revision of the Turritellidae', *Proc. malac. Soc. Lond.*, **32**, (4), 144–166.

87c. MOORE, D. R. 1962. 'The Systematic Position of the Family Caecidae' (Mollusca: Gastropoda)', *Bull. Mar. Sci. Gulf & Caribbean*, **12**, no. 4, 695–701.

88. MOORE, R. C. (ed.) 1960. *Treatise on Invertebrate Paleontology: Part I Mollusca 1*, II–I351, fig. 1–216, (Kansas University Press).

89. PALLA, P. 1967. 'Gasteropodi pliocenici della bassa Val d'Elsa (Toscana occidentale)', *Riv. ital. Paleont.*, **73**, no. 3, 931–1020, pl. 71–75.

89a. PILSBRY, H. A. and OLSSON, A. A. 1954. 'Systems of the Volutidae', *Bull. Am. Paleont.*, **35**, no. 152, 1–36, pl. 1–4.

90. SCHENCK, H. G. 1926. 'Cassididae of Western America', *Bull. geol. Univ. Calif.*, **16**, 69–98, pl. 12–15.

91. SCHILDER, F. A. 1932. *Fossilium Catalogus: I. Animalia: Pars 55. Cypraeacea*, 1–276 (W. Junk: Berlin). [Schilder's own numerous papers on Cypraeidae from 1922 on are given in the bibliography.]

92. SCHILDER, F. A. 1939. 'Die Genera der Cypraeacea', *Arch. Molluskenk.*, **71**, 165–201, pl. 7–8.

93. STEWART, R. B. 1926. 'Gabb's California Type Gastropods', *Proc. Acad. nat. Sci. Philad.*, **78**, 287–447, pl. 20–32. [See also (72) *supra*.]

94. VOORTHUYSEN, J. H. VAN. 1944. 'Miozäne Gastropoden aus dem Peelgebiet (Niederlande) (Rissoidae-Muricidae, nach Zittel's Einteilung 1924)', *Meded. geol. Sticht.*, (C), **4**, (1), no. 5, 1–116, pl. 1–13.

95. VREDENBURG, E. W. 1927. 'A review of the genus *Gisortia* . . .', *Mem. geol. Surv. India Palaeont. indica*, N.S., **7**, no. 3, 1–124, pl. 1–32. [Taxonomy later partly revised by Schilder.]

96. WENZ, W. 1923–30. *Fossilium Catalogus: I. Animalia: Pars 17. Gastropoda Extramarina Tertiaria*, 4 vols (W. Junk: Berlin). [A most complete and model catalogue, with almost unnecessarily full references, full synonymy, an index to all generic and trivial names including synonyms, and another to fossil localities. Stratigraphical classification rather different from that here used, Ypresian being included in the Palaeocene, so that Lutetian becomes Lower Eocene and Auversian Middle Eocene, while Montian is excluded from the Tertiary.]

443

97. WENZ, W. 1938–44. 'Gastropoda: Allgemeiner Teil und Prosobranchia (Amphigastropoda u. Streptoneura)', *Handbuch der Palaozoologie*, **6**, Teil 1, 1–1639 (Gebrüder Borntrager: Berlin). [This, together with Zilch (see **106**), constitutes an excellent descriptive and illustrated coverage of practically all known gastropod shells above the rank of species, up to the dates of publication.]

98. WOODWARD, H. 1885. 'Recent and Fossil Pleurotomariae', *Geol. Mag.*, **2**, (3), 433–439, pl. 11.

99. WOODRING, W. P. 1957. 'Geology and Paleontology of Canal Zone and Adjoining Parts of Panama: Geology and Description of Tertiary Mollusks (Gastropods: Trochidae to Turritellidae)', *Prof. Pap. U.S. geol. Surv.*, **306–A**, 1–145, pl. 3–23.

100. WOODRING, W. P. 1959. 'Geology and Paleontology of Canal Zone and Adjoining Parts of Panama: Description of Tertiary Mollusks (Gastropods: Vermetidae to Thaididae)', *Prof. Pap. U.S. geol. Surv.*, **306–B**, 147–239, pl. 24–38.

101. WOODRING, W. P. 1964. 'Geology and Paleontology of Canal Zone and Adjoining Parts of Panama: Description of Tertiary Mollusks (Gastropods: Columbellidae to Volutidae)', *Prof. Pap. U.S. geol. Surv.*, **306–C**, 241–297, pl. 39–47.

102. WRIGLEY, A. 1927–30. 'Notes on English Eocene Mollusca', *Proc. malac. Soc. Lond.*: [I. See **45**]; II. 'The Fusinidae', **17**, 216–249, pl. 33–35; III. '*Ficus*', **18**, 235–251, pl. 15–16; IV. 'The Muricidae', **19**, 91–115, pl. 9–10.

103. WRIGLEY, A. 1932. 'The English Eocene species of *Sassia*, with a note on the Morphology of the Cymatiidae and the Bursidae', *Proc. malac. Soc. Lond.*, **20**, 127–140, pl. 10–11.

104. WRIGLEY, A. 1934. 'English Eocene and Oligocence Cassididae, with notes on the nomenclature and morphology of the family', *Proc. malac. Soc. Lond.*, **21**, 108–130, pl. 15–17.

105. WRIGLEY, A. 1953. 'English Eocene *Siphonalia* and *Pseudoneptunea*', *Proc. malac. Soc. Lond.*, **30**, (4–5), 121–130, figs. 1–15.

106. ZILCH, A. 1959–60. 'Gastropoda: Euthyneura', *Handbuch der Paläozoologie*, **6**, Teil 2, 1–834 (Gebrüder Borntrager: Berlin).

Chapter V

TERTIARY OSTRACODA

REFERENCE-LETTERS on the figures, following the figure numbers:
a, right valve external.
b, left valve external.
c, right valve internal.
d, left valve internal.
e, dorsal view.
f, radial pore canals (enlarged).
g, muscle scar pattern.

The subclass Ostracoda consists of small crustaceans which live inside a bivalved shell called the *carapace*, the two elements of which articulate dorsally along a *hinge line*. They include some forms which have a marine habitat and some which have a fresh-water habitat. Although some of the marine forms live in deep waters, the majority live in the shallow waters of the continental shelves.

Although the soft and chitinous parts of the animal are much used in the classification of living ostracoda, the classification of fossil forms depends partly on their presumed relationships to living forms but, for the palaeontologist, primarily on the nature of the *muscle scars*, the type of *hingement*, the characters of the *duplicature*, and the general *form* and *ornament*.

The *adductor muscle scar* consists of a series of impressions on the valve interior for attachment of muscle used for closing the valves, usually found just in front of mid-length. The *antennal muscle scar* is an impression on the valve interior for attachment of muscle joined to the antenna; it is located in front of and generally above the adductor muscle scar. The *mandibular muscle scar* is one where muscle leading to the mandibular appendage was attached; it is usually in front of the adductor scars and below and sometimes in front of the antennal muscle scar.

In the living animal the body is covered anteriorly, ventrally and posteriorly by a thin *inner lamella*, chitinous except for calcified marginal parts forming a *duplicature* in some groups. The *outer lamella* is the relatively thick mineralized shell layer enclosed by thin chitinous layers. In some groups there is a space (termed *vestibule*) between the duplicature and the outer lamella. The shape and size of the vestibule, if present, and the number and arrangement of the *radial pore canals* at each end are of importance in determining genera.

FIG. 888 TYPES OF OSTRACOD HINGEMENT

Dorsal views of right valves

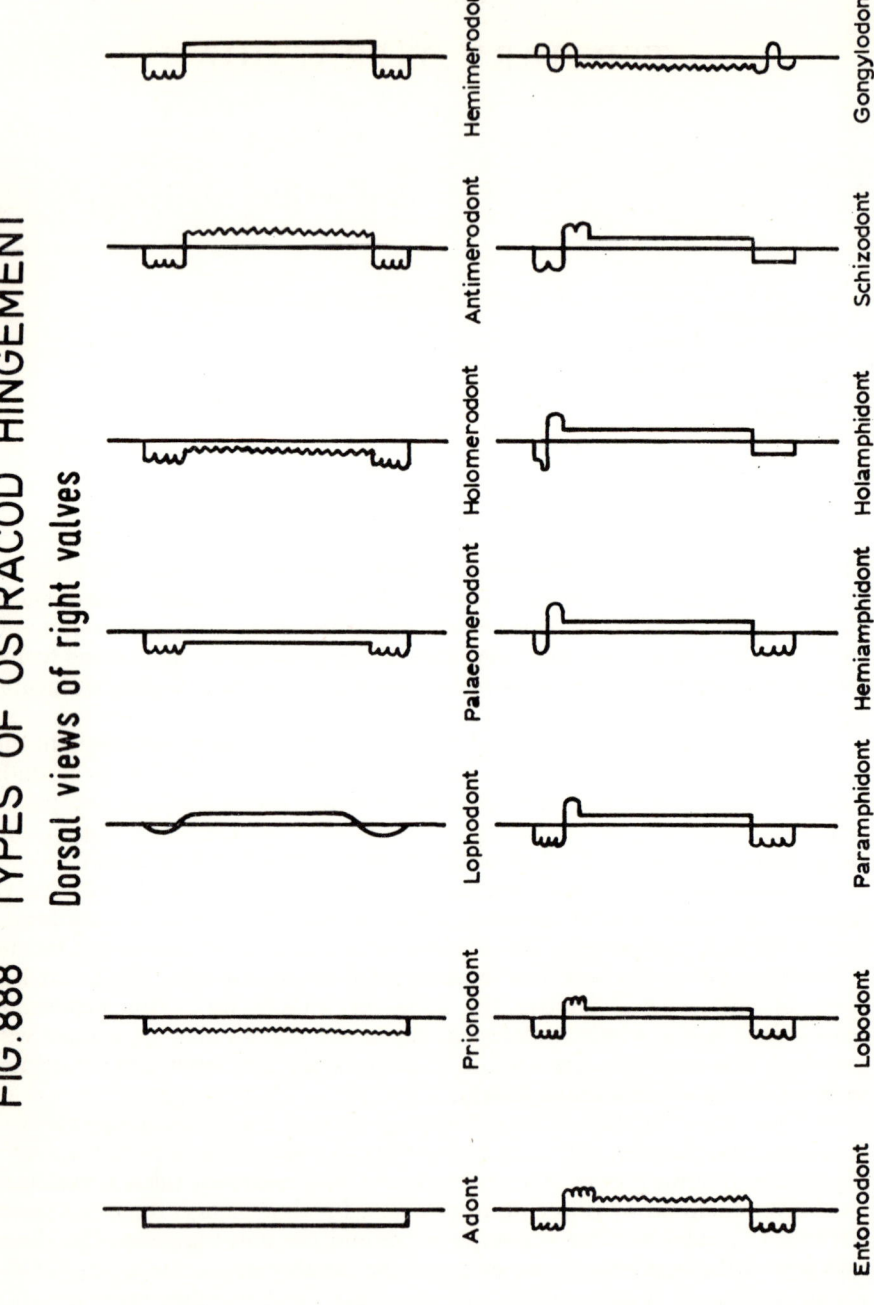

Adont

Prionodont

Lophodont

Palaeomerodont

Holomerodont

Antimerodont

Hemimerodont

Entomodont

Lobodont

Paramphidont

Hemiamphidont

Holamphidont

Schizodont

Gongylodont

446

The fourteen principal technical terms applied to different types of hinge-ment are illustrated in Fig. 888; in each case they are dorsal views of the right valve (the left valve being complementary), elements to the right of the guide line being median groove and terminal sockets, those to the left of the guide line being median bar and terminal teeth.

Some ostracod species show sexual dimorphism, adult female carapaces being relatively more inflated and less elongate than male carapaces.

Stratigraphical use. There has been such great development in the study of Tertiary ostracoda during the last twenty years or so that their use in strati-graphy has very considerably increased, especially in the case of marine sediments. The normal method of preparation of material for study (washing and picking) is exactly the same as for foraminifera. The evidence of ostracoda forms a useful supplement to that of foraminifera, and, when no foraminifera are present, it may even be just as useful.

Order PODOCOPIDA

Dorsal margin curved, or, if straight, shorter than maximum length of carapace; duplicature narrow or wide; adductor muscle scar pattern a circular group of many scars in more primitive forms (Metacopina); number of individual scars reduced in others, either grouped or biserial (Platycopina), or discrete and variously arranged (Podocopina); hinge margin undifferentiated in Platycopina and some Metacopina, usually differentiated into three or more elements in Podocopina. Possibly a polyphyletic order. (Ord.-Rec.)

Suborder PODOCOPINA

Carapace with valve margins incurved in the middle of the ventral margin; muscle scar pattern of discrete scars, those of the adductors being distinct from those of muscles attached to appendages; duplicature usually wide, with or without a vestibule; free margin with a selvage which overlaps that of the other valve; hinge often differentiated into three or four elements, any or all of which may be denticulate. Habitat marine and fresh-water. (Ord.-Rec.)

Superfamily BAIRDIACEA

Dorsally convex; wide duplicature; anterior vestibule wide, posterior vestibule narrow or wide; muscle scar pattern of discrete scars in a more or less circular area; hinge rabbeted or (e.g. *Macrocypris*) of five elements; in some forms denticles are present along contact margins. Marine. (Ord.-Rec.)

Family BAIRDIIDAE

Usually with asymmetrical, angulated, convex and concave rounded and acuminate 'bairdian' shape in lateral view; ends acuminate in dorsal view; left valve larger than right valve, overreaching and overlapping it; hinge rabbeted; distinct duplicature and vestibule. (Ord.-Rec.)

Bairdia (Figs 889, 890): dorsal margin arched, posterodorsal margin

447

slightly concave, anterodorsal margin less so, anterior end rounded, ventral margin nearly straight but upturned at the ends; owing to the greater concavity of the posterodorsal margin the posterior end is often like a snout; because of the strong but varying degree of overlap the two valves are often not quite the same shape, the smaller right valve being straighter dorsally and ventrally; surface smooth or punctate; anteroventral and posteroventral margins sometimes denticulate; hinge rabbeted, the bar in the right valve; wide duplicature and a vestibule present. Ord.-Rec. Cosmop. [This is the longest-ranging known ostracod genus.] *Bairdoppilata:* like *Bairdia*, but with short series of teeth and sockets in anterodorsal and posterodorsal positions in selvage of right valve and selvage groove of left valve. Cret.-Tert. N.Amer., S.Amer., N.Africa. *Triebelina:* like *Bairdia*, but surface coarsely pitted and with two or three straight or arched ridges, and right valve with weak terminal tooth at each end of hinge. U.Eoc.-Rec. Cosmop. *Bythocypris* (Fig. 891): reniform in outline, dorsal margin arched, ventral margin gently concave, posterior end more sharply rounded than the anterior; although lacking the asymmetry of *Bairdia*, the overlap, hinge, duplicature, vestibule and muscle scar are similar. ?Ord.-Rec. Cosmop.

Family MACROCYPRIDIDAE

Elongate, dorsal margin arched; wide vestibules anteriorly and posteriorly; hinge of five elements. (?Ord.-?Mio.-Plio.-Rec.)

Macrocypris (Fig. 892): smooth, compressed, elongate, anterior end rounded, posterior end pointed, ventral margin straight or concave; right valve larger than left valve, overreaching it everywhere except anteriorly; radial pore canals straight, crowded at anterior and posterior ends; muscle scar pattern consisting of a rosette of about nine scars, with three more close above and two more anterodorsally. ?Ord.-?Mio.-Plio.-Rec. Cosmop. [The

---

Figs 889–905. OSTRACODA: BAIRDIACEA AND CYPRIDACEA
Figs 889–891 after Shaver, in Moore; Fig. 892 after Sylvester-Bradley, in Moore; Figs 893–905 after Swain, in Moore.
889a. *Bairdia oklahomensis* Harlton, U.Carb. Illinois. × 20.
890c. *Bairdia formosa* Brady, Rec. Medit. × 25.
891a, *Bythocypris reniformis* Brady, Rec. N.Atl. × 25.
892b, d. *Macrocypris minna* (Baird), Rec. Norway. T. × 15.
893b. *Eucypris virens* (Jurine), Rec. Eur. T. × 15.
894c, d. *Protoargilloecia minor* (Jones and Hinde), Cret. England. T. × 25.
895d, e. *Cypridopsis vidua* (O. F. Müller), Rec. Eur., N.Amer. T. × 35.
896b. *Potamocypris fulva* (Brady), Rec. British Is. T. × 40.
897c, f. *Herpetocypris reptans* (Baird), Rec. British Is. T. c × 10, f enlarged.
898b. *Cyclocypris globosa* (Sars), Rec. Eur. T. × 25.
899b, *Cypria exculpta* (Fischer), Rec. Eur. T. × 30.
900b, g. *Candona candida* (O. F. Müller), Rec. Eur. b × 20, g × 65.
901b, e. *Ilyocypris gibba* (Ramdohr), Rec. N.Eur., N.Amer. T. × 30.
902b, f. *Cyprois marginata* (Strauss), Rec. Eur. T. b × 15, f × 60.
903d. *Paracypris polita* Sars, Rec. Norway. T. × 20.
904a. *Pontocyprella harrisiana* (Jones), Cret. England. T. × 30.
905d, e. *Argilloecia cylindrica* Sars, Rec. Norway. T. × 35.

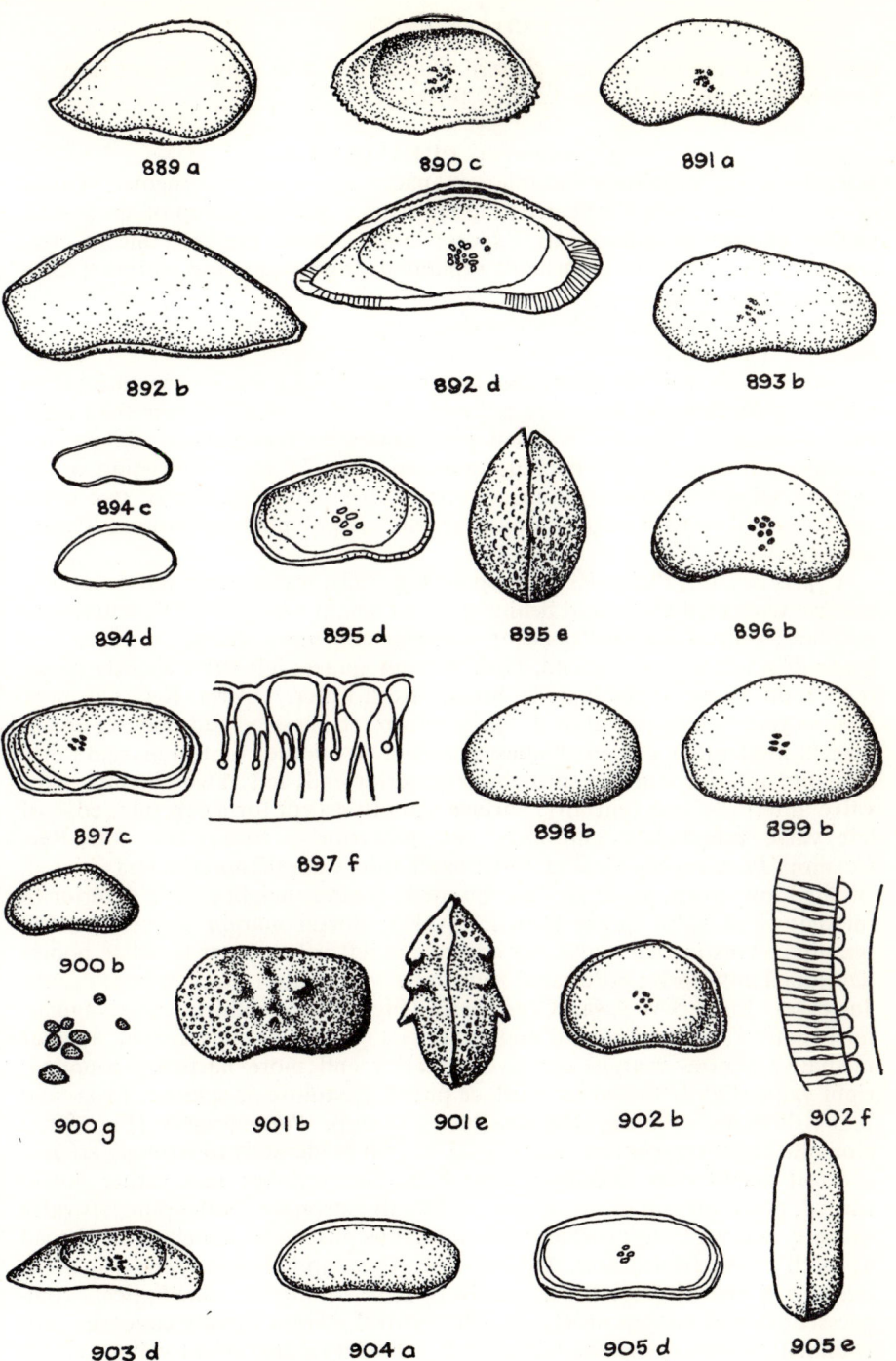

889 a

890 c

891 a

892 b

892 d

893 b

894 c

894 d

895 d

895 e

896 b

897 c

897 f

898 b

899 b

900 b

900 g

901 b

901 e

902 b

902 f

903 d

904 a

905 d

905 e

characters of the muscle scar pattern, duplicature and hinge do not seem to have been confirmed in pre-Tertiary forms.]

Superfamily CYPRIDACEA

Size, shape and ornament variable; carapace calcareous or corneous; ventral margin concave; hinge rabbeted; muscle scars a median group of spots with usually additional anteroventral spots; inner lamella present; line of concrescence and inner margin usually separated; fresh-water and marine. (?Sil.-?Perm., Trias.-Rec.)

Family CYPRIDIDAE

Usually subovoid to subtriangular, greatest height median or anterior to middle, posterior end the more pointed; left valve usually larger than right valve; ornament usually of small pits, knobs or reticulations; hinge line usually, but not always, curved, rabbeted; muscle scars consisting of an anteromedian or median group of four or five spots and one or two extra anteroventral spots. Fresh-water and marine. A polyphyletic group. (?Perm., Jur.-Rec.)

Cypris is ?Jur., Pleist.-Rec. *Eucypris* (Fig. 893): transversely suboval, fairly convex, truncated above and behind greatest height which is a little anterior to mid-line; ventral margin concave; posterior end more sharply rounded and lower down than anterior end, both without spines; left valve slightly larger than right valve; inner lamella broad; fresh-water. U.Cret.-Rec. Cosmop. *Heterocypris:* elongate-suboval, fairly convex, thickest behind middle; dorsal margin moderately convex, highest posterior to middle; ventral margin fairly straight; anterior end not so high as posterior end; left valve overlaps right valve anteriorly and ventrally; surface smooth except for a few pits; edge of left valve tuberculate anteriorly and posteriorly; fresh-water. Pal.-Rec. Cosmop. *Protoargilloecia* (Fig. 894): small, thin, smooth; anterior end pointed, situated low down, posterior end rounded; greatest height a little anterior to middle; right valve larger than left valve; dorsal margin convex, ventral margin convex in right valve but concave in left valve; inner lamellae poorly developed anteriorly; pore canal zone narrow, with thin, straight pore canals. Jur.-Mio. Eur., S.W.Asia. *Cypridopsis* (Fig. 895): reniform-subtriangular, fairly convex, thickest behind middle; dorsal margin strongly arched, angular medially; ventral margin concave; anterior end more narrowly rounded; right valve slightly the larger; surface pitted, pustulose or spinose; fresh- and brackish-water. ?Perm., U.Cret.-Rec. Cosmop. *Potamocypris* (Fig. 896): elongate-reniform, compressed; dorsal margin moderately to strongly arched, greatest height near mid-line; ventral margin concave; ends rather downturned, posterior end more narrow; right valve strongly overlapping left valve dorsally and ventrally; surface with numerous pits; inner lamella fairly broad at both ends; fresh-water. U.Cret.-Rec. Cosmop. *Herpetocypris* [*Candona auctt. non* Baird] (Fig. 897): relatively large, elongate, subelliptical, compressed; dorsal margin nearly straight, ventral margin slightly concave; ends rounded, posterior end the higher; left valve larger than right valve; surface smooth; inner margin broader anteriorly; radial pore canals complex; fresh-water. Tert.-Rec. Cosmop.

Family CYCLOCYPRIDIDAE

Small, subovate-subtriangular, maximum height near middle; dorsal margin arched, ventral margin almost straight; ends rounded, the anterior one usually narrower; maximum length below mid-height; compressed to strongly convex, thickest behind middle; subequivalve to strongly inequivalve; surface smooth, pitted, reticulated or with lines. Fresh-water. (U.Jur.-Rec.)

Cyclocypris (Fig. 898): subovate, swollen, maximum height median; dorsal margin strongly convex, ventral margin straight; anterior end truncated above, narrow; right valve a little the larger; surface smooth, weakly pitted or reticulated. Tert.-Rec. Eur., Asia, N.Amer. *Cypria* (Fig. 899): subovate, compressed, greatest height a little behind middle; dorsal margin strongly arched, ventral margin fairly straight; anterior margin the narrower; left valve a little the larger; surface smooth or punctate. Tert.-Rec. Cosmop.

Family CANDONIDAE [EUCANDONIDAE]

Fairly long, subreniform, compressed to fairly convex, maximum height usually behind middle; dorsal margin arched, ventral margin somewhat concave; anterior margin the more narrowly rounded; equivalve or inequi-valve; surface smooth or pitted; inner lamellae typically broad. (?Perm., Trias.-Rec.)

Candona [*Eucandona*] (Fig. 900): elongate-subreniform, fairly convex, maximum height behind middle; dorsal margin strongly arched to almost straight, ventral margin concave; anterior end the more narrowly rounded; left valve slightly the larger, but right valve may overlap it ventrally; surface smooth or finely punctate; inner lamellae broadest anteriorly; radial pore canals few, simple, widely spaced; fresh-water. ?Perm., Trias.-Rec. Cosmop.

Family ILYOCYPRIDIDAE

Subquadrate, compressed, thickest behind middle, with one or more dorsomedian sulci; dorsal margin straight, ventral margin straight to slightly concave; anterior end the more broadly rounded; left valve larger than right valve; surface pitted, tuberculate or spinose; inner lamellae fairly narrow. Fresh-water. (Trias.-Rec.)

Ilyocypris (Fig. 901): transversely subquadrate, compressed, bisulcate; no anteroventral marginal notch; marginal compressed rim present; pitted, pustulose or tuberculate; dimorphic. ?Trias., U.Jur.-Rec. Cosmop.

Family NOTODROMADIDAE

Subovate, inflated, with prominent anterodorsal eye spots; dorsal margin arched, ventral margin flattened and carinate; subequivalve; smooth or pustulose. Fresh-water. (Pal.-Rec.)

Notodromas is Recent only. *Cyprois* (Fig. 902): subovate, fairly convex; dorsal margin strongly arched, straighter on anterior slope; ventral margin

451

slightly concave; posterior end more sharply rounded, a little produced; anterior and ventral margins compressed; left valve slightly the larger; smooth; inner lamellae broad anteriorly; radial pore canals numerous anteriorly, fewer posteriorly. Pal.-Rec. Eur., N.Amer.

Family PARACYPRIDIDAE
Smooth, elongate; duplicature wide; anterior and posterior vestibules large; radial pore canals usually branched. Marine and fresh-water. (?Sil., Jur.-Rec.)
Paracypris (Fig. 903): elongate, wedge-shaped, posterior end pointed; left valve the larger; inner lamellae very broad; radial pore canals bifurcate. Resembles *Macrocypris* in general form, but is smaller, the dorsal margin is usually less convex, and the muscle scar spots are fewer; marine. ?Sil., Jur.-Rec. Cosmop. *Pontocyprella* (Fig. 904): small, transversely subquadrate or bean-shaped, dorsal and ventral margins nearly parallel, posterior end pointed; left valve overlaps right valve; hinge rabbeted, ridge in right valve; inner lamellae feebly developed anteriorly only; pore canal zone rather broad, with straight, closely spaced canals. Jur.-L.Mio. Asia, Eur.

Family PONTOCYPRIDIDAE
Elongate, dorsal margin moderately arched; posterior end the more pointed; nearly equivalve; usually smooth. Marine and fresh-water. (?Dev., Trias.-Rec.)
Pontocypris, a Recent form, is only doubtfully known fossil (?Dev.). *Argilloecia* (Fig. 905): subelliptical, sides flattened, posterior margin truncated; right valve a little the larger; inner lamellae usually very broad; marked sexual dimorphism. Cret.-Rec. Cosmop.

Superfamily DARWINULACEA
Elongate-ovate, narrower and less inflated anteriorly; usually smooth; right valve the larger or the smaller; hinge simple, a right valve overlap; muscle scar consisting of several radially arranged spots, but total outline flatter on upper side. Fresh-water or estuarine. (?Ord., U.Carb.-Rec.)

Family DARWINULIDAE
Characters of the Superfamily. (?Ord., U.Carb.-Rec.)
Darwinula (Fig. 906): elongate, slightly higher posteriorly; right valve much the larger; no calcified inner lamellae. ?Ord., U.Carb.-Rec. Cosmop.

Superfamily CYTHERACEA
Mostly defined by characters of soft parts which cannot be fossilized; adductor muscle scars usually a vertical row of four, with one or two antennal scars and usually three mandibular scars in front; there are a few adont families, but most have a compound hinge usually divided into three or four elements, any or all of which may be dentate or crenulate. Mainly marine. (M.Ord.-Rec.)
This is one of the most important superfamilies of Tertiary Ostracoda,

many genera and species being of stratigraphical significance. Characters of use in classification are the characters of the hinge, the general shape, the pattern of the muscle scars (adductor, antennal and mandibular), the duplicature, the presence or absence of a vestibule and an eye tubercle, and the nature of the radial pore canals.

Family CYTHERIDAE

Subquadrate, subreniform or ovate; subequivalve, right valve slightly overlapping left dorsally, but overlapped by left valve ventrally; smooth or reticulate, with large, scattered, usually sieve-like normal pore canals; hinge usually antimerodont, but some elements may be smooth; muscle scars in an almost vertical row of four adductors, one or two antennal scars in front, below which is a mandibular scar; marginal areas fairly broad, with a few straight or wavy radial pore canals; sometimes a small vestibule at ends. (Jur.-Rec.)

Cythere is Recent only. *Cnestocythere* (Fig. 907): subquadrate in side view, upper part of posterior end somewhat caudate; no strong overlap; hinge antimerodont, with strong teeth; ornament of strong longitudinal ridges, reticulate in between; strong eye tubercle; inner margin and line of concrescence coincide everywhere; anterior margin with about five simple, straight, radial pore canals; one rather long antennal scar. Mio. Eur. [The Oligocene species *C. reticulata* Moyes (27, p. 26) has only ventral and posterior ridges and is otherwise reticulate, and there is no prominent eye tubercle; it seems to belong elsewhere.] *Loxocythere:* subquadrate to subtriangular in side view, with ventral ridge and reticulate ornament; hinge antimerodont, with weakly crenulated median element; small vestibules at each end; no eye tubercle; radial pore canals few, straight; sexes distinct. Olig.-Rec. N.Z.

Family BRACHYCYTHERIDAE

Inflated, especially ventrally, smooth or with reticulate ornament; often with a ventral ridge or ala; end view subcircular to strongly triangular; usually with an eye tubercle; hinge amphidont, varying from paramphidont to hemiamphidont or holamphidont, often with accommodation groove; adductor muscle scars in a vertical row of four, modified by subdivision of the upper pair and fusion of the lower pair; antennal scar single or double; radial pore canals tending to be bulbous at mid-length; no vestibule. (U.Trias.-Rec.)

Brachycythere (Fig. 908): subtriangular to subovate in side view, rather inflated ventrally where there is often a small carina; flanks smooth or reticulate, ventral surface usually longitudinally striated; prominent eye tubercle; hinge hemiamphidont; radial pore canals numerous. U.Cret.-Rec. Eur., N.Amer., Africa, M.East (unpublished). *Alatacythere:* subrectangular to subovate in side view, ventrally rather inflated and with strong pointed alae, dorsally with a sharp ridge; ends spinose; flanks usually smooth; hinge hemiamphidont; marginal areas with eight to ten groups of two to four radial pore canals (medially bulbous), each group leading to a spine. U.Cret.-Olig. N.Amer., Eur. *Pterygocythereis:* like *Alatacythere*, but hinge usually

453

holamphidont and flanks sometimes with tubercles or spines. Eoc.-Rec. Eur., N.Amer., Carib., S.Amer. *Pterygocythere:* like *Alatacythere*, but with thicker shell material and broader accommodation groove in left valve. U.Cret.-Eoc. N.Amer. *Bosquetina:* shaped like *Brachycythere*, but without eye tubercle and with weakly holamphidont hinge; ventral carina present; lateral surface smooth; dimorphous. Olig.-Rec. Eur.

Family BYTHOCYTHERIDAE

Mostly with short caudal process, and many forms with a median dorsal sulcus; hinge lophodont, with long median element, and variously modified; adductor muscle scars usually in an arcuate group of five or more scars. (Dev.-Rec.)

Bythocythere: subquadrate, with straight dorsal margin and short caudal process, smooth, ventrally inflated or alate; no median sulcus; hinge lophodont, no accommodation groove; moderate anterior vestibule; radial pore canals few and straight; adductor muscle scars forming an arcuate group of six. Olig.-Rec. Cosmop. *Monoceratina* (Fig. 909): usually elongate, with long, straight dorsal margin ending in caudal process; median sulcus extending half way down carapace, usually surrounded by a crescentic swelling sometimes bearing one or more spines; hinge modified lophodont, anterior elements reduced or absent; adductor muscle scars in an arcuate group of four or five scars. Dev.-Rec. Cosmop.

Family CYTHERETTIDAE

Valves thick, with obliquely rounded ends, the posterior end the narrower; surface smooth, longitudinally pitted, or reticulate, some forms with three longitudinal ribs; no external submedian tubercle; hinge strongly holamphidont, the left valve anterior socket being only weakly enclosed ventrally and the median element being smooth or weakly crenulated; marginal area broad, with irregular embayments, not parallel to outer margin; small anterior vestibule usually present; radial pore canals long, sinuous or bulbous;

Figs 906–918. OSTRACODA: DARWINULACEA AND CYTHERACEA
Fig. 906 after Swain, in Moore; Figs 907–916 after Howe, in Moore; Fig. 909 after Sylvester-Bradley and Kesling, in Moore; Figs 917–918 after Reyment, in Moore.
906a, e, g. *Darwinula stevensoni* (Brady and Robertson), Rec. Eur. T. a, e × 40, g × 1000.
907b, e. *Cnestocythere lamellicosta* Triebel, Mio. Vienna B. T. × 30.
908a, f. g. *Brachycythere sphenoides*(?) Reuss, U.Cret. Texas. a × 22·5, f, g × 90.
909b. *Monoceratina ventralis* Roth, U.Carb. Oklahoma. T. × 25.
910b, d. *Cytheretta subradiosa* (Roemer), Rec. Medit. T. b × 35, d × 70.
911b. *Cytheridea muelleri* (Münster), U.Olig. Germany. T. × 18.
912b. *Eucythere declivis* (Norman), Rec. N.Atl. T. × 50.
913c. *Krithe papillosa* (Bosquet), Mio. Aquitaine. × 37·5.
914c. *Neocytherideis subulata fasciata* (Brady and Robertson), Rec. England. T. × 45.
915a, c. *Pontocythere grosjeani* (Keij), M.Eoc. Belgium. × 37·5.
916a, d. *Schuleridea acuminata* Swartz and Swain, Jur. Louisiana. T. × 30.
917a. *Cytherura gibba* (O. F. Müller), Rec. Holland. T. × 40.
918b. *Cytheropteron aureum* (Hornibrook), Rec. N.Z. × 50.

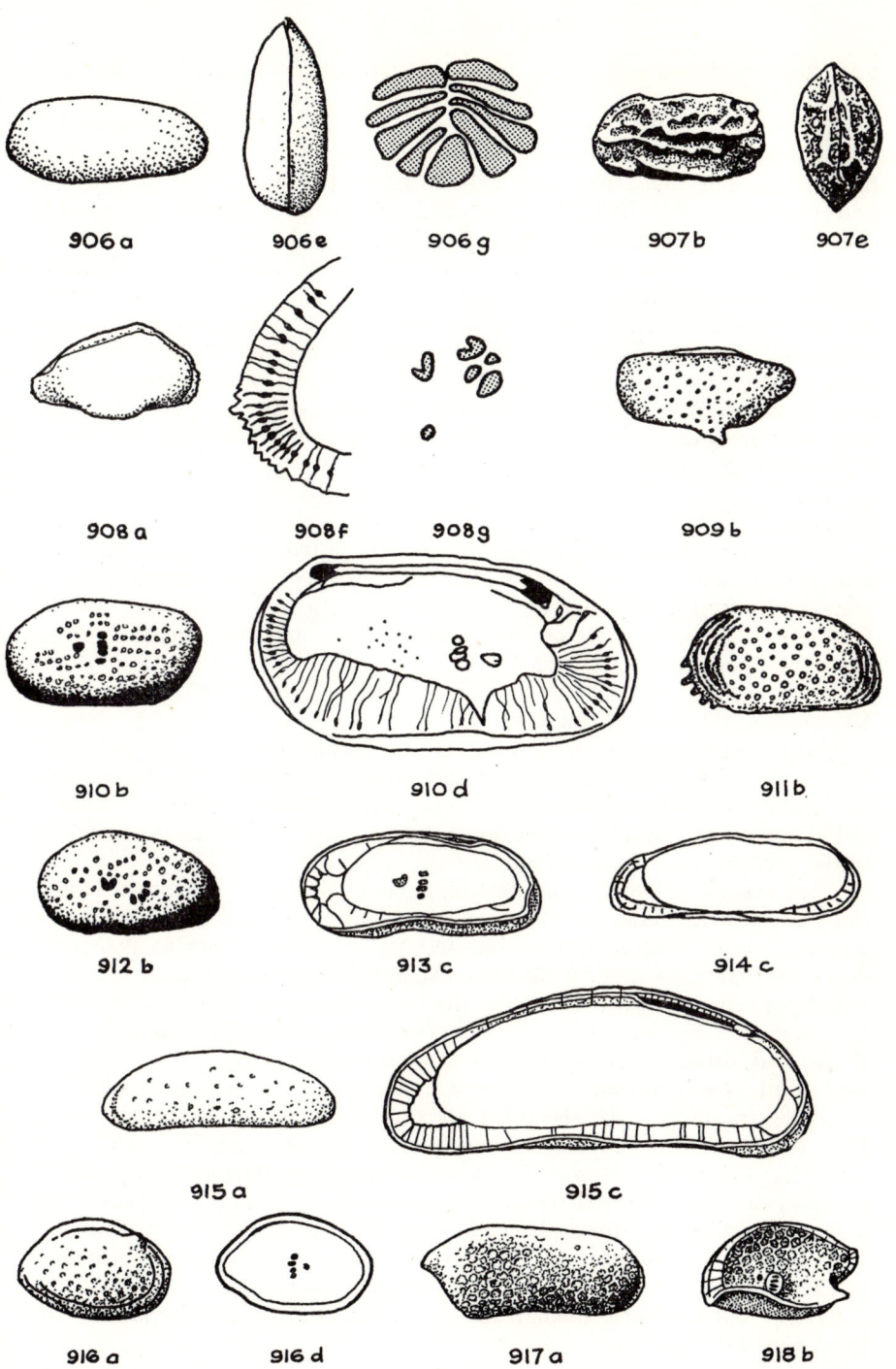

906 a   906 e   906 g   907 b   907 e

908 a   908 f   908 g   909 b

910 b   910 d   911 b

912 b   913 c   914 c

915 a   915 c

916 a   916 d   917 a   918 b

adductor muscle scars in a vertical row of four; antennal scars fused into a V. (U.Cret.-Rec.)

Cytheretta (Fig. 910): ovate in side view, left valve larger and shorter than right valve; surface smooth or pitted (the pits sometimes in longitudinal rows); hinge solid; marginal areas irregularly broad; radial pore canals long, curved and branching. Eoc.-Rec. Eur., Medit., N.Amer., Africa (unpublished). *Paracytheretta:* posterior end slightly caudate, and left valve overlapping right valve antero-dorsally; surface reticulate, with three longitudinal ridges; internally like *Cytheretta*. U.Cret.-Rec. Eur., N.Amer., S.Pac. *Protocytheretta:* elongate-ovate, inflated, ends rounded, but posterior end slightly caudate and with small spines; surface longitudinally reticulate and with three longitudinal ridges; internally like *Cytheretta*. Olig.-Rec. N.Amer. *Flexus:* more elongate and tapering than *Paracytheretta*, the narrow posterior end with small denticles; smooth or reticulate, with three very long ridges; internally like *Cytheretta*. Olig.-Mio. Eur.

Family CYTHERIDEIDAE

Ovoid to pyriform, surface usually smooth, pitted or reticulate; adductor muscle scars in a vertical or inclined row of four, usually with one, two or more scars in front; hinge adont to merodont, smooth or crenulated. The sub-families are defined mainly on the bases of hinge characters, marginal areas, and radial pore canals. (Perm.-Rec.)

Subfamily CYTHERIDEINAE

Hinge antimerodont or holomerodont (some species, e.g. in *Haplo-cytheridea*, with hinges reversed in the two valves); marginal areas widest anteriorly; sometimes with small anterior and smaller posterior vestibules; radial pore canals usually rather numerous; usually two muscle scars in front of the vertical row of four. (Perm.-Rec.)

Cytheridea (Fig. 911): ovate to triangular in side view, both ends tending to be finely denticulate, posterior end often bluntly pointed ventrally; surface punctate; hinge antimerodont, but left valve median element posteriorly depressed; marginal areas widest at ends where there may be small vestibules; radial pore canals numerous, rather thickened medially, anteriorly tending to be grouped. Olig.-Rec. Eur., ?N.Amer., Carib. *Clithrocytheridea:* form of *Cytheridea*, usually more strongly ornamented and tending to have ridges as well as pits; hinge antimerodont; marginal area fairly broad, ends with small vestibules; radial pore canals numerous; one V-shaped antennal scar. U.Cret.-Rec. N.Amer., Eur. *Cyprideis* (*s.s.*): ovate, surface sometimes with a few tubercles; hinge as in *Cytheridea*; marginal areas regular, without vestibules at ends; radial pore canals rather numerous, straight; antennal muscle scar V-shaped; markedly dimorphic; habitat usually brackish-water. Mio.-Rec. Eur., Carib. *Haplocytheridea:* like *Cytheridea*, but hinge holomerodont. U.Cret.-Rec. N.Amer., Carib., S.Amer., Eur. *Neocyprideis* [*Goerlichia*]: like *Cyprideis*, but left valve median hinge element is an undifferentiated crenulated bar. Pal.-Plio. Eur.

Subfamily EUCYTHERINAE

Hinge lophodont, antimerodont or lobodont, sometimes almost edentulous; marginal area broad anteriorly, with or without vestibules. (Jur.-Rec.)

Eucythere (Fig. 912): ovate-triangular, rather pointed posteriorly; smooth, but large normal pore canals visible; hinge lophodont; with an anterior and usually also a posterior vestibule; radial pore canals few, straight; antennal muscle scar U- or V-shaped. U.Cret.-Rec. Eur., N.Atl.

Subfamily KRITHINAE

Elongate, reniform, postero-ventral region somewhat more pointed; smooth; posterior margin usually inturned, closed valves in dorsal view showing a re-entrant V; marginal areas broad anteriorly; usually with large anterior and small posterior vestibules; radial pore canals not very frequent; hinge essentially adont, larger valve with a smooth or crenulated furrow; reversal of hingement and valve size occurs; one antennal muscle scar. (U.Cret.-Rec.)

Krithe (Fig. 913): postero-dorsal margin usually inturned; large, widely spaced normal pore canals; left valve hinge with median furrow for reception of dorsal edge of right valve; one U-shaped antennal muscle scar (sometimes divided); anterior vestibule mushroom-shaped; radial pore canals not numerous, rather short. U.Cret.-Rec. Eur., N.Amer., Asia, Carib.

Subfamily NEOCYTHERIDEIDINAE

Usually more elongate than Cytherideinae, surface smooth, pitted or reticulate or even noded; hinge usually lophodont; marginal areas variable in width; vestibule at one or both ends; one or two antennal muscle scars. (Jur.-Rec.)

Neocytherideis [*Sahnia*] (Fig. 914): elongate-oval, with low anterior end; hinge lophodont, weak; marginal areas narrow; large anterior vestibule; radial pore canals few, short, straight; surface smooth. Mio.-Rec. Eur.

Pontocythere [*Cytherideis auctt., Hemicytherideis*] (Fig. 915): elongate-subcylindrical, anterior end higher than in *Neocytherideis*; surface smooth or very finely pitted; hinge lophodont; anteriorly with a vestibule having a serrate margin, several radial pore canals developed from each serration; one large antennal muscle scar. Jur.-Rec. N.Amer., Carib., Eur; Africa (unpublished). *Cushmanidea:* like *Pontocythere*, but with reticulate ornament, especially well developed on lower half of valves. Mio.-Rec. N.Amer., Eur., Africa (unpublished).

Subfamily PERISSOCYTHERIDEINAE

A dorsal swelling on the posterior half has a sulcus in front of it; ventral ridge present; hinge antimerodont; radial pore canals few and widely spaced, some not reaching outer margin; sieve-type pore canals. (Mio.-Rec.)

Perissocytheridea: inflated-pyriform, dorsal margin nearly straight, ventral margin convex; surface with pits and ridges or smooth; marginal areas broad;

457

narrow terminal vestibules; strongly dimorphic; brackish-water. Mio.-Rec. N.Amer., Carib., S.Amer.

Family SCHULERIDEIDAE

Subovoid to oblong; surface smooth or pitted, rarely with ridges; hinge hemimerodont to antimerodont or more or less adont. (Jur.-Mio.)

Subfamily SCHULERIDEINAE

Hinge hemimerodont to antimerodont; one antennal muscle scar; tendency for radial pore canals to be arranged fan-like around anterior margin. (Jur.-Mio.)

Schuleridea (*s.s.*) (Fig. 916): subovoid, posterior end more narrowly rounded, greatest length below mid-line; left valve distinctly the larger; smooth or finely pitted; eye tubercle strong on right valve, weaker on left valve, with a slight depression behind them; hinge paleaomerodont; one antennal muscle scar. Jur.-Mio. N.Amer., Eur.

Subfamily CUNEOCYTHERINAE

Ovate, posterior end sometimes rather pointed; left valve the larger; hinge adont or nearly so; marginal areas broad, especially anteriorly; radial pore canals usually numerous and long; one antennal muscle scar. (Jur.-Mio.)

Cuneocythere: ovate, thick-shelled; surface pitted to reticulate; sometimes with a small anterior vestibule; left valve median groove with a small knob at its anterior end. Eoc.-Mio. Eur.; N.Africa, E.Africa (unpublished).

Family CYTHERURIDAE

Usually small and with a caudal process; ventral surface sometimes flattened, even becoming alate ventrolaterally; usually, but not always, well ornamented; hinge modified entomodont; with or without eye tubercle; marginal zones usually wide; one or two antennal muscle scars; dimorphism variable. (Jur.-Rec.)

Cytherura (Fig. 917): transversely subquadrate, with subdorsal caudal process; ornament variable, punctate, reticulate, or ribbed; eye tubercles absent or weak; right valve hinge with two smooth to weakly crenulated terminal teeth separated by a furrow that enlarges at each end; at least two antennal muscle scars; no vestibules; radial pore canals few, thin, simple, long; dimorphic, but the male atypically the more inflated. Cret.-Rec. Cosmop. **Cytheropteron** (Fig. 918): ovate, with somewhat upturned caudal process and ventrolateral pointed wing; smooth or rather weakly ornamented; no eye tubercles; right valve hinge with terminal crenulated teeth joined by a crenulated furrow; narrow anterior vestibule; radial pore canals few, straight, simple; two antennal muscle scars, one of which is V-shaped. ?Jur., Cret-Rec. Cosmop. **Eucytherura:** like *Cytherura*, but surface tuberculate or reticulate and with prominent eye tubercle; one kidney-shaped antennal muscle scar. Cret.-Rec. Eur., N.Amer., Carib. **Paijenborchellina** (Fig. 919): elongate-pyriform, with long caudal process which is below mid-height; left valve slightly the larger; ventral ridge present; surface pitted to reticulate with weak ridges, with a shallow furrow descending from dorsal margin to middle of flank; right valve hinge with two terminal teeth joined by a crenulated furrow; no vestibules; radial pore canals few, straight; eye tubercles weak. Cret.-Eoc. Eur., W.Africa, N.Africa, M.East (unpublished). *Paracytheridea:* similar to *Cytherura,* with alate ventral process, surface reticulate with various knobs and ridges; hinge weak, right valve with terminal crenulated teeth and crenulated median furrow; radial pore canals medially swollen. ?Carb., Cret.-Rec. Eur., N.Amer.

Family HEMICYTHERIDAE

Ovate to subrectangular or rather almond-shaped, rather truncated posteriorly, the angulation being at or close to the end of the ventral margin and sometimes produced into a caudal process; margin usually concave above the angulation; surface smooth, pitted, reticulate or with longitudinal ridges; ventral margin often with low keel; except in one Recent genus the hinge is merodont in youth but holamphidont in adults; vestibules rare; radial pore canals numerous; median scars of the four adductor muscles variously divided; two or three mandibular-antennal muscle scars. (Eoc.-Rec.)

Hemicythere (Fig. 920): inflated, subrectangular, postero-dorsal margin gently concave; surface pitted or reticulate; hinge heavy; second adductor muscle scar (from top) divided; two antennal muscle scars. Eoc.-Rec. Cosmop. *Aurila:* ovate to almond-shaped, rounded in front and pointed behind; surface strongly pitted, with distinct eye tubercles; hinge holamphidont, but with median element finely crenulated; small anterior and posterior vestibules; the two middle adductor muscle scars divided; three antennal muscle scars. L.Mio.-Rec. Eur., Carib. *Caudites* (Fig. 921): small, elongate- subtriangular, not inflated, with postero-ventral caudal process; surface with one or more longitudinal ridges and usually also one near and parallel to posterior end; hinge as in *Hemicythere*. L.Eoc.-Rec. N.Amer., Carib., Eur; N.Africa, E.Africa (unpublished). *Pokornyella:* egg- to kidney-shaped in side view, with small eye tubercles and distinct posteroventral caudal process; surface pitted or reticulate; radial pore canals anteriorly less than half the number present in *Hemicythere* (25 instead of 60–80); no vestibules; all four adductor muscle scars undivided; two antennal muscle scars; hinge holamphidont. M.Eoc.-Mio. Eur.

Family LEGUMINOCYTHEREIDIDAE

Elongate-ovate in both side and dorsal views; surface smooth to reticulate, sometimes with a ventrolateral ridge, but without the dorsal and median ridges of trachyleberids; submedian node and small eye tubercles sometimes present; hinge modified holamphidont, anterior socket of right valve elongate longitudinally; marginal areas a little wider anteriorly, where there is a small vestibule; radial pore canals fairly numerous, straight; two antennal muscle scars (fused into a V in *Basslerites*). (Eoc.-Rec.)

Leguminocythereis (Fig. 922): rather bean-shaped in side view, with dorsal and ventral margins subparallel; surface reticulate, with an oblique element in upper half; weak submedian node usually present. Pal.-Mio. N.Amer., S.Amer., Eur. **Basslerites:** ovate, with dorsal and posterior margins almost forming a right angle; surface smooth or nearly so; hinge strong; radial pore canals numerous; antennal muscle scars fused into a V. Eoc.-Rec. N.Amer., S.Amer., Eur., W.Africa.

Family LEPTOCYTHERIDAE

Relatively small, elongate to subquadrangular in side view, with distinct posterior angle; surface almost smooth to strongly ornamented; marginal area broad, with characteristic polyfurcated radial pore canals; hinge modified entomodont; one antennal muscle scar. (?Jur., Tert.-Rec.)

Leptocythere is Recent only. **Callistocythere** (Fig. 923): ornament of reticulation and wavy ridges; at least three terminal teeth of left valve median hinge element enlarged; vestibule almost nil. ?Jur., Tert.-Rec. Cosmop.

Family LIMNOCYTHERIDAE

Subequivalve, weakly or strongly calcified, with smooth, noded, or reticulate surface; marginal area fairly broad, sometimes forming small vestibules; radial pore canals straight; some of the four adductor muscle scars sometimes divided; antennal muscle scar somewhat crescentic; hinge usually adont, sometimes with terminal teeth in right valve; sometimes strongly dimorphic; fresh- to brackish-water. (Jur.-Rec.)

Limnocythere (Fig. 924): carapace thin, horny; surface reticulate, tuberculate or spinose; radial pore canals numerous. Jur.-Rec. Cosmop.

Family LOXOCONCHIDAE

Small, reniform to subrectangular in side view, with nearly smooth, pitted,

---

Figs 919–934. OSTRACODA: CYTHERACEA, PLATYCOPINA AND THAUMATOCYPRIDACEA

Figs 919 and 933 after Reyment, in Moore; Figs 920–922, 924–925, 928 and 931–932 after Howe, in Moore; Figs 923 and 927 after Hanai, in Moore; Fig. 926 after Sylvester-Bradley and Howe, in Moore, Figs 929–930 and 934 after Sylvester-Bradley, in Moore.

919b. *Paijenborchellina ijuensis* Reyment, Eoc. Nigeria. × 62·5.
920a, g. *Hemicythere villosa* (Sars), Rec. N.E.Atl. T. a × 40, g × 90.
921a. *Caudites medialis* Coryell and Fields, Mio. Panama. T. × 60.
922b. *Leguminocythereis scarabaeus* Howe and Law, Olig. T. × 40.
923b, d. *Callistocythere littoralis* (Müller), Rec. Italy. T. × 50.
924a, d. *Limnocythere inopinata* (Baird), Rec. N.W.Eur. T. × 40.
925b. *Loxoconcha rhomboidea* (Fischer), Rec. N.E.Atl. T. × 45.
926c. *Paradoxostoma variabile* (Baird), Rec. Holland. × 40.
927a, c. *Pectocythere quadrangulata* Hanai, Plio. Japan. T. × 52·5.
928a. *Schizocythere tessellata* (Bosquet), M.Eoc. Paris B. × 40.
929a. *Trachyleberis scabrocuneata* (Brady), Rec. Japan. T. × 60.
930a, c. *Buntonia shubutaensis* Howe, Eoc. Louisiana. T. × 60.
931a, d, e. *Xestoleberis aurantia* (Baird), Rec. N.E.Atl. a, e × 50, d × 72·5.
932b. *Falunia sphaerolineata* (Jones), Mio. France. T. Enlarged.
933b, e. *Cytherella abyssorum* Sars, Neogene. British Is. × 19.
934b. *Thaumatocypris echinata* G. W. Müller, Rec. Indian Ocean. T. ×. 15.

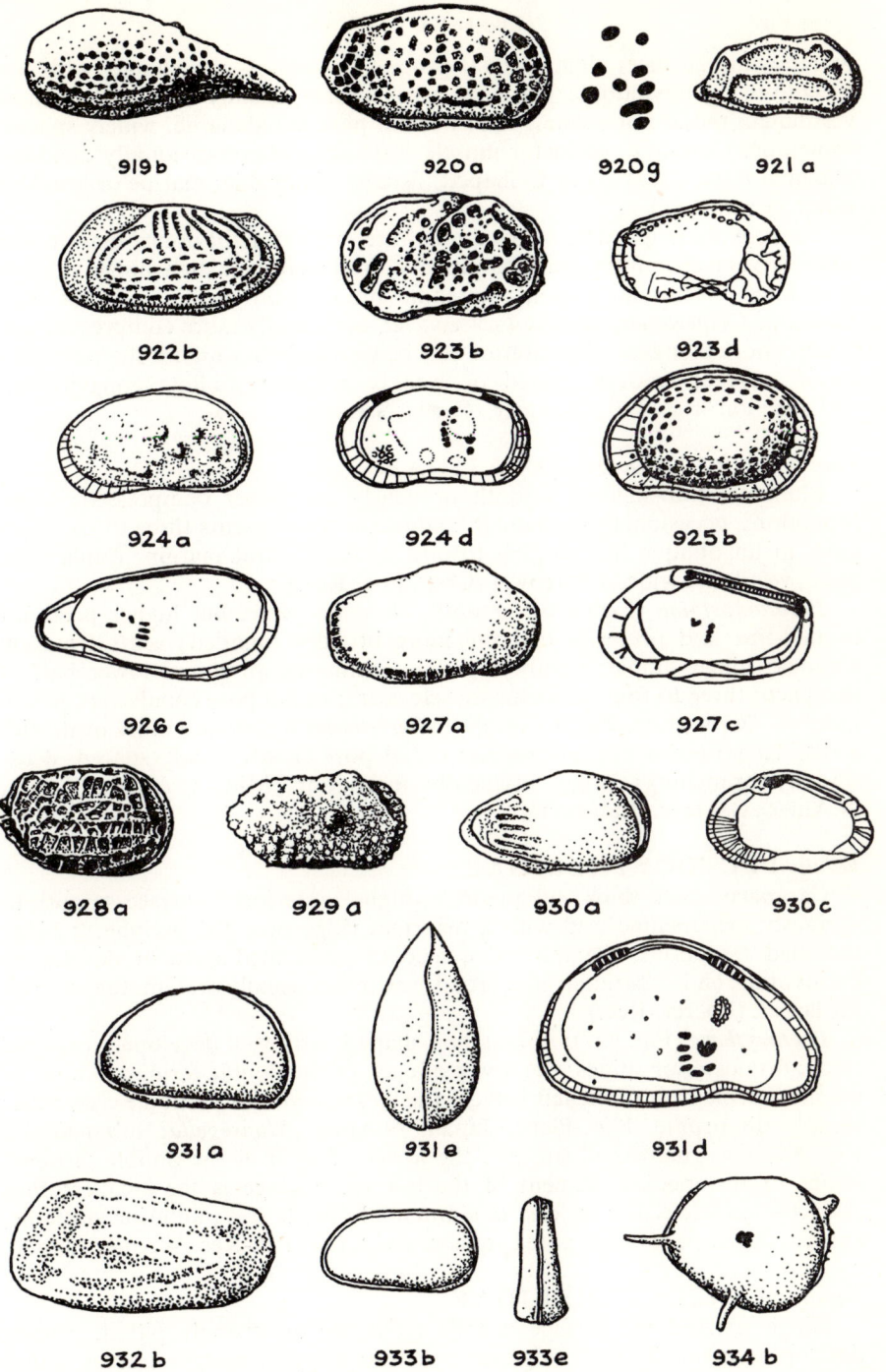

919 b

920 a

920 g

921 a

922 b

923 b

923 d

924 a

924 d

925 b

926 c

927 a

927 c

928 a

929 a

930 a

930 c

931 a

931 e

931 d

932 b

933 b

933 e

934 b

or reticulate surface; slight caudal process in some forms; hinge, with few exceptions, gongylodont; marginal areas broad, usually with rather small vestibules; radial pore canals few; normal pore canals large, widely spaced, sometimes sieve-like; adductor muscle scars elongate (occasionally divided); antennal muscle scar U- or C-shaped; usually dimorphic; marine or brackish water. (Cret.-Rec.)

Loxoconcha (Fig. 925): oval-subquadrate, with straight dorsal margin and sinuous ventral margin; surface pitted or reticulate; median hinge element crenulated; vestibules present; antennal muscle scar crescentic. Cret.-Rec. Cosmop. *Cytheromorpha:* like *Loxoconcha*, but usually more compressed and a little more elongate, the anterior end being somewhat higher than the posterior; surface smooth, pitted or weakly reticulate; hinge gongylodont. ?M.Jur., Pal.-Rec. Eur., N.Amer., Carib., S.Amer.

Family PARADOXOSTOMATIDAE

Elongate, thin-shelled, smooth or nearly so, rather compressed; hinge lophodont, occasionally denticulate; adductor muscle scars three to six, elongate, in an oblique line sloping toward antero-ventral margin; duplicature wide; vestibules wide or narrow. (?Cret., Eoc.-Rec.)

Paradoxostoma (Fig. 926): smooth, elongate-ovate, but higher posterior to mid-line and posterior end the more broadly rounded; ventral margin sinuous; hinge simple; vestibule wide, continuous all round lower half of carapace; three to four adductor muscle scars; radial pore canals very few in number. ?Cret., Eoc.-Rec. Cosmop. *Pellucistoma:* higher and more oval, with a distinct posterior caudal process; radial pore canals widely spaced, occasionally branching; hinge atypically merodont. Mio. C.Amer., Carib., N.Amer., S.Amer., Indonesia.

Family PECTOCYTHERIDAE

Carapace small, thick, subquadrate, higher anteriorly; surface smooth to variously ornamented and with a marginal ridge near the periphery; hinge modified merodont, the left valve median crenulated element developing knobs at the ends; marginal area broad; vestibules usually present, the anterior the larger. (L.Cret.-Rec.)

Pectocythere (Fig. 927): oblong box-shaped, with well developed marginal ridge; no posterior spines; the lower of each of the double terminal knobs of the median element of the left valve is the larger; anterior vestibule crescentic; slightly dimorphic. Plio.-Pleist. Japan, N.Amer. *Munseyella:* subquadrate, with heavily ornamented surface; the lower of each of the double terminal knobs of the median element of the left valve hinge is the smaller; four adductor muscle scars; at least one antennal muscle scar; anterior vestibule, at least, developed; radial pore canals straight, few. Pal.-Rec. N.Amer., Japan.

Family SCHIZOCYTHERIDAE

Hinge schizodont; surface usually reticulate or strongly pitted, usually developing a longitudinal ridge near ventral margin; marginal areas rather broad; radial pore canals few; four adductor muscle scars; one or two antennal muscle scars. (U.Cret.-Rec.)

Schizocythere (Fig. 928): ovate-subquadrate, with small posterior caudal process; surface pitted to strongly reticulate; distinct eye tubercle present; hinge schizodont with median element crenulated; normal pore canals few, widely spaced, sieve-like; one rounded antennal muscle scar. Eoc.-Plio. Eur. *Paijenborchella* (*s.s.*): elongate, with long caudal process (often with a serrated end) near lower posterior end; surface smooth, with an almost vertical median sulcus, with median and ventral longitudinal ridges, the latter forming a small spine; hinge schizodont, with crenulated median element; very few radial pore canals; one antennal muscle scar. M.Olig.-Rec. Indonesia, Eur., N.Amer. The subgenus *Eopaijenborchella* is like *Paijenborchella*, but the surface is reticulate, there are three longitudinal ridges, and the pointed caudal process is shorter. U.Cret.-Rec. Cosmop. *Neomonoceratina:* like *Paijenborchella*, but the short, pointed, upturned caudal process is above the mid-line. U.Eoc. E.Africa (unpublished); L.Mio.-Rec. Indonesia, Eur.

Family TRACHYLEBERIDIDAE

Subrectangular (with dorsal and ventral margins parallel or slightly convergent towards posterior end) to pyriform or subtriangular; anterior end rounded, posterior end subtriangular to gently caudate below mid-line; with or without a submedian tubercle; surface with strong spines or ridges or both, sometimes reticulate, or even smooth; hinge in post-Jurassic forms strongly amphidont, but juveniles merodont; duplicature of moderate width; vestibule narrow or absent; normal pore canals large, of sieve-type; radial pore canals numerous, sometimes crossing and sometimes widening medially; usually four adductor muscle scars (occasionally some subdivided); usually one U-shaped or two oval antennal muscle scars; sexual dimorphism common. (Jur.-Rec.)

From the Cretaceous onward species of this family are abundant in shallow water and littoral zones, some extending into deeper water. The family is distinguished from the Hemicytheridae by often having a submedian tubercle, not having such a distinct caudal process, and only rarely having one of the adductor muscle scars divided.

Subfamily TRACHYLEBERIDINAE

Subrectangular; submedian tubercle present. (Jur.-Rec.)

Trachyleberis (Fig. 929): ornament of scattered spines, tubercles or blades; submedian tubercle distinct; surface sometimes with reticulations as well; hinge holamphidont. Pal.-Rec. Cosmop. [It does not seem that the few Danian records can be substantiated.] *Bradleya:* subquadrate, without caudal process; smooth to reticulate, with dorsal and ventral ridges, sometimes with a few small tubercles; no vestibules; hinge hemiamphidont, with crenulated median element; two antennal muscle scars; radial pore canals simple or bifurcate, with median swellings. U.Cret.-Rec. Cosmop. *Costa:* outline of *Trachyleberis*, but surface reticulate and with three longitudinal ridges and an anterior rim;

caudal process spinose below; hinge holamphidont; one V-shaped antennal muscle scar. U.Eoc. Carib. (unpublished); Olig.-Rec. Eur., Asia, Africa, ?N. Amer. *Henryhowella* [*Howella*]: like *Costa*, but anterior end with small tubercles, and three spinose ridges on posterior half. Mio. N.Amer., Carib.; also (unpublished evidence) U.Eoc.-Plio. Eur., N.Africa, M.East, E.Africa, C.Amer., S.Amer. *Hermanites:* subquadrate, reticulate, with dorsal and ventral longitudinal ridges and strong submedian tubercle; one crescentic antennal muscle scar. Eoc.-Mio. N.Amer., Carib., Eur. *Quadracythere:* like *Bradleya*, but a little shorter and with distinct caudal process. Eoc.-Rec. Cosmop. *Trachyleberidea:* like *Costa*, but hinge hemiamphidont and two antennal muscle scars present; submedian tubercle strong, the median longitudinal ridge being discontinuous. Pal.-L.Mio. Eur., N.Amer. *Veenia:* like *Costa*, also with holamphidont hinge, but median ridge only slightly convex upward, all ribs less well defined; fairly strongly dimorphic. U.Cret.-Pal. Eur., N.Amer.

Subfamily BUNTONIINAE

Pyriform to subtriangular in outline; no submedian tubercle. (Jur.-Rec.)

Buntonia (Fig. 930): plump, subtriangular, with narrowly rounded posterior end; surface smooth, punctate, or with weak longitudinal ridges; hinge holamphidont, with crenulated median element; one antennal muscle scar. U.Cret.-Rec. N.Amer., Eur., E.Africa, Carib., S.Amer. *Protobuntonia:* like *Buntonia*, but posterior end more pointed, eye tubercles prominent, and with two antennal muscle scars. U.Cret.-Pal. N.Africa, W.Africa. *Echinocythereis:* ovate-subquadrate, with more rounded posterior end; surface finely reticulate and covered with small spines; prominent, small eye tubercle; two antennal muscle scars. U.Cret.-Rec. Cosmop.

Family XESTOLEBERIDIDAE

Ovate-reniform, plump, more inflated behind mid-line; surface smooth or pitted; internally usually with a reniform scar behind eye region; hinge adont or merodont; antennal muscle scar more or less arcuate; marginal areas broad, with anterior vestibule; radial pore canals short. Habitat marine. (Cret.-Rec.)

Xestoleberis (Fig. 931): ovate, with left valve the larger; hinge merodont, the median element smooth or finely crenulated; anterior end lower and narrower, with a vestibule; radial pore canals short, straight; antennal muscle scar arrow-shaped. Cret.-Rec. Cosmop. *Uroleberis:* bairdiiform in outline, swollen ventrally, surface smooth or pitted; hinge merodont, with smooth median element; antennal muscle scar V-shaped. Eoc.-Rec. Eur.

The family placing of the following two genera in the Cytheracea is uncertain:

Falunia (Fig. 932): subquadrate, anterior end only slightly the higher; surface with fine longitudinal ridges and reticulations; hinge merodont, with finely crenulated median element; marginal zones narrow, with poorly developed and few radial pore canals. U.Eoc.-Rec. Eur.; also Medit., N.Africa,

E.Africa, M.East (unpublished). *Ruggieria:* rather like *Cytheretta* in outline, with small spines anteriorly and posteriorly and a distinct eye depression; surface partly or wholly ornamented with reticulations or longitudinal ridges with distinct pits in between them, and with a small posteroventral spine; hinge hemiamphidont, but with a crenulated median element; marginal area fairly broad; no vestibules; radial pore canals rather numerous, simple, wavy, widened medially; antennal muscle scar U-shaped. Mio.-Rec. Eur.

Suborder PLATYCOPINA

Ovate with rounded ends to subquadrate; right valve typically overlapping left valve; hinge rabbeted; muscle scar pattern a biserial aggregate of small scars; duplicature narrow or lacking; dimorphism recognized by posterior swelling; compared with the Podocopina, there is no well defined inner calcareous lamella. (Jur.-Rec.)

Family CYTHERELLIDAE

Usually oval to subquadrate in side view; no radial pore canals. (Jur.-Rec.)

Cytherella (Fig. 933): transversely ovate to subquadrate, smooth or practically so, anterior half a little the less inflated; posterior end may be weakly denticulate in some forms, and is usually slightly more sharply rounded than the anterior end. Jur.-Rec. Cosmop. *Cytherelloidea:* like *Cytherella*, but with a few longitudinal ridges which may join at either end; usually less inflated than *Cytherella*. Jur.-Rec. Cosmop. *Platella:* like *Cytherella*, but surface ornamented with numerous distinct pits; usually with a slight median sulcus. Jur.-Rec. Cosmop.

Order MYODOCOPIDA

Subequivalve, smooth or ornamented, with or without an anterior rostrum; dimorphic. Marine. (Ord.-Rec.) [Includes most planktonic ostracodes.]

Suborder MYODOCOPINA

Dorsal margin straight or curved; anterior margin usually with a rostrum; may attain 2 or 3 cms in diameter. (Ord.-Rec.)

Superfamily THAUMATOCYPRIDACEA

Subcircular in outline, without rostrum, but with a few projecting spines near the margin. (M.Jur.-Rec.)

Family THAUMATOCYPRIDIDAE

Characters of the Superfamily. (M.Jur.-Rec.)

Thaumatocypris (Fig. 934): the only genus. M.Jur.-Rec. Cosmop.

Suborder CLADOCOPINA

Subcircular in side view, without gape, usually rather small (less than 1 cm.); muscle scar pattern of three scars placed close together. (?Dev., Carb.-Rec.)

465

Family POLYCOPIDAE

Characters of the Suborder. (?Dev., Carb.-Rec.)

Polycope (Figs 935, 936): subcircular in side view, with slight indication of cardinal angles; surface smooth, punctate or reticulate; left valve hinge anteriorly and posteriorly with a short ridge above a groove; anterior margin not serrate. ?Dev., Jur.-Rec. Cosmop.

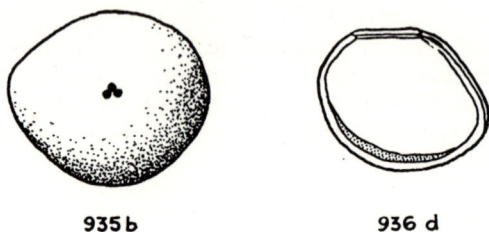

935 b   936 d

Figs 935–936. OSTRACODA: CLADOCOPINA

Figs 935–936 after Sylvester-Bradley, in Moore.

935b. *Polycope orbicularis* Sars, Rec. Norway. T. × 35.

936d. *Polycope sublaevis* Sars, Rec. Atl. × 35.

SHORT GLOSSARY OF TECHNICAL TERMS APPLIED TO OSTRACODA

Accommodation groove. Furrow above median element of hinge for reception of dorsal edge of opposite valve when carapace is opened.

Adductor muscle scar. Impressions on valve interior of muscle used for closure of valves, usually found just in front of mid-length.

Adont. Hinge without teeth, interlocking merely by a ridge in one valve and a groove in the other.

Ala. Ventral wing-like lateral extension of valve.

Amphidont. Type of hinge in which right valve has a posterior tooth, a median groove followed by a socket and then an anterior tooth; left valve arrangement complementary.

Antennal muscle scar. Impression on valve interior of attachment for muscle joined to antenna, located in front of and generally above adductor muscle scar.

Anterior. That part of the carapace in which antennae and antennules are located; the front end.

Anterior cardinal angle. The junction between the anterior margin and the hinge line.

Antimerodont. Type of merodont hinge in which the right valve has a crenulate posterior tooth, a crenulate median groove and a crenulate anterior tooth; left valve complementary.

Carapace. Protective covering of ostracod soft parts and appendages, forming two nearly symmetrical valves joined together by hinge along dorsal margin.

Cardinal tooth. Major projection at or near one or both extremities of the hinge, fitting into a socket in the opposite valve.

Caudal process. Posterior projection of valve margin.

Dimorphism. Development within a species of two shapes of adult carapaces, that of females being moderately to greatly different from that of the males.

Dorsal. The upper part when in normal position, i.e. that containing the hinge, eyes, antennae and antennules.

Duplicature. That part of the border, usually narrow, in which the calcareous peripheral portion of the inner lamella is in contact with the outer lamella or separated from it by a vestibule.

Entomodont. Type of hinge in which the right valve has a crenulate posterior tooth, a smooth or crenulate median groove, and a crenulate anterior socket followed by a crenulate tooth; left valve complementary.

Eye tubercle. Polished transparent knob in anterodorsal region of valve forming lens of eye.

Free margin. Anterior, ventral and posterior parts of margin, excluding the hinge.

Gongylodont. Type of hinge with a posterior socket bounded by two rounded teeth, a median crenulate bar, and an anterior rounded tooth bounded by two sockets; left valve complementary.

Hemiamphidont. Type of hinge in which the right valve has a crenulate posterior tooth, a smooth median groove and a simple anterior socket followed by a simple tooth; left valve complementary.

Hemimerodont. Type of hinge in which the right valve has a crenulate posterior tooth, a smooth median groove and a crenulate anterior tooth; left valve complementary.

Hinge. That part of the valves along and near the dorsal margin serving for articulation.

Holamphidont. Type of amphidont hinge in which the right valve has a smooth

467

posterior tooth, a smooth or finely crenulate median groove, and an anterior socket followed by a stepped (not crenulate) tooth; left valve complementary.

Holomerodont. Type of merodont hinge in which the right valve has a crenulate posterior tooth, a crenulate median bar and a crenulate anterior tooth; left valve complementary.

Inner lamella. Thin layer covering body in anterior, ventral and posterior parts of carapace, chitinous except for calcified marginal parts forming duplicature.

Instar. One of several successive moults.

Line of concrescence. Proximal line of junction of duplicature with outer lamella, coinciding with inner border of chitinous adhesive strip.

Lobodont. Type of hinge resembling entomodont except that anterior socket in right valve is lobate, not crenulate.

Lophodont. Type of hinge in which right valve has short, smooth anterior and posterior teeth and a smooth median furrow; left valve complementary.

Mandibular muscle scar. Impression on interior of valve where muscle leading to mandibular appendage was attached, usually in front of adductor scars and below and sometimes in front of antennal muscle scar.

Merodont. Type of hinge in which right valve has crenulate anterior and posterior teeth and a smooth or crenulate furrow or bar in between; left valve complementary.

Moult. Carapace cast off in moulting.

Muscle scar. Impression on interior of valve for attachment of a muscle.

Normal pore canal. Tubule piercing the valve almost at right angles, usually with enlarged proximal part.

Outer lamella. Relatively thick mineralized shell layer enclosed by thin chitinous layers; protects soft parts of body and appendages.

Overlap. Closure of valves so that contact margin or selvage of one valve extends over that of the other valve.

Palaeomerodont. Type of hinge in which right valve has crenulate terminal teeth and a smooth median bar; left valve complementary.

Paramphidont. Type of hinge in which right valve has a crenulate posterior tooth, a smooth or weakly crenulate median furrow followed by a round socket and then a crenulate anterior tooth; left valve complementary.

Pore canal. Minute tubule passing through shell.

Prionodont. Type of hinge in which right valve has a crenulate bar and left valve a crenulate furrow.

Rabbeted. Articulation in which one valve has a recess for receiving edge of opposite valve.

Radial pore canal. Tubule extending through adhesive strip from inner to outer surface of duplicature.

Schizodont. Type of hinge in which right valve has crenulate posterior tooth, smooth median furrow, and bifid anterior socket followed by bifid tooth.

Selvage. Middle ridge of contact margin, the principal ridge of the duplicature, sealing valves when closed.

Sieve-type pore canal. Wide normal pore canal partly closed by an internal perforate plate.

Socket. Pit in hinge area of one valve for reception of tooth of the other valve.

Submedian tubercle. Prominent anteromedian node on outer surface of some podocopids (e.g. trachyleberids), corresponding with muscle scar pit on interior.

Tooth. Projection on hinge area of one valve, fitting into socket on other valve.

Vestibule. Space between duplicature and outer lamella.

SELECT BIBLIOGRAPHY OF TERTIARY OSTRACODA

1. ALEXANDER, C. I. 1934. 'Ostracoda of the Midway (Eocene) of Texas', *J. Paleont.*, **8**, 206–237, pl. 32–35.
2. APOSTOLESCU, V. 1955. 'Description de quelques ostracodes de Lutétien du Bassin de Paris', *Cah. géol. Thoiry*, no. 28–29, 241–279, 7 pls.
3. BLAKE, D. B. 1950. 'Gosport Eocene Ostracoda from Little Stave Creek, Alabama', *J. Paleont.*, **24**, (2), 174–184, pl. 29–30.
4. BOLD, W. A. VAN DEN. 1946. *Contribution to the study of Ostracoda with special reference to the Tertiary and Cretaceous microfauna of the Caribbean region*, 1–167, pl. 1–18 (J. H. de Bussy: Amsterdam).
5. BOLD, W. A. VAN DEN. 1950. 'Miocene Ostracoda from Venezuela', *J. Paleont.*, **24**, 76–88, pl. 18–19.
6. BOLD, W. A. VAN DEN. 1957. 'Oligomiocene Ostracoda from southern Trinidad', *Micropaleontology*, **3**, (3), 231–254, 4 pls.
7. BOLD, W. A. VAN DEN. 1958. 'Ostracoda of the Brasso Formation of Trinidad'. *Micropaleontology*, **4**, (4), 391–418, pl. 1–5.
8. BOLD, W. A. VAN DEN. 1960. 'Eocene and Oligocene Ostracoda of Trinidad', *Micropaleontology*, **6**, (2), 145–196, pl. 1–8.
9. BOLD, W. A. VAN DEN. 1963. 'Upper Miocene and Pliocene Ostracoda of Trinidad', *Micropaleontology*, **9**, (4), 361–424, pl. 1–12.
10. BOLD, W. A. VAN DEN. 1965. 'Middle Tertiary Ostracoda from northwestern Puerto Rico', *Micropaleontology* **11**, (4), 381–414, pl. 1–7.
11. BOSQUET, J. A. H. 1852. 'Description des entomostracés fossiles des terrains Tertiaires de la France et de la Belgique', *Mem. Sav. Acad. r. Sci. Belg.*, **24**, 1–142, pl. 1–6.
12. BUTLER, E. A. 1963. 'Ostracoda and correlation of the Upper and Middle Frio from Louisiana to Florida (with a discussion of the Gulf Coast Miocene-Oligocene Boundary)', *Geol. Bull. La Geol. Surv.*, **39**, 100 pp., 6 pls.
13. ELLIS, B. F. and MESSINA, A. R. 1952. *Catalogue of Ostracoda*, (Am. Mus. Nat. Hist.). [Supplements later.]
14. GOERLICH, F. 1953. 'Ostrakoden der Cytherideinae aus der Tertiären Molasse Bayerns', *Senckenbergiana*, **34**, 117–148, pl. 1–9.
15. GREKOFF, N. 1956. '*Guide pratique pour la determination des Ostracodes post-paleozoiques*', 1–95, 16 pls (Maçon & Cie.: Paris).
16. HORNIBROOK, N. DE B. 1952. 'Tertiary and Recent marine Ostracoda of New Zealand', *Palaeont. Bull. Wellington*, **18**, 1–82, pl. 1–18.
17. HOWE, H. V. and CHAMBERS, J. 1935. 'Louisiana Jackson Eocene Ostracoda', *Geol. Bull. La*, no. 5, 1–65, pl. 1–6.
18. HOWE, H. V. and LAW, J. 1936 'Louisiana Vicksburg Oligocene Ostracoda', *Geol. Bull. La*, no. 7, 1–96, pl. 1–6.
19. JONES, T. R. 1857. 'A monograph of the Tertiary Entomostraca of England', *Palaeontogr. Soc. Monogr.* (*1856*) *1857*, 1–68, pl. 1–6.
20. KEIJ, A. J. (in Drooger, C. W., Kaasschieter, J. P. H. and Keij, A. J.). 1955. 'The Microfauna of the Aquitanian-Burdigalian of southwestern France: Pt 4. Ostracoda', *Verh. K. ned. Akad. Wet.*, (1), **21**, (2), 101–136, pl. 14.
21. KEIJ, A. J. 1957. 'Eocene and Oligocene Ostracoda of Belgium', *Verh. K. Belg. Inst. Natuurwetensch.*, no. 136, 1–210, pl. 1–33.
22. KINGMA, J. T. 1948. *Contribution to the knowledge of the Young-Caenozoic Ostracoda from the Malayan region*, 1–118, pl. 1–11 (Kemink en Zoon N.V.: Utrecht).

23. KOLLMANN, K. 1960. 'Cytherideinae und Schulerideinae n. subfam. (Ostracoda) aus dem Neogen des ostlichen Osterreich', *Mitt. geol. Ges. Wien.*, **51**, 89–195, 21 pls.

24. LIENENKLAUS, E. 1894. 'Monographie der Ostrakoden des nordwestdeutschen Tertiärs', *Z. dt. geol. Ges.*, **46**, 158–268, pl. 13–18.

25. MOORE, R. C. (ed.) .1961. *Treatise on Invertebrate Paleontology: Part Q Arthropoda 3: Crustacea: Ostracoda*, Q1–Q442, fig. 1–334 (Kansas Univ. Press).

26. MORKHOVEN, F. P. C. M. 1962. *Post-Palaeozoic Ostracoda: their Morphology, Taxonomy and Economic Use: 1. General*, 1–204, 8 pls (Elsevier Publishing Co.: Amsterdam).

27. MOYES, J. 1965. *Les Ostracodes du Miocene Aquitain*, 1–338, pl. 1–13 (Imprimerie E. Drouillard: Bordeaux).

28. OERTLI, H. J. 1956. 'Ostrakoden aus der oligozänen und miozänen Molasse der Schweiz', *Abh. schweiz. paläont. Ges.*, **74**, 1–119, pl. 1–16.

29. OERTLI, H. J. 1961. 'Ostracodes du Langhien-Type', *Riv. ital. Paleont. Stratigr.*, **67**, (1), 17–44, pl. 1–4.

30. REUSS, A. E. 1850. 'Die fossilen Entomostraceen desöster reichischen Tertiär-Beckens', *Naturw. Abh. her. v. Haidinger*, **3**, (3), 41–92, pl. 8–11.

31. SWAIN, F. M. 1962. 'Emendation of Candoninae, Eucandoninae, *Candona* and *Eucandona* as published in *Treatise on Invertebrate Paleontology: Q*', *J. Paleont.*, **36**, (4), 838–839.

Chapter VI

OTHER TERTIARY INVERTEBRATA

Class SCAPHOPODA

A small group of marine molluscs, bilaterally symmetrical, with an external elongate, tubular, tapering calcareous shell which is open at both ends and usually somewhat curved, the convex side being ventral. The aperture is at the larger end, and the posterior opening may be simple or variously slit or notched, sometimes with a small terminal tube. They are liable to be confused with some of the tubicolous Annelids, from which they differ in the greater regularity of form and in the microscopic structure of the shell (which is composed of three layers, while those of annelids consist of only two). (?Ord., Dev.-Rec.)

Family DENTALIIDAE

Shell usually regularly tapering, smooth or ornamented, greatest diameter at aperture. (?Ord., Dev.-Rec.)

Dentalium (*s.s.*) (Fig. 937): shell a curved and tapering cylinder (like an elephant's tusk); with distinct longitudinal threads or costellae, especially towards the apex which is polygonal; aperture circular; small end simple or with a slight notch or slit. U.Cret.-Rec. Cosmop. The subgenus *Antalis* is less strongly ornamented than *Dentalium*, and the apex, which is not polygonal in section, has a V-shaped notch and a plug with a small central tube. Trias.-Rec. Cosmop. *Fustiaria* (*s.s.*): slender and slightly curved, circular or nearly circular in section; no longitudinal ornament, only regular annular grooves; apical end simple or with a long, narrow slit on convex side. L.Cret.-Rec. Eur., Asia, Indonesia, Africa, S.Amer. The subgenus *Laevidentalium* is completely smooth and slightly curved, circular or slightly oval in section; the apical end has one or two short notches. Trias.- Rec. Cosmop.

Family SIPHONODENTALIIDAE

Shell usually small and smooth, the apertural area commonly constricted. (?Trias., L.Cret.-Rec.)

Siphonodentalium (Fig. 938): smooth, moderately to strongly curved, usually circular in section; apex cut into lobes. Pal.-Rec. Cosmop. *Cadulus* (*s.s.*): usually smooth, distinctly swollen medially; apical end simple. U.Cret.- Rec. Cosmop. The subgenus *Gadila* (Fig. 939) is like *Cadulus*, but is distinctly more slender. L.Cret.-Rec. Cosmop. The subgenus *Polyschides* has its maximum swelling further back from the aperture than in *Gadila*; apex with four or more notches. M.Eoc.-Rec. Cosmop. The subgenus *Dischides* is slender, with the maximum diameter near the aperture, and there are two deep apical slits. M.Eoc.-Rec. Cosmop.

SHORT SELECTED BIBLIOGRAPHY FOR SCAPHOPODA

1. MOORE, R. C. (ed.). 1960. *Treatise on Invertebrate Paleontology: Part I. Mollusca 1*, I37–I41 (Kansas Univ. Press).
2. EMERSON, W. K. 1962. 'A Classification of the Scaphopod Mollusks', *J. Paleont.*, **36**, (3), 461–482, pl. 76–80.

Class CEPHALOPODA

The Cephalopoda, so important in the Mesozoic, have relatively few survivors in the Tertiary. The Ammonoidea made their last appearance in the Maastrichtian (*Indoceras*, the last Ammonite, and *Baculites*) and were already extinct in the Danian. The Nautiloidea, on the contrary, continue as abundant as in the Mesozoic and even seem to make a feeble attempt at throwing off new lines (Hercoglossidae, Aturiidae), reproducing some of the features of the early goniatites. The Coleoidea are really the most flourishing branch, but as they tend increasingly towards a soft-bodied condition their fossil representatives appear scanty and sporadic: they show, however, most interesting transitions from the belemnites to modern forms. (Cambr.-Rec.)

Subclass NAUTILOIDEA

Small to large, chambered shells, always planispirally coiled (in the Tertiary) but with wavy septal sutures, and with siphuncle either approximately central in the septum or approaching the internal (dorsal) edge. (Cambr.-Rec.)

The other two nautiloid subclasses of the Cephalopoda (Endoceratoidea and Actinoceratoidea) are Palaeozoic only.

Order NAUTILIDA

Shells curved to coiled. (Dev.-Rec.)

Superfamily NAUTILIACEAE

Involute, usually smooth, some forms with sinuous folds or costae; whorl

---

Figs 937–943. SCAPHOPODA AND NAUTILOIDEA
Fig. 938 after Ludbrook, in Moore; Figs 940–943 after Kummel, in Moore; Figs 937 and 939 original.
937. *Dentalium elephantinum* Linné, L.Plio. S.France. T. × 1.
938. *Siphonodentalium lobatum* (Sowerby), Rec. Atl. T. × 1.
939. *Cadulus (Gadila) clarae* Maury, M.Mio. Oak Grove, Florida. × 6.
940a, b. *Nautilus pompilius* Linné, Rec. S. W. Pac. T. × 0·35.
941a, b. *Eutrephoceras laverdei* Durham, L.Cret. Columbia. × 0·2.
942. *Cymatoceras pseudoelegans* (d'Orbigny), Cret. Eur. × 0·2.
943a, b. *Hercoglossa orbiculata* (Tuomey), Pal. Alabama. T. × 0·4.

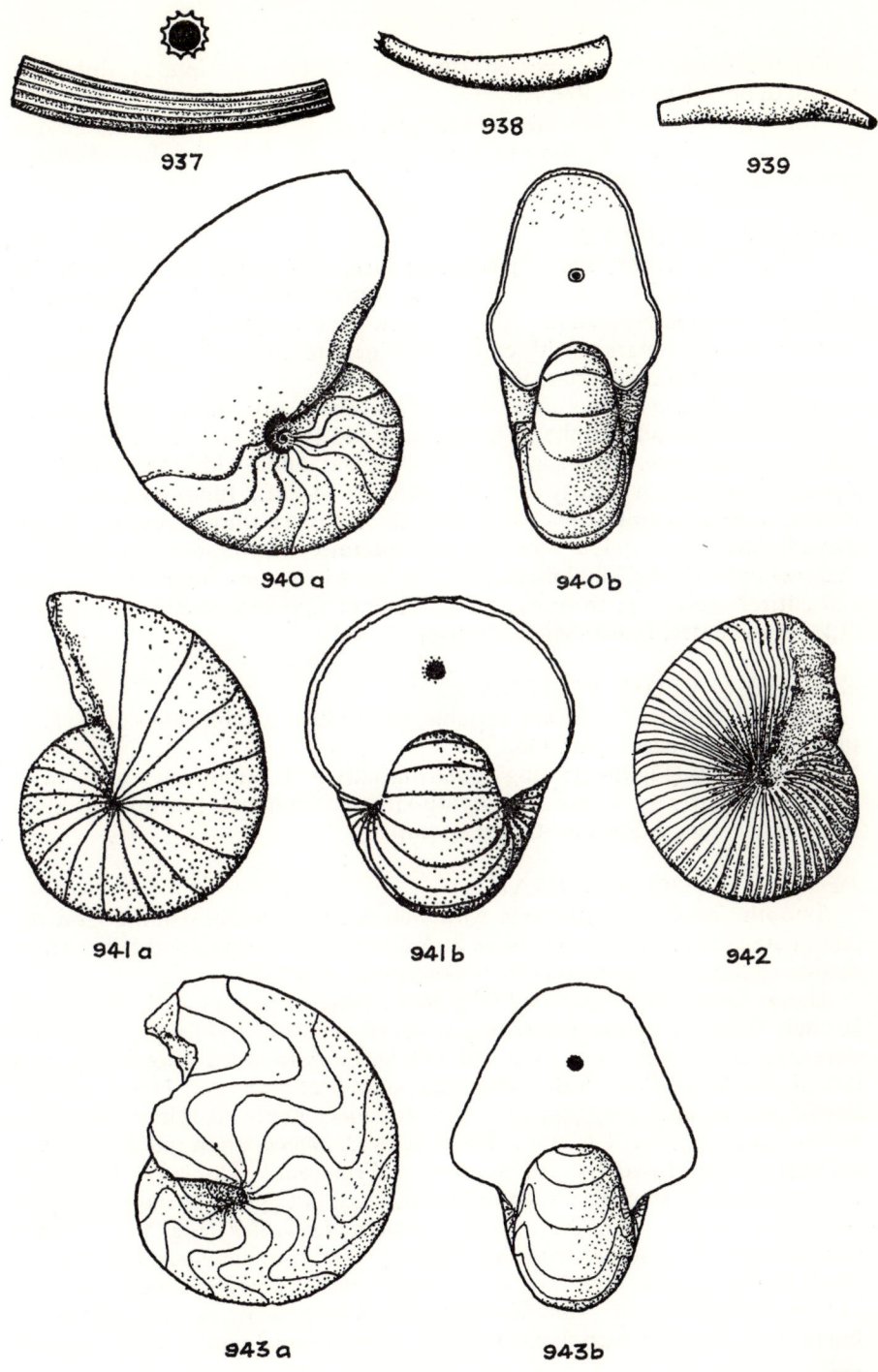

937

938

939

940 a

940 b

941 a

941 b

942

943 a

943 b

section low to laterally compressed; sutures straight to very sinuous; siphuncle central to dorsal. (Trias.-Rec.)

Although Spath (**15, 16**) referred the three well marked types of Nautiloidea that are found in the Tertiary to three families, the following four Tertiary families are now recognized.

Family NAUTILIDAE

Surface (in Tertiary species) smooth; form more or less completely involute; whorl section rounded or subquadrate; suture line comparatively straight to moderately sinuous, without a conspicuous lateral saddle; siphuncle central; apertural margin with convex outline laterally and shallow hyponomic sinus peripherally. (Trias.-Rec.)

Nautilus (Fig. 940): smooth, involute or nearly so; suture line sinuous, having a low, broad peripheral saddle, a broad lateral lobe descending well below the guide-line, and a small but well-marked saddle within the umbilicus; siphuncle central or nearly so. Olig.-Rec. Indo-Pac., Indonesia, Austral., Russia. *Eutrephoceras* (Fig. 941): smooth, inflated; whorl section kidney-shaped, broadly rounded ventrally and laterally; hyponomic sinus broad, shallow and rounded; umbilicus more or less nil; suture line nearly straight and entirely above (in front of) the guide-line; siphuncle small, central or a little below centre. U.Jur.-Mio. Cosmop.

Family CYMATOCERATIDAE

Involute, costate; form very variable, suture less variable. Only one genus survived the Mesozoic. (Jur.-Olig.)

Cymatoceras (Fig. 942): subglobular, usually with rounded whorl section; involute or nearly so; suture only slightly sinuous; siphuncle position variable; sides and venter costate. Jur.-Olig. Cosmop.

Family HERCOGLOSSIDAE

Smooth, involute; suture line with well marked saddle, definitely on the lateral area (not in the umbilicus as in *Nautilus*), and deep lateral lobe; whorl section rounded or triangular (*Deltoidonautilus*). (Jur.-Olig.)

Hercoglossa [*Enclimatoceras*] (Fig. 943): more or less inflated, whorls not strongly rounded laterally, narrowly rounded ventrally and deeply impressed dorsally; suture with broad ventral saddle, deep rounded lateral lobe and lateral saddle, shallow, broad lobe near umbilical wall, and broad internal lateral saddle and deep dorsal lobe; siphuncle never marginal. ?L.Cret., U.Cret.(Danian)-Eoc. Cosmop. [The species *H. danica*, often misidentified, is actually a zone fossil for the uppermost Cretaceous (Danian).] *Cimomia*: suture line much smoother than in *Hercoglossa*, but agrees in the position of the lateral saddle. Jur.-Olig. Cosmop. *Deltoidonautilus* (Fig. 944): whorl section triangular; suture with rather acute ventral saddle, large lateral lobe, small lateral saddle and lobe on umbilical wall; siphuncle near dorsum. U.Cret.-Olig. Cosmop. *Aturoidea* [*Paraturia*] (Fig. 945): rather compressed laterally; suture line with distinct, broad, blunt ventral saddle, deep, narrow,

474

asymmetrical, linguiform lateral lobe, broad, rounded, asymmetrical lateral saddle, broad, rounded lobe near umbilical seam, broad, rounded saddle on side of impressed zone, and U-shaped dorsal lobe; siphuncle a little nearer dorsal side, without the special features of that of *Aturia*. U.Cret.-Eoc. Eur., N.Amer., S.Amer., Africa, Asia, Austral.

Family ATURIIDAE

Shell smooth, compressed, discoidal, completely involute; suture line with wide, rounded lateral saddle, deep, narrow linguiform lateral lobe, flat peripheral saddle, broad lobe on umbilical slope to dorsal area, and broad saddle on dorsal area; siphuncle internal (dorsal), consisting of a series of wide, funnel-like extensions of the septa. (Pal.-Mio.)

Aturia (Fig. 946): characters of the Family (the only genus). Pal.-U.Mio. Cosmop.

Subclass COLEOIDEA [DIBRANCHIATA]

Shell, when present, internal (except for the unchambered external shell of the female of *Argonauta*); two gills; eight or ten arms. (Palaeozoic-Rec.)

Much caution is advisable before accepting statements of the discovery of belemnites of Tertiary age, for two reasons. In the first place, they are fossils which will stand much wear and tear (witness their survival along with oysters in glacial deposits) and may therefore very well occur as *derived* fossils. Secondly, one of the Octocorallia (*Graphularia*—see p. 484) could easily be mistaken for a belemnite. Most of the alleged Tertiary belemnites have been accounted for in one of these two ways. Nevertheless, there remains a small number of fossils of undoubted Tertiary age which, if found in Mesozoic strata, would without hesitation be called 'belemnites' in the broad sense of the term, and there is a larger number which are undoubted offshoots from the Belemnitidae, some of them of special interest as transitional to modern types of cuttlefish.

Order BELEMNITIDA

Shell consisting of three parts: a rostrum (cigar-shaped), a phragmocone (conical), and a proostracum (projecting, but rarely preserved); ten arms. (Palaeozoic-Rec.)

Family BELOPTERIDAE [NEOBELEMNITIDAE]

Rostrum somewhat reduced, often with two deep lateral furrows; phragmocone often somewhat curved. (?Jur., U.Cret.-L.Olig.)

Beloptera (Figs 947, 948): resembles a short belemnite guard, constricted in the middle and bearing in this constricted part a pair of flat projecting wings; a boreal form. Pal.-L.Olig. Eur. *Belopterina:* like *Beloptera*, but without projecting wings, merely lateral crests. L.Eoc. Eur. *Bayanoteuthis* (8, 13):

like a long, cylindrical Mesozoic belemnite with two ventro-lateral grooves; alveolus oval in cross-section and very deep (53 mm. in the largest specimen, a fragment 70 mm. long but probably less than half the full length). M.Eoc. Eur. *Styracoteuthis* (**1**) (Fig. 949): guard stout and cylindro-conical, about 75 mm. long and 18 mm. thick, oval to subtriangular in cross-section, with two deep ventro-lateral grooves running for the greatest part of the length, and a deep and acute alveolus. Pal.-Eoc. Arabia, W.Pakistan. *Vasseuria* (**9**) (Fig. 950): much smaller than *Bayanoteuthis*, about 5 cm. long and 6 mm. wide; the alveolus, elliptical in cross-section, is very deep and wide, the guard only forming a thin sheath around the phragmocone; except for its straightness, the guard has much the aspect of a *Dentalium*; the siphuncle is stated to resemble that of *Spirula*. M.Eoc. Eur.

Order SEPIIDA

Shell internal, tending to be bent ventrally, the phragmocone constituting an important part of it; ten arms. (Jur.-Rec.)

Family SEPIIDAE

Siphon considerably enlarged, chambers not of normal shape; tissue formed of numerous pillars; rostrum reduced or absent. (Jur.-Rec.)

Sepia (cuttle-fish): shape of skeleton much as in *Belosepia*, but guard reduced to a small spike, and phragmocone to a spongy mass (by development of vertical pillars between the septa). U.Eoc.-Rec. Widespread.

Family BELOSEPIIDAE

Phragmocone small; ventral curvature feeble and thickness great; no pillars or intercalary septa; rostrum strong, bent dorsally. (Eoc.)

Belosepia (Fig. 951): appearing transitionally between *Belemnites* (*s.1.*) and *Sepia*; guard greatly reduced in proportion to phragmocone, which is curved, with very oblique septa and a very wide siphuncle; proostracum thick, wide, with rugose and tuberculate surface; as a rule only the guard and adjacent part of phragmocene are preserved. Pal.-U.Eoc. Widespread.

Figs 944–951. NAUTILOIDEA AND COLEOIDEA

Figs 944–946 after Kummel, in Moore; Figs 947–948 after Cossmann and Pissarro; Fig. 949 after Crick; Fig. 950 after Cossmann; Fig. 951 after R. B. Newton.

944a, b. *Deltoidonautilus triangularis* (de Montfort), U. Cret. France. × 0·3.

945a, b. *Aturoidea olssoni* Miller, Eoc. Peru. × 0·3.

946a, b. *Aturia angustata* (Conrad), Mio. Washington. × 0·3.

947a, b. *Beloptera belemnitoides* Blainville, M.Eoc. Paris B. T. × 0·75. a, ventral view; b, side view.

948. *Beloptera curta* Cossmann, M.Eoc. Paris B. × 0·94. (Anterior view showing alveolus).

949. *Styracoteuthis orientalis* Crick, Eoc. Oman, Arabia. T. × 0·6.

950. *Vasseuria occidentalis* Munier-Chalmas, M.Eoc. Bois-Gouët, Brittany. × 0·75.

951a, b, c. *Belosepia sepioidea* (Blainville), M.Eoc. Hampshire. T. × 0·56, a, ventral view; b, side view; c, dorsal view.

476

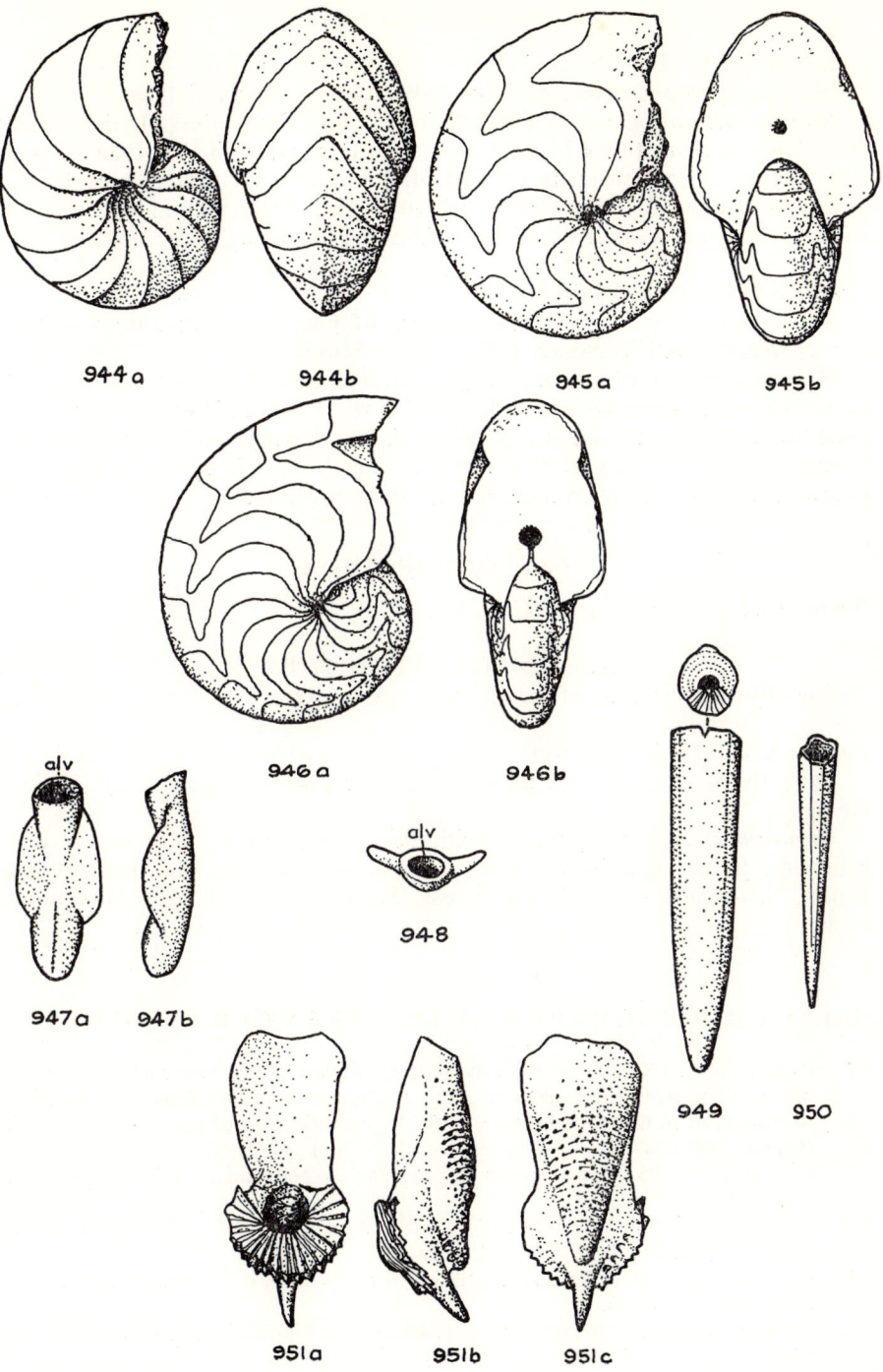

944a

944b

945a

945b

946a

946b

947a

947b

alv

948

949

950

951a

951b

951c

477

Family SPIRULIDAE

No guard; phragmocone rolled up into a loose plane-spiral. (Rec.)

Spirula, the only genus, is not definitely known fossil. Possibly descended from *Spirulirostra*. It lives in deep oceanic waters and until 1921 (when many were caught in deep water in the Atlantic) had only very rarely been found alive, but its light, gas-filled shell is drifted to great distances and is cast up on shore in many parts of the world, sometimes in great abundance.

Family SPIRULIROSTRIDAE

Strong tendency for ventral curvature of phragmocone; antero-ventral part of rostrum well developed. (?Cret., Eoc.-Mio.)

Spirulirostra: guard curved, sharply pointed behind, compressed in front so as to have a somewhat triangular outline when viewed laterally; phragmocone long, strongly curved at first, less so later, with siphuncle along concave ventral margin; curved part of phragmocone embedded in compressed part of guard. Of rather rare occurrence. Eoc.-Mio. Eur., Austral.

Order OCTOPODIDA

No internal shell; eight arms. (Cret.-Rec.)

Suborder POLYPODOIDEA

Defined by the characters of the soft parts. (Mio.-Rec.)

Family ARGONAUTIDAE

Only the female excretes an external, unchambered spiral shell. (Mio.-Rec.)

Argonauta: shell compressed, with flattened periphery and costae and tubercles, rather reminiscent of the Jurassic *Cosmoceras*. Mio.-Rec. Styria, Japan, N.Africa, Italy, N.Z., warm seas. ***Kapal*** is an allied form, Mio. Indonesia.

SELECT BIBLIOGRAPHY OF TERTIARY CEPHALOPODA

1. CRICK, G. C. 1904. 'On a Dibranchiate Cephalopod, *Styracoteuthis orientalis*, n.g. et n.sp., from the Eocene of Arabia', *Proc. malac. Soc. Lond.*, **6**, 274–278.
2. FLOWER, R. H. and KUMMEL, B. 1950. 'A classification of the Nautiloidea', *J. Paleont.*, **24**, 604–616.
3. KUMMEL, B. 1953. 'The ancestry of the family Nautilidae', *Breviora*, no. 21, 1–7, 1 pl.
4. KUMMEL, B. 1956. 'Post-Triassic nautiloid genera', *Bull. Mus. comp. Zool. Harvard*, **114**, no. 7, 324–484, pl. 1–28.
5. MILLER, A. K. 1947. 'Tertiary nautiloids of the Americas', *Mem. geol. Soc. Am.*, **23**, 1–234, pl. 1–100.
6. MILLER, A. K. and THOMPSON, M. L. 1933. 'The Nautiloid Cephalopods of the Midway Group', *J. Paleont.*, **7**, 298–324, pl. 34–38.

7. MOORE, R. C. (ed.). 1964. *Treatise on Invertebrate Paleontology: Part K Mollusca 3: Nautiloidea*, K383–K466 (Kansas Univ. Press).

8. MUNIER-CHALMAS, P. 1872. [*Bayanoteuthis*, gen. nov.]. *Bull. Soc. géol. Fr.*, (2), **29**, 530.

9. MUNIER-CHALMAS, P. 1880. [*Vasseuria*, gen. nov.]. *Bull. Soc. géol. Fr.*, (3), **8**, 291.

10. NEWTON, R. B. 1894. 'British Eocene Cephalopoda', *Proc. malac. Soc. Lond.*, **1**, 119–131, pl. 10.

11. PALMER, K. VAN W. 1937. 'The Claibornian Scaphopoda, Gastropoda and Dibranchiata Cephalopoda of the southern United States', *Bull. Am. Paleont.*, **7**, no. 32, 1–730, pl. 1–90.

12. PIVETEAU, J. (ed.). 1952. *Traité de Paléontologie*, **2**, Dibranchiata, 689–755, (Masson & Cie.: Paris).

13. SCHLOENBACH, U. 1868. 'Ueber *Belemnites rugifer* Schloenb. sp. nov., aus dem eocenen Tuffe von Ronca', *J. K.-K. geol. Reichsanst.*, *Wien*, **18**, 455, pl. 11, fig. 1a–h.

14. SHIMANSKIY, V. N. 1957. 'Systematics and phylogeny of the order Nautilida', *Bull. Moskov. Obshch. Ispyt. Prir.*, **32**, 105–120.

15. SPATH, L. F. 1927. 'On the Classification of the Tertiary Nautili', *Ann. Mag. nat. Hist.*, (9), **20**, 424–428.

16. SPATH, L. F. 1927. 'Revision of the Jurassic Cephalopod Fauna of Kachh (Cutch)', *Mem. geol. Surv. India Palaeont. indica*, N.S., **9**, no. 2. (Deals also with Tertiary Nautiloidea, pp. 19–26.)

17. STOLLEY, E. 1928. 'Die Belemniten des alpinen Eozäns', *Zentbl. Miner. Geol. Paläont.*, (B), no. 2, 110–124.

18. TEICHERT, C. 1940. 'Contributions to nautiloid nomenclature', *J. Paleont.*, **14**, 590–597.

19. TEICHERT, C. and GLENISTER, B. F. 1952. 'Fossil nautiloid faunas from Australia', *J. Paleont.* **26**, 730–752, pl. 104–108.

20. WAGNER, J. 1938. 'Die Dibranchiaten Cephalopoden der Mittel-Oligozänen (Rupélien) Tonschichten von Kiscell und neue Sepiinae aus dem ungarischen Eozän', *Annls hist.-nat. Mus. natn. hung.*, **31**, 179–199.

Phylum BRACHIOPODA

The Brachiopoda play a much smaller part in the palaeontology of the Caenozoic era than they did in earlier eras, and consequently they are not dealt with in detail here. The following list of references should be a guide to more detailed investigations.

SELECT BIBLIOGRAPHY OF TERTIARY BRACHIOPODA

1. ALLAN, R. S. 1931. 'Descriptions of Tertiary Brachiopoda from New Zealand', *Trans. N.Z. Inst.*, **62**, 1–5, pl. 20–22.

2. ALLAN, R. S. 1932. 'Tertiary Brachiopoda from the Chatham Islands, New Zealand', *Trans. N.Z. Inst.*, **63**, 11–23, pl. 4–6.

3. ALLAN, R. S. 1937. 'Tertiary Brachiopoda from the Mount Brown Beds of Mount Donald, Weka Pass District, New Zealand,' *Rec. Canterbury Mus.*, **4**, no. 3, 131–137, pl. 18.

4. ALLAN, R. S. 1937. 'Tertiary Brachiopoda from the Forest Hill Limestone (Hutchinsonian) of Southland, New Zealand', *Rec. Canterbury Mus.*, **4**, no. 3, 139–153, pl. 19–20.

5. ALLAN, R. S. 1939. 'Studies on the Recent and Tertiary Brachiopoda of Australia and New Zealand', *Rec. Canterbury Mus.*, **4**, no. 5, (1), 231–248, pl. 29–31.

6. ALLAN, R. S. 1940. 'Studies on the Recent and Tertiary Brachiopoda of Australia and New Zealand', *Rec. Canterbury Mus.*, **4**, no. 6, 277–297, pl. 35–37.

7. COOPER, G. A. 1959. 'Genera of Tertiary and Recent rhynchonelloid brachiopods', *Smithson. misc. Collns*, **139**, 1–90, 22 pls.

8. DALL, W. H. 1903. 'Contributions to the Tertiary fauna of Florida', *Trans. Wagner free Inst. Sci. Philad.*, **3**, (6), 1219–1620, pl. 48–60.

9. DAVIDSON, T. 1852. 'A monograph of the British fossil Brachiopoda: pt 1, Tertiary, *Palaeontogr. Soc. Monogr.*, 1–23, pl. 1–2.

10. DAVIDSON, T. 1870. 'On Italian Tertiary Brachiopoda', *Geol. Mag.*, 7, 359–370, 399–408, 460–466, pl. 17–21.

11. DAVIDSON, T. 1874. 'On the Tertiary Brachiopoda of Belgium', *Geol. Mag.*, dec. 2, **1**, 150–159, pl. 7–8.

12. ELLIOTT, G. F. 1940. 'Deux brachiopodes nouveaux de l'Auversien du Bassin de Paris', *Bull. Soc. géol. Fr.*, (5), **9**, 539–598.

13. ELLIOTT, G. F. 1954. 'New Brachiopoda from the Eocene of England, France and Africa', *Ann. Mag. nat. Hist.*, (12), **7**, 721–728, pl. 15.

14. ELLIOTT, G. F. 1960. 'Brachiopodes tertiaires d'Arabie et de Syrie', *Bull. Soc. géol. Fr.*, (7), **2**, no. 2, 152–155, pl. 4.

15. HATAI, K. M. 1940. 'The Cenozoic Brachiopoda from Japan', *Sci. Rept. Tohoku Imper. Univ.*, (2, Geol.), **20**, 1–413, 12 pls.

16. HAYASAKA, I. 1922. 'On some Tertiary Brachiopoda from Japan', *Sci. Rept. Tohoku Imper. Univ.* (2, Geol.), **6**, no. 2, 139–163, pl. 7–8.

17. HERTLEIN, L. G. and GRANT, U. S. 1944. 'The Cenozoic Brachiopoda of western North America', *Calif. Univ. Publ. Math. Phys. Sci.*, 3, 1–172, 21 pls.

18. MOORE, R. C. (ed.). 1965. *Treatise on Invertebrate Paleontology: Part H Brachiopoda*, **1–2**, H1–H927, (Kansas Univ. Press).

19. MUIR-WOOD, H. M. 1938. 'Notes on British Eocene and Pliocene Terebratulas', *Ann. Mag. nat. Hist.*, (11), **2**, 154–181.

20. PAJAUD, D. 1965, 'Remarques sur les Thecideidae (Brachiopodes) tertiaires. Sur une nouvelle forme du Miocène rhodanien: *Glazewskia demarcqui* nov. gen., nov. sp.', *Bull. Soc. géol. Fr.* (1964), (7), **6**, no. 2, 258–261, pl. 12a.

21. THOMSON, J. A. 1916. 'Additions to the knowledge of the Recent and Tertiary Brachiopoda of New Zealand and Australia', *Trans. N.Z. Inst.*, **48** (1915), 41–47, pl. 1.

22. THOMSON, J. A. 1927. 'Brachiopod morphology and genera (Recent and Tertiary)', *Man. N.Z. Board Sci. Art*, no. 7, 1–338, pl. 1–2.

23. VINCENT, E. G. 1893. 'Contribution à la paléontologie des terrains tertiaires de de la Belgique: Brachiopodes', *Ann. Soc. malac. Belg.*, **28**, 38–64, pl. 3–4.

The following are some of the more common generic names encountered in the literature: *Lingula* (?Ord., Sil.-Rec.), *Discinisca* (?Trias., Jur.-Rec.), *Crania* (?Carb., Cret.-Rec.), *Hemithiris* (Mio.-Rec.), *Terebratula* (Mio.-Plio.), *Gryphus* [*Liothyris*] (?Eoc., Olig.-Rec.). *Liothyrella* (Mio.-Rec.), *Cancellothyris* (Mio.-Rec.), *Terebratulina* (U.Jur.-Rec.), *Megathiris* (U.Cret.-Rec.), *Argyro-*

theca (U.Cret.-Rec.), *Platidia* (Eoc.-Rec.), *Dallina* (?Eoc., Mio.-Rec.), *Terebratalia* (Olig.-Rec.), *Terebratella* (Olig.-Rec.), *Magellania* (Olig.-Rec.).

Phylum BRYOZOA (POLYZOA)

The scope of this work does not allow of any treatment of this important group. The following list of references will, however, serve as a guide to intending students.

SELECT BIBLIOGRAPHY OF TERTIARY BRYOZOA

1. BALAVOINE, P. 1959. 'Bryozoaires du Lutétien de Bois-Gouët (Loire-Atlantique)', *Bull. Soc. géol. Fr.*, (7), **1**, no. 3, 245–251, pl. 6, 7a.
2. BOBIES, C. A. 1958. 'Die Crisiidae (Bryozoa) des Tortons im Wiener Becken', *Jb. geol. Bundesanst. Wien*, **101**, Heft 1, 147–165, pl. 13–15.
3. BROWN, D. A. 1952. *The Tertiary cheilostomatous Polyzoa of New Zealand*, 1–399, fig. 1–296 (Brit. Mus. (N.H.): London).
4. BUSK, G. 1859. 'A monograph of the fossil Polyzoa of the Crag', *Palaeontogr. Soc. Monogr.*, 1–136, pl. 1–22.
5. CANU, F. 1907–10. 'Bryozoaires des terrains tertiaires des environs de Paris', *Annls Paléont.*, **2**, 57–89, 137–160, 8 pls; **3**, 61–104, 2 pls; **4**, 101–140, 4 pls; **5**, 89–112, 4 pls.
6. CANU, F. 1907–19. 'Les bryozoaires fossiles des terrains du Sud-Ouest de la France', *Bull. Soc. géol. Fr.*, (4), **6**, 510–518, pl. 12–13; **8**, 382–390, pl. 6–7; **9**, 442–458, pl. 15–18; **10**, 840–855, pl. 16–19; **11**, 444–445, pl. 7–8; **12**, 623–630, pl. 20–21; **13**, 298–303, pl. 4–5; **14**, 465–474, pl. 14–15; **15**, 320–334, pl. 3–4; **16**, 127–152, pl. 2–3; **17**, 350–361, pl. 12–13.
7. CANU, F. and BASSLER, R. S. 1920. 'North American early Tertiary Bryozoa', *Bull. U.S. natn. Mus.*, **106**, 1–162, fig. 1–279.
8. CANU, F. and BASSLER, R. S. 1923. 'North American later Tertiary and Quaternary Bryozoa', *Bull. U.S. natn. Mus.*, **125**, 1–302, pl. 1–47.
9. CANU, F. and BASSLER, R. S. 1929. 'Bryozoaires Éocènes de la Belgique', *Mém. Mus. r. Hist. nat. Belg.*, no. 39, 1–69, pl. 1–5.
10. CANU, F. and BASSLER, R. S. 1931. 'Bryozoaires Oligocènes de la Belgique', *Mém. Mus. r. Hist. nat. Belg.*, no. 50, 1–27, pl. 1–4.
11. CANU, F. and LECOINTRE, G. 1925–30. 'Les bryozoaires cheilostomes des faluns de Touraine et d'Anjou', *Mém. Soc. géol. Fr.*, **4**, N.S., 1–130, pl. 1–25.
12. CIPOLLA, F. 1921. 'I briozoi pliocenici di Altavilla presso Palermo', *G. Sci. nat. econ. Palermo*, **32**, 1–185, pl. 1–8.
13. DARTEVILLE, E. 1933. 'Contributions a l'étude des bryozoaires fossiles de l'Éocène de la Belgique', *Annls Belg. Soc. Roy. Zool.*, **63**, 55–116, pl. 2–4.
14. DAVID, L. 1949. 'Quelques Bryozoaires nouveaux du Miocène du Gard et de l'Hérault', *Bull. Soc. géol. Fr.*, (5), **19**, 539–544, pl. 20.
15. DAVIS, A. G. 1934. 'English Lutetian Polyzoa', *Proc. Geol. Ass.*, **45**, 205–245, pl. 13–15.
16. DUVERGIER, J. 1921–24. 'Note sur les bryozoaires du Néogène de l'Aquitaine', *Act. Soc. linn. Bordeaux*, **72**, 145–181, pl. 1–4; **75**, 145–190, pl. 1–6.
17. FAURA Y SANS, M. and CANU, F. 1917. 'Sur les bryozoaires des terrains tertiaires de la Catalogne', *Treb. Inst. catal. Hist. nat.*, **2**, 59–193, pl. 1–9.

18. FURON, R. and BALAVOINE, P. 1959. 'Les Bryozoaires aquitaniens de Qoum (Iran)', *Bull. Soc. géol. Fr.*, (7), **1**, no. 3, 294–303, pl. 12–14.
19. GREGORY, J. W. 1893. 'On the British Palaeogene Bryozoa', *Trans. zool. Soc. Lond.*, **13**, 219–279, pl. 29–32.
20. LAGAAIJ, R. 1952. *The Pliocene Bryozoa of the Low Countries*, 1–233, pl. 1–29. (E. van Aelst: Maastricht).
21. LAGAAIJ, R. 1963. '*Cupuladria canariensis* (Busk)-Portrait of a Bryozoan', *Palaeontology*, **6**, (1), 172–217, pl. 25–26.
22. LAGAAIJ, R. 1963. 'New additions to the bryozoan fauna of the Gulf of Mexico', *Publs Inst. marine Sci., Texas*, **9**, 162–236, pl. 1–8.
23. MACGILLIVRAY, P. H. 1895. 'A monograph of the Tertiary Polyzoa of Victoria', *Trans. R. Soc. Vict.*, N.S., **4**, 1–166, pl. 1–22.
24. MCGUIRT, J. H. 1941. 'Louisiana Tertiary Bryozoa', *Bull. La geol. Surv.*, **21**, 1–177, pl. 1–31.
25. MAPLESTONE, C. M. 1898–1913. 'Further descriptions of the Tertiary Polyzoa of Victoria', *Proc. R. Soc. Vict.*, N.S., **11**, 14–22, pl. 1–2; **12**, 1–13, pl. 1–2; **12**, 162–169, pl. 17–18; **13**, 1–9, pl. 1–2; **13**, 183–190, pl. 23–24; **13**, 204–213, pl. 34–35; **14**, 65–74, pl. 6–8; **15**, 17–27, pl. 1–2; **16**, 140–147, pl. 16–17; **21**, 233–239, pl. 7–8; **23**, 266–284, pl. 37–47; **24**, 355–356, pl. 27.
26. MOORE, R. C. (ed.). 1953. *Treatise on Invertebrate Paleontology: Part G Bryozoa*, G1–G253, fig. 1–175 (Kansas Univ. Press).
27. REUSS, A. E. 1874. 'Die fossilen Bryozoen des osterreichisch-ungarischen Miocäns', *Denkschr. Akad. Wiss. Wien*, **33**, 141–190, pl. 1–12.
28. SOUAYA, F. J. 1965. 'On the Bryozoa of Gebel Gharra (Cairo-Suez Road) and other Miocene sections in Egypt', *J. Paleont.*, **39**, no. 6, 1129–1144, pl. 135–139.
29. VIGNEAUX, M. 1949. 'Révision des bryozoaires néogènes du Bassin d'Aquitaine', *Mém. Soc. géol. Fr.*, **60**, N.S., 28, 1–155, pl. 1–11.
30. WATERS, A. W. 1881–87. Several papers on the Tertiary polyzoa of Australia and New Zealand. See *Q. Jl geol. Soc. Lond.*, **37**, 309–347, pl. 14–18; **38**, 257–276, pl. 7–9; **38**, 502–513, pl. 22; **39**, 423–443, pl. 12; **40**, 674–697, pl. 30–31; **41**, 279–310, pl. 7; **43**, 40–72, pl. 6–8; **43**, 337–350, pl. 18.

The following are some of the more common generic names encountered in the literature: *Crisia* (Eoc.-Rec.), *Diastopora* (Jur.-Rec.), *Spiropora* (Jur.-Rec.), *Tubulipora* (Eoc.-Rec.), *Crisisina* (Cret.-Rec.), *Proboscina* (?Ord., Jur.-Rec.), *Entalophora* (Jur.-Rec.), *Diaperoecia* (Cret.-Rec.), *Actinopora* (Cret.-Rec.), *Hornera* (Eoc.-Rec.), *Heteropora* (Trias.-Rec.), *Ceriopora* (Trias.-Mio.), *Lichenopora* (Cret.-Rec.), *Membranipora* (Mio.-Rec.), *Acanthodesia* (Eoc.-Rec.), *Conopeum* (Cret.-Rec.), *Cupuladria* (Pal.-Rec.) (*teste* Lagaaij, *verb.*), *Electra* (Eoc.-Rec.), *Hincksina* (Eoc.-Rec.), *Callopora* (Cret.-Rec.), *Membraniporidra* (Cret.-Rec.), *Onychocella* (?Jur., Cret.-Rec.), *Floridina* (Cret.-Rec.), *Micropora* (Cret.-Rec.), *Lunulites* (Cret.-Rec.), *Calpensia* (Plio.-Rec.), *Steginoporella* (Eoc.-Rec.), *Thalamoporella* (Olig.-Rec.), *Cellaria* (Eoc.-Rec.), *Nellia* (Pal.-Rec.) (*teste* Lagaaij), *Canda* (Eoc.-Rec.), *Porina* (Cret.-Rec.), *Beisselina* (Cret.-Eoc.), *Schizoporella* (Eoc.-Rec.), *Hippoporina* (Cret.-Rec.), *Microporella* (Mio.-Rec.), *Mucronella* (Eoc.-Rec.), *Margaretta* [*Tubucellaria*] (Eoc.-Rec.), *Retepora* (Eoc.-Rec.), *Adeona* (Tert.-Rec.), *Cellepora* (Eoc.-Rec.), *Holoporella* (Eoc.-Rec.).

482

Class ANTHOZOA (Corals)

The same remarks apply as under Bryozoa. The following is a short bibliography:

1. ACHIARDI, A. D'. 1875–76. 'Coralli eocenici del Friuli', *Atti Soc. tosc. Sci. nat.,* **1**, 70–86, 115–124, 147–222, pl. 1–2, 6–19.
2. ALLOITEAU, J. 1952. 'Sous-classe des Alcyonaria'; in Piveteau, J., *Traité de Paléontologie*, **1**, 408–417. (Masson & Cie.; Paris).
3. ALLOITEAU, J. 1952. 'Madreporaires post-Paléozoïques', in Piveteau, J., *Traité de Paléontologie*, **1**, 539–684, pl. 1–10 (Masson & Cie.; Paris).
4. DAINELLI, G. 1915. *L'Eocene Friulano*, 204–344, pl. 27–41 (M. Ricci: Firenze).
5. DENNANT, J. 1899–1904. 'Descriptions of New Species of Corals from the Australian Tertiaries', *Trans. R. Soc. S. Aust.*, **23** (1899), 112–122, pl. 2–3, and 281–287, pl. 9–10; **25** (1901), 48–53, pl. 2; **26** (1902), 1–6, pl. 1, and 255–264, pl. 5–6; **27** (1903), 208–215, pl. 1–2; **28** (1904), 52–76, pl. 22–25.
6. DUNCAN, P. M. 1863–68. 'On the fossil corals of the West Indian islands', *Q. Jl geol. Soc. Lond.*, **19** (1863), 406–458, pl. 13–16; **20** (1864), 20–44, 358–374, pl. 2–5; **24** (1868), 9–33, pl. 1–2.
7. DUNCAN, P. M. 1864–76. Several papers on the Tertiary Corals of Australia and Tasmania in *Ann. Mag. nat. Hist.*, (3), **14** (1864), 161–168, pl. 5–6; **16** (1865), 182–187, pl. 8; and in the *Q. Jl geol. Soc. Lond.*, **21** (1865), 394–395; **26** (1870), 284–318, pl. 19–21; **31** (1875), 673–674, pl. 38A, and 677–678, pl. 38C; **32** (1876), 341–348, pl. 22.
8. DUNCAN, P. M. 1866. 'A Monograph of the British Fossil Corals, 2nd series', pt. 1. *Palaeontogr. Soc. Monogr.*, 66 pp., 10 pls.
9. DUNCAN, P. M. 1880. 'A Monograph of the Fossil Corals and Alcyonaria of Sind', *Mem. geol. Surv. India Palaeont. indica*, (14), **1**, 110 pp., 28 pls.
10. DURHAM, J. W. 1942. 'Eocene and Oligocene coral faunas of Washington', *J. Paleont.*, **16**, 84–104, pl. 15–17.
11. FELIX, J. 1884. 'Korallen aus ägyptischen Tertiärbildungen', *Z. dt. geol. Ges.*, **36**, 415–453, pl. 3–5.
12. FELIX, J. 1915–20. 'Jungtertiäre und quartäre Anthozoën von Timor und Obi', *Pal. von Timor*, **2**, 1–48, pl. 37–38; **8**, 1–40, pl. 128.
13. FELIX, J. 1921. 'Fossile Anthozoën von Borneo', *Pal. von Timor*, **9**, 1–64, pl. 141–144.
14. GERTH, H. 1921. (in MARTIN, K). 'Die Fossilen von Java-Anthozoa', *Samml. geol. Reichsmus. Leiden*, N.S., **1**, 387–445, pl. 55–57.
15. GERTH, H. 1923. 'Die Anthozoenfauna des jungteriärs von Borneo', *Samml. geol. Reichsmus. Leiden*, (1), **10**, 37–136, pl. 1–9.
16. GERTH, H. 1933. 'Neue Beiträge zur Kenntnis der Korallenfauna des Tertiär von Java: I. Die Korallen des Eocän und des älteren Neogen', *Wet. Med. Dienst Mijnb. Ned. Oost-Indië*, no. 25, 45 pp., 5 pls.
17. GREGORY, J. W. 1925. 'The Collection of Fossils and Rocks from Somaliland,, etc: Part IV Fossil Corals', *Monogr. geol. Dep. Hunter. Mus.*, **1**, 22–45, pl. 4–7. (Oligocene Corals, 28–42, pl. 5–7). [Really Lower Miocene.]
18. GREGORY, J. W. 1930. 'The Fossil Fauna of the Samana Range and some Neighbouring areas: Part vii the Lower Eocene Corals', *Mem. geol. Surv. India Palaeont. indica*, N.S., **15**, 80–128, pl. 11–16.
19. KÜHN, O. 1933. 'Das Becken von Isfahan-Saidabad und seine altmiocäne Korallenfauna', *Palaeontographica*, **79**, (A), 143–218, pl. 17–19.

20. MILNE-EDWARDS, H. and HAIME, J. 1850. 'A Monograph of the British Fossil Corals: Part 1. Introduction; corals from the Tertiary and Cretaceous formations', *Palaeontogr. Soc. Monogr.*, 71 pp., 11 pls.

21. MONTANARO, E. 1929. 'Coralli Tortoniani di Montegibbio (Modena)', *Boll. Soc. geol. ital.*, **48**, 107–137, pl. 3.

22. MOORE, R. C. (ed.). *Treatise on Invertebrate Paleontology: Part F. Coelenterata*, F1–F498, 358 figs (Kansas Univ. Press).

23. PREVER, P. L. 1921–22. 'I coralli oligocenici di Sassello nell'-Appennino Ligure: Pte. I. Corallari a calci confluenti, *Palaeontogr. italica*, **27** (1921), 53–100, pl. 7–15; **28** (1922), 1–40, pl. 1–7.

24. REUSS, A. E. 1868–1872. 'Paläontologische Studien uber die älteren Tertiärschichten der Alpen', *Denkschr. Akad. Wiss. Wien*, **28** (1868), 129–184, pl. 1–16; **29** (1869), 215–298, pl. 17–36; **33** (1872), 1–60, pl. 37–56.

25. ROEMER, F. A. 1863, 'Die Polyparien der norddeutschen Tertiär-Gebirges', *Palaeontographica*, **9**, 199–245, pl. 35–39.

26. UMBGROVE, J. H. F. (in RUTTEN, L. and HOTZ, W.). 1924. *Geological, Petrographical and Palaeontological Results . . . in the island of Ceram: Report on Pleistocene and Pliocene Corals from Ceram, 2nd ser., Palaeontology, No. 1,* 1–23, pl. 1–2.

27. VAUGHAN, T. W. 1900. 'The Eocene and Lower Oligocene Coral Faunas of the United States with a few doubtfully Cretaceous species', *Monogr. U.S. geol. Surv.*, **39**, 1–263, pl. 1–24.

28. VAUGHAN, T. W. 1919. 'Fossil corals from Central America, Cuba, Porto Rico, with an account of the American Tertiary Pleistocene and Recent coral reefs', *Bull. U.S. natn Mus.*, **103**, 189–524, pl. 68–152.

29. VAUGHAN, T. W. and WELLS, J. W. 1943. 'Revision of the suborders, families, and genera of the Scleractinia', *Spec. Pap. geol. Soc. Am.*, **44**, 1–363, pl. 1–51.

The following are some of the more common generic names encountered in the literature: *Actinastrea* (U.Trias.-Rec.), *Stylocoenia* (Eoc.-Mio.), *Astrocoenia* (Eoc.-Mio.), *Stylophora* (Eoc.-Rec.), *Madracis* (U.Cret.-Rec.), *Acropora* (Eoc.-Rec.), *Trochoseris* (Cret.-Olig.), *Cyathoseris* (U.Cret.-Mio.), *Agaricia* (Mio.-Rec.), *Siderastrea* (Cret.-Rec.), *Cyclolites* (Cret.-Eoc.), *Cycloseris* (Cret.-Rec.), *Fungia* (Mio.-Rec.), *Micrabacia* (Cret.-Rec.), *Stephanophyllia* (Eoc.-Rec.), *Actinacis* (Cret.-Olig.), *Goniopora* (Cret.-Rec.), *Porites* (Eoc.-Rec.), *Favia* (Cret.-Rec.), *Favites* (Eoc.-Rec.), *Goniastrea* (Eoc.-Rec.), *Platygyra* (Eoc.-Rec.), *Hydnophora* (Cret.-Rec.), *Montastrea* (U.Jur.-Rec.), *Antiguastrea* (U.Cret.-Mio.), *Astrangia* (Eoc.-Rec.), *Oculina* (Cret.-Rec.), *Madrepora* (Eoc.-Rec.), *Trochosmilia* (Eoc.), *Mussa* (Plio.-Rec.), *Caryophyllia* (U.Jur.-Rec.), *Trochocyathus* (M.Jur.-Rec.), *Paracyathus* (Eoc.-Rec.), *Ceratotrochus* (Cret.-Rec.), *Stephanocyathus* (Eoc.-Rec.), *Turbinolia* (Eoc.-Olig.), *Sphenotrochus* (Eoc.-Rec.), *Parasmilia* (Cret.-Rec.), *Euphyllia* (Eoc.-Rec.), *Flabellum* (Eoc.-Rec.), *Balanophyllia* (Eoc.-Rec.), *Dendrophyllia* (Eoc.-Rec.).

The octocoral *Graphularia* (London Clay of England) is straight and rod-like in form, and shows radial, and sometimes concentric, structure like that of a belemnite, but it shows very little tapering, however, and no cavity that could be taken for an alveolus.

Phylum ANNELIDA (worms)

(Proterozoic-Rec.)

Class POLYCHAETIA

Mostly marine. (Cambr.-Rec.)

Order SEDENTARIDA

Living in tubes or burrows. Marine. (Cambr.-Rec.)

Family SERPULIDAE

Builds a calcareous tube that is circular, triangular or polygonal in cross section, externally ornamented or not; operculum calcareous or horny; usually attached, sometimes free. (Cambr.-Rec.)

The following are the three commonest Tertiary genera:

Serpula (*s.s.*): tube tapering irregularly, twisted, lower end attached; fine concentric ornament; operculum horny. Sil.-Rec. Eur., N.Amer. **Ditrupa:** tube tapering, open at both ends, gently arcuate. Tert.-Rec. Cosmop. **Tubulostium** [*Rotularia, Spirulaea*]: spirally coiled, part of last whorl detached, aperture somewhat contracted; smooth or concentrically wrinkled, sometimes with a few longitudinal keels. U.Cret.-U.Eoc. Cosmop.

Chapter VII

TERTIARY VERTEBRATA

(Contributed by Dr R. J. G. Savage)

Fossilized vertebrate remains are in most cases readily distinguishable from invertebrates and if in doubt a simple acid test will usually differentiate; the calcium carbonate of invertebrates is attacked by dilute (about 10 per cent) acetic acid whereas the calcium phosphate of vertebrates is unaffected. Since, however, vertebrates tend to fossilize as isolated bones, scales and teeth, their taxonomic identification is difficult for the non-specialist in vertebrate palaeontology. The aim of this chapter is to enable a geologist to identify remains sufficiently to make stratigraphic and environmental deductions. Given reasonable material, he should be able to place a fossil in its class and order. Most fossils are preserved in water-laid deposits, but most reptiles, birds and mammals live on land, so their mode of preservation may be only an indirect clue or even no clue as to the environmental interpretation of the facies. Nonetheless the information afforded is no less than that for invertebrates. For instance, sharks are almost all exclusively marine; crocodiles are tropical and subtropical; mammoths occur only in the Pleistocene.

Perhaps the biggest problem in vertebrate identification is the number of skeletal parts—vertebrae, ribs, girdle bones, limb bones, foot bones, jaws, skulls, teeth, etc. To attempt to describe all of these for each order would be futile. Teeth are usually the most diagnostic parts and fortunately also the most frequently preserved. Hence emphasis will be placed on dentitions and the geologist should remember this when he finds vertebrate remains.

All vertebrate classes save some primitive fish groups are present in the Tertiary and these are:

Class CHONDRICHTHYES	Sharks and Rays.
Class OSTEICHTHYES	Bony Fish.
Class AMPHIBIA	Frogs and Newts.
Class REPTILIA	Reptiles.
Class AVES	Birds.
Class MAMMALIA	Mammals.

Throughout the chapter all major taxa will be mentioned and only a few minor groups omitted where the fossil record in the Tertiary is either absent or very sparse. To include all orders and families living in the Tertiary would lengthen the chapter unduly and cause confusion. For example, the class Agnatha is represented today by lampreys and hagfish; they have fossil ancestors in the Palaeozoic, but there are no known Mesozoic or Caenozoic fossils; some may be discovered but their identification would be a task for the specialist. The extinct mammalian order Xenungulata is poorly known from only a handful of scrappy specimens, referred to one genus, and has been found only in Upper Palaeocene deposits in South America.

Class CHONDRICHTHYES

Characteristics. Class comprises sharks and rays; size may be very large (Recent and Tertiary species 14 m. long); no true bone or scales present; skeleton entirely cartilaginous and rarely fossilized; some elements may be secondarily ossified, especially vertebrae, and may thus fossilize. Teeth are of two kinds (a) pointed biting teeth of carnivorous sharks and (b) flat dental plates of bottom-feeding, crustacean- and mollusc-crushing sharks, rays and skates. Teeth vary considerably according to position in jaw, so each individual has teeth of several slightly differing shapes.

Occurrence. Almost exclusively marine. Worldwide, more common and larger in tropics. Teeth abundant in many Tertiary marine deposits. Little evolutionary change during the Tertiary and hence of little stratigraphic value.

Classification. In each of the two orders only those families which have a reasonable Tertiary record are listed; these with their examples, however, serve to give some idea of the range of variation in the dentitions.

Order SELACHII

Family HETERODONTIDAE
Port Jackson sharks.
Heterodontus (Fig. 952): U.Jur.-Rec.

Family HEXACANTHIDAE
Six-gilled sharks.
Notidanus (Fig. 953): L.Jur.-Plio.

Family CHLAMYDOSELACHIDAE
Frilled sharks.
Chlamydoselache (Fig. 954): Mio.-Rec.

Family CARCHARIIDAE
Sand sharks.
Odontaspis (Fig. 955): L.Cret.-Rec.

Family ISURIDAE
Mackerel sharks.
Isurus (Fig. 956): Cret.-Rec. *Lamna* (Fig. 957): L.Cret.-Pleist. *Carcharodon* (Fig. 958): Pal.-Rec.

Family ORECTOLOBIDAE
Nurse sharks.
Ginglymostoma (Fig. 959): U.Cret.-Rec.

Family SCYLLIORHINIDAE
Dogfish.
Scylliorhinus (Fig. 960): U.Cret.-Rec.

Family CARCHARARHINIDAE
Ground sharks.
Galeocerdo (Fig. 961): Eoc.-Rec.

Family SQUALIDAE
Spiny dogfish.
Squalus (Fig. 962): U.Cret.-Rec.

Order **BATOIDEA**

Family PRISTIDAE
Sawfish.
Pristis (Fig. 963): Eoc.-Plio.

Family MYLIOBATIDAE
Eagle rays.
Myliobatis (Fig. 964): U.Cret.-Rec.

Class **OSTEICHTHYES**

Characteristics. There are two subclasses of bony fish, the Actinopterygii and the Sarcopterygii. The latter comprise only the coelacanth and lungfish, both rare today but lungfish occasionally found in the African Tertiary (***Protopterus*** (Fig. 966). Eoc.-Rec.). So in practice remarks can be limited to the Actinopterygii. The bones are lightly built, the vertebrae are amphicoelus (biconcave) and without processes. Skull bones are loosely knit and readily fall apart. Teeth are variable, and include both biting and crushing dentitions. The scales

---

Figs 952–965. VERTEBRATA: PISCES
Fig. 953 after Smith Woodward; Figs 952, 954, 956 and 959–962 after Romer; Fig. 965 after British Museum (Natural History); Figs 955, 957, 958, 963 and 964 original.
952. *Heterodontus* sp. × 0·5. (Anterior (top) and posterior teeth.)
953. *Notidanus serratissimus* Agassiz, L.Eoc. Sheppey, Kent. × 1.
954. *Chlamydoselache* sp. × 0.5.
955. *Odontaspis elegans* Agassiz, M.Eoc. Hampshire. × 0·75.
956. *Isurus* sp. × 0·5.
957. *Lamna obliqua* (Agassiz), L.Eoc. England. × 0·7.
958. *Carcharodon megalodon* Agassiz, Mio. Malta. × 0·5.
959. *Ginglymostoma* sp. × 0·5.
960. *Scylliorhinus* sp. × 0·5.
961. *Galeocerdo* sp. × 0·5.
962. *Squalus* sp. Slightly enlarged.
963. *Pristis* sp., U.Eoc. Libya. × 0·5. (Fragment of jaw).
964. *Myliobatis* sp., M.Eoc. Hampshire B. × 0·5. (Tooth battery).
965. *Acipenser* sp. (sturgeon), L.Olig. Hampshire B. × 1. (Lateral scute).

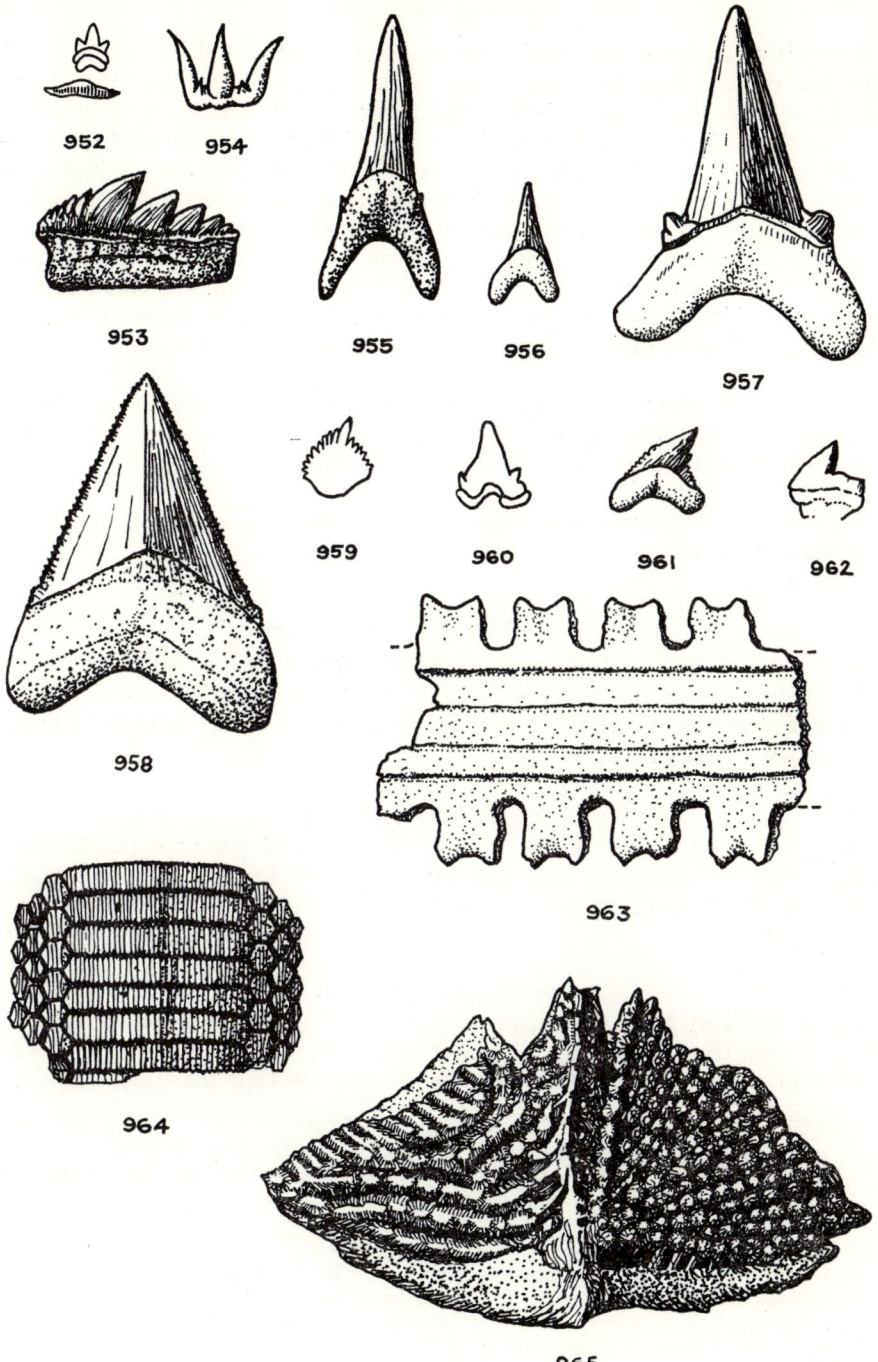

952   954

953

955   956

957

958

959   960   961   962

963

964

965

of primitive bony fish are small and rhomboid (true of Palaeozoic taxa and the few present day survivors). Later, scales become larger and cycloid or ovoid, as in the carp, then thin and ctenoid or comb-shaped as in the herring, and finally lost in more advanced forms as plaice and sole. Without the complete fish identification to generic level is difficult; details of the fins and skull bones are required and these are often badly crushed. A few have characteristic structures, as the crushing teeth of ptycodonts and the swordfish with its very elongate premaxillary bones which form a sword. In the ear region there are minute bones known as otoliths (Fig. 970); these have highly characteristic shapes, are identifiable and are of stratigraphic value. If a sample is washed for foraminifera or other microfossils, otoliths too may be found.

Occurrence. Worldwide, virtually all aquatic environments. Rapid and diverse radiation in the Tertiary; over 20,000 described species. Genera mostly long-lived but otoliths of considerable stratigraphic value.

Classification. The actinopterygian fishes can be grouped into three infraclasses as follows:

Subclass ACTINOPTERYGII

Infraclass CHONDROSTEI

Living sturgeon (*Acipenser* (Fig. 965): U.Cret.-Rec.) and bichir of African rivers (*Polypterus*: Eoc.Rec.). Most primitive forms in this infraclass; living descendants specialized.

Infraclass HOLOSTEI

Living bowfin (*Amia* (Fig. 967): U.Cret.-Rec.) and gar-pike (*Lepidosteus* (Fig. 968): U.Cret.-Rec. *Pycnodus* (Fig. 969): U.Jur.-Eoc.). Intermediate between primitive Chondrostei and advanced Teleostei; scales thick and rhomboid.

Infraclass TELEOSTEI

Almost all living bony fish. Scales composed of ganine, thin and often lost. Best known from 'fish-beds' such as Green River in Wyoming and Monte Bolca in northern Italy, both Middle Eocene. Within the Teleostei eight

---

Figs 966–972. VERTEBRATA: PISCES AND AMPHIBIA
Fig. 969 after Egerton; Fig. 972 after Swinton; Figs 967, 968 and 970 after British Museum (Natural History); Figs 966 and 971 original.
966. *Protopterus* sp., L.Olig. Libya. × 1. (Dental plate).
967. *Amia* sp., L.Olig. Hampshire B. × 1. (Opercular plate).
968. *Lepidosteus suessionensis* Gervais, L.Eoc. London B. × 1. (Scale).
969. *Pycnodus* 'pachyrhinus' Egerton, L.Eoc. Sheppey, Kent. × 0·75.
970. *Albula eppsi* White and Frost, L.Eoc. London B. × 1. (Otolith).
971. *Rana pueyoi* Navas, L.Plio. Spain. × 0·5.
972. *Andrias scheuchzeri* Cuvier, U.Mio. Germany. × 0·1. ('Homo diluvii testis').

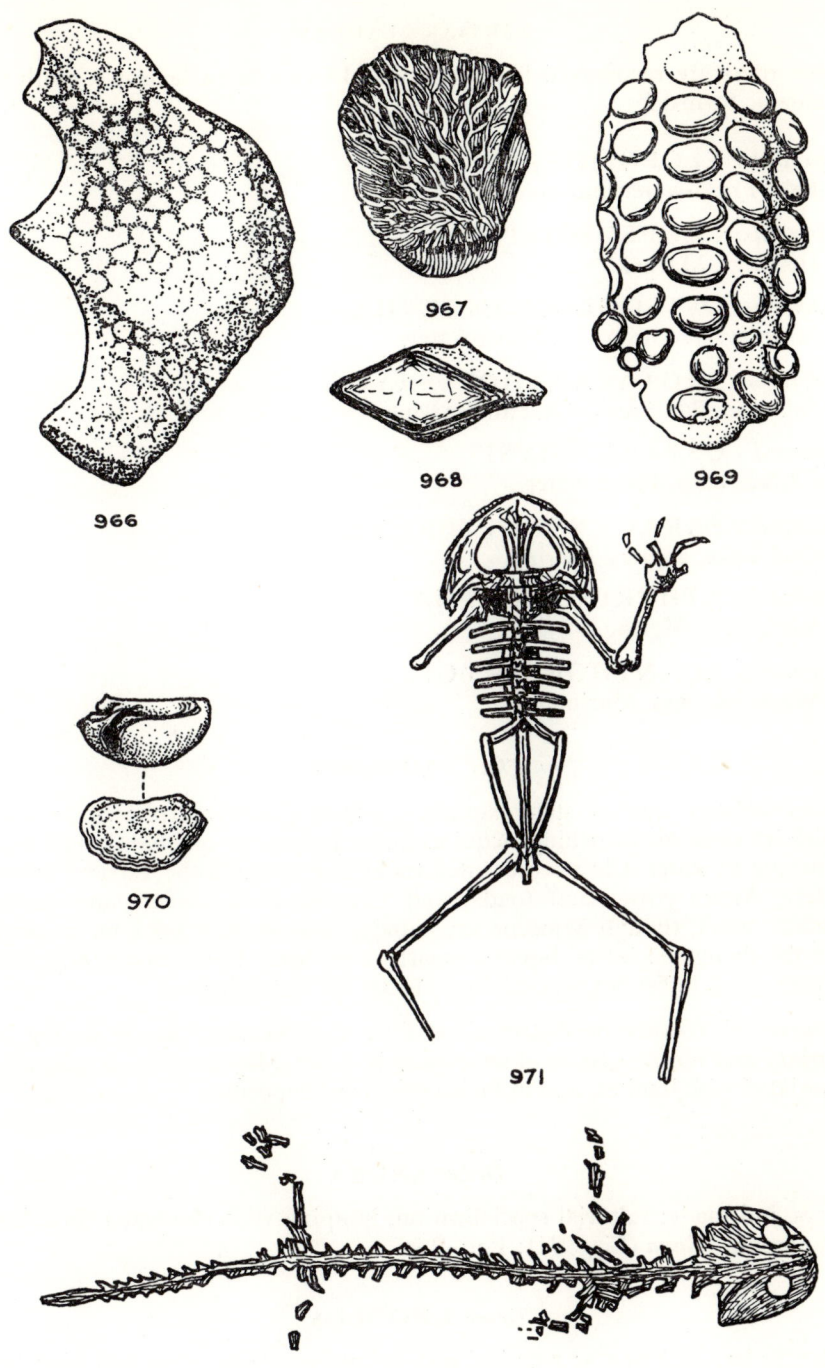

966

967

968

969

970

971

972

superorders are recognized for Tertiary and living fish (there is a ninth for Mesozoic forms).

Superorder ELOPOMORPHA
Eels. Fresh-water and marine.

Superorder CLUPEOMORPHA
Herrings. Marine.

Superorder OSTEOGLOSSOMORPHA
Mooneys, featherbacks. Fresh-water.

Superorder PROTACANTHOPTERYGII
Salmon, pike. Fresh-water and marine.

Superorder OSTARIOPHYSI
Catfish, carp. Fresh-water.

Superorder PARACANTHOPTERYGII
Cod, hake, anglers. Marine.

Superorder ATHERINOMORPHA
Silversides. Marine.

Superorder ACANTHOPTERYGII
Mackerel, bass. Marine.

Class AMPHIBIA

Characteristics. Larval stage aquatic (tadpole); metamorphoses to adult which becomes air-breathing, acquires limbs and moves on land though often returning to water. Class represented today and through the Tertiary by two orders, Anura (frogs and toads) and Urodela (salamanders and newts). Usually small, though Miocene salamander *Andrias* reached 1 m. in length and the living species of Japan is nearly 2 m. long. Bones are lightly built, tubular, with indistinct articulations. Teeth are very small.

Occurrence. Worldwide distribution; little evolutionary change during the Tertiary and too rare to be of stratigraphic value. May occur in ponds, often associated with lignites, and in Pleistocene cave deposits.

Classification:

Order ANURA

Frogs and toads; saltorial specialization, jumping with elongated hind legs; tail absent. **Rana** (Fig. 971): Eoc.-Rec.

Order URODELA

Salamanders and newts; running and swimming forms; fore and hind legs about same size; long tail. **Andrias** (Fig. 972): Olig.-Rec.

492

Class REPTILIA

Characteristics. Although the Mesozoic is the 'Age of Reptiles' and many spectacular forms become extinct before the beginning of the Caenozoic, crocodiles and turtles are abundant in many Tertiary deposits. Reptiles have well ossified skeletons; limb bones have poorly defined articulations when compared with mammals. There is no larval stage and the egg hatches out as a small reptile which continues to grow throughout life. Scales are usually present but rarely fossilized; crocodiles, however, have bony scutes covering their backs and these fossilize well. Turtles have a shield usually composed of fused bony plates; the upper part is the carapace and the lower side the plastron; the shield usually breaks after death but its parts fossilize. Teeth of reptiles are simple, conical and rootless or single-rooted; they vary in size in the jaws of any one individual and are continually replaced during life. Turtles and tortoises have no teeth, but instead a horny beak. The reptilian skull breaks up more readily after death than does the mammalian; it articulates with the vertebral column by one condyle (two in mammals) and the brain is small. Behind the eye socket can usually be seen two openings in the skull, except in turtles and tortoises where the area is deeply indented. The mandible or lower jaw is composed of several bones (only one in mammals).

Occurrence. Worldwide, more and bigger in tropics both during Tertiary and Recent. Land, fluviatile and marine habitats. Little evolutionary change during Tertiary; of limited stratigraphic value.

Classification. Four orders of reptiles are present in the Tertiary:

Order CHELONIA

Turtles and tortoises. Turtle carapace and plastron remains common in fluviatile and lagoonal facies, especially in warm temperate and tropical environments; may be very large. (Pliocene *Colossochelys* of India over 2 m. long).
    Trionyx (Fig. 973): U.Jur.-Rec.

Order RHYNCHOCEPHALIA

The tuatara (*Sphenodon*) of New Zealand the only living survivor of Mesozoic order. Lizard-like; unknown between Jurassic and Recent.

Order CROCODILIA

Crocodiles and alligators. Common in Tertiary shallow water deposits, both non-marine and near-shore; scutes, skull and jaw bones have reticulate pattern; teeth conical, with longitudinal fluting, set in sockets in jaws.
    Crocodilus (Fig. 974): U.Cret.-Rec. *Tomistoma* (Fig. 975): Eoc.-Rec.

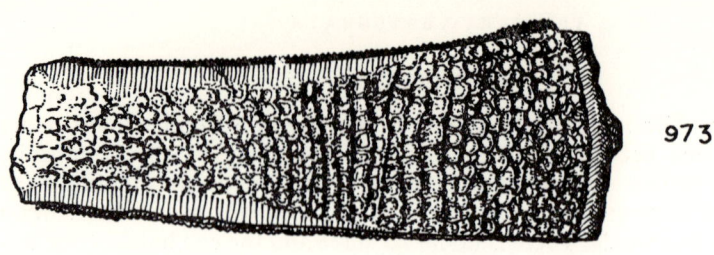

973

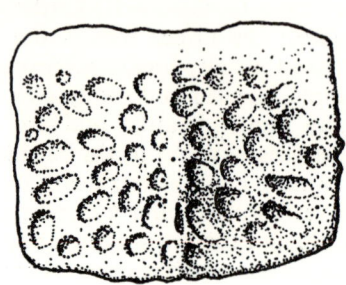

974

975

976

Figs 973–976. VERTEBRATA: REPTILIA
Fig. 973 after British Museum (Natural History); Figs 974–976 original.
973. *Trionyx circumsulcatus* (Owen), U.Eoc. Hampshire B. × 0·5 (Costal bone).
974. *Crocodilus* sp., L.Mio. Libya × 0·6. (Scute).
975. *Tomistoma* sp., U.Eoc. Libya. × 0·3. (Part of mandible).
976. *Palaeophis typhaeus* Owen, M.Eoc. Hampshire B. × 1. (Vertebra).

494

Order SQUAMATA

Lizards and snakes. Snakes are the last major stock of reptiles to appear in late Mesozoic; bones very fragile and rarely fossilized; vertebrae with complex articulation processes. Lizards never abundant in fossil record; usually much smaller than crocodiles; bones without ornament; no bony scutes; teeth often lanceolate and fused to jaw margin; some lizards marine, as *Palaeophis* (Fig. 976): Eoc.

Class AVES

Characteristics. The features in which birds differ from reptiles are nearly all ones associated with the adaptation to flight, though some birds are secondarily flightless. They are warm-blooded for greater muscle efficiency, their respiration is much improved on the reptilian pattern; they have feathers which are lighter than scales, retain body heat and are used in flight control. Vision is usually good and the brain is relatively larger than in reptiles, with a particularly well developed cerebellum, the centre concerned with balance and the coordination of movements. The bones are pneumatic with thin walls and large air-filled cavities which lightens them. The tail is usually short. The strong pelvis and stout hind limbs support the animal in standing and running. The fore limb is modified as a wing, or may be vestigial in running birds. The ribs carry unciform processes not found in reptiles; in flying forms the sternum often has a high keel for attachment of flight muscles. Skull bones are thin, have beaks and no teeth.

Occurrence. Worldwide; major radiation in Tertiary but fossil record not good and identification difficult. Most fossil birds are wading and aquatic forms. Eggs very rare and difficult to distinguish from those of reptiles.

Classification. Aside from the Mesozoic birds, all others are either ratites (running birds) or carinates (flying birds).

Superorder PALAEOGNATHAE [RATITAE]

Ostriches and emus. Non-flying birds; limited to southern hemisphere in Recent; some very large in Tertiary and Pleistocene (*Dinornis* (Fig. 977) of New Zealand about 4 m. high); almost unknown in N.America, Europe and Asia north of the equator.

Superorder NEOGNATHAE [CARINATAE]

All flying birds, and some non-flyers (e.g. penguins). Skeletons all very similar; some Miocene forms very large (*Phororhacos* (Fig. 978) of S.America, over 2 m. high with skull 60 cm. long).

Class MAMMALIA

Characteristics. There are about a thousand living genera of mammals and about twice as many extinct genera. Of some thirty-five orders about half are extinct. Sufficient variety, however, remains for them to be familiar and living mammals are readily distinguishable from reptiles. Mammals give birth to their young alive and are not egg-laying with the exception of a few Australian monotremes. The body temperature is constant and usually above that of the environment, which enables mammals to function more efficiently than cold-blooded reptiles. The heat is retained by body covering of hair, though this may be lost in some (whales and elephants, for instance), replaced by spines (as in hedgehog and porcupine) or scales (as in pangolin) or even by shell (as in extinct glyptodon). Mammals have limited period of growth and not the life-long size increase of reptiles; their brains are relatively large and complex. They are usually very active animals, though some forms hibernate.

Bone characteristics found in mammals and not in reptiles are:

(i) Epiphyses present. For example, a mammalian limb bone has three parts, the shaft or diaphysis and the two terminal elements or epiphyses. The epiphyses are joined to the shaft by cartilage in juveniles (and would thus separate if fossilized) but fused to the shaft in adults. Reptilian limb bones lack these separate ends and the articular surfaces are less complex.

(ii) Mandible or lower jaw composed of one bone only, the dentary. In reptiles there are several bones present in the mandible.

(iii) Middle ear has three ossicles, whereas only one is present in reptiles.

(iv) Skull has large braincase, a single nasal opening, and it articulates with the vertebral column on two condyles. In reptiles the braincase is small, the nasal opening is double, and vertebral articulation is by one condyle.

(v) Dentition is differentiated along the jaw from front to back; in reptiles teeth are usually similar all along the jaw. Mammalian teeth are replaced only once in life, the milk dentition succeeded by the permanent; in reptiles the teeth are continuously replaced.

Mammalian dentitions. Teeth preserve better than any other parts of the mammalian skeleton. They are highly adapted for functional efficiency and constant through a species: thus they are valuable taxonomic characters as well as offering much information on the evolution and habits of their owner.

Structurally, teeth comprise a crown with root or roots; the crown is composed of dentine with a capping of enamel, and may be multicusped. Teeth are confined to the jaw margins and are grouped as incisor, canine, pre-

---

Figs 977–978, 980–981. VERTEBRATA: AVES AND MAMMALIA
Fig. 980 after Piveteau; Figs 977–978 after Swinton; Fig. 981 original.
977. *Dinornis maximus* Owen, Pleist. N.Z. × 0·06.
978. *Phororhacos longissimus* Ameghino, L.Mio. Patagonia. × 3·5.
980. *Ptilodus wyomingensis* Jepsen, Pal. N.Amer. × 2. (Right mandible).
981. *Peratherium* sp., U.Eoc. Hampshire B. × 8. (Lower molar).

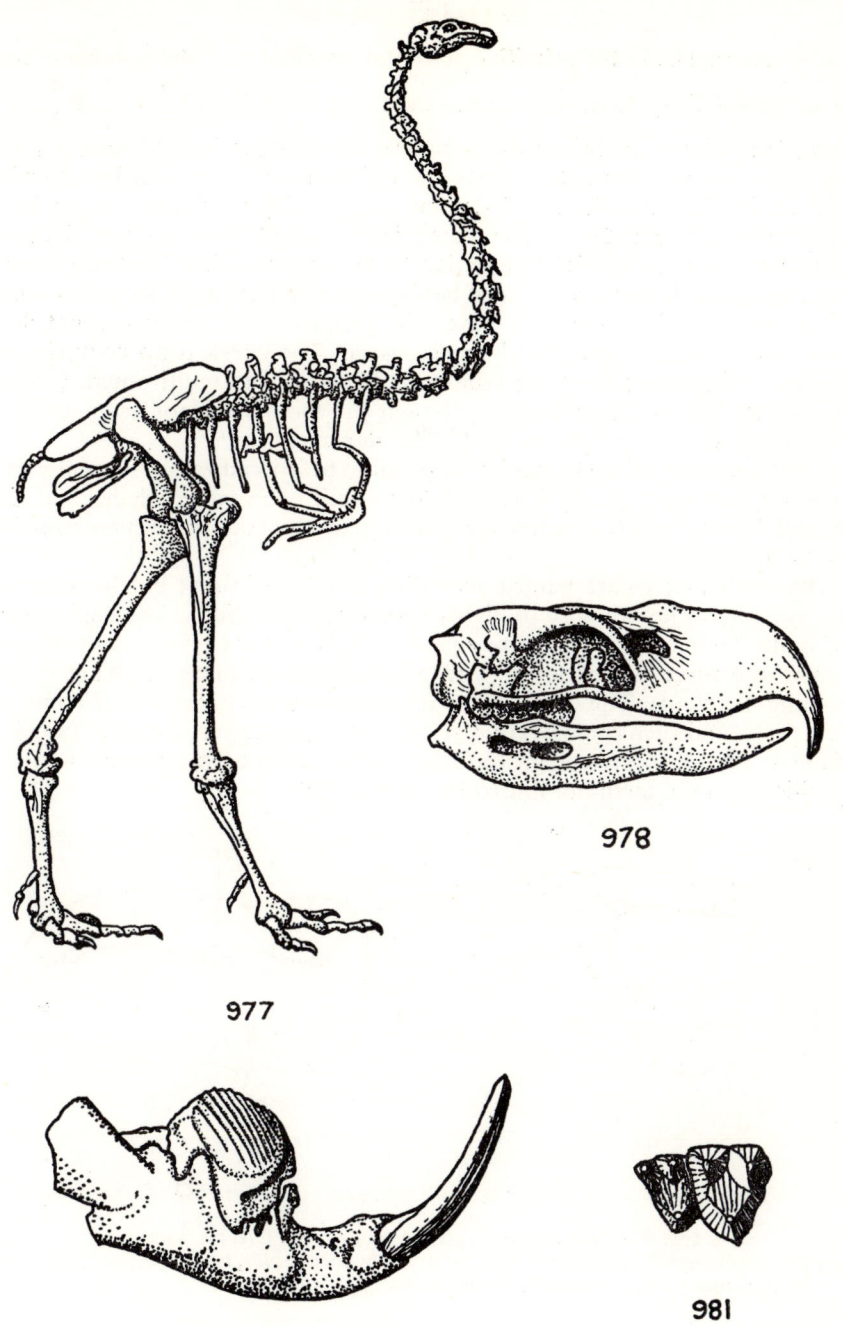

977

978

980

981

molar and molar. In the primitive placental mammal the total number of teeth in an adult is forty-four, expressed in the dental formula as $I\frac{3}{3}$, $C\frac{1}{1}$, $P\frac{4}{4}$, $M\frac{3}{3}$. Since individuals are bilaterally symmetrical, only the half number is given; the upper figures refer to the upper jaw and the lower figures to the lower jaw. The canine has usually a well defined tusk or dog-tooth-like shape; in the upper jaw it is situated at the premaxillary-maxilla bone suture; the lower canine occludes immediately anterior to the upper canine. Teeth in front of the canine are incisors and those behind are often grouped as 'cheek' teeth. Both incisor and canine teeth have milk precursors to the permanent dentition; all are single-rooted and single-cusped. The cheek teeth comprise premolars and molars; these are usually multicusped and multirooted. Only the premolars have milk precursors, though the molars do not erupt until the milk teeth are being replaced by the permanent ones.

The number of teeth varies according to specialization—usually it is reduced, but occasionally it exceeds forty-four as in some whales and marsupials. Terms used to describe the general pattern of cheek teeth are explained in the glossary.

Individual cusps are named according to a set of rules, but these are not always easy to homologize in highly specialized dentitions. The basic pattern of molar teeth in fairly primitive placentals is a four-cusped squarish upper molar and three-cusped lower with posterior tail or talonid, giving roughly rectangular shape. The cusps of the upper molars are given names composed of a prefix and such roots as cone, style, conule; the lower molars are similarly named, but all terms end in '-id'. The more common cusp terms and their positions on the teeth are illustrated in Fig. 979.

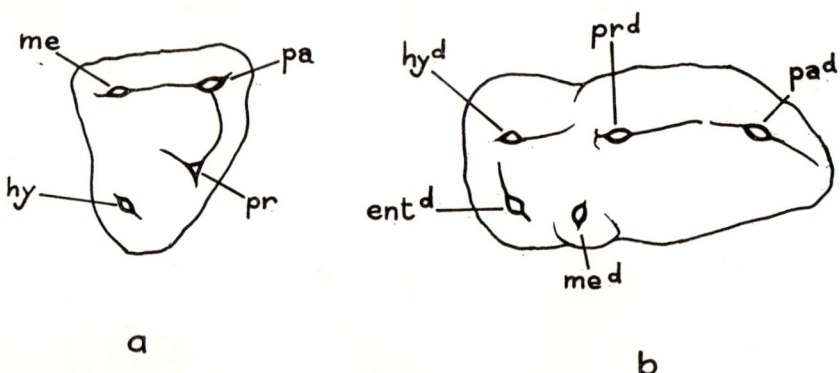

Fig. 979. DIAGRAMMATIC MAMMALIAN CHEEK TEETH.
a, right upper; b, left lower. *ent*^{d}, entoconid; *hy*, hypocone; *me*, metacone; *me*^{d}, metaconid; *pa*, paracone; *pa*^{d}, paraconid; *pr*, protocone; *pr*^{d}, protoconid.

Occurrence. Worldwide distribution throughout the Tertiary, though as yet many gaps in the record. Marine mammals rare compared with land mammals.

498

Mesozoic record poor but Tertiary record often abundant with much evolutionary diversity. Mammals of stratigraphic value, especially small forms.

Classification. The class Mammalia comprises three subclasses. The Prototheria (with Australian egg-laying platypus and spiny ant-eater) have almost no fossil record. The Allotheria contain only multituberculates, a taxon of small- to medium-sized rodent-like mammals which were diverse in the Cretaceous and persisted into the early Tertiary. The third subclass Theria contains the marsupials and the placentals, that is, almost all living and Tertiary mammals. Not all of the twenty-nine orders of placental mammals will be detailed; six are very poorly known in the fossil record (Tillodontia, Dermoptera, Pholidota, Tubulidentata, Xenungulata and Embrithopoda) and hence omitted.

Subclass ALLOTHERIA

Order MULTITUBERCULATA

A Mesozoic order that survived into the Eocene when their place was taken by the rodents. Small- to medium-sized gnawing mammals with chisel-like incisor teeth followed by a diastema; the cheek teeth are multicusped giving a grinding battery; the first lower premolar is very much the largest tooth in the series. Very rare in the Tertiary. (Pal.-Eoc. N.Amer., Eur.).
Ptilodus (Fig. 980): Pal. N.Amer.

Subclass THERIA

Order MARSUPIALIA

Pouched mammals. Restricted today to Australia and South America; fossils known from North America and western Europe.

Marsupials, fossil and recent, fall fairly readily into six major taxa, the grouping of which into superfamilies and higher taxa is too contradictory to be attempted here. Marsupials are recorded from the early Tertiary of western Europe, but none from Asia and Africa. In N.America the record goes back into the Cretaceous, and into the Palaeocene in S.America. In Australia the great majority of fossil marsupials are surprisingly late in the Tertiary, but one record at least appears to go back into the Oligocene.

Family DIDELPHIDAE
Opossums, known from Upper Cretaceous onwards, and recorded from North and South America and Europe. Molar teeth tricuspid; usually small, insectivore-like animals.
Peratherium (Fig. 981): Eoc.-Mio. N.Amer., Eur.

Family BORHYAENIDAE
An extinct family from the Palaeocene to Pliocene of S.America. These

499

marsupials parallel the true carnivores in size and dental adaptations, even evolving a sabre-toothed genus.

Thylacosmilus (Fig. 982): Plio. S.Amer.

Family DASYURIDAE

Pliocene of Australia onwards. Range in size from that of a shrew to that of a wolf; basically insectivorous, with shearing dentition; mostly small, but several medium-sized carnivores.

Sarcophilus: Plio. Austral.

Family PERAMELIDAE

The bandicoots of Australia; few fossils.

Family CAENOLESTIDAE

Small rodent-like marsupials of S.America, known from Palaeocene times onwards.

Family PHALANGERIDAE

Comprises the kangaroos, wallabies, koalas and wombats, together with a number of fossil forms, occurring in Australia from Oligocene times onwards. Cheek teeth bilophodont; diastema usually present; incisors protudent and cutting.

Prionotemnus (Fig. 983): large, kangaroo-like. Plio. Austral.

Order INSECTIVORA

Small mammals, known from Cretaceous to Recent; include living shrews, moles and hedgehogs, and from early Tertiary taxa most other placental orders can be derived. Teeth highly trenchant, usually tricuspid. Fossil record not rich, but becoming increasingly important as more micro-mammals are found by washing and sieving sites.

Lantanotherium (Fig. 984): Mio. Eur., N.Amer., Africa.

Order CHIROPTERA

Essentially flying insectivores. Fore limb modified for flight by elongation of fingers to support wing membrane. Two suborders. Fossil record poor; already fully developed flight in Palaeocene.

Figs 982–985. VERTEBRATA: MAMMALIA

Figs 982–983 after Piveteau; Fig. 984 after Thenius; Fig. 985 after Hoffstetter, in Piveteau.
982. *Thylacosmilus atrox* Riggs, Plio. Argentine. × 0·5. (Skull and mandible).
983a, b. *Prionotemnus palakarinnicus* Stirton, U.Tert. S.Austral. × 1. a, part of maxilla and upper right dentition; b, part of mandible and lower left dentition.
984. *Lantanotherium longirostre* Thenius, M.Mio. Austria. × 4·5. (Upper dentition).
985. *Panochthus tuberculatus* Owen, Pleist. S.Patagonia. × 0·5. (Right mandible).

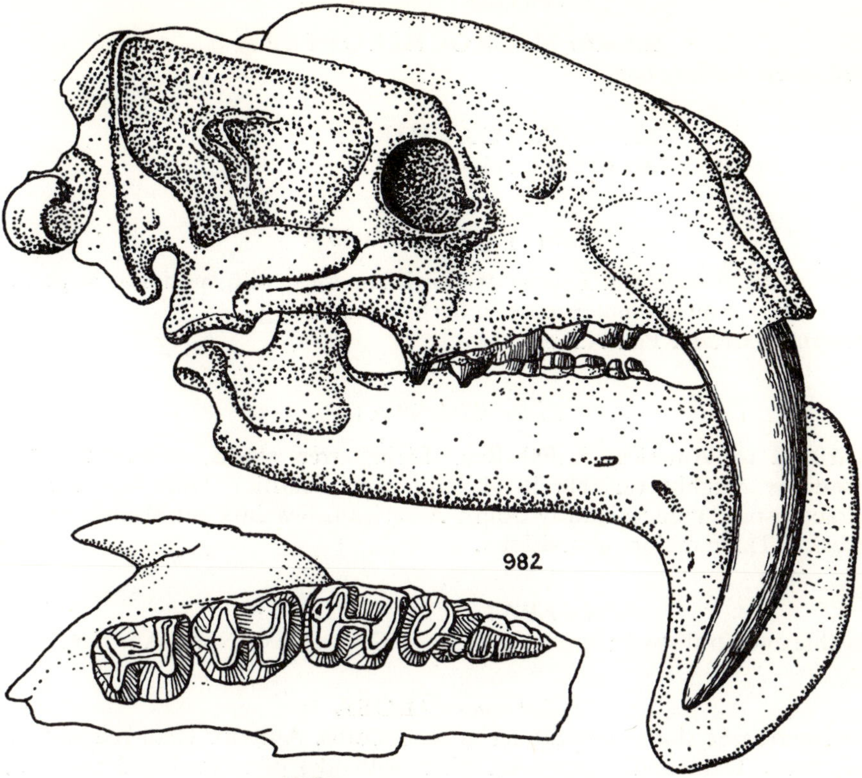

982

983a

983b

984

985

Suborder MICROCHIROPTERA
Small insect-feeding bats.

Suborder MEGACHIROPTERA
Larger fruit-eating bats of the tropics.

Order TAENIODONTA
Order of Palaeocene and Eocene of North America, with species as large as pigs, earliest mammalian exercise in herbivorous living.
Psittacotherium: Pal. N.Amer.

Order EDENTATA
Restricted to America, U.Pal.-Rec. Living tree sloths, ant-eaters and armadillos. Highly variable in skeleton and dentition with few overall characteristics. Predominantly South American, few in Central and North America. Three major subdivisions:

Suborder PALAEONODONTA
Possible ancestral stock; late Palaeocene to Oligocene of North America.

Suborder PILOSA
Ant-eaters, tree sloths and extinct ground sloths. Anterior teeth reduced or lost; cheek teeth undifferentiated, often with loss of enamel and rootless; browsing Pleistocene ground sloths up to 7 m. long; ant-eaters with long edentulate jaws.

Suborder CINGULATA
Armadillos and extinct glyptodonts. Armoured bony carapace; anterior teeth lost; cheek teeth undifferentiated but increased in number to give grinding battery.
Glyptodon: about 3 m. long. Pleist. S.Amer. *Panochthus* (Fig. 985): Pleist. S.Amer.

Order PRIMATES
Medium-sized mammals, mainly in tropical regions. Not common in the fossil record; importance greatly exaggerated due to inclusion of man in the order. Usually primitive five-toed feet; relatively large brain; omnivorous dentition; teeth usually quadrate and low-crowned, relatively unspecialized. The three major subdivisions are Lemuroidea, Tarsioidea and Anthropoidea.

Suborder LEMUROIDEA
Fossil and living lemurs of Madagascar, closely allied lorises of S.E.Asia and

502

the bush babies of equatorial Africa; fossils in the Palaeocene and Eocene of Europe and N. America.

Megaladapis: very large, skull 30 cm. long. Pleist. Madagascar.

Suborder TARSIOIDEA

Small. Essentially Palaeocene and Eocene of Europe and N.America, with the living *Tarsius* in S.E.Asia.

Microchoerus: Eoc. Eur.

Suborder ANTHROPOIDEA

Comprises four families:

Family CEBIDAE

S.American monkeys. Miocene onwards. Have three premolars; lower premolars usually single-rooted.

Family CERCOPITHECIDAE

Old World monkeys. Oligocene onwards. Have two premolars; lower premolars birooted.

Family PONGIDAE

Apes, represented today by gibbons and orangutans in Asia and by chimpanzees and gorillas in tropical Africa. Quadrupedal, with well developed canines.

Dryopithecus (Fig. 986): Mio.-Plio. Eurasia, Africa.

Family HOMINIDAE

Basically man (genus *Homo*) and fossils believed to be of common ancestry with man. Upright posture, enlarged brain, dental arcade without diastema and with reduced canines. Due to the antics of emotionally charged hominologists, nomenclature is in an indescribable muddle. Probably Pliocene to Recent.

Australopithecus: Pleist. Africa.

Order CREODONTA

Basically carnivores of the Palaeogene. These fossils have variously been included with true carnivores, grouped with insectivore-like taxa in the order Deltatheroidea, and split up among several other orders; the attribution used here has at least the elegance of Ockham's razor and may even truly reflect phylogeny. Small to very large mammals. Palaeocene to Pliocene; found in Europe, Asia, Africa and N.America; they parallel modern carnivores in their radiation. Feet with cleft claws; brain small; dog-like canines, two or three pairs of shearing molar teeth (only one pair in true carnivores); individual teeth difficult to distinguish from those of true carnivores.

503

Hyainailouros: skull over 60 cm. long. Mio. Eurasia, Africa. *Hyaenodon* (Fig. 987): Eoc.-Mio. Eur., Asia, Africa, N.Amer.

Order CARNIVORA

Probably originated independently of the Creodonta in the late Palaeocene. Worldwide (Australia by introduction); wide radiation in Neogene to carnivorous habitats. Large and small; ground, arboreal, running, swimming, etc; very active; sight usually good; brain large compared with creodonts. Comprises dogs, stoats, bears, raccoons and pandas, civets, hyaenas, cats, seals, sea-lions and walruses. Aquatic seals and sea-lions with simple, undifferentiated peg-like teeth; walruses with large upper canines forming tusks and flattened cheek teeth. Land forms with biting incisors, tearing canines and shearing cheek teeth; two cusps of P^4 and M_1 specialized as shearing blade in most genera, but shear lost in bears, raccoons and pandas. Crushing molars present behind the carnassial shearing teeth in dogs, stoats and civets, but lost in cats and hyaenas. Hyaenas with very powerful bone-crushing premolars.

Smilodon (Fig. 988): sabre-toothed cat, with very enlarged canine teeth. Pleist. N.Amer., S.Amer. *Amphicyon* (Fig. 989): Olig.-Plio. Eur., Asia, N.Amer.

Order CETACEA

The whales have worldwide distribution and are known from early Eocene times onwards. They mostly occur in marine facies and occasionally in nearshore deposits. They can reach over 30 m. in length and this great size alone may be sufficient to identify isolated bones. Most common remains are isolated teeth and ear bones (Fig. 990). Teeth usually simple reptile-like ones, often large. Ear bones dense spherical or cup-shaped. There are three suborders:

Suborder ARCHAEOCETI

Extinct, almost exclusively Palaeogene. Fully aquatic; may reach 20 m. in length; the jaws bear relatively undifferentiated teeth with serrated cusp.

Suborder ODONTOCETI

Toothed whales, dolphins and porpoises. Dentition is composed of numerous undifferentiated simple peg-like teeth; these whales are carnivores, and found from late Eocene onwards.

Figs 986–990. VERTEBRATA: MAMMALIA
Figs 987–989 after Piveteau; Figs 986 and 990 original.
986. *Proconsul africanus* Hopwood, Mio. Kenya. × 1. (Lower dentition).
987a, b. *Hyaenodon cruentus* Leidy, Olig. U.S.A. × 0·5. (Upper and lower dentition).
988. *Smilodon californicus* Bravard, Pleist. Calif. × 0·25. (Skull and mandible).
989. *Amphicyon major* Blainville, Mio. Eur. × 0·6. (Upper dentition).
990. Tympanic bone of Whale, U.Plio. Suffolk. × 0·5.

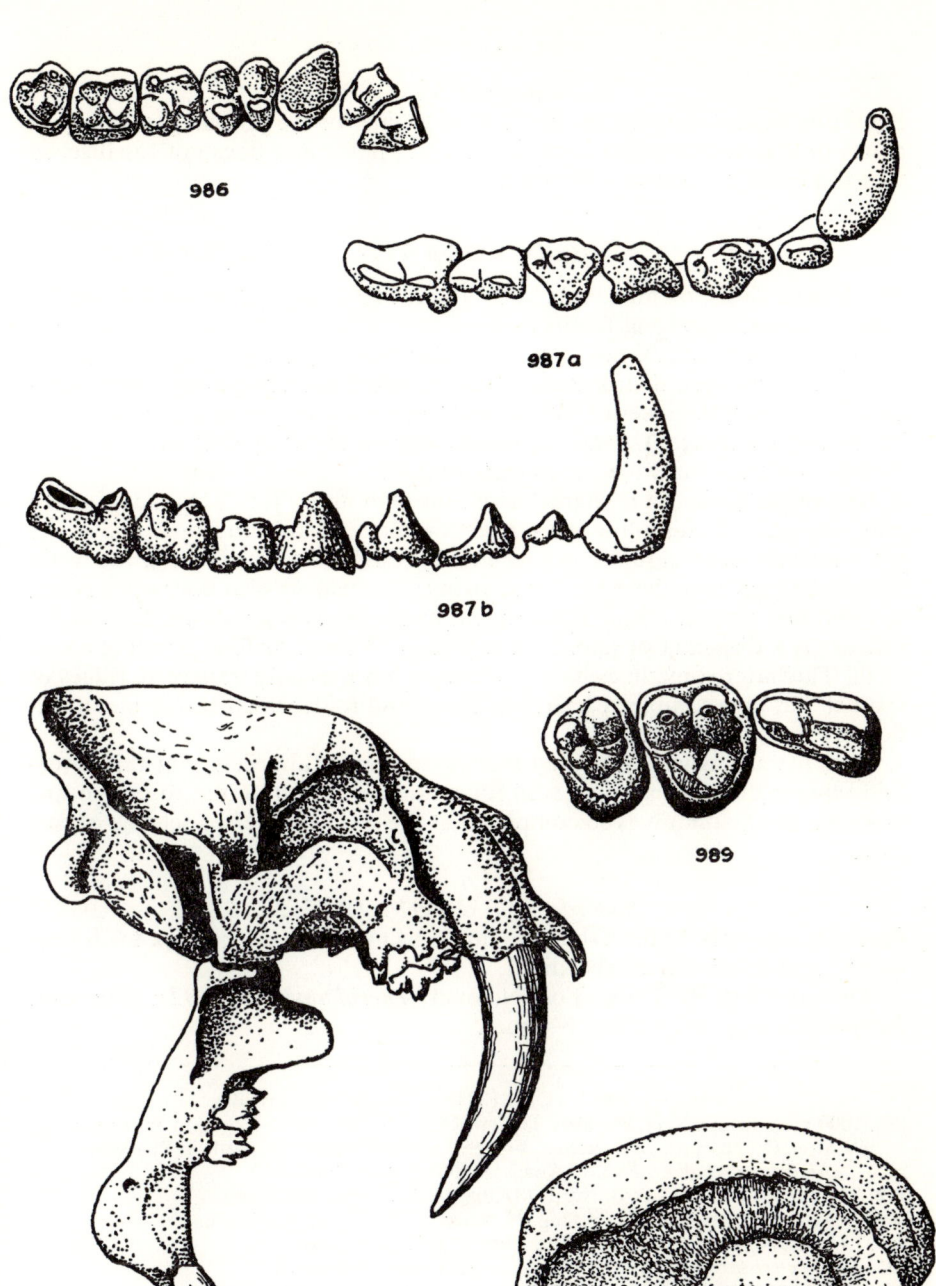

986

987 a

987 b

988

989

990

505

Suborder MYSTICETI

Plankton feeders. Include the great Blue Whale. In the jaw hangs an array of plates of whalebone which is chitinous and pliable and does not fossilize, so that the fossil jaw appears toothless.

Order RODENTIA

The largest order of mammals, with about 350 living genera and about 300 extinct genera. Known as fossils throughout the Tertiary from late Palaeocene onwards. Mostly small, few medium-sized; ground, burrowing, arboreal or aquatic. Worldwide distribution; fossil remains not commonly found until matrix is sieved, then teeth and jaws may be locally very abundant; becoming increasingly important stratigraphically due to rapid evolution, short time duration of taxa and widespread distribution.

The order Rodentia is characterized throughout by the dental adaptation; only one pair of chisel-like incisor teeth is present in both upper and lower jaws, with the enamel (often orange in life) restricted to the front of the teeth, so that a sharp edge develops for gnawing; the incisors also have open roots, so continual growth allows replacement as the tip is worn down. Behind the incisors is a diastema or gap, then a battery of three or four grinding cheek teeth. The latter sometimes have low cusps, more usually transverse ridges of enamel that give complex patterns of loops and folds which enable identification, even of isolated teeth.

Rodents are usually classified into three major suborders, but this threefold scheme does not account well for a number of minor and fossil groups; however, the taxonomy is so complex that the simple threefold scheme is used here.

Suborder SCIUROMORPHA

Basically squirrels. Often arboreal; small- to medium-sized; cheek teeth low-cusped and relatively unspecialized.

Paramys (Fig. 991): Pal.-Eoc. N.Amer. **Theridomys** (Fig. 992): Olig. Eur. **Ischyromys** (Fig. 993): Olig. N.Amer.

Figs 991–1001. VERTEBRATA: MAMMALIA

Fig. 1000 after Romer; Fig. 998 after Tobien; Figs 991–996 after Schaub, in Piveteau; Fig. 997 after Dechaseaux, in Piveteau; Figs 999 and 1001 after Lavocat, in Piveteau.

991. 'Paramys' sp. Bridger, Eoc. N.Amer. × 2·5. (Upper dentition).
992. Theridomys lembronicus Bravard, M.Olig. × 5. (Upper dentition).
993. Ischyromys typus Troxell, M.Olig. N.Amer. × 3. (Upper dentition).
994. Cricetodon gaillardi Schaub, Mio. Eur. × 5. (Upper dentition).
995. Hystrix cristata Linné, Rec. Africa. × 3. (Upper dentition).
996. Sciamys latidens (Scott), L.Mio. S.Amer. × 10. (Lower dentition).
997. Palaeolagus haydeni Leidy, Olig. N.Amer. × 6. (Lower dentition).
998. Lagopsis verus (Hens), Mio. Eur. × 7. (Lower dentition).
999. Phenacodus primaevus Cope, Eoc. U.S.A. × 0·8. (Upper and lower dentition).
1000. Diadiaphorus sp., Mio. S.Amer. × 1. (Upper molars).
1001. Theosodon sp., Mio. Patagonia. × 0·8. (Upper dentition).

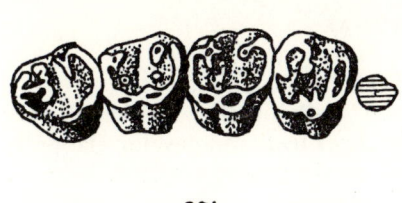

991

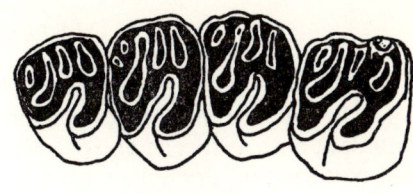

992

993

994

995

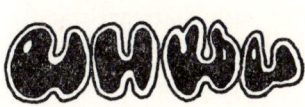

996

997

998

999

1000

1001

Suborder MYOMORPHA

Basically rats and mice, voles and lemmings. Usually small or very small; very abundant both as species and as individuals; cheek teeth can develop elaborate fold patterns; no premolars, only three molars, the largest anterior.

Cricetodon (Fig. 994): Olig.-Plio. Eur., N.Amer.

Suborder CAVIMORPHA

True porcupines of Eurasia and Africa, tree and New World porcupines of N. and S.America, and a vast array of S.American rodents—guinea-pigs, pacas, coypu, capybara, etc. Often medium-sized, occasionally large—capybara reaches 1·5 m. in length; cheek teeth usually hypsodont with fully developed lophs or transverse ridges of enamel; only one upper premolar present; mandible with characteristically flared posterior angle.

Hystrix (Fig. 995): Plio.-Rec. Asia, Eur., Africa. *Sciamys* (Fig. 996): Mio. S.Amer.

Order LAGOMORPHA

Rabbits and hares. The order parallels the rodents in some ways, but the two are of quite distinct origin. Lagomorphs may date back to the Palaeocene of C.Asia. Like rodents, the lagomorphs have gnawing incisors and grinding cheek teeth separated by a large diastema; but two pairs of upper incisors are present, I^2 behind I^1, and the enamel is continuous around the teeth. Molars are transversely elongate, hypsodont and often bilophodont.

Palaeolagus (Fig. 997): Olig. N.Amer. *Lagopsis* (Fig. 998): Mio. Eur.

Order CONDYLARTHRA

An extinct order of primitive ungulates with full placental dental formula, 3. 1. 4. 3; usually quadrate cheek teeth, brachyodont and bunodont. Common in Palaeocene and Eocene of N.America; rarer in S.America and Europe and very rare in Asia. Contains seven families from which can be derived most if not all of the ungulate orders of placental mammals.

Phenacodus (Fig. 999): Pal.-Eoc. N.Amer., Eur.

Order LITOPTERNA

An extinct order of exclusively S.American mammals ranging from late Palaeocene to Pleistocene. Origin from Condylarthra, and containing two families that have evolved forms closely paralleling horses and camels. Molar teeth tend to selenodonty with W-shaped loph on upper molars.

Diadiaphorus (Fig. 1000): Mio.-Plio. S.Amer. *Theosodon* (Fig. 1001): Olig.-Mio. S.Amer.

Order NOTOUNGULATA

A large extinct order, abundant in S.America from Palaeocene to Pleistocene. Hoofed three-toed feet; incisors chisel-like; molars triangular and lophid. Four major subdivisions.

Suborder NOTIOPROGONIA

The only stock outside S.America, in N.America and China. Primitive and small.

Suborder TOXODONTA

Abundant in Oligocene, Miocene and Pleistocene. Dentition without diastema; molars triangular, lophid and inwardly curved. Many genera closely resemble genera in Old World orders.

Toxodon: rhinoceros-like. Pleist. *Nesodon* (Fig. 1002): tapir-like. Mio. *Homalodotherium:* chalicothere-like. Mio.

Suborder TYPOTHERIA

Small- to medium-sized, with close resemblances to Caviomorph rodents; dentition with diastema. (Plio.-Pleist).

Typotherium: Pleist.

Suborder HEGETOTHERIA

Small-sized, closely parallel lagomorphs; dentition with diastema. (Eoc.-Pleist).

Pachyrukhos (Fig. 1003): Olig.-Mio.

Order ASTRAPOTHERIA

An extinct order of S.American ungulates beginning in the Eocene, mostly found in Oligocene and Miocene. Tend to be large and very specialized; skull with very reduced premaxillary bones and no upper incisors; lower incisors small, probably used for cropping like cattle; upper and lower canines big tusks; diastema behind canines, then large molars, triangular and lophid; probably carried proboscis like tapirs; feet small.

Astrapotherium (Fig. 1004): Olig.-Mio. S.Amer.

Order PYROTHERIA

An extinct order from S.America, Eocene and Oligocene. Reached size of elephants, which they probably resembled closely, or even more the African Eocene *Barytherium*; two pairs of chisel-like upper incisors enlarged as tusks; molars bilophodont.

Pyrotherium (Fig. 1005): Olig. S.Amer.

509

Order PANTODONTA

An extinct order of the Palaeocene to Oligocene of Eurasia and N.America. Tend to increase in size until as big as pigmy hippopotami which they resemble; clawed feet; canines enlarge to give tusks; molars brachyodont with some lophodonty.

Coryphodon (Fig. 1006): Pal.-Eoc. Eur., N.Amer.

Order DINOCERATA

An extinct order of the Palaeocene and Eocene of N.America and Asia. Large-sized rhinoceros-like mammals with bone swellings on forehead and very large upper canines in the males, the roots of which project upwards to give the appearance of two anterior horns on the face; molars V-shaped, lophodont.

Uintatherium (Fig. 1007): Eoc. N.Amer.

Order PROBOSCIDEA

Mastodons and elephants. Fossil record from Eocene time onwards; world-wide distribution except Australia. Tend to grow gigantic; smallest in Eocene, about the size of a pigmy hippopotamus; in the Pleistocene some stood nearly 5 m. at the shoulder. Incisors reduced to one pair, which enlarge to form tusks; cheek teeth reduced in number and enlarged in size, hypsodont and lophodont. Within the order five major groups can be conveniently recognized.

Suborder MOERITHERIOIDEA

Restricted to the Eocene of N.Africa. About pigmy hippopotamus size; almost full dentition with incipient upper tusks (I^2); quadrate bunodont molars becoming lophodont.

Moeritherium (Fig. 1008): Eoc.-Olig. N.Africa.

Suborder MASTODONTOIDEA

A term used here to include two families, Gomphotheriidae and Mammutidae; the taxonomy within these families is complex and confusing to the non-specialist. Range from Oligocene to Pleistocene; occurring in Africa, Eurasia and N.America. Tending to large size; tusks often in both upper and lower jaws; always two or more cheek teeth in each jaw; cheek teeth multicusped,

---

Figs 1002–1006. VERTEBRATA: MAMMALIA

Figs 1002–1006 after Lavocat, in Piveteau.

1002. *Nesodon imbricatus* Owen, Mio. Patagonia. × 1. (Upper and lower dentition).
1003. *Pachyrukhos* sp., Mio. Patagonia. × 2. (Upper dentition).
1004. *Astrapotherium magnum* Owen, Mio. Patagonia. × 0·13. (Skull and mandible).
1005. *Pyrotherium romeri* Ameghino, Olig. Patagonia. × 0·25. (Mandible).
1006. *Coryphodon testis* Cope, Eoc. U.S.A. × 1. (Upper molars).

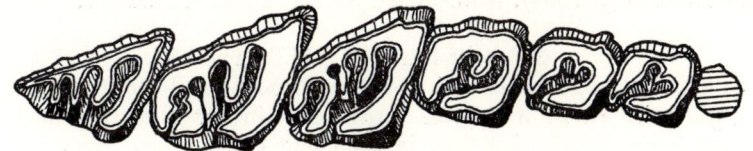

1002

1003

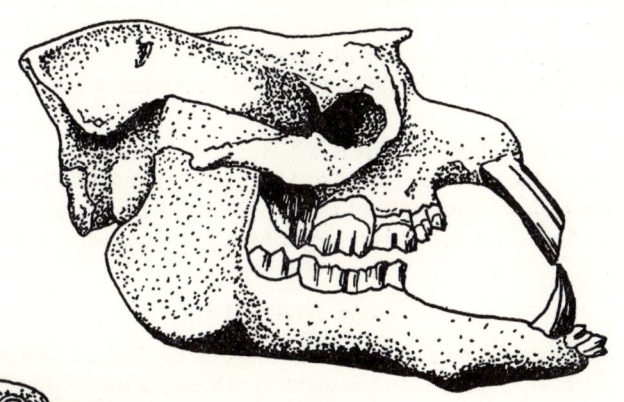

1004

1005

1006

the main cusps arranged in pairs transversely and with tendency to unite to give loph; usually numerous accessory cusps present between and around the main cusps.

Mastodon (Fig. 1009): Mio.-Pleist. Africa, Eurasia, N.Amer.

Suborder ELEPHANTOIDEA

The family Elephantidae is limited to the Pliocene and Pleistocene, with two living genera in Africa and India. Tusks only in upper jaws, never in lower; only one cheek tooth at a time in each half of each jaw; cheek teeth very hyposdont and lophodont; the number of lophs may reach thirty in the last molar to erupt.

Mammuthus: Pleist. Eurasia, Africa, N.Amer.

Suborder DEINOTHERIOIDEA

Only one genus, *Deinotherium* (Fig. 1010): basic change through time is increase in size but usually larger than contemporary mastodons; never any tusks in upper jaw, one pair of downward- and backward-turned tusks in lower jaw; cheek teeth bilophodont. Mio.-Pleist. Africa, Eurasia.

Suborder BARYTHERIOIDEA

Only one genus, *Barytherium*: size of African elephant, two pairs of tusks in upper and lower jaw; tusks chisel-edged, followed by diastema; cheek teeth bilophodont, almost indistinguishable from deinothere teeth. Eoc. N.Africa.

Order SIRENIA

The sea-cows, comprising today the dugongs of the Indian Ocean and manatees of the rivers that issue into the Atlantic Ocean. Fully aquatic marine and fluviatile mammals without hind legs, swimming with their tails. Fossil record from M.Eocene times onwards, mainly in Atlantic and Tethys (=Mediterranean). Ribs are the most commonly found parts and most easily recognized; sirenia ribs are the only mammalian ribs to exhibit pachyostosis, i.e., dense thick bone without a spongy core; ribs look like large fossilized bananas, and are usually found in near-shore facies. Skulls and teeth rare; anterior dentition with one pair of upper incisor tusks, other incisors reduced or lost; cheek teeth essentially bilophodont.

Halianassa (Fig. 1011): Mio. Eur., N.Amer.

Figs 1007–1011. VERTEBRATA: MAMMALIA

Fig. 1011 after British Museum (Natural History); Fig. 1007 after Piveteau; Figs 1008–1010 original.

1007. *Uintatherium mirabile* Marsh, M.Eoc. U.S.A. × 0·5. (Upper and lower dentition).
1008. *Moeritherium* sp., U.Eoc. Egypt. × 0·5. (Upper dentition P^2–M^3).
1009. *Mastodon angustidens* Cuvier, L.Mio. Germany. × 0·33. (Upper molar M^3).
1010. *Deinotherium* spp., Plio. Eur. (Mandible (× 0·09) and lower molar).
1011. *Halianassa fossile* (de Blainville), M.Mio. Eur. × 1. (Lower left M^3).

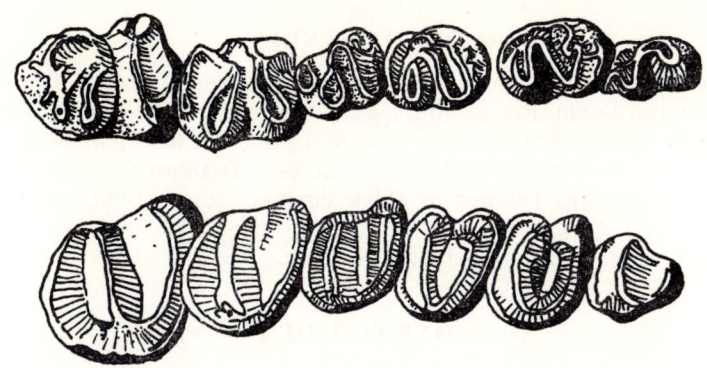

1007

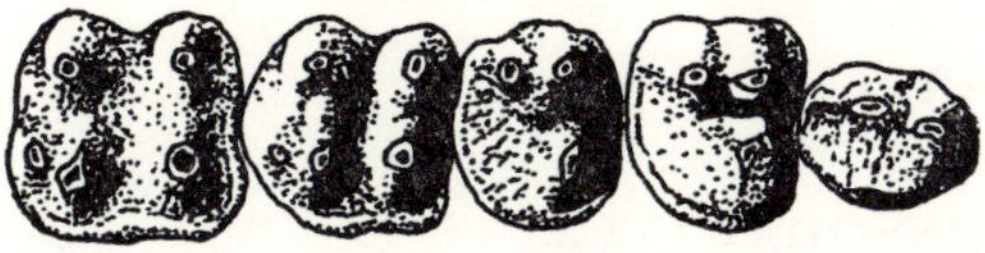

1008

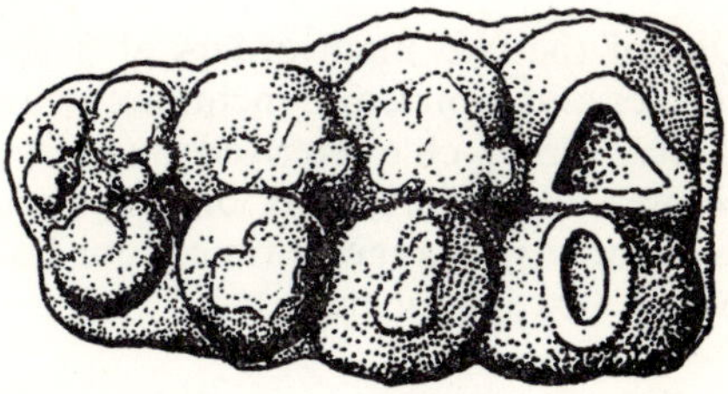

1009

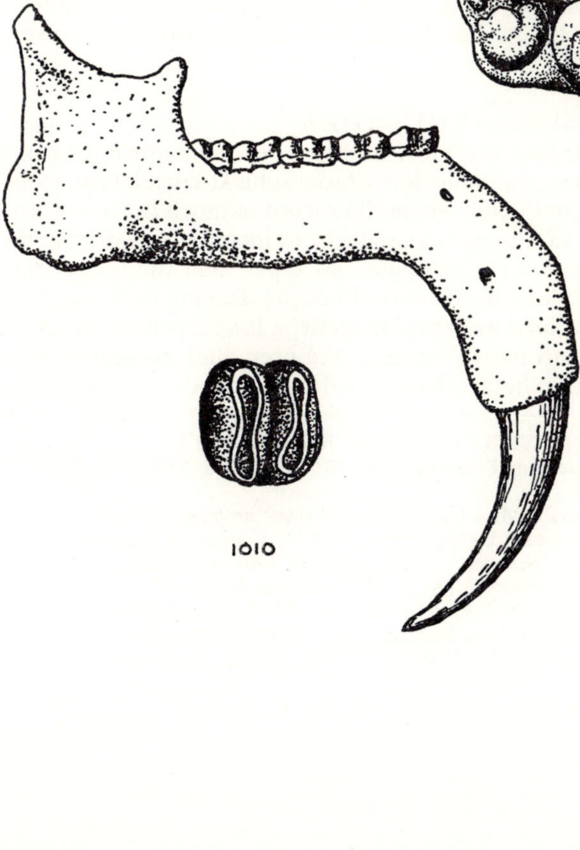

1010

1011

Order DESMOSTYLIA

An extinct order of aquatic mammals from the Miocene of the Pacific coasts of N.America and Japan. Retain hind legs and probably not closely related to Sirenia, merely parallel adaptations; skull horse-like in *Palaeoparadoxia*, and with elongate snout—bearing short incisor tusks in *Desmostylus*; cheek teeth composed of numerous tubules of thick enamel, quite unlike any other mammal.

Desmostylus (Fig. 1012): Mio. Japan, N.Amer.

Order HYRACOIDEA

The hyraxes or 'conies' of Africa. Only three living genera, but more abundant in the Tertiary of Africa and a few in the Pliocene of Eurasia. Small- to medium-sized herbivores; teeth and foot characters suggest parallels if not affinities with perissodactyls.

Megalohyrax (Fig. 1013): Olig. Africa.

Order PERISSODACTYLA

One of the two main orders of herbivores, comprising today horses, rhinoceroses and tapirs, and including two extinct stocks, brontotheres and chalicotheres. Order characterized by uneven-toed condition; three toes in rhinoceroses and tapirs, only one in living horses (compare with even two-toed artiodactyls). Range in time from Palaeocene onwards and worldwide distribution (except Australia).

Suborder EQUOIDEA

The horses have one of the best and most complete records known for fossil mammals and are extremely valuable for stratigraphical correlation in N. America, less so in other continents where the record is much less complete. In the course of their evolution there is a tendency to increase in size from the dog-size of the early Eocene. Legs elongate and taper, and the lateral toes become reduced and finally lost in the early Pliocene, leaving only one-toed horses. The incisors are retained as cropping teeth; a long diastema develops behind these; the cheek teeth form a battery, the premolars becoming large and fully molariform. Early horses have quadrate, bunodont four-cusped

---

Figs 1012–1018: VERTEBRATA: MAMMALIA
Figs 1012–1013 after Dechaseaux, in Piveteau; Figs 1014-1016 and 1018 after Viret, in Piveteau; Fig. 1017 original.
1012. *Desmostylus hesperus* Marsh, Mio. Calif. × 0·7 (Lower molars).
1013. *Megalohyrax niloticus* Schlosser, L.Olig. Egypt. × 0·75. (Upper right molars).
1014. *Hyracotherium* spp., L.Eoc. N.Amer. × 2. (Upper and lower dentition).
1015. *Hipparion matthewi* Abel, L.Plio. Turkey. × 0·75. (Upper dentition).
1016. *Brontops brachycephalus* Osborn, L.Olig. U.S.A. × 0·25. (Upper dentition).
1017. *Lophiodon remensis* Lemoine, L.Eoc. France. × 0·5. (Upper dentition).
1018. *Hyracodon nebrascensis* Leidy, Olig. N.Amer. × 0·66. (Upper dentition).

1012

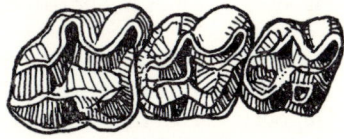

1013

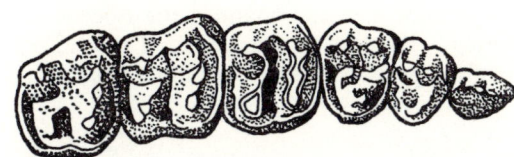

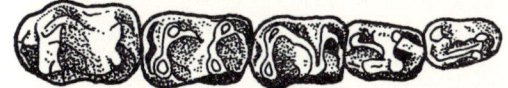

1014

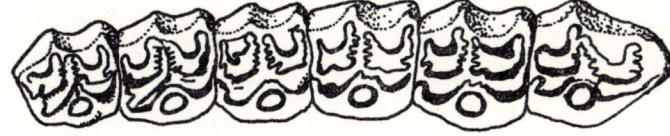

1015

1016

1017

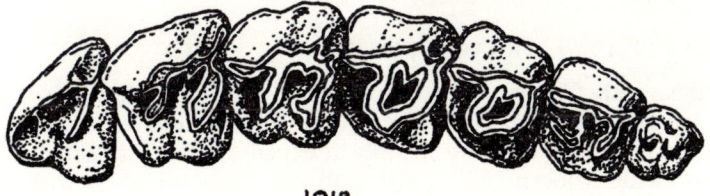

1018

515

molars; they soon become selenodont and by Miocene times become hypsodont when they change from browsing habit to grazing on grasses. The complexity of the enamel pattern increases as more is incorporated in the crown to
resist wear. In Europe there was the true horse *Hyracotherium* [*Eohippus*]
(Fig. 1014) in the early Eocene, and a side-line, the Palaeotheres. In Miocene
times the Anchitheres migrated to Europe and Asia from America. In Pliocene
times the *Hipparion* (Fig. 1015) fauna encircled the northern hemisphere. By
Pleistocene times the horses had become worldwide (excepting Australia).

Suborder BRONTOTHERIOIDEA

The extinct titanotheres reached a gigantic size and spanned the Eocene and
Oligocene of Eurasia and N.America. The skull is characterized by the presence on the snout of a pair of horn-like growths, sometimes branched. Molar
teeth with W-shaped loph.
   Brontops (Fig. 1016): Olig. N.Amer.

Suborder CHALICOTHERIOIDEA

A rare extinct group which ranged from Eocene to Pleistocene and whose
remains have been identified in N.America, Eurasia and Africa. Remains
usually very fragmentary; cheek teeth similar to Brontotheres; feet with cleft
claws—the most readily recognizable elements.

Suborder TAPIROIDEA

Medium-sized; three-toed; a relatively conservative stock. Living tapirs of
S.E.Asia and C.America; fossil record from Eocene onwards from Eurasia
and N.America. Reduced upper canines; small diastema develops; molariza-
tion of premolars; molars brachyodont and bilophodont. All browsers and no
grazers.
   Lophiodon (Fig. 1017): Eoc. Eur.

Suborder RHINOCEROTOIDEA

Very varied in fossil record that extends from the Eocene and ranges through
Eurasia, Africa and N.America. Size medium to gigantic (*Baluchitherium*
from the Oligocene of Asia stood 6 m. at the shoulder); feet three-toed, with
slow and fast running forms, and some amphibious; incisors with or without
tusk development; upper molars have π pattern, lower molars with two
crescents.
   Hyracodon (Fig. 1018): Olig. N.Amer. *Trigonias* (Fig. 1019): Olig. N.Amer.

Order ARTIODACTYLA

Even-toed ungulates, with four or two toes on each foot, the 'cleft-hoof'
mammals. Very varied and numerous; worldwide distribution excepting
Australia. Few in the Eocene, increasing steadily at expense of perissodactyls,
explosive radiation in the Miocene with grazing forms. Early members with
complete dentition, bunodont browsers; later lose upper incisors, diastema

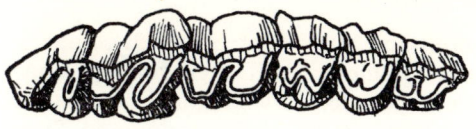

1019

1020

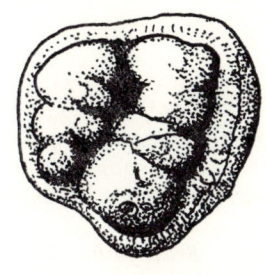

1021

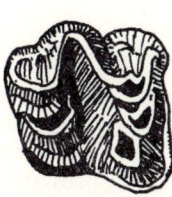

1022

1023

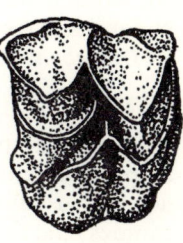

1024

1025

1026

1027

Figs 1019–1027. VERTEBRATA: MAMMALIA

Figs 1020, 1024 and 1027 after Viret, in Piveteau; Figs 1019, 1021–1023 and 1025–1026 original.

1019. *Trigonias* sp., L.Olig. U.S.A. × 0·5. (Upper dentition).
1020. *Dichobune* cf. *robertiana* Gervais, M.Eoc. Eur. × 2. (Upper dentition).
1021. *Entelodon deguilhemi* Répelin. M.Olig. France. × 0·5. (Upper right molar).
1022. *Brachyodus borbonicus* (Gervais), M.Olig. France. × 1. (Upper right M³).
1023. *Hexaprotodon* sp., U.Plio. E.Africa. × 0·5.
1024. *Cainotherium* sp., Aquit. Eur. × 4. (Upper molar).
1025. *Merycoidodon culbertsonii* Leidy, M.Mio. N.Amer. (Wyoming). × 1. (Upper dentition P⁴–M³).
1026. *Dichodon simplex* Kowalsky, M.Eoc. Eur. × 1·5. (Upper dentition).
1027. *Pliauchenia magnifortis* Gregory, L.Plio. N.Amer. × 0·33. (Upper dentition).

517

develops, cheek teeth become selenodont and hypsodont. Stomach compartmented and very efficient in digestion of cellulose. Skull often develops bone growths, horns or antlers. There are four suborders of which the Pecora is the largest and most important, the others being Palaeodonta, Suina and Tylopoda.

Suborder PALAEODONTA

An extinct, ancestral stock; mainly Eocene and Oligocene of Europe and N.America; teeth quadrate, bunodont.

Dichobune (Fig. 1020): Eoc.-Olig. Eur.

Suborder SUINA

Namely pigs, peccaries and hippopotami, and two extinct stocks, anthracotheres and entelodonts. Basically heavily built, short-legged, some semiaquatic, usually four-toed. Dentition bunodont (four, five or many cusps); canines tend to enlarge. Rooters and ground-browsers.

Entelodon ('giant pigs', Fig. 1021): Olig. Eur., N.Amer. *Brachyodus* (Fig. 1022): analogues of entelodonts in the Old World; molars bunoselenodont, usually with fifth cusp. Olig.-Mio. Eurasia. *Hexaprotodon* (Fig. 1023): a hippopotamus; molars with four trifoil cusps. Plio.-Pleist. Eurasia.

Suborder TYLOPODA

Camels and three extinct related stocks. Feet four- or two-toed; canines reduced, cheek teeth selenodont; stomach with three compartments.

Family CAINOTHERIIDAE

Mainly Oligocene of Europe; small hare-like gnawers; long slender legs with four toes; dentition without diastema, molars with five cusps.

Cainotherium (Fig. 1024): Olig.-Mio. Eur.

Family MERYCOIDODONTIDAE

The oreodonts, very abundant in the Oligocene of N.America; pig-like in build; dentition without diastema; enlarged upper canine; molars brachyodont to selenodont, four-cusped.

Merycoidodon (Fig. 1025): Olig. N.Amer.

Family XIPHODONTIDAE

A primitive stock, possibly ancestral to camels; two-toed; dentition complete; molars selenodont with five cusps. *Dichodon* (Fig. 1026): Eoc.-Olig. Eur.

Family CAMELIDAE

Camels have fossil record mostly in N.America, from Eocene times onwards. Represented today in Asia and by introduction in Africa, and in S. America by llama, alpaca, guanaco and vicuna. Dentition without upper canine; cheek teeth biselenodont with four cusps; legs and neck long; feet splayed, two-toed; increase in size from rabbit-size in Eocene.

Stenomylus: gazelle-like. Mio. N.Amer. *Oxydactylus:* giraffe-like. Mio. N.Amer. *Pliauchenia* (Fig. 1027): Plio. N.Amer.

Suborder PECORA

The largest and most important division of the artiodactyls, with chevrotains, deer, giraffes, antelopes and cattle. Two-toed; four-compartmented stomach; often with skull outgrowths. There are three superfamilies, Traguloidea, Cervoidea and Bovoidea.

Superfamily TRAGULOIDEA

Living chevrotains. Small; tropical; gazelle-like; molars selenodont with four cusps; extinct Protoceratidae of N.America with elongate face and nasal bony growths, often branching.

Synthetoceras (Fig. 1028): L.Plio. N.Amer.

Superfamily CERVOIDEA

Deer and giraffes, and the extinct ancestral family Palaeomerycidae from the Oligocene and Miocene of N.America and Eurasia. Giraffidae have ossicones or bony outgrowths of skull which appear in both male and female, may or may not branch, are not shed, nor covered in horn as in bovids. While deer are limited to the northern temperate and tundra zones, giraffes occupy a similar niche to the south in the tropical and savannah zones of Africa with fossils in S.Europe and Asia.

Palaeotragus (Fig. 1029): Mio.-Plio. Eurasia, Africa.

Superfamily BOVOIDEA

True grazers, dentition without upper incisors, upper canine reduced, molars hypsodont and selenodont. Miocene onwards. Two families.

Family ANTILOCAPRIDAE

Prong-buck, antelope-like, of N.America, but horns forked and shed annually.

Family BOVIDAE

Many and varied, few in the Miocene, major radiation in the Pliocene. Essentially Old World, few in N.America, none in S.America or Australia. Horns persistent, with bony core and horny covering, unbranched, present in both sexes; these more useful than teeth in identification. Includes cattle, sheep, goats, antelopes, gazelles, bison, etc.

Palaeoreas (Fig. 1030): Mio.-Plio. Eurasia, Africa.

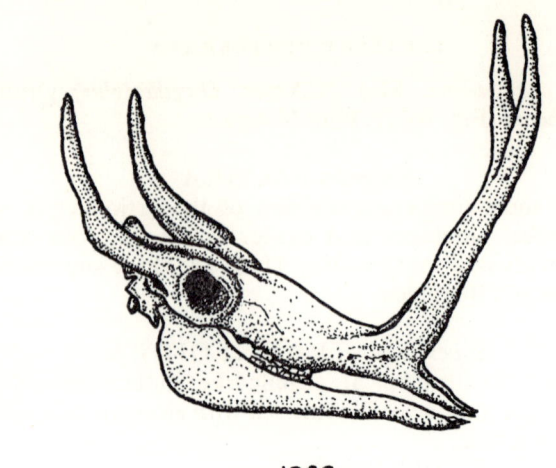

1028

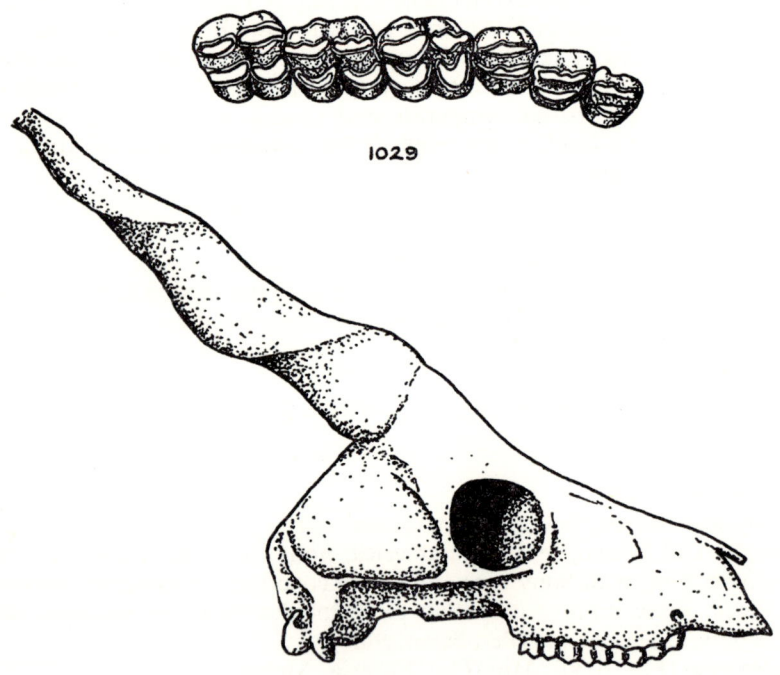

1029

1030

Figs 1028–1030. VERTEBRATA: MAMMALIA

Fig. 1028 after Romer; Figs 1029–1030 after Viret, in Piveteau.

1028. *Synthetoceras* sp., Plio. N.Amer. × 0·08. (Skull and mandible).
1029. *Palaeotragus roueni* Gaudry, L.Plio. Pikermi (Greece) × 0·5. (Upper dentition).
1030. *Palaeoreas lindermayeri* Gaudry, L.Plio. Pikermi (Greece). × 0·25. (Skull).

520

SHORT GLOSSARY OF TECHNICAL TERMS APPLIED TO VERTEBRATA (CHIEFLY MAMMALIA)

Amphicoelous. Vertebra with concave facets on the centrum.

Brachyodont. Tooth with low crown, not capable of long-continued wear and tear.

Bunodont. Tooth bearing a number of low and rounded tubercles on the crown.

Bunoselenodont. Teeth with both blunt tubercles and crescentic ridges.

Canines. The front teeth of the maxilla bone, and corresponding teeth of lower jaw.

Carapace. The continuous dorsal bony covering of a Chelonian formed partly by upgrowths from vertebrae and ribs, partly from dermal ossifications.

Carnassial. Adapted for flesh-eating.

Carnivore. An animal feeding on the flesh of other animals.

Cement. A bony deposit on the crowns of teeth, chiefly filling hollows between enamel ridges.

Centrum. The ossified notochord in a vertebra.

Cerebellum. Part of the brain concerned with balance.

Cheek-teeth. Teeth posterior to the canines.

Conid. A conical elevation on the occlusal surface of a lower molar.

Crown. The exposed part of a tooth, free of jaw bone and gums, usually covered with enamel. [This is not the dentists' sense of the term, which is equivalent to what is here called the occlusal surface.]

Ctenoid. Ovoid fish scale with comb-like edge.

Cusp. An elevation of the crown of a tooth.

Cycloid. Oval-shaped fish scale.

Denticle. A pointed cusp at the side of the tooth in some fishes.

Dentine. Calcareous tissue forming the main part of any tooth.

Diastema. Gap in dentition between front teeth and cheek teeth.

Enamel. A very hard calcareous material usually forming the surface of the crown of a tooth.

Epiphysis. A separate centre of ossification at each end of a long bone, separated from the main ossification by cartilage, which permits of growth without distortion of shape.

Herbivore. An animal subsisting on vegetable food other than fruits and seeds, either browsing on soft leaves or grazing on hard grass etc.

Hypsodont. High-crowned tooth.

Incisors. Teeth borne by the premaxilla in the upper jaw, and the corresponding teeth of the lower jaw.

Loph. Any enamel ridge in an upper jaw cheek tooth.

Lophid. Any enamel ridge in a lower jaw cheek tooth.

Lophodont. Tooth in which lophs or lophids are the dominant feature.

Marsupial. Pouched mammal (e.g. kangaroo).

Maxilla. Facial bone which carries cheek dentition.

Molar teeth. The hinder cheek-teeth, which have no milk predecessors.

Monotreme. Primitive egg-laying mammal.

Occlusal. The surface of a tooth which comes in contact with the corresponding surface of the tooth opposite it in the other jaw (equivalent to crown in the dentists' sense).

Omnivorous. Unspecialized feeding habit, including both plant and animal sources.

Pachyostosis. Thickening of bone found in some aquatic vertebrates.

521

Pelvis or *pelvic girdle*. The series of bones within the trunk to which the bones of the hind limbs are attached.

Placental. Mammal in which foetus is enclosed in a sheath (placenta).

Plastron. The ventral bony shield of a Chelonian, formed of dermal bones only.

Premaxillary. Bone in upper jaw which carries incisor teeth.

Premolars. The front cheek-teeth, which have milk predecessors.

Root of a tooth. The portion embedded in gum and jaw bone, and not covered with enamel.

Saltorial. Jumping gait.

Selenodont. Cusps of tooth in occlusal view have a crescent shape.

Sternum. Breast bone.

Trenchant. Teeth with sharp cutting facets and steep sides.

Unciform. Hook-shaped (process on the ribs of birds).

Ungulate. Hooved, having the horny epidermis on the tips of fingers and toes developed into thick flat hooves.

Vertebra. Any one of the long series of similar bones constituting the backbone or vertebral column: the vertebrae differ to some extent in the five regions of the column: cervical (neck), thoracic (back), lumbar, sacral and caudal (tail).

SELECT BIBLIOGRAPHY OF TERTIARY VERTEBRATA

1. AMEGHINO, F. 1889. 'Contribución al conocimiento de los mamiferos fósiles de la República Argentina', *Actas Acad. nac. Cienc. Córdoba*, **6**, 1–1028, 98 pls. [Description of many South American taxa.]
2. ANDREWS, C. W. 1906. *A Descriptive Catalogue of the Tertiary Vertebrata of the Fayum, Egypt*, 1–324, 26 pls (British Museum: London). [Description of early proboscideans, *Arsinoitherium*, hyracoids and sirenians.]
3. BENEDEN, P. J. VAN and GERVAIS, P. 1880. *Ostéographie des cétacés vivants et fossiles*, 1–634, 64 pls (A. Bertrand: Paris).
4. BOHLIN, B. 1926. 'Die Familie Giraffidae mit besonderer Berücksichtigung der fossilen Formen aus China', *Palaeont. sin.*, (C), **4**, no. 1, 1–179, 12 pls.
5. BRODKORB, P. 1963–64. 'Catalogue of fossil birds', *Bull. Fla St. Mus., biol. Sci.*: part 1, 'Archaeopterygiformes through Ardeiformes', **7**, (4), 179–293; part 2, 'Anseriformes through Galliformes', **8**, (3), 195–335.
6. CRUSAFONT PAIRÓ, M. 1952. 'Los jiráfidos fósiles de España', *Mem. Comm. Inst. geol. Barcelona*, **8**, 1–239, 67 pls.
7. DAVIS, D. 1964. 'The giant panda: a study of evolutionary mechanisms', *Fieldiana Zool. Mem.*, **3**, 1–339, 1 pl.
8. ERDBRINK, D. P. 1953. *A review of fossil and Recent bears of the Old World*, 1–597, 18 pls (Drukkerij Jan de Lange: Deventer).
9. FLOWER, W. H. 1885. *An Introduction to the Osteology of the Mammalia*, 3rd ed., 1–383 (Macmillan & Co.: London). [Still the best account of the subject.]
10. FLOWER, W. H. and LYDEKKER, R. 1891. *An Introduction to the Study of Mammals, Living and Extinct*, 1–763 (Adam & Charles Black: London). [Still a useful anatomical account.]
11. GREGORY, W. K. and HELLMAN, M. 1939. 'On the evolution and major classification of the civets (Viverridae) and allied fossil and Recent Carnivora: a phylogenetic study of the skull and dentition', *Proc. Am. phil. Soc.*, **81**, 309–392, 6 pls.
12. HÜRZELER, J. 1936. 'Osteologie und Odontologie der Caenotheriden', *Abh. schweiz. palaont. Ges.*, **58**, 1–88; **59**, 91–112, 8 pls.
13. KELLOGG, R. 1936. 'A review of the Archaeoceti', *Publs Carnegie Instn*, no. 482, 1–366, 37 pls.
14. LAMBRECHT, K. 1933. *Handbuch der Palaeornithologie*, 1–1024, 4 pls (Gebrüder Borntraeger: Berlin). [A well documented account.]
15. MATTHEW, W. D. 1910. 'The phylogeny of the Felidae', *Bull. Am. Mus. nat. Hist.*, **28**, 289–316.
16. MATTHEW, W. D. 1930. 'The phylogeny of dogs', *J. Mammal.*, **11**, 117–138.
17. MILNE-EDWARDS, A. 1867–71. *Récherches Anatomiques et Paléontologiques pour servir à l'Histoire des Oiseaux Fossiles de la France*, **1**, 475 pp., 96 pls.; **2** 632 pp., 104 pls (Masson & Cie.: Paris).
18. OSBORN, H. F. 1918. 'Equidae of the Oligocene, Miocene and Pliocene of North America, iconographic revision', *Mem. Am. Mus. nat. Hist.*, N.S., **2**, (1), 1–330.
19. OSBORN, H. F. 1929. 'The titanotheres of ancient Wyoming, Dakota and Nebraska', *Monogr. U.S. geol. Surv.*, no. 55, **1**, 1–702, 42 pls.; **2**, 703–953, 194 pls.
20. OSBORN, H. F. 1936–42. *Proboscidea: A Monograph of the Discovery of Evolution, Migration and Extinction of the Mastodonts and Elephants of the World*, **1**

(Moeritherioidea, Deinotherioidea, Mastodontoidea), 1–804, 12 pls; **2** (Stego-dontoidea, Elephantoidea), 805–1676, 18 pls (American Museum of Natural History: New York).

21. PEARSON, H. S. 1927. 'On the skulls of early Tertiary Suidae, together with an account of the otic region in some other primitive Artiodactyla', *Phil. Trans. R. Soc.*, (B), **215**, 389–460.

22. PEARSON, H. S. 1928. 'Chinese fossil Suidae', *Palaeont. sin.*, (C), **5**, (5), 1–75, 4 pls.

23. PETERSON, O. A. 1909. 'A revision of the Entelodontidae', *Mem. Carneg. Mus.*, **4**, 41–156, 9 pls.

24. PILGRIM, G. E. 1932. 'The fossil Carnivora of India', *Mem. geol. Surv. India Palaeont. indica*, N.S., **18**, 1–232, 10 pls.

25. PILGRIM, G. E. 1939. 'The fossil Bovidae of India', *Mem. geol. Surv. India Palaeont. indica*, N.S., **26**, (1), 1–356, 8 pls.

26. PILGRIM, G. E. and HOPWOOD, A. T. 1928. *Catalogue of the Pontian Bovidae of Europe in the Department of Geology of the British Museum*, 1–106, 9 pls (British Museum (N.H.): London).

27. PIVETEAU, J. (ed.). 1952–. *Traité de Paléontologie*, 7 vols (Masson & Cie.: Paris). [Written mainly by French palaeontologists. **4–7** give a comprehensive up-to-date account of fossil vertebrates.]

28. RADINSKY, L. B. 1963. 'Origin and early evolution of North American Tapi-roidea', *Bull. Peabody Mus. nat. Hist.*, **17**, 1–106, 4 pls.

29. REYNOLDS, S. J. 1913. *The Vertebrate Skeleton*, 2nd ed., 1–535 (Cambridge University Press: Cambridge).

30. ROMER, A. S. 1956. *Osteology of the Reptiles*, 1–772 (University of Chicago Press: Chicago).

31. ROMER, A. S. 1966. *Vertebrate Paleontology*, 3rd ed. 1–468 (University of Chicago Press: Chicago). [Exhaustive bibliography.]

32. SIMPSON, G. G. 1932. 'Fossil Sirenia of Florida and the evolution of the Sirenia', *Bull. Amer. Mus. nat. Hist.*, **59**, 419–503.

33. SIMPSON, G. G. 1940. 'Review of the mammal-bearing Tertiary of South America', *Proc. Am. phil. Soc.*, **83**, 649–709.

34. SIMPSON, G. G. 1945. 'The principles of classification and a classification of mammals', *Bull. Amer. Mus. nat. Hist.*, **85**, 1–450. [Standard work on mam-malian classification with complete generic lists.]

35. STEHLIN, H. G. and SCHAUB, S. 1951. 'Die Trigonodontie der Simplicidentaten Nager', *Schweiz. paläeont. Abh.*, **67**, 1–385.

36. STIRTON, R. A. 1940. 'Phylogeny of North American Equidae', *Univ. Calif. Publs Bull. Dep. Geol.*, **25**, (4), 165–198.

37. SWINTON, W. E. 1958. *Fossil Birds*, 1–63, 11 pls (British Museum (N.H.): London).

38. THENIUS, E. 1959. *Tertiar: Wirbeltierfaunen*, 1–328, 10 pls–Band III, 2 Teil of *Handbuch der Stratigraphischen Geologie*, ed. Fr. Lotze (Ferdinand Enke Verlag: Stuttgart).

39. THENIUS, E. and HOFER, H. 1960. *Stammesgeschichte der Säugetiere: eine Uebersicht über Tatsachen und Probleme der Evolution der Säugetiere*, 1–322 (Springer-Verlag: Berlin).

40. VAN VALEN, L. 1966. 'Deltatheridia: a new order of mammals', *Bull. Amer. Mus. nat. Hist.*, **132**, no. 1, 126 pp., 8 pls.

41. WOOD, A. E. 1962. 'The early Tertiary rodents of the family Paramyidae', *Trans. Am. phil. Soc.*, N.S., **52**, (1), 1–261.

42. WOOD, H. E., CHANEY, R. W., CLARK, J., COLBERT, E. H., JEPSEN, G. L., REESIDE, J. B., JR. and STOCK, C. 1941. 'Nomenclature and correlation of the North American continental Tertiary', *Bull. geol. Soc. Am.*, **52**, 1–48, 1 pl.
43. YOUNG, J. Z. 1957. *The Life of Mammals*, 1–820 (Clarendon Press: Oxford).

SYSTEMATIC INDEX

Names of genera and subgenera are printed in ordinary type, those of higher grades in small capitals. All names regarded as synonymous (for whatever reason) are printed within brackets.

Abertella, 133
ABERTELLIDAE, 133
Abida, 412
Abra, 241
Acamptochetus, 362
Acanthechinus, 109
Acanthina, 354, 353, 364
Acanthinula, 413
ACANTHINULINAE, 413
Acanthocardia, 226, 228
Acanthodesia, 482
ACANTHOPTERYGII, 492
Acar, 182
ACAVACEA, 425
ACAVIDAE, 425
Acervulina, 74
ACERVULINIDAE, 74
Acesta, 202
Achatina, 419
ACHATINACEA, 419
ACHATINIDAE, 419
Acila, 177
Acipenser, 490
ACLIDIDAE, 398
Aclis, 398
Acmaea, 291
ACMAEIDAE, 291
Acme, 309
ACMEIDAE, 309
ACOCHLIDIACEA, 392
Acrilla, 324
ACRILLINAE, 324
ACROLOXIDAE, 409
Acroloxus, 409
Acropora, 484
Acroreia, 406
ACROREIIDAE, 404
[Acroria], 406
[Actaeon], 393
[ACTAEONIDAE], 392
Acteocina, 395, 396
ACTEOCINIDAE, 395

Acteon, 393, 392
ACTEONIDAE, 392
Acteonina, 393
ACTEONINAE, 393
ACTEONININAE, 393
Actinacis, 484
Actinastrea, 484
ACTINOCERATOIDEA, 472
Actinopora, 482
ACTINOPTERYGII, 490
Actinosiphon, 76
Adacna, 230
[ADELOMELONINAE], 375
Adelosina, 33
Adeona, 482
[ADEORBIDAE], 309
[Adeorbis], 309
[ADESMACEA], 265
Admete, 385
ADMETINAE, 385
Adusta, 337
Adytaster, 144
Aeolopneustes, 109
Aequipecten, 195
Africoterebellum, 332
Agaricia, 484
Agaronia, 377
Agasoma, 355
Agassizia, 148
[Agina], 261
AGNATHA, 486
Akera, 401
AKERIDAE, 401
Aktinocyclina, 78
Alabamina, 84
ALABAMINIDAE, 84
Alatacythere, 453, 454
Albea, 430
Albinula, 413
Alcithoe, 374
ALCITHOINAE, 374
Alectrion, 365

529

GENERAL INDEX

Technical terms not found below should be looked for in the Glossaries at the end of Chaps. I, II, III, IV, V and VII. The following abbreviations are used below: F, Foraminifera; E, Echinoidea; B, Bivalvia; G, Gastropoda; O, Ostracoda; P, Pisces; AM, Amphibia; A, Aves; R, Reptilia; M, Mammalia; V, Vertebrata.

abactinal, (E) 161
aboral, (E) 161
Abrard, R., 59
accessory apertures, (F) 22, 88
accommodation groove, (O) 467
acervuline, (F) 88
actinal, (E) 161
actinal furrows, (E) 161
actinodont, (B) 174, 275
adapical, (E) 101, 161
adducter, (B) 175, 275
adductor muscle scar, (O) 445, 467
adont, (O) 446, 447
adoral, (E) 101, 161
adradial, (E) 101, 161
adventitious, (F) 88
agglutinating, (F) 20
ala, (O) 467
alar prolongations, (F) 56, 88
alligators, (R) 493
alpaca, (M) 518
alveolus, (F) 88
amb, (E) 161
ambitus, (E) 101, 161
ambulacrum, (E) 161
amphicoelus, (P) 488, 521
amphidetic, (B) 172, 275
amphidont, (O) 467
amphisternous, (E) 161
anagenesis, (F) 59
anal fasciole, (E) 161, (G) 434
anchitheres, (M) 516
anglers, (P) 492
anisomyarian, (B) 175, 275
annular chambers, (F) 88
anomphalous, (G) 281, 434
ant-eater, (M) 499, 502
antecurrent, (G) 284, 436
antelopes, (M) 519
antennal muscle scar, (O) 445, 467
anterior, (E) 161, (G) 434, (O) 467
anterior canal, (G) 284
anterior cardinal angle, (O) 467
anthracotheres, (M) 518

antimerodont, (O) 446, 467
aperture, (F) 21, 88, (G) 281, 434
apetaloid, (E) 161
apes, (M) 503
apical system (disk), (E) 100, 161
apophysis, (B) 265, 275
arbacioid type, (E) 112
areal aperture, (F) 88
arenaceous, (F) 20
areole, (E) 161
Aristotle's lantern, (E) 102, 161
armadillos, (M) 502
arrangement of chambers, (F) 21
articulate, (G) 285, 434
auricles, (E) 161
auriform, (G) 434
axial, (F) 88, (G) 434
axial ornament, (G) 285
axial section, (F) 56, 88
axis, (F) 88

bandicoots, (M) 500
base, (G) 281, 434
basicoronal, (E) 161
bass, (P) 492
bats, (M) 502
beak, (B) 169, 275
bears, (M) 504
Bernard, F., 174
bichir, (P) 490
biconical, (G) 283, 434
bidentate, (E) 161
bigeminate, (E) 161
bilamellar, (F) 88
birds, (V) 486
biserial, (F) 21, 88
bison, (M) 519
bivium, (E) 161
blue whale, (M) 506
body whorl, (G) 281
bony fish, (P) 486, 488
boring habit, (E) 102, (B) 260
boss, (E) 161
bourrelets, (E) 136, 161

563